Big Science in the 21st Century

Economic and societal impacts

Online at: https://doi.org/10.1088/978-0-7503-3631-4

Big Science in the 21st Century

Economic and societal impacts

Edited by
Panagiotis Charitos
CERN, Switzerland

Theodore Arabatzis
National and Kapodistrian University of Athens, Department of History and Philosophy of Science, Greece

Harry Cliff
University of Cambridge, United Kingdom

Günther Dissertori
ETH Zürich, Switzerland

Juliette Forneris
Danish Technological Institute, Denmark

Jason Li-Ying
Technical University of Denmark, DTU, Denmark

IOP Publishing, Bristol, UK

ISBN 978-0-7503-3631-4 (ebook)
ISBN 978-0-7503-3629-1 (print)
ISBN 978-0-7503-3632-1 (myPrint)
ISBN 978-0-7503-3630-7 (mobi)

DOI 10.1088/978-0-7503-3631-4

Version: 20231201

IOP ebooks

British Library Cataloguing-in-Publication Data: A catalogue record for this book is available from the British Library.

Published by IOP Publishing, wholly owned by The Institute of Physics, London

IOP Publishing, No.2 The Distillery, Glassfields, Avon Street, Bristol, BS2 0GR, UK

US Office: IOP Publishing, Inc., 190 North Independence Mall West, Suite 601, Philadelphia, PA 19106, USA

In loving memory of Steven Weinberg (1933–2021)

Contents

Part II Economic and innovation impact of Big Science projects [Editor: Jason Li-Ying]

Preface

Last year platoons of scientists discovered exoplanets orbiting distant stars, successfully launched the James Webb Space Telescope, our latest state-of-the-art 'window' into deep space, finished sequencing the human genome, publicly offered generative AI tools, while accelerated innovation in vaccine technologies and promising therapeutics for other diseases pushed into new territories. Answering big questions often requires scientists and institutions to pool resources and data, whether the research involves detecting gravitational waves in deep space, understanding the interactions between elementary particles that are the fundamental constituents of matter, or investigating the genetics of brain development.

Impressive by any standard, such achievements nonetheless raise questions about the direction of 21st century science. The 20th century saw the rise of Big Science with its roots traced back to the First World War. Subsequently, the Second World War led to the Manhattan Project's development of the atomic bomb, which is widely considered to be the dawn of the Big Science era. Following the end of the war, intensive globalisation led to the foundation of international laboratories such as CERN in Switzerland, as well as many international collaborations led by NASA, ESA, ESO, EMBL and other Big Science organisations worldwide. Big Science projects were often perceived as the key to pushing back the frontiers of science as well as to a nation's prestige. Over the past years, Big Science has also changed phase, becoming not simply larger in scale but even more interdisciplinary and international. The open questions and mind-bending concepts explored by big scientific instruments in particle physics and astronomy also serve to draw bright young minds into science, even if some of them will later work in other areas. Fundamental science has to make its case not only on the basis of the cultural value of fundamental knowledge, but also in terms of the socioeconomic benefits that it brings to society. The impact of Big Science on economic prosperity becomes more evident if we consider its cumulative contributions to the 21st-century knowledge economy, which relies heavily on research and innovation. However, Big Science has also drawn criticism due to its long timelines and consumption of significant resources, as well as the view that large projects may reduce innovation or lead to 'group think'.

These tensions, articulated at different levels within the scientific community and the public sphere, have been transforming the scope and the characteristics of the Big Science model. Capturing these changes, and how they are manifested in different areas, is one of the motivations for this volume. The volume is organised in five parts. The first looks at lessons from Big Science organisations and best practices in increasing the return of benefits to society. We have strived to include a variety of Big Science facilities, at different levels of maturity, offering an overview of the challenges linked to planning, building, and operating these facilities. The second part presents the view of economists working in relevant fields and includes case studies that illustrate previous socioeconomic impact assessments for large-scale research infrastructures, offering new insights into the impact of Big Science organisations. The third part takes a historical perspective on Big Science, tracing its development in the aftermath of the Second World War. The essays, extending over several thematic and geographical boundaries, discuss the ways and whys that shaped the evolution of Big Science in different disciplines (e.g., physics, biology and engineering) and offer valuable insights into the modern debate about the challenges and merits of Big Science. The fourth part explores the wider cultural and educational impact of Big Science projects over the past decades. The essays discuss the potential of Big Science projects to inspire and alter how we look at our world through different media, from documentaries to Hollywood's blockbusters, and from museums to the classroom. The last part of the book offers a broader and more global perspective, by widening the discussion to cases beyond North America and Western Europe. The essays in this part examine the different challenges that Big Science projects face in other parts of the globe, while reminding us of the synergies and collaborations developed between laboratories and academia at a global scale.

Undeniably, Big Science has transformed the world in which we live and will continue to do so even if its character is transformed. The different viewpoints in this book demonstrate the ways in which Big Science has delivered intellectual, cultural and economic advancement, while reminding us that this path is not always linear. Furthermore, the blend of historical and contemporary perspectives reminds us that science is not only about discoveries but is also about the collection, replication and corroboration of results with new and more powerful instruments. A finding that seems mundane or trivial may become immensely important years later, when a parallel discovery contextualises or clarifies its implications, or by adding more precision that reveals small deviations compared to our present theoretical models. It's like exploring an unknown land; we can never know in advance whether over the next hill lies a new ocean or just another hill.

We would like to thank the authors who kindly accepted our invitation and joined us in this amazing journey. Balancing different projects and approaches and striving for an inclusive representation of the many facets of Big Science has been an arduous task for us as editors. This task was further complicated by the global pandemic, as many of us faced lockdowns and shared stories about losses and grief. As we went through different phases of the pandemic certain authors had to withdraw while delays were introduced. We are also grateful to the Institute of Physics (IOP) team, who encouraged us from the first moment and steered the production of this volume smoothly and efficiently. We would also like to thank the external reviewers who evaluated our original proposal for this book, providing invaluable feedback and further guidance. We dedicate this work to the memory of Steven Weinberg, who had kindly supported the work that became this volume. Weinberg, who died in July 2021, was not only a Nobel laureate physicist but also a prolific public intellectual who greatly supported the development of Big Science.

This volume aspires to offer a comprehensive overview of the different ways in which Big Science has impacted our society and, consequently, to open new perspectives on how to evaluate its societal benefits. It offers a glimpse of the complex realities that characterise the development and present status of Big Science. A central theme of the volume is that the impact of Big Science should be considered beyond its economic benefits to industry and innovation, or the opportunities that its facilities offer for smaller experiments, but also via its interaction with society at large, including the training of new scientists and broader cultural benefits for new generations. We hope that these essays will encourage a dialogue with other scientific communities and bring new connections between people working in science and technology studies and economists and historians.

Editor biographies

Panagiotis Charitos

Panos holds a bachelor's and a master's degree in astrophysics from Imperial College London and an MSc in Media and Communications from the London School of Economics. He joined CERN in 2011 as a member of the ALICE Outreach Group and since 2013 he has also served as co-editor of CERN's Physics Department newsletter. From 2015 to 2020 he was in charge for the communications of the Future Circular Collider project exploring the feasibility of a next generation of circular colliders at CERN. Before that, Panos had worked for major daily newspapers and publishing houses in London and Athens, and was the founder of ROPI Publications, a niche publishing house specializing in the history and philosophy of science.

Theodore Arabatzis

Theodore Arabatzis holds a Diploma in electrical engineering from Aristotle University of Thessaloniki, and an MA and a PhD in history of science from Princeton University. He was awarded fellowships from Princeton, MIT, and the Max Planck Institute for the History of Science. His research has focused on the history of modern physical sciences and on historical philosophy of science. He has published numerous articles in these areas and the monograph *Representing Electrons: A Biographical Approach to Theoretical Entities* (University of Chicago Press, 2006). He has served as co-editor of the journal *Metascience* (2010–2014). In 2017 he was awarded the IUHPST Essay Prize in History and Philosophy of Science by the International Union of History and Philosophy of Science and Technology. He was president of the European Society for the History of Science (Sept 2020–Sept 2022). More detailed information about his work can be found at http://scholar.uoa.gr/tarabatz/home.

Harry Cliff

Harry Cliff is a particle physicist at the University of Cambridge's Cavendish Laboratory working on the LHCb experiment, a huge particle detector buried 100 metres underground at CERN near Geneva. He is a member of an international team of around 1400 physicists, engineers and computer scientists who are using LHCb to study the basic building blocks of our universe, in search of answers to some of the biggest questions in modern physics. His research is focussed on the very rare decays of exotic fundamental particles called beauty quarks, which have the potential to reveal evidence of hitherto undiscovered particles or forces.

Cliff is also an active communicator of science. From 2012 to 2018 he held a joint post between Cambridge and the Science Museum in London, where he curated two major exhibitions: Collider (2013) and The Sun (2018). He gives a large number of public talks, including two at TED and the Royal Institution that have each been viewed millions of times, alongside appearances on television, radio and podcasts. His first popular science book, *How To Make An Apple Pie From Scratch*, was published in August 2021 and a second, Space Oddities, will be published in 2024.

Günther Dissertori

Günther Dissertori studied Physics at the University of Innsbruck. From there, he moved to work as a doctoral student at CERN in Geneva, where he began his research career. In 2001, he was offered an assistant professorship at ETH Zurich, and six years later was made a full professor. At CERN, he was strongly involved in the CMS experiment, one of the two experiments at the Large Hadron Collider that discovered the Higgs boson. With his group, Dissertori also developed new positron emission tomography (PET) devices, including a cost-efficient brain PET scanner, which was brought to the market by the spin-off company Positrigo AG. In February 2022, he took up the role of Rector of ETH Zurich.

Juliette Forneris

Juliette Forneris holds a PhD in Geology from Arizona State University. From doctorate and post-doc studies focused on the thermodynamics of minerals, she transitioned into the fascinating world of Big Science through subsequent positions at Danish firms supplying goods and services to Big Science organizations. From 2010 to 2019 she was network manager for BigScience.dk, a publicly financed initiative to promote Danish companies towards the Big Science market. Throughout that period, she served as Industrial Liaison Officer for CERN and ESO, played a central role in the first Big Science Business Forum in 2018 in Copenhagen, and also chaired CERN's ILO forum from 2015 to 2018. In 2019, Juliette became the director of the International Centre at the Danish Technological Institute, while still keeping in touch with the world of Big Science through collaborative R&D projects.

Jason Li-Ying

Jason Li-Ying is a well-recognized figure in the international research society of innovation management. His research and teaching interests include technology & innovation management, innovation and organizational learning, strategic management, and technology transfer. Besides his excellence in research and education, Jason also has extensive experience in innovation strategy and business development for SMEs by acting as strategic advisors for a number of innovative companies. He also has an excellent track record in teaching, coaching, and supervising EMBA students and teaching courses at Master's and PhD level programmes in product design innovation, innovation management, and business strategy at Technical University of Denmark (DTU). Jason is also visiting professor at University of Management and Economics (ISM) in Lithuania, a faculty member of DTU Executive School of Business, and a faculty member for the 'Eurotech Universities' with respect to the Innovation Economics PhD program. His recent running research projects focus on innovation and corporate entrepreneurship at Big Science facilities (Research Infrastructures); the relationship between technology searching (through licensing) and innovation and digital business model innovation for SMEs.

List of contributors

Paolo Acunzo
Chair of the 2019–2022 Steering Board, Pan-European Research Infrastructures ILO Association (PERIIA), Italy

Daniel Adams
Chief Director, Basic Sciences and Infrastructure Department of Science and Technology, South Africa

Yuki Akimoto
Chiba University, Japan

Giovanni Anelli
Head of Knowledge Transfer Group, CERN, Switzerland

Graham Appleby
Business Development Manager, ISIS Neutron and Muon Source, UK

Theodore Arabatzis
Department of History and Philosophy of Science, National and Kapodistrian University of Athens, Greece

Rosalind Attenborough
University of Edinburgh, UK

William S Bainbridge
National Science Foundation, USA

David Baneke
University of Utrecht, Netherlands

Andrea Bastianin
University of Milan, Italy

Anne-Gentil Beccot
CERN, Switzerland

Leonardo Biagioni
Deputy Chief Financial Officer, F4E, Spain

Sukanta Bose
Professor, Inter-University Centre for Astronomy and Astrophysics (IUCAA, India) and Washington State University (USA), and Project Coordinator, LIGO-India, India

Caroline Boudou
Institut Laue-Langevin (ILL), France

Alison Boyle
Science Museum, London, UK

Jelena Brankovic
CERN, Switzerland

Gaia Cantelli
Strategy Officer, EMBL, Germany

Panos Charitos
CERN, Switzerland

Fahmida Chowdhury
National Science Foundation, USA

Harry Cliff
University of Cambridge, UK

Simon Connell
University of Johannesburg, South Africa

Katharina C Cramer
University of Bonn, Germany
and
Lund University, Sweden

Robert Crease
Stony Brook University, USA

Riccardo Crescenzi
London School of Economics, UK

Sunje Dallmeier-Tiessen
CERN, Switzerland

Matias del Campo
Taubman College of Architecture and Urban Planning, University of Michigan, USA

Ana Belen Del Cerro Gordo
Member of the 2019–2022 Steering Board, Pan-European Research Infrastructures ILO Association (PERIIA), CDTI, Spain

Erika Susan Dietrichson
Technical University of Denmark (DTU), Denmark

Guenther Dissertori
ETH Zurich, Switzerland

David Eggleton
University of Sussex, UK

Sibel Erduran
Professor of Science Education, University of Oxford, UK

Jan Fell
Institute for Entrepreneurship and Innovation, Vienna University of Economics and Business (WU), Austria

Barbara Ferreira
Media Manager, ESO, Germany

James Fleming
Charles A Dana Professor of Science, Technology, and Society, Colby College, ME, USA

Massimo Florio
University of Milan, Italy

Pamfilos Fokianos
CERN, Switzerland

Juliette Forneris
Danish Technological Institute, Denmark

Kostas Gavroglou
Department of History and Philosophy of Sciences, National and Kapodistrian University of Athens, Greece

Ralph Gilles
Heinz Maier-Leibnitz Zentrum (MLZ), Technische Universität München, Germany

Florian Gliksohn
Executive Director, ELI ERIC, Czech Republic

Steve Goldfarb
University of Melbourne, Australia

Jean-Baptiste Gouyon
University College London, UK

Chen Guang
Chinese Academy of Sciences, China

Ute Gunsenheimer
Designated Secretary General, EOSC Association, Belgium

Stefan Gustafsson
Commercialisation Officer at European Space Agency

Stephan Haid
Director of Administration, CTAO, Germany

Anna Hall
Member of the 2019–2022 Steering Board, Pan-European Research Infrastructures ILO Association (PERIIA), Big Science Sweden, Sweden

Olof Hallonsten
Lund University School of Economics and Management, Sweden

Yuko Ikkatai
Faculty of Human Sciences, Kanazawa University Institute of Human and Social Sciences, Japan

Laurent James
Member of the 2019–2022 Steering Board, Pan-European Research Infrastructures ILO Association (PERIIA), CNRS - INSU, France

Oliver James
Chief Scientist, DNEG, UK

Paul Josephson
Professor of History, Colby College, ME, USA

Peter Keinz
Associate Professor, Institute for Entrepreneurship and Innovation, Vienna University of Economics and Business (WU), Austria

Ariane Koek
Curator and writer, founder and former Head of Arts at CERN, UK

Alexander Kohls
CERN, Switzerland

Helge Kragh
Professor Emeritus, Niels Bohr Institute, University of Copenhagen, Denmark

Linn Kretzschmar
CERN (previously at Vienna University of Economics and Business (WU)), Austria

Frederick M Kronz
National Science Foundation, USA

Leong-Chuan Kwek
National Institute of Education, Nanyang Technological University, Singapore
and
Quantum Science and Engineering Centre (QSEC), Nanyang Technological University, School of Electrical and Electronics Engineering, Block S2.1, 50 Nanyang Avenue, Singapore 639798
and
Center for Quantum Technologies, National University of Singapore, 3 Science Drive 2, Singapore 11754

Anne-Flore Laloë
Information Management Coordinator, EMBL, Germany

Fiona Larner
Head of Evaluation, Science and Technology Facilities Council, UK Research and Innovation (STFC-UKRI), UK

Sune Nordentoft Lauritsen
Managing Director, ESA Business Incubation Centre Denmark, Denmark

Frank Lehner
DESY, Germany

Rodrigo Liscovsky Barrera
University of Edinburgh, UK

Jason Li-Ying
Professor, Technical University of Denmark (DTU), Denmark

Luciano Maiani
Sapienza Universita e INFN, Roma I, (former Director General CERN), Italy

Sandra Manninger
Taubman College of Architecture and Urban Planning, University of Michigan, USA

Klaus Marhold
Institute for Entrepreneurship and Innovation, Vienna University of Economics and Business (WU), Austria

Joseph D Martin
Department of History, Durham University, UK

Michela Massimi
University of Edinburgh, UK

Charlotte Matthieu
Economist, ESA, France

Johanna McEntyre
Associate Director, EMBL-EBI Services, UK

Ed Mitchell
European Synchrotron Radiation Facility, France

Cyrus Mody
Maastricht University, The Netherlands

Micha Moskovic
CERN, Switzerland

Sekazi Mtingwa
TriSEED Consultants, USA

Kamran Naim
CERN, Switzerland

Prosper Ngabonziza
Department of Physics and Astronomy, Louisiana State University, Baton Rouge, LA, USA
and
Department of Physics, University of Johannesburg, Johannesburg, South Africa

Poul Zimmermann Nielsen
Space Solutions Manager, ESA Business Incubation Centre Denmark, Denmark

Markus Nordberg
Head of Resources Development, CERN, Switzerland

Grigoris Panoutsopoulos
Department of History and Philosophy of Sciences, National and Kapodistrian University of Athens, Greece

Antonios Papadopoulos
CERN, Switzerland

Wonyong Park
Southampton Education School, University of Southampton, UK

Gabriele Piazza
London School of Economics, UK

Stephen Pompea
University of Arizona, USA

Valangiman Subramanian Ramamurthy
Director, National Institute of Advanced Studies, India

Michael Riordan
Adjunct Professor of Physics, University of California, Santa Cruz, and Lecturer, Stanford University, CA, USA

Veronica Riquer
CERN, Rome University, and INFN, Rome, Italy

Mark Robinson
SOAS, University of London, UK

Pedro Russo
University of Leiden, The Netherlands

Alejandro Sánchez
Associate Director for the Industrial Liaison Office, ALBA Synchrotron Light Source, Spain

Alexandra Schmidli
Senior Communications Officer, ELI ERIC, Czech Republic

Robert Seidel
Professor Emeritus, University of Minnesota, USA

Elizabeth Shotton
Head of Industrial Liaison Group, Diamond Light Source, UK

Dinesh Srivastava
Homi Bhabha Chair Professor, National Institute of Advanced Studies, India

Simon Stewart
Director of the Centre for European and International Studies Research, Department of Sociology, University of Portsmouth, UK

Adrian Tiplady
Deputy Managing Director, Strategy and Partnerships, South African Radio Astronomy Observatory (SARAO), South Africa

Martin Townsend
Member of the 2019–2022 Steering Board, Pan-European Research Infrastructures ILO Association (PERIIA), CCFE, UK

Emanuel Tsesmelis
CERN, Switzerland

Simone Turchetti
University of Manchester, UK

Thomas Turnbull
Max Planck Institute for the History of Science, Germany

Lenka Unge
European Spallation Source ERIC, Sweden

Toon Verhoeven
Member of the 2019–2022 Steering Board, Pan-European Research Infrastructures ILO Association (PERIIA), Big Science Netherlands, Netherlands

Niki Vermeulen
University of Edinburgh, UK

Silvia Vignetti
Centre for Industrial Studies, Milan, Italy

Lucia von Bredow
Bioethics Manager, Science and Society Lead, EMBL, Germany

Ting Wang
Associate Professor, Institutes of Science and Development, Chinese Academy of Sciences (CASISD), Beijing, China

Allen Weeks
Director General, ELI ERIC, Czech Republic

Stéphanie Willekens
Economist, ESA, France

Andrew Williams
External Relations Officer, Executive Office of the Director General, ESO, Germany

John Womersley
Former Director General, European Spallation Source ERIC, UK

Hiromi M Yokoyama
Deputy Director and Professor at Kavli IPMU, University of Tokyo, Japan

Nikolaj Zangenberg
Member of the 2019–2022 Steering Board, Pan-European Research Infrastructures ILO Association (PERIIA), DTI, Denmark

Milosz Artur Zielinski
CERN, Switzerland

Part I

Voices from European Big Science Organizations [Editor: Juliette Forneris]

IOP Publishing

Big Science in the 21st Century
Economic and societal impacts
Panagiotis Charitos, Theodore Arabatzis, Harry Cliff, Günther Dissertori, Juliette Forneris and Jason Li-Ying

Chapter 1

Introduction to part I: Voices from European Big Science Organizations

Juliette Forneris

Big Science projects represent the scientific flagships of the twentieth and twenty-first centuries, created by the combination of intense international research collaboration and massive investment. Whether located on Earth, in space, or the virtual world, they symbolize the incredible human endeavours to understand the world around us and to solve the challenges of our times.

The common traits of Big Science projects are the level of scientific excellence, the enormous budgets to plan, build, and upgrade facilities sometimes compared to giant 3D puzzles, the huge amount of scientific, technical, managerial, and administrative staff, the engineering skills required to make the science happen, and the complexity of the project management.

There is no consensus, neither in this book nor in the existing literature, about the terminology used to describe Big Science projects and the organizations behind them. The terms Big Science projects or facilities, research infrastructures (RI), large scale research infrastructures (LSRI), and analytical research infrastructures (ARI) are some of the names that are used, sometimes interchangeably, to describe them. This is likely because they have their origins in different scientific communities but also because Big Science is a concept that has undergone evolution and diversification from the middle of the twentieth century until now:

1. *Scientific domains*: The early Big Science facilities had a strong focus on physics and astronomy. They were for the most part driven by a quest for advancing our understanding of the Universe. They are still often referred to as ‘Big Physics’. Towards the end of the twentieth century, a diversification in scientific domains took place, giving rise to Big Science facilities for the life sciences, ICT, and social sciences.
2. *From fundamental to applied research*: Although many currently running Big Science facilities are still focused on producing beyond state-of-the-art

doi:10.1088/978-0-7503-3631-4ch1

fundamental research, more and more facilities have been built for the purpose of applied research in a broad range of scientific domains. These Big Science facilities are characterized by a large user community encompassing both academics and industrial users.
3. *From single-sited to distributed and virtual facilities*: As an effect of the increasing size and complexity of Big Science projects, together with advances in data science, distributed and even virtual Big Science facilities have emerged.

For the first part of this book we reached out to a range of both mature and newer European Big Science Organizations (BSOs) to give the reader an overview of the challenges linked to planning, building, and operating Big Science facilities, including the ever-growing demand from their stakeholders for documenting the socio-economic impacts. Several of these BSOs have ongoing collaborations established, for example, via EIROforum[1], ESFRI[2], LENS[3], LEAPS[4], CERIC-ERIC[5], or bilateral agreements. However, in 2018, the Big Science Business Forum (BSBF2018) was the first large public event to showcase the existence of a consolidated Big Science market and to create a stronger collaboration amongst BSOs. Under the leadership of the Danish Ministry of Higher Education and Science, nine BSOs (CERN, EMBL, ESA, ESO, ESRF, ESS, European XFEL, F4E, and ILL) joined forces to present future investments worth billions of euros (BEUR) for the Big Science industry. During the preparation of the event, the interest for other BSOs to participate grew and an additional nine BSOs (ALBA, DESY, ELI-NP, ENEA, FAIR, MAX IV, SCK•CEN–MYRRHA, PSI, and SKA) joined BSBF2018 as affiliated BSOs. During the three day event, over 1000 participants from more than 500 businesses and organizations representing 29 countries gathered to listen to over 100 speakers from these 19 BSOs and discuss how they could do business out of the over 12 BEUR in future investments presented. The success of this first event led to a pledge for a follow-up event in 2020, which due to the COVID-19 pandemic, only took place in 2022, in Granada. The idea of bringing some of these organizations together for part I of this book 'Voices from European Big Science Organizations' is also one of the direct impacts of BSBF2018.

Part I is structured around four overarching themes:

1. *The efforts to assess the socio-economic impacts of Big Science projects* are at the core of the essays from Anelli, Nordberg, and Charitos for CERN, Mathieu, Laurich, and Willekens for ESA, and Williams and Ferreira for ESO. Due to the maturity of these three organizations, a large amount of data is available to identify the long-term impacts measured after years of operation of their facilities. The conceptual frameworks and methodologies

[1] https://www.eiroforum.org/
[2] https://www.esfri.eu/
[3] https://lens-initiative.org/
[4] https://leaps-initiative.eu/
[5] https://www.ceric-eric.eu/

for impact assessment are discussed by the authors specifically for each of their organizations and Big Science projects. The essays cover a broad range of benefits from Big Science projects: scientific, technological, strategic, societal, and economic. Socio-economic impact assessments are increasingly important to BSOs and their funding bodies in light of future investment needs, but are complex analyses requiring important efforts and a systematic approach.

2. *Developing business cases for new Big Science facilities and measuring short-term impacts.* Not all Big Science projects are currently at the operation stage. Nevertheless, investigating the shorter-term impacts resulting from the construction of a Big Science facility before it even starts producing scientific results illustrates other benefits to society, as argued by Biagioni (Fusion for Energy (F4E)) in the case of ITER. The question of how to develop business cases for new Big Science facilities is addressed from different angles by Womersley for the European Spallation Source (ESS) and by Weeks, Gliksohn and Schmidli for the Extreme Light Infrastructure (ELI ERIC). While the science case is still at the heart of the decision-making process to build a new Big Science facility, key requirements are placed on the funding, governance, project management, stakeholder engagement, and cost–benefit analysis.
3. *Big Science engaging with industry.* Big Science facilities are typically planned by and built for scientists but much of the engineering work behind the science is done in collaboration with industrial suppliers (the Big Science industry). This covers a broad range from unique products based on state-of-the-art R&D work (e.g. detectors, superconducting cables, and neutron guides), to the supply of high-tech products (e.g. high-precision mechanical components, special coatings, and power supplies) to completely off-the-shelf products. Verhoeven *et al* explore how industrial liaison officers act as key intermediaries between BSOs and their industrial suppliers to ensure the best match between the demand and the supply. In their essay 'Analytical research infrastructures and industry engagement: drivers, challenges, and impact' Appleby *et al* explore the complex ecosystems for industry engagement around ARIs focusing on three paths of engagement: industry as a user, industry as a technology partner, and enterprises and research technology organizations (RTOs) as intermediaries.
4. *Big Science: constantly morphing to adapt to scientific and societal challenges.* In addition to being measured against their socio-economic impacts, BSOs are also under constant pressure to reinvent themselves to stay at the forefront of scientific developments. A prime example of this long-term sustainability is illustrated by Haid and Lehner through the almost 40 years of operation and three 'lives' of the DORIS storage ring at DESY in Hamburg. In their essay von Bredow *et al* investigate how the European Molecular Biology Laboratory (EMBL) and by extension other BSOs can affect and also respond to changing societal, scientific, and cultural norms in areas such as bioethics, data privacy, environment policies, and research integrity.

Part I, ‘Voices from European Big Science Organizations’, was not conceived to present an exhaustive catalog of the largest European BSOs. The purpose was instead to give the reader an idea of the scientific and technological complexity combined with the socio-economic accountability that comes with investments in Big Science. We hope that we have succeeded in giving these insights and that this work will inspire further studies encompassing more ESFRI landmarks and projects[6], and potentially exploring the specificities of the growing number of virtual infrastructures in the context of the socio-economic impacts.

[6] https://roadmap2021.esfri.eu/media/1251/rm21-part-2.pdf

Chapter 2

CERN: the study of the infinitesimally small and the rise of Big Science

Giovanni Anelli, Markus Nordberg and Panagiotis Charitos

2.1 The rise of Big Science: a brief account

During the twentieth century, particle physics has become a truly global activity, with experimental collaborations such as those of the Large Hadron Collider (LHC) experiments at CERN bringing together thousands of researchers from all over the world, to design, develop, and operate this enormous research infrastructure. The LHC, the world's largest particle collider, and its detectors have performed admirably, recording the mountain of data flowing in from the collisions with staggering precision and efficiency. When the 2013 Nobel Prize in Physics was awarded to Peter Higgs and François Englert for the discovery of the Higgs boson, CERN, the European Organization for Nuclear Research, was praised for its role in confirming the new fundamental particle. It was reported (Höne and Korbalija 2018) that CERN's LHC was described by the Swedish academy as 'probably the largest and the most complex machine ever constructed by humans'. A detailed account of the steps that led to the creation of the LHC can be found in the essay by Gavroglou and Panoutsopoulos in the third part of this book.

The creation of CERN in 1954 was based on an effort to establish a laboratory on European soil comparable to those that had been established in the USA during the Second World War. CERN would provide a new stage for physics, helping to attract European scientists back to the continent and stemming the tide of scientific migration westwards across the Atlantic. CERN's history is, to a certain extent, intertwined with that of the Manhattan Project, a programme that is often considered to have served as a stepping stone towards the rise of Big Science, despite the ethically questionable results it produced. There exists a rich literature on the history of CERN (Herrmann *et al* 1990, Martin 1992, Krige 1996) and in his essay in part III of this volume, Schopper (2009). Beyond its specific scientific

doi:10.1088/978-0-7503-3631-4ch2

achievements, CERN has also been characterised as an early example of institutionalised scientific cooperation (Engelen 2012, Copeland 2016), a paradigmatic example for the promotion of peace through science by bringing individuals and countries together and overcoming historical and political fractures (Quevedo 2013, Ruffini 2017), and an important element of the process that led towards European integration (Stein 2002).

Big Science remains a contested notion, one that challenges any attempts at a universal definition. For essays on the birth of Big Science, see Derek J de Solla Price's work *Little Science, Big Science … And Beyond* (de Solla Price 1986) as well as the book *Big Science: The Growth of Large-Scale Research* edited by Peter Galison and Bruce Hevly (Galison and Hevly 1992). The use of increasingly larger and more complex scientific instruments is linked to certain characteristics which are evident in the history of CERN and key when discussing its impact. First, the rise of Big Science has been characterised by an increase in the number of collaborating scientists, as well as in the number of scientific disciplines involved in the development and construction of these projects. It has also been followed by a marked shift in the way in which researchers collaborate—they needed, after all, to adapt to working within massive, inter-disciplinary groups, comprising those who develop the various technologies, those with an expertise in running the complex infrastructures, and those with the know-how of collecting and analysing the results. As early as 1943, Arthur Holly Compton, one of the Manhattan project leaders, recognised the potential of putting chemists, physicists, metallurgists, engineers, and machinists together, under one roof, and urged the president of his university to invest in a new model that would change the way in which research is conducted (Compton 1965). Today, CERN is home to more than 11 000 scientists from all around the world, who contribute to the construction and operation of new scientific instruments and devise experiments in order to glean insights into the fundamental laws of nature. A number of scientists of such a scale also presents certain challenges with regard to the sharing of available resources, requiring the development of a robust infrastructure and various tools for the global distribution and analysis of data. When discussing CERN's broader impact, it is important to keep in mind the opportunities and challenges generated by the transformation of scientific collaborations in the era of Big Science.

Furthermore, the rise of Big Science is connected to a shift in public perception regarding the role of scientists, and the need to devise new funding models which challenge all pre-existing private and public funding hybrids. As Gian Giudice notes in his essay on Big Science and the LHC (Giudice 2012):

> The Manhattan Project was only an episode that accelerated an already inevitable evolutionary process. Well before the Second World War, rapid scientific and technological progress had propelled science beyond its academic borders. On the one hand, science was having an increasingly crucial impact on society; on the other hand, science was requiring financial resources that could be found only outside of the limited world of universities and research institutions.

A unique period in the rise of Big Science is the period of the Cold War, marking the closer intertwining of science with technology, politics, and national pride. On 12 April 1961, Yuri Alekseyevich Gagarin became the first man to journey into space, causing a massive reaction in the USA, and thus leading to the launch of the Apollo programme, one of the largest scientific programmes in world history. On 25 May 1961, President John F Kennedy addressed the US Congress and nation with the famous words: 'I believe that this nation should commit itself to achieving the goal, before this decade is out, of landing a man on the Moon and returning him safely to Earth'. Public opinion was firmly in his favour and Congress displayed no hesitation, approving, almost unanimously, this colossal project, estimated to have cost between 20 and 40 billion dollars.

The construction of increasingly more advanced and costly instruments has gradually become decisive for the pursuit of progress across many scientific fields. Historically, in the aftermath of the Second World War, new government agencies in the USA, such as the Atomic Energy Commission (which gradually transformed into what is known today as the Department of Energy) and the National Science Foundation (NSF) were established in order to channel federal funding into research aimed at furthering US interests. This model was also reproduced by other nations' funding agencies, although a detailed comparison goes beyond the scope of this essay. It should be noted, however, that these arguments are still being heard, despite being seemingly at odds with the transnational nature of Big Science projects. The new funding mechanisms have also given birth to certain administrative bodies, charged with overseeing the allocation of budgets and resources. These bodies often lie wholly outside of the academic environment, and thus their role usually invokes a criticism often levelled against the methodology of Big Science in general, which is perceived as the diminishment of scientific methodology in favour of PR and communication. As Alvin Weinberg (1967) notes: 'Unfortunately, science dominated by administrators is science understood by administrators, and such science quickly becomes attenuated if not meaningless.'

Before closing this historical account, we will refer to a third characteristic of Big Science that, in our view, is relevant in the discussion of CERN's overall benefit to society. This concerns the link between scientific research and the rise of new, socially beneficial practical applications as well as the intensification of the relation between fundamental science and technological innovation. Historically, many US economists have considered scientific research and technological innovation as the keys to constant economic growth, which produces jobs and, therefore, prosperity. As a consequence, it can also be implied that they represent a solution to the social problems of the poorer classes, as well as a pre-emptive cure for all kinds of political instabilities. Vannevar Bush vigorously asserted these links in his report 'Science: The Endless Frontier' which he presented to President Harry S Truman on July 5, 1945: 'The simplest and most effective way in which the Government can strengthen industrial research is to support basic research and to develop scientific talent' (Bush 1990). Bush thus identified basic research as the decisive factor in promoting progress, claiming that technology is an inescapable consequence of leading-edge science. The goal of the government is thus to support and nurture the most advanced research institutions, without paying a lot of attention to the aspects of technological innovation. According to Bush, 'basic research is the pacemaker of technological progress', and his report also urged for the creation of what, in 1950, became the National Science Foundation. This view is also reflected in the modern debate on the role of public funding and its ability to drive innovation. With some variation, authors such as Mazzucato (2018) and Bozeman (2002) and Bozeman and Crow (2021) echo the original doctrine of Vannevar Bush, i.e. that basic research has a substantial and positive impact on socio-economic innovation via direct and indirect mechanisms. Interestingly, however, more recent literature has argued that, while it is commonly believed that Bush maintained an unquestioning faith in an integrated and linear model of innovation, his notion was more sophisticated and involved symbiotic cross-fertilisation. While Bush understood that basic research and applied research benefit and complement each other, they also succeed by working as separate systems, or stacks. Consequently, the goal of modern scientific and economic policy mechanisms should be to coordinate these two systems, allowing each to operate on the basis of its own logic and success criteria, yet simultaneously cultivating specific points where they can nurture each other (Leyden and Menter 2018).

We will conclude this section by acknowledging certain criticism of publicly funded Big Science projects, doubts that have accompanied its rise not only in society, but also within the scientific community itself. The most authoritative dissenting voices within said community included the physicists Merle A Tuve, Alvin M Weinberg, and Philip W Anderson, and the astrophysicist Fred Hoyle. In 1961 Alvin Weinberg wondered whether Big Science was ruining science, identifying in the process certain issues that are still worth discussing today:

> In the first place, since Big Science needs great public support it thrives on publicity. The inevitable result is the injection of a journalistic flavour into Big Science which is fundamentally in conflict with the scientific method… The spectacular rather than the perceptive becomes the scientific standard.

Weinberg was referring to the mid-twentieth century's space programme, but his words still seem relevant today.

2.2 The birth of CERN

In the late 1940s Europe was struggling to emerge from the ruins of the Second World War. While the US had played a vital role in the conflict, and had undoubtedly incurred heavy losses, it had been left relatively unscathed from a material perspective. The rise of fascism and the relentless war that had ensued had depleted most of the continent's resources, and many of its top scientists had fled to the United States. A shining vision of life across the Atlantic therefore represented a beacon of hope for millions of Europeans living in austerity. CERN was first conceived in the late 1940s at the beginning of the post-war era.

A consensus, led by several Nobel laureates, emerged regarding the need for a world-class facility dedicated to nuclear research, one imbued with a purely scientific and fundamental character. French physicist Louis de Broglie first proposed a multinational European lab in 1949. This call was matched by the 1950s European movement of reconciliation after the Second World War. During the global conflict, many eminent European scientists that had fled the rise of Nazism in Europe had been drawn into the Manhattan Project at Los Alamos. Post-war, US science remained pre-eminent. Anxious to stem a 'brain drain' of talent, farsighted pioneers saw that Europe was in need of a comparable scientific focus. This was the seed of an idea for a European centre for atomic research. Forward thinking scientists, such as Pierre Auger (French), Raoul Dautry (French), Lew Kowarski (Russian born, naturalised French), Edoardo Amaldi (Italian), and Niels Bohr (Danish), put aside national interests and strove to build bridges between nations in education, science, and culture.

One of the organisations established in the wake of the Second World War to help promote world peace and cooperation was the United Nations Educational, Scientific and Cultural Organisation (UNESCO). At the UNESCO General Conference in Florence, Italy, in June 1950, the idea for a European scientific laboratory still lay dormant. Among the US delegation in Florence was Isidor Rabi, who had won a 1944 Nobel Prize for his work on the magnetic properties of nuclei. Rabi was lobbying for years in favour of a high-energy physics 'global laboratory'. Leading a United States delegation to the newly established UNESCO, he settled on the idea that this laboratory should be based in Europe. However, on arrival in Florence he was disturbed to find that there was no mention of this idea on the agenda. Two Europeans, Pierre Auger (then UNESCO's director of exact and natural sciences) and Edoardo Amaldi, who was to be a constant driving force behind the project, helped Rabi through the intricacies of European committee formalities, and so the European seed was fertilised and, within a few years, CERN was born.

The timely cooperation between the physicists and the politicians provided an opening and thus 12 European nations came together to sign the convention, officially forming CERN in 1954. Nations that just ten years earlier had been at war were now bound together in peaceful research—a dramatic improvement in international relations that the community had worked together to achieve.

It is worth noting that the establishment of CERN signalled not only the start of a strong collaboration between European and US researchers, but also one between pure experimentalists, theoreticians, and accelerator physicists, as well as instrumentation specialists. Marking a long-standing tradition of international collaboration, two members of the Brookhaven team—John and Hildred Blewett—moved to CERN and contributed to the commissioning of the Proton Synchrotron (PS). The first high-energy protons circulated in the PS in November 1959. In the early 1970s, CERN specialists travelled to Fermilab to help build and commission the new, enormous US synchrotron. A few years later, US machine physicists visited CERN when the comparable Super Proton Synchrotron (SPS) was getting under way. In the mid-1970s, Burt Richter of SLAC, while spending his sabbatical at CERN, worked to set the scale for CERN's LEP electron–positron collider, helping to minimise the losses caused by synchrotron radiation from the circulating beams of electrons and positrons.

This US–Europe collaboration continues to this day, both at the LHC and in the design of future machines such as the Electron–Ion Collider (EIC) in the US and the Future Circular Collider (FCC) project in Europe. The desire for greater coordination also manifests itself in the bottom-up approaches of the US Particle Physics Community Planning Exercise ('Snowmass') in the US and the updates of the European Strategy for Particle Physics. It can be argued that this trend emerged within a context of rising costs and a growing number of projects, accompanied by a general drop in state budgets following the financial crisis of 2008. It also stems from increased competition on a global level, with both Japan and China having put forward plans to host their own Big Science facilities in particle physics. Pooling together resources would allow their most efficient utilisation and the development of new research infrastructures, enabling the widest possible scientific programme. Proposals for the current LHC began in the early 1980s, but the machine was not actually completed until 2009, when the first collisions took place. So, while the LHC experiments will continue to run until around 2030—with an intermediate upgrade to the High Luminosity LHC—it is the right time to start talking about its successor, with a wide consensus agreeing that it should be a Higgs factory, allowing us to scrutinise the newly discovered particle.

The methodological cosmopolitanism that we have outlined in this section is also in line with the complexity behind the decision-making mechanisms of such projects. It also accommodates the fact that the approval of large projects is not necessarily in conflict with a robust programme of scientific diversification. As a matter of fact, in the past, large and small projects have always risen and fallen together. For example, there is no evidence that, following the cancellation of the Superconductor Super Collider (SSC), the scientific fields whose representatives openly opposed its continuation enjoyed any financial benefit. If a distinction must be made, it is better to make it between projects and directions of research that drive real advancement in knowledge and those that lead to blind alleys or propose repetitive experiments of little scientific value. Science needs different investigational methodologies to create opportunities favourable for making progress. As Freeman Dyson noted:

> [T]here is no illusion more dangerous than the belief that the progress of science is predictable. If you look for nature's secrets in only one direction, you are likely to miss the most important secrets, those which you did not have enough imagination to predict. (Dyson 1988)

Dyson's perceptive words, however, do not necessarily undermine the reasons for big scientific enterprises. Even such an adamant Small Science advocate and Big Science critic as Dyson concurs that the extraordinary progress in astronomy and particle physics during the last 60 years was made possible only by a wise admixture of large and small projects.

We will close this brief account on the rise of Big Science and the creation of the organisation, by attempting to outline the mission of CERN, as it stands today. This mission revolves around four pillars: those of technology, research, education, and collaboration, and its scope extends much farther than any single experiment, technology, or breakthrough. As an organisation, CERN strives to look ahead to the future—focusing not only on the possibilities inherent in new theories, new machines, and new experiments, and any and all possible applications for the benefit of society, but also in the field itself, and the individuals that constitute it. Therefore, its greater role as an institution is to serve as a vivarium for the next generations of scientists, engineers, and innovators—nurturing them and ensuring stability and continuity, even as the world in general faces an increasingly uncertain future—whilst also striving to make the inner workings of the scientific process as open and as accessible as possible, both to scientists outside of it and to the general public as well.

2.3 CERN's knowledge transfer mechanisms

2.3.1 A diverse hub for scientific knowledge: benefits to the scientific community

CERN is one of the world's leading research centres for fundamental physics. It provides the necessary infrastructure so that more than 11 000 scientists from around the world can gain a greater understanding about the smallest constituents of matter, their interactions, and the origin and evolution of the Universe.

Particle colliders have played a crucial role both in validating many of the Standard Model's predictions and in guiding further theoretical developments. Experimentally, in recent decades thanks to the collider physics programme we have discovered a host of fundamental particles, most notably the W and Z in 1983 at CERN's SPS, the top-quark in the 1990s at the Tevatron, and the Higgs boson at the LHC. Moreover, experiments at LEP have been crucial for confirming the existence of a third generation of leptons and for narrowing the mass of the Higgs boson down to a range that was reachable at the LHC. These developments led to the 2012 discovery of a Higgs boson by the ATLAS and CMS experiments at CERN. Ever since, the LHC's experiments have been studying the properties of the new boson and its interactions with the other known particles, while also putting stringent limits on new physics models proposed to address some of the questions left open by the Standard Model. Testing the boundaries of the Standard Model requires more experimental data and consequently novel detection/experimental techniques.

Technological developments in accelerator and detector technology occurring within the frameworks of large-scale projects have often assisted smaller experiments as well, which, as a rule, profit from the available research infrastructure, even if the latter is beyond their scope. In the past, the resources (both human and financial) required for setting up large-scale facilities led to the emergence of a dichotomy between small- and large-scale experiments. Such a distinction is far less relevant today; thinking in terms of this dichotomy ignores the symbiotic relationship between large and small-scale experiments, with regard to the infrastructure they share, as well as the flow of data between them. Adopting a form of methodological cosmopolitanism could shed more light on the inner workings of modern scientific inquiry—and CERN is a fertile ground for such an approach. To understand the importance of this cross-fertilisation between Big Science projects and smaller experiments, one needs to look no further than the steep rise in the number of smaller experiments that were running or being approved at CERN during the construction of the LHC. The development and operation of this massive project enabled a diverse experimental programme that benefited from the accelerator complex, the available workshops, and beams for testing as well as the—exceptionally crucial—software and IT infrastructure for receiving, distributing, and analysing results. As an example of the latter, the framework developed at CERN to host and distribute application software stacks for LHC users, the CernVM File System (CERN 2021), is also used by the LIGO collaboration studying gravitational waves. In parallel to the LHC, other planned facilities such as DUNE in the US and HyperK in Japan also profit from testing certain technologies (e.g. proto-DUNE) at CERN's facilities and from detailed simulations of topics of common interest.

The construction of the LHC highlighted a number of benefits for small-scale projects and helped to expose physicists to new techniques, some of which found unprecedented new applications. FASER, one of the proposals regarding the search for long-lived particles within the LHC, serves as an apt example. Motivated by the lack of experimental evidence for new particles, physicists put forward ideas for particles that may decay long after the detector has ceased recording events. FASER is one of the detectors proposed to be installed during Run 2 of the LHC. It uses recycled components from the ATLAS and LHCb experiments and moreover demonstrates that the LHC—like any collider—is, in reality, quite a versatile facility, that allows for occasional out-of-the-box thinking.

Hardware and software synergies during the development and realisation of Big Science projects enable a wider research programme and thus make the boundaries of the laboratory more porous and more inclusive; a manifestation of the methodological cosmopolitanism in favour of which we argue. Beyond these synergies, there is also a profound epistemological synergy at play between data obtained from CERN's experiments (collider and non-collider) and other fields beyond high-energy physics. For example, more accurate data on particle interactions or nuclear structure can be important in analysing and understanding astrophysical and cosmological phenomena. Another example is the CLOUD experiment at CERN that studies aerosol and cloud formation in an ultra-pure chamber, shedding light on information that could help shape policies for the reduction of air pollution. Finally, one should not forget the value of the so-called unknown unknowns. Data can also reveal new phenomena and raise new questions, as is the case of the collective structure revealed during asymmetric collisions of protons with heavy-ions at the LHC.

Before we close this section, we should note that the precise quantification of the impact of knowledge gained through these complex scientific instruments remains an open challenge for economists. The long-term value of discoveries stemming from fundamental research often eludes precise estimation, due to its unpredictable nature and the often long timescales it involves. The wider impacts on science, the economy, and society are also typically difficult to forecast and to assess and:

> they can even be difficult to evaluate retrospectively, since science typically advances over a broad front, with many separate but interlinked discoveries producing overall societal change. (OECD 2014)

The recent cost–benefit analysis (CBA) of LHC/HL-LHC (Bastianin and Florio 2018) focused mainly on the impact that scientific results have on the career development of scientists and introduced a measure derived on the basis of the number of citations in scientific publications. Using this methodology, they accounted for 2% of the total socio-economic benefits of the LHC/HL-LHC programme; a figure that as the authors note 'is related to the number of publications and citations and to their content'; a statement hinting at the limitations of including the direct scientific methodology in a socio-economic impact analysis.

A more detailed account would go beyond the scope of this chapter; however, we hope that the above may help shine a light on the rich mosaic of relations that have evolved around CERN and Big Science projects in general, which may be better understood by adopting a new methodological approach. The relation between small and big experiments is much more complex than that implied by a 'one or the other' approach and a failure to acknowledge this diversity leads to a misestimation of the impact that CERN and other Big Science facilities have as research hubs. In brief, large-scale, long-term projects facilitate strong partnerships between universities, laboratories, and industries that lead to scientific and technological breakthroughs—with industrialisation assisting with applications outside high-energy physics (HEP)—while small and big experiments profit from technology exchanges that often reduce costs and facilitate the realisation of even more experiments. This cross-fertilisation of experimental techniques and technologies is one of the foremost characteristics of Big Science in the beginning of the twenty-first century and should be taken into account when assessing the wider impact of CERN as well as other Big Science projects.

2.3.2 From idea creation to knowledge transfer: the path from CERN to society

While CERN's research is primarily motivated by curiosity, it also has a significant impact on society and everyday life. Reaching ambitious scientific objectives requires the development of advanced instruments and new technologies, making CERN and the collaborating institutes and laboratories drivers of innovation. The active dissemination of CERN knowledge and technology is rather focused on all of the collective knowledge and technology required to execute the high-energy physics research. To design and build experiments at the scale of the LHC, significant technological developments are required in the field of magnets, radio-frequency cavities, superconductivity, detectors, ICT, and many other domains, and these can all have a direct positive impact on society.

Beyond technological innovation, CERN is also a catalyst for intellectual and social innovation. The latter refers to the design and implementation of new solutions that imply conceptual, process, product, or organisational change, which ultimately aim to improve the welfare and wellbeing of individuals and communities (OECD 2006). This process is reflected in the number of users and the diversity behind the collaborations that drive these large scientific endeavours, adopting new technological solutions and devising new partnerships to deliver the tools required to advance the construction of large-scale detectors. Moreover, the complex of social relations developed around big scientific projects often acquires an independent status as a driver of technical developments and innovation. As Autio (2014) writes:

> Big Science facilities provide a platform for global research networks; this social capital has been shown to provide an excellent setting for combining ideas leading to innovations such as the World Wide Web.

The scientific exchanges in the lifetime of Big Science collaborations can generate a type of interlaced knowledge.

Sociologists have long been studying the structure of scientific organisations, several with CERN as a guinea pig. They find that particle physics collaborations uniquely engage in 'participatory collaboration', which brings added value in knowledge generation, technology development, and innovation. This approach, along with the presence of global experiments hosted by our laboratory, brings high value to CERN's knowledge transfer activities. In the book *Collisions and Collaboration: The Organization of Learning in the ATLAS Experiment at the LHC* (Boisot *et al* 2011), it is noted that

> a kind of 'interlaced' knowledge emerges, distributed across the different members of the collaboration to become a collective possession. Interlaced knowledge gradually produces an infrastructure of 'boundary objects' through which much of the organizational coordination across heterogeneous teams, practices, and languages can then take place.

The social and intellectual innovations that develop around Big Science facilities contribute greatly towards technological innovation. The history of the web represents what is, perhaps, the most often scrutinised example of this process. In the summer of 1991, researchers at CERN released a set of protocols called the World Wide Web, that ran on top of the existing Internet protocols, and allowed for a very flexible and general-purpose access to material stored on the Internet in a variety of formats. It was this feature of access across formats, machines, operating systems, and standards that allowed the Web to become so popular, so rapidly. In addition to developing the Web, Berners-Lee together with his colleagues also developed a program that allowed easy access to the software from a personal computer. The social structures at CERN allowed Berners-Lee to develop his ideas, and the culture of openness allowed CERN's management to make the Web available to all for free. One could argue that the Web could have come from anywhere, and indeed the time was right for such an innovation, yet it is no accident that it ended up coming from CERN. Few areas of human intellect have an ecosystem that combines cutting-edge technology, intellectual rigour, and a culture of openness to the same degree as fundamental research. This spirit allowed CERN to utilise a touchscreen-operated control system for the SPS in the 1970s, and to foster the development of technologies in fields ranging from aerospace to cultural heritage.

The first step in a scientific journey is the birth of a new idea. This fragile, albeit invaluable, moment is what CERN's IdeaSquare (ID2) tries to preserve and nurture by offering the required infrastructure as well as access to a large network of experts. From its birth in 2014, CERN's IdeaSquare has been bringing together researchers, engineers, and people from different organisations, including industry and young students, and encouraging them to come up with new ideas that are useful for society, inspired by CERN's ongoing physics and detector detector R&D projects. ID2 was initially created as a prototype or test-bed for the EU-funded ATTRACT[1] initiative, which aims to create a new type of innovation ecosystem in Europe, built around detection and imaging technologies. ID2 coordinates this initiative, engaging six European research infrastructures and two innovation and entrepreneurship-driven universities and business schools. A dedicated student programme, called Challenge Based Innovation (CBI), was put in place. Inspired by CERN and design thinking (DT), CBI incorporates user aspects or experience into the innovation process at an early stage, involving CERN technologies whenever it is meaningful. In addition to its integration in ATTRACT, several CBI-like student projects run in parallel within ID2 as part of a wider network of inter-disciplinary, DT-driven programmes called Design Factory Global Network (DFGN).

In terms of the innovation value chain, one can argue that IdeaSquare precedes knowledge transfer (KT) activities, as the former connects CERN researchers with inter-disciplinary students working on societal challenges, still at the 'fuzzy front end' of the conceptual phase. Once the initial ideas develop further and the conceptual prototypes become more solid, possibly supported by CERN IP, the KT phase can then take over. Naturally, KT works in other, parallel ways as well: continuously seeking and identifying knowledge and technology transfer opportunities arising from the daily activities in the laboratory. A constant effort is made to map the (new) technologies and know-how developed at CERN (visit https://kt.cern/technology-portfolio), analysing the value that they can represent to external partners and in particular to industry. These partners are the necessary vehicle for technology transfer from CERN to society, and often have a crucial role in understanding how CERN's technologies can solve challenges in other fields of application.

CERN started technology transfer activities in 1988, when the CERN Industry and Technology Liaison Office was founded to stimulate interaction with industry and to assist in issues related to CERN's intellectual property. The field and objectives of the activities have changed over the years, as has the general structure of CERN, and thus the knowledge and technology transfer office at CERN has gone through many iterations before reaching its current form as the CERN Knowledge Transfer Group. An objective that was included in the inception of these activities at CERN was revenue creation, providing extra income. However, even though there is still an objective to gain a fair return on the transfer of knowledge, considerations such as contributing back to society and strengthening CERN's role in society have priority. Today the mission of CERN's Knowledge Transfer Group is to 'maximise the technological and knowledge return to the Member States and promote CERN's image as a centre of excellence for technology'; in addition, the CERN IP policy

[1] The ATTRACT project has received funding from the European Union's Horizon 2020 research and innovation programme under grant agreements No. 101004462 and No. 777222.

(https://kt.cern/activities-services/intellectual-property-management) states that impact having priority over revenue creation is a steering principle for the activities. The current policy at CERN, focusing on maximising impact, makes it clear that using only commercial modes of knowledge transfer, such as licensing and paid consultancy, is not the only path to choose. A main part of the objective of the CERN Knowledge Transfer Group is therefore to evaluate different dissemination methods and select the appropriate mode to fulfil this mission. CERN is therefore increasingly sharing its knowledge through a variety of different commercial and non-commercial modes.

Moreover, CERN is investing in the development of an entrepreneurship culture among its staff members and its collaborators. CERN supports personnel and entrepreneurs in creating spin-off companies. Since 2012, CERN has created a network of ten Business Incubation Centres (BICs), to assist in taking innovative CERN technologies from technical concept to market reality. As of today, about 40 start-ups and spin-offs use CERN technologies, with applications in domains as diverse as biotechnology, medical applications, and material science.

2.4 Procurement activities and industry

Innovation, in any field, calls for significant investments in technological R&D as well as the early involvement of industry. This is much more so in the case of Big Science where the industry's involvement enables the development of intricate prototypes, and makes the large-scale production required by the construction of massive research instruments possible. A long-term vision, of the kind that Big Science projects entail, provides, in turn, ample motivation for the further expansion of technological limits—i.e. beyond what is currently needed for market applications.

The construction of large-scale projects—arguably the LHC presents such an example—helps technologies to mature from a laboratory-prototype to industrial scale production. A key factor in many of CERN's procurement activities is that each new accelerator or upgrade brings about even more challenging requirements and therefore pushes industry to its limits. In the case of the LHC, the large production volumes were an incentive for potential suppliers to invest heavily in both time and resources. These projects also provide significant potential for technological learning and innovation (see chapter 8 by Erkko Autio *et al* in Boisot *et al* 2011). Unlocking this potential hinges on two related issues: first, how the technological specifications of Big Science procurement projects are produced; and, second, how the procurement process itself is organised. CERN, for example, has operated as a 'specification factory', often undertaking significant technological research to create exact specifications for how deliveries should be designed (Unnervik 2009). This preparatory research (i.e. research that preceded the issue of tenders) was often undertaken in collaboration with industrial suppliers and member countries' universities. This approach provided the opportunity for member countries' universities and industrial companies to leverage CERN's R&D resources for the development of technological competence. Large parts of the cost of new large-scale projects translate into opportunities for specific R&D projects, in partnership with industrial and academic partners in a collaborative fashion. In addition, one should also take into account the industrialisation of these technologies and large-scale manufacturing. This is important for a more effective transfer of new knowledge to areas beyond particle physics, such as medical applications or industrial manufacturing processes. Finally, for smaller markets, the industry often hesitates to invest the time and money required, which often leads CERN to designing its own prototypes or working alongside industry for many years to ensure that firms build up the necessary competence and skills.

The various benefits for industry from its involvement with CERN's projects emerge gradually, as each project moves from the initial stage of design and development all the way up to the operations phase. Benefits also occur upstream, in the creation and development of construction and engineering capabilities and supply chain innovations for delivering large and complex scientific facilities and equipment. Furthermore, they extend downstream through the commercialisation and industrial applications of innovative technologies and outcomes that emerge from the frontier of science. Finally, these benefits are also tied to activities undertaken midstream, through the myriad of collaborations and solutions provided by industry that are necessary for the realisation of large-scale experiments. Valuable skills that hold potential for industry are developed at all stages of Big Science projects.

Further benefits stem from the procurement activities of Big Science facilities. The different ways in which these benefits are realised make it rather challenging to quantify and estimate the return-of-investment (ROI) of Big Science projects. Financial multipliers ranging from 1.9 (CERN) to 3.7 (ESA) have been found by previous studies (Autio *et al* 2003), meaning that each euro invested in industry by Big Science generates a two- to four-fold return for society. A study on LHC-related procurement and contracting, published in Autio *et al* (2003), demonstrated clear benefits for CERN suppliers: some 38% had developed new products, 44% had improved technological learning, 60% acquired new customers thanks to the CERN contracts, and all firms questioned stated that they had derived great value from using CERN as a marketing reference. A more recent study by Florio and Sirtori (2016) provides evidence of a positive and statistically significant correlation between LHC procurement and supplier R&D effort, innovation capacity, productivity, and economic performance. Moreover, it documents that this also depends on the nature (middle- or high-tech) of the procurement contracts. Specifically, firms supplying innovative and high-tech products, reported, in turn, greater learning benefits and economic performance.

Beyond the direct procurement activities, lies the less documented effect of the experience that technological firms gain when working with Big Science projects. Contracts with CERN also help industries drive their innovation as well as their business model. The benefits for industries involved in CERN's procurement activities range from technology and product development to organisational changes in the companies, thereby allowing their internationalisation and resulting in a bigger circle of market operation. As Autio *et al* (2014) note:

> benefits vary extensively among suppliers, learning and innovation benefits tend to occur together: one type of learning is positively associated with other types of learning. Technological learning stands out as the main driver. Indeed, 44% of respondents indicated that they had acquired significant technological learning through their contract with CERN.

Suppliers find benefits from their involvement in CERN projects in terms of organisational effects, such as improvements to their manufacturing capability, quality-control systems, and R&D processes. This picture has been recently confirmed by a second independent survey by Castelnovo *et al* (2018), which demonstrates that the benefits from procurement activities spill-over across the entire supply chain to subcontractors too, thus highlighting the extent of CERN's impact on the industrial landscape in CERN's member states. The study revealed that the expected present value of the economic benefits for high-tech suppliers over the entire LHC lifetime amounts to approximately 2 billion EUR (at 2008 prices), covering approximately 15% of the total cost for the LHC construction and operation until 2025. Moreover, the study affirmed that the closer the collaboration between CERN and its suppliers, the better the innovation and learning benefits enjoyed by supplier firms, and the better their future economic performance. In other words, industry benefits both from improving their products and developing new products as well as through expanding their market reach by developing new networks. Linked to the latter are the positive reputational effects enjoyed by most suppliers, including smaller and non-high-tech firms, which improves their business model and helps them to expand in new markets. In other words, CERN provides a prestigious customer reference that firms can use for marketing purposes. As a result of their interaction with CERN, 42% of the respondents have increased their international exposure, and 17% have opened a new market (figure 2.1).

Beyond the raw benefits of procurement activities, be they direct or indirect, recent studies such as those by Zimmermann and Thomä (2019), Florio *et al* (2016), Florio (2019), Bastianin *et al* (2022) have revealed five major impact pathways: training, the value of science projects, industrial spillovers, open software and systems, cultural goods, and the public good value (figure 2.2).

A detailed discussion of these results is presented in the essay by Bastianin and Florio in chapter 16 of this book. As a brief comment, we can conclude this section by noting that the application of the CBA framework in Big Science has also been met with a certain level of scepticism. A previous study by OECD (1994) concluded that

> no rigorous cost/benefit analysis is possible. Research into the fundamental workings of the Universe, such as that performed by particle physics, enriches our culture and helps to stimulate other scientific fields.

In 2016, CERN's former Director General Herwig Schopper (2016) published a warning on the breadth that CBA analysis methods can offer when applied in large-scale scientific projects. He pointed out certain difficulties in integrating costs in the CBA method such as, for example, the costs related to the administration or the

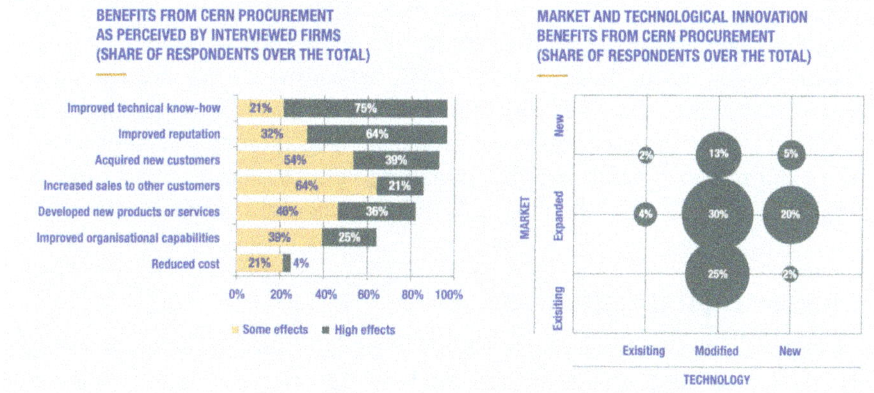

Figure 2.1. Measuring the economic benefits for CERN suppliers revealed the different benefits that firms acquire through their participation in CERN's procurement process. The different types of benefits and their effects are shown in the left figure. All interviewed firms confirmed that collaborating with CERN contributed to several important outcomes, ranging from the acquisition of technical know-how and the development of new products to improved organisational capabilities and reputational effects. The right figure shows the relation between the technological innovation that firms introduced as part of the procurement process in relation to their access to larger market share. The figure shows that the majority of firms modified existing technologies allowing them to expand their present market (30%) or even enter new markets (13%). Only firms for which CERN has a valid e-mail address were invited to the survey. The full CERN procurement database counts some 4200 suppliers with orders larger than 10 000 CHF. Reprinted from Sirtori *et al* (2019).

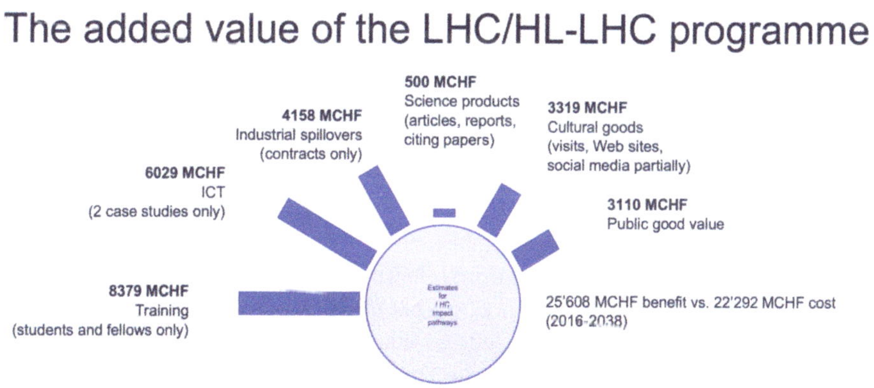

Figure 2.2. The added value of the LHC/HL-LHC programme. Reprinted from Sirtori *et al* (2019).

value of previous experience in developing these projects. Schopper also pointed out cultural elements that can escape the strict scope of a CBA analysis. He notes:

> CERN was founded with the objective to promote science but also to foster the understanding between nations. An evaluation of the social value of a project like the LHC cannot be complete without considering this aspect of 'science for peace'. The election of CERN as an advisory

organisation to the UN is only one recognition of these achievements. Another element is the establishment of a new style of global international cooperation, both by the LHC experiments (in reality large independent projects), but also how CERN as a whole is managed. In a world where the complexity of global actions increases continuously, the exploration of new ways of managing new challenges has a considerable social value independent of CERN's scientific success.

2.5 Human capital formation

Human capital, an intangible yet absolutely essential asset that refers to the value of the experience and skill sets of professionals, is nurtured through a wide range of training programmes at CERN. It manifests itself in a multitude of ways, from scientists devising new algorithms in order to refine their analyses, to developing new materials and designs for electronics to cope with the special requirements of high-energy physics experiments. The open questions and mind-bending concepts explored by particle physics and astronomy also serve to draw bright young minds into science, even if individuals go on to work in other areas. Skills acquired in research laboratories are highly valued both in the academic and non-academic fields and extend beyond the realm of hard (i.e. technical and scientific) skills, to 'soft' ones, that are gained through professionals from different cultures and domains of knowledge via the networking afforded by conferences and workshops.

This fact was confirmed in a study performed in conjunction with students from the LEP's DELPHI experiment (Camporesi 2000). They reported that the private sector was interested in specific skills acquired during this experience, such as the ability to work effectively in large and diverse teams, the drive and motivation to solve complex problems, the exposure to cutting-edge technologies in the electronics and computing domain, and a familiarity with software techniques related to the handling of large quantities of data and the performance of sophisticated modelling.

In 2014, a survey performed by CERN's KT group among past CERN fellows (Bruzzi and Anelli 2014) aimed to measure the 'knowledge transfer' promoted by CERN through its unique international learning environment, in an attempt to address the shortcomings of previous attempts to measure the socio-economic impact of CERN. The study, based on 831 fellows who worked at CERN during the period from 2007 to 2012, identified that among the 288 (35%) who had left CERN:

- 85% had found a new position, 5% had started their own company, 2% had returned to their old employer in a new position, 2% had returned to their old position, and 5% were unemployed since they had ended their fellowship;
- of those currently employed, 41% were in the industry/services sector, 26% in other research organisations, 30% in universities, and 3% in other sectors;
- 73% gave a positive response concerning the importance of the fellowship in being able to secure their current position; and
- 55% reported that they held a position in a different country than the country where they had received their last diploma, indicating that CERN is a strong contributor to international mobility.

The study demonstrates the important role that the skills the fellows acquired by participating in CERN's fellowship programme played in their entrance into the labour market, while also highlighting their high overall rate of satisfaction from their fellowship. According to the authors:

> a link between basic research and the labour market exists, confirming that the investment in human capital made by CERN through its Big Science facilities generates positive externalities that we called as a 'fertilisation effect' to the benefit of all the global economic system.

This fertilisation effect becomes even more apparent when one considers that a large fraction of early-career researchers, including the 2400 PhD students registered at CERN, leave the field of HEP and move onto new roles in the public and private industry, including in the field of entrepreneurship.

More recent surveys among the members of CERN's Alumni Network (Catalano *et al* 2018) confirmed that their stay at CERN offered them a variety of skills, considered to be important in their post-CERN career. These skills are related to software technologies, working in an international group, data analysis, logical thinking, and communication; furthermore, there is a distinct correlation between the diversification of the aforementioned skills and the time spent at CERN. The survey also identified that performing research and acquiring skills at CERN (particularly at the LHC/HL-LHC programme and related experiments) leads to a salary premium of around 5%–13% compared to the normal return to education in the relevant fields; the exact percentage depends on the acquisition of additional competencies—beyond the scope of the technical domain of the project—such as data analysis and software development skills for physicists and engineers. CERN's training programme generates human capital with significantly higher qualifications which, in turn, translates into a salary premium compared to individuals who enter the labour market right after the conclusion of their graduate studies. In their paper, Catalano *et al* (2018) concluded that:

> The salary premium effect is either direct, valuing the additional skills and experience or indirect, if the person enters a position that is characterised by a higher salary because it includes managerial or leadership tasks.

A comparative study of the impact on human capital formation is shown in table 2.1 (Catalano *et al* 2018).

In addition to the fellowship programme and the EU-funded Marie Curie actions, there are many other student programmes. CERN's summer student programme trains nearly 300 students from 70 different countries every year, from both CERN's member and non-member states. The programme combines theoretical lectures and practical work through integration with existing research teams. CERN also runs a programme for technical students, PhD students, and administrative students, as well as various schemes for teaching and training engineers, scientists, and other

Table 2.1. CERN human capital premium: results from primary data.

Estimated salary premium	Data source	Year of the survey
9%–12%	Survey with 384 CERN current and former students and additional estimation	2014–2015
5%–12%	Survey with 384 CERN current and former students	2014–2015
9%–13%	Survey with the 384 original and 54 newly collected questionnaires from the CERN Alumni community	2018
5%–12%	Survey with 322 team leaders	2018

professionals. These programmes all contribute towards knowledge transfer by spinning off human capital from CERN and further highlight the additional benefits generated by the presence of a large-scale research infrastructure which inspires and attracts talent from around the world.

CERN also reaches out to members of the public, teachers, and school students through a variety of programmes. The laboratory welcomes some 100 000 visitors every year, with the role of presenting them with a basic introduction to particle physics and to the mission and research of CERN Building on this tradition, CERN has recently launched the Science Gateway, a new facility for scientific education and outreach. The goal is to create a hub of scientific education and culture that could welcome about 300 000 visitors per year, inspiring them with the beauty of science and informing them of how CERN collaboratively builds new instruments and advances the frontier of human knowledge.

One of the flagship education programmes of CERN is the High School Teacher Programme which features an intensive three-week summer session, funded primarily by CERN, but with some support in individual countries by other organisations. Stipends are granted for travel and living expenses, and on-site housing is provided. The programme accepts applications from individual teachers from CERN Member and Observer States, while there is also some support for teachers from non-member states. Applicants are required to provide some evidence of their motivation. The curriculum—which is presented in English—focuses on particle physics, but with excursions into domains such as cosmology and relevant applied topics such as medical applications and superconductivity. The National Teacher Programmes are held across all the CERN member countries, with instruction given in the local language. CERN provides all of the materials as well as administrative and technical support, with the individual countries being responsible for the travel and expenses of the teachers. Typically, these instructional programmes take place over several days. It is noteworthy that much of the instructional content is created by scientists who are members of the CERN staff or of the experimental teams. In this way, CERN leverages its explicitly international character to provide added value to member countries at no incremental cost. Founded in 2014 at the time of CERN's 60th birthday, the Beamline for Schools competition inspires and trains the youngest researchers in science. Since Beamline for Schools was launched in 2014, almost 10 000 students from 84 countries have participated.

International Master Classes have accounted for a major part of the overall activity of the International Particle Physics Outreach Group (IPPOG) supported by CERN's Education, Communication and Outreach group during the LHC era. Through IPPOG, the masterclasses profit from scientists taking an active role, conveying the fascination of fundamental research and thereby reaching young people. IPPOG offers a reliable and regular discussion forum and information exchange, enabling world-wide participation. All of the LHC collaborations involved acknowledge the potential—and the success—of educational programmes that communicate the important discoveries of the LHC to high school students, by providing large samples of the most recent experimental data. Other physics fields also utilise the masterclasses as a model for physics education and science communication. For example, in the UK, nuclear-physics masterclasses cover nuclear fusion and stellar nucleosynthesis. Astroparticle physics has also joined the masterclass scene. For example, in Germany, the Netzwerk Teilchenwelt hosts masterclasses that use data from the Pierre Auger Observatory to reconstruct cosmic showers or energy spectra, or data on cosmic muons that the students take themselves using Cherenkov or scintillation detectors. Since 2012, students at the Notre Dame Exoplanet Masterclass in the US have used data and tools from the Agent Exoplanet citizen science project run by the Las Cumbres Observatory Global Telescope Network to measure the characteristics of exoplanets from their effects on the light curves of the stars they orbit. New international masterclasses have been established on the search for very high-energy cosmic neutrinos at the IceCube Neutrino Observatory. Like other benefits stemming from Big Science facilities, evaluating the aspect of human capital creation calls for refined methodologies and for the design of data collection systems. This would further advance a cultural approach within the facilities of Big Science but would also allow the definition of new metrics for measuring the impact of Big Science infrastructure on human capital formation.

2.6 Lessons and future steps

In section 2.3 we briefly referred to the follow up of the CBA analysis carried out during 2015–2020 for the LHC programme. In 2020, the CERN Council announced that it has updated the strategy that will guide the future of particle physics in Europe within the global particle physics landscape. It highlights the need to pursue an electron–positron collider acting as a 'Higgs factory' as the highest-priority facility after the LHC while also stating that Europe, in collaboration with the world-wide community, should undertake a technical and financial feasibility study for a next-generation hadron collider at the highest achievable energy, with an electron–positron collider as a possible first stage. To address this challenge, CERN has launched the FCC feasibility study. As part of this effort, ongoing studies aim to understand the potential impact of the FCC and ways to maximise it.

Together with academic partners, CERN has launched a number of studies including the regional industrial development potentials for superconducting radio-frequency development and production. This study also benefits from previous data from the construction of Big Science projects such as the European XFEL and will explore the generation of regional impacts. In addition, it seeks to understand the potential impact on local employment as well as the study of local development opportunities and synergies. Furthermore, in line with the principles of a circular economy and to address the question of sustainability, the FCC collaboration, with the support of the EU-funded FCC Innovation Study (FCCIS), has launched the international competition 'Mining for the Future' seeking potential solutions for the reuse of molasse excavation materials. The goal is to find solutions that can be applied in sectors beyond particle physics and estimate the industrial added value of reusing excavated materials that are encountered in other construction projects. Other areas of interest when assessing the socio-economic impact of a post-LHC collider lie in understanding the value of open software developed at CERN (Indico, Zenodo) and used world-wide in areas beyond the focus of a Big Science project. Finally, an important aspect is the economic impact generated by the number of scientists and tourists to a large-scale facility and the surrounding visiting points. Work is needed in the future to understand the ways in which impact is created as well as designing a sustainable model for science tourism in the area.

The work carried out for the socio-economic impact investment of the LHC established a methodology that led to socio-economic impact analysis as a repeatable process offering a baseline on which monitoring and continuous improvement can be built. In addition, it revealed certain challenges for providing an accurate evaluation of Big Science's impact related to the available data and resources as well as the establishment of mechanisms for data gathering. Furthermore, one has to acknowledge the challenges related to the implementation of a CBA analysis and specifically the selection of data for assessing the socio-economic impact of Big Science facilities. A second difficulty in the implementation of a CBA concerns the definition of a counterfactual scenario. The challenges of implementing a CBA analysis could result in very conservative estimates or fairly large uncertainties. Despite its challenges, CBA offers a solid way to assess the impact of Big Science facilities often in line with the results from different approaches. Further refining of the CBA tools along with complementary approaches could perhaps offer a more holistic understanding of the socio-economic impact of Big Science facilities. Finally, an important open question revolves around finding ways of quantifying the socio-economic impact of the new knowledge created in such Big Science projects—with the broader epistemological horizon that they offer.

Beyond efforts to develop new tools for studying the different aspects of the socio-economic impact generated by Big Science projects at CERN, it is also important to refer to the shifting role of Big Science projects as spaces for co-innovation with the industry. Based on previous lessons, a new role for the relation between industry and Big Science facilities is possible. According to Bozeman (2000) a new 'cooperative technology' model of technology transfer emerges in which public funded laboratories and research infrastructures play an important role in technology innovation and economic growth.

In this setting, CERN is a fecund ground for innovation, with concrete examples of international knowledge transfer. CERN's knowledge transfer activities are shifting from technology push to market pull with the main focus on five areas in which CERN's expertise and technology competence can have an impact. The areas identified are: environment and energy, medical, digital, quantum, and aerospace.

The new strategy also focuses on open innovation and aims to maximise the dissemination of our technologies and know-how. This means we aim to increase value for all, rather than focus on revenue. In that context, patent volume is not a relevant measure of successful intellectual property management. Since we focus on dissemination, when patents are the best tool for that to happen, we use them, but there is a wide range of intellectual property tools useful for knowledge transfer. Instead, the CERN Knowledge Transfer group measures the number of new technology disclosures (14 in 2021), number of contracts and agreements with external partners and industry (47 in 2021, and more than 400 since 2000). Moreover, six EU-funded projects with a strong knowledge transfer component started in 2021, namely ATTRACT2, AIDAinova, HITRIplus, I.FAST, PRISMAP, and RADNEXT. We monitor how our culture of collaboration and entrepreneurship translates into spin-off and start-up companies—there are currently about 40 using CERN technology, some hosted directly in CERN's network of BICs. Together with the impressive breadth of application fields of CERN technologies, we believe these are clearer measures of successful impact.

The recently launched EU-funded ATTRACT project, coordinated by CERN, is scaling up this approach and tries to implement a new innovation ecosystem in the field of detection and imaging technologies. Imaging and detection technologies could have a core function in almost all technically sophisticated commercial products representing an estimated annual market of over $100 billion (Frost & Sullivan 2015). These domains include, among others, medical device and imaging technology, advanced manufacturing, automation, microelectronics, materials and coatings, environment and sustainability, and information and communication technology. ATTRACT is designed to directly address the ineffectual transition—or disconnection—between the technology-push mechanisms and the market-pull instruments in many ways (Auerswald and Branscomb 2003).

In this respect, ATTRACT represents a departing moment from previous funding instruments, such as the Future and Emerging Technologies Programme (FET). In the case of ATTRACT, both research infrastructures and industrial players constitute focal actors with equal weights. The adopted approach also takes into account existing relationships between research infrastructures and their industrial suppliers; in particular, the highly specialised small and medium-sized enterprises that contribute to the engineering, construction, and operation of some of the world's most sophisticated technologies. Specifically, for projects involving European research facilities and industrial organisations, the most immediate use of their technologies is in sight. In this sense, a first 'internal market' can be imminent. This 'internal market' paves the way for industry to target other applications and new commercial opportunities.

The approach described above is consistent with recent calls by Mazzucato (2017), arguing that the government can go beyond its role as a regulator or fixer of markets towards an entrepreneurial role, absorbing the risks in strategic sectors until technologies have reached a sufficiently mature state to be attractive to private and venture capital. ATTRACT sets an example of how to accomplish this coordination task. Indeed, it can be perceived 'an experiment in innovation policy' that could be adopted in other fields for the development of large-scale scientific projects that can profit by acknowledging the novel constellation of actors, resources, design, and governance as well as it providing novel channels for cross-fertilisation of technology and entrepreneurial options.

2.7 Conclusion

Looking at just the financial contributions to CERN from member states in 2018, for example, more than 40% of the CERN budget was returned to industry in its member state countries through the procurement of supplies and services, generating corollary benefits such as opening new markets.

The overview presented shows how CERN actively disseminates its knowledge in an attempt to consistently maximise the benefits that it brings to society. The grand ideas and open questions in particle physics and cosmology already inspire many young people to pursue STEM studies, while several related technological spin-offs, including—but not limited to—medical treatments, Big Data handling, and radio-frequency technology have entered the sphere of public discourse. Less well known however, are the significant, but harder-to-quantify, economic benefits of Big Science. Evaluating and putting in place measures on how to manage, measure and stimulate this diffusion of knowledge remains an open challenge.

A number of previous studies highlight the benefits for early-stage researchers as the single most important societal benefit. CERN could invest more in activities facilitating the transition to the international job market. Similarly, cooperative relations with suppliers of technology and the development of innovative software, data storage, networking, and computing solutions are strategic levers that CERN could utilise to boost its social benefits. Finally, cultural effects, in particular those related to on-site visitors and social media, have great potential for generating societal benefits, hence outreach and communication strategies are important.

Studies have quantified the impact of CERN's activities in a number of areas ranging from collaboration with industry on R&D activities, procurement actions, training programmes, and educational initiatives among others. A recent cost–benefit analysis of the LHC, conducted by the University of Milan (Bastianin and Florio 2018), concluded with 92% probability that benefits exceed costs, even when attaching the very conservative figure of zero to the value of the organisation's scientific discoveries. The Milan group conservatively estimates for the HL-LHC that every €1 of costs returns €1.2 to society, while a similar study concerning the FCC estimates the benefit/cost ratio to be even higher, at 1.8.

However, as discussed in this chapter, evaluating the scientific impact of Big Science facilities calls for a new methodological approach. This should acknowledge the diversity of research programmes resulting from Big Science projects as well as the cross-fertilisation of technologies and know-how between Big Science and smaller projects. CERN's founding convention states that knowledge dissemination is part of its *raison d'être*. Beyond its core scientific missions, CERN works to generate a path for sharing the knowledge created during the design, construction, and operation phases of its projects.

Future research should focus on creating metrics for evaluating and comparing the impact of all of these different modes of knowledge transfer. The design, construction and use of Big Science facilities span decades, and funds must be provided in advance for their maintenance, operation, and eventual decommissioning. In parallel, a well-designed data collection mechanism should be integrated in the design of a research infrastructure to evaluate its socio-economic impact and identify ways to further increase the resultant benefits for society.

In addition to the above, it should be noted that there is an impressive list of SMEs in member states built on CERN know-how, underlining the success of spin-out from CERN Finally, being a member state of CERN will generally enhance the country's skilled workforce and brings many benefits for teaching and education.

References

Auerswald P E and Branscomb L M 2003 Valleys of death and Darwinian Seas: financing the invention to innovation transition in the United States *J. Technol. Transfer* **28** 227–39

Autio E 2014 Innovation from Big Science: Enhancing Big Science impact agenda (Imperial College London) https://dera.ioe.ac.uk/19649/1/bis-14-618-innovation-from-big-science-enhancing-big-science-impact-agenda.pdf

Autio E, Bianchi-Streit M and Hameri A P 2003 Technology transfer and technological learning through CERN's procurement activity, CERN - Education and Technology Transfer Division Geneva https://cds.cern.ch/record/680242/files/CERN-2003-005.pdf

Autio E, Kenney M, Mustar P, Siegel D and Wright M 2014 Entrepreneurial innovation: the importance of context *Res. Policy* **43** 1097–108

Bastianin A and Florio M 2018 Social cost–benefit analysis of HL-LHC *FCC Document Report* No. CERN-ACC-2018-0014 CERN

Bastianin A *et al* 2022 Big science and innovation: gestation lag from procurement to patents for CERN suppliers *J. Technol. Transf.* **47** 531–55

Boisot M, Nordberg M, Yami S and Nicquevert B 2011 *Collisions and Collaboration: The Organization of Learning in the ATLAS Experiment at the LHC* (Oxford: Oxford University Press) ch 8

Bozeman B 2000 Technology transfer and public policy: a review of research and theory *Res. Policy* **20** 627–55

Bozeman B 2002 Public-value failure: when efficient markets may not do *Public Adm. Rev.* **62** 145–61

Bozeman B and Crow M 2021 *Public Values Leadership: Striving to Achieve Democratic Ideal* (Baltimore, MD: Johns Hopkins University Press)

Bruzzi S and Anelli G 2014 Basic research, knowledge transfer and labor market: evidence from CERN's fellowship programs *EuroAtlantic Union Rev.* **115** 12

Bush V 1990 *Science: The Endless Frontier. A Report to the President* (Washington, DC: National Science Foundation) p 16 (reprint of 1945 original)

Camporesi T 2000 Physics acts as a careers stepping stone *CERN Courier* 21 September

Castelnovo P, Florio M, Forte S, Rossi L and Sirtori E 2018 The economic impact of technological procurement for large-scale research infrastructures: evidence from the large Hadron Collider at CERN *Res. Policy* **47** 1853–67

Catalano G *et al* 2018 The Value of Human Capital Formation at CERN *Future Circular Collider Publication* CERN-ACC-2018-0025 https://cds.cern.ch/record/2635864/files/CERN-ACC-2018-0025.pdf

CERN 2021 CernVM File System http://cernvm.cern.ch/fs

Compton A H 1965 *Arthur Holly Compton Papers* (Chicago, IL: Hanna Holborn Gray Special Collections Research Center)

Copeland D 2016 Science diplomacy *The Sage Handbook of Diplomacy* ed C M Constantinou, P Kerr and P Sharp (Thousand Oaks, CA: Sage), pp 628–41

de Solla Price D J 1986 *Little Science, Big Science … and Beyond* (New York: Columbia University Press)

Dyson F 1988 Six cautionary tales for scientists *From Eros to Gaia* (New York: Pantheon) pp 11–28

Engelen J 2012 The Large Hadron Collider project: organizational and financial aspects (of physics at the terascale) *Philos. Trans. Royal Soc.* A **370** 978–85

Florio M 2019 *Investing in Science. Social Cost–Benefit Analysis of Research Infrastructures* (Cambridge, MA: MIT Press)

Florio M, Forte S, Pancotti C, Sirtori E and Vignetti S 2016 Exploring cost–benefit analysis of research, development and innovation infrastructures: an evaluation framework arXiv: 1603.03654

Florio M and Sirtori E 2016 Social benefits and costs of large scale research infrastructures *Technol. Forecast. Soc. Change* **112** 65–78

Frost & Sullivan 2015 Sensor technologies that will have the highest impact in 2015 https://store.frost.com/2015-top-technologies-in-sensors-control-technical-insights-19841.html

Galison P and Hevly B (ed) 1992 *Big Science: The Growth of Large-Scale Research* (Stanford, CA: Stanford University Press)

Giudice G F 2012 Big Science and the Large Hadron Collider *Phys. Perspect.* **14** 95–112

Herrmann A, Weiss L, Pestre D, Mersits U and Krige J (ed) 1990 Building and running the laboratory, 1954–1965 *History of CERN* vol 2 (Amsterdam: North-Holland)

Höne J and Korbalija J 2018 Accelerating basic science in an intergovernmental framework: learning from CERN's science diplomacy *Glob. Policy* **9** 67–72

Krige G J 1996 *History of CERN* vol 3 (Amsterdam: Elsevier)

Leyden D P and Menter M 2018 The legacy and promise of Vannevar Bush: rethinking the model of innovation and the role of public policy *Econ. Innov. New Technol.* **27** 225–42

Martin B R 1992 'Big history' for big science? Critical review of history of CERN *Scientometrics* **24** 371

Mazzucato M 2017 Mission-oriented innovation policy: challenges and opportunities *Working Paper Series* IIPP 2017-1 UCL Institute for Innovation and Public Purpose https://ucl.ac.uk/bartlett/public-purpose/sites/public-purpose/files/moip-challenges-and-opportunities-working-paper-2017-1.pdf

Mazzucato M 2018 *The Entrepreneurial State* (London: Penguin)

OECD 1994 Symposium on International Technology Co-operation *Working Group on Innovation and Technology Policy* OCDE/GD(94)88 Organisation for Economic Co-operation and Development https://oecd.org/officialdocuments/publicdisplaydocumentpdf/?cote=OCDE/GD(94)88&docLanguage=En

OECD 2006 *Conf. on Social Innovation in the 21st Century: A Dialogue (New South Wales, Australia, 28–30 April)* https://www.oecd.org/cfe/leed/conferenceonsocialinnovationin-the21stcenturyadialoguenewsouthwalesaustralia28-30april2006.htm

OECD 2014 *Report on the Impacts of Large Research Infrastructures on Economic Innovation and on Society: Case Studies at CERN* (Paris: OECD)

Quevedo F 2013 The importance of international research institutions for science diplomacy *Science Diplomacy* 7 January https://sciencediplomacy.org/perspective/2013/importance-international-research-institutions-for-science-diplomacy

Ruffini P-B 2017 *Science and Diplomacy. A New Dimension of International Relations* (New York: Springer)

Schopper H 2009 *LEP—The Lord of the Rings at CERN 1980–2000* (Heidelberg: Springer)

Schopper H 2016 Some remarks concerning the cost/benefit analysis applied to LHC at CERN *Technol. Forecast. Soc. Change* **112** 54–64

Sirtori E, Catalano G, Giffoni F, Pancotti C, Caputo A and Florio M 2019 Impact of CERN procurement actions on industry : 28 illustrative success stories (Geneva: CERN) http://cds.cern.ch/record/2670056?ln=en

Stein J A 2002 Science, technology and European foreign policy: European integration, global interaction *Sci. Public Policy* **29** 463–77

Unnervik A 2009 Lessons in Big Science management and contracting *The Large Hadron Collider: A Marvel of Technology* ed L R Evans (Lausanne: EPFL Press) pp 38–56

Weinberg A 1967 *Reflections on Big Science* (Cambridge, MA: MIT Press)

Zimmermann V and Thomä J 2019 Interactive learning or R&D: how do small and medium-sized enterprises generate innovations? *Focus on Economics* No. 264 KfW Research 28 August

IOP Publishing

Big Science in the 21st Century
Economic and societal impacts
Panagiotis Charitos, Theodore Arabatzis, Harry Cliff, Günther Dissertori, Juliette Forneris and Jason Li-Ying

Chapter 3

Assessing the socio-economic impacts of Big Science: case studies from the science and exploration programmes of the European Space Agency (ESA)

Charlotte Mathieu, Laurent Laurich and Stéphanie Willekens

The development of space science and research is one of the European Space Agency's (ESA) cornerstone activities. Science has a particular place in the agency, as the main mandatory programme for its member states. Scientific purposes are also pursued in Earth observation missions as well as through ESA's space exploration programme which involves a wide variety of scientific communities. This chapter aims to highlight the increasing importance for ESA and its member states to assess the socio-economic benefits brought by public investment in space science and exploration. First, the methodological approaches developed specifically for ESA science and exploration programmes are outlined, presenting the list of socio-economic impact indicators and the methodology used to assess them either quantitatively or qualitatively. Second, a deep-dive is performed through the results of two socio-economic impact assessments: the first on a sample of eight ESA scientific missions and the second on the agency's investment in the International Space Station (ISS). The studies were conducted by ESA in the framework of its council meeting at the ministerial level in 2019 and 2016, respectively. The large array of economic, strategic, scientific, and social benefits stemming from the programmes are presented. Finally, the chapter concludes by emphasizing the importance of investing in research and innovation, while presenting upcoming ESA Big Science projects.

doi:10.1088/978-0-7503-3631-4ch3

3.1 Introduction

The European Space Agency (ESA) is an intergovernmental organisation established in the 1970s which now has 22 member states. Its purpose is to promote cooperation between member states, in space research and technology as well as in space applications. The agency's budget was €6.5 billion in 2021, with almost €2 billion coming from the budget of the European Union, for the Galileo (navigation) and Copernicus (Earth observation) programmes and from EUMETSAT[1] for meteorological satellites. The rest of the budget comes directly from the contributions of the member states.

ESA is one of the few space agencies in the world to combine responsibility in nearly all areas of space activity. Space science is a mandatory programme, i.e. all member states contribute to it according to their GDP[2]. All other programmes including human and robotic exploration are optional[3], funded on a voluntary basis by participating states. Science and human spaceflight and exploration together account for about 20% of the budget.

Space is more and more part of our daily lives and has become an engine of economic growth and innovation with impacts that go well beyond its industrial sector. ESA's member states wish to ensure maximum benefits to the economy and society from their investments in space. The European Space Agency has therefore been measuring the impact of its activities on the European space industry as well as on the European economy and society since the 1990s, providing evidence that investing in space creates jobs and supports a competitive European economy, while developing strategic tools to implement sovereign policies.

Over the past few years, ESA has been increasingly conducting socio-economic impact assessment studies of its programmes to inform decision-makers. The agency has also been putting significant effort in further consolidating and improving the methodological approach to conduct these so-called socio-economic impact assessments, continuously gathering best practices from the experience of member states and relevant international bodies such as the Organisation for Economic Co-operation and Development (OECD) to find the best approaches applicable to the wide variety of its space activities.

In 2016 [1] and 2019 [2], [3] [4], ESA conducted socio-economic impact assessments of its participation in the International Space Station (ISS) programme, its exploration programme and its science programme, respectively, to provide ESA's member states with evidence of their generated impacts. These studies were conducted using specific methodological approaches to estimate the various benefits deriving from the two programmes since their launch, over 20 years ago.

The methodological approaches and the studies' results are introduced in the sections below.

---

1 European Organisation for the Exploitation of Meteorological Satellite.

2 Gross domestic product.

3 The eight Science missions assessed were Hubble, SOHO, XMM-Newton, Rosetta, JWST, SMILE, JUICE, and ARIEL.

4 Studies available on https://space-economy.esa.int/restricted-area. Complete electronic library accessible upon free registration.

The present chapter has the editorial choice to highlight the socio-economic benefits generated by the science and exploration programmes of ESA. A very complementary example is provided in the essay 'Space technology entrepreneurship drives societal and economic benefits' in chapter 20, which focuses on the ESA Business Incubation Centres (BIC) network and its impact as an international powerhouse for innovative space entrepreneurship.

3.2 Methodological approach

3.2.1 Objectives and principles

The purpose of socio-economic impact assessment studies is to identify and evaluate either ex-post or ex-ante the impacts (positive or negative) of the investment decision taken by Member States in ESA programmes. The methodological approach developed by ESA for socio-economic analysis of its programmes has relied on extensive consultation and continuous exchange with both national space agencies in ESA member states and reference institutions such as the OECD. The approach covers:

- The types of benefit indicators to be measured (e.g. revenue generation, job creation, publications in peer-reviewed journals, support to students take-up of STEM subjects).
- The methodologies to be applied (e.g. cost–benefit analysis, economic modelling, multi-criteria analysis).
- The data collection approach (identifying both the data sources and the way to collect data).

3.2.2 Understanding the specifics of science and exploration

Representing an essential part of ESA's mandate and activities, investments in science and exploration have been the subject of several economic analyses conducted by the agency. Public spending in science and exploration missions has been demonstrated to generate a wide variety of benefits to both the economy and society, from scientific knowledge production to technological demonstration, from industry competitiveness to inspiration.

The agency recognises the need to tailor the methodology applied for measuring the positive effects of science and exploration programmes to the specific nature of such activities, considering the complete value chain of the programmes.

For example, the benefits of a science mission need to be evaluated across the multiple steps required for its end-to-end implementation which include (a non-exhaustive list):

- Preliminary technology development activities required for a given mission.
- The design, development and integration of the mission elements, including scientific payloads and instruments.
- Scientific objectives definition.
- The involvement of the relevant scientific communities.
- The potential technology transfers and spill-overs further generated by the mission.

In addition to the diversity of the activities conducted and the stakeholders involved, both science and exploration missions have particularly long lead cycles, increasing the challenges for the benefit quantification.

Finally, science and exploration programmes generally require significant budgets to meet the challenging requirements of state-of-the-art technological innovation underlying astronomy, planetary science, astrophysics research, and exploration. Many space science and space exploration missions are implemented through international partnerships, involving different space agencies, and financial contributions from various sources and of various natures (e.g. the European share in the common operating costs for the International Space Station are offset via the provision of European Service Modules for the NASA Artemis missions). The multilateral nature of such missions requires specific methodological attention to identify and isolate the benefits from ESA's participation (question of attribution).

3.2.3 Pragmatic approach and further investigation

Past studies measuring the benefits of ESA member states' investment in Science and Exploration have adopted a progressive approach, starting from *traditional* and *space-specific* benefit indicators and further investigating the inclusion of more challenging types of benefits to capture the specifics of science and exploration activities and their impacts on the economy and society.

The so-called *traditional* and *space-specific* benefit indicators have been identified and selected for the agency's socio-economic impact assessment studies through:

- A thorough literature review of authoritative references in socio-economic impact assessments such as the European Commission Impact Assessment Guidelines (2009) which includes, for example, the generation of revenues and related gross value added (GVA), the creation of jobs, the support to small and medium enterprises (SMEs), and the generation of intellectual property.
- The identification of space-specific indicators developed for space programmes in particular, to account for benefits from technological advancements (e.g. an increase of the technology readiness level, achievement of in-orbit demonstration or in-orbit validation), or strategic benefits such as European non-dependence (ensuring e.g. European capabilities in critical and highly innovative technological areas).

In addition to this basis, further research has been (and keeps on being) performed in identifying complementary indicators and approaches to capture a greater portion of the value generated by space science and exploration missions. For example in terms of scientific benefits, this effort includes more sophisticated analysis of publication volume and quality, knowledge transfer, or the cooperation of international scientists. The impact in terms of public awareness and inspiration is also being investigated better, tracking benefits such as the number of related PhD theses or using student awareness surveys.

Table 3.1 presents a summary description of the scientific, strategic, societal, and economic impact categories used in a previous socio-economic impact assessment of the agency's science programme, which are then further presented in section 3.3.

Table 3.1. Socio-economic impact categories and indicators from ESA science mission assessment [4].

Impact category	Summary description from previous socio-economic impact assessment of the agency's science programme
Scientific impacts	The scientific impacts of ESA's science programme were evaluated through indicators such as the *scientific production volume and quality* which provide information on the *number of refereed papers*, but also the *number of paper citations, prizes*, and *nominations*. The *interest from the scientific community* towards ESA science missions was also measured by estimating the *size of said community* and their *accessibility to the missions' data*. Science is impactful when its results and knowledge can be shared, understood, and disseminated beyond the research framework. For this reason, *knowledge transfer* was assessed towards both the scientific communities at large and industry. Finally, due to the high involvement of international scientific communities in such large scale science projects, it was also relevant to measure the *level of international cooperation between scientists*.
Strategic impacts	Strategic impact indicators from ESA science programme focused on the *increase in international collaboration* between stakeholders (e.g. space agencies, governments, industry) identifying *future collaborations generated thanks to the science missions*. They also contributed to reinforcing European non-dependency, by developing critical technologies or capabilities (e.g. that had so far been imported from outside Europe).
Societal impacts	Societal impacts were estimated by assessing the *outreach and awareness* related to the science missions, and how they inspire the younger and future generations through education. In addition, and while some science mission discoveries happen in outer space, their results may *contribute to better preserve our environment*, which is why *scientific contributions to space weather programmes and environment-friendly spin-offs* were relevant indicators to observe.
Economic impacts	Economic impacts consider traditional indicators such as the *generation of gross value added (contribution to GDP)*, the *change in employment* and the interest in related jobs as well as the economic benefits brought by the level innovation and technology transfers through *spin-offs and spill-overs* (also identified as technological impacts).

Future studies will be an opportunity to continue improving and extending the methodological approach applied by ESA to measure the socio-economic benefits of its programmes, including the science and exploration programmes. The agency will continue to rely on the exchanges with an ever-growing network of institutional partners and research centres, such as CERN, who are willing to demonstrate and fully capture the value generated by Big Science.

3.3 A closer look at the socio-economic impacts of the science and exploration programme

3.3.1 The socio-economic impacts of ESA's exploration programme

Space exploration is about science, knowledge, understanding mankind's place in the Universe, and expanding human presence into the Solar System. It is about mankind's origin and future, but also about technology and engineering applicable to earthly issues. Scientific knowledge is on one hand the result of exploration but on the other hand also an enabler for human exploration. Answers to these questions and challenges are so important that multilateral cooperation between space agencies world-wide has become a standard in this field, in parallel with international competition between both public and private actors. Human and robotic exploration have always been great sources of inspiration for our generation, and will keep attracting generations to come. ESA's human and robotic exploration (HRE) programme is an optional programme[5] and represents 10% (€672 million) of the ESA budget in 2021.

The European Space Exploration Envelope Programme (E3P), also called *Terrae Novae*, secures Europe's central role in global space exploration and aims to deliver new results in both basic and applied science and to offer a compelling vision of global endeavour enriching society and inspiring the coming generations. It comprises work on exciting missions serving the three European exploration destinations: low Earth orbit (LEO), the Moon, and Mars. Activities and missions, operational or under development include the following:

- European astronauts continue to perform science on the International Space Station, complemented with innovative microgravity science on non-ISS platforms and space environment analogues.

[5] ESA's programmes fall into two categories—'mandatory' and 'optional'. All member states contribute to mandatory programmes on a scale based on their gross national product (GNP). The other programmes, known as 'optional', are only of interest to some member states, who are free to decide on their level of involvement. Optional programmes cover areas such as Earth observation, telecommunications, satellite navigation, and space transportation. Similarly, the ISS and microgravity research are financed by optional contributions.

- The European Service Modules (ESMs) providing NASA's Orion spacecraft its main power source and propulsion mechanism. Europe is responsible for a critical component to get humans travelling to the lunar vicinity, for the first time since the Apollo programme.
- Contribution to the NASA led Lunar Gateway (a small space station around the Moon also supporting human and robotic exploration of the lunar surface), ESA leading the International Habitation module and providing critical technologies (communication, refuelling, etc).
- Sustained operation of the ExoMars Trace Gas Orbiter (TGO), exploring the atmosphere of Mars to understand its habitability, and providing a vital telecommunications relay capability for future Mars missions of both the ESA and other agencies.
- Contribution to the NASA–ESA joint Mars Sample Return campaign, the most advanced robotic planetary exploration campaign so far conceived. It is composed of several missions that will collect Martian samples, send them into Martian orbit and bring them back to Earth.
- Activities aimed to develop future exploration technology and new projects in human and robotic exploration activities.
- Commercial partnerships will support European industry to step up commercial exploration services and secure their market share.

In order to support policy decision making and to communicate towards the participating ESA member states, the wider stakeholder community and the general public, the agency launched its two aforementioned studies on the European contribution to the International Space Station [5] (including activities such as cargo transport using automated transfer vehicles (ATVs)) and the European Exploration Envelope Programme Exploration Roadmap [6].

In addition to these two studies, it is worth noting that ESA is running a benefit management process to monitor, optimise, and communicate on the impacts of the European Exploration Envelope Programme (E3P).

The objective of the 'Assessment of the socio-economic impacts of the ESA participation to the ISS Programme' study from 2016 was to estimate the benefits of ESA's participation to the ISS programme for ESA participating member states, from 1995–2015.

The 'Benefits of the ESA Exploration Roadmap in Socioeconomics (BEERS)' study was performed in 2019 and focused on evaluating the impacts deriving from selected E3P missions since their respective launch, and in the future with a five-year projection (2020–25).

The benefits and impacts to be presented for both studies include:

- Scientific impacts.
- Technological impacts.
- Societal and inspirational impacts.
- Strategic impacts and international cooperation.
- Economic impacts, including contribution to GDP and employment.

3.3.1.1 Scientific impacts

One of the main reasons to participate in the ISS programme was the execution of research in microgravity (μg) conditions. In fact, the reduced gravity conditions in the ISS enable the observation of phenomena that cannot be observed on Earth due the stronger gravity force, yet produce effects relevant on human beings, flora, fauna, and even industrial processes.

Over the 2001–17 period, around 900 research experiments were conducted on the ISS. As shown in figure 3.1, the annual number of European experiments executed on the ISS has increased considerably over 15 years.

The step-increase in 2008 was due to the integration of the European 'Columbus' Laboratory module on the ISS the same year. It marked a radical change in terms of experiment locations, by hosting about 50% of the total number of experiments, the other half being hosted by the US laboratory 'Destiny' (34%) and the Russian segment (16%).

The data collected by ESA on European investigations performed on the ISS also indicates that European research activities are manifold and diverse, with life science representing 47% of the European research activities performed on board the ISS between 2001 and 2017, including human physiology (31%) and biology (16%).

The life science research fields are very important as they give the possibility to conduct experiments with the direct support of astronauts in space, resulting in many different applications both in space and on Earth. These research activities contribute to:

- Supporting the preparation of future exploration missions, through a better understanding of how the human body adapts to a microgravity environment.
- Developing new treatments for diseases and improving the quality of life through the understanding of the basic mechanisms regulating human physiology.

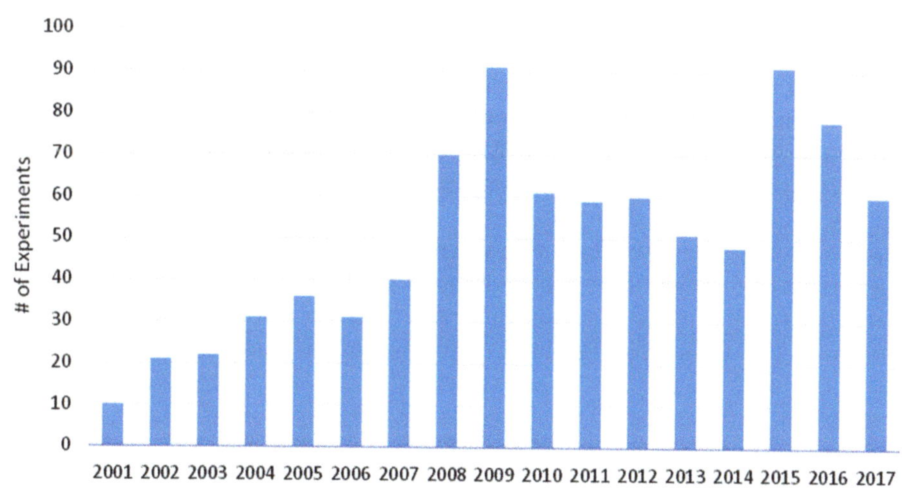

Figure 3.1. Distribution of the European led experiments on the ISS implemented from 2001 to 2017. Adapted from [7].

The remaining 53% of European research activities include:

- Technology demonstration (19%).
- Physical science (13%), in particular material science and fluid physics.
- Space science (11%), in particular radiation physics, and solar and plasma physics.
- Educational activities (8%).

The conducting of these experiments on the ISS involved a large number of European researchers. In fact, ESA fundamental and applied research on the ISS involved around 340 team coordinators with an additional 1900 researchers involved in different experiments, thus enhancing collaboration among the European and global science communities.

Regarding the robotic exploration missions, ExoMars 2016 and 2020 are creating long-term benefits in scientific fields such as life-detection, water detection, geology, and climate evolution. As of 2020, over 2800 scientific publications mentioned the Exomars mission.

The ESM-1 mission will support the lunar mission and create broad benefits for advancing geology, Solar System science, new materials, the origin of the Moon, and its impact upon Earth.

3.3.1.2 Technological impacts

The collaborations between European researchers and global science communities generate benefits along the entire research value chain, including the learning and exchange of best practices for space projects as well as the development and spin-off of innovative technologies.

During the first step of the research value chain, the stakeholders need to define the objective of the research activities by exchanging with scientific and industrial stakeholders, the ESA, and national space agencies. This collaboration helps create human capital and knowledge transfer. Once the research objectives are determined, the experimental facility must be designed in order to fulfil the scientific and operational requirements for the research. This process leads to new technology creation and technology transfers, such as the miniaturisation of components, the maximisation of the re-use of resources (e.g. power, water, and oxygen), and the reduction of waste. Many involved SMEs were able to leverage this experience by commercialising spin-off solutions on Earth, for commercial or humanitarian purposes.

Spin-off solutions are very common when it comes to ISS-developed technologies. In fact, most of these technologies have found purposes in other space projects and attracted new investments towards European space companies. For example, the European ATV[6] technology has built confidence in the European industry to such

[6] From 2008 to 2015, Europe's ATV has been indispensable to the International Space Station, delivering equipment and spare parts as well as food, air, and water for the crew. ATVs delivered more than 31 500 kg of supplies but it also boosted the ISS to raise its orbit numerous times and similarly moved it out of the way of space debris.

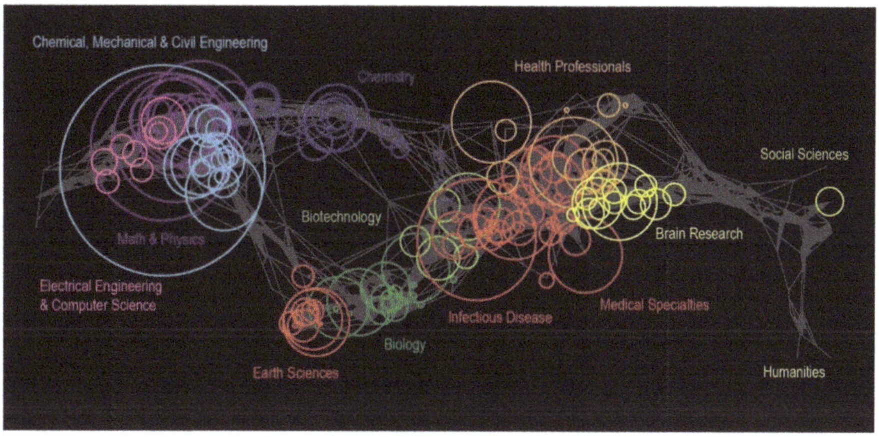

Figure 3.2. ISS map of science. (Source: ESA.) Reprinted from [9].

an extent that NASA trusted ESA to provide the ESMs for Orion. In fact, the European space industry will also supply 50% of the Lunar Gateway elements.

In figure 3.2 the very broad range and research focus of the ISS can be seen on the map of science, mapping the twelve different scientific disciplines in which ISS research has published scientific articles. In 2019, for example, 2571 publications were identified, representing the work of over 4000 scientists, including 1380 unique European science participants and 268 science coordinators representing over 1000 ESA ISS experiments. This confirms the large array of technological fields benefitting from research on the ISS.

Regarding the ExoMars mission, the development of the Schiaparelli lander enhanced the technological expertise of ESA member states and European industrial partners and paved the way to testing and evaluating technologies such as entry descent, landing systems, sensors, and scientific instruments. This opportunity is very valuable as these new technologies could be used in future Mars missions and collect data that could improve models of the Martian atmosphere.

3.3.1.3 Societal, inspirational, and educational impacts

The presence of astronauts in space has always fascinated people and, consequently, raised interest in spaceflight and STEM disciplines. This outreach has been drastically improved with the increasing use of social media in the past decade, creating new opportunities and means of communication for ESA, and giving citizens a direct exchange platform with the astronauts on board the ISS. For this reason, the inspirational value created by the ISS is very high.

An analysis of social media, introduced alongside the more traditional communication tools, shows that astronauts and ISS catch the interest of users more than any other ESA space programme. The number of video views on ESA's YouTube channel were analysed, collecting all videos posted by ESA related to all types of ESA missions from human spaceflight to planetary exploration. Among the 180 million YouTube views, the most popular videos posted by ESA were ISS related, demonstrating the importance of the astronauts' outreach activities and the level of attractiveness of human spaceflight activities.

An analysis of ESA's Facebook page and Twitter account was also carried out, revealing a strong interest for ESA space activities, with respectively 4 million friends and 6 million followers.

ESA astronauts also receive a lot of interest on social media, with Samantha Cristoforetti, Luca Parmitano, Alexander Gerst, Tim Peake, Andreas Mogensen, Matthias Maurer, and Thomas Pesquet cumulating over 5 million followers on Twitter and 3.5 million friends on Facebook. The inspirational dimension of their activities has been reflected in the 2021 ESA astronaut selection campaign, with over 22 500 applications, a 270% increase in applications since the last selection campaign in 2008.

Finally, traffic was analysed on the ESA ISS Tracker which allows anyone to track the station in real-time. Regularly used world-wide by many institutions as well as normal citizens, it has been viewed about 350 million times between 2014 and 2020, with an average of 160 000 views per day.

Regarding ExoMars and the ESM, the missions generated societal benefits by contributing to a better understanding of clean atmosphere management, contamination avoidance, robotic development, and new forms of waste management. These domains are particularly important due to the growing interest in new human habitats in harsh environments on Mars, which is starting to create inspiration in several public spheres.

The human and robotic exploration programme has also had strong educational impacts in 2019, through its numerous educational initiatives, such as school education projects[7], space-related educational challenges, and even animated cartoons destined for children. At the school and university levels, over 47 000 teachers were trained in order to further communicate about space exploration and its inspiring dimensions, and over 2.2 million pupils and students were reached through such initiatives.

3.3.1.4 Strategic impacts and international cooperation

In the beginning of the ISS utilisation phase, ESA was cooperating closely with Russia to learn about the conducting and executing of experiments on the ISS.

Since the integration of the Columbus Laboratory module in 2008, about 75% of the European experiments were implemented with European funds and have involved European research organisations (both institutional and commercial). The remaining 25% have been implemented in co-funding with other international entities such as ISS international partners (NASA, Roscosmos, JAXA) and European national space agencies. These co-funded projects created interest and synergies among the wider ISS community, attracting additional funding from international and European partners, and enabling research on scientific and technological topics of common interest between stakeholders. The ISS has brought

[7] Astro Pi Challenge (https://astro-pi.org), Moon Camp Challenge (https://mooncampchallenge.org), Mission X: Train like an Astronaut (www.stem.org.uk/missionx).

together 15 international partners and, as of summer 2019, has been visited by 237 astronauts from 18 countries working together in international crews.

Not only did the ESA's research activities on the ISS increase scientific knowledge, they have also created value along the entire research value chain, increasing expertise and enabling different types of spin-offs.

Regarding human spaceflight, the Agency's ESM-1 that was delivered to NASA in 2018 is putting Europe on the critical path of the US exploration programme. New cooperation is now on-going on the Lunar Gateway (with European IHAB and ESPRIT) and on Mars Sample Return (with the European Earth Return Orbiter, Sample Fetch Rover, and Sample Transfer Arm).

3.3.1.5 Economic impacts

While ESA's funding of the ISS Programme was around €8 billion over 1995–2016, the estimated value added[8] generated over the same period was of €14.6 billion. Also, ESA's investment in the ISS programme was estimated to have sustained around 210 000 jobs, of which 110 000 were in the space sector. This means that with every 100 additional jobs created in the space sector in the countries studied, 90 additional jobs are supported in the wider economy, thus impacting employment well beyond the space sector.

In addition, around €7 billion was generated in governmental revenues over 1995–2016, meaning that around 88% of the contributions of the member states to the ISS Programme have been retrieved by governments through various taxes and social contributions.

Positive economic impacts were also estimated for the on-going European investments in the ExoMars 2016 and 2020 missions, and the ESM-1. With an initial ESA investment of €1.8 billion across all three missions, the estimated value added expected to be generated at completion is €4.4 billion. In terms of employment created, ESA investments in ExoMars 2016, ExoMars 2020, and ESM-1 have created 6500 jobs. However, around 19 000 jobs are expected to be sustained in the following years, meaning that with every 100 additional jobs created for these missions, 210 additional jobs will be supported in the space sector and in the wider economy.

3.3.2 The socio-economic impacts of ESA's science programme

Space science addresses major questions about the origin of our Universe and our Solar System, the origin of life, and whether we are alone in the Universe. The science programme is the backbone of ESA and is a mandatory programme created in 1984.

[8] GVA is measured to assess the contribution of a given space programme to GDP in the ESA member states. It is estimated through a input–output modelling approach which derives direct, indirect, and induced economic impacts (including GVA, employment, and government revenues).

These are the main questions to which space scientific missions try to provide answers, by better understanding the celestial objects of our Universe and by providing observations and data supporting the research. Representing 9% (€557 million) of ESA's budget in 2021, this development and research still represents one of ESA's cornerstone activities, aiming to contribute to the creation and enhancement of knowledge about our Universe.

Science is the underpinning theme of ESA, as it unifies all elements of the agency's activities. It fosters innovation and therefore technological advances which lead to progress and economic development. It also drives inspiration and contributes to fascinating and educating new generations of scientists and engineers.

Science missions are ambitious and require significant investment, with results available in the medium to long term. The majority of science missions therefore operate over decades. ESA conducted a socio-economic impact assessment of its science programme in 2019 [8], to estimate the various benefits deriving from it over its past 30 years. The study focused on eight science missions, of which four were ongoing/past and four were future missions. These missions were selected in order to cover various mission profiles for a more realistic assessment. The present socio-economic impact assessment focused on evaluating the benefits brought by the science programme for each of the eight missions separately and at an aggregated level, through four *impact categories*:

- *Scientific impacts* that constitute a core objective of scientific missions.
- *Strategic impacts* for ESA and the other European stakeholders involved in these missions.
- *Societal impacts* derived from these missions and contributing to raise public awareness, support and promote education, and protect the Earth environment.
- *Economic impacts*, such as contribution to GDP and employment.

3.3.2.1 Scientific impacts

In terms of scientific results, high-quality and large volumes of scientific publications have always been a constant indicator of good productivity across the four past missions. The scientific community also shows interest in data quality, which further feeds scientific research.

For example, Hubble[9] is considered to have one of the highest scientific impacts of all the missions ever flown in space. Not only did the observations from the telescope contribute to the Nobel Prize in Physics in 2011 for the discovery of the accelerating expansion of the Universe through observations of distant supernovae (Saul Perlmutter, Lawrence Berkeley National Laboratory and University of California, and Adam Riess, John Hopkins University, and Brian P Schmidt, Australian National University), it also has led to over 16 000 refereed papers and 800 000 citations, produced by almost 20 000 authors mainly publishing on

[9] The Hubble Space Telescope is a joint ESA/NASA project launched in 1990 in LEO, with the objective to enable astronomers to observe planets, stars and galaxies.

astronomy and astrophysics. European authors had a major role in the amount of publications, representing over 60% of Hubble-related papers thanks to ESA's role in Hubble. Hubble's enabling potential is international and sets science as a common language.

High volumes and high-quality papers are also good indicators of scientific value. However, scientific knowledge is most valuable when it can be shared and transferred. It can be transferred to researchers from the same scientific domain, from a different domain or to industry.

In this way, SOHO[10] had a very strong scientific impact with around 5400 publications and over 150 000 citations, and led to significant knowledge transfer to meteorology, atmospheric science, aerospace and aeronautics, optics, and nuclear and particle physics.

ESA's science programme develops and operates a portfolio of missions, which ensure that the European scientific community maintains its scientific leadership position in terms of discovery and innovation across all of space astronomy, heliophysics, planetary science, and fundamental physics. It also supports the development of new world-leading hardware and software solutions within European industry, reducing dependence on outside suppliers.

3.3.2.2 Strategic impacts

Because of their international dimension, complex space science missions foster strong collaboration between research teams from different countries, disciplines, and fields of expertise. Some of these missions also generate significant benefits for the European industry competitiveness.

Due for launch in 2024, the SMILE[11] mission is expected to have strong strategic impacts, with positive outcomes at the programme level, at institutional level, and at market level. The mission's activities are expected to help improve efficiency toward US export control regulations on American made systems, sub-systems, and components. At the institutional level, an improved relationship between ESA and the Chinese Academy of Sciences (CAS) is foreseen, as well as between European and Chinese universities and research institutions. These improved relations are expected to impact markets positively by nurturing future commercial collaboration between the two respective aerospace industries.

3.3.2.3 Societal, inspirational, and educational impacts

Not only do most of these scientific missions have a strategic value, they also substantially benefit society at large. Scientific missions raise public awareness by being our eyes in space and sharing spectacular images of celestial bodies.

Communicating effectively about these missions also helps to promote education, by stimulating interest in science, technology, engineering, and mathematic

[10] SOHO is a joint ESA/NASA project launched in 1995 to study the Sun's activity. Initially planned for two years, it has been operational for over 25 years.

[11] SMILE is a future joint ESA/CAS mission that will provide understanding of the interaction between the solar wind and Earth's magnetosphere.

disciplines. Past assessments of ESA science missions have always demonstrated a strong impact on education, which is very important, given that hundreds of PhD theses were produced based on the four past missions.

The societal impacts of the on-going SOHO mission over the period 1995–2015 were very important due to its cross-sectorial contribution to space weather programmes. Leveraging on an open data policy and the development of the Internet, SOHO was one of the first scientific missions to publish results and images in near real-time, thus boosting its outreach.

The Rosetta mission[12] (2004–16), on the other had, operated with a radical change in communication and had a strong societal impact in terms of education. At the same time, more than 50 PhD students graduated with a thesis involving Rosetta, while many master's students and undergraduates took part in data exploitation activities.

3.3.2.4 Economic impacts

Overall, the economic impacts of the scientific space missions were high, with important GDP and employment multipliers.

In fact, for a total investment of €4.4 billion (from 1984 up to 2016), the GDP impact (GVA) was of €7 billion, demonstrating an important economic benefit for ESA member states. In terms of employment level, ESA spending on the eight science missions is estimated to generate 16 500 jobs in the space sector and 18 000 jobs in the wider economy over 1984–2030, showing strong job creation supported by the ESA's initial investment and related contracts passed to the European space industry.

On top of these direct effects on the economy, the missions under scrutiny had and/or are expected to have significant technology transfer implications. Past missions have demonstrated important positive externalities in terms of technology re-use or adaptation, human resources contributing to other space missions later on thanks to their expertise and accumulated know-how, or involvement of ESA and/or European institutes in other scientific missions that present similar challenges or synergies. This implies a large number of economic activities that have an origin in the first mission developed.

3.4 Conclusion

The studies presented in the current chapter showcased how Big Science projects are at the very core of ESA activities—activities which have a strong impact on economy and society and generate important transversal benefits for all actors on the value chain including governments, companies, universities, research, citizens, and last but not least for Earth's environment.

Research, innovation, and technology progress are widely recognized as the main drivers of long-term productivity growth. Public spending in such domains hence

[12] Rosetta was launched in 2004 and terminated in 2016. It was a mission focused on studying the origin of comets, the relationship between commentary and interstellar material, and its implications with regard the origin of the Solar System.

fosters sustainable, inclusive, and smart growth. According to the European Commission, innovation contributed to over 65% of total European productivity growth between 2010 and 2016[13]. Investing in Big Science hence contributes to generating such benefits, only at a larger scale. It can change the scientific and industrial landscape through its discoveries and demonstrate strong impacts for the future of humanity.

These studies were presented to ESA member states for the council at the ministerial level that took place in 2019, together with seven other socio-economic impact assessment studies of ESA programmes[14]. By doing so, ESA tried to bring more and robust arguments to its member states in favour of sustained and increased investment in European space activities. The agency also paved the way towards a systematic impact assessment approach for its programmes and missions looking forward and will consistently inform decision-makers of the impacts of its upcoming activities. In fact, ESA conducted a set of 15 new studies in view of the past council at the ministerial level, held in 2022.

References

[1] PwC 2016 Assessment of the socio-economic impacts of the ESA participation to the International Space Station (ISS) Programme

[2] PwC 2019 Socio-economic impacts assessment of ESA's Science Programme May

[3] Open University 2019 Benefits of the ESA Exploration Roadmap in Socioeconomics (BEERS)

[4] PwC 2016 Assessment of the socio-economic impacts of the ESA participation to the International Space Station (ISS) Programme

[5] PwC 2016 Assessment of the socio-economic impacts of the ESA participation to the International Space Station (ISS) Programme

[6] Open University 2019 Benefits of the ESA Exploration Roadmap in Socioeconomics (BEERS)

[7] PwC 2014 Assessment of the socio-economic impacts of the ESA participation to the International Space Station (ISS) Programme

[8] PwC 2019 Socio-economic impacts assessment of ESA's Science Programme May

[9] Diallo O, Ruttley T and Tate-Brown J 2018 *Annual Highlights of Results from the International Space Station* (NASA)

[13] European Commission, 'Science, research and innovation performance of the EU 2020, 2020'—innovation is broadly defined as the aggregate of total factor productivity (TFP) and intangible investments.

[14] All executive summaries of the socio-economic impact assessments prepared for ESA councils at the ministerial level are available publicly on the ESA Space Economy website https://space-economy.esa.int/.

IOP Publishing

Big Science in the 21st Century
Economic and societal impacts
Panagiotis Charitos, Theodore Arabatzis, Harry Cliff, Günther Dissertori, Juliette Forneris and Jason Li-Ying

Chapter 4

Bringing value from Big Science projects to society: the view from ESO

Andrew Williams and Bárbara Ferreira

The European Organisation for Astronomical Research in the Southern Hemisphere, also known as the European Southern Observatory (ESO), is an intergovernmental science and technology organisation in astronomy. By designing, building, and operating world-class observatories on the ground, ESO enables scientists worldwide to discover the secrets of the Universe for the benefit of all. ESO also plays a leading role in promoting and organising European and international cooperation in astronomical research. This chapter describes the outcome of a study undertaken by ESO to understand its broader societal benefits. The overall objective of the activity was to provide a factual and compelling narrative of ESO's cumulative, broad societal benefits across a range of domains, focusing on impact rather than performance. The term 'societal benefits' intends to convey the combination of direct benefits that ESO brings to the public, industry, government, and scientific communities, in addition to broader and indirect benefits that are at the level of society. First, key findings on the societal benefits of ESO are described across each impact element: science and engineering, economy and innovation, talent development, education and outreach, and international collaboration and policy. Second, the authors offer some reflections on the nature of societal benefits and socio-economic impact evaluation in Big Science facilities.

4.1 Introduction

Humankind has always looked to the heavens in search of meaning and to reveal the workings of the Universe. The unimaginable vastness of the Cosmos has spurred our curiosity to explore and to develop the science of astronomy in order to do so. The quest to understand our Universe has had wide-ranging impacts on our society. We lie not at the centre of the Heavens but orbit a relatively common star in the outer

doi:10.1088/978-0-7503-3631-4ch4

regions of a typical galaxy, one out of countless billions in the known Universe. The fundamental laws of physics are probably the same everywhere, but our current understanding can only explain approximately 5% of the matter in the Universe. There are as many planetary systems as grains of sand on the Earth, if not more, some of which include habitable planets. Astronomy has influenced our understanding of nature, time, and the change of seasons, providing the tools to enable early sailors to navigate the seas. Astronomical knowledge about our place in the cosmos has yielded several transformational paradigm shifts, with sometimes dramatic societal consequences, by giving us an objective picture of our existence, location, and importance. In the near future, ground and space observatories will be built that allow direct imaging of exoplanets, and detailed analysis of the chemical make up of exoplanet atmospheres, which could yield the answer to the ultimate question—are we alone?[1]

From its relatively humble origins of small, independently funded, and private observatories in the late nineteenth and early twentieth century, astronomy has more recently emerged as a 'Big Science', partly spurred by the increasing need for ever larger and more sensitive telescopes to probe further into the Cosmos. In recent decades, many governments around the world have devoted substantial financial and human resources to the development of ever more complex observatories in ground and space, indicating that society still values this pursuit of knowledge. The European Southern Observatory (ESO) represents a classic case of this interest. Founded in 1962, ESO is an intergovernmental science and technology organisation in astronomy. By designing, building, and operating world-class observatories on the ground, ESO enables scientists worldwide to discover the secrets of the Universe for the benefit of all. ESO also plays a leading role in promoting and organising European and international cooperation in astronomical research. As an intergovernmental organisation, states ratify an international treaty at the highest level of government in order to join ESO as a member state. From its humble origins of five European founding countries desiring to build jointly a small 1 m size telescope, ESO now counts 16 member states and a ten-year strategic partnership with Australia and is constructing and operating 'Big Science' facilities in Chile, ESO's host state.

The Chilean Atacama Desert, a marvellous place with unique conditions to observe the sky, hosts all ESO telescopes. The organisation operates three observing sites: La Silla, Paranal, and Chajnantor. At Paranal, ESO designed, constructed, and now operates the Very Large Telescope (VLT) and its Very Large Telescope Interferometer (VLTI), as well as two survey telescopes, VISTA working in the infrared and the visible-light VLT Survey Telescope. Also at Paranal ESO will host and operate the Cherenkov Telescope Array South, the world's largest and most sensitive gamma-ray observatory. Together with international partners, ESO operates APEX and ALMA on Chajnantor, two facilities that observe the skies in the millimetre and submillimetre range. ESO is constructing the Extremely Large Telescope (ELT)—the 'world's biggest eye on the sky' and a landmark infrastructure

[1] https://spacewatch.global/2021/05/spacewatchgl-opinion-five-reasons-why-astronomy-is-important to our future-in-space/

in the European Strategy Forum for Research Infrastructures (ESFRI) Roadmap—on Cerro Armazones, but fully integrated into the Paranal Observatory. The ELT is a shining example of a science-driven European enterprise, full of technical challenges and seeds of innovation for European industry. When complete in the late 2020s, this revolutionary telescope will answer some of the most pressing scientific questions and will keep Europe at the forefront of ground-based astronomy for many years to come (figure 4.1).

ESO sustains a large, global community of astronomers. Each year, about 2000 research proposals are made for the use of ESO telescopes, requesting between four and six times more nights than are available. In 2021 alone, over 1100 peer-reviewed papers were published based on data from ESO's telescopes, making ESO one of the most productive ground-based astronomical observatory in the world[2]. Furthermore, ESO's telescopes are outfitted with a variety of instruments constructed in partnership with large consortia of research institutes and industry in the

Figure 4.1. An artist's impression, based on the final design for the telescope, of ESO's Extremely Large Telescope (ELT). (Credit: ESO/L Calçada.)

[2] https://www.eso.org/public/announcements/ann22004/

member states. The astronomy community inherently understands the value of ESO in terms of enabling scientific results, but what about the impacts of ESO beyond the production of amazing scientific results?

In 2020 ESO decided to undertake an exercise to understand its broader societal benefits[3]. The overall objective of the activity was to provide a factual and compelling narrative of ESO's cumulative, broad societal benefits across a range of domains, focusing on impact rather than performance, and employing novel methods that move away from narrow economic-based assessments. The term 'societal benefits' intends to convey the combination of direct benefits that ESO brings to the public, industry, government, and scientific communities, in addition to broader and indirect benefits that are at the level of a society. The results of the activity aimed to provide ESO's key governmental, scientific, and industry stakeholders with a comprehensive picture of ESO's societal benefits. The outputs aimed to serve as a mechanism to increase engagement and deepen relationships with member states by stimulating interactions with delegates, government representatives, visiting diplomatic staff, and senior scientific representatives of member state communities, and supporting ESO strategic communications and communications products in the long term.

ESO commissioned the European Future Innovation System (EFIS) Centre, in partnership with NEBS Sprl Consultancy, to conduct the societal benefits project in 2020. The project encompassed four activities: 1. A conceptual framework of ESO's impacts was developed, drawing from a range of interviews with ESO's key stakeholders (see figure 4.2 for an example). 2. Evidence from internal documents, reports, articles, and further interviews was gathered to illustrate the conceptual

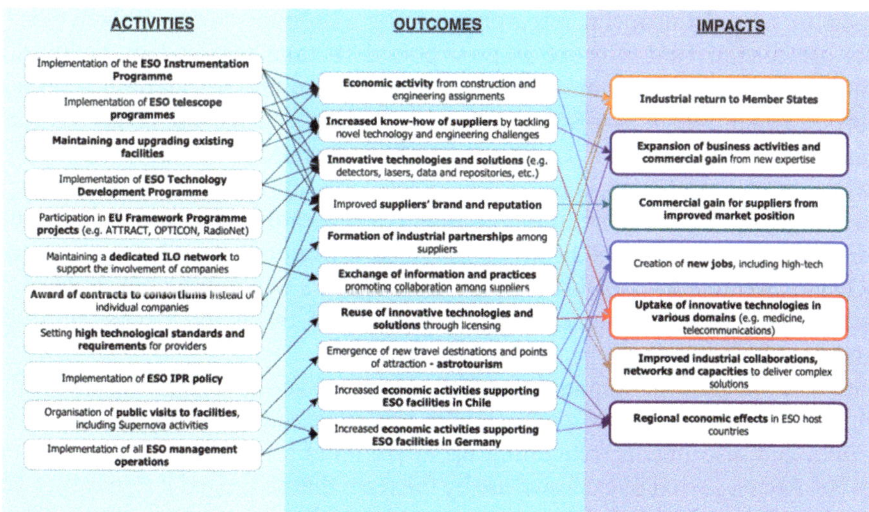

Figure 4.2. An example of the conceptual impact framework developed for the societal benefits study. Credit: ESO / European Future Innovation System Centre ABSL.

[3] https://www.eso.org/public/about-eso/societal-benefits/

impact framework. 3. Several case studies were developed to illustrate the impact area in greater detail. 4. A strategic communication kit was created to support ESO's Communications Department. After the study, ESO's communications team then launched a communications campaign and promotional materials[4]. Some of the descriptions in this chapter are drawn directly from these materials.

This chapter provides an overview of the societal benefits exercise and its results. First, key findings on the societal benefits of ESO are described across each impact element: science and engineering, economy and innovation, talent development, education and outreach, and international collaboration and policy. Second, the authors offer some reflections on the nature of societal benefits and socio-economic impact evaluation in Big Science facilities. This work focuses primarily on the benefits gained in the European member states, rather than in the host country of Chile, which accrues a similar but distinct set of impacts by virtue of the physical location of the observatories and the unique Chilean–ESO relationship. A separate study was made on the societal benefits in Chile[5].

4.2 Scientific and engineering benefits

ESO stands as a testament to the possibilities and achievements of European scientific collaboration and is seen as a cornerstone in the European astronomy community. By pooling resources in a stable legal structure, ESO's member states have created a large portfolio of telescopes and instruments, opening up a range of scientific areas. The activities and collaborations that serve this vibrant scientific programme have allowed scientific and technical communities to flourish, develop world-leading capability, and define scientific excellence, and ultimately are a core component of the European research area for astronomy and astrophysics. The results are unequivocal: Europe is now poised to take the lead in building and operating the next generation of 25–40 m class optical and infrared telescopes that will produce unimaginable discoveries and offer unparalleled opportunities for the communities of scientists and engineers in ESO's member states and beyond.

ESO's most significant benefits to society are in the areas of science and engineering. By building and maintaining world-leading telescopes, as well as enabling access to them, ESO has transformed astronomy and increased scientific knowledge and excellence. Observations done with its telescopes contribute to over 1000 refereed publications each year, making ESO the world's most productive ground-based observatory.

ESO serves a growing number of astronomers and access to its telescope is highly sought after. More than 2500 proposals are submitted each year for the use of ESO telescopes and instruments, requesting up to six times as much observing time as is available. After a period of exclusive use by the team that requested it (typically 1 year), ESO makes data available to astronomers worldwide through its petabyte-sized science archive, opening up more research opportunities to more people (figure 4.3).

[4] https://www.eso.org/public/products/brochures/brochure_0076/

[5] https://www.eso.org/public/chile/products/brochures/brochure_0078/

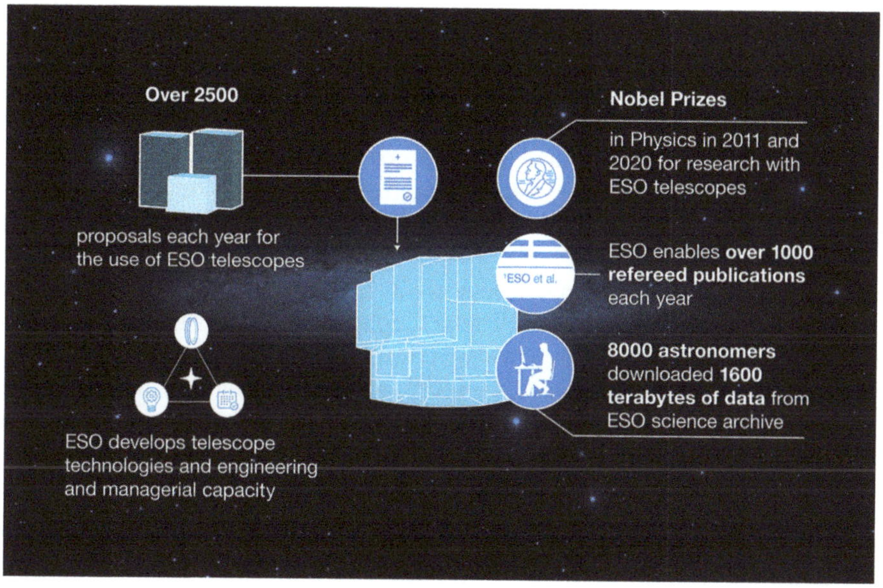

Figure 4.3. ESO's scientific metrics. Credit: ESO.

The increase in scientific excellence is evident in the breakthrough discoveries made possible with ESO telescopes. The 2020 Nobel Prize in Physics was awarded for research with ESO telescopes on the Milky Way's supermassive black hole, and ESO telescopes also contributed to the discovery of the accelerating Universe, recognised with a Nobel Prize in Physics in 2011 that was awarded to Saul Perlmutte, Brian P Schmidt, and Adam G Riess. In addition, two of the winners of the 2019 Nobel Prize in Physics, namely Michel Mayor and Didier Queloz, conducted much of their exoplanet research using ESO telescopes.

ESO pushes the frontiers of technology to enable astronomers to make great strides in advancing our knowledge of the Universe. ESO staff developed active optics, a key technology that allowed ESO—and other observatories—to build bigger and optically accurate primary mirrors for its telescopes. ESO's contributions were also indispensable in developing adaptive optics, a technology that allows ground-based observatories to build telescopes able to obtain extremely sharp images by using deformable mirrors, lasers, and sensors to correct for the blur caused by Earth's atmosphere. To correct for this blurring effect, telescopes require guide stars—bright stars that are affected by turbulence in the same way as the object under study. However, in most cases, natural stars that can be used as guides are not available in the regions of the sky astronomers are investigating. Instead, light from sodium lasers on the telescope is used to excite sodium atoms in the Earth's atmosphere, artificially creating a bright, guide star. ESO, partially supported by EU grants, worked with Toptica AG in Germany to develop compact 20 W lasers and later licensed the technology to Toptica and Canada-based MPB Communications. This was the first transfer of patented technology from ESO to industry and provides opportunities both for industry to market the technology and for ESO to have a supplier for future requirements. Laser systems are currently in operation on ESO's VLT, and will be used on the ELT in future.

Other telescopes worldwide also employ the technology. In addition to their application in astronomy, the high-powered lasers have uses in the fields of space situational awareness and optical satellite communications. Without active and adaptive optics, as well as other game-changing technologies developed at ESO in collaboration with industry, large telescopes such as the VLT and the ELT would not be possible.

By operating and providing access to a broad portfolio of telescopes and instruments, equipped with state-of-the-art technology, ESO offers astronomers a unique toolbox with which to explore the cosmos over a wide range of wavelengths and to look deeper into specific objects of study. Regular upgrades to its observatories and instruments open up new areas of scientific inquiry, providing more opportunities to tackle astronomical challenges and broadening the potential of ESO telescopes to make unexpected and fascinating discoveries.

4.3 Economy and innovation benefits

ESO's main economic impact comes from investments in telescope and instrumentation development, as well as in construction and technology development. In 2019, for example, ESO spent about 60% of its ~€300 million budget on the design and construction of telescopes and instruments. Over 90% of that was invested in high-tech innovation led by European industry and research institutions in the member states. For its newest telescope, the ELT, over 80% of the €1.3 billion development construction budget is being invested in contracts with industry. During the construction of the VLT in the 1990s, 91 contracts were awarded to 76 companies in 12 countries. Contracts are primarily distributed across ESO's member states, providing a significant industrial return to these countries and creating new high-skilled jobs. A network of national industry liaison officers (ILOs) helps to reach into national industrial ecosystems and connect reliable and capable partners. The growth of suppliers in both existing, new, and niche areas also grows employment in these sectors. While ESO contracts on a primarily competitive basis, without a *juste-retour* system in place like ESA, supply contracts are monitored to ensure a reasonable balance between countries. The network of suppliers also involves a substantial amount of sub-contracting between member states and also third countries, catalysing a wide range of international industrial partnerships.

More crucial than the direct financial impact of ESO's activities in industry are the innovation benefits that collaboration with ESO brings to European companies. ESO's investments promote the emergence of new technologies and expertise, leading to new markets and opportunities, as well as to the creation of new jobs and industrial collaborations. ESO drives forward technology development and capabilities in Europe.

The high standards and ambitious requirements of ESO projects also foster the formation of industrial partnerships that pool existing capabilities and promote the exchange of know-how to address complex engineering and technological challenges. ESO maintains a dedicated network of industrial liaison officers to promote the exchange of information and best practices and to facilitate cooperation between partners from different member states.

By collaborating with ESO, industries can improve their market position and expand their business activities and commercial gain. ESO has stringent standards for its telescopes and instruments, which often require companies themselves to make R&D developments; in addition, contracted companies often benefit from ESO's technical and project management expertise. Companies benefit in the long term, with collaborations leading to improved reputation and technical expertise. For example, in 2020, French company CILAS[6] won a multi-million Euro contract with the US-based Thirty Meter Telescope International Observatory, thanks in part to technology developed jointly with ESO. Likewise, ESO's collaboration with German company Etalon on a technology for high-precision measurements is a success story and an example of how ESO's benefits to industry bring direct economic impacts. While the technology (now called Absolute Multiline Technology) was not originally developed for telescopes, ESO engineers extensively tested, verified, and validated the measuring system, showing it could be employed in telescopes. This opened up a new market for the Absolute Multiline Technology, allowing the company to sell this product for application in telescopes at ESO and elsewhere. The technology will be used on the US-led Giant Magellan Telescope and is also employed by French company Safran-Reosc, in the production of the segments that make up the primary mirror of ESO's ELT.

Another important benefit that results from ESO's collaboration with industry is the take-up of innovative technologies in various domains. Technologies developed at ESO have found applications in areas such as optics, intercontinental data transfer, medicine, and imaging, sensor, and detector technology. For example, through the EU-funded ATTRACT initiative, the German institute innoFSPEC is bringing astronomical technology to bear in the fight against cancer, by transferring techniques from the MUSE integral field spectrograph on ESO's VLT to cancer diagnostics[7].

4.4 Talent development benefits

ESO plays a pivotal role in attracting and developing the talent base required to maintain European astronomy research and related instrumentation development at world-leading levels. Through its competitive studentship and fellowship programmes, ESO contributes to career development in astronomy and increases the availability of highly skilled engineers and technical experts. As of November 2021, ESO has 471 international staff members (ISMs) and 164 local staff members. These categories of personnel are denoted as 'established' and may be contracted indefinitely. Non-established members of personnel, who are engaged only on temporary contracts, include fellows, paid associates, unpaid associates, students,

[6] https://www.eso.org/public/announcements/ann20004/

[7] https://innofspec.de/en/news/news-events/79-kick-off-for-eu-attract-projects; https://www.eso.org/public/news/eso1909/

and guest professors. ISMs are all hired initially on fixed term contracts of up to three years. As from a second contract and the fourth year of service, these contracts may become of indefinite duration according to demonstrated skills, competence, business need, and budget. Local staff members are initially hired on a fixed term contract of two years, which may then become of indefinite nature thereafter based on the same criteria. No nationality is in principle excluded, however, recruitment preference is given to nationals of ESO's member states, host state Chile, and strategic partner, Australia.

Over the past decade, ESO has trained over 200 students in science and engineering and has hosted over 150 postdoctoral fellows. With students coming from more than 40 countries and fellows from over 30 (in the past decade), the impact of ESO on talent generation and development goes beyond European borders. ESO empowers its students and fellows to acquire skills that see application in careers beyond astronomy research or engineering. Students and fellows participate in ESO projects and can support operations, which gives them the opportunity to gain transferable skills in areas such as communication, project management, or data analysis. They also benefit from ESO's international environment, which fosters openness to other cultures and multilingual competencies, and provides access to the global network of astronomers with connections to ESO.

> ESO was the key to the success of my career and gave me the opportunity to be involved in building cutting-edge instrumentation and access to a wide community. In the end, 90% of my papers are related to ESO.
>
> —Didier Queloz, Nobel Prize Laureate in Physics 2019

ESO's internship programme is another key example of how ESO supports career development in a variety of occupations, helping transfer know-how and develop talent. Since 2010, ESO has hosted over 90 people training in science writing, graphic design, and administration, in addition to astronomy and engineering.

In general, people who study or work at ESO go on to have successful careers in astronomy and engineering, as well as in operations, information technology, education, business development, programme and project management, and media and communications. When they go on to work in member states, in industry or academia, after their time at ESO they enrich national systems with the technical skills and collaborative approach they acquired.

ESO's impact on talent development goes beyond the people it trains directly. Research carried out at ESO finds its way into teaching materials, mostly in astronomy—an inspiring discipline that encourages students to pursue careers in science and technology. ESO also facilitates collaboration with national astronomy organisations, which form part of a broader network in European teaching and competence building, thus increasing the number of all-level students studying astronomy in ESO member states.

ESO plays a pivotal role in attracting and developing the talent base required to maintain European astronomy research and related instrumentation development at world-leading levels. The impact on talent generation and development is global and the skill sets developed are much broader than those required to carry out astronomy or astrophysics research. ESO mobilises significant numbers of people in specialised (niche) engineering and computing (data analytics) fields. People who work at or with ESO gain not only a kaleidoscope of technical know-how but also a strong set of soft skills that enable them to manage, design, and develop complex projects in an international and cross-cultural environment. ESO is not so much a place of work as a community of practice in which know-how is transferred formally and informally between people and across institutes, companies, and countries.

4.5 Education and outreach

Astronomy is inspiring. ESO's discoveries of new planets, observations of distant galaxies, and colourful images of stars and nebulae have genuine public appeal and excite the public about new scientific findings and the technology required to make them. ESO's communication, education, and outreach activities harness this excitement and help increase public support for astronomy, science, and technology, as well as to promote scientific literacy in society. They also help place planet Earth into the context of the Universe and instill a sense of humility and the fragility of life that makes people live more consciously and interact with our planet more responsibly. ESO's public engagement activities have served as a benchmark for other astronomy organisations.

One way in which ESO fosters the public's inspiration and fascination about astronomy and science is through its vast database of visuals: by offering more than 18 000 astronomical images and videos for free[8], ESO ensures audiences everywhere, from teachers to journalists and astronomy enthusiasts, can experience the wonders of the cosmos.

ESO also plays a pivotal role in astronomy education and in fostering engagement of young people in STEM fields and promoting STEM careers. With activities such as organising astronomy summer camps for teenagers, contributing content to online science education platforms, and running an exciting outreach programme though the ESO Supernova Planetarium and Visitor Centre, ESO directly inspires children and teenagers and helps teachers engage their students. The ESO Supernova was designed as the first open-source digital planetarium in the world to reach audiences beyond the centre's immediate vicinity. By developing and offering planetarium shows free of charge, the ESO Supernova provides planetariums worldwide with access to high-quality shows that they can use to inspire local audiences.

Every year ESO's public and media visits to its observatories in the Chilean Atacama Desert attract some 8000 national and international astronomy enthusiasts and have helped promote Chile as the world's capital of astronomy. ESO has also helped the public to develop a sensibility towards the appreciation of dark skies as a cultural heritage, by raising awareness of the impacts of light pollution.

The reach of ESO's communication and outreach activities goes beyond its member and host states. Through its social media channels, ESO raises awareness of

[8] https://www.eso.org/public/images/; https://www.eso.org/public/videos/

the latest astronomy discoveries and the newest technology advancement to over half a million followers worldwide. ESO press releases are translated into over 15 languages, increasing the reach of ESO news; in the last 10 years, ESO has been mentioned in over 40 online press articles per day on average.

ESO has also contributed to arts and culture, in particular through the partnerships it has established with artistic organisations; musicians have produced music videos at ESO observatories and artists regularly use ESO's sites and image archive as a source of inspiration for their work. By making sure a comprehensive approach to communication and outreach activities targets a wide variety of people, ESO ensures that astronomy can truly be meant for and be enjoyed by everyone.

4.6 International collaboration and policy benefits

ESO offers a platform for international collaboration and science diplomacy that actively encourages political and cultural understanding between different nations. It brings countries together to create the scientific, technical, and political capacity needed to develop ambitious technical projects that allow nations to further knowledge about our Universe, spur innovation, and raise society's awareness about the cosmos. ESO has come to symbolise a unified, European approach to astronomy and has greatly contributed to making Europe a leader in astronomy and astrophysics research. ESO has evolved into an internationally representative body that is able to negotiate and enter into international partnerships on behalf of multiple ESO member states. ESO leverages the strengths and opportunities that stem from combining efforts and resources to create a scientific and political capacity for development that is beyond the reach of an individual country.

A key project where ESO significantly fosters international collaboration is the ELT, the telescope of the future, which brings together individuals, organisations, and industries in ESO member and host states. For example, to develop instruments for the telescope, ESO encouraged universities and research institutes to work together, forming consortia. Instrument development for the ELT involves over 50 institutes from 20 countries. Didier Queloz, 2019 Nobel Prize winner in Physics, summarised this status[9]:

> ESO has provided three key benefits to the European and global community: first, a world-class observatory system that generates many efficiencies for national astronomy communities in terms of its operations and options for instrumentation, and possibilities to host national projects in the observatory; second, a deep body of experience in the management of megaprojects in astronomy; and third, a stable framework that has allowed countries to fairly pool resources to enable these world-leading projects. A project like the ELT is only possible in Europe with ESO.

[9] Interview with the author (Williams), 2 July 2020.

ESO is a founding member of the European Intergovernmental Research Organisation forum (EIROforum), which brings together eight European research infrastructures to exchange experiences, cooperate on strategic projects, and thereby foster international cooperation. ESO also maintains formal cooperation agreements with the European Space Agency and CERN.

Through its influential role, ESO helps promote the interests of European astronomy, shapes the priorities of European research strategies, and, more broadly, impacts science policy. For example, ESO has driven the development of open data standards in astronomy and ethical review processes in science, and contributes to the concept of a European Open Science Cloud. On the global stage, ESO has a seat in the United Nations Committee on the Peaceful Uses of Outer Space and has made key policy contributions in areas such as satellite mega-constellations[10] and the protection of the dark sky, and is a member of the UN-mandated International Asteroid Warning Network, which monitors the skies for potentially threatening asteroids[11].

By bringing countries together to develop challenging projects, ESO has become a role model in R&D collaboration. It has accumulated numerous best practices, knowledge and expertise that have been propagated across the landscape of astronomy and Big Science facilities. For example, the Square Kilometre Array (SKA), an international project with 14 member countries, benefited from ESO as a model on how to set up an international organisation and how to implement successful operations. In the US, the National Science Foundation's National Optical-Infrared Astronomy Research Laboratory (NOIRLab), which brings together four observatories under one umbrella, has also drawn inspiration from ESO's organisational model.

> Through focused, consistent investment in world-class facilities, forefront technologies, and crucially in human-capital, the European community has enabled ESO to be a world force in ground-based astronomy.
>
> —Matt Mountain,
> President of the Association of Universities for
> Research in Astronomy (AURA), USA

4.7 Closing thoughts: the future of understanding the value of Big Science

This chapter has described the multifaceted benefits of ESO beyond its immediate scientific outputs. The evaluation framework was qualitatively derived from the key stakeholder interviews, but the elements of the framework are consistent with standard models recently created by the OECD[12] and others[13]. A comprehensive

---

[10] https://www.eso.org/sci/publications/messenger/archive/no.184-sep21/messenger-no184-3-7.pdf

[11] https://www.eso.org/public/blog/hunting-for-neos/

[12] OECD 2019 Reference framework for assessing the scientific and socio-economic impact of research infrastructures *OECD Science, Technology and Industry Policy Papers* No. 65 (Paris: OECD Publishing) https://doi.org/10.1787/3ffee43b-en

[13] For example, the RI-Paths project funded by Horizon 2020 (https://ri-paths-tool.eu/en)

societal benefits study internal report was presented to ESO Council in December 2020, with an accompanying public communications materials posted on the ESO website[14]. The results provided a set of arguments to ESO's member states to support continued engagement and investment in the development of ESO's world-leading astronomy facilities. While individual and aggregated data on member state performance is presented to the ESO council on a regular basis, until this study ESO had not conducted an evaluation of its comprehensive benefits to broader society, in addition to those benefits to the astronomy community. The results of this activity will therefore support increased engagement with member state delegates and funding agencies, and also inform ESO's communications strategy in general.

The societal benefits exercise also offers the opportunity to present some broader thoughts on the nature of socio-economic impact evaluation and the evaluation of the benefits of Big Science facilities, particularly as the need for socio-economic evaluations is increasing. Indeed, European nations and particularly the European Union (EU) generally require formal evaluations of public facilities, in particular where large financial investments are made. The increasing investments made in large scientific facilities across Europe has likewise resulted in a growing expectation to conduct socio-economic or other impact assessments for research infrastructures. Many funding agencies, including the European Commission (EC), also require *ex-ante* socio-economic impact assessments as part of funding applications. Furthermore, the creation of the ESFRI and its recent EU Council mandate[15] to conduct performance monitoring of the facilities on its strategy roadmap has further added to the demand and expectation to conduct such evaluations.

There are some challenges, however, to the evaluation of large multipartnership facilities, particularly for pan-European research facilities. The drivers of scientific excellence and scientific competition naturally lead to internationalisation, where only by combining resources across many countries can scientific communities gain access to world-leading facilities that match those of large countries such as the United States and Japan. This creates a complexity in evaluating the impacts of multi-partner facilities, whereby each partner's contribution—and therefore socio-economic return—in a facility may be relatively small, but the combined impact can be very large. Impact pathways from industrial effects to scientific community impacts may play out in very different ways across different nations and communities. Focusing on the standard metrics of economic impact may therefore miss important areas of non-economic benefit which escape the attention of funding agencies and governments. This underlines the importance of the approach taken by ESO to emphasise a narrative approach of societal benefits captured from key stakeholder perceptions in addition to quantitative data.

While many other research infrastructures and facilities have conducted socio-economic impact assessments, often drawing on quantitative economic models, metrics, and cost–benefit analyses, the extent to which these apply in the case of ESO

[14] https://www.eso.org/public/about-eso/societal-benefits/

[15] EU Competitiveness Council, 29 May 2018. Council conclusions on 'Accelerating knowledge circulation in the EU' https://data.consilium.europa.eu/doc/document/ST-9507-2018-INIT/en/pdf

deserves scrutiny. The support of the member states is already generally assured and ESO is recognised as a vital component of the astronomy industry. In other words, European partners recognise that without pooling their resources, a world-leading facility would unlikely be achievable by any one member alone. Thus this singular value transcends all the economic arguments. While economic arguments are paramount in supporting specific investment decision purposes, the 'return on investment' thinking is a rather narrow lens through which to appreciate the complex system of an intergovernmental organisation's interaction over time through multiple countries' scientific, industrial, and political communities. While ESO's key stakeholders already know and appreciate some or even all of these economic and scientific components, an important objective of this study is to address the general perception of ESO and ensure that all key stakeholders are fully aware of ESO's benefits in a comprehensive and integrated way.

IOP Publishing

Big Science in the 21st Century
Economic and societal impacts
Panagiotis Charitos, Theodore Arabatzis, Harry Cliff, Günther Dissertori, Juliette Forneris and Jason Li-Ying

Chapter 5

The low hanging fruit—the economic boost from the construction of large science infrastructure in the twenty-first century

Leonardo Biagioni

Considerate la vostra semenza: fatti non foste a viver come bruti, ma per seguir virtute e conoscenza

—Dante Alighieri, *Inferno*, Canto 26, v. 117–20

5.1 Summary

The positive social and economic impact from the construction of a large scale scientific and technological undertaking has been addressed by many authors in past decades. Indeed, since the dawn of modern 'Big Science' (during the second half of the twentieth century) this theme has been one of the main pillars for defining, promoting, and evaluating scientific projects in Western countries.

Past attention has mostly focussed on the long term benefits generated by the new scientific infrastructure once completed and operating. This is embodied by new knowledge and scientific breakthroughs, the development of the skills of scientists, the engagement of highly qualified young researchers, spin-offs, local economic benefits, etc.

While important and extremely valuable, in particular in today's global competition for the 'knowledge economy', these positive impacts take a long time to materialize. Many modern Big Science projects need several decades to go from the initial definition to the design and construction phases. At the same time, the pace of social, economic, and technological transformation in a globalized world is accelerating, making it difficult to predict the future impact from a large scientific infrastructure and in which areas it may generate significant return.

On the other hand, as the typical size (several billion euros, spent over a timespan of one or more decades) of scientific infrastructures has increased during the past

doi:10.1088/978-0-7503-3631-4ch5

century, a second type of positive economic feedback has become increasingly apparent—one deriving from the construction of the facility itself. This positive impact occurs as soon as the project begins, decades before it starts scientific operations, and can be analysed and understood as it develops. The characteristics of large science infrastructures (often complex, high technology, multidisciplinary undertakings pushing the limits of industrial state-of-the-art in many dimensions and requiring a flexible and adaptable industrial organizations) are such that the construction itself becomes a highly challenging endeavour, in which industrial and scientific communities must work together with an unprecedented degree of collaboration and integration. The effect is extremely positive for both communities, the increased competitiveness, economic benefits, and development of new knowledge and skills being the 'low hanging fruit' that immediately benefits society.

Using quantitative information and examples from the European contribution to the ITER international fusion project (EU-ITER), one of the largest world-wide Big Science projects currently under construction, this work analyses how this immediate positive impact materializes and which of its dimensions can be best monitored and reinforced. The EU-ITER case will also be used to identify good practice, intended to enhance and better direct the short-term impact from large science construction projects, with the objective to maximize the benefits that modern societies can derive from these.

5.2 Big Science and the development of modern industrial infrastructure

The expression 'Big Science' is understood today to comprise large scale installations or projects dedicated to scientific and technological investigation, requiring the collaboration of thousands of workers over many years, significant funding (mostly public), entailing high risk and often leading to advancements in human knowledge and understanding. The main objective of these installations is to facilitate scientific and/or technological breakthroughs, although in most cases this is coupled with political, economic, and social targets that are proportionate to the size of the undertaking. Most Big Science projects also aim to promote international collaboration ('science diplomacy', of which the ITER project is a prime example [1]) and to establish the prestige, influence, and legacy of the funding nation(s).

Big Science is not a modern concept. Indeed, one may argue that Big Science-type objectives were pursued as early as 3000 BC with the construction of Stonehenge in the United Kingdom[1] (figure 5.1) and were a decisive element for Egyptian pyramids in 2500 BC. For both these examples, large (for the time) construction activities were promoted by the respective rulers as a demonstration of power and prestige (i.e. the science diplomacy aspect) and according to many studies to then operate as astronomical observatories or timekeeping devices (i.e. the knowledge advancement aspect). However, the expression 'Big Science' came into use during the 1960s [2, 3]

[1] Incidentally the diameter of Stonehenge's sandstone circle (the Sarsen Circle, 33 m) is almost equal to the diameter of the ITER tokamak core (30 m).

Figure 5.1. Stonehenge: Big Science, circa 3000 BC. (Photo from [https://unsplash.com/photos/U9ZXRF7LA8I]. Credit: Cajeo Zhang, Unsplash, 2020.)

to describe the coordinated, international projects which became a dominant feature of the world's science and technology landscape during the second half on the twentieth century.

Although modern Big Science has been traditionally linked to large, concentrated installations devoted to the development of the physical and engineering sciences, more recently the meaning has broadened to include distributed, non-material facilities in which large scale coordination is implemented through a highly connected network of resources which are geographically far away (e.g. the Human Genome Project). Consequently, the understanding of Big Science has enlarged to include projects stemming from all fields of modern science and technology [4].

Regardless of the field of application or the nature of the facility (physical or immaterial), Big Science projects are invariably born out of a symbiotic relationship between scientific, technological, and industrial ambitions. To be truly innovative, Big Science projects need to push the boundaries of the technological state-of-the-art, which requires a close collaboration of scientists with the industrial communities. In turn, the challenging objectives allow (or even force) industry to undertake risky developments which would in most cases not be pursued under a purely market-driven logic.

In particular, during the twentieth century, Big Science has been a key driver of social, industrial, and economic development in the most advanced part of the world: landmark developments such as nuclear energy, the World Wide Web, wired and wireless digital information networks, DNA sequencing, spaceflight, etc, were born largely through and thanks to Big Science. Less visibly but not less importantly, the industrial infrastructure and capability of the countries undertaking these projects has been shaped by them, with an enormous impact on their innovation potential, competitiveness, and technological leadership. The enormous pay-off was quickly understood by governments [5], which generously funded Big Science projects during the second half of the century and received proportionally large benefits. Several twentieth century Big Science projects were so large that they accounted for almost 1% of the total funding countries' GDP at their peak.

The driving technological role, the socio-economic benefits, and the enormous impact of twentieth century Big Science were clearly evidenced, measured, and demonstrated during that timeframe. Although a lively debate took place about the role of Big Science in the broader scientific development framework and about its prioritization in the allocation of public funding [6], the general consensus remains positive about the fact that most of these projects delivered net benefits to the nations undertaking them [7].

5.3 A changing context

Now, at the beginning of the twenty-first century, Big Science projects are developing in a quickly changing socio-economic environment. In absolute terms the funding levels have remained relatively constant over the years (figure 5.2). For example, European countries are supporting several Big Science projects for a total expenditure in the order of 10 billion EUR per year [8], a figure which has been roughly constant since the year 2000. However, the size of the underlying economies has grown significantly and today even the largest Big Science projects account for not more than 0.01%–0.05% of the underlying GDP. Big Science is no longer one of

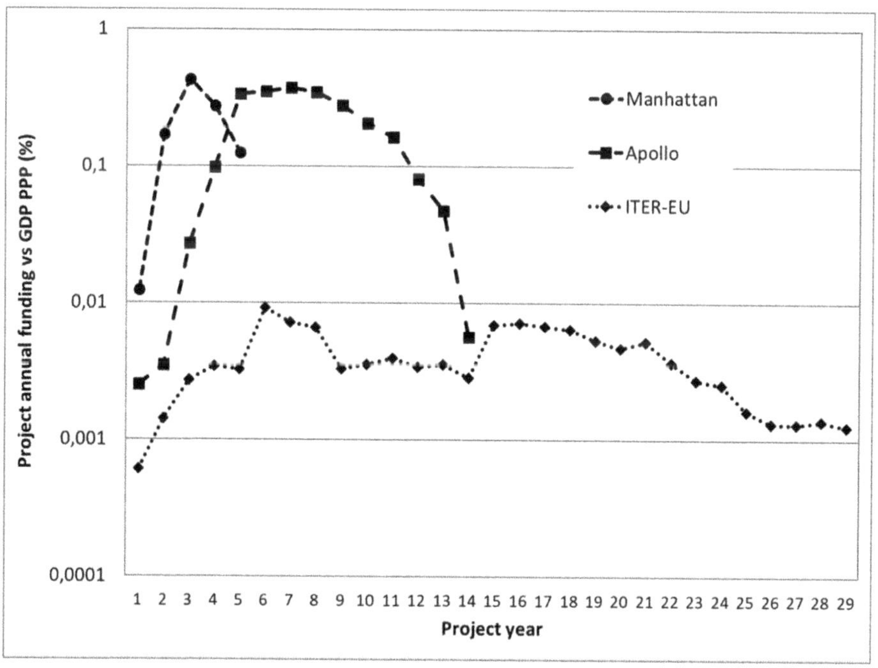

Figure 5.2. Big Science project annual funding versus underlying GDP (produced on the basis of funding data from [9, 10], GDP data from the US Bureau of Economic Analysis and EUROSTAT).

the dominant drivers for industrial and technological development, being displaced by other social and economic macro-trends (e.g. wireless information exchange, globalization and decentralization of manufacturing processes, etc).

As the complexity of Big Science projects has increased, so has the timespan of their lifecycle. The largest twentieth century Big Science projects covered a time from inception to conclusion of the order of 10 years. At the beginning of the twenty-first century, the largest Big Science projects under construction or being proposed cover a timespan of more than 30 years [11], which means that these projects will require two generations of scientists and engineers, and several government funding cycles, in order to reach their operational stage.

In addition to the very practical challenges that these expanding timespans imply (e.g. how to transfer the know-how from one generation to another, how to manage flexible scientific objectives in consideration of possible changes in priorities, how to establish and maintain a stable and long term supply chain, how to ensure long term political backing, etc) a different type of question emerges. To which socio-economic environment will these facilities contribute once they reach their operational phase, 30 years in the future? How can we estimate the benefits obtained through their operation, when in today's quickly changing, globalized world we can barely imagine how society and economy will look like in 5 or 10 years' time?

The intrinsic value of increased scientific knowledge for the long term development of humankind and the lasting impact of major science and technology undertakings remain unchallenged. However, it is much more difficult today to estimate the economic return that Big Science facilities will generate when operating and producing scientific results many years into the future [12]. As a consequence, an important debate has started within scientific, political, and industrial communities about the role that Big Science will have in the twenty-first century and on what basis decisions will be taken about which projects get funded and which unfortunately do not.

Among the many threads in this debate, an important one is about the indicators used to assess the socio-economic impact of Big Science projects, to define their objectives and to evaluate their performance. New aspects of the project lifecycle must be brought into focus in this context—while past cost–benefit analysis and performance evaluations focused on the facility impact during its operation [13, 14], the present understanding is that a significant portion of beneficial effects is already present during the development/construction phase of a large scale, high technology facility. This is not a new concept *per se*, e.g. the economic impact of setting up and growing the Apollo programme's supply chain has been studied and analysed using several approaches in the past [15]. However, this was done *a posteriori* (after the programme conclusion) and, while this aspect was indeed considered in the process leading to the decision to start the programme [16], it was not among its explicit objectives.

5.4 Which fruit is ripening first?

The development and construction of a Big Science facility generates immediate positive socio-economic impact through several mechanisms. In this section the three main ones will be discussed: increased industrial competitiveness, concurrent technology transfer, and spending in higher value added industries. These areas are among those 'low hanging fruits' which ripen first and can therefore provide short-term positive impact during the early years of a major project.

A modern Big Science project requires a high level of partnership and integration among the scientific and industrial communities. The scale, system complexity, and technological difficulty of a large science infrastructure very often exceed the state-of-the-art, requiring that the two communities merge and interoperate in unprecedented ways. This has profound and beneficial effects on both.

To successfully drive large projects to completion, scientists and technologists are required to develop skills that are not often part of their education and training: project management, system engineering, financial planning, as well as political and economic negotiation skills. These skills are sometimes acquired, sometimes inherited or injected from other projects, but always bring permanent changes in the receiving community and in its capacity to conceive and manage future developments.

On the other hand, industry needs to operate in an environment in which large scale production occurs in a new, unmapped and highly uncertain environment. New processes, methods, and technologies must often be adopted with a high degree of uncertainty on their outcome; rapidly evolving designs imply that most items produced are in fact first-of-a-kind; small manufacturing batches mean that the staff almost constantly work in the steep part of the learning curve. All these aspects push industry to move out of their comfort zone (low risk, recurring business) and into areas requiring a high adaptability which, particularly in the present fast-moving economy, brings higher competitiveness and agility. Like athletes training at high altitude, exercising through Big Science projects pushes industry to their limits, which in turn enable them to innovate and grow in their core business and to open new markets.

This effect will be referred to as type I in the following sections. It can be observed shortly after the start of a large project's industrial phase (3–5 years) and can be measured and analysed through direct field data collection.

Exploring nature in new and unmapped directions requires adopting novel methods, i.e. technologies, in the broadest meaning of the word[2]. This is invariably true, regardless of the field of exploration, be it physical, geographical, biological, psychological, or something else [17]. When the scale of this exploration increases, so does the extent and ambition of the new technologies. Science projects require technologies which are beyond the state-of-the-art: detectors, imaging equipment, remote handling and robotics, cryogenic, data handling and transmission, and thousands more technologies are developed further by the research and industrial communities to allow for novel scientific infrastructure.

Joint development beyond the state-of-the-art has long been a characteristic of Big Science. For example, ocean exploration during the seventeenth to eighteenth centuries was a true Big Science undertaking, with large government funded

[2] From the Greek word *techne*, meaning art, craft.

activities being pursued to answer strategic needs through the expansion of human knowledge. In this context, clock artisans with little or no prior involvement in sea navigation undertook significant research and development to produce novel high accuracy watches suitable for sea use, following a 1714 'call for tender' by the British Parliament and eventually leading to the invention of the compact chronometer [18]. Similarly to their twenty-first century descendants, this eighteenth century technological development quickly became available for broader applications impacting every-day human activities.

This process is known as technology transfer and has long been recognized as the most powerful source of long term benefit from Big Science projects [19]. For twentieth century Big Science, the shorter duration of project development activities and longer technology adoption times meant that technology transfer efforts often emerged in the decades after the conclusion of the project. A typical example is provided by the Apollo programme, for which the most evident technology transfer cases occurred up to 20 years after the conclusion of the programme in 1975 [24]. However, the faster technology adoption cycles characterizing twenty-first century economy mean that differently to the past new technologies are in fact already transferred during the development and construction phases of a Big Science project. Cross-breeding between different applications of a new technology still under development eventually benefits both sides of the transfer: the originating Big Science project and the receiving commercial application.

This effect will be referred to as type II or 'technology transfer' in the following sections. It can be observed several years after the start of a large project's industrial phase (5–10 years) and can be measured and analysed through direct field data collection. Its effects may take more than 10 years to become relevant for society; still, this is the effect that can be most easily communicated to the general public.

Compared to other types of public infrastructure projects, Big Science engages a proportionally much larger fraction of the spending/investment in areas which entail complex technological processes. Even without considering leading-edge technology developments, these projects require extensive manufacturing and fabrication activities, digital equipment, software, etc, as well as specialized human resources. These industrial areas are well known to economists as generating large added value to the economy of modern countries.

Economists generally agree that public spending propels economic growth, at least up to a certain fraction of public expenditure versus GDP. However, the effect of public spending in different areas has shown to generate significantly different returns [20]. Anecdotal evidence shows that public spending on the type of activity necessary for Big Science projects generates a higher than average return for the total public investment [25]. This trend is confirmed for the ITER case, as detailed in the following section.

This effect will be referred to as type III in the following sections. It can be observed almost immediately after the start of a large project's industrial phase (2–3 years) and can be measured and analysed through indirect methods (e.g. econometric models).

5.5 The European ITER case

The ITER project is a global collaboration among seven parties (China, India, Japan, Russia, South Korea, the United States, and the European Union) intended to demonstrate the scientific feasibility of fusion on a scale needed for a future energy source[3]. The project is organized through in-kind contribution from the parties, providing goods and services for the construction of a single facility (the largest ever superconducting tokamak) in southern France, operated by the ITER International Organisation (IO). The ITER programme dates to 1986, with design, development, and prototyping activities starting in 1992. However, the actual construction started in 2007 and will progress, though several construction and operation phases, until at least 2035 (figure 5.3).

ITER is the epitome of twenty-first century Big Science, with significant political implications, unprecedented magnitude, wide adoption of new technologies, and multi-generational timescale.

The scale (in industrial, technological, and financial terms) of the European contribution to the ITER project (EU-ITER) has brought significant attention to the effect that these activities have on the European socio-economic framework. For this reason, since 2012 Fusion for Energy[4] (F4E, the organization through which the European contribution to ITER is provided) and the European Commission have started recording and analyzing data about the real-time (or short-term) impact of the design, manufacturing, and research contracts awarded to European industry and research institutions.

Figure 5.3. Aerial view of the ITER site during construction. (Source: ITER Organisation http://www.iter.org/, May 2021.)

[3] The ITER project, www.iter.org.

[4] The European Joint Undertaking for ITER and the Development of Fusion Energy.

In 2017 and 2020, the European Commission sponsored two econometric studies [26, 27] that evaluated the impact that ITER-related design and construction activities are generating (and will generate) on the economy of European members. The studies utilize different economic models and methodologies, however, they both use as input an extensive dataset developed by F4E and more recently by IO, and yield as outputs the impact of the ITER construction contracts on the European gross value added (GVA) and employment. Thus, the two studies address quantitatively type III effects, although both discuss types I and II in qualitative terms.

Detailed discussion about the methods followed and the results are contained in the final reports from the studies. In the following paragraphs additional considerations stemming from those results will be presented.

In spite of the different econometric models used, both studies confirm that in addition to the direct GVA (which can be considered as a proxy for the gross margin from involved companies) the ITER construction activities generate significant indirect and induced GVA across the European member states. When evaluated from the start of construction until 2019 (the latest year covered by the studies), the ratio between the direct and indirect GVA impact is estimated at 1.2—this means that for each euro of value added activities performed under ITER contracts, the economy overall benefited by more than 1 euro in indirect and induced effects. Also of relevance, this ratio increased from about 0.5 in the early years (when the project was therefore generating little indirect and induced impact) and is expected to remain around 1 over the whole duration of the construction, i.e. until 2035. This led the study authors to a first important observation, in that '[…] while the impact […] had been dominated by funding for ITER itself in the earlier years […], the economic benefits of ITER spending has spread from the ITER plant into the wider economy and the supply chain […]' as the project progressed.

When looking at the ratio between jobs created directly by ITER activities and the indirect job creation effect, the trend is similar: over the 2007–19 timeframe the job ratio is evaluated at 1.2, implying that more than one indirect job is created on top of each job directly linked to ITER contracts.

The GVA and job ratios for different economic sectors are, as expected, significantly different. The ITER project is found to have the largest positive effect on the industrial sector—in this case the sectorial GVA ratio during the duration of the project is estimated to be almost continuously increasing, peaking at 2.4 in 2035. This is a very similar ratio to what found for other industry-intensive high technology programmes in Europe, such as the European Space Agency's launch vehicle developments. Similarly, the industry job multiplier will reach 2.9 in 2035, signalling that by the end of the construction '[…] for every job created in direct relation to ITER's industrial activities, almost three more jobs are created in the supply chain of ITER and the wider industry of Europe's economy'.

In addition to the analysis of the gross GVA (which constitutes an indicator of the difference between ITER spending and no spending), both studies also attempt to provide a first answer to the question: what if the ITER funds were spent in a different way by the public sector? The net impact evaluations quantify the difference between the ITER spending scenario and an alternative, reference scenario where funds would have been transferred to non-ITER-related investments and triggered different effects in other parts of the economy.

Again, the two econometric models agree with the conclusion that ITER spending generates overall a positive net value added (NVA) with respect to the reference scenarios. In other words, public funds allocated to ITER generate a larger benefit for the EU economy (and in particular to its industrial sector) than generic, non-targeted public expenditure. The ITER NVA is estimated by both econometric models to be positive by more than 5% of the base figures. In particular the 2020 study, benefiting from a longer data time series, concludes that while direct effects are almost equivalent (hence a direct NVA close to zero) the indirect and induced NVA from ITER activities is 104 million euro during 2008–19. This indicates that the ITER construction generated over this timeframe a more than 12% higher indirect and induced GVA than the reference scenario (871 versus 767 million euro). The NVA from ITER construction over the full 2008–35 period is estimated to be, respectively, 586 million euro (2017 study) and 332 million euro (2020 study, which even considers the negative effect of the COVID-19 crisis). The large estimated NVA for ITER confirms previous observations about the higher proportional positive impact from 'mission-oriented' R&D projects and policies [21], of which Big Science projects are the extreme case.

Similarly to the GVA impact, the NVA becomes positive a few years after the start of the project (figure 5.4). In the ITER case, the NVA remained slightly

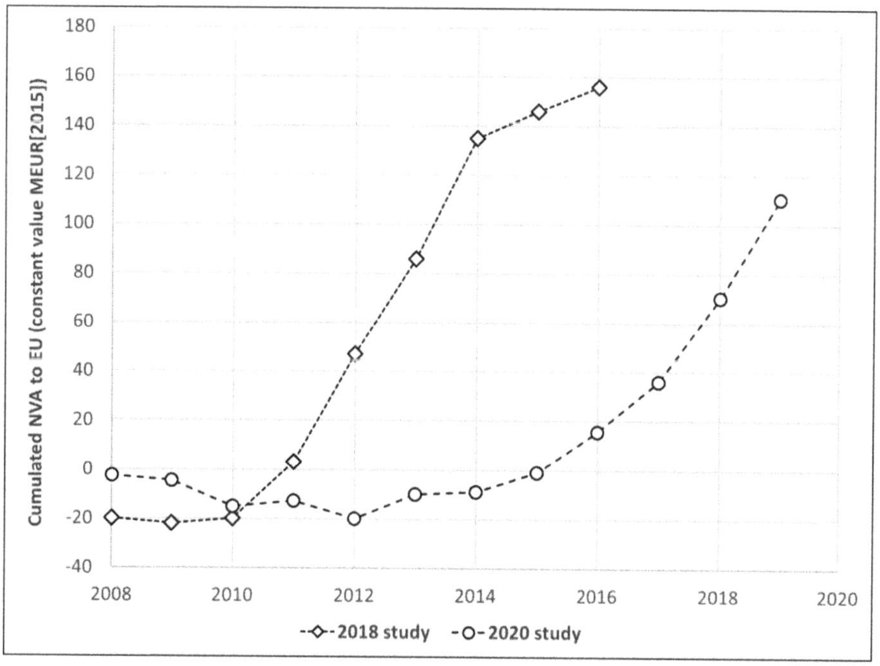

Figure 5.4. ITER net added value with respect to the reference scenario. (Source: Trinomics, 2018, IHS Markit, 2020.)

negative until 2010–11 and only began to grow significantly as of 2013. This was caused by the fact that ITER started as a 'green field' project, therefore initial project expenses were mostly low-added value site preparation ones. After 2013 the NVA remains positive with an increasing trend. This confirms the previously discussed general characteristic of twenty-first century Big Science projects—due to the longer construction timescales, the positive net effect of the activities starts being significant well before the end of the infrastructure completion date. The time lag between start of the funding ramp-up and significant positive economic effect is estimated at around 5 years for the ITER project.

Neither of the two studies includes the effect of technology transfer in their estimations. However, they both provide anecdotal evidence of the fact that the process already started (providing actual examples, qualitatively described). The 2017 study estimates that this additional contribution can generate an increase of the GVA output over the whole construction phase by 15%–60%.

When comparing this large, unexplored potential (15%–60%) to the more modest econometric results (12%) it can be easily recognized that accelerating the technology transfer processes can quickly become the driving element in the socio-economic impact of a Big Science project such as ITER. While this has long been an important element in the strategy of Big Science organizations, the results from the ITER project confirm previous observations [22] and further highlight that to maximize the benefits to it is important to start promoting and managing technology transfer at the very beginning of the project.

5.6 Industrial policies for Big Science

The decision to proceed with a large infrastructure project, be it scientific (Big Science) or otherwise, is primarily driven by long term strategic considerations regarding social, economic, scientific, and technological progress, development opportunities and sometimes national prestige.

The decision-making process can be made easier by recognizing the fact that modern Big Science facilities can have a more than proportional short-term benefit deriving from the construction itself. This benefit starts to materialize well before the facility becomes operational and helps to boost the return of investment balance during the critical construction years, when uncertainty about the socio-economic environment in which the facility will operate is at its peak.

Short-term (or immediate) benefit from development and construction activities occurs naturally in any large undertaking. The experience from the ITER project and other major twenty-first century Big Science projects highlights that these benefits materialize through several mechanisms, and that some of these can be designed by the stakeholders in the project objectives in a way which increases the benefits. The policies supporting a new Big Science project can be designed in order to generate early return, in certain cases even its design and scientific objectives could be adapted to promote a more sustainable return of investment and help broaden acceptance by stakeholders (including the general public).

Often the focus on broader socio-economic impact for a Big Science project is channelled by stakeholders through an industrial policy, containing a strategic set of objectives intended specifically to ensure that the projects find a trade-off between the purely operational objectives (performance of the facility, timely completion, etc) and those related to its interaction with the supply chain and the economic system. A proper, well thought-through industrial policy or strategy is an essential element to ensure that significant short-term socio-economic impact materializes from a Big Science project. Its implementation should be monitored constantly by the institutional stakeholders, through specific and measurable objectives and should influence all important decisions in the project.

A few examples of relevant topics are discussed in the following paragraphs.

Among the aspects covered by a proper industrial policy, it is impossible to overestimate the importance of putting in place a strong technology transfer programme in the early stages of the project. Evidence from previous studies [23] shows that a proportionally large fraction of potential spin-offs is generated during the initial phases of a project—to capture this opportunity, focus on technology transfer should not be delayed. As discussed above, the results of type II effects take significant time to become evident (between 5 and 10 years) but they can generate by far the largest and most durable positive effect of Big Science spending. These effects can be favoured by appropriately allocating ownership and rights of the intellectual property (IP) generated through the project activities, For example, in many cases the Big Sciences organization only needs the right to use this IP, but ownership can be left to industry, thus promoting a faster valorization and exploitation. In addition, the design choices in a Big Science system or components should whenever possible favour the use of technologies which are not too specific and sectorial, or the potential for transfer can be significantly reduced.

Several examples of top-driven technology transfer programmes have been started by European Big Science organizations. Notable in this area are the efforts by ESA and CERN that have reached maturity and have become a stable feature in the performance evaluations for these organizations.

A second area of concern for industrial policy, should be designing procurement and contracting practices with a view to maximize type I effects, i.e. stable improvements in the industrial organization of companies involved.

In this regard it is important to notice that more often than not the key industrial actors in a Big Science project are neither small nor large companies, but medium-sized ones. These companies have the production and financial capacity to undertake contracts on the scale and complexity required by a large scientific infrastructure; have the appetite for novelty, entrepreneurial spirit, and adaptability which is often missing in risk-averse large companies; and have the financial stability and long term perspective which small companies have not yet achieved.

From the above considerations it follows that Big Science industrial policies should target medium-sized companies as their first layer of strategic suppliers and promote the progression of small companies to medium size through organic growth or aggregation. This can be achieved, for example, by opposing the often-seen trend of releasing scope to industry in a fragmented, piecemeal approach and rather structuring contracts that cover a comprehensive and multidisciplinary scope.

Finally, as the boundary conditions within which a Big Science project operates changes constantly, the industrial policy should be seen as a living document that evolves following the developments occurring around the project. The driving dynamics and business models in industrial markets, and more specifically the project's supply chain, will seldom remain constant over a timescale of decades. Therefore, the way in which the different beneficial impacts are pursued (i.e. its industrial policy) should be continuously re-evaluated and if necessary updated.

A successful Big Science project is an entity living within and in symbiotic connection with the social and economic fabric. The project receives from it the resources (material and immaterial) needed to develop its potential and returns the knowledge and inspirational fruits which in turn help society to advance and prosper.

Acknowledgements and disclaimer

The author wishes to acknowledge the interesting discussions and ideas brought on this subject by several colleagues at Fusion for Energy and the European Commission, in particular but not limited to: Victor Saez Lopez-Barrantes, Raymond Monk, Luigino Petrizzi, Christopher Ibbott, Carles Dedeu Fontcuberta, and Laban Coblentz.

The opinions expressed in this work are the author's own and should not necessarily be interpreted as the view of Fusion for Energy, the ITER Organisation, or the European Commission.

References

[1] Harding T K, Khanna M J and Orbach R L 2012 International fusion energy cooperation: ITER as a case study in science and diplomacy *Science Diplomacy* **1** March 2012

[2] Weinberg A M 1961 Impact of large-scale science on the United States *Science* **134** 161–4

[3] De Solla Price D J 1963 *Little Science, Big Science* (New York: Columbia University Press)

[4] Vermeulen N and Penders B 2007 Big Science *Encyclopedia of Earth* ed C J Cleveland and K Vranes (Washington, DC: National Council for Science and the Environment)

[5] Bush V 1945 *Science: The Endless Frontier* (Washington, DC: US Office of Scientific Research and Development)

[6] Gisler S 2008 'Bubbles in Society': the example of the United States Apollo Programme arXiv:0806.0273

[7] Wareham J 2020 Big Science infrastructures, technology spinoffs and social impact *Do Better* 25 February https://dobetter.esade.edu/en/social-impact

[8] Big Science Sekretariatet 2014 Vejen til en Big Science industri i Danmark *BigScience.dk* https://bigscience.dk/media/1084/vejen_til_big_science_industri_danmark_2014.pdf

[9] Stine D D 2009 The Manhattan Project, the Apollo Program, and Federal Energy Technology R&D programs: a comparative analysis *CRS Report for Congress* RL34645 Congressional Research Service

[10] European Commission 2017 The EU contribution to a reformed ITER project *Communication from the Commission to the European Parliament and the Council* COM (2017)319 European Commission

[11] Benedikt M, Blondel A, Janot P, Mangano M and Zimmerman F 2020 Future circular collider succeeding the LHC *Nat. Phys.* **16** 402–7

[12] Hallonsten O 2016 *Big Science Transformed* (London: Palgrave Macmillan)

[13] Glasson J, Weston A and Chadwick J 1995 *The Joint European Thorus Project Impact on the Local Economy* (Oxford: Oxford Brookes University Press)

[14] OECD 2014 *The Impact of Large Research Infrastructures on Economic Innovation and on Society: Case Studies at CERN* (Paris: OECD)

[15] Johnson S B 2006 *The Secret of Apollo* (Baltimore, MD: John Hopkins University Press)

[16] Sorensen T C 2008 *Counselor: A Life at the Edge of History* (New York: Harper)

[17] Vermeulen N 2009 Supersizing science: on building large-scale research projects in biology *PhD Thesis* Maastricht University

[18] Sobel D 2007 *Longitude* (London: Bloomsbury)

[19] Scarra D and Piccaluga A 2020 *The Impact of Technology Transfer and Knowledge Spillover from Big Science: A Literature Review, Technovation* (Amsterdam: Elsevier)

[20] Mandl U, Dierx A and Ilzkovitz F 2008 The effectiveness and efficiency of public spending *Economic Papers* 301 European Commission

[21] Mazzucato M 2015 *Mission Oriented Research and Innovation in the European Union* (Brussels: European Commission)

[22] Autio E 2014 Innovation from Big Science: enhancing the Big Science impact agenda *UK Department for Business Innovation and Skills Report* BIS/14/618

[23] Bach L, Cohendet P and Schenk E 2002 Technology transfers from the European Space Programs: a dynamic view and comparison with other R&D projects *J. Technol Transfer* **27** 321–38

[24] NASA 2004 *Benefits from Apollo* NASA FS-2004–07–002-JSC

[25] PwC 2014 Socio-economic impact assessment of access to space in Europe: an ex-post analysis of the Ariane 5 and Vega programmes *ESA Contract* Ref. 4000 110 988/14/F/MOS

[26] Rademaekers K *et al* 2018 Study on the impact of the ITER activities in the EU *Trinomics* https://trinomics.eu/project/iter-impacts/

[27] Sertaç E *et al* 2021 Follow up study on the economic benefits of ITER and BA projects to EU industry *LGI and HIS Markit, European Commission* https://op.europa.eu/en/publication-detail/-/publication/3db11048-6a89-11eb-aeb5-01aa75ed71a1/language-en

Chapter 6

Developing a business case for international research infrastructures: the European Spallation Source

John Womersley

In this contribution I will explore how to make a strong business case for investment in an international research facility, using the European Spallation Source (ESS) as a case study and example. ESS is a three billion euro project to build the world's most intense neutron scattering facility on a green-field site in Sweden. At the time of writing, ESS is roughly 75% complete and science experiments should begin towards the middle of the decade.

I will start, somewhat unashamedly, from the viewpoint that scientific and technical innovation is essential if we are to address the biggest problems we face. These include the global challenges of energy, climate, environment, and healthcare—it is not enough to roll out existing technologies and demand societal adjustment, our toolkit is simply not sufficient. The flagship report *Net Zero by 2050* by the International Energy Agency [1] makes this point very clearly—something like half of the reductions in CO_2 emissions needed for a carbon neutral future by 2050 depend upon technologies that do not yet exist. Scientific and technological innovation is also essential to address the economic and societal challenges in developed economies driven by deindustrialisation, stalled productivity, and long term wage stagnation—combined now with a likely post-Covid economic shock that we have yet to fully experience. These are complex problems and I do not believe anyone really has the answer—but we do know that innovation and STEM skills will position economies and societies dramatically better to weather these challenges by generating higher-value knowledge-based employment. And of course, the experience with Covid-19 during 2020 and 2021 has served to dramatically re-emphasise the importance of scientific capacity and capability in dealing not just with the challenges we know about, but with the unexpected.

6.1 Research infrastructures

Scientific and technical innovation takes place in a complex ecosystem that requires strong support for higher education, for university and laboratory-based research, for business, and for knowledge exchange between these actors. Here we will focus on research infrastructures, by which I mean national and international scale investments in shared research capability that go beyond what any single institution (or even nation) can support. Such projects have a key role to play in the overall success of the ecosystem because they dramatically increase scientific reach, address research questions of long duration requiring pooled effort, and promote collaboration, interdisciplinarity, and interaction with industry. Figure 6.1 attempts to show schematically how research infrastructures complement and extend the reach of university and industry based research teams and fill a unique and complementary role in the ecosystem. However, such billion-€/$ scale investments naturally attract a lot of scrutiny and therefore require a robust investment case and pooling of resources.

In 2016 The European Strategy Forum for Research Infrastructures (which I chaired at the time) reviewed the implementation progress of the projects on its roadmap [2], in particular those which had been added in 2006 but which had still not proceeded to implementation ten years later. We found that inadequate stakeholder engagement and the lack of a credible funding plan were the biggest barriers—much more so than any weakness of the science case. This points to important areas of the investment case that must not be neglected.

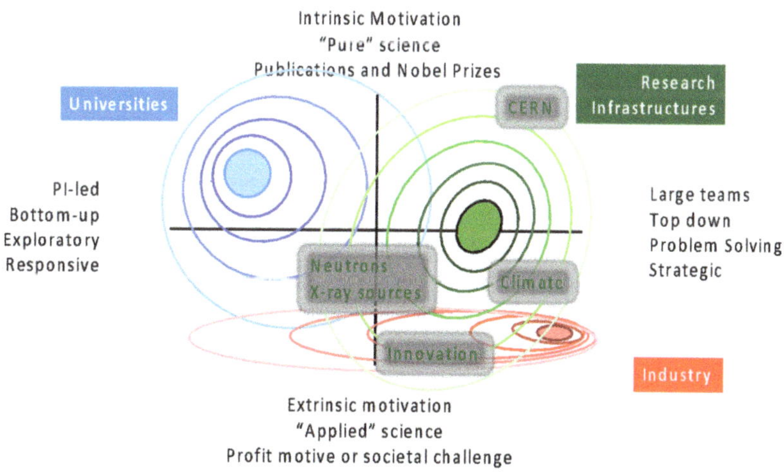

Figure 6.1. Schematic representation of the motivations for research (vertical axis, from 'pure' science motivated by intrinsic rewards to more 'applied' research motivated by profit or the desire to solve problems external to science) and the style of work (horizontal axis, from individual researchers writing bottom-up proposals to large teams addressing strategic challenges). The space typically occupied by university and industry based research is shown in blue and red, respectively, and that made possible by large research infrastructures in green.

An extreme example of this situation is the Superconducting Super Collider, a project to build what would have been the world's most powerful particle accelerator in Texas in the United States. The science case and the technical design were robust and had the project been completed it would have discovered the Higgs boson long before CERN did so. However, stakeholder support in the US Congress was not deeply grounded. When cost and schedule issues started to accumulate, and a new administration did not make it a priority, Congress cancelled the project in 1993 after several billion dollars had already been spent [3]. It is no exaggeration to say that particle physics in the United States has never fully recovered from this blow.

The key requirements in building sustainable support for any science megaproject are listed below:

- A strong science case, endorsed by acknowledged experts in the field.
- Technical R&D carried out to the level where the remaining risks are understood and cost estimates can be made, including appropriate contingency at perhaps the 30% level.
- A project management plan following international best practices.
- A credible funding and governance plan.
- Stakeholder engagement—support for the project from a wide range of actors.
- A compelling investment case—including the anticipated societal benefits of the project.

Since the end of the Second World War, there have been three broad waves of investment in research infrastructures for Big Science:

1. The first wave grew out of the technologies that had been developed during the war, and focused on accelerators for nuclear and particle physics, large optical and radio telescopes, and space projects. This is still what most people think of when they think of 'Big Science'. Examples in Europe include CERN, FAIR, ESO, and ESA.
2. The second wave took technologies such as nuclear reactors and particle accelerators and exploited them as user facilities for very large numbers of scientists typically from university-based research groups, interested in the structures of materials and molecules. Examples in Europe include ILL, ESRF, and ESS. In many cases such facilities have been located so as to exploit and re-purpose existing national laboratories created in the first wave—examples include Brookhaven, SLAC, and LBNL in the United States, and DESY and the Rutherford Appleton Laboratory in Europe.
3. The third and most recent wave is focused on access not to facilities but to data. A large number of new pan-European research infrastructures have been established over the past decade bringing together national centres into European-wide groupings and allowing access to digital datasets spanning the continent. This has expanded the reach of 'Big Science' into areas such as bioinformatics, climate, the environment, social sciences, and humanities. Examples in Europe include the European Social Survey, CLARIN, and ELIXIR. The European Open Science Cloud is the logical next step in this development.

6.2 The evolving decision-making context

Since the end of the cold war, there has been a pronounced shift in emphasis in the priorities that governments use in judging the returns from Big Science.

Figure 6.2 shows, on the left, the sort of values that science investments were broadly expected to promote in roughly the period 1950–90. There was an emphasis on science as an exemplar of national values, whether they be openness and democracy, or the superiority of the socialist planned economy. Of course there was also an element of hypocrisy here, as governments also expected Big Science to demonstrate industrial and technical capacity relevant to defence, particularly in areas such as nuclear physics and space. Since roughly 1990, however, the pendulum has swung much more to the right. Big science is now expected to deliver tangible economic and societal benefits such as jobs and growth, or solutions to grand challenges such as climate. It is not enough to show excellence in scientific enquiry as an end in itself. In some ways the science community is happier on the left of this chart—it is closer to scientists' own value system and a lot simpler to be judged on one's cultural value than on one's contribution to job creation. Nonetheless, the right-hand side is where decisions are made, and any successful project must find a way to engage with these priorities in constructing its business case.

6.3 The European Spallation Source

Let us now take a closer look at the European Spallation Source (ESS) to see how a successful business case for a €3B new investment in scientific capability can be justified.

ESS is a neutron scattering facility—it is a machine that directs intense beams of subatomic particles at samples to answer the question 'where are the atoms and what do they do?' This kind of research has a long and successful track record in Europe. It is key to the development of new materials, new drugs, new chemical processes, food technology, engineering, and new energy capture and storage technologies. Two very topical examples of neutron research that are highly relevant to today's challenges and being carried out right now include understanding the structure of

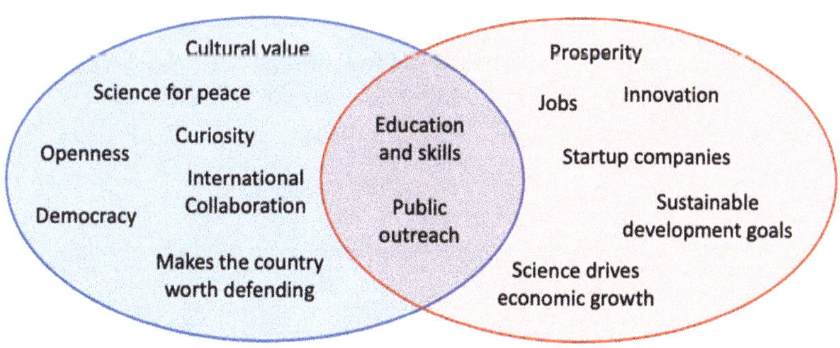

Figure 6.2. More traditional (in blue) and current (in pink) justifications for investments in 'Big Science' in terms of its broader return to the nation. The shift from blue to pink corresponded roughly with the end of the Cold War around 1990.

lipid nanoparticles used as a delivery mechanism for mRNA therapy, as used for Covid-19, and understanding materials for future higher energy-density electric vehicle batteries as part of the European Union's Battery 2030 initiative.

Figure 6.3 shows an aerial view of the facility under construction. Its layout follows its function. The long straight building extending towards the top of the picture traces the path of the underground superconducting particle accelerator, which will be the world's most powerful (5 MW beam power) when fully complete. It will accelerate protons to roughly 2 GeV in energy, or about 90% of the speed of light. This beam of protons will then be directed on to a neutron production target housed in the tall building in the centre. 'Spallation' is simply the scientific term for the nuclear physics process by which the proton beam liberates neutrons from the tungsten atoms in the target. These neutrons are then captured and cooled in a liquid hydrogen 'moderator' before being directed along beam guides to 15 scientific instruments located in three large experimental halls, one on either side of the target building and one fan-shaped structure to the lower left. Each instrument will have distinct and complementary capabilities and these are where the scientific experiments will be carried out. The other buildings dotted around the campus contain offices, cafeteria, lecture hall, maintenance laboratories, and utilities including electrical power, cooling water, and a cryogenic liquid helium system to maintain the particle accelerator close to absolute zero in temperature. Each of the experimental stations at ESS will be on average twenty times more sensitive than today's best anywhere in the world. The facility will deliver around 800 scientific experiments per year, each of a few days' duration, serving many thousands of European scientists.

ESS is being constructed on a green-field site just outside Lund in southern Sweden, roughly 45 minutes travel from Copenhagen and its international airport, and next to the Swedish national light source MAX IV. The roughly €3B investment covering the construction, commissioning, and initial operations of ESS up to 2025

Figure 6.3. Aerial view of the European Spallation Source in spring 2021. The project was roughly 75% complete at this point; all the major buildings were complete and installation and commissioning of technical equipment was underway. Credit: Perry Nordeng/ESS.

has been made by a partnership of 13 European nations. The two host countries, Sweden and Denmark, are contributing 47% of the construction cost; the remainder is covered by the Czech Republic, Estonia, France, Germany, Hungary, Italy, Norway, Poland, Switzerland, Spain, and the United Kingdom. The project is legally constituted as a European Research Infrastructure Consortium or ERIC, with these 13 countries as members. The ERIC framework allows an international organisation to be established relatively simply and gives it certain privileges under EU law, including exemption from VAT and the ability to set its own procurement rules.

6.4 Scientific impact

Of course the science impact of any facility is the key deliverable. ESS's vision is to build and operate the world's most powerful neutron source, enabling scientific breakthroughs in research related to materials, energy, health, and the environment, and addressing some of the most important societal challenges of our time. Since the science impact of any facility cannot be measured in advance, the ESS business case was based on the relevance of ESS research to key challenges (energy, environment, life sciences), current European strength in these areas, and the ability for ESS to do ten to twenty times better, as shown simply and graphically in figure 6.4. The data plotted here are accurate, but this is very much an advocacy plot and not a scientific plot, as there are many other factors to be considered in judging the scientific capability of each of these facilities. Figure 6.4 also helpfully makes the point that other regions outside Europe, notably Japan (J-PARC), the United States (SNS), and China (CSNS), are investing heavily in similar facilities, emphasising the need

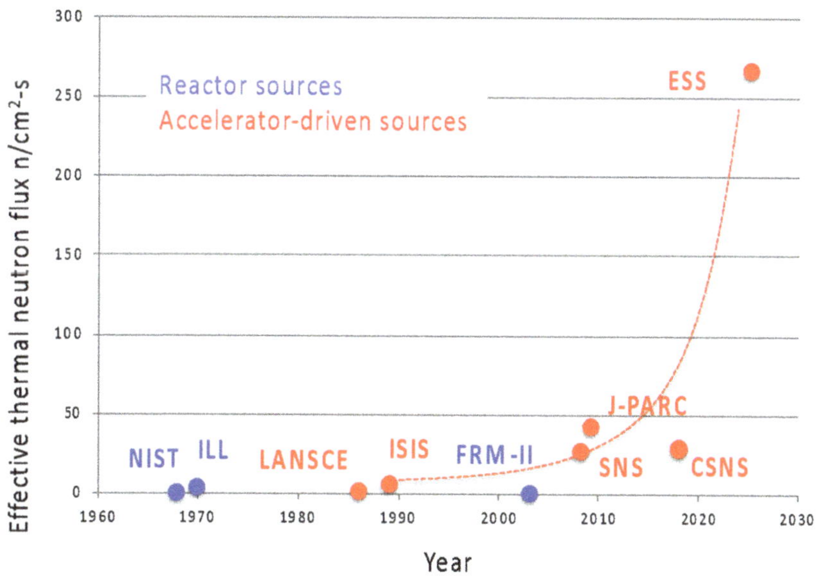

Figure 6.4. Effective thermal neutron flux from a selection of neutron sources plotted as a function of the first year of operation. Reactor sources are shown in blue and accelerator based sources in red.

for Europe to protect is current strong position. It was also important to involve key industry partners and persuade them to act as advocates—so that it was not just scientists making the case for more investment in science.

6.5 In-kind contributions

The business case for ESS in the partner countries was strengthened by the central role foreseen for in-kind contributions from the partners. By in-kind contributions, we mean significant technical work-packages that were procured or constructed in the partner countries rather than centrally from Lund. This means that the partner holds the budget and spends the money locally, but they also have to manage the risk. This sort of arrangement is a project management challenge, particularly when the time comes for installation and integration of components from this very diverse supply chain, but in many cases it is a political necessity. It exploits the strengths of national laboratories and industrial supply chains in the partner countries, especially important in the case of a facility being constructed in a green-field site without an established engineering function. It also spreads the economic benefit of the work carried out, and avoids a situation where the host region benefits from everyone else's investments. In the case of ESS, the target was for contributions from the 11 non-host countries to be 70% in-kind and 30% in cash, while the host countries contributed only in cash. This means that the total in-kind fraction was about 35% overall, which is just about manageable. In-kind contributions to ESS included the particle accelerator, much of the target system and the instruments; while the cash contributions paid for the conventional facilities construction contract and staff salaries—see figure 6.5. Some other big projects, such as the ITER fusion reactor in Cadarache, have a much higher fraction of in-kind contributions (up to 90%) and this can become very hard to manage; there is not enough cash available centrally to deal with the unexpected challenges which always arise in any large project.

Modern expectations for a research infrastructure must also be addressed. In the case of ESS that has meant environmental concerns and working to minimise the CO_2 footprint of the facility. I expect this to be a growing concern for new projects in the coming decades. At ESS, the office buildings meet modern energy-use criteria

Figure 6.5. Typical contributions to ESS by the host states, showing construction of the conventional facilities underway (left), and by in-kind partners, showing a superconducting accelerator cryomodule under test after delivery to ESS (right). Credit: Ulrika Hammarlund/ESS.

(BREEAM certification), the construction site is 'green' with 100% biodiesel-fuelled machinery and no waste taken to landfill. The project is committed to purchase all electrical power from renewable sources, which is relatively straightforward in Sweden, and the waste heat from the electrical machines and accelerator power supplies is recovered into the Lund hot-water district heating system.

ESS is committed to make its data open and FAIR (in the sense of Findable, Accessible, Interoperable and Reusable) through the European Open Science Cloud. Users of the facility will have an exclusive access period of three years after which their data will be made open for anyone to use. ESS will provide full support for computing, software, and analysis of ESS data, and its long term curation, at the ESS Data Management and Software Centre (DMSC) in Copenhagen.

ESS will also facilitate the application of its results through engagement with industry and support for innovation and spin-outs. A science and innovation campus is foreseen for the area between ESS and the MAX IV light source.

6.6 Exploring the socio-economic impact of ESS

Up to this point we have mainly been describing ways of promoting the future impact of the ESS project. But with ESS three quarters complete, we are now in a position to have some quantitative data on the impact so far. This I am sure will serve as examples that can be quoted by other projects of what is achievable.

These data generally cover the period 2013–18 and derive from a study carried out in 2019–20[4]. The motivation was twofold: first to start to capture the impacts of the construction of the facility, while the procurement and supply chains were still in place to be queried; and second to start to prototype and develop methodologies and metrics for use during operations. The work was carried out by Professor Jason Li-Ying's group at the Centre for Technology Entrepreneurship in the Technical University of Denmark (DTU).

Some lessons learned from this exercise are worth noting:

- Start collecting data early—five years of construction is already a significant period of data to mine. If one waits until construction is complete, both key staff and suppliers will have moved on and data collection will become much harder.
- Collect a wide range of metrics—one does not know in advance which areas will prove to be of most interest or most fruitful.
- Use interviews and surveys to capture the impact on the supply chain. The engagement rate to such surveys is high and most of ESS's suppliers were enthusiastic to share their experiences, both good and bad.

6.6.1 Publications

Scientific data and publications will of course be one of the key metrics of success once the facility is in operation. But even during construction, we would want to see the innovative techniques and developments that were required published and disseminated. ESS scientists should also continue to be active researchers and build collaborations with the future user community. Over the period considered,

654 papers were published. Of these, 64% were co-authored with scientists from other research facilities, 54% with universities, and 8% with industry. These papers had been cited 4671 times by the end of 2018. 42% are freely available through Open Access—perhaps less than ideal, but still higher than the global average.

6.6.2 Communications and outreach

The next aspect of dissemination is through broader communications and outreach activities. During the five year period, ESS hosted 19 465 visitors to the site. These visits included everything from school groups, fellow scientists, and future ESS users up to and including science ministers, HM the King of Sweden, the Governor-General of Canada, and the United Nations Security Council. ESS featured in 11 293 online articles with a potential readership of over 7 million.

6.6.3 Impacts of constructing the facility

Any large construction project will have a significant impact on the local and regional economy. What we need to capture is the extra value that is added by the scientific character of the investment. This extra value includes the role of ESS in attracting and training skilled staff, in upskilling suppliers, and in driving innovation.

Over the period considered, 214 highly qualified people were attracted to ESS from over 45 countries. The ESS staff are roughly 40% Swedish, 40% from the rest of Europe, and 20% worldwide, with scientists and engineers coming from as far away as Australia, New Zealand, China, Japan, the United States, and Brazil. Of the staff 32% are female, which is not too bad for a highly technical workforce. This ability to attract highly skilled people to the region was cited as part of the original case for building ESS—promoting a 'Nordic brain gain'.

Between 2013 and 2018, ESS placed 192 large contracts (with values over 50k€). On average two to three tenders were received per contract. ESS operates its own procurement rules (a privilege of ERIC status) but generally aims to follow the spirit of EU procurement practices. A supplier survey was carried out for the impact report, with 284 respondents. Of these, 37% reported experiencing increased overall profitability and 20% having entered a new market as a result of their work with ESS. Around 10% had either started a completely new business unit or established a new R&D unit. The survey showed that 36%–40% of the contracts were for new or innovative products or services, 53% required a high degree of customisation for ESS, and only 16%–18% could be met by off-the shelf components. Among the 208 respondents who had not provided off-the-shelf components, 50% reported improved technical know-how and many derived other innovation benefits from their work with ESS. These impacts were spread widely across all ESS member states, with positive impacts reported by commercial suppliers in Sweden, Denmark, France, Italy, Germany, Spain, Switzerland, and the UK. Procurements cover areas including electrical and electronic equipment, power supplies, utilities, cryogenic systems, and a wide range of services from design to installation to logistics.

6.6.4 In-kind contributions

As noted earlier, ESS relies strongly on in-kind contributions for many of its most innovative technical components. What of the impacts here? A total of 33 in-kind partners joined the project during the 2013–18 period, with at least one in each member country other than the host states, as shown in figure 6.6. Achieving this broad geographic spread was important, as it helps secures the best possible anchoring of support for ESS in all its members. It also means that as well as exploiting existing expertise in countries like the UK, Switzerland, France and Germany, ESS is helping to establish new centres of excellence in some of the newer European Union members such as the Czech Republic, Poland, Hungary, and Estonia. The first formal in-kind agreements were signed in 2015 and by the end of 2018 twenty-two agreements were in place with six more in preparation. Sweden and

Figure 6.6. Locations of ESS in-kind partners and collaborating institutes across Europe. ESS member countries are shaded blue in the map.

Denmark do not have in-kind partners, but academic institutions in those countries are making contributions through collaboration agreements in the same spirit.

Just as for contracts, the in-kind partners report high levels of innovation in their contributions. A survey yielded 27 responses. Nearly three quarters of the in-kind contributions required either highly customised technologies and services, or entirely new and advanced products and services.

6.6.5 Economic benefit to the host region

Something that ESS does not over-emphasise, but which is nonetheless true, is that there remains a huge economic benefit to Sweden and the local region from hosting ESS, despite the efforts to spread the impact widely through in-kind contributions and Europe-wide procurement. Over 80% of commercial contracts (and 66% of the high value contracts) went to Swedish suppliers. This includes a single 520M€ construction contract which was awarded to Skanska (see figure 6.6). The total number of suppliers to ESS from Sweden is 2283, over eight times more than any other country. Of course this benefit is reflected in large part by Sweden's paying a large fraction of the total construction cost (37.5% = 645M€) and all in cash rather than in-kind, so that it can justifiably be said that Sweden's own contribution more than covers the value of the contracts placed with Swedish companies. It can also be argued that the host state contracts are in general focused on less innovative and lower-technology areas such as construction and administrative services rather than high technology procurements.

6.6.6 Attracting new members

How can a project like ESS broaden its membership and attract new national partners? Clearly the construction of the facility and the benefits that it brings are already allocated to the existing members. What is the business case for a new partner to join? First and foremost is the ability to access ESS's science capabilities. The ESS access policy is clear that use of the facility will be prioritised for scientists from member countries, and this is a central reason why the existing members chose to be part of the project. Any new member country would also be encouraged to add to the scientific capability of the facility for the benefit of all, for example by building an additional scientific instrument or adding to the capability of the existing fifteen. The same in-kind model would of course apply so that much of the investment could be made 'at home'. In order to reduce any perceived barriers to participation by non-European nations who may not wish to become part of the ERIC, ESS has created a category of associate membership which can be negotiated very flexibly.

6.7 Concluding remarks

Investment in research capability is essential if we are to address the huge challenges facing society and the economy. International research infrastructures are a key part of the solution, but as large investments they attract a lot of scrutiny and require a robust investment case. ESS offers a good example of how such a case can be successfully constructed.

The ESS business case was developed over ten years ago. Ground was broken in 2015 and the project is now 75% complete. It is interesting therefore to ask how the business case has withstood the test of time. The science case is still strong, perhaps even stronger than it was a decade ago. Key application areas such as materials, energy storage, and drug development remain as topical as ever. The recent commitments by many countries to becoming carbon neutral in the next two to three decades and the heightened awareness of the risk of novel infectious diseases have only re-emphasised the need for scientific innovation in these areas. Neutron scattering facilities are being invested in heavily in the USA, Japan, and China. Like any big infrastructure, ESS has experienced budget and schedule challenges. These have been due to a post-Fukushima tightening of radiation safety requirements, to issues with the in-kind supply chain, and most recently to delays caused by Covid-19 restrictions across Europe. Despite these challenges the financial commitments from the partners remain strong. An additional 800M€ commitment to commissioning and initial operations up to 2025 is now in place, showing the continued support of the partners, and in-kind commitments are being delivered.

The original case can now be augmented with impact data from the first five years of construction, which will also serve as an example for other projects to cite in terms of what can be delivered. We can now show that construction of a large facility like ESS has a real and positive impact in attracting a talented workforce and in driving innovation in the high-tech supply chain. And of course the major science impacts of the investment are still to come!

References

[1] IEA 2021 *Net Zero by 2050* (Paris: IEA) https://www.iea.org/reports/net-zero-by-2050

[2] ESFRI 2016 ESFRI Roadmap 2016 https://www.esfri.eu/roadmap-2016

[3] See for example ed Appell D 2013 *Scientific American* 15 October https://www.scientificamerican.com/article/the-supercollider-that-never-was/

[4] European Spallation Source 2020 Exploring the socio-economic impact of ESS 2013–2018 https://ess.eu/article/2020/12/16/anchoring-ess-society. This work was supported within the BrightnESS^2 project by the European Union Framework Programme for Research and Innovation, Horizon 2020, under grant agreement 823867

IOP Publishing

Big Science in the 21st Century
Economic and societal impacts
Panagiotis Charitos, Theodore Arabatzis, Harry Cliff, Günther Dissertori, Juliette Forneris and Jason Li-Ying

Chapter 7

Debating the impact of Big Science in the twenty-first century. ELI ERIC: from local to global—addressing the multiple dimensions of impact

Allen Weeks, Florian Gliksohn and Alexandra Schmidli

The Extreme Light Infrastructure ERIC (ELI ERIC) is an international laser facility offering new interdisciplinary research opportunities using extreme light from the highest peak power laser sources currently available and dedicated to the purpose of research. These lasers and the secondary radiations derived from them will enable unprecedented discoveries across a broad range of scientific disciplines as well as societally relevant applications. With this variety in research activities, the ELI facilities are expected to deliver significant societal impact on a multi-dimensional and multi-faceted scale.

7.1 Introduction

To meet the societal challenges of our time, new technologies are needed to discover and answer the fundamental questions of our Universe. The field of photonics has been identified as one of the key enabling technologies (KET) of the twenty-first century by the European Commission (EC) (European Commission 2018).

The Extreme Light Infrastructure (ELI) is the world's leading laser-based research infrastructure (RI) offering the most diverse collection of state-of-the art, advanced, laser and secondary sources for external users. This international user facility accommodates some of the most intense lasers in the world which will contribute to new discoveries and pave the way to the advent of compact, reliable, and affordable laser-driven sources of x-rays, plasma, and accelerated particles for a wide range of applications.

ELI is expected to have a significant impact on science, society, and economy on a local, regional, European, and also on a global level. The ELI facilities provide

doi:10.1088/978-0-7503-3631-4ch7

employment and business opportunities that drive Europe's competitiveness and open up unprecedented possibilities in particle physics, nuclear physics, high-energy beam science, nonlinear field theory, and ultra-high-pressure physics. Experiments at the ELI facilities will spur discoveries and innovations in a range of fields that have a global reach and impact, from nuclear materials and radioactive waste management, industrial tomography and gamma radiography, pharmaceutical radioisotopes, medical imaging, and radiation and proton cancer therapy, among many others still to be explored.

Emerging from a coordinated effort of a multi-national scientific laser community, ELI is the first, multi-site ESFRI Landmark (ESFRI 2006) constructed in the Central Eastern European Member States with three complementary facilities located in the Czech Republic (ELI Beamlines), Hungary (ELI ALPS), and Romania (ELI NP). ELI relies on an innovative funding model combining European Regional Development Fund (ERDF) and co-funding from the hosting countries for construction, and contributions to a European Research Infrastructure Consortium (ERIC) for operations (EU 2021).

7.1.1 A new era for laser science and applications

The technology used at ELI is enabled by techniques that decrease the laser pulse duration to timescales on the order of a billionth of a billionth of a second (attoseconds). This technology would not be possible without the invention of chirped-pulse amplification (CPA), for which Gérard Mourou and Donna Strickland were awarded the Nobel Prize in Physics in 2018. This technology enables the production of extreme laser powers in the order of petawatts (10^{15} Watt), which is a thousand times more than the power of all of the power stations in the world combined (Mourou *et al* 2011).

The idea of CPA boils down to making optimal use of the concept that power is defined as energy divided by time. Mourou and Strickland demonstrated that by stretching a laser pulse in time by separating its component colors, amplifying the elongated pulse, and then compressing it again to very short timescales while re-combining the different colors, peak powers of a million times a billion watts (petawatts) or even a thousand times higher (exawatts) could be achieved (see figure 7.1).

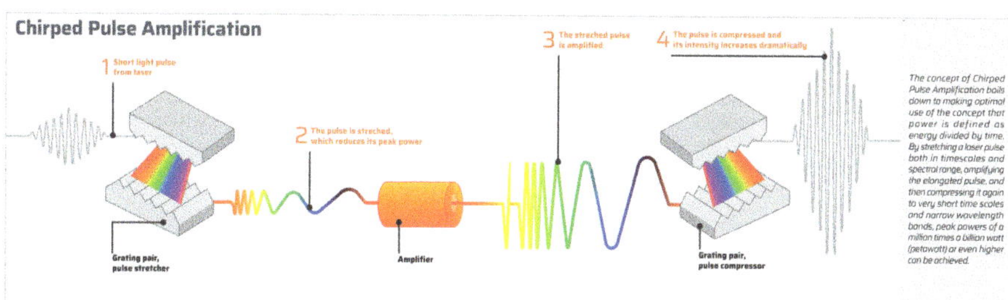

Figure 7.1. Diagram of the chirped-pulse amplification (CPA) concept.

7.1.2 The revolution of high-power, high-repetition lasers

CPA has opened the door to unprecedented laser intensities, the current state of the art being about 10^{22} W cm^{-2} (see figure 7.2). The next generation of high-power lasers will increase this number by orders of magnitude with multi-10 petawatt and sub-exawatt lasers expected to reach 10^{26} W cm^{-2}. At intensities of this order charged particles traversing a laser wavelength gain huge electromagnetic energies such that even protons start travelling at almost the speed of light, or in technical terms, become relativistic.

ELI's ultra-high-power lasers with focusable intensities and average powers reaching far beyond existing laser systems will open new frontiers in fundamental science. Over the past decades, lasers have been extremely effective in improving our understanding of the molecular and atomic structure of matter and the associated dynamical events. As the first infrastructure dedicated to the fundamental study of laser–matter interaction in the ultra-relativistic regime, the scientific research at ELI will challenge the vacuum critical field, as well as provide a new avenue to ultrafast attosecond studies of laser–matter interaction.

ELI will open the possibility of taking snapshots of the electron dynamics in atoms, molecules, plasmas, and solids. The facilities will enable new investigations in particle physics, nuclear physics, gravitational physics, nonlinear field theory, ultra-high-pressure physics, astrophysics, and cosmology. In addition to its fundamental physics mission, a paramount objective of ELI is to provide ultra-short energetic particle (10–100 GeV) and radiation (up to few MeV) beams produced from compact laser plasma accelerators. These will be used for the benefit of industry and society. For instance, the secondary sources provide x-ray technologies that can help clarify the complete time history of reactions such as protein activity and protein folding, radiolysis, monitoring of chemical bonds, and catalysis processes. This will lead to a better understanding and control of key events during chemical bond formation and destruction. Since these processes play a major role in, for

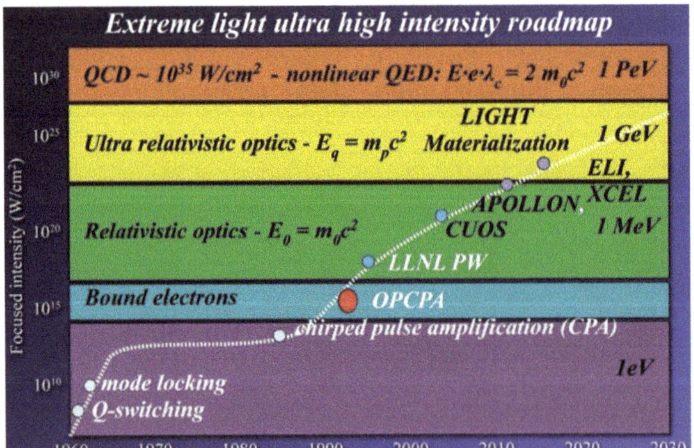

Figure 7.2. The development of laser powers over time. Adapted with permission from [7], CC BY 4.0.

example, the creation of new drugs, high-power lasers will have a high impact on industry and society.

7.1.3 ELI technology and application areas

The ELI ERIC facilities consist of two facilities, ELI ALPS and ELI Beamlines, each with a specific focus in high-power, high-repetition-rate lasers. A third ELI facility, ELI NP, is expected to join ELI ERIC in the near future.

The ELI Beamlines located in Dolní Břežany (near Prague), Czech Republic (figure 7.3), focusses on the development of short-pulse secondary sources of radiation and particles, and on their multidisciplinary applications in molecular, biomedical, and materials sciences, the physics of dense plasmas, warm dense matter, and laboratory astrophysics. In addition, the facility will utilise its high-power, high-repetition-rate lasers for high-field physics experiments investigating exotic plasma physics, and nonlinear quantum electrodynamic effects.

The ELI Attosecond Light Pulse Source (ELI ALPS) in Szeged, Hungary (figure 7.4), provides light sources in the form of ultra-short pulses with high repetition rates between the terahertz (10^{12} Hz) and x-ray (10^{18}–10^{19} Hz) frequency ranges. ELI ALPS will be dedicated to extremely fast dynamics by taking snapshots in the attosecond scale (a billionth of a billionth of a second) of the electron dynamics in atoms, molecules, plasmas, and solids. Research with ultra-high intensity lasers is also pursued.

The ELI Nuclear Physics (ELI NP) in Magurele, Romania (figure 7.5), is focused on laser-based nuclear physics. ELI NP hosts a very high-intensity laser, where beams from two 10 PW lasers are coherently added to get intensities of the order of 10^{23}–10^{24} W cm^{-2}. The very intense, brilliant gamma beam obtained by incoherent Compton back scattering of a laser light off a brilliant electron beam from a

Figure 7.3. The ELI Beamlines facility, Czech Republic. (Source: The ELI Beamlines facility.)

Figure 7.4. The ELI ALPS facility, Hungary. (Source: The ELI-ALPS facility.)

Figure 7.5. The ELI NP facility, Romania. (Source: ELI NP.)

conventional linear accelerator is under construction. Applications include frontier fundamental physics, new nuclear physics, astrophysics, nuclear materials, and radioactive waste management.

7.1.4 ELI's institutional background and funding model

The Extreme Light Infrastructure project started as a bottom-up initiative by the European scientific laser community and the network of national laser facilities Laserlab-Europe. In 2006 ELI was added to the European Strategy Forum on Research Infrastructures (ESFRI) roadmap (ESFRI 2006). As the first international RI completely located in Central and Eastern Europe and the newer Member States of the European Union, ELI applied a novel funding model. The overall investment volume exceeds €850 million. with on average 85% stemming from the European Regional Development Fund (ERDF), and 15% from co-funding from the national budgets of the host countries. This novel funding approach was implemented through three individual projects monitored separately by national managing authorities in the host countries. The investment split of the construction budget is about 30% in buildings, 60% in technology, and 10% in personnel and services.

The ELI facilities were implemented by three independent local 'beneficiaries', the Institute of Physics of the Czech Academy of Sciences for ELI Beamlines, the ELI-HU non-profit Ltd for ELI ALPS, and the Horia Hulubei National Institute for R&D in Physics and Nuclear Engineering for ELI NP. An international non-profit association under Belgian law, the ELI Delivery Consortium (ELI-DC), was established in 2013 with the purpose to facilitate coordination between the three entities in charge of ELI's implementation, promote and preserve the European character of ELI and pave the way towards future integrated and sustainable operations of ELI by supporting the negotiations on the establishment of a European Research Infrastructure Consortium (ERIC).

The legal status of the ERIC was granted to ELI by the European Commission on 30 April 2021 (EUR-Lex 2021). The ERIC is a specific legal form designed to facilitate the establishment and operation of RIs of European interest. As a main statutory mission, ELI ERIC is responsible for making the ELI facilities, initially ELI Beamlines and ELI ALPS, available to the scientific community as a single international organisation, with unified governance and management. The Czech Republic and Hungary are joined by Italy and Lithuania as founding members of ELI ERIC, while Germany and Bulgaria are founding observers with ambitions toward future membership. The third ELI facility, ELI NP, is completing construction in Romania and expected to complement the two other ELI ERIC facilities in the future. Altogether, the ELI facilities represent nearly 85 000 sqm on three sites and already staff over 600 scientists, technicians, and support staff charged with user support and administration.

The integration of the ELI ERIC facilities into a single organisation is underway and is expected to take place over the next two-to-three years. As of 1 January 2023 the ELI Beamlines facility has legally integrated into ELI ERIC with the integration of ELI ALPS expected to take place on 1 January 2024. During that time, management, technical, and scientific procedures in the different facilities are being harmonised. The merger is being facilitated by a €20 million grant from the European Commission under the Horizon 2020 programme known as 'IMPULSE' (https://impulse-project.eu).

For all major investment projects (over €50M) applying for assistance of ERDF, the EU Cohesion Policy regulations require a cost-benefit analysis (CBA) to assess the relative merit of such large-scale investments. The CBA is a well-established economic evaluation tool commonly used in the public sector to make transparent investment decisions. With a CBA it is possible to assess the relevance of an investment (and to compare it with other investment options) by computing the net present value of its financial and economic costs and benefits over a certain reference period (i.e. weighing them over time). By applying for national investment and ERDF funding, ELI was required to conduct a CBA and in doing so pioneered its use for RIs. Each facility conducted separate CBAs initially between 2010 and 2012. Those were reviewed and updated in 2015–17 when the projects were extended into the following programming period (Gliksohn 2020).

Structure and objectives of CBA

A CBA consists of two interconnected parts, a financial analysis and a socio-economic analysis, to measure the variation in the level of welfare among a population of reference, generated in and out of markets (i.e. with or without a price mechanism). The benefits and costs that should be taken into account in a socio-economic analysis are those that are generated directly by the project. In the context of a grant application, the objective of the CBA is not to provide an all-inclusive evaluation of the socio-economic impacts, but to demonstrate that the economic net present value is positive and that the economic rate of return is higher than the socio-economic discount rate.

ELI's experience

For the purpose of the CBA, key indicators need to be defined, quantified, and monetised. The main indicators used by ELI were closely connected to the resources it uses (high-skilled human capital) and to its output as a research organisation:

- Knowledge (e.g. the number of publications from the facility and its scientists).
- Development (e.g. the number of patents granted).
- Education and training (e.g. the number of students and graduates, MA and PhD level).
- Employment (e.g. the number of newly created jobs).
- Knowledge transfer and collaborations (e.g. the number of collaborators and strategic partners).

The monetisation of these impact indicators was based on guidelines provided by the managing authorities of the host countries, which are derived from accepted standards in the economic literature.

In the case of ELI Beamlines and ELI NP, the impact of the future provision of access to external users was also considered in the CBA. As no standard valuation approach exists in the current literature, ELI built on the notion of 'willingness to pay' to assign a value to access units and thus quantify the socio-economic impact of access. In practice, this approach consisted in estimating the price users would be willing to pay to access the ELI facilities taking into consideration the costs of operations of other laser facilities and the competitive edge of ELI.

While the usefulness of the CBA as a tool to assess investment opportunities is well established, its use in the context of an RI proves complex due to the very nature of the outputs of the organisation (new knowledge), which are by definition difficult to quantify. The exercise is therefore somewhat 'artificial' as it focuses more on the quantification of impact than on the understanding of the mechanisms and pathways driving and maximising impact. For that reason, a CBA should always be complemented by a clear strategy supporting delivery and regional appropriation of impact.

However, the CBA had a very tangible impact on ELI's implementation as many of the indicators used in the CBA are used by the managing authorities to monitor that the ELI facilities are delivering on their promise as investment projects.

7.2 The various types and dimensions of ELI's expected socio-economic impact

ELI will have impact on science, society, and economy on different levels, ranging from the local and regional level, to the Member States and as such the European level, as well as on a global level. These different geographies of impact have specific outputs that can overlap and complement each other. The first section of this chapter will briefly explore these dimensions, whereas the following sections elaborate further by type of impact.

7.2.1 Dimensions of impact

7.2.1.1 Local and regional impact

A key reason for investing in and constructing ELI in Central and Eastern European countries is to leverage and catalyse the impact of a high-value research and technology centre on the host regions. A direct economic effect of the project is the creation of high-skilled job opportunities not only for researchers but also for support staff. Given the international character of the organisation, these contribute on the one hand to attracting foreign staff, but also to reversing the brain drain that has long affected the host regions.

The ELI ERIC facilities will bring considerable new economic resources to the region. As a secondary effect, it is expected that the project increases the demand for and raises the standards of public and private services. Increasing incomes, business turnovers, and consumption means growing tax revenues that enable the improvement of public services and living conditions for the regions overall.

The ELI facilities are located in or near cities with links to universities. The universities have established dedicated programmes which train and educate more students in the field. The proximity and access to the ELI centres leads to improved education and training opportunities and develops people and knowledge to build human capital. Through international research co-operations, exchange programmes, and the activities of guest researchers and lecturers, the city, the university, and the facilities strengthen the knowledge production of the region which will in turn attract external interest and talent. Furthermore, ELI's potential for multidisciplinary applications will expand the scientific and academic communities beyond the fields of laser and photonics.

Local and regional industries will also benefit from the opportunities to sell products or even by adapting a product offer to meet the needs of the facilities and surrounding infrastructure. By establishing an industry liaison office, information sharing between the facilities and industry will be facilitated. In addition, having a network of qualified local suppliers cuts transportation time and costs, and facilitates joint development projects.

7.2.1.2 European impact

As a whole, ELI will increase the competitiveness of Europe and the European Research Area by building, retaining, and attracting scientific expertise and talent while increasing the innovation potential of the European industry. ELI will intensify scientific research cooperation across disciplinary and national borders and will provide educational and training opportunities for new scientists and engineers in photonics and laser-enabled areas of research and beyond. In particular, it will ensure that Europe remains the front-runner in high-peak power laser facilities. Similar to the local level, Europe will benefit from the job creation, knowledge production, training, and education opportunities that the ELI facilities will offer.

Member countries in particular benefit from the opportunity to contribute to and guide the ELI ERIC governance and with that the overall scientific direction of the research. Membership also gives the national scientific community access to the ELI facilities with the opportunity for close collaboration on projects and research and the possibility to influence future investments into and upgrades of the ELI facilities but also with a focus on training and education opportunities. This in turn has the potential to strengthen national expertise in specific fields related to ELI and drive national knowledge production.

7.2.1.3 Global impact

On a global level, ELI will have a significant impact as a driver for scientific breakthroughs which will benefit global society as whole. Furthermore, the scale and ambitions of operating ELI ERIC demands a global effort with long-term sustainable investments. It is anticipated that, by collaborating with partners and working with industry, facilities can lower the costs of operations and increase availability of access time for users. This offers the opportunity to realise unique economies of scale and scope, enabling significant technological advances in a shorter time period. Such advances will resonate globally throughout the laser industry and scientific community.

7.2.2 Scientific impact

Collectively operated under ELI ERIC, the ELI facilities are the most multifunctional of all existing and projected laser facilities. With a collection of the most powerful and shortest-pulsed systems in the world, ELI is also among the infrastructure providing the most extreme intensities. More precisely, ELI is dedicated to the fundamental study of laser–matter interaction in the ultra-relativistic regime with intensities exceeding 10^{24} W cm^{-2}. Furthermore, the facilities are testing a new generation of innovative, very compact accelerators delivering energetic particle and radiation beams of femtosecond (10^{-15} s) to attosecond (10^{-18} s) duration.

ELI ALPS in Hungary harbours lasers that combine high repetition rates, broad spectral ranges, and extremely short pulses with high average power and high peak intensity. ELI ALPS will propose new methods to investigate exotic states of matter, breaking new ground in photonics, plasmonics, semiconductor physics, and plasma physics. This new knowledge makes it possible to manipulate processes at extremely short timescales and by doing so create new materials and technologies.

The unprecedented intensity of the ELI ALPS laser systems enables, for example, the control of molecular dynamics in such a way that chemical reactions can be pushed in desired directions. The high repetition rate makes the ELI ALPS facilities extremely interesting for, amongst others, the acceleration of electrons for time-sensitive experiments, such as those concerning biological samples which often degrade very fast. Furthermore, it facilitates experiments to understand ultrafast carrier dynamics, which will provide valuable input for the rational design and selection of materials, morphology, nanostructures, heterojunctions, and interfaces. This will result in improved performance in solar cells, solar fuel generation, and organic electronics.

In the Czech Republic, ELI Beamlines focuses on the development of short-pulse secondary sources of radiation and particles. The unique feature of ELI Beamlines is that the optical, x-ray, and particle beams the facility provides are perfectly synchronised due to their generation process. This enables pump-probe investigations in a very broad range of energies for photon (eV–MeV) and particle beams (eV–GeV) with very high accuracy.

ELI Beamlines focuses on multidisciplinary applications in molecular, biomedical and material sciences, the physics of dense plasmas, warm dense matter, and laboratory astrophysics. In addition, the facility utilises its high-power, high-repetition-rate lasers for high-field physics experiments, investigating exotic plasma physics and nonlinear QED effects.

The instruments at ELI Beamlines enable researchers to follow in real time how the geometry changes when one molecule binds to other molecules, and to see how the molecule changes shape on a femtosecond timescale. In the field of plasma and astroparticle physics, ELI Beamlines is able to produce the extreme dense states of matter that are present in the cores of stars.

The facility also focuses on the further development of laser-driven accelerators that accelerate either ions or electrons. Since using lasers for particle accelerators drastically reduces the size and cost of these systems, this development makes, amongst others, accelerator-based applications such as proton therapy accessible for more hospitals.

7.2.3 ELI as a platform for innovation

In terms of ELI's potential impact on innovation, multiple aspects should be considered: innovation through technological development required by ELI, innovation through access to the ELI facilities, and innovation through knowledge transfer to industrial actors.

ELI's technology contracts drove the growth and the technological development of many European companies. A few technological innovations have already been made available to the market thanks to the developments for the ELI contracts, such as:

- Optical parametric chirped-pulse amplification (OPCPA) enabling shorter, more powerful pulses.
- The world's highest peak power diode arrays.
- A pump laser generating up to 200 joules at a 10 Hz repetition rate.
- A gas-cooled short-pulse titanium-doped sapphire amplifier.
- A dual chirped-pulse-amplification high-contrast short-pulse front end.

The market for lasers in research and science is naturally highly dependent on the technology push of research initiatives such as ELI and is innovation-driven: the developments designed and realised for ELI are leading references and have already been adopted by the laser industry and the other laser laboratories around the world. A long-term relation with industrial partners will ensure sustainability of the operations through continuous technological developments.

A few examples of areas where further technological innovation is foreseen are in: key optical components for the high-power/high-repetition-rate lasers that are able to withstand high-energy and high-peak-power pulses; short pulses, large beams metrology to measure the performances of the systems on short pulses and large beams, and complete control for complex systems.

In addition to technology development as a response to the needs and requirements of ELI, industry will also benefit through the possibility of proprietary access to the facility, as well as collaborations on projects with universities and the research community resulting in knowledge transfer to the industry. It is expected that the principal interest of industrial users will be in the sectors of life sciences (pharmaceutical, diagnostics methods), chemistry and femto-chemistry (ultrafast processes), advanced materials (microelectronics, semiconductors), energy research (energy storage, batteries, nuclear fission/fusion power plants), food and agriculture industries, steel and construction materials, and automotive and aerospace industries.

7.2.4 Multi-faceted impact as a driver of the scientific community

The scientific success and enthusiasm in the field of high-power laser systems has led to an increase in European installations in the last decade (see figures 7.6 and 7.7).

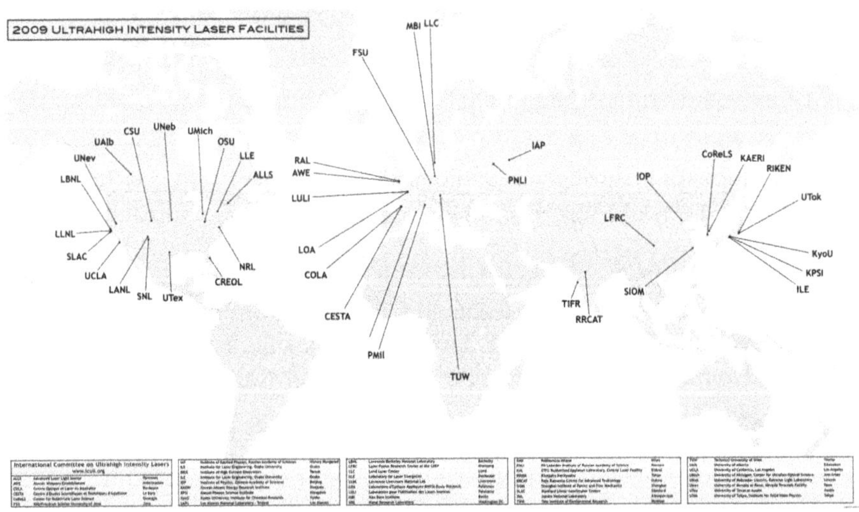

Figure 7.6. World map by the International Committee for Ultra-Intense Lasers (ICUIL) showing the 2009 global extent of PW-class laser facilities. (Source: Professor C P J Barty, University of California, Irvine.)

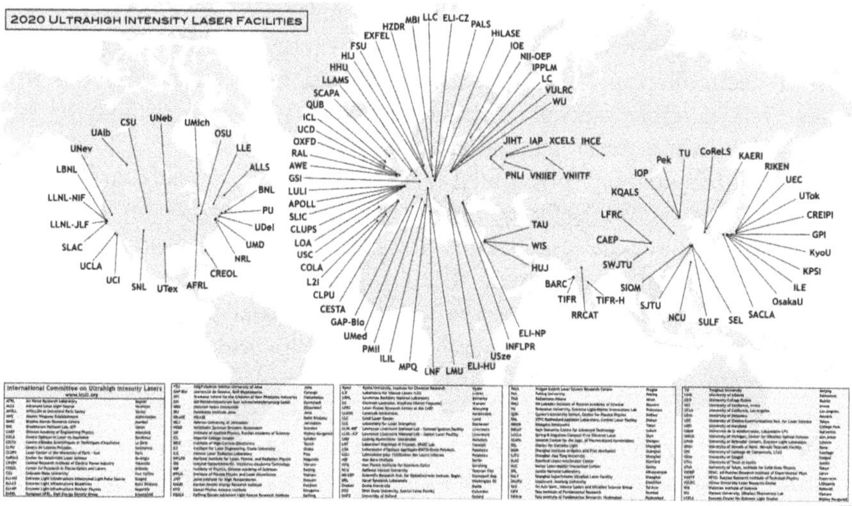

Figure 7.7. World map by the International Committee for Ultra-Intense Lasers (ICUIL) showing the 2020 global extent of PW-class laser facilities. (Source: Professor C P J Barty, University of California, Irvine.)

Much of the support for ELI was generated by the early success of the Laserlab-Europe consortium, a network involving over 35 European laser laboratories and institutes (www.laserlab-europe.net) collaborating on the provision of access to users and joint research activities. Based on the success of those laboratories, the expertise, knowledge, and proofs-of-concept for ELI were developed. A community of users has matured and significant investment has taken place, also supporting the development of European companies to build ELI and other PW-class facilities.

Europe has taken a clear competitive lead relative to the rest of the world in this field. Much of the global, aggregated distribution of peak power that is currently either operational or under construction is concentrated in the three ELI facilities. Continued cooperation in the community is critical for Europe's sustained competitiveness and leadership in this field.

ELI is expected to build on the scientific and technical developments and scientific community in a multi-faceted manner by driving and initiating change on an institutional and organisational level. Constructing facilities as large as the ELI ERIC facilities involves a number of operational challenges that entail a change of paradigm in how the facilities are operated and how users and industries interact with them. In this area, ELI is expected to innovate in a number of ways, ranging from the development of new laser technologies, control systems, and metrology devices to the development of new operational management concepts and data management procedures, solutions, and services.

While some of these developments take place within the ELI framework, others require collaboration with academic and industrial partners to define standards. Substantial impact on the community, other laboratories, and facilities is expected from such developments through knowledge transfer and best practice exchanges.

The main conditions for this type of impact to materialise require openness in the dissemination of ELI's results, and association—wherever possible—of institutional partners in technological developments, co-definition of standards, and training measures. The tangible aspects of these impacts may be the availability of new technologies, software solutions, and standards.

As such, ELI is uniquely positioned to have long-term impacts on the scientific field and community. While the community involves a number of large facilities, these national facilities are significantly smaller than ELI in terms of capital and operational investment. They rely on shorter-term financial cycles and have limited capacity on their own to have initiatives that have potential community-wide effects. In comparison, ELI is a truly multi-national effort of a scale allowing initiatives of such ambition to emerge.

The European laser community is organised differently from that of synchrotrons and neutron sources. Based on its critical mass ELI has the capacity to steer long-term initiatives involving the laser community in a coordinated fashion in collaboration with Laserlab-Europe. An initiative to support the definition and implementation of joint strategies and standards at community level in a number of areas such as technology development, access provision, operational standards, data management, etc, has great potential. Such coordination has the potential to enable a more efficient use of resources, effective operations, and ultimately contributes to increased competitiveness. The involvement of the community of users, national facilities, and laboratories is essential for the sustainability of ELI because it promotes internal organisation within the community and positions individual institutions and national communities in a consistent manner around jointly defined challenges and horizons.

7.3 ELI's delivered impact

The ELI ERIC facilities have completed the construction phase. Their systems are now gradually coming online and entering operations. As such ELI is already generating a wide range of meaningful impacts. To capture this, a set of key indicators have been selected to demonstrate the current impact on a local, national, European, and global level (figure 7.8).

The following data are limited to the two ELI ERIC facilities, ELI ALPS and ELI Beamlines, due to the fact that ELI NP is not operating within the framework of the ELI ERIC at this stage.

7.3.1 Employment

The ELI ERIC facilities have already impacted the local and regional economy significantly by creating jobs, not just in the scientific and technical sector but also in administration and maintenance. As previously noted, the increase in employment opportunities is a boost for the local and regional economy and contributes to the improvement of the quality of life as employees support the industry through spending, consumption, and overall business, which in turn makes the region more competitive, attracting new business and talent.

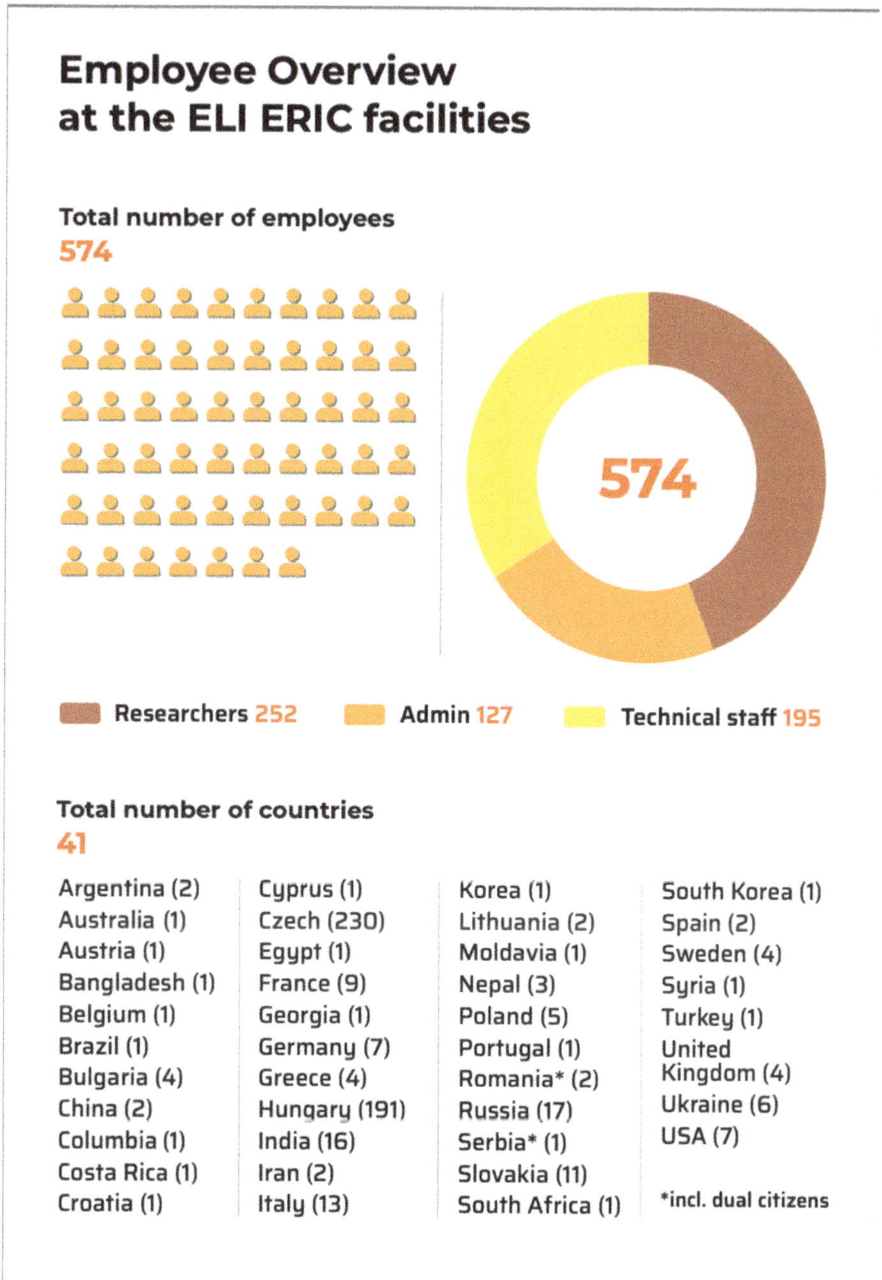

Figure 7.8. ELI ERIC employment figures. (Source: ELI ERIC.)

The focus in the data presented lies on the direct impact on local job creation, however, it is worth noting that jobs are not only created on a local and regional level but beyond that through collaborations and partnerships with other institutions and organisations.

7.3.2 Collaborations and partnerships

ELI ERIC's main mission is the output of new, excellent science. This will result from the scientific experiments performed by its users, the development of innovations, new technologies, and the expansion of capabilities needed to be at the cutting edge and enable scientific and innovative breakthroughs. ELI already achieves this through a variety of cooperation with partners across borders (figure 7.9). The

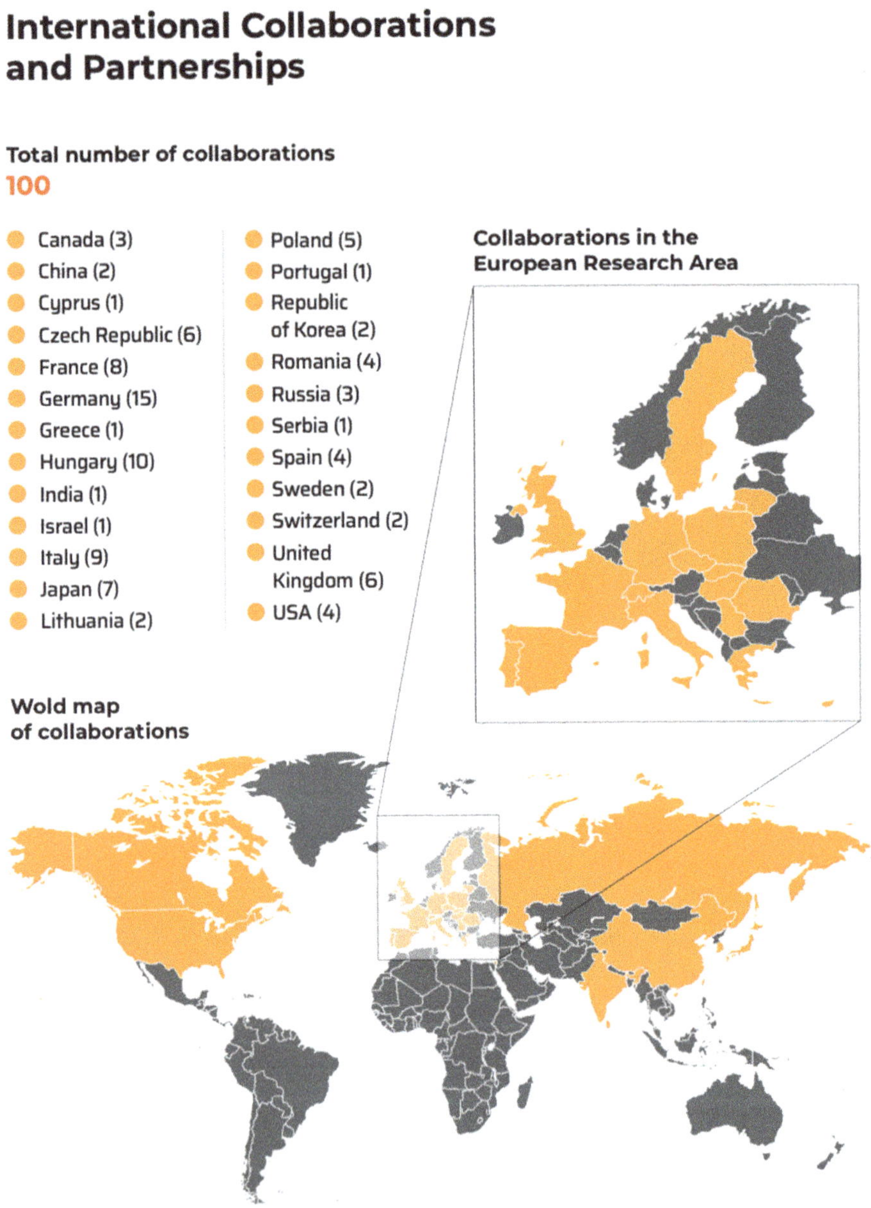

Figure 7.9. ELI ERIC partnerships and collaborations. (Source: ELI ERIC.)

benefit of such collaborations is the exchange of knowledge and skills that drives new scientific and technological discoveries and leads to innovation. Additionally, the international collaborations and partners of course also have a direct benefit to access the most powerful and advanced laser systems of the world. A diverse network of collaboration and partnerships is conducive to fruitful exchanges and advantageous for all partners involved.

7.3.3 User groups for commissioning experiments

The rich exchange and value of collaborations is also evident in the context of the user experiments for commissioning. The ELI ERIC facilities have invited a limited number of user groups to commission the facilities and equipment (figure 7.10). Such exchanges during commissioning experiments establish close relationships with a wider scientific community which can lead to future collaborations and cooperation across borders. Additionally, working with leading experts in their respective fields ensures that the facilities are at the cutting edge of research and even has the potential for identifying new scientific communities and applications to work with.

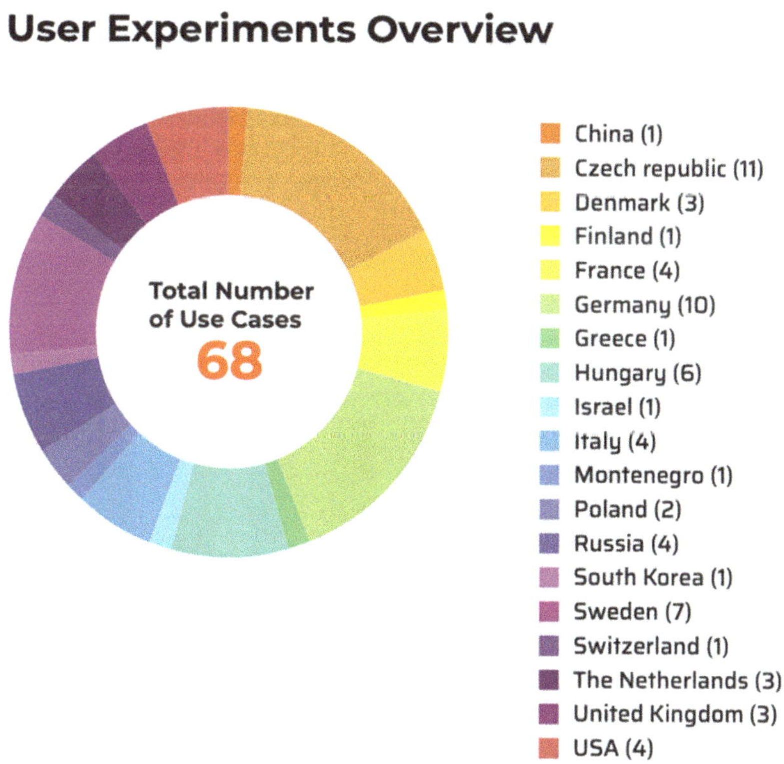

Figure 7.10. ELI ERIC Commissioning user experiments. (Source: ELI ERIC.)

7.3.4 Publications

The scientific community relies on publications for the exchange of information and knowledge. The scientific and technical developments at ELI generate knowledge and enhanced skills. In order to share this knowledge with a wider community of peers the information and data is made public through articles, papers and report which are presented in specialised journals or platforms, presented at conferences and other events for critical review discussions (figure 7.11). Further dissemination through citations results in reaching a wider audience. Such sharing of information also acts as a quality control of the data. This kind of open knowledge exchange leads to better science, and better discoveries, and impacts society positively.

7.3.5 Industry

The advancement of laser technology into more powerful and ultrafast systems has broadened the possibilities for industrial application. This has led to wide range adoption of lasers by industry, from innovative systems for medical imaging, radio- or proton-therapies, advanced process for manufacturing, new generation of compact accelerators and related light sources, and very short wavelength light sources for security and cargo inspection, to name a few. The access to these application markets with high-peak laser technology requires the development of integrated systems, up to high-readiness levels, from the component validation in a laboratory environment to the demonstration in an operational environment.

In parnership with industry, ELI pushes the advancement of technology such as diode-pumped laser systems, light pulse compression, or amplification techniques, to a level where new products and innovations can be developed. The integration of such new technologies further expands the range of applications into new sectors such as gamma-ray machines for security, ion or electron based systems for

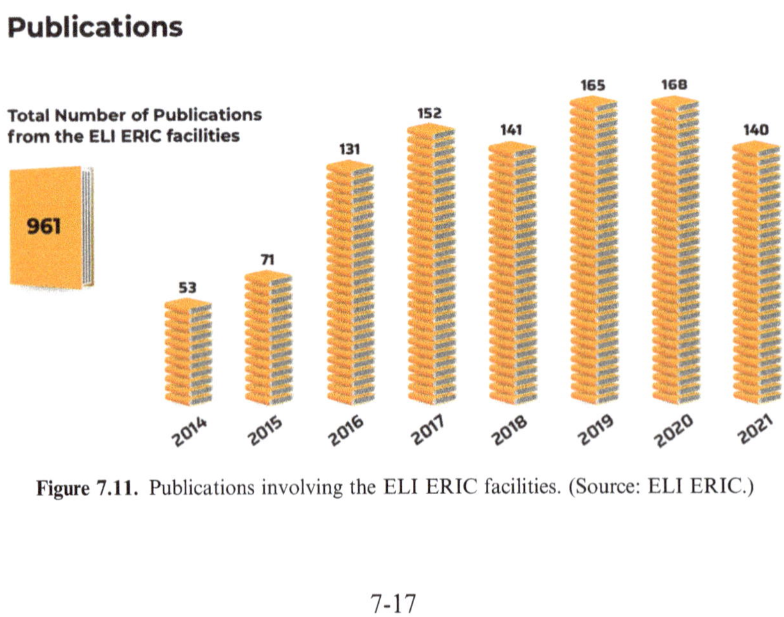

Figure 7.11. Publications involving the ELI ERIC facilities. (Source: ELI ERIC.)

materials science applications, x-ray imaging machine for biomedical applications, XUV systems for lithography, etc.

Beyond technological advances, ELI also facilitates the strategic development of the resources needed to achieve industrial impact, from tools to track ongoing success, review of technical tactical plans and targets, support to industry to achieve international positioning and sales, and of course the training and cultivation of highly qualified personnel. ELI also benefits industry with its capability for technology risk assessment, for example by reviewing emerging trends, regulatory issues, IP protection, and compliance. As such, ELI has emerged as a platform for high-power laser science and a driver of laser science technologies in collaboration with industry.

ELI has already had a profound impact on innovation and industrial technologies during the first stages as the three ELI facilities have strongly involved European industry in the construction of the facility. Combined, the ELI projects have contributed significantly to the competitiveness of European industry with more than €488 million in contracts to companies in 19 European countries. Of this budget almost €200 million concerns technology contracts outside the host countries[1].

7.3.6 Education and training

Education and training are a core mission of ELI ERIC as these activities are essential for building the scientific community to use the facilities in the future and meet the demand for the skilled professionals needed to design and operate the equipment. ELI aims to expand the community that will exploit the fundamentally novel research opportunities deriving from the world-leading technologies and experimental capabilities available at the ELI facilities. Education and training is key to ensure the sustainability of the scientific community of users, the operators of the facilities, and industrial actors in the relevant fields, both in terms of recruiting students and young scientists to enter the field and to building and developing the knowledge of researchers, paving the way towards emerging technologies, innovation, and scientific breakthroughs.

Education and training measures are being implemented at all ELI facilities and on various levels. Each of the ELI facilities have internship and student training programmes while also offering PhD and post-doctoral positions. The proximity and relation to universities of all ELI facilities supports the recruitment of new students and many of the staff members hold teaching positions. The yearly ELI Summer School (ELISS) is a key outreach and training event, already in its tenth edition. In 2021 ELISS welcomed more than 300 participants from 29 countries for a week-long programme focused on high-power, high-intensity lasers, among many other topics relevant for ELI.

To broaden the already existing training and education measures is developing a European Masters and Graduate Programme within which young scientists being

[1] Figures based on data from 2019.

educated at universities and institutes across Europe would have the opportunity to apply their theoretical knowledge directly at the ELI facilities. The programme will provide exchange opportunities and mobility to participating students and strengthen ties between the ELI facilities and the academic community. Additionally, such a programme has the potential to expand and build potential new user communities.

7.3.7 Communication and outreach

The ELI ERIC facilities and ELI ERIC engage in a variety of communication and outreach activities to initiate an informed dialogue with relevant stakeholders and key audiences, including external and internal actors. Communication and outreach are important tools to demonstrate the return on investment and to minimise misunderstanding. At the same time communication and outreach activities are a platform to promote the work of the scientists and the facility. Dissemination to the broader scientific community supports the scientific community building, capacity building, and networking efforts. Outreach measures also contribute to engage potential partners and stakeholders to ensure sustainable operations of the facilities in the future.

While the scientific and technical staff engage in many conferences, workshops, seminars, etc, to present, promote, and exchange their work, there are a number of ELI specific events that are essential to interact with the scientific community and stakeholders, such as the ELI User Meeting which takes place annually and gathered nearly 500 participants in 2021. This is the main event to inform the user community about new developments and the scientific opportunities which ELI has to offer.

The ELI ERIC and ELI ERIC facility websites are the main channels for communication. These are supported by social media channels including Twitter, LinkedIn, Facebook, Instagram, and YouTube. Twitter is used for daily updates to present news and events while serving as a way to connect with the scientific community as well as other projects and organisations, with over 1200 total followers. LinkedIn has become one of the key platforms for engagement about job opportunities with over 3000 followers between the accounts. Facebook is also used to connect with the community, share news and updates, and connect with other projects and network. With a combined number of nearly 5000 followers, it is the largest platform.

In addition to these outreach measures, ELI also engages in a number of other activities geared at the general public through events such as researcher's nights and science fairs, music and cultural festivals, etc. Both ELI ERIC facilities also offer visitor programmes with interactive features to help the audience better understand the science behind ELI. For example, ELI ALPS recently opened a dedicated visitor space 'Focus' to welcome external guests.

7.4 Summary

As the world's largest and most advanced high-power laser infrastructure and a global technology and innovation leader, ELI provides the international scientific community with access to leading experimental capabilities that will enable ground-breaking scientific discoveries, new technologies, and a broad range of research applications.

Beyond enabling excellent science, which is the core of its mission, ELI has the ambition to deliver a wide range of impacts supporting local and regional economies, but also Europe's competitiveness and society at large. The net creation of over 600 highly-skilled jobs, the development of training and education opportunities, and the promotion of attracting international talents are the most immediate but also lasting and tangible representations of ELI's impact. This human capital coupled with ELI's technological resources are the foundations for longer-term impacts in the many application areas and technological directions pursued by ELI. The development of sustainable innovation partnerships and ecosystems with technological and industrial players will be key in this respect.

ELI emerged from a European community effort driven by the ambition to explore the many scientific applications of a new generation of ultra-high-power short-pulse lasers, but, due to the pioneering use of ERDF for the construction of the ELI facilities, ELI was also conceived early on as an opportunity to generate socio-economic returns and contribute to the development of the hosting regions. While construction has already generated substantial impacts, the establishment of ELI ERIC and the gradual transition of the ELI facilities to operations open up a new cycle in the life of ELI that comes with impact opportunities. As a consortium of European member states, ELI ERIC represents a sustainable framework to nurture the development of European-wide initiatives and partnerships supporting innovation and education. Operating as an integrated organisation, the challenge and ambition of ELI is now to implement the strategies that will turn these opportunities into long-lasting impacts.

Acknowledgements

This contribution has been developed with support from a number of individuals. In particular, the authors would like to acknowledge the kind contributions from the ELI ERIC Facility Directors, Roman Hvězda of ELI Beamlines and Gábor Szabó of ELI ALPS and their entire staff for providing the necessary information and input to compile this chapter. Furthermore, Federico Canova and Jean-Claude Kieffer supported this work with input but also by reviewing the article. Finally, Sonja Knols-Jacobs made a valuable contribution supporting the initial compilation and editorial work. We would also like to thank the editors of the volume for the invitation to contribute and their feedback on the article.

References

EU 2021 Setting up the Extreme Light Infrastructure—Extreme Light Infrastructure European Research Infrastructure Consortium (ELI ERIC) *Commission Implementing Decision* 2021/960 of 30 April 2021, notified under document C(2021) 2923 https://eur-lex.europa.eu/eli/dec_impl/2021/960

European Commission 2018 Re-finding industry: defining innovation *Directorate-General for Research and Innovation* Publications Office https://data.europa.eu/doi/10.2777/475890

ESFRI 2006 European roadmap for research infrastructures *Report* https://www.esfri.eu/sites/default/files/esfri_roadmap_2006_en.pdf

EUR-Lex 2021 *Statutes of the Extreme Light Infrastructure ERIC* 2021/C 230/01 https://eur-lex.europa.eu/legal-content/EN/TXT/?uri=uriserv:OJ.C_.2021.230.01.0001.01.ENG

Gliksohn F 2020 ELI Case Study *Workshop on Impact Assessment, Evaluation and Monitoring of Research Infrastructures; ACCELERATing Europe's Leading Research Infrastructures* http://accelerate2020.eu/wp-content/uploads/20201216_Workshop-on-impact-assessment_-ELI-protocol-Florian-Gliksohn.pdf

Mourou G A *et al* (ed) 2011 *Whitebook ELI—Extreme Light Infrastructure; Science and Technology with Ultra-Intense Lasers* (Berlin: Thoss) https://eli-laser.eu/media/1019/eli-whitebook.pdf

Tajima T and Mourou G 2002 Zettawatt-exawatt lasers and their applications in ultrastrong-field physics *Phys. Rev. Accel. Beams* **5** 031301 https://journals.aps.org/prab/abstract/10.1103/PhysRevSTAB.5.031301

Chapter 8

Industrial liaison officers, key intermediaries between Big Science organisations and their industrial suppliers

Toon Verhoeven, Martin Townsend, Ana Belen Del Cerro Gordo, Paolo Acunzo, Anna Hall, Laurent Jammes and Nikolaj Zangenberg

8.1 Introduction

The construction of Big Science facilities requires a strong and consistent engagement from companies in order to supply state-of-the-art components. In order to ensure a strong relation with the national industry in the member states (MSs) of a Big Science organisation (BSO), most organisations and MSs appoint industrial liaison officers (ILOs). The function of an ILO may vary from BSO to BSO and from one MS to another. The ENRIITC project [1] works with the definition given by the European Commission:

> Industry Liaison Officers (ILO) are officially appointed by the Member States and Associated Countries to stimulate the collaboration among the national industry and the international research infrastructures, providing advice on business opportunities, R&D collaborations, call for tenders and industrial services.
>
> (From the European Commission Horizon 2020 Framework Programme [2].)

The ENRIITC project further elaborates:

> ILOs, typically motivate the respective national industry to participate in tenders issued by research infrastructures, identify business opportunities at research infrastructures, promote them to national industry and support national industry in the development of bids if necessary.

ILOs are typically employed and financed by the MSs with the overall purpose of maximising the industrial return from the BSO to the MS. Thus, the ILOs see it as their main task to keep national business/industry aware of opportunities and developments in BSOs and in so doing, help businesses to prepare for and participate in commercial call-for-tender (CFT). The ILO activity is particularly important for high-tech small and medium sized enterprises (SMEs) who rarely can afford the cost or time to support a cross-sector business scanning capability and hence may not see opportunities outside their own marketplace. Future success for BSOs is intrinsically linked to the development of breakthrough technologies, in which the ILO network plays a key role in the promotion of co-development between BSOs and industry and the subsequent optimisation of knowledge transfer by means of spin-ins and spin-outs.

Because the above interaction between BSOs and companies is supported by ILOs, the ILOs have a rather rare and unique oversight into the industrial supply lines. This helps to give BSOs a broad and thorough overview of the capabilities, capacity, and technology readiness of the industry. For BSOs this is useful to overcome the technical challenges they encounter and to succeed in the construction and operation of such complex facilities.

Over the past few years, the term 'Big Science market' has emerged. This market recognises that a supplier to one BSO may also supply to other BSOs since many of the underlying technology requirements are the same, leading to the specifications and required competences from the companies often being very similar. The ILOs have been instrumental in defining this as a common market by motivating the national companies to offer their products and services across several BSOs.

8.2 The interaction between Big Science and industry

Big Science is a significant breeding ground for scientists and engineers meet. They challenge each other to strive beyond the current capabilities/performances of scientific instrumentation. Human desire to understand, challenge, and pose questions has no bounds, solving these questions increasingly requires pushing technological developments beyond the current state of the art and subsequently often leads to new questions that are even more challenging, creating a never-ending cycle. This is the natural cycle of Big Science—from scientific inspiration to technological realization, answering questions and defining new questions. Not only is it feeding our curiosity, but it offers the possibility to better understand, challenge, and solve other global societal challenges, in e.g. healthcare, climate change, and sustainable energy production.

Continuous innovation is fundamental to success and high-tech industry is invaluable in this cycle. There is an increasing need to involve industry for the development of the required complex break-through technologies and to start this process at the initial steps of the technological development. From the perspective of the industry, the involvement in Big Science is motivated by the opportunities to improve capabilities and to introduce new technologies in existing or new markets. But this is not without risks—projects are getting ever more complex with the associated very long lead times and there may not be a short-term return on investment. Technologies for Big Science often require significant investments to realize commercially viable returns giving an even more difficult barrier for SMEs. To keep the cycle running we need ways to manage these risks through appropriate risk sharing strategies and still achieve the gains at the end of the road, both for Big Science and for the industrial partners.

In this perspective it is important to develop a sustainable European market for the BSOs in order to optimize industrial investment and promote technology transfer from one sector/BSO to another. This common approach will be relevant also for improving SME participation in specific activities of those Big Science projects. ILOs have been seen to constitute an important 'grease' in the gears of this ecosystem mixing innovation with commercial interests in a high-tech market.

The way the ILO networks are organized within the different European countries are quite diverse, the following approaches can be identified:

- ILO networks directly hosted and financed by a ministry (e.g. Spain). In this way, a very close collaboration between the different ILOs in a country is facilitated.
- ILOs financed by different public entities and aligned to the national facilities but with an overall coordinating function (e.g. the Netherlands). This allows for a strong interaction of the different ILOs with their home research institutes where they are usually employed.
- ILOs financed by different public entities and aligned to the national facilities but without an overall coordinating function (e.g. Italy, United Kingdom). The ILOs play a key role in making the links between national and pan-European programs. Nevertheless, these countries do have ways to organize national events incorporating all the BSOs.
- ILO networks financed by a ministry but outsourced to an RTO or university or combination (e.g. Denmark, Switzerland). This enables specialists with different background to work together and introduces a stronger link to companies.

8.3 Best practices for the interactions between BSOs and industrial suppliers

The Danish national ILO-network, BigScience.dk, was formed in 2010 and thus was among the earliest countries to analyze and try to formulate the common characteristics of the 'Big Science market'. Many of their observations and recommendations were presented in the report: 'Best practices for the interactions between Big Science organisations and industrial suppliers' [3]. The report recognizes that the main purpose of the BSOs is to provide scientific excellence for the partners/MSs, either for fundamental or applied research. The report does, however, also highlight that the BSOs are under pressure to demonstrate a broader industrial and societal impact—or 'return on investment'—for the MS that are financing them. This impact is often monitored by considering industrial return, technology transfer opportunities, as well as educational opportunities for scientists, engineers, and technicians.

The report highlights one of the main challenges in the high-tech marketplace:

> Industrial contracts with BSOs are a quantifiable measure of the rate of return on investment but are also seen as a source of innovation and knowledge transfer to the involved companies. However, there is an inherent conflict between the need for long-term relationships with suppliers, who can provide innovative solutions, and the desire for openness and transparency expressed in the fact that many European BSOs are purchasing by following public procurement rules. [3]

This dilemma has also been discussed academically over recent years in the context of innovation procurement which, in principle, can be done as so-called 'competitive dialogue' under EU-law [4, 5].

The BigScience.dk report [3] made a series of observations relevant for new companies, BSOs and ILOs that are presented in the following sections.

8.3.1 Barriers for entering the Big Science market

A new company who wants to enter the Big Science marketplace will face challenges:

> The [Big Science] market is characterised by being relatively conservative in procedures and practices, while pushing the bounds of technology and innovation. The barriers for new companies are often high and there are clear trends of consolidation, and in some cases, monopolies have formed.
>
> Thus, access to the market for new companies can often be subject to some degree of national political intervention, for example through the activities of ILOs and launching of national R&D programs with industrial participation. When a company has acquired the first contract, it is easier to achieve the following orders, since references are in place and it is no longer a risk for the individual buyer at a research facility to place the next order at the company. [3]

The market includes several large companies that are also active in other demanding and national political importance, e.g. aerospace, defence and civil nuclear technologies. These areas also include both technical and organisational knowhow that enables the companies to become significant players in the Big Science marketplace as well. These companies are sometimes seen to have a national, strategic role and interact with public authorities. It is a good advice to companies with an interest in becoming suppliers to the Big Science market, to engage with one of these 'prime contractors' as sub-suppliers.

The Big Science market has evolved from the mid-50s when the establishment of national and international BSOs took off. Some companies have played a significant role during the full life cycle of the establishment of new BSOs, from the initial concept design, through construction and during the operation and maintenance of a BSO.

For conventional commercially off the shelf (COTS) products or services, the market is more normal and competitive, being generally characterised by open tenders and price competition. The collaboration and corporation with research facilities is often an advantage, but not a necessity.

Several structural factors can make the Big Science market difficult to access for especially smaller companies.

The report [3] also highlights five mainly administrative challenges for a company:

Many different procurement procedures and a wide variety of supplies

Experience from supplying to a Big Science facility cannot always be transferred to another one since each organisation works with its own procurement rules. The legal and administrative procedures are often difficult and burdensome; especially for companies without own legal services. Big Science organisations typically purchase a wide variety of products via quite diverse procurement procedures. This often makes it difficult for companies to find out which opportunities are a good match for their skills.

Price-based competition

Procurement rules where open competitive calls are awarded on the basis of the lowest technically compliant bidder are typically appropriate for standard off-the-shelf products and solutions. However, in the case of high-tech procurements, this can be a source of problems since they often involve an R&D process with an industrial supplier. Many industrial suppliers (particularly SMEs working with a shorter-term horizon than large companies) do not see a viable path for capitalising on their investment in the R&D phase if the following contract is opened to competition and solely based on price. They may therefore be hesitant to participate in an R&D process with a Big Science organization.

Time consuming procurement processes

Many tendering procedures are quite time consuming for companies. Tenders that are very large and/or spreading across different disciplines often exceed the financial and technical capacity of a single company. This requires access to a network that can make it easier to find national and international partners. Even though ILOs are there to help with this task, companies have to invest a lot of time in building their contacts within the Big Science organisations and participating in industry days and other relevant events. As a result, the business system favours the participation of large companies with Big Science as a main market. Smaller companies trying to find their way into the market have to make a large investment in time, with very little guaranty for success.

High demand for documentation/quality control, standards, and references

The demand from Big Science organisations for this type of documentation and references varies quite a lot and is not always a barrier for new companies. However, BigScience.dk has experienced over the years many examples of tenders requiring compliance with certain national standards (e.g. in the nuclear field), when equivalent European norms would result in a more transparent tendering procedure rather than giving a competitive advantage to companies familiar with national standards. In the same way, requirement for very specific references can disqualify very competent companies, who have references from other market segments but have not yet worked so many years with the Big Science market.

National participation in certain programs may exclude some companies
As previously mentioned, considerations regarding in-kind and 'fair return' can, in some cases, have adverse effect towards the industrial participation in Big Science projects. Some qualified companies may be excluded from tender procedures for example because their country does not have an in-kind contribution or because they come from a well-balanced member state in an organisation using 'fair return'. It is a balancing act for Big Science organisation to lift poorly balanced member states, while not discouraging companies from well-balanced member states to participate [3].

Individual companies have only limited capacity and access to BSOs to handle these barriers. Here, the ILOs play an important role to initiate a dialogue between the BSO and a company and to mediate the dialogue which may even lead to improved handling of procurement at the BSO since the ILOs represent the partners/MSs responsible for the BSO.

8.3.2 Best practices and recommendations

Combining the recommendations from [3] and the experiences from the PERIIA board, a series of recommendations directed to the BSOs are formulated below:

1. *Supplier databases*: The ILOs can help BSOs ensure that their databases are updated and, in their national work, encourage new potential suppliers to register in their databases. In order to create and update such databases, it is very useful to arrange company visits from the technical staff of the BSOs or organize promotional visits to the BSO. For both types of visits, ILOs should take an active role and support the involved BSOs by identifying and establishing contacts for the most relevant companies. The recent COVID-19 pandemic has changed the interaction between BSOs, ILOs, and industry and actually made it easier to establish a first contact via a virtual visit, meeting, or webinar. This is particularly an advantage for SMEs with restricted budget for travels.
2. *Arrange industry events*: Several BSOs arrange larger industry events (for example ESA Industry Space Days, ITER Business Forum or CERN HiLumi Industry Days) with the purpose of presenting a specific Big Science project to a large industrial audience. These industry events usually give companies the opportunity to present their skills to representatives from the BSOs and/or to other industrial suppliers during prearranged individual meetings.
3. *Join industry fairs*: Key technical staff members from the BSOs should participate in recurring industrial fairs such as the Hannover Messe (Germany), Elmia Subcontractor (Sweden), and Precision Fair (Netherlands) and other similar events in Europe, to find and network with potential suppliers. Furthermore, having dedicated persons at BSOs focused on the interaction with industry and ILOs has demonstrated to be very useful in building long term relationships.

8.4 Creating a consolidated European Big Science marketplace

The idea of a consolidated European Big Science marketplace is not new. The consequences of the lack of such a common market have been named in several publications evaluating the challenges and impact of building Big Science facilities. It has been addressed in the framework of the EIROforum [1], particularly by heads of procurement through the EIROforum Working Group on Procurement. From 2015 the topic was also discussed in the Horizon 2020 project BrightnESS, a collaboration of 18 European partners under the leadership of the European Spallation Source [6], which proposed to establish a common tender platform for business opportunities at BSOs. Many ILOs have taken it upon themselves to stimulate interest from the industry towards the Big Science market by organising events with participation of several BSOs, thereby showing the cross-organisational nature of some of the suppliers.

Those efforts could bc intensified and coordinated in the future to increase the attractiveness of the Big Science market towards the European industry. Big Science facilities and suppliers could both benefit from the development of several tools that could increase transparency across different facilities and different industrial sectors. A common vocabulary for systems and components could be a starting point for standardization of call codes. This could also be the basis for a common register of suppliers on system/component level.

An important step in the direction of a consolidated European Big Science marketplace was to organise a joint Big Science Business Forum (BSBF) with the participation of several BSOs. The main purpose of the event was to enabling BSOs to engage directly with suppliers and potential supplier companies from all over Europe. The event also presented the opportunity to connect national and international policy makers coupled with key decision makers in the BSOs in order to work towards developing successful and sustainable boundary conditions for the Big Science market.

The first BSBF was held in Copenhagen 2018 and featured a series of technology themed sessions, where the respective heads of subject from the BSOs presented their program opportunities and challenges under each theme. The themes included: data acquisition, robotics, precision engineering, diagnostics, and more. One of the objectives of the activity was to help industry to understand the requirements of the different programs, the chairs of the various sessions were challenged to sum up and explore commonality in challenges. The applications were very different, but, interestingly, the underlining technology challenges were surprisingly similar. For example, the presentations showed that in robotics, the common issues that ITER, CERN, ESS, ESA, etc, face are extreme environments, reliability, radiation hardness, latency, and human machine interfaces.

The event was a huge success, as can be seen in figure 8.1.

[1] EIROforum (https://www.eiroforum.org/) is the network of many of the largest Big Science Organisations in Europe.

1037	Participants in total
530	Businesses and organisations were represented
500	Participants for conference dinner
250	Participants for site-visits at ESS and MAX-IV
120	Speakers and presentations
71	Exhibition spaces including 207 exhibiting organisations
38	Journalists and media representatives present
29	Countries and 39 nationalities
16	Parallel session tracks
12	Billion EURs in investments presented
9	Organising Big Science organisations
9	Affiliated Big Science organisations
5	Side events
1	Cross-border, cross-organisational ILO summit with 49 ILO's present

Figure 8.1. Fact box about BSBF 2018 (numbers supplied by the organiser).

The Forum featured more than 120 presentations from BSOs about their future purchasing needs and collaboration opportunities. It attracted more than 1000 participants representing 530 companies from 29 countries. One of the most impressive figures extracted was the overall estimated value of the Big Science market for the four year period 2018–21 of no less than € 38 billion.

The impressive turnout was the result of a long campaign from the organising BSOs and using the network of ILOs which had been organised the year before in the Pan-European Research Infrastructures ILO Association (PERIIA). As seen in figure 8.2, one-third of the participants learned about the event from the ILOs. The survey of the participants also revealed a satisfactory rating of 95%, and 98% of the respondents indicated that they would join the next event.

The second edition of the Big Science Business Forum event, being organised by CDTI, is scheduled for 4–7 October 2022, in Granada, Spain, after being postponed twice due to the COVID-19 pandemic. BSBF2022 will present the main technological opportunities in the Big Science market through 16 parallel sessions on specific topics. The International Organising Committee consisting of the largest BSOs participating in the event estimate that for the period 2022–25, the market worth is more than € 37 billion. A string of new sessions will be introduced, including the research infrastructures in Horizon Europe, SME aspects, and affiliated BSOs. Furthermore, in this second edition the event is broadening its initial scope, paying special attention to the importance of high-technology SMEs in this market, and the opportunities for technology transfer. Although this second edition has been postponed several times due to the COVID-19 pandemic, this has also been an opportunity to consolidate the forum, thanks to the organisation of

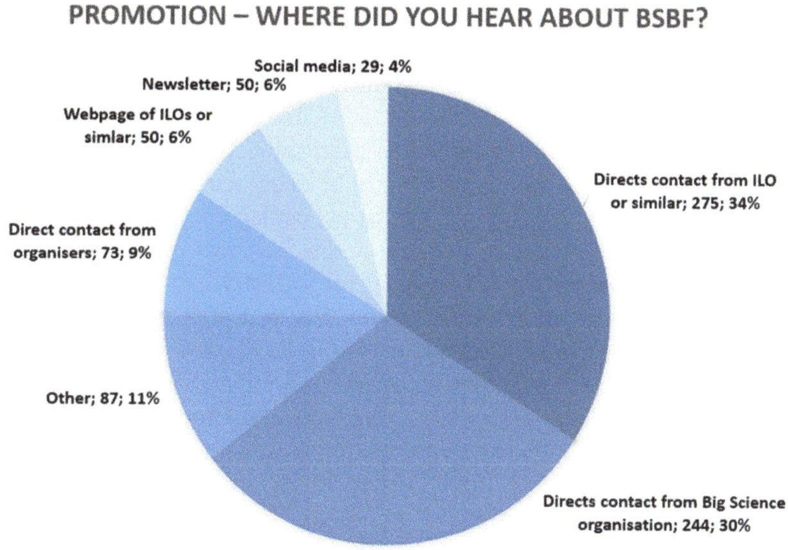

Figure 8.2. Survey response from BSBF 2018 participants (poll numbers supplied by the organiser).

multiple webinars, devoted to short-term opportunities in Big Science market, the impact of COVID-19, technology transfer, or future perspectives in astronomy, fusion, or particle physics. All these activities have contributed not only to keeping companies updated in the international Big Science market in a time with many travel restrictions, but also to facilitate a lot of discussions between industry, ILOs. and BSOs. Thus, it has strengthened the paramount importance of BSBF as a meeting point for the Big Science market stakeholders.

This type of event increases European companies' awareness of market opportunities in the Big Science market and actively engages private industry with a view to succeed in constructing, upgrading, and operating Big Science facilities. This type of joint effort also provides a stepping-stone towards a more integrated collaboration between the involved BSOs.

8.5 Future perspectives of the Big Science market

8.5.1 Different perspectives on the Big Science market

The Dutch ILO-Net took the initiative to map the relevant issues in relation to the Big Science market during a working conference in Brussels in 2017 [2]. Speakers from BSOs, industry, and government agencies were invited to present the most important issues from their perspective. The presentations were followed by a working session during which separate working groups addressed the specific challenges that were identified and tried to find ways to tackle them. The keynote speeches discussed the challenges that Big Science will be confronted with in the next twenty years from the

[2] The presentations from the workshop held on 8 June 2017 may be found at https://bigscience.nl/en/documents/presentations.

relevant perspectives; the Big Science facilities, the industry, and science. The findings and perspectives were collected in the 'Position paper on the future of Big Science' [7]. It includes several interesting quotes:

> What could be called Big Science has been subject to an evolution over the last 50 years. For 21st century Big Science activities trends of less favorable conditions can be noticed, in particular decreasing infrastructure spending which leads to only moderate business opportunities and less industrial engagement. The appeal for industry can be improved by introducing more attractive business models and coordination between various Big Science programmes, creating a larger, more accessible and attractive market with inherent business continuity.
>
> (Leonardo Biagioni, Head of Contracts and Procurement, Fusion for Energy 'Big Science in a changing business context' [7, p 3].)

> Collaboration, Innovation and Technology Transfer are the enabling elements that stimulate the return of Big Science for all parties involved. The LHC experiments at CERN can be seen as an example in which all these elements were actively pursued and where CERN tries to realize an 'ecosystem' including relevant activities. This is done using the current IdeaSquare initiative at CERN and the new EU initiative ATTRACT (http://www.attract-eu.org/) as implementations.
>
> (Markus Nordberg, Head of Resources Development of the Development and Innovation Unit at CERN: 'How to organize and support large collaborations in Big Science endeavours during many years, maintaining the innovation chain, and assuring spin-off and technology transfer along the way?' [7, p 3].)

> The cross-cutting Science & Technology segment within VDL, being a larger high-tech lead industry, has various goals which fit well with participation in Big Science projects. However, the large number of projects and tenders with a high degree of inherent complexity requires priority setting. Strong partnering with Big Science agencies, national institutes and specific innovative high-tech industries is a pre-requisite. There are pros and cons of SME participation in BS. This is thought to be generally difficult unless in collaboration with larger lead industries.
>
> (Hans Priem, Business Manager Science & Technology at VDL ETG: 'VDL & Big Science relevance' [7, p 3].)

> A new trend in Big Science is data intensive astronomy, requiring large scientific collaborations and new scientific (often global) infrastructures including large and complex central and regional Data Centers. The SKA initial data archive will be roughly 300 Petabytes, posing substantial challenges. Realizing these new infrastructures will stimulate

technology development, partnerships with industry, global networks and initiation of world-wide scientific collaboration.

(Michael Wise, Head of the Astronomy Group at ASTRON: 'The Square Kilometre Array—Big Science for Global Astronomy' [7, p 3].)

8.5.2 Concerns and recommendations

During this Brussels conference several barriers for successful collaboration were identified. In each of these areas, which are obviously interrelated, recommendations were formulated.

Sustaining the relatively high public funding seen in Big Science during the twentieth century is becoming very challenging due to competing demands for investment funding, i.e. infrastructure, social healthcare, and education. Therefore, the Big Science market gets less attractive to enter for industry, especially for high-tech innovative SMEs. The risks associated with (long-term) investments are relatively high. To promote industrial participation and to lower existing barriers the following recommendations should be considered:

- Develop specific funding programs for SMEs to enter Big Science projects and to promote low TRL development (e.g. SBIR).
- Create a shared vision on Big Science among the involved government agencies, based on an innovation chain approach. Make strategic choices and align funding sources along the chain, involving science driven technology development, innovation, and societal challenges.
- Improve the information flow between science and industry on both a national and on a European level, for instance by creating a European database for finding technological solutions and suppliers, a common on-line forum for suppliers and Big Science programs for matchmaking.
- Promote the creation of long-term roadmaps for the development of key enabling technologies for Big Science on a European scale.
- Promote the establishment of European databases in which information on required skills and competences can be easily accessed to facilitate matchmaking between the BSOs, research organizations, and industry.
- Develop procurement strategies that enable SMEs to minimize management load, and mitigate financial and technical risks.
- Standardize the codes used for call for tenders by different BSOs.
- Moreover, cultural differences still complicate enduring sustainable forms of collaboration between science and industry. Therefore, establishing exchange programs between scientists and industry professionals could be considered.
- Promote transition in science/industry cooperation approach emphasizing collaboration and a more explicit 'target-orientated' approach.

8.5.3 The innovation ecosystem for Big Science

Societal relevance of science has become increasingly important. Curiosity driven (excellent) science goals are no longer sufficient to evaluate the performance of research institutes or in many cases to justify the investment. Research institutes must contribute to the creation of national value, creation of know-how, support technology transfer, and create/develop sustainable business growth (creating jobs). National funding schemes have an increasing focus on (national) strategic programs, which is affecting the flexibility to support long term curiosity driven research in an international environment. This is an outcome of a team building session in June 2020 by the Dutch ILO-net where the following arguments were collected from a science perspective [8].

Existing 'ecosystems' in Europe are highly motivated by specific societal requirements (such as health and climate) largely driven by short term return on investment, through roadmaps that are led by industry. Science is in a subordinate position and must find its niche in selected ecosystems. An ecosystem for Big Science would restore science as a strong and exclusive driver for innovation, contributing to long-term European competitiveness. Such an ecosystem does, however, require a long term perspective and continuity, based on excellent (basic) science to start the innovation chain. Being part of an ecosystem will enable research institutes to raise funding much more easily for their missions, since the synergy between the relevant interests will already be explicit, and can be easily recognized and supported by dedicated funding instruments.

An ecosystem for Big Science would be a dedicated ecosystem in which an added value (multiplier) can be derived from the continuous interaction between geo-return, co-development, and technology transfer, driven by the ambition to push the frontiers of scientific knowledge. The ecosystem is characterized by long-term roadmaps, the development of breakthrough (key) technologies, and high risk/high gain.

Research institutes that act as the national 'home base' for large BSOs have the capability of creating unique science/technology combinations in their endeavor to push the frontiers of scientific knowledge. They maintain a continuous cycle of scientific inspiration and technological realization, thereby challenging high tech industry.

Industrial liaison officers (ILOs) are already acknowledged for their role as intermediaries between science and industry. Their focus can be different depending on the specific mission of their research institutes, but ranges from technology transfer, initiating co-development to raising geo-return. An ILO-network, such as PERIIA, encompasses all these capacities and can serve as a powerful lubricant in a Big Science ecosystem. Looking (for instance) at the developments around a new Horizon Europe program, and the European Innovation Council (EIC) there is a definite trend towards the support of developing breakthrough or 'key (enabling) technologies' which are particularly crucial for the curiosity driven sciences.

BSOs are pivotal in sustaining and enhancing the competitiveness and world-class excellence of European science and have significant value for and impact on European industry, society and economy. This is so concluded in a recent EIROforum paper [9].

European BSOs:

- harness scientific expertise that drives discoveries and creation of knowledge.
- offer access to researchers from Europe and beyond to the best and in some cases, unique, state-of-the-art facilities.

- enable integration of research communities from all countries in Europe.
- maintain the world-class excellence of European science.

BSOs are also drivers of innovation as they need to develop advanced instruments and a variety of cutting-edge technologies. These developments are often done in close collaboration with industry, and lead to both incremental and breakthrough innovation. Knowledge and technology transfer from BSOs to industry via collaborative R&D activities, procurement of goods, equipment and services, and the creation of spin-off companies is of significant benefit for European industry. BSOs now become key drivers for the continuous cycle of technology developments, which can feed the cycle of innovation well beyond their respective scientific areas. BSOs provide numerous benefits to European society and economy, either directly, through addressing societal challenges or industrial needs, or indirectly through training and education, as well as knowledge and technology transfer.

8.5.4 The formation of PERIIA

As can be seen from the previous sections, both in Denmark and the Netherlands (as well as several other places around Europe) the ILOs were becoming organised in national groups and discussing the challenges and barriers. Since it was also evident, that these challenges were common it was recognised that some form of ILO coordination across borders would be beneficial, and the ILOs created a Pan-European Research Infrastructure ILO Association (PERIIA, https://periia.eu/). This association aims to progress and develop the interaction between BSOs and business/industry to create a more integrated and sustainable Big Science market and in doing so help support business continuity and growth. The first steps toward the creation of PERIIA were initiated at a parallel meeting during the ITER Business Forum in 2017 in Avignon (IBF/17). The formal decision to create PERIIA was taken at a meeting the day before the start of the BSBF2018 in Copenhagen. During IBF/19, the founding General Assembly for the PERIIA Network was held and the board of PERIIA was elected.

PERIIA provides a forum for the exchange of ideas/best practises from across the wide network, strives to enhance cohesion and functionality across the different BSO ILO networks and furthermore acts as a consultative reference body for European institutions and other stakeholder groups to contribute to establish a common market of Big Science to support research, innovation, and economic development in Europe.

The importance of the PERIIA network has increased during the time of the COVID-19 pandemic. The free association of ILOs, working with the major BSOs in Europe and coming from 30 different European countries, has become the ideal place to compare national needs and suggest common responses to new industrial problems due to the COVID-19 pandemic restrictions.

One important action of PERIIA was to analyse the effects and economic impact of COVID-19 on the Big Science market and present proposals for supporting industries and SMEs involved in the sector. In fact, the Big Science market must be managed as an important sector for innovation and economic development for the European competitiveness in the world. For this reason, it needs specific measures to be adopted in synergy by different stakeholders such as BSOs, EU institutions, national governments, and many other international stakeholders.

Based on a questionnaire submitted to all ILOs and discussed at the ENRIITC seminar 'Industrial impact on the Big Science market after one year of the COVID pandemic' held June 2021, PERIIA suggested the following priorities in different areas in order to archive the main goal to develop and structure a common market of Big Science in Europe:

Regarding procurement:

- Identify activities to maximize SMEs access into the Big Science market.
- Simplified procurement process for low value orders.
- Develop financial support mechanisms to support companies involved in Big Science in Europe.
- Subsidise the investments necessary to prepare large and complex procurement activities for industry in Big Science.
- Encourage harmonization of Big Science organizations tender codes, through the establishment of a Codes Standardization Committee.

Regarding market:

- Structuring a forum for sharing success stories and exchange of experiences among national ILOs working with different BSOs.
- Increase the universal visibility of the otherwise rather fragmented Big Science market.
- Encourage harmonization of BSOs purchasing practices, processes, and procurement rules.
- Make available a universal and free software package to enable online matchmaking events for business.
- Define specific requirements of Next Generation EU and other European funds in order to support Big Science industry.

8.5.5 Recommendations from the ENRIITC project

The ideas about establishing a more formal platform for exchange of training material, best practices, and ideas concerning the interaction between BSOs and industry has been matured in the European ENRIITC project (the European Network of Research Infrastructures and Industry for Collaboration) [1]. It is a cooperation of 11 partners and 60+ associates working together to establish a pan-European network of industrial liaison and contact officers (ILOs/ICOs) to improve the cooperation between BSOs and industry and to boost the innovation ecosystem in Europe. ICOs are defined as employees at the BSOs in charge of the non-supplier interactions with industry, e.g. usage of the facility or technology transfer.

The coordination between ILOs and ICOs enables mutual learning. It further maps the collaboration potential between BSOs and industry, develops and refines strategies and best practices to foster these collaborations, raises awareness among industry for collaboration opportunities at research infrastructures, and demonstrates impact.

While the roles of ILOs and ICOs are very different, some of the activities they perform can overlap. ENRIITC aims to provide advice to both by developing strategies for innovation, training, and outreach, and formulating policy recommendations to encourage efficient interactions between ILOs, ICOs, and industry. A total number of six focus groups were set up to participate in this work. PERIIA plays an important role in ENRIITC, and its board members are present in almost all of the focus groups.

8.6 Conclusions

In this chapter we have illustrated the role of ILOs as key intermediaries between Big Science organisations and their industrial suppliers, particularly for SMEs. Combining forces of all ILOs, organized within PERIIA framework and developing policy recommendations via EU projects such as ENRIITC, are essential steps towards the creation of a single consolidated Big Science marketplace. A supplier to one BSO is often very well equipped to supply to other BSOs, therefore, the appeal for industry can be improved by introducing more attractive business models and better coordination between various Big Science programs, creating a larger, more accessible, and attractive market with inherent business continuity.

Research in the field suggests that there is significant benefit to be gained if a shared vision of Big Science could be created among the involved government agencies, perhaps based on an innovation chain approach. Strategic choices should be made, and funding sources could be aligned along the chain, involving science driven technology development, innovation, and societal challenges.

An ecosystem for Big Science would be the dedicated ecosystem in which an added value can be derived from the continuous interaction between geo-return, co-development, and technology transfer, driven by the ambition to push the frontiers of scientific knowledge. The ecosystem is characterized by long-term roadmaps, the development of breakthrough (key) technologies, characterized as high risk/high gain.

All of these considerations led, as a first step of the consolidation process, to the organization of the Big Science Business Forum (BSBF). The first version of BSBF in February 2018 in Copenhagen was such a great success that now, despite the 2 years of delay because of the COVID pandemic, we are pleased that the second version of BSBF will go ahead in October 2022 in Granada, Spain.

References

[1] ENRIITC The ENRIITC Project https://enriitc.eu/

[2] European Commission Horizon 2020 Network of research infrastructure industrial liaison and contact officers *Framework Programme* Topic ID: INFRAINNOV-02-2019 https://ec.europa.eu/info/funding-tenders/opportunities/portal/screen/opportunities/topic-details/infrainnov-02–2019 (Accessed: 24 November 2021)

[3] Jensen A, Forneris J, Nielsen L P, Zangenberg N, Nielsen R and Korsholm S B 2017 Best practices for the interactions between Big Science organisations and industrial suppliers *Report* BigScience.dk https://www.bigscience.dk/media/1078/best_practices_big_science_industrial_suppliers_17022017.pdf (Accessed: 24 November 2021)

[4] European Union 2014 EU procurement directive *Official Journal of the European Union* 2014/24/EU

[5] Li-Ying J, Forneris J, Korsholm S B, Jensen A and Zangenberg N 2021 How European Big Science organizations and suppliers innovate through public procurement *Res.-Technol. Manag.* **64** 46–56

[6] Cordis EU Research Results 2020 Building a research infrastructure and synergies for highest scientific impact on ESS *BrightnESS project* https://cordis.europa.eu/project/id/676548

[7] Cornet G, van Dijk H, Boom E, Klöpping R, Visser J and Verhoeven A G A 2017 Position paper on the future of Big Science *Position Paper* Dutch ILO-net https://bigscience.nl/files//rapporten/Position%20Paper%20on%20the%20future%20of%20Big%20Science%202018_def_.pdf (Accessed: 24 November 2021)

[8] Cornet G, Boom E, Klöpping R, Visser J and Verhoeven A G A 2020 *Position paper of the Dutch ILO-net* https://www.bigscience.nl/files//rapporten/ilo-net-position_paper_2020.pdf

[9] EIROforum 2020 Research infrastructures: value and impact for European science, industry and society *Position Paper* https://www.eiroforum.org/wp-content/uploads/eiroforum-paper-value-of-RIs-may-2020.pdf (Accessed: 24 November 2021)

IOP Publishing

Big Science in the 21st Century
Economic and societal impacts
Panagiotis Charitos, Theodore Arabatzis, Harry Cliff, Günther Dissertori, Juliette Forneris and Jason Li-Ying

Chapter 9

Analytical research infrastructures and industry engagement: drivers, challenges, and impact

Graham Appleby, Caroline Boudou, Ralph Gilles, Ed Mitchell, Alejandro Sánchez and Elizabeth Shotton

Large-scale analytical research infrastructures (ARIs) provide unique capabilities to study materials and living matter at levels far beyond those of conventional techniques. The x-ray and neutron analytical research infrastructures (synchrotron, free-electron laser, nuclear reactor, and spallation neutron sources) are at the centre of the ARI family and provide a vast range of techniques and skills made available to researchers in academia and industry alike to study, probe, and monitor materials under static as well as real manufacturing and working conditions. The unique probes (mostly non-destructive) allow deep vision in materials, making inner structures visible across broad length- and time-scales. These capabilities mean that such facilities are increasingly on the radar of industry as sophisticated means to provide a competitive edge to their research and development (R&D) needs. This becomes even more relevant with materials development demanding ever more detailed views on structure to improve performance, sustainability, and reusability. In the 'make–measure–model' cycle of materials development, the analytical research infrastructures can therefore play an important, if not in fact a critical role. However, their use by industry, directly and indirectly, remains under-exploited and significant opportunities are, as yet, largely unexplored and underused.

This essay explores the routes that the European ARIs are using to engage and work with industry, laying out the overall ecosystem with its key drivers and current challenges and impact, with examples, and then looks ahead to building more and stronger links with industry to use these facilities to drive industrial innovation.

9.1 Introduction

Dorothy, in the *Wonderful Wizard of Oz* (Baum 1900), realises very quickly that she has been transported to another realm, remarking to Toto, her dog, 'I've a feeling we're not in Kansas any more. We must be over the rainbow!' Being at a large-scale analytical research infrastructure (ARI) and working with industry is not dissimilar. One rapidly realises that working with clients and researchers from the industry sector comes with a very different perspective to that of the traditional academic user community, and that paving the equivalent of the yellow brick road to the Emerald City of stronger and better engagement with industry brings its own challenges and drivers, but also rewards and impact, and excellent science.

Just after synchrotron light was first observed in the 1940s, an 'X-rays in Industry' conference was held in Melbourne with the proceedings published in the *British Journal of Applied Physics* (Ackland *et al* 1950) commenting that 'Although the synchrotron is still in the developmental stages, there is every indication that it may become a useful industrial x-ray generator'. While 'man's first self-sustaining nuclear chain reaction' took place in 1942 in Chicago (OSTI 1982) under the supervision of Nobel Prize winner Enrico Fermi, thermal neutrons were used to obtain a diffraction pattern a few years later in 1945. The industrial sectors that first showed an interest in neutron reactors were energy (thermonuclear power plants) and nuclear medicine (radioisotopes). Since these early activities, synchrotron light, neutron, and, most recently, free-electron laser ARIs have emerged and matured on the research landscape, with their state-of-the-art scientifically excellent instrumentation and professional support for researcher access. The facilities were developed initially for fundamental science, but, as foreseen, their usefulness for industrial applications was realised with industrial application as research tools. As early as the mid-1980s, three industrial companies (ICI, BP, and Shell) identified the value of synchrotron radiation for industrial research and joined together as 'The Industrial Consortium' for commercial access to the Synchrotron Radiation Source in Daresbury (UK).

However, the landscape has changed very much and particularly so over the last ten years, with industrial engagement moving from a 'nice to have' to a 'must have' status for new and ongoing investments in such capital-heavy large-scale research facilities. This is reflected in the strong expectation of stakeholders when making investments, and particularly so perhaps of the light and neutron sources which are well adapted to advanced materials characterisation and whose mission is to provide access to routine excellence for both academic and industrial researchers. This expectation has grown with the transformation of the scientific community towards a model of contributing to a richer multimodal mix of fundamental, applied, and commercial research. Stakeholders in the ARIs demand stronger engagement, delivering on more immediate economic interests and impact, as well as the traditional longer-term blue-sky research opportunities. In the context of economic impact and working with industry, which this essay focuses upon, ARI support is provided across pre-competitive (i.e. publishable) research, collaborative projects, and competitive (i.e. confidential proprietary) studies, as well as industry benefiting from the ARI technology developments through knowledge and technology transfer.

Contact from industry to ARIs starts with an unsolved problem in industry. The reasons can be manifold, e.g. not the right equipment or technique available in industry to perform the required experiment, insufficient knowledge in the experimental field, or no tools to evaluate and interpret the measured data. Work with industry is spread across the technology readiness levels (TRLs) and covers issues in supporting baseline innovation, improving manufacturing processes, failure analysis, reliability, and ageing processes. ARIs also contribute to the arising challenges of our society, as recently highlighted with priority access for important coronavirus disease COVID-19 experiments for both academics and industrial researchers.

This essay sets out a perspective on the landscape of how the light and neutron sources open their doors for industry, creating opportunity and paving the way for innovation, and looks ahead as to how ARIs and industry could extend their relationship in the future.

9.2 Industry and ARIs: a rich land of opportunity

ARIs of all flavours across the world strive to provide 'accessible excellence' to state-of-the-art capabilities which are precise, sensitive, fast, and provide insights far beyond those of conventional techniques for materials and life sciences. Traditionally used by academic researchers, ARIs are also providing actual competitive advantages to those companies which are aware of the potential value of ARI capacities for their research and innovation. The advanced techniques provided by light and neutron sources are powerful tools for multi-scale material and process characterisation down to atomic detail. As such, they are useful for a wide range of industry sectors (figure 9.1) for developing new advanced materials, new innovative products, and more efficient manufacturing processes. Pharmaceuticals, biotechnology, medical technology, isotope production, cosmetics and consumer

Figure 9.1. Industrial sectors addressed by light and neutron sources gathered, respectively, in the League of European Accelerator-based Photon Sources (LEAPS) and League of Advanced European Neutron Sources (LENS) consortia.

products, metallurgy, additive manufacturing, batteries, energy storage, photovoltaics, environment, agro-food, energy and mining, packaging, semiconductor, aerospace, building materials, advanced instrumentation, polymers, and textiles is a non-exhaustive view of industry sectors already benefiting from using ARIs.

The rich landscape of opportunity provided by the ARIs is characterised by facilities missioned to engage with industry, supporting research and providing a direct return on the public investments made in the large-scale instruments. With the growth of synchrotron light sources and neutron sources (figure 9.2), the ecosystems set around the facilities and in which the facilities are embedded have extended and become more complex and gradually more active in partnering and collaborating together. This section explores and highlights the versatility and depth of the mechanisms in place for industry and the ARIs to work together.

The traditional access to ARI instruments and expertise is provided via a peer-review process, by which researchers submit proposals that are reviewed by external expert panellists with access granted by subjective 'scientific excellence'. Industry typically requires quick and confidential results to solve their problems, to develop innovative products or to provide new solutions. To that end, fast and proprietary access, requiring neither peer-review nor publication of results or data, has been put in place by most ARIs. For peer-review access, the research performed at light and

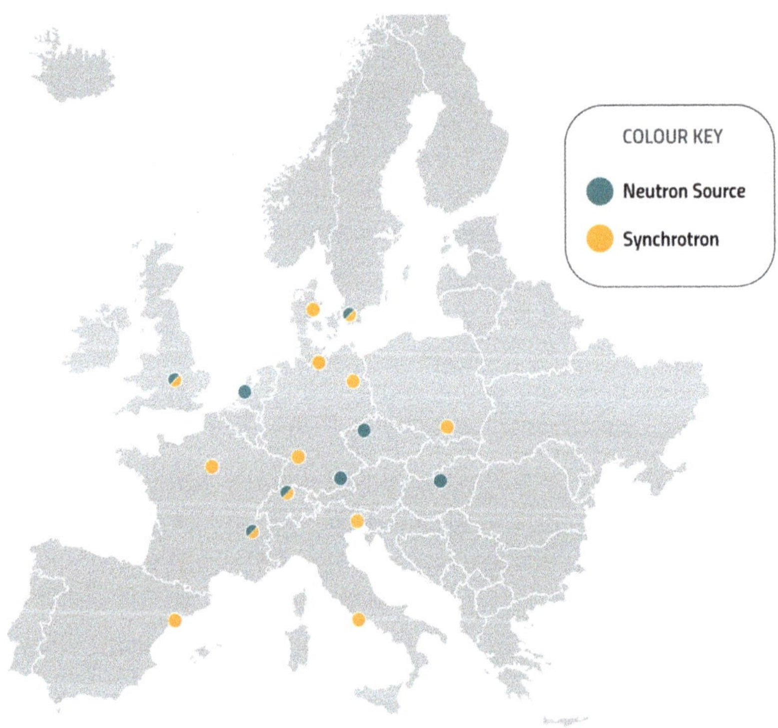

Figure 9.2. Map of light and neutron sources established in Europe. List of organisations and networks can be found on the webpages of Analytical Research Infrastructures in Europe (ARIE) www.arie-eu.org.

neutron sources normally encompasses TRLs from 1 to 2 and pushing to level 4. However, the proprietary route extends user research activity across all TRLs from 1 to 9 with experiments ranging from those supporting basic innovation within companies, to those seeking data to resolve manufacturing and operational issues. It is important to note that the ARI's own internal R&D on advanced instrumentation goes all the way to TRL 9, producing operational equipment for the facility itself, some of which can be commercialised rapidly via technology transfer and licensing agreements with technology providers.

In working with industry, the commercial access and industry-related activities enhance the impact of ARIs, demonstrating their role with industry and creating a modest but important virtuous cycle of growing industrial access and engagement—though at times creating tension with the traditional peer-review access and internal development priorities as capacity can limit timely access for academics and industry alike.

9.2.1 Pushing open the ARI doors for industry

No longer a 'could have', nor 'nice to have', industry is now a 'must have' component of the ever widening ARI missions, with industry an ARI community in its own right and having its own place and interests (Mitchell *et al* 2011, Malbet-Monaco *et al* 2013, Petry 2015). Stakeholders and facility managers are one of the driving forces in a top down approach, encouraging facility-industry engagement, but there is also a bottom up push with facility staff also wishing to work with industry in a stronger application of their beamlines and techniques to real world challenges.

Over time, the ARIs have acknowledged the importance of industry by setting up internal offices with dedicated staff to deal with industrial clients, such as the so-called 'industry contact officers' (ICOs), and by opening access avenues which are parallel to, but completely different from, the traditional academic peer-review process to be in line with the response times and needs expected by industry. These offices are increasingly professionalised, being proactive in outreach and marketing (no longer a forbidden word!) towards prospective commercial users and collaborators. Typically, therefore, the ICO works as the main contact point with the industrial clients providing them with advice and guidance on the scientific techniques and identifying the most relevant instrument to be used for better solving the problem posed by the client. The ICO teams are responsible for the administrative tasks, outreach and marketing, training, defining the commercial offer, preparing and issuing quotes; specific ICOs (who may be based in the industry offices or in beamline groups according to facility policy) and who are dedicated to focussed scientific areas can also support elements of the technical side with instrument configuration and of course measurements, data acquisition, analysis and interpretation, and customer follow-up. ICOs may also take on a role wider than being a reactive 'shop', and proactively stimulate collaboration with industry, joint work and engagement with the commercial world through external funding such as national and European framework programmes. The role and value of the ICOs has grown to such an extent that the European Commission has funded a project ENRIITC (ENRIITC n.d.) to build a permanent network linking ICOs (and the procurement equivalent of industry liaison officers, ILOs) from all types of the major ARIs across Europe. Complementing these offices may also be industry or innovation committees, or industry user representatives which provide advice to ARI management on industry engagement strategies.

Although there has been a remarkable growth within the light and neutron sources of industry dedicated staff and offices, internal structuring to liaise with industrial users of the facilities or of the technology varies amongst facilities as it depends on the institute's own strategy and ambition regarding socio-economic relevance and the level of expected (or demanded) commercial revenues, as well as maturity of the facility (Thiry and Boudou 2020). In this latter case, newer facilities may have an operational model with industry fully 'built in' from the outset compared to older facilities which have incrementally constructed an industry strategy as industrial engagement has become 'must have'. This ambition and actual activity can and does therefore vary widely amongst facilities and is not always simply based on the commercial income targeted or generated, but other key performance indicators can play a role such as publications with industrial authors, collaboration agreements signed, PhD students or research staff funded, technology licences generated, and so on (figure 9.3).

The industrial activities are usually very relevant for the ARIs as an important element of demonstrating relevance, and the ICOs typically report to a director of the facility and have direct contact with all of the instrument and beamline scientists.

It is important to note that internal structuring goes far beyond simply setting up and letting loose an industry office with an intrapreneurial mission. The culture of 'industry' needs to encompass the beamline and instrument staff, instrumentation engineers and technicians, administrative staff, and also the facility operational schedule and technology developmental projects with the flexibility (or not) to incorporate industry needs and interests. Without the whole facility being on board to work with industry, challenges, tensions, and mis-matched expectations can quickly mount up.

Industrial users address the facilities with questions for which they seek answers, usually in a very timely manner matching industrial research or production cycle times. As mentioned earlier, facilities provide access to the state-of-the-art

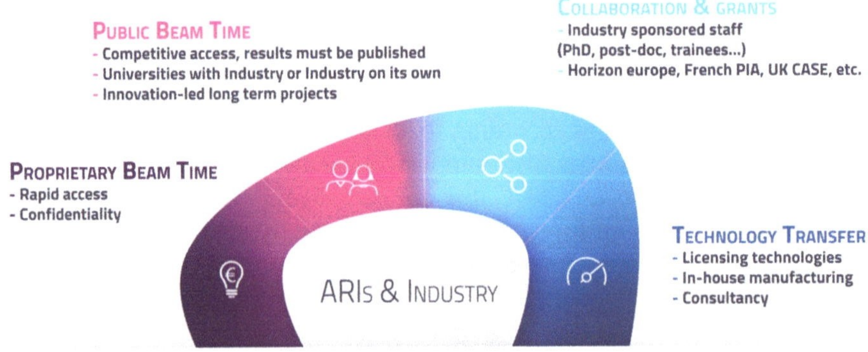

Figure 9.3. ARIs working with industry and creating value.

instruments and support to data acquisition and reduction but most, built on a model of 'expert' users, do not provide the analysis and/or interpretation of the results for the peer-review user base (though automation for selected routine services has dramatically changed this outlook and which may see further developments with the application of artificial intelligence and machine learning). Further support for data reduction and analysis for proprietary users varies, with some facilities providing a full 'project level' support through dedicated staff, others focussing on their core business of data production, with some developing a mixed approach of providing analysis for selected techniques. Industry is also adapting with some companies investing in-house and learning how to handle and work with some types of raw data which are of particular longer-term importance to them. Importantly, private enterprise is also filling this gap where higher added value exists, with the emergence of intermediaries (as described in a dedicated sub-section below).

Experience has demonstrated that having a specific office or team missioned to work with industry, identify opportunities, and work with an intrapreneurial objective leads to a stronger industry engagement. Over a lead time of a few years, the offices and staffing typically more than pay for their costs and make a positive return to the facility in many ways—cash, impact, visibility, and excellent science.

Across the European light and neutron sources, commercial service provision is typically based upon cost recovery. The aim of the facilities is not to make a cash 'profit' but to enable their advanced facilities to be exploited by industry, helping to push forward commercial developments and having a more direct impact on the economy and the quality of life of stakeholder citizens. The commercial income generated is usually seen as an additional income for the facility. The internal policies of facilities may vary in the details, but often there is a policy of returning at least a fraction of the commercial income to the scientific or engineering groups or teams generating the income, where it can be used to provide a better overall service to all users by purchasing better equipment and hiring more staff. In the case of technology transfer or inventions by staff, there may be a policy of rewarding staff directly according to the national legal obligations in place for the host country of the ARI.

When it comes to industrial users involved in pre-competitive research, the (usually long lasting) collaboration with ARI scientists becomes fruitful, often in terms of co-authored publications which benefit the ARI staff as quantities of papers remain an important metric for career progression, and sometimes in terms of exploitable intellectual property which can be protected before the collaborative research is published according to facility rules for peer-review access (ALBA 2020, Catalano *et al* 2021).

All in all, facilities are investing resources and putting efforts into internal structuring to enable working with industry to be as effective as possible, whilst respecting their statutes, missions, and overall user community demands.

9.2.2 Versatility and ecosystems for industry engagement

One size does not fit all and ARIs have come to realise this in working with industry, striving to open their doors to industry as effectively as possible by appreciating the different contexts and drivers from industry partners, activating different modes and growing flexibility within the remits and scope of the ARI statutes and governance. In some cases, these statutes may limit how an ARI can behave with industry, be it in terms of being a fully public institute, or by a particular governance framework.

Table 9.1 summarises the methods most commonly made available by the light and neutron sources to work with industry—the table is intended as an overview and is not exhaustive. Section 9.3 then provides more details and examples.

Table 9.1. Overview of approaches to work at/with ARIs for industrial researchers and engineers.

Approaches to work at/with ARIs for industrial researchers and engineers	Complementary information
Pre-competitive research through peer-reviewed access for beam time on instruments Beam time is granted (free of charge) after review and ranking of all proposals received at a given ARI. Industry can benefit from trans-national access support of Europe or ARI own travel support schemes on an equal footing to academics.	Results must be published and facility data policies mean raw data is released after an embargo period (usually three years). This route is open for industry to use. Beam time application forms often include innovation value or impact as an indicator for evaluation. Industry is treated equally with academics in the process, though there yet, even now, can remain a risk that industry driven work may be treated with a certain jaundiced eye by heavily academic review committees. A few facilities have access routes which are more industrial-friendly and based on innovation driven (yet still pre-competitive) R&D needs. But these remain very modest and not mainstream access.
Competitive research through proprietary (paid-for) beam time access and service provision. Beam time provided as rapidly as possible without formal review and application. The client company pays all costs related to the access and confidentiality is granted.	Commercial access is available at almost all facilities via the ICO offices as described above. Access is aimed to be rapid, and is confidential. Typically a few percent of all access at a facility is commercial, though it varies widely according to technique. Limits to commercial work may apply according to the facility statutes and/or the level of commercial income which may be generated.
Competitive research via owned or jointly owned beamlines and instruments	In some facilities, industry has invested in long-term ownership of beamlines with the aim of guaranteed access at cost effective rates offset against the original investment.

	The involvement of companies as 'investors' can allow a facility to speed up beamline build-out and demonstrates a strong industry interest and commitment.
Technology transfer (TT), knowledge transfer (KT) By their nature ARIs are pools deep with advanced instrumentation which have value in the ARI world and, albeit rarely, beyond. The technologies concerned are mainly detectors, optical devices, electronics, sample environments, and accelerator/insertion devices.	TT or KT occurs via collaborations with other facilities, with balanced contributions, and via technology licences with technology providers. Patents, largely due to their cost and administratively heavy nature, seem to be rarely used by European ARIs—though this may vary if ARIs are integrated within a university or as a co-managed component of a large research-based campus, perhaps enabling them to make a better use of patents as tool for intellectual property (IP) rights valorisation and management. Facility policies usually mean that software is typically made available as 'open source', with its use not tracked, via download from open source warehouses such as github. Open source policies are the norm as software is an element of support to user researchers and its availability outside of the facility is seen as an important asset to the user communities which need to process, visualise, and handle data after their experiments upon return to home laboratories.
Hardware and people investment	Long-term collaborations are often characterised by a strong involvement of the parties involved, who are counting on years in order to gain competence/expertise or to adapt and develop techniques and methodologies of analysis to fit their purposes. See section 9.3.1.4.
Collaborative engagement This route encompasses a wide range of approaches including funded PhD/post-doc projects, grants such as EU Framework projects, bi- or multi-lateral agreements, long-term innovation based access, and co-development.	EU framework programmes and national funding for research and innovation increasingly demand industrial inclusivity and provide opportunity for facilities and industry to work together. This can range from modest-scale access programmes to support small and medium enterprises (SMEs), to larger-scale projects to drive forward open innovation.

(*Continued*)

Table 9.1. *(Continued)*

Approaches to work at/with ARIs for industrial researchers and engineers	Complementary information
Collaboration is a growing and creative area as industry becomes more open to working with academic research institutes, and as barriers lower and R&D culture adapts for such opportunities.	PhD and post-doc funding via national schemes such as 'CIFRE'[1] in France, Germany's BMBF[2] or BMWI[3] funding or the UK's 'CASE'[4], or European scale such as the Marie Skłodowska-Curie Actions can also provide a collaborative carrot for ARIs and industry.
Production related needs going beyond pure R&D and supporting industry with material or expertise for immediate manufacturing requirements	ARIs encompass a wide range of specialised support laboratories and facilities for their own development needs and operation, but which can be valuable to industry, such as magnet measurement labs, alignment, metrology, and other specialised equipment. This can be considered especially helpful for SMEs who cannot afford nor risk the capital expenditure, but can benefit from an ARI's panoply of equipment for their own manufacturing needs or extended service provision. The reactor neutron sources have a particularity of high value in the production of a range of rare medical radioisotopes for use in therapy and very homogeneous wafer doping which can be only obtained by irradiation in the high-flux nuclear reactors at the source of the neutron beams used for the user programmes. Facilities may also manufacture and sell specialised components for integration in systems. This is usually only feasible where the ARI has set up a dedicated lab or machinery to supply its own in-house needs and can supply excess production to other facilities and industry.

[1] CIFRE: Conventions industrielles de formation par la recherche, a PhD funding scheme supported by France's MESRI (Ministry of Higher Education, Research and Innovation).
[2] BMBF: Federal Ministry of Education and Research (Germany).
[3] BMWI: Federal Ministry for Economic Affairs and Energy (Germany).
[4] Cooperative Awards in Science and Technology from the UK Research Councils provide support for PhD students to work in collaboration with a non-academic partner.

9.3 Examples of industry engagement

In this section, further details, examples, and success stories, mostly from Europe, of industrial work at the light and neutron source ARIs are provided and presented according to the type of access route or approach.

As mentioned above, within Europe, and indeed across much of the rest of the world, funding agencies see ARI industry engagement as a key strategic activity, leveraging value from the ARI investments. Towards this, the European Framework funding programmes have been and continue to be drivers of this engagement, providing the resources which allow the ARIs to go beyond basic missions to better work with demanding user communities like industry.

Many recent Horizon2020 projects have encompassed specific networking activities with industry and between the ARIs, allowing ICOs to build strong outreach programmes and to share knowledge and experience amongst themselves. Examples of such integrating activity projects are SINE2020[5], NFFA[6], CALIPSO[7], and CALIPSOplus[8]. Recently, the ARIs have become also involved in European supported projects targeting industry more directly such as the BIG-MAP[9] project on accelerated battery developments, the open innovation test bench (OITB) project TeesMAT[10] on providing services for battery research and testing, and EASI-STRESS[11] on developing quality based routine measurements for engineers, including at ARIs

The European support extends beyond industry as a user of the ARIs and also encompasses co-development and collaboration with industry in terms of developing advanced instrumentation. The recent ATTRACT and ATTRACT2[12] projects have set out to encourage industry and academia and ARIs to work together in the domain of detection and imaging technologies. The first round funding resulted in 170 breakthrough concepts being supported with the hottest projects going on in the second round with larger funding scales aiming to de-risk the technology. Starting in April 2021, the LEAPS Innovation project[13], is bringing together the light sources with industry in different configurations to pilot new ways of working together to co-create, co-develop new instrumentation required at advanced light sources. Developed in Europe, this new instrumentation may then be exploited by industry as suppliers.

9.3.1 Industry as a user of ARI facilities

9.3.1.1 Traditional peer-review access

The traditional access to ARI instruments and expertise is provided via a peer-review process, by which researchers submit proposals that are reviewed by external

[5] www.sine2020.eu
[6] www.nffa.eu
[7] https://cordis.europa.eu/project/id/730872
[8] www.calipsoplus.eu
[9] www.big-map.eu
[10] www.teesmat.eu
[11] www.easi-stress.eu
[12] www.attract-eu.com
[13] www.leaps-innov.eu

expert panellists. Beam time (i.e. access to instruments) for the measurement is granted based on the scientific excellence of the proposed project. This is a highly competitive process—most facilities receive two to three times more requests for beam time than can be supplied. However, access is granted for free and the results of measurements are required to be published. The facility data policies also mean that the experimental data are released after an embargo period (usually three years).

Perhaps surprisingly, this is still the main route for experiments which involve industry in some way. Between 15% and 40% all peer-reviewed access has some degree of industrial collaboration. Numbers vary amongst infrastructures and instruments. For some instruments, such as protein crystallography, stress–strain, small-angle scattering, computed tomography, and microscopy with strongly applied capabilities, this number can be at the high end of the range. For a few very specialised techniques there may well be almost none or no industry-related measurements taking place.

The facilities encourage this access route for industry-supported research, as this pre-competitive and publishable industrial research is often high impact work and it also encourages industry to be involved and see for themselves the value of ARI data. Data analysis is generally performed by the academic partner and the results are published in scientific articles which include company staff in the author lists. The science is generally of very high quality and usually of course highly relevant. Staff of major companies, benefiting from large R&D operations, such as Rolls Royce, AstraZeneca, P&G, Unilever, Johnson Matthey, General Electric, and many other large international companies, are frequent co-authors on these types of publications. The ARIs are then able to produce industrial case-studies from these publications to demonstrate societal impact, without the concern for confidentiality usually linked with the proprietary access route. A selection of papers with industry authors is given as examples:

- AstraZeneca and their academic collaborators published a study dealing with the improvement of subcutaneous administration of mRNA. Neutron scattering appeared to be the only available technique to probe the lipid distribution and resolve the surface due to the labelling of hydrogen atoms to the respective molecules (Yanez Arteta *et al* 2018, Davies *et al* 2021, MLZ 2021a, Sebastiani *et al* 2021).
- BASF together with scientists from the Universitat Politècnica de Catalunya (UPC) and the ALBA Synchrotron have determined how clays and superplasticizers interact in cement pastes. These results pave the way for improving the design of new superplasticizers more resilient to the clays that usually accompany the sands used in concrete as aggregates (ALBA 2019, Borralleras *et al* 2019).
- Using a range of neutron scattering techniques at ISIS Neutron and Muon Source to investigate how naturally occurring surfactants interact with their synthetic analogues, Unilever scientists study materials with application in more eco-friendly products (de Laune 2020, Tucker *et al* 2020)

- Rolls Royce are frequent users of neutron diffraction at ISIS Neutron and Muon Source, to understand formation and precipitates in nickel-base superalloys, the material of choice for the hottest sections of gas turbine engines (Collins *et al* 2020).

However, the competitive nature of the peer-review process can lead to difficulties in the industry-supported proposals being granted beam time, as fundamental science proposals can often be ranked with a higher scientific merit than more routine, although technologically valuable, industrial measurements which are essentially of economic value rather than being scientifically ground-breaking. Furthermore, as mentioned above, good industrial proposals may be bumped out of peer-review procedures and pushed to pay for access due to potentially out-dated reflections on what merits access. Further limitations arise from the duration of these processes—it often takes six to twelve months from the submission of the proposal to the beam time actually taking place. Thus the peer-review access route is suitable and beneficial for long-term collaborations between academic researchers and their industrial collaborators, but for fast access to facility infrastructure without a requirement to publish potentially sensitive intellectual property, alternative access routes have therefore been developed for industrial users.

The gradual recognition from stakeholders and the scientific community that innovation based access proposals have a legitimate place amongst peer-review access is leading to more beam time being provided for such work. Innovation impact is now often included in access forms and assessment procedures, and specific access paths for such work have even been set up. This again shows the increasingly diverse exploitation of ARIs, with user communities growing beyond traditional fundamental science applications to be more real world relevant and having a role to play in important mission-driven science and technology programmes.

As an alternative to the highly competitive peer-review access route, the UK's ISIS Neutron and Muon Source has established a unique 'Industrial Collaborative R&D (ICRD)' access route by which an industrial proposal is reviewed based on its potential economic benefit, rather than scientific merit. In ten years of running this scheme (2011–20) 46 companies have submitted 69 proposals, resulting in an average of 48 days per year of additional industrial use of ISIS (Frost and Appleby 2020).

9.3.1.2 Routine commercial services and products

Some synchrotron or neutron based techniques have proven to be highly amenable to become routine for both academic users and industrial clients alike—creating a 'menu' of readily available services. Routine services are provided with the existing equipment available at the ARIs and do not require specific developments according to a client's needs to carry them out. Nonetheless, significant instrumentation and software developments have certainly been made at the ARI, on its own or in a co-design approach with industry, to allow the services to become routine and reliable, and possibly high-throughput and automated. Although the foundations of such routine services may be initially designed led by industry driven needs, they are often then taken up by academics via the peer-review access.

ARI routine services usually take one to two months from request to supply and are based on a pay per service (often on a 'per sample' basis). However, access can also be extremely rapid, e.g. less than a week turnaround in some cases, according to capacity, flexibility (notably enabled by automated sample and data pipelines), and prioritisation. Furthermore, access through mail-in service provision eliminates the need for clients to travel to the ARI. These mail-in services are usually provided under full-service conditions, although remote access, particularly for protein crystallography at synchrotrons, is very popular. Under full-service conditions, facility staff run the experiments, with the costs usually reflecting the increased service provision—although offset against automation streamlining the procedures. In the remote mode, facility staff put samples on the beamline, with the client researchers running the beamline and data collections from the comfort of their own home laboratory. The gain for the client generally far outweighs the cost by releasing the industry scientists to focus on exploiting the data rather than performing a routine measurement—and the highly experienced facility experts (or the smart automated data collection pipelines) are likely to obtain better quality data. This requires a strong element of trust and confidence from the industry client in the ARI and its staff and is an important element of the ARI–client relationship.

Routine services such as protein crystallography, supporting drug discovery, and computed tomography, supporting a huge range of industries with remarkable data, are part of the services which usually include default automated or computer assisted workflows with the service deliverable being reduced or processed data—though structural biology workflows can go as far as delivering a final protein structure (if the commercial client allows it). It is anticipated that data reduction or first level analysis will increasingly be part of a default ARI service offer for industrial and academic researchers alike, enabled by modern artificial intelligence and machine learning processes. The growing European Open Science Cloud (EOSC) may also have an important role to play in enabling these enhanced services. If required, data analysis services may also be offered depending on expertise available at each ARI or added in from third-party collaborators.

A routine service used extensively by the pharmaceutical industrial sector is synchrotron x-ray protein crystallography (structural biology)—as mentioned above, a very valuable technique for drug discovery and design. Atomic structural insight of new drugs and their interactions with specific targets (e.g. viral proteins, DNA, etc) allows large screening of drug candidates and shortens time-to-market. Pharma companies have become, as a result, significant commercial users of light sources—even extending to taking stakes in beamlines (see section 9.3.1.4) and daring in a few cases to stop using lab-based x-ray sources and rely on light sources entirely for their supply of pipelined x-ray crystallographic data. Standardisation across Europe of sample holders and their transport has been central to democratising the use of these services, as well as a joint approach amongst facilities to developing control software and sample tracking/data tracking tools. These services have become more sophisticated, particularly in recent years, with extensions to so-called 'fragment based drug discovery' with screening services part of some light source service offers—either on their own or in partnership with other experts.

High-performance cryo-electron microscopy has also joined the drug discovery party, with light sources using their experience in operating user-access platforms to invest in cryo-EM instruments for academic and commercial access.

Neutron sources facilitate the testing of radiation hardness (and thus reliability) of semiconductor components (ISIS-STFC n.d.). A specific irradiation service (PAC-G n.d.) is being implemented through support provided by investments via the French IRT-Nanoelec involving both the Institut Laue-Langevin (ILL) neutron and the European Synchrotron Radiation Facility (ESRF) x-ray ARIs based in Grenoble and benefiting from the co-location of these two facilities with other local French facilities. Chip irradiation at the UK's ISIS Neutron and Muon Source is the main commercial service in terms of income with studies mimicking, for example, atmospheric neutron radiation effects in an accelerated way and in the desired specific areas of the components, which provides invaluable information to industries for designing novel hardening strategies against errors. Such work will be critical, for example, in helping to bring driverless cars towards reality. For decades now, routine neutron radiographies have been part of the control step of spacecraft parts (at ILL, the Swiss Spallation Neutron Source SINQ, at FRM II and Laboratoire Léon Brillouin before its closure), and more recently for fuel element components (ILL).

In addition to the core advanced characterisation techniques provided by ARIs, specialised laboratories have been deployed and ancillary technologies developed by the ARIs to cover their own specific operational needs. Some have turned out to be useful for industry as well. One such case uses the magnetic measurement laboratories or 'benches' which are available to perform accurate magnetic measurements of long magnet structures used in ARI accelerators (e.g. dipoles, quadrupoles, sextupoles, insertion devices) with a high spatial accuracy and precision. Similarly optics and metrology laboratories[14] are used to characterise optical elements with very high accuracy, providing unique systematic error suppression methods and very accurate optimisation models for mirror benders. This interest for industry, in particular SMEs, can even extend to what may be considered 'basic' laboratory equipment but which SMEs cannot afford to purchase. As stated in the table earlier, beyond the x-ray and neutron beamlines, ARIs can therefore also be seen as equipment-rich pools and which can be opened for industry—perhaps a rather under-exploited support at this time.

The reactor pool of a number of neutron facilities serves routine production needs in the semiconductor industry (homogeneous doping of Si ingots (MLZ Silicon doping)) and for medical applications. Focussing on the latter, radioactive isotopes are produced in neutron sources (in the reactor pool) via neutron capture by stable isotopes which serve for medical applications such as imaging and diagnostics (ILL n.d., MLZ Industry and medicine). These radioactive isotopes are used by biotechnology and radiopharmaceutical companies dedicated to the development, production and global supply of targeted diagnostic and therapeutic radiopharmaceuticals (e.g. ITM Isotopen

[14] https://www.albasynchrotron.es/en/technology/additional-services

Technologien München AG[15]). The neutron source efforts in the supply chain of such isotopes is clearly of strong societal value and is in fact a strong activity at several neutron sources (e.g. FRM II, ILL, research reactor in Řež, Czech Republic).

9.3.1.3 Tailor-made measurements and analysis

As both neutron and synchrotron radiation provide very deep penetration depths and are mainly non-destructive, *in situ* and *operando* measurements are often in the focus of industry requirements. Samples from industry can be measured with regular set-up configurations on dedicated instruments providing the right length- or time-scale for monitoring or with specific sample environments, specially developed for the purpose or related topics to solve the open questions from industry. The latter tailor-made access can be realised in two ways. First, by installing an experimental set-up supplied by the industrial partner on the beamline and the changes of the incoming and outcoming beam to obtain insights into the sample or apparatus. Second, by developing a brand new sample environment dedicated to the studies of an industrial client or an industrial sector (Chang *et al* 2019).

To visualise this approach an example is mentioned here. For investigations of combustion engines in gas turbines it is important to understand the changes in the microstructure of the alloy at very high temperatures and under strong forces. Such information is essential for alloy development with the goal to reach higher service temperatures and lower CO_2 emissions. ARIs developed special testing machines (Kümmel *et al* 2021) to emulate these working conditions with a specific furnace and a device to load and compress the specimens (figure 9.4), e.g. with the company VDM Metals[16] (MLZ 2021b).

Examples from the energy sector involve the charging/discharging processes of batteries under e.g. temperature variation, because measurements using neutrons are very sensitive to the light element lithium, and are able to distinguish neighbouring elements in the periodic table. Such experiments and others (Gilles 2020) allow the study of lithium intercalation process in the anode or the influence of nickel, manganese, and cobalt variation in the cathode material as part of a collaboration between Volkswagen and VARTA Microbattery, scientists from the Heinz Maier-Leibnitz Zentrum (MLZ 2017). Typical engineering topics are strain and stress of manufactured components as railway tracks, automotive components with high mechanical demands (cranked shafts), welding seams[17], etc.

Synchrotron radiation has a wide demand for *in situ* work and tailored access matching the precise needs of industry. Imaging is rapidly growing in this area, with synchrotron x-ray computed tomography and radiography bringing unique insights. One example is the ultra-high speed diagnostic imaging of the functionality of an inhaler for asthma with PLM Medical (ESRF 2017). The x-ray imaging showed the flow of drug particles with the aim to improve the design of the inhalers with better drug delivery. The company has made use of the ESRF for this work via both

[15] https://itm-radiopharma.com/home

[16] https://www.vdm-metals.com/

[17] Work carried out for OHB Systems, https://www.ill.eu/neutrons-for-society/industry/transport

Figure 9.4. Testing machine (FRM II) for alloy development to improve gas turbines. (Source: R Gilles.)

commercial and peer-review access routes. Also structure based drug design, drug delivery for therapy, and vaccination or interaction with pharmaceutical products, e.g. BioNTech and Pfizer research at DESY regarding next generation RNA drugs against COVID-19 (DESY 2020), cosmetics, and topics in life science which need very high time resolution are in the focus.

9.3.1.4 Long-term collaborations

Medium- or long-term collaborations (from 1 to 3 or 5 years) between companies and ARIs take several forms and can be found in almost every ARI. These collaborations are characterised by a strong involvement of the parties involved, who are counting on years in order to gain competence/expertise or to adapt and develop techniques and methodologies of analysis to fit their purposes.

It is relatively common to have staff from industry or (co)funded by industry located on site at the ARIs: PhD students, post-docs, engineers, senior researchers, or a combination of these. For example, Johnson Matthey's (JM) involvement in the Harwell Campus (UK), having a large JM R&D centre located just a few miles away, is significant in this context. From being regular users of synchrotron methods for many years, JM took the strategic decision to install one of their high-end electron microscopes at Diamond rather than on their home site. The JM team based within Diamond is working closely with the beamline teams to exploit the complementarity between the electron and x-ray methods.

Building on this, Johnson Matthey also part-employs industrial research fellows at ISIS Neutron and Muon Source, and the Central Laser Facility, both also co-located on Harwell Campus, indicating how serious they are about benefiting from the use of the ARIs.

A doctoral programme supported by ESRF and ILL (InnovaXN[18]) and co-financed by the European Commission's MSCA COFUND started in 2020. This programme consists of hosting 40 PhD students (in two cohorts over five years) with research topics using synchrotron x-rays and/or neutrons and driven by industry partner interests with researchers from the industry partners intimately involved and co-supervising the students. A key aim of the programme is for the students to experience industry driven pre-competitive work exploiting the ARIs and seeing both sides of academic and commercial work. In addition, the MSCA GNeuS project started during 2021 with financing for 45 post-doc grants and which includes nine industry partners for post-doc secondments.

Researchers fully funded by Henkel were hired by ALBA Synchrotron (ALBA n.d.a) and based at its premises within the framework of a long-term collaboration for carrying out research on new materials and for developing innovative products, in particular for a new generation of adhesives, using synchrotron techniques.

Long-term collaboration can also take shape through the ARI's instruments and beamlines that may be owned, co-invested, co-funded and/or co-operated by industry (see table 9.2 for selected representative examples).

9.3.1.5 A focus on SMEs

SMEs are well known as the drivers of breakthrough innovation. As such, ARIs are available to support and work with such companies, but SMEs bring their own particular challenges. As small companies, they can be highly subject to cash flow limitations, very focussed on work which only enhances their bottom-line, and unable to work on more blue-sky research like the large multinationals. Within Europe alone there are vastly more tech-rich SMEs which could benefit from ARI support compared to the household-name companies with dedicated R&D divisions. The numbers alone make it hard to reach out to SMEs and make them aware of how the ARIs can help their business and innovation needs. Due to this, the ARIs may most often interact with SMEs in their geographically close ecosystems.

Fortunately, Europe allows SMEs to benefit from support. This has opened the way for European R&D Framework Programme funding to subsidise access by SMEs to ARIs, conturning budget issues, supporting a modest outreach, and allowing the SMEs to keep results confidential whilst benefiting from trans-national access. The 'TamaTA' programme[19] of CALIPSOplus, a Horizon2020 project, has piloted such an SME access system, enabling over 28 SMEs from across Europe and beyond to benefit from using synchrotron x-ray beamlines from different European facilities, characterising materials or testing instrumentation to boost innovation. The strong collaboration among the European synchrotron and neutron facilities to provide SME support is quite unique in the world and it is also reflected in the

[18] https://www.innovaxn.eu/

[19] http://www.calipsoplus.eu/taa-tamata

Table 9.2. Examples of co-owned or co-exploited beamlines and instruments with industry.

RI (country) type of ARI	Industry partner(s)	Type of collaboration	Collaboration summary
NIST (USA) Neutron	As of 2019, in alphabetical order: Amgen, Aramco, Corning, Dow, Genetech, L'Oréal, P&G, Pfizer, Regeneron, Solvay, Toyota	Co-funded, co-exploited	nSOFT (soft materials such as biodegradable plastics, membranes for energy storage, composites for transportation, protein based biopharmaceuticals, additive systems for water management in soils, and a range of soaps, lotions and transdermal products)
Swiss Light Source at the Paul Scherrer Institute (Switzerland) Synchrotron	Actelion, Boehringer, Hoffmann La Roche, Ingelheim, Novartis, Proteros	Joint funding providing guaranteed access	The industrial partners co-funded two protein crystallography beamlines at SLS providing a routine access and capacity to serve the commercial needs
ESRF and ILL (France) Synchrotron and neutron	Programme d'Investissements d'Avenir financed investments on BM05 (ESRF) and D50 (ILL) for the microelectronics industry with local partners Schneider Electric, SOITEC, and STMicroElectronics	Investment providing enhanced services and access provision	Part of France's overall investment into technology, this work at ESRF and ILL is part of the Technology Research Institute 'NanoElec' programme in Grenoble and specifically of the Characterisation Programme and 'Platform for Advanced Characterisation—Grenoble' (PAC-G) which serves the microelectronics industry
SPring8 (Japan) Synchrotron	Quite common at SPring8. An example: Toyota	Owned by Toyota Central R&D Labs	Toyota beamline (BL33XU) Dynamics of chemical reactions

(Continued)

[20] https://irtnanoelec.fr/characterisation-programme-irt-nanoelec/

Table 9.2. (*Continued*)

RI (country) type of ARI	Industry partner(s)	Type of collaboration	Collaboration summary
APS (USA) Synchrotron	A number of 'Collaborative Access Teams' with direct industry support, for example: IMCA-CAT ('Industrial Macromolecular Crystallography Association') with Abbvie, Bristol Myers Squibb, Merck, Novartis, Pfizer, Evotec and Relay Therapeutics, DND-CAT ('DuPont—Northwestern—Dow' for materials science) with The Dow Chemical Company and DuPont de Nemours, Inc.	(Co)-funding Exploitation	IMCA, established in 1990, is the association of pharmaceutical companies that supports the operation and development of the IMCA-CAT facility at the Advanced Photon Source. IMCA contracts with Hauptman-Woodward Medical Research Institute to manage the IMCA-CAT facility. The Northwestern Synchrotron Research Center manages DND-CAT on behalf of its partner institutions
MAX-IV (Sweden) Synchrotron	ForMAX beamline is funded by the Knut and Alice Wallenberg Foundation, while the operation costs will be covered by the forest industry via Treesearch	Co-funded, co-exploited	ForMAX—biocomposites, nanocellulose, modification of wood, the pulping process, fibre ultrastructure and fibre-fibre bonding
ALBA (Spain) Synchrotron	In partnership with ESA and to be operated by the Dutch company cosine	Co-funded, co-exploited	ESA-ALBA Fixed-Energy beamline (MINERVA) for characterising and aligning x-ray optics (mirror modules) of the ATHENA telescope to be launched by the European Space Agency (ESA)

creation of the LEAPS[21] and LENS[22] organisations bringing together light sources and neutron sources in Europe. It would be a worthy ambition to pursue the pilot and extend to a wider range of ARIs, or even RIs beyond the analytical ones,

[21] https://leaps-initiative.eu/
[22] https://www.lens-initiative.org/

with a comprehensive marketing campaign to the SME innovation power houses of Europe.

9.3.2 Industry as technology partner

9.3.2.1 Co-development

RIs are important drivers of new advanced technology and instrumentation. In contrast to the previously described in-house engineering (see section 9.3.1.3), many ARIs also collaborate directly with technology providers, such that the ARI describes its requirements for a new technology, which is then developed by an industrial supplier, often with continued support from the ARI. The supplier is thereafter often able to sell its newly developed technology to other consumers—usually with a return to the initiating ARI.

There are typically four models by which ARIs and industry co-develop technology (Appleby 2018)—these are largely differentiated by the role of each partner and the ownership of the IP:

- The ARI provides the IP, the company provides the engineering—by which an ARI licences its IP to a company which develops the commercialisation. Example: PSI jointly developed a component of their on-site proton therapy centre with an industrial supplier. The supplier now owns the licence to this new technology, and can sell it commercially to other proton therapy centres, whilst PSI receives royalties from these sales.
- The ARI proposes an evolution of a commercial product implemented at the facility—by which the company can evolve its product as an option or as a standard, and rewards the ARI with royalties. Example: European XFEL staff modified the design of an instrument that was being built for them by an industrial supplier. The supplier then upgraded this instrument free of charge based on the designs of European XFEL, and provided the improved instrument to European XFEL.
- The ARI proposes an idea, the company develops the technique and the technology to realise a new product/service. Example—a contract from European XFEL requiring a particularly advanced vacuum environment allowed a supplier to significantly expand its own expertise and capabilities, to produce advanced vacuum systems that can be of value to other research facilities.
- The ARI and a company create a so-called 'Joint Lab' with shared resources in terms of work forces, financial support, and equipment—ensuring development of a successful long-term strategy and producing state-of-the-art outcomes that will fit the market—e.g. the IMPULSE Joint Lab between Amplitude Technology and LaserlabEurope LIDYL laboratory of CEA (France)

The role of ARIs in development of technology has been recognised in European funding with projects such as the European-funded ATTRACT project which brings ARIs together with industry and investors in the domain of imaging and detection technologies, and LEAPS Innovation project which brings the light sources together with industry in different configurations to develop new instrumentation.

9.3.2.2 Knowledge and technology transfer

As drivers of new advanced technology and instrumentation, ARIs have highly experienced in-house engineering groups which design much of the new equipment required for the sources and the beamlines: from advanced source instrumentation such as undulators for light sources, photon and neutron optics, high-performance control electronics and control software, as well as specialised detectors or beam monitors.

The central driver for these in-house technology projects is to enable the ultimate in performance via often unique one-off tailored instrumentation. Nonetheless, in some circumstances the instrumentation can serve other research infrastructures and may spillover into other research and industry domains where such high-performance instrumentation is required. When such opportunities are identified, industry can enter the picture to exploit the developments and commercialise the technology or the knowledge that has been generated by the facility. The market for these products is often niche—largely other advanced analytical research infrastructures, but industry may also be able to innovate beyond the original design objectives and spot other openings, exploiting the technology in a wider domain.

The transfer of knowledge and technology is handled usually by the light source itself with the internal industry/business or technology transfer offices. Although if a light source is itself part of a larger structure, the structure may handle the transfer procedures. Licences for the technologies are made with a royalty payment to the facility concerned which is generally recycled into the operations budget. In some cases, a spin-off company may be formed based upon the technology in hand.

It is interesting to note that the Swiss Light Source (SLS) has co-funded a company—SLS TechnoTrans AG[23]—dedicated to the transfer of technology, but also development of specific beamlines to serve the requirements of industry and supervision of industry research activities—effectively outsourcing the 'ICO' role.

A few examples of light and neutron source technologies and their transfer and exploitation by industry are provided below.

1. **Spin-off companies**
    - DECTRIS[24], a spin-off created in 2006 from the Paul Scherrer Institut (PSI) in Switzerland, has developed to become a major supplier of state-of-the-art pixel detectors to light sources around the world. It exploits technology developed at PSI and tested and matured on the Swiss Light Source.
    - KIUTRA[25], a spin-off from the Physics Department of TUM and FRM II established in 2018. The focus is on low temperature equipment.
    - KYMA is a spin-off company formed in 2007 and owned by Elettra Sincrotron Trieste, as existing suppliers could not meet its requirements for undulator magnetic components. Kyma has expanded into other

[23] https://synchrotron-analysis.ch/about-us
[24] www.dectris.com
[25] www.kiutra.com

high-performance components and now supplies its products to many other light sources in Europe, Asia and the USA.

- Quantum Detectors (www.quantumdetectors.com) is a spin-out created by the UK's Science and Technology Facilities Council (STFC) and Diamond Light Source—and is based on the Harwell Campus which hosts the UK's light and neutron sources. It exploits new detector technologies and readout systems developed at STFC and Diamond, as well as high-performance voltage to frequency converters developed at the European Synchrotron Radiation Facility (ESRF). The company was founded in 2007, and has grown to about 30 employees, with an ever increasing range of products installed at ARIs worldwide.
- SwissNeutronics[26] company for neutron optics, neutron guides, and focussing devices (PSI spin off but alumni colleagues of FRM II)
- TM Isotopen Technologien München AG[27], founded in 2004 and located around FRM II, is a privately owned biotechnology and radiopharmaceutical group of companies dedicated to the development, production, and global supply of targeted diagnostic and therapeutic radiopharmaceuticals and radioisotopes for use in cancer treatment. The company frequently performed experiments at the FRM II (e.g. Lu177) and has around 300 employees and is active worldwide.
- XENOCS[28] was spun out of the ILL in 2020 to develop special x-ray optics based on knowledge developed originally for neutron optics. XENOCS then started to develop sub-systems and full-systems for material characterisation at the nano-scale. The company has now 50 employees from 12 different nationalities and operates worldwide.

2. **Knowledge transfer**
   - SEF Technologies[29] is an SME based in Toulouse (France) and is a supplier of electromagnets used in accelerators such as light sources. The ESRF is working with SEF Technologies on a knowledge and material transfer for high-performance magnetic measurements from its Insertion Device team, enabling SEF Technologies to expand its business by providing added value services.
   - A team of researchers at the Paul Scherrer Institute (Switzerland) has developed and patented a radioisotope embedded in a molecule that could potentially fight a particular form of cancer (Goldhahn 2018). In 2018, after two years of technical and commercial discussions, the PSI molecule belongs to the Lausanne-based pharmaceutical company Debiopharm[30], for which PSI receives royalties.

[26] www.swissneutronics.ch/
[27] www.itm-radiopharma.com
[28] https://www.xenocs.com/
[29] www.sef-technologies.com
[30] https://www.debiopharm.com/

3. **Technology transfer**
    - Alibava Systems[31], a spin-off company of the Spanish National Research Council (CSIC) located in Barcelona (Spain), is supplying full mechanical and electronic system integration and radiation detectors. ALBA Synchrotron (ALBA n.d.b) in collaboration with Institute of Microelectronics of Barcelona (IMB-CNM CSIC) developed a device based on a thin transmissive photodiode to detect and characterise x-rays while absorbing only a small fraction of the incoming intensity. ALBA signed a licence agreement with CSIC and Alibava to allow the commercialisation of the device.
    - The open electronics standard MicroTCA.4, which the German synchrotron DESY has been instrumental in developing, was adapted for controlling the European XFEL, and is now being used in other research institutions and by numerous industrial companies around the world. The MicroTCA Technology Lab was recently set up at DESY in order to give the industry access to the development of further applications, which has been hailed as an outstanding example of technology transfer from fundamental research to broader practical applications.
4. **Open source software**
    - Tango Controls[32] is an open source software controls solution for highly heterogeneous environments. The concept of Tango Controls was developed at the European Synchrotron Radiation Facility about 20 years ago. Since then the work on the core of the toolkit has begun. Although Tango Controls has proven itself as a mature toolkit with many users, as an open source the Tango Controls toolkit is always being improved. The toolkit is now used widely in research infrastructures and beyond, and has spawned its own commercial service provider ecosystem for support and development, including companies such as Atos Worldgrid and Byte Physics.

9.3.3 Enterprises and research and technology organisations as intermediaries

In a world increasingly driven by innovation and complex materials, a wider range of companies, large and small, could benefit from insights provided through ARIs. 'Intermediary' or 'facilitator' companies and well known research and technology organisations are actors which can help to draw in industry to use the power of ARIs for commercial and technological research.

Typically the ARI scientists, be they beamline staff or industry dedicated staff, provide service to the industrial clients. However, in case of resource limitations or for those ARIs without a dedicated team of scientists supporting industrial users, working with third parties is a way to provide a comprehensive offering to industrial

[31] https://alibavasystems.com/

[32] www.tango-controls.org

researchers, drawing on their resources to provide some or all of sample preparation, data analysis, and reporting as well as support for data collection. Data collection is usually straightforward for the ARI to provide as data production is their core mission; it is the value-added services around the data itself which have been traditionally more difficult. This largely stems from the fact that the user communities of ARIs have been expert, with training since PhD level in using the facilities and exploiting the scientific data. This means that resources for data analysis are not part of the core mission of ARIs. However, as an aside, this model is perhaps gradually changing as ARIs become more integrated into 'daily' scientific discovery and with some techniques becoming routine in use. This in turn means the user communities expand with fewer experienced researchers wishing to integrate ARI data into their research, but with their level of experience (and perhaps motivation to learn) required to use the facility lowered. Automation in sample changing and expert systems for data reduction and analysis are driving this trend further. This can be expected to further open the ARIs, democratising their use—although intermediaries supporting sophisticated commercial use will have a long lasting impact.

9.3.3.1 Specialised intermediary service companies

Entrepreneurs and academic scientists have spotted the gap between ARIs and possible commercial clients. This has resulted over the last decade in a number of 'micro' specialised companies being set up with the specific goal to hold the hand of industry in exploiting sophisticated analytical ARI techniques. Examples include Novitom[33], Excelsus[34], Finden[35], 3Dmagination[36], Expose[37], and Xploraytion[38]. These companies usually specialise in one technique and bring very high quality science experience which can add considerable value for industry bringing complex research challenges—such as Finden with strong knowledge in the study of catalysts using synchrotron radiation.

Intermediaries, and indeed larger contract research organisations (CRO), prevalent in the health sector, which offer structural biology services to pharmaceutical clients have a close connection to synchrotrons, accessing the beamlines on a regular basis to collect crystallography data. Examples are numerous and include Evotec[39] in the UK, Saromics[40] in Sweden, and NovAliX[41] in France. Whilst these companies undoubtedly offer similar services to those offered by facilities themselves, they also offer a much broader spectrum of integrated research services to

[33] www.novitom.com/en/
[34] www.excelsusss.com
[35] www.finden.co.uk/
[36] www.3dmagination.uk
[37] www.expose-crystals.com
[38] www.xploraytion.com/
[39] www.evotec.com/en/execute/drug-discovery-services/hit-identification
[40] www.saromics.com/technologies/crystallization/
[41] novalix.com/biology/protein-x-ray-crystallography/

industry (e.g. expertise in biochemistry, modelling/prediction, pre-clinical and clinical stage research).

9.3.3.2 Research and technology organisations (RTOs) working with the ARIs
Intermediaries and multipliers are not limited to the micro-companies or CROs which help to bridge ARIs and industry, but also include much larger-scale organisations such as the RTOs (e.g. CEA in France, the Fraunhofer Institutes in Germany, DTI in Denmark, and RISE in Sweden) and universities or clusters supported by public funds such as LINX[42] in Denmark. Such institutes are essentially driven by applied research with strong collaborations with industrial partners. However, some RTOs are aware of the power of the analytical ARIs to provide research-critical data and exploit ARIs in their research programmes, particularly if the RTO is physically located close to an ARI. These valuable relationships can bring mature technology networks to the ARIs, growing use of ARIs for industry and technology development.

As the academic and industrial R&I worlds draw ever closer together, this potentially very rich area of collaboration can increasingly draw upon ARI facilities for advanced characterisation and experiments which are required within industry driven partnerships with research institutes and universities. This culture is growing with the specific industry–academic–RI triple helix becoming more important in using ARIs for applied and commercial R&D.

9.4 A future perspective

As we have outlined above, the ARIs are open to work with industry as a user, partner and collaborator. However, it is nonetheless time to enable more and to move on from the period in the late twentieth and early twenty-first centuries when most of the user-based light and neutron facilities were established in the European landscape, and look ahead to the ARI research narrative of 2025 and beyond.

It has become a cliché to say that humanity is facing colossal challenges. Scientific and technological research, whether fundamental, applied, or industrial, grows humanity's knowledge and provides the means to advance the quality of life and to protect our planet. ARIs have a mission to provide the underlying data to provide answers to these technological and societal challenges of our time. Moving beyond the image of being ivory towers of elite science and devoting significant time, access and resources for industry can only help in drawing ARIs into the full value chain from fundamental understanding to applied research and to support innovation. In this context, governments, stakeholders and funding agencies can and must continue to support and promote collaboration with and use of ARIs by industry.

For this the ARIs themselves need to continue to open fully to technology-based and industry driven research, by bringing systems up to speed for such impact based access with evaluation criteria being inclusive rather than scientifically exclusive. Outreach and education needs to be across the board, working with young students

[42] https://www.nbi.ku.dk/english/industrial-collaboration-at-nbi/cases/linx/

from university stage (thereby engaging with the future generations of industry researchers and engineers) to life-long learning in industry on ARI application and exploitation as key innovation enablers.

Trans-ARI thematic platforms are starting to show their impact (e.g. the NFFA H2020 project), and the new open innovation test benches, which include ARIs as key elements of materials and device characterisation, are a further nascent example. These scintillating constellations of ARIs, grouped to provide a critical mass of support and access to industry-based research are likely to become more common in mission-led activities (e.g. metallurgy, energy storage, or drug development). But to work, these platforms need to be inclusive, with industry-friendly access and support procedures—for which the current European schemes are not yet fully compatible.

Across all of these areas the role and proactive engagement of new innovation driven European entities such as the European Innovation Council is absolutely critical where it can act as assembler, catalyser, and funder, drawing in the ARIs as a first-thought in supporting technology development and not as an after-thought. Associations, clusters, and conveners bringing together industry must also be implicated to efficiently draw in industry, large and small. It is important though to realise that a self-sustaining model of working with industry is an unlikely objective; supporting the system to enable industry-ARI engagement and innovation is and will always be required, and will always demonstrate its longer-term return on such investments in innovation, science and skills development for a technology-based economy.

The ARIs are open for industry, and are prepared to do business, supporting commercial R&D and enabling innovation, bringing polychrome facets to the ARIs with new challenges and exciting science. The over-the-rainbow Emerald City of the introduction thus has its foundations in place and is ready to support considerably enhanced work with industry. The objective over the coming few years is to assure that the walls and roof are constructed and a polished and quality engagement between ARIs and industry is enabled and supported, generating more and better economic impact with industry.

Acknowledgements

The authors gratefully acknowledge the ICOs and collaborators at neutron and light source facilities, as well as our industrial colleagues, for fruitful exchanges of views over the past years, both formally and less formally, and the insights that this has brought to this essay. Special thanks to Steph Monfont (ILL) for figures 9.1–9.3.

References

Ackland R G, Garrod R I and Richardson J F 1950 Summarized proceedings of conference on 'X-Rays in Industry'—Melbourne, Australia, November, 1949 *Br. J. Appl. Phys.* **1** 217–24

ALBA 2019 BASF, UPC and ALBA propose a methodology for producing better additives for concrete technology https://www.cells.es/en/media/news/basf-upc-and-alba-propose-a-new-methodology-for-producing-better-additives-for-concrete-technology

ALBA 2020 CSIL report release: from experiment to innovation impact of ALBA https://www.cells.es/en/media/news/alba-csil-report-release-from-experiment-to-innovation-impact-of-alba

ALBA n.d.a ALBA and Henkel sign a R&D collaboration agreement https://www.cells.es/en/media/news/alba-synchrotron-and-multinational-company-henkel-sign-a-r-d-collaboration-agreement

ALBA n.d.b ALBA, CSIC and ALIBAVA develop an x-ray detector https://www.cells.es/en/media/news/alba-csic-and-alibava-develop-an-x-ray-detector

Appleby G 2018 Synergy and innovation potential of EUCALL. Zenodo *Deliverable* 3.2 https://zenodo.org/record/1219079

Baum F 1900 *The Wonderful Wizard of Oz* (Chicago: G M Hill) https://archive.org/details/wonderfulwizardo00baumiala/mode/2up

Borralleras P, Segura I, Aranda M A G and Aguado A 2019 Influence of experimental procedure on D-spacing measurement by XRD of montmorillonite clay pastes containing PCE-based superplasticizer *Cem. Concr. Res.* **116** 266–72

Catalano G, García López G, Sánchez A and Vignetti S 2021 From scientific experiments to innovation: impact pathways of a synchrotron light facility *Ann. Public Coop. Econ.* **92** 447–72

Chang S-Y *et al* 2019 A versatile liquid-jet/sessile droplet system for operando studies of reactions in liquid dispersions and solutions by x-ray absorption spectroscopy *React. Chem. Eng.* **4** 679–87

Collins D M, D'Souza N, Panwisawas C, Papadaki C, West G D, Kostka A and Kontis P 2020 Spinodal decomposition versus classical Γ' nucleation in a nickel-base superalloy powder: an *in-situ* neutron diffraction and atomic-scale analysis *Acta Mater.* **200** 959–70

Davies N *et al* 2021 Functionalized lipid nanoparticles for subcutaneous administration of MRNA to achieve systemic exposures of a therapeutic protein *Mol. Ther. Nucleic Acids* **24** 369–84

de Laune R 2020 ISIS beamlines help Unilever develop biodegradable products *ISIS-STFC* https://www.isis.stfc.ac.uk/Pages/SH20_ISIS_beamlines_help_Unilever_develop_biodegradable_products.aspx

DESY 2020 DESY's x-ray source PETRA III points possible ways to better RNA vaccines https://desy.de/news/news_search/index_eng.html?openDirectAnchor=1977&two_columns=0

ENRIITC n.d. The ENRIITC https://enriitc.eu/ (Accessed: 28 June 2021)

ESRF 2017 Improving asthma and other respiratory diseases https://esrf.fr/home/Industry/industry-news/content-news/esrf-news-list/improving-asthma-and-other-respiratory-diseases.html

Frost C and Appleby G 2020 *ISIS Industrial Collaborative R&D (ICRD) Scheme—Review 2020* (Chilton: ISIS Neutron and Muon Source, STFC) https://isis.stfc.ac.uk/OtherFiles/20210610%20ICRD%20Review.pdf

Gilles R 2020 How neutrons facilitate research into gas turbines and batteries from development to engineering applications *J. Surf. Invest.* **14** S69–74

Goldhahn S 2018 *Licence Agreement with Swiss Pharma Firm for Development of a Cancer Drug* (Würenlingen: Paul Scherrer Institute) https://psi.ch/en/media/our-research/licence-agreement-with-swiss-pharma-firm-for-development-of-a-cancer-drug

ILL n.d. Isotopes production for medicine https://www.ill.eu/neutrons-for-society/neutron-technology/isotopes-production-for-medicine (Accessed: 28 June 2021)

ISIS-STFC n.d. ChipIr, Instrument for the irradiation of microelectronics https://isis.stfc.ac.uk/Pages/Chipir.aspx (Accessed: 12 July 2021)

Kümmel F, Kirchmayer A, Solís C, Hofmann M, Neumeier S and Gilles R 2021 Deformation mechanisms in Ni-based superalloys at room and elevated temperatures studied by *in situ* neutron diffraction and electron microscopy *Metals* **11** 719

Malbet-Monaco S, Leonard G A, Mitchell E P and Gordon E J 2013 How the ESRF helps industry and how they help the ESRF *Acta Crystallogr.* D **69** 1289–96

Mitchell E P, Boller E, Gordon E J, Leonard G A and Monaco S 2011 The industrial research program at the European synchrotron radiation facility: the past, the present, and challenges for the future *Synchrotron Radiat. News* **24** 18–23

MLZ 2017 The lost lithium: neutrons explain the aging in battery cells used in stationary storage systems https://www.frm2.tum.de/en/frm2/news-media/article/the-lost-lithium-neutrons-explain-the-aging-in-battery-cells-used-in-stationary-storage-systems/

MLZ 2021a Research with neutrons for better MRNA medicines https://www.frm2.tum.de/en/frm2/news-single-view-en/article/research-with-neutrons-for-better-mrna-medicines/

MLZ 2021b Pressure and heat: Testing alloys for gas turbine https://mlz-garching.de/englisch/news-und-press/news-articles/pressure-and-heat-testing-alloys-for-gas-turbines.html

MLZ Industry and medicine https://www.frm2.tum.de/en/frm2/industry-medicine/ (Accessed: 28 June 2021)

MLZ Silicon doping https://www.frm2.tum.de/en/frm2/industry-medicine/silicon-doping/ (Accessed: 28 June 2021)

OSTI 1982 The first reactor [40th anniversary commemorative edition] *Report DOE/NE* 0046

PAC-G n.d. PAC-G platform for advanced characterisation Grenoble https://irtnanoelec.fr/pac-g/

Petry W 2015 Neutrons for industry *EPJ Web Conf.* **104** 01001

Sebastiani F *et al* 2021 Apolipoprotein E binding drives structural and compositional rearrangement of MRNA-containing lipid nanoparticles *ACS Nano* **15** 6709–22

Thiry M and Boudou C 2020 Business model for scope, access and IP for 'Neutrons and Industry' SINE2020 *Deliverable* D1.6. SINE2020 https://cordis.europa.eu/project/id/654000/results (Accessed: 28 June 2021)

Tucker I M, Burley A, Petkova R E, Hosking S L, Thomas R K, Penfold J, Li P X, Ma K, Webster J R P and Welbourn R 2020 Surfactant/biosurfactant mixing: adsorption of saponin/nonionic surfactant mixtures at the air–water interface *J. Colloid Interface Sci.* **574** 385–92

Yanez Arteta M *et al* 2018 Successful reprogramming of cellular protein production through MRNA delivered by functionalized lipid nanoparticles *Proc. Natl. Acad. Sci.* **115** E3351

Chapter 10

The socio-economic impact of DORIS

Stephan Haid and Frank Lehner

When the DORIS storage ring—located at DESY in Hamburg—started in 1974, it was used mainly as a particle physics collider with substantial scientific contributions towards the establishment of the Standard Model. Later it became an indispensable x-ray source that pioneered many applications in synchrotron radiation science to the benefit of numerous diverse scientific fields. The case of DORIS very impressively demonstrated how a major research infrastructure could be successfully adapted to meet the developing requirements in science, thus providing research opportunities and long-term sustainability to scientific communities over many decades until its shut-down in 2013. After describing the science-historical context of DORIS this chapter addresses the different dimensions of socio-economic impacts. Apart from scientific impacts that exemplify the importance of new scientific instrumentation and the creation of new scientific communities, the impact on people and the outstanding value of education at large-scale research facilities is discussed on the basis of a survey among former DORIS graduate students. The economic impact is assessed by an input–output study looking at the economic 'chain reaction' of the spending for DORIS. Finally, the regional impact of DORIS, its role on the DESY campus and its importance for the networking and knowledge transfer with other research institutions in the Hamburg metropolitan region is illustrated.

10.1 Introduction

The 'Doppel-Ring-Speicheranlage' DORIS at DESY in Hamburg was built [1] as a multi-GeV storage ring more than 50 years ago as the first of its kind. During its entire lifetime between 1974 and 2013 it went through three distinct phases [2] and has generated a tremendous variety of scientific findings and applications in the fields of particle physics, accelerator science and technology, photon science, and in the life sciences. Generations of researchers have used DORIS to generate new knowledge and to pioneer, develop, and test innovative experimental methods,

doi:10.1088/978-0-7503-3631-4ch10

instrumentation, and related technologies that have subsequently been applied at other research laboratories all over the world and in industry. Hundreds of doctoral students and postdoctoral researchers earned their first merits at DORIS. Countless scientific cooperations were forged between national, international, and often interdisciplinary teams to carry out joint experiments in Hamburg and many of these connections still exist today and have since then been firmly strengthened on institutional grounds, building the core of the multidisciplinary science and innovation campus around DESY in Hamburg-Bahrenfeld.

The nearly 40 years of operation of DORIS represent a unique opportunity for a lifetime analysis of a multidisciplinary facility impacting various scientific and technological fields, promoting the building of new scientific communities and creating strong networks that still exist today. Only very few major research facilities in the world can look back at such a long, versatile, and successful history in the world of science. In this report we will attempt to capture the most relevant impacts of DORIS and will illustrate also the structural developments and institutional clustering effects on the campus.

10.2 Science-historic background

DORIS was originally designed for particle physics—at a time when quarks were known to be only speculative point-like objects. The novel and ambitious idea at the end of the 1960s was to construct a new type of accelerator known as a storage ring with colliding electron and positron beams (figure 10.1). The plan was controversial

Figure 10.1. Photograph of the construction of DORIS starting in 1969 on the DESY campus in Hamburg. Copyright DESI.

since several experts did not believe in the scientific potential of electron–positron collisions at multi-GeV energy [3]. The start-up of this *first phase* of DORIS as a collider accidentally fell in 1974, the year in which a revolutionary change in particle physics occurred due to the spectacular discovery of the new charm quark in the US. Immediately after the discovery, DORIS made important contributions by independently confirming the existence and by detecting and analyzing new excited bound charm quark states [4]. Later, in 1977, additional acceleration sections were installed to be able to reach higher energies to study the next heavier quark, the b-quark. The experiments at DORIS produced the most important findings to establish the Standard Model of particle physics, which was still new at that time.

The use of synchrotron radiation (SR), which was originally seen as a by-product of the accelerator's operation, opened up a plethora of new scientific opportunities at DORIS with truly impressive applications in physics, materials/nano sciences, geoscience, and chemistry, and with profound impacts in biology/life sciences. From the very beginning, researchers at DORIS were able to make use of the intensely collimated x-ray radiation generated by the accelerator and interest in SR applications grew quickly [5]. Most remarkable in this context was also the strong growth of SR applications in the life sciences at DORIS. The early success of pioneering experiments at the original DESY synchrotron triggered the decision to incorporate an SR beamline for hard x-rays with focus on biological applications (figure 10.2).

In a *second phase* of DORIS (DORIS II) a new lab, the so-called Hamburg Synchrotron Radiation Laboratory (HASYLAB) was opened in 1981 with the goal to meet the increasing demand and to better organize SR-related research at DORIS. Soon, it became an internationally renowned facility and one of the incubators of modern and successful research with SR worldwide [6].

In this period, particle physicists and HASYLAB researchers shared the same storage ring in the 1980s. The ring was improved with special focal elements to enhance the collision rates at the interaction points of particle physics detectors [7]. During two thirds of the time DORIS operated in a mode optimized for particle physics. The x-ray beams could then only be used 'parasitically' during these intervals, so applications were restricted. One third of the time was reserved for HASYLAB experiments, with the beam parameters adjusted to ensure it would generate as much radiation as possible. In this mode, DORIS II was the most

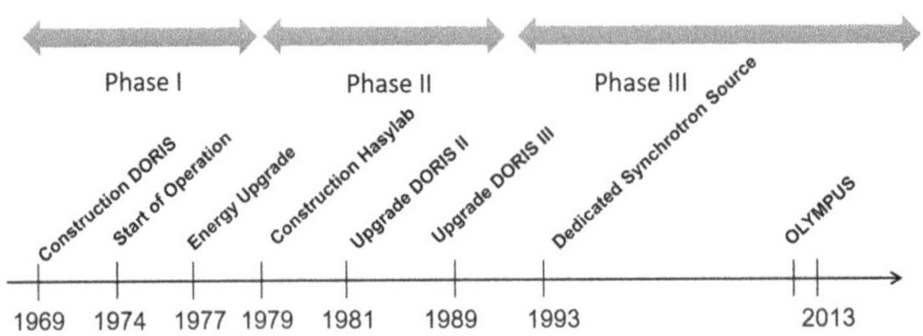

Figure 10.2. The scientific life of DORIS in three distinct phases.

brilliant source of x-rays in Europe—until 1994, when the European Synchrotron Radiation Facility (ESRF) began operating in Grenoble, France.

By the end of 1980s a further upgrade to the facility was proposed in order to accommodate the increasing demand for SR users and to stay competitive with other international facilities. In this *third phase* of DORIS one of the two straight sections was replaced with a curved section that would be fitted with integrated special magnets known as wigglers and undulators [8]. This enabled the number of measuring stations to be increased and also improved the quality and intensity of the x-ray beams considerably. In 1993, almost 20 years after the first beams in DORIS, particle physics stopped and the ring became a dedicated SR facility which attracted more than 2000 researchers per year from all over the world for their experiments: physicists, chemists, and materials scientists, as well as biologists, geoscientists, and engineers from industry (figure 10.3).

In 2012 a refurbished detector from MIT was installed at one of the former interaction zones of DORIS to study a long-standing puzzle in electron–proton

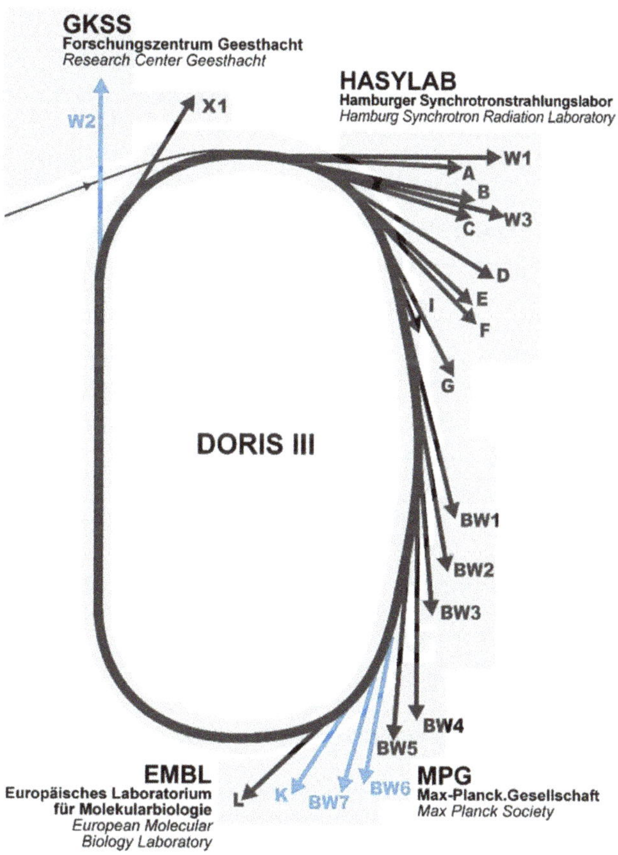

Figure 10.3. Layout of DORIS III as dedicated SR facility with the characteristic curved section and the co-location of research institutions EMBL, MPG, and GKSS. Copyright DESI.

interactions. The experiment (OLYMPUS) took place until the storage ring was finally shut down in January 2013.

10.3 Scientific and technological impacts

Research infrastructures such as DORIS were built and operated with a primary scientific mission and with the goal to contribute to the national and international stock of knowledge and to its dissemination and diffusion into other areas (Fact box 1). It is well beyond the scope of this paper to record the complete set of scientific impacts that materialized over the 40 years of the lifetime of DORIS. Instead, we have chosen a pragmatic approach to focus on a few selected examples giving sufficient evidence for the various types and pathways how research at DORIS led to scientific impacts.

10.3.1 Particle physics

The golden times for particle physics at DORIS and at other accelerators in the world came basically in two eras. A remarkable short and exciting period in the years between 1974 and 1978 leading to a plethora of new results and to a true paradigm shift in particle physics—two new, heavier quark particles, 'charm' and 'bottom', and a new third generation lepton particle, the 'tau-lepton' a heavy sister of the electron and muon, were suddenly discovered in the US and needed to be explained. Subsequent key measurements at DORIS complemented these results and were instrumental in firmly validating the quark model as one of the major building blocks of the famous Standard Model in particle physics. Although DORIS was not the main 'discovery engine' it was one of the two or three major places in the world that contributed with many key results significantly to the precise scientific understanding of the underlying fundamental physics and thus to the establishment of what is known today as the Standard Model of particle physics [9].

In a second golden era, in the 1980s, the improved DORIS collider with the ARGUS experiment explored the world of bound states composed of one bottom quark with another light quark, the so-called B mesons. This led 1987 to the discovery of mixing of B mesons at DORIS, a fundamental process in physics, which paved the way [10] to a completely new scientific field in particle physics and to new

Fact box 1:

Scientific harvest of DORIS (in a nutshell) between 1974 and 2013

- More than 12 000 publications in peer reviewed journals.
- More than 20 000 researchers used DORIS for their research either in collaboration or as SR users.
- More than 1200 PhD students finished their thesis.
- The user community at DORIS came from more than 300 institutes from >40 countries.

generations of dedicated facilities in US and Japan to study CP-symmetry breaking phenomena which is still today at CERN's LHC and at the KEK Belle II detector in Japan a subject of enormous research interest.

10.3.2 Accelerator science and technology

DORIS was commissioned in 1974 and considered as one of the most advanced collider facilities of its time. Many modern technologies and innovations were used in its construction—some of which were extremely relevant for industry, such as vacuum and magnet technologies including early large-scale applications of superconductors [11].

Significant contributions to the fundamental understanding of modern electron–positron colliders at high energies and intensities were done at DORIS in the early years. The careful studies of unexpected effects in the accelerator that were observed in the beginning of the operation of DORIS led to a much better understanding of underlying phenomena and to new knowledge in beam physics [12]. The success of these investigations also relied on the rapid advancement of computing power and the possibility of developing numerical software for non-linear beam dynamics and particle tracking. The knowledge and proper understanding of many of the effects encountered at DORIS as well as the advent of numerical programs in accelerator design were important contributions to the global expertise in accelerator physics of electron storage rings, and today's powerful accelerators and synchrotron light sources largely benefitted from these studies.

Examples include the understanding of non-linear effects in accelerator beam optics and dynamics, the tackling of beam instabilities, the development of higher mode damping antenna in accelerator cavities which have now become a standard in RF accelerator technology, and the investigation of beam–beam interactions and crossing angles which were a major breakthrough in understanding colliding-beam designs and which had significant impacts on all subsequent electron–positron colliders in US, Europe and Asia. Moreover, compact strong-focusing quadrupoles ('mini-beta scheme') were implemented at DORIS II close to the interaction points allowing an extremely strong focusing of the beams at the interaction points and correspondingly higher luminosities [6]. Such a scheme was later used successfully at many other colliders. DORIS was also one of the first facilities in the world that used sophisticated correction technologies based on fast-feedback beam position monitors and control systems to improve the photon beam stability for SR users.

10.3.3 Photon science

Soon after DORIS started up, the SR produced by the storage ring as a by-product proved to be a valuable research instrument for numerous researchers from many scientific fields and disciplines. In the early 1990s, DORIS ultimately became one of the leading x-ray sources in the world and developed and established a wide spectrum of SR relevant scientific applications and techniques that are now found in almost all SR labs across the world [5, 13]. The impact that DORIS were so broad in scope and ranged from studies of the structure of atoms, molecules, clusters, nanomaterials, solids, surfaces and liquids, of polymers, biological samples and complex organic structures such as proteins and enzymes, to applications in chemistry and the geo and environmental sciences, to the development of diagnostic methods for medicine and to technological studies, including elemental analysis of a wide variety of objects, for instance for research on cultural and natural heritage [2]. In almost all research areas work was carried out at the DORIS storage ring with great success and international recognition by many national and international research groups. The DORIS user community published a total of about 12 000 publications, many of them in high-impact journals. Key users at DORIS obtained many highly prestigious awards for their research including the Nobel prize in Chemistry 2009 to Ada Yonath for the deciphering of the ribosome (figure 10.4).

In the following we give a few examples of methodological breakthroughs and highlights in SR research at DORIS knowing that this is by no means a complete overview:

- 1975: Early research and developments on x-ray lithography methods were performed raising strong interest by industry; these studies and results were later used in industrial applications for semiconductors.
- 1976: Imaging of living cells with unprecedented resolution using SR x-ray microscopy techniques with zone plates [14]. This work started the close institutional cooperation between DESY and U Göttingen in biological imaging.
- 1984: Pioneering experiment on Mössbauer spectroscopy with SR [15]. After many years of tedious preparatory work at DORIS, the first Mössbauer spectrum ever obtained with SR was recorded. Mössbauer spectroscopy with

Figure 10.4. The Nobel Laureate Ada Yonath in front of the new experimental hall (extension hall east) of the current SR-source PETRA III at DESY in 2016. Due to her significant research on the Ribosome structure at DORIS III, the new experimental hall at PETRA III was named after her. (Copyright: © DESY/Gesine Born.)

SR is now carried out with great success at many laboratories worldwide. Because of the sharpness of the line, measurements can be made with unprecedented precision and in numerous applications.

- 1986: Phonon spectra determined by high-resolution inelastic x-ray scattering —for the first time research groups at DORIS [16] succeeded in measuring the excitation of individual phonons (lattice vibrations) in solids with SR. This made it possible to carry out investigations of the elastic properties of materials that were previously only possible at nuclear reactors with neutron scattering. Another large field of materials research with synchrotron radiation was thus opened up.
- 1987: Circular dichroism as a powerful tool for studies of magnetic properties of matter. At DORIS the first spin-dependent absorption measurements of magnetic materials were done to allow the investigation of magnetic properties of solids and thin films on an atomic scale [17]. Today, this technique is widely used for the element-specific and quantitative determination of local magnetic spin and orbital moments providing new insights into magnetic, electronic, and geometric structures and phenomena.
- 1999: Experiments on quantum cutting phenomena on rare earth doped systems to develop new luminescent materials [18]. Quantum cutting processes to efficiently convert vacuum ultraviolet radiation to visible light materials are an active field of research and has been placed on many optical technology roadmaps worldwide in the quest for mercury-free fluorescent lamps and improved plasma display panels.
- 2002: A large clinical study on human intravenous coronary angiography with synchrotron radiation took place at DORIS [19].

10.3.4 Life sciences

A profound and most spectacular imprint of SR has been in the life sciences and, in particular, in macromolecular crystallography. The key challenge here is to reveal the three-dimensional molecular structures of biologically interesting large objects, such as proteins, lipids, or nucleic acids, in order to deepen the knowledge of the molecules of life and of their biological function at atomic levels. The advent of much better x-ray sources provided by SR laboratories and associated technologies has transformed this field over the past 40 years and led to tremendous progress in our understanding of the complexities of living organisms and cells. A major driver in this development was the early presence of the European Molecular Biology Laboratory (EMBL) at DESY which started from early on its own outstation in order to precisely analyze biomolecules [20] and later on the establishment of three permanent research groups in molecular biology of the Max Planck Society (MPG), with one of them led by Ada Yonath.

DORIS and the associated labs have played then a key role in all these SR developments and promoted the growth of SR applications in life sciences from a small effort to a worldwide mainstream activity. This boom in SR experiments for life sciences was mainly created in the late 1970s and 1980s, with continuous upgrades of equipment and beamline technologies increasingly providing access to international scientists for protein crystallography, small-angle scattering, and high-resolution x-ray spectroscopy.

The leadership role in life science research at DORIS in these days is impressively underlined by the number of structural biology publications associated with SR in the early 1990s. Almost 40% of the published papers in this field in 1992–93 refer to DORIS beamlines, with a high proportion of the publications appearing in high-impact journals such as *Nature*, *Science*, or *Cells* [21]. Altogether, the DORIS and EMBL life science beamlines contributed tremendously to the determination of protein structures. In the late 1990s, for instance, approximately 15% of all protein structures worldwide were decoded at DORIS [22].

Among the most visible and important impacts of the various research efforts at DORIS throughout the decades is the structural determination of the ribosome. For this work, Ada Yonath, the leader of the MPG ribosome structure group at DESY, was awarded the 2009 Nobel Prize in Chemistry together with Thomas A Steitz and Venkatraman Ramakrishnan.

A very few selected methodological breakthroughs and milestones are illustrated as examples:

- 1970: First diffraction studies using SR at biological samples, in that case muscles (at DESY).
- 1986: Start of the work of Ada Yonath at DORIS on the ribosome.
- 1987: Development of an image plate detector, later the first online imaging plate scanner for protein crystallography was produced, commercialised, and widely used.
- 1990s: Pioneering research by Eva-Maria and Eckhard Mandelkow on the decisive role of the tau protein in connection with the development of Alzheimer's disease [23].
- 1998: The three-dimensional structure of titin was revealed—a giant protein that plays an important role in muscle development and contraction [24].
- 1999: Starting from the end of 1990s, EMBL Hamburg developed the program package ATSAS [25]. It became the world most comprehensive program suite for small-angle scattering data analysis from biological macromolecules, with currently over 7500 users. Over 60% of biological solution scattering papers worldwide employ and reference ATSAS

10.4 Footprint of DORIS on the research system

Over the 40 years DORIS has greatly contributed but also benefited from the surrounding research, education, and innovation system. As a research facility it has served as an anchor for national and international collaboration, provided access to university research groups, acted as platform for the training and education of young researchers, and enabled as a vehicle multidisciplinary scientific research and technology development through various communities. This improved the advancement of knowledge and related technologies in an effective way and generated many impacts on the research landscape *per se*:

- The footprint of DORIS on other research facilities, which were inspired by the pathfinding work and breakthrough results of DORIS was enormous. For example, the result on B meson mixing of ARGUS at DORIS (figure 10.5) immediately paved the way for the development of dedicated B-facilities in the US and Japan. In photon science, there was a huge impact by DORIS on the establishment and growth of synchrotron radiation facilities, in particular at BESSY in Germany in 1977 and then later at the ESRF, the European Synchrotron Radiation Facility in Grenoble. In both cases, key personnel from DORIS were either driving designs and developments and/or took over important positions in the establishment of the new facilities.
- DORIS as a user facility was also a major driver to shape numerous collaboration and relationships with (national and international) universities and research institutes, from which DESY still benefits today. A large number of institutional collaborative frameworks were formed including dedicated professorships at several German Universities for DESY research topics that also helped to strengthen the research profiles at the academic institutions and faculties.
- The close interactions with (national) policy makers and funding agencies improved the way how science and research activities as well as access to the facility was organized. For instance, in the early phase of HASYLAB a new funding scheme ('Verbundförderung') was introduced by the German Ministry which granted a special budget line for research groups from German universities. Funding from this budget has greatly enabled universities to build and install additional experimental stations and instruments at

Figure 10.5. The ARGUS collaboration with the detector in the background. The discovery of the B meson mixing started the construction of new facilities in the US and Europe. Copyright DESI.

the facilities, which was very supportive of the further development and one key element to create a strong and active national photon science community, close links to universities and a large number of institutional agreements.

10.5 Impact on people

Research facilities in basic science such as DORIS are not only promoting scientific knowledge and stimulating economic growth, but also strongly influencing the people working there and enhancing their capacities. During the service of DORIS (1974–2013), several thousands of people developed, maintained, and scientifically exploited the facility. Next to the administrative, technical, and scientific staff of DESY, these included external users from a wide range of disciplines.

A study by the German Physical Society (Deutsche Physikalische Gesellschaft (DPG)) revealed the great economical value of physicists [26]. Thanks to the transferable skills acquired during their education, physicists are found to easily adapt to different industrial sectors and fields of activity. This makes them much sought-after by the labour market.

Against the background of a society that changes to a post-industrial, knowledge-based economy, highly educated people with technical, professional, and social skills and knowledge become more and more important.

To visualise the career paths of doctoral degree recipients who worked at DORIS and to analyse the acquirement and transfer of skills into other sectors of employment, a survey among all known former DORIS graduate students was conducted by DESY. These included PhD students employed by DESY, EMBL, and several universities in Germany and abroad. The idea of the study (carried out in 2015) was to serve as an explicit proxy to track the transfer and utilisation of skills and knowledge created at DORIS in all fields of employment and all sectors of the economy.

Between 1973 and 2014 about 1200 students completed their PhD on a topic related to DORIS, 17% of them female. About half of them could be contacted via e-mail, and 144 participated in the survey. The results of the survey prove the outstanding value of education and training at large-scale research facilities such as DORIS for science, economy, and society in general. The answers of the participating former PhD students clearly show that their skill set gained at DORIS goes far beyond scientific expertise. In the following, some of the main findings of the survey are summarized:

- The training at large research infrastructures such as DORIS enables students to follow careers also outside of public science and research. Even in their first job after the PhD, 22% started a career in the private sector. 10% of the students became entrepreneurs, of which 67% had employees.
- Regarding the PhD students in the private sector, the data show major accumulations in the sectors information technology (IT) and healthcare, which will both gain increasing importance in the future due to the megatrends related to the technological and demographic change of society.

- To all appearances, the PhD students did not only gain specific scientific skills and knowledge related to their research area, but also non-specific knowledge and soft skills that are even more valuable for the private sector. Among the five most important skills that the former students acquired during their PhD work, only one refers to science (research methodologies/techniques). Instead, problem solving, team working, creative thinking, and independent working were considered the most useful for their subsequent careers.
- Working at the research facility laid the foundations for successful professional careers of the students within or outside the commercial world. Only five years after graduation, more than half of the former DORIS PhD students were responsible for their own budget (55%) and 45% led their own staff. In the long run (current position), 71% managed a budget and 69% had personal staff, clearly indicating an executive position.

Although fundamental science sometimes seems to be vague and hardly application-oriented, the skills that people gain are all the more applied—they solve concrete problems. In addition to being trained in scientific methods and techniques, the PhD students at DORIS (and the same most certainly applies to other large-scale research facilities) learn how to manage projects and find creative and innovative solutions in international and interdisciplinary teams (figure 10.6).

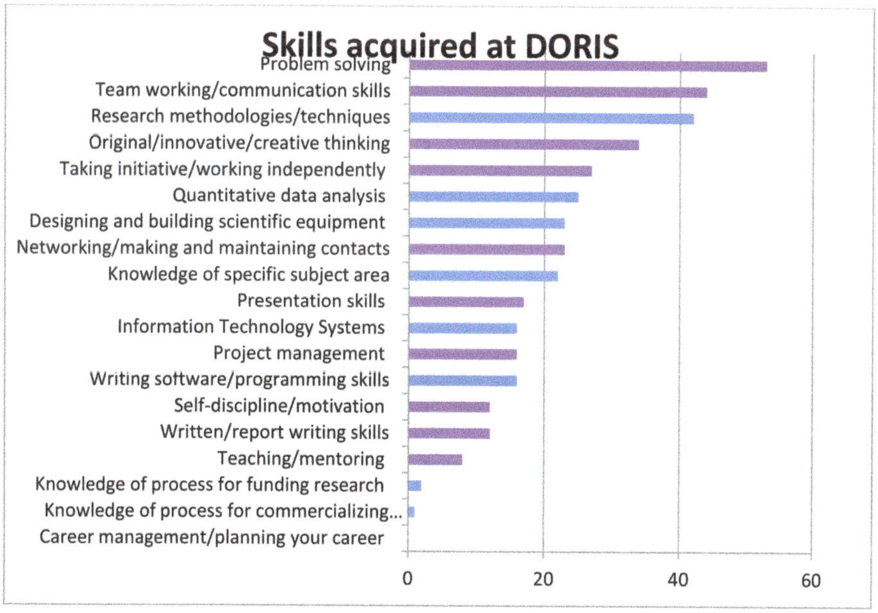

Figure 10.6. Result of a survey of former PhD students at DORIS on the rank of the skills acquired during their PhD work that are seen useful for their current job. The horizontal axis is given in % of the respondents. Purple coloured bars indicate soft skills, blue coloured bars indicate knowledge-based skills.

10.6 Impact on economy

The economic effects of research facilities are manifold. Here we distinguish between direct and indirect economic added value and the effect of research infrastructures on economic innovation.

The construction and development of DORIS demanded major financial investments which generated economic impacts mainly to the metropolitan area of Hamburg and northern Germany, but also beyond. A reasonable method to elaborate these effects is a multi-stage, regionalized input–output-analysis [27], in which financial effects are decomposed into regional and sectoral effects. Essentially one would like to understand the economic 'chain reaction' of investments considering that money is re-spent again and again in the economy, creating jobs and income for businesses and other workers.

There are three different aspects of the financial impact to be considered. First, the primary input triggers the so-called direct effects. They include on the one hand revenues of the direct suppliers (direct sales effects) and on the other hand the jobs that were directly generated at the facility (direct employment effects). Second, indirect effects are generated. Since every supplier and subcontractor obtains certain inputs from other companies, there is an economic effect in the supply chain of every direct supplier. Furthermore, the induced effects capture the money flows, which are created by salaries, both directly at the facility and at other companies involved in the value chain. The induced effects can be put on a level with consumer spending of all employees, which are affected directly or indirectly by the facility's spending.

Even though the results of such analysis have to be taken with care—given a scarce data basis of 40 years of investments—it gives an impression of the benefits of the investment in DORIS for the region and which economic sectors profited most.

It is estimated by the authors that the overall DORIS investments throughout its 40 years of lifetime summed up to approximately 1.4 billion euros (inflation-adjusted) including costs for (re)construction and operations (figure 10.7).

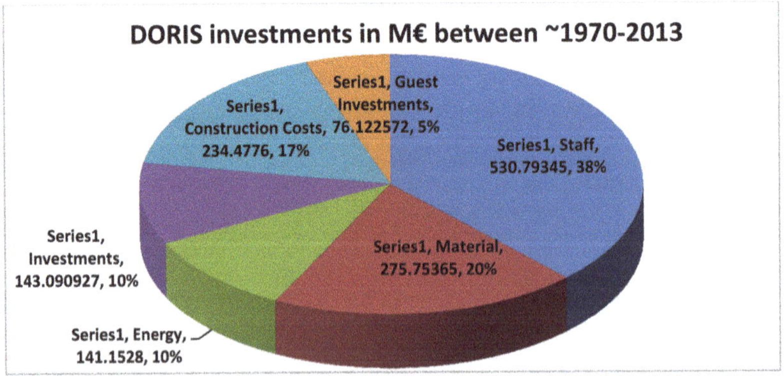

Figure 10.7. Distribution of the estimated investment in DORIS construction and operation between ~1970 and the end of operation of DORIS in 2013.

This number was derived by taking as an educated guess an appropriate fraction of the overall budget of DESY during the different phases of the lifetime of DORIS. However, third-party funding, particularly from universities invested in DORIS could be only partially tracked. The spending of the guest scientists is subject to assumptions as well—an approximation was done based on the number of scientists per year and a fixed amount of spending.

From the input–output-analysis we receive an indirect multiplier of 1.83 and an induced multiplier of 1.90. This means that one euro spent on DORIS produced as a result of this direct spending an additional 0.83 € sales volume through indirect effects from business–business transactions in the supply chains and additionally 0.90 € through induced effects from increased personal income through direct and indirect effects. Fact box 2 summarizes the short term economic impact of DORIS.

Even though the knowledge of these demand effects can help to gain political support, basic research is not and should not be regarded as a 'business cycle' program to boost employment. The demand effects are socio-economic 'side effects'. The following effects on economic innovation reflect more the nature of research infrastructures like DORIS.

A large impact has been undoubtedly achieved through all those advancements in computer and network technology, and in electrical engineering, vacuum technology, radiology, and several other fields that have followed in the wake of the pushing of technical and scientific limits in accelerator and detector construction at DESY and other particle physics laboratories globally. Similarly, the collected global effort in advancing technologies and research in the area of synchrotron radiation has also led to advances that have had direct benefits in the development of more efficient batteries, LCD displays, and data storage devices in consumer electronics, enhanced and completely novel drug treatments, as well as software development, to name just a few.

Fact box 2:

Short term economic impact of DORIS

- Total budget spent (over the DORIS lifetime): 1.4 bn €
- Direct sales effects: 700 M€
- Indirect sales effects: 590 M€
- Induced sales effects: 1.2 bn €
- Indirect (induced) multiplier: 1.83 (1.9)
- Additional jobs created in Germany: 12 500 FTE; which equals 300 FTE per year
- Effects in Hamburg: 200 direct jobs per year at DESY plus 450 million euros triggering about 1000 FTEs (= 25 FTE per year)

10.7 Impact on the region

The regional impact of a research infrastructure is manifold. As described in the previous chapters, large scientific facilities and research labs have characteristic imprints on the regions which are hosting these facilities in terms of scientific output, higher education services, economic innovation, but also in terms of regional collaboration and networks. Moreover, research infrastructures often represent path-shaping investments and carry an institutional resilience that means continued and repeated investments in similar projects locally.

DORIS is a prime example for such structural socio-economic effects in the Hamburg metropolitan region. Already during or after the times of DORIS many other and even larger science investments accompanied by the co-locations and aggregation of cooperating partner institutions and by auxiliary infrastructures followed—and all of them have made their own specific contributions to the local, regional, national, and international economy.

- *Research infrastructures*: Many experimental methods and technologies that were pioneered at DORIS have now become standards and led the way to further infrastructure investments. The continued improvements and the innovative atmosphere together with a remarkable collaborative spirit led later to the construction and operation of unique photon science infrastructures on the campus such as PETRA III, one of the most brilliant x-ray source in the world, and FLASH, the world's first free-electron laser in the x-ray range, and lastly the European XFEL as an international flagship facility for FEL science in the hard x-ray regime.
- *Aggregation of research institutions*: Already in 1975, EMBL established an outstation on the DESY campus to use the intense light for the investigation of biomolecules. In 1986, the MPG added another outstation at DORIS for three research group units followed by other external research organisations such as the GKSS-Forschungszentrum Geesthacht, now the Helmholtz-Zentrum Hereon. The University of Hamburg has strongly increased the presence of its research institutions and interdisciplinary research centers on site and decided recently to also move the complete chemistry department to the campus. MPG founded in 2014 an own institute on structural dynamics (with own research building) on the campus.
- *Interdisciplinary centers*: The concentration of world-leading research infrastructures and competences to decode the structure and dynamics of matter led to the establishment of further scientific and multidisciplinary institutions as collaborating platforms such as the Center for Free-Electron Laser Science (CFEL), the Centre for Structural Systems Biology (CSSB) and the Center for X-ray and Nano Science (CXNS).
- *Innovation*: Even if DORIS was not a major driver of innovation and transfer it is seen as a starting point for a further dynamic development with the recently opened start-up laboratory Bahrenfeld as a new place for hi-tech and science entrepreneurship and start-ups from the physical and biological sciences.

DORIS as a pioneering facility opened the way to the formation of new competence clusters on the campus, mainly in nano, bio, laser, and engineering materials. This marked also the transition of the campus from a previously mono-disciplinary site, into a science and innovation ecosystem with a rich and versatile multidisciplinary landscape characterized by world-leading research infrastructures, co-location of renowned research institutes, and a stimulating innovation environment from which the Hamburg Metropolitan area greatly benefits today.

The development continues by the DESY plan to build a new key research infrastructure PETRA IV, an ultimate and unique 3D x-ray microscope, and the establishment of the so-called Science City Hamburg-Bahrenfeld [28], presently the largest and most ambitious future urban planning project of the City of Hamburg for the coming decades.

10.8 Conclusions

The case study of DORIS very impressively demonstrates that a major scientific facility can be continuously adjusted and upgraded to meet the changing requirements in science. DORIS went through various phases and different scientific missions, showing that the facility and its related services was sustainably utilized over many decades. The facility has been a pioneer in several ways, in accelerator science and technology, in particle physics and, most prominently, in photon and life sciences. It paved the way towards modern colliders and third generation SR machines and developed numerous methods and instrumentation that are now being used worldwide.

In addition to the long-term sustainability and the enormous scientific impact, the case of DORIS also demonstrates the role and added value of research infrastructures as a focal point for collaboration among a multitude of scientific actors. The strong network of universities and research organizations was crucial and highly beneficial for the further development. Together with these partners the transformation of DESY from a single discipline laboratory into a multidisciplinary research center was enabled.

This transformation of the campus triggered by research at DORIS had and still has a significant impact on the regional development of the larger Hamburg metropolitan area. Not only that the demand effects of the major financial investments in the construction and operation of research infrastructures that followed DORIS such as PETRA III, FLASH, and European XFEL, generated a large economic impact, but also the multidisciplinary research at DORIS was the origin of the formation of new competence clusters, the begin of an aggregation of research institutes and the start of the growth of an innovation ecosystem on campus.

Finally, DORIS played an important role in capacity building. Young researchers and technical personnel were trained at this research instrument and gained knowledge that goes beyond scientific methods and techniques. Above all, they learned how to work in international and interdisciplinary teams on innovative solutions for complex questions. Most likely, the development of this enormous work force of skilled people for the industry is probably the largest economic impact that DORIS left.

During the analysis of the case of DORIS it has quickly become clear that the separation of one research infrastructure from the rest of the research center DESY and often from the whole environment of research organizations, universities, and other research infrastructures is hardly possible. The impacts are often not only linked to one but to a variety of research infrastructures and instruments. It is the collaboration between these multiple actors around research infrastructures and the open access to these tools that generates their enormous added value for society.

References

[1] Jentschke W *et al* 1967 *Vorschlag zum Bau eines 3 GeV Elektron–Positron-Doppelspeicherrings für das Deutsche Elektronen-synchrotron* (Hamburg: DESY)

[2] Mundzeck T 2012 The three lives of DORIS: from charm quarks to cell biology *CERN Courier* 27 November https://cerncourier.com/a/the-three-lives-of-doris-from-charm-quarks-to-cell-biology/

[3] Lohrmann E and Söding P 2009 *Von schnellen Teilchen und hellem Licht: 50 Jahre Deutsches Elektronen-Synchrotron DESY* (Weinheim: Wiley)

[4] Wolf G 1976 Elektron-Positron-Wechselwirkungen *Phys. Bl.* **32** 692–703

[5] Kunz C, Skibowski M and Sonntag B 2015 How it all started at DESY in 1964 *Synchrotron Radiat. News* **28** 16–19

[6] Heinze T, Hallonsteen O and Heinecke S 2015 From periphery to center: synchrotron radiation at DESY, part II: 1977–1993 *Hist. Stud. Nat. Sci.* **45** 513–48

[7] Wille K 1981 DORIS II/III—a 5.8 GeV $e^+ e^-$ storage ring with high luminosity *DESY preprint* DESY-81-047

[8] Nesemann H *et al* 1995 DORIS III as a dedicated source for synchrotron radiation *Proc. Particle Accelerator Conf.* (Piscataway, NJ: IEEE) S 195–7

[9] Wegener D 2013 A fructibus eorum cognoscetis eos from charm to beauty *Talk at the DORIS Days, DESY*

[10] Lehner F and Faverot-Spengler S 2007 Argus-Fest. 20 years of B meson mixing 1987–2007 *DESY Conf.-Proc.*

[11] Horlitz G *et al* 1975 PLUTO, example of a large superconducting magnet *Siemens-Z.* **49** 764–8

[12] Piwinski A 1977 Satellite resonances due to beam–beam interaction *IEEE Trans. Nucl. Sci.* **24** 1408–10

[13] Gehrke R and Flegel I 2005 Meeting reports: forty years with synchrotron radiation at DESY *Synchrotron Radiat. News* **18** 20–2

[14] Schmahl G 1983 X-ray microscopy *Nucl. Instrum. Methods Phys. Res.* **208** 361–5

[15] Gerdau E *et al* 1985 Nuclear Bragg diffraction of synchrotron radiation in yttrium iron garnet *Phys. Rev. Lett.* **54** 835

[16] Burkel E, Peisl J and Dörner B 1987 Observation of inelastic x-ray-scattering from phonons *Europhys. Lett.* **3** 957

[17] Schütz G *et al* 1987 Absorption of circularly polarized x rays in iron *Phys. Rev. Lett.* **58** 737

[18] Wegh R T *et al* 1999 Visible quantum cutting in $LiGdF_4:Eu^{3+}$ through downconversion *Science* **283** 663

[19] Dix W R *et al* 2003 Comparison of intravenous coronary angiography using synchrotron radiation with selective coronary angiography *J. Synchrotron Radiat.* **10** 219

[20] Brochure 2014 *40 Years EMBL* (Heidelberg: European Molecular Biology Laboratory)

[21] Ealick S E and Walter R L 1993 Synchrotron beamlines for macromolecular crystallography *Curr. Opin. Struct. Biol.* **3** 725–36

[22] Jiang J and Sweet R M 2004 Protein Data Bank depositions from synchrotron sources *J. Synchrotron Radiat.* **11** 319–27

[23] Mandelkow E M and Mandelkow E 1998 Tau in Alzheimer's disease *Trends Cell Biol.* **8** 425–7

[24] Mayans O *et al* 1998 Structural basis for activation of the titin kinase domain during myofibrillogenesis *Nature* **395** 863–9

[25] Manalastas-Cantos K 2021 ATSAS 3.0: expanded functionality and new tools for small-angle scattering data analysis *J. Appl. Crystallogr.* **54** 343–55

[26] Deutsche Physikalische Gesellschaft 2010 Physikerinnen und Physiker im Beruf—Arbeitsmarktentwicklung, Einsatzmöglichkeiten und Demografie https://www.dpg-physik.de/veroeffentlichungen/publikationen/studien-der-dpg/studie-arbeitsmarktentwicklung

[27] Hoppe H C and Pfähler W 2001 Ökonomie der Grundlagenforschung und Wissenschaftspolitik *Perspekt. Wirtschaftspol.* **2** 125–44

[28] Science City: Wissenschafft Stadt https://sciencecity.hamburg/

IOP Publishing

Big Science in the 21st Century
Economic and societal impacts
Panagiotis Charitos, Theodore Arabatzis, Harry Cliff, Günther Dissertori, Juliette Forneris and Jason Li-Ying

Chapter 11

A case study of Big Science at an IGO: how does EMBL respond to changes in societal norms?

Lucia von Bredow, Gaia Cantelli, Anne-Flore Laloë and Johanna McEntyre

11.1 Introduction

The European Molecular Biology Laboratory (EMBL) was initiated on the basis that the nascent discipline of molecular biology could only advance if several countries pooled their resources to do so. Indeed, by the 1960s it had become clear that studying life at the molecular level would require large infrastructures that were too big for a single country to fund or operate. In addition, Europe was becoming an increasingly united and collaborative space both politically and scientifically: the Treaty of Rome had been signed in 1957 and the European Organization for Nuclear Research (CERN) was established in 1954. It is in this context that the EMBL was devised and established as an intergovernmental organisation (IGO) by the European Molecular Biology Organization.

What is an intergovernmental organisation (IGO)?

An IGO is an organisation that is made up of several sovereign member states or other IGOs to meet specific missions as outlined by an establishing charter. IGOs and their employees have certain privileges and immunities that are given to them to ensure that IGOs can deliver their missions effectively, independently of the influence of any one of its component members. This principle is known as 'functional necessity' and means that IGOs are generally immune from the jurisdiction of national courts. The supervision and legal accountability of IGOs should instead be assured by internal mechanisms.

Multilateralism lies at the very core of IGOs: they bring together the governments of several national entities to work together to address a defined challenge or to solve a shared problem. The legal framework which allows IGOs to deliver their missions reflects this multilateralism, ensuring that they can operate within their own individual regulatory and, often, cultural environments, independent of those of their members. This operational framework enables IGOs to act effectively, and therefore deliver their missions efficiently, be it the distribution of humanitarian aid, the setting of international standards, or the acceleration of science.

In 1973 EMBL's ten founding member states enshrined EMBL's mission as being to 'promote co-operation among European States in fundamental research, in the development of advanced instrumentation and in advanced teaching in molecular biology as well as in other areas of research essentially related thereto, and to this end shall concentrate its activities on work not normally or easily carried out in national institutions'. [1]

Big Science is a specific term that describes scientific work which inherently relies on large infrastructures, and therefore also large budgets and high staff numbers. While the term was originally applied to astronomical or physical sciences, it is applicable to the life sciences, for example in the case of genomics research or imaging and microscopy facilities. These words from EMBL's Establishing Agreement encapsulate several ideas about the relationship between Big Science and societal norms which this chapter will address. By focussing on science that is too complex and too expensive for any one entity to carry out, EMBL places Big Science at its core.

By approaching molecular biology in a multilateral manner and through its IGO status, EMBL can be independent from the national science priorities of its member states, while still ensuring that the results of its work are fed back to the public, thus adding value to national science activities [2]. EMBL, as Europe's only life sciences IGO, is ideally placed to provide research and services for the life sciences in an international and intergovernmental context. This is overseen by its governing body—EMBL Council [3]. Therefore, EMBL has both the remit and the infrastructure necessary to carry out truly international life sciences research and answer Big Science questions. EMBL Council is composed of representatives of each of EMBL's member states, associate member states, and observers who set the agenda for EMBL's scientific, technical, and administrative activities[1].

The nature of EMBL's status as an IGO and its streamlined governing organs mean that EMBL is able to rapidly respond to changing societal, scientific, or cultural norms. With limited bureaucracy, internal policies or guidelines can be approved relatively quickly, compared with the pace of national law-making bodies, while still ensuring adequate oversight. Nevertheless, several important questions need to be asked: how should this flexibility be used, does EMBL have a duty to respond to societal norms, and how do these pertain to the life sciences? Conversely, should EMBL be permitted to act according to its mandate, and in so doing, exceed its mission? This chapter will explore EMBL's activities in the context of Big Science and society.

[1] Article 6 of the agreement establishing EMBL.

What are societal norms?

Norms are social phenomena that propagate themselves through a group of individuals. They can be codified, for example through laws, or they can be informal. They are generally behavioural and can be public and social, for example, queuing to enter a shop, or private, as in a family's preferred way of setting a table.

Norms become norms over time, as they are first introduced, then adopted widely, and then internalised (or normalised) until they are no longer questioned. Social norms can include things such as attitudes towards data protections or the facilitation of work–life balance in professional life. Specifically in academia, a normalisation of open science is certainly being seen in some disciplines, both as attitudes towards publishers change and as institutions legislate in favour of open access.

11.2 Does EMBL have a duty to respond to changes in societal norms?

11.2.1 EMBL's mission and programme

EMBL has five core missions: (i) to perform excellent fundamental research in molecular biology; (ii) to offer vital services to scientists in the member states and the world; (iii) to train scientists, students, and visitors at all levels; (iv) to actively engage in technology development, transfer, and industry relations; and (v) to coordinate and integrate European life science research. The notion of Big Science, and the prospect of multilateral, multi-faceted, and international collaboration, is fundamental to achieving these goals.

EMBL's activities are planned in five-year cycles and outlined in a document called the EMBL Programme. Every five years, the member states which fund and govern EMBL agree a funding plan called the Indicative Scheme, which will support the EMBL Programme. The EMBL Programme is developed with input from people across the organisation, including scientific faculty and technical, service, and administrative staff. It outlines how EMBL aims to push the boundaries in research, services, training, and technology development and transfer, developing EMBL's position as an international role model for research institutions.

The past half century has seen an incredible acceleration of knowledge and technologies in the life sciences, particularly in the field of molecular biology. Things can change rapidly at the frontiers of biomedical research. In just five years, previously unheard of methods and technologies can become essential research tools. New discoveries can spark whole new fields which scientists may want to access or need to be trained in. To retain EMBL's ability to adapt to such rapid and disruptive changes, the EMBL Programme serves as a broad guideline rather than a detailed directive. While the EMBL Programme is not prescriptive, it provides direction, informing decisions from strategic hires and investments in technology, to establishing partnerships with other institutes.

The EMBL Programme is mostly developed internally, but is reviewed by EMBL's Scientific Advisory Committee [4]. In general, SAC, which is composed of external scientific experts, is tasked with advising the EMBL Council specifically on the realisation of the EMBL Programme. Moreover, by also being involved in its elaboration, it can provide input on the Programme's plans, feasibility and relevance. The eventual approval by the EMBL Council ensures that it is aligned with the collective wishes of EMBL's member states.

11.2.2 Societal impact

In addition to following this comprehensive and iterative planning process, which involves all of the stakeholders of EMBL governance structure, EMBL has programmes specifically designed to create a bridge between EMBL and society—such as the Science & Society and ELLS Programmes (which will be discussed in section 11.2.4), and site-specific outreach initiatives such as the Heidelberg Forum [5], the EMBL-EBI public engagement strategy [6][2], and the World of Molecular Biology Exhibition [7]. This ensures that EMBL's scientific activities and its delivery of Big Science do not operate in a vacuum, whether scientific or societal.

EMBL's scientific and societal impact: a selection of statistics from 2020

- *Scientific service users*: 64M daily requests on data resources, and over 5400 annual users of core facilities.
- *Scientific papers*: 729.
- *Training and conferences*: 9900 participants to courses and conferences, with 101 countries represented.
- *The European Learning Laboratory for the Life Sciences* worked with 245 high-school teachers and 1298 students from 39 different countries.
- *Collaborations with industry*: four new companies, 20 tech development projects in collaboration with industry, 30 inventions disclosed.

The organisation also has a number of endeavours designed to create connections between EMBL and industry. For example, EMBLEM [8], the wholly owned commercial subsidiary and exclusive technology and knowledge transfer partner of the EMBL, are responsible for identifying, protecting, and commercialising intellectual property and associated technologies and materials developed in the EMBL-world.

In addition, the EMBL-EBI Industry Programme is aimed at companies that make significant use of the data and resources provided by EMBL-EBI as a core part of their R&D. The programme is unique, providing strategy meetings and expert-level workshops on topics prioritised by the members. It also serves as an interface between industry-focussed initiatives at EMBL-EBI and organisations

[2] 30 outreach events and 2443 interactions, involving 48 staff.

such as the Innovative Medicines Initiative (IMI), the Pistoia Alliance, the Clinical Data Interchange Standards Consortium (CDISC), and more [9].

An independent impact review conducted by the Technopolis Group[10] in 2021 reflected that EMBL Experimental Services is

> helping academic users to deliver wider social impacts in a range of thematic areas beyond health, such as industrial biotechnology, environment and climate change, and food security and sustainable agriculture. Notably, EMBL-enabled research facilitated a new enzymatic recycling approach to lower the environmental impact of plastic waste. Academic users maximise the societal impact of their research through public engagement activities. A smaller proportion of academic users also generates economic impact through commercial activities, including research contracts and consultancy for industry and creating spin-off companies [11].

In addition, the Beagrie Report 'Data-driven discovery' [12], which focused on EMBL Data Services, also completed in 2021, highlighted that 'EMBL-EBI's managed data resources have been making a major contribution to science throughout the COVID-19 Pandemic. Some 32% of respondents said they valued EMBL-EBI data resources more as a result of the pandemic and 25% said they had used them more. It also found that 'during the last year the use of EMBL-EBI managed data resources contributed to the wider realisation of research impacts conservatively estimated to be worth some £2.2 billion annually (almost 140% higher than 2015–16), or up to £15 billion over 30 years in net present value'.

It is clear, therefore, that EMBL can not only demonstrate societal impact, but that it also has the legal and organisational ability to respond to changes in societal norms. But which kind of norms may be relevant to a research, service, and training IGO, and whose interpretation of norms should be considered? This will be examined in section 11.2.3, and will help us assess whether EMBL has a duty to respond to any changes in such norms—and, if so, how its response should be constructed and implemented.

11.2.3 Relevant norms

This chapter will explore the concept of Big Science at an IGO in more detail and, using two case studies, will examine how EMBL has previously chosen to respond to societal changes. The case studies chosen are EMBL's Scientific Services—and in particular the open biological data resources hosted by one of EMBL's six sites, the European Bioinformatics Institute (EMBL-EBI)—and EMBL's ongoing work to improve the accessibility of scientific research outputs within the open science movement.

Although this chapter will examine in detail only two specific case studies, it is worth noting that there are a multitude of areas at EMBL which may have an impact on and indeed be influenced by societal norms, from its research projects to its working practices.

While it is true that the impact of a basic research institute such as EMBL can often be hard to estimate, the ethical, legal, and social implications of EMBL's research have the potential to affect the whole of society, rather than just the scientific community, touching almost every area of everyday life. Furthermore, the impact of the work carried out at EMBL is not limited to citizens of member states, but rather affects the global community. EMBL affects the international community through impacts beyond its research, services, and training activities. In fact, EMBL's influence extends in a number of other areas, including:

- *Bioethics*, including human, animal, and environmental ethics. When EMBL researchers develop or contribute to the development of emerging technologies or new protocols, they have a real or perceived impact on the wider world. For instance, research carried out at EMBL can have a direct influence on advances in fields such as personalised medicine, combatting climate change, or in infectious disease management. Furthermore, when communicating about its research, EMBL must be mindful that it has the capacity to shape the societal perception of science.
- *Data privacy* and the protection of sensitive, personal data. Data protection is an important aspect of society in general and scientific research on or with human subjects specifically. How personal data is managed for individuals, for their relatives and for society raises a number of questions relating to public health, epidemiology, and genomic medicine. Data privacy considerations are becoming increasingly relevant to the bioethics debate, as recent scientific and technological advances are allowing an ever greater amount of sensitive data to be gathered. For instance, researchers can gather information on an individual's genome, their metabolism, their microbiome, their ancestry, and their predisposition to thousands of medical conditions.
- *Environmental policies*. These are closely linked to the biological sciences, and EMBL's research and institutional actions can influence societal movements. Foremost, the link between EMBL's overarching goal of understanding the molecular basis of life covers a wide spectrum of biology, including cell biology, developmental biology, tissue and organ biology, and microbiology. These considerations, alongside EMBL's vision of applying expertise in molecular biology to study life in its natural context, ensure that EMBL's research is in fact closely linked to the environmental sciences. In addition, the new 'Green EMBL' initiative is part of the organisation's response to the need to tackle climate change. Set up in 2019, the initiative centres on three main pillars: ensuring EMBL research is environmentally responsible; ensuring EMBL carries out research relevant to environmental and climate science; and sharing the knowledge it gains in environmental research with the public.
- *Research integrity* and communication. These are important to ensuring societal trust in scientific research and its outputs is maintained. The manner

in which EMBL carries out research and seeks to uphold principles of research accountability and transparency in science sends an important message to its member states and to society as a whole. While the IGO model allows a great degree of flexibility to research organisations, it is still paramount that IGOs themselves demonstrate a deep culture of research ethics and integrity of processes.

Administrative matters in IGOs and societal norms

Because of their legal status, IGOs often manage many administrative matters which would otherwise fall under the remit of national laws and infrastructures. This includes social issues such as pensions, medical insurances, staff development, leave policy, and more. This sometimes leads to differences between the employment conditions of employees of IGOs and those of non-IGO employees who live in the same country.

While it could be said that this discrepancy places IGOs outside wider societal norms, it is important to note that, as a consequence, IGOs can also find themselves in a position to react faster on certain issues. The IGO's administrative norms, therefore, approved through its multilateral instruments, are a reflection of that IGO's societal perspective.

11.2.4 Whose interpretation of norms prevails?

When working with and for many member states, all of whom have different historical, legal, and cultural approaches to said norms, an infinite number of approaches of interpretation or implementation can result on any given subject. In some cases, EMBL might not wish to take a particular position on an issue, but ultimately, the manner in which EMBL acts might result in it being viewed as a proxy for the laboratory's ethos and values. In this way, any organisational actions, or the voicing of any views or opinions externally, will effectively be perceived as the IGO position on the issue in question.

This is why, in the instances where there is no clear interpretation of which societal norms should prevail, EMBL has sought to take a more measured approach—such as facilitating dialogue on the ethical, legal, and social implications of its research through the Science & Society Programme [13], while remaining impartial throughout. This approach is also underpinned by other outreach and engagement initiatives such as the European Learning Laboratory for the Life Science [14] (ELLS), EMBL's science education department, which is open to European school science teachers and young people of all backgrounds aged between 10 and 19 years. The ELLS programmes cover complex, cutting-edge topics in life science research in an exciting and insightful way, fostering the discovery of current research trends, the scientific method, and scientific career paths.

Furthermore, EMBL (on a rotational basis with other EIROs) hosts the EIROforum *Science in School* online magazine to support teachers in delivering STEM curricula in order to foster a positive attitude towards science that attracts students to careers in these fields. Other informal learning activities invite people of all ages and backgrounds to connect with science and scientists and to be inspired by life science research. These include open days, visits to the EMBL site(s) and local science fairs. All these initiatives help further multi-disciplinary discussions on European life sciences.

However, it is clear that in some cases, specific societal issues require a proactive approach, demanding that an IGO raises its head above the parapet, and articulates its position clearly. These sorts of issues are often linked with prescribed regulatory or legal changes, such as essential due diligence processes in animal experimentation or human subject research, which at EMBL are implemented via the EMBL Bioethics Office.

A further example of particular relevance to the new frontiers of research being explored by scientists at EMBL is the implementation of the Nagoya Protocol, which concerns the fair and equitable sharing of benefits arising out of the utilisation of genetic resources, thus contributing to the conservation and sustainable use of biodiversity. In some cases, the EMBL position on issues is easily defined by its adherence to international directives. For instance, when EMBL states that it will adhere to a relevant EU Directive, it is a given it will remain within appropriate boundaries of that Directive, thus defining precisely its position on the topic.

It is the issues which fall outside of this category which are perhaps more interesting. It may be fair to say that, when there is no pre-defined institutional position, the prevailing views at a Big Science IGO such as EMBL tend to be simply those which align with the organisational core values. This is not to say that IGOs should not take a role as an expert stakeholder in shaping discussions about issues in the life sciences (such as recent discussions by the Convention on Biological Diversity, CBD, on regulating synthetic biology [15]). In the right circumstance, such positions are indeed crucial. Rather, it is submitted that there are relatively few occasions where IGOs are required to respond to changes in societal norms through internal policy changes or through external advocacy. When they do have such a duty, it is taken very seriously.

Interpretation of norms: key issues concerning Big Science at an IGO

- *Multilateralism* has huge benefits, but also means dealing with a plethora of different views from different collaborators.
- *Where no clear interpretation of norms prevail* an IGO like EMBL can act as a neutral facilitator and create a platform for dialogue and debate.
- *What kind of external change triggers this debate?* Legal, societal, technological and issues of *force majeure* may all trigger the conversation about an IGOs response to a change in societal norms. But does the organisation have a duty to respond?

11.2.5 When do IGOs have the duty to respond to changes in societal norms?

IGOs may prioritise, and thus may have a duty to respond to changes in societal norms that align with their core mission. Some possible issues with this framework are explored below through two landmark external events—the COVID-19 pandemic and social movements for equality (such as Black Lives Matter, Pride, disability awareness, women in science, and so on).

As the pandemic emerged, EMBL stepped into the fight against SARS-CoV-2 through its research, service provision, and training arms. Research into the virus by EMBL groups, for instance, has brought us new understanding of how the SARS-CoV-2 virus may protect itself from the immune system in the gut, while EMBL's training teams have organised a number of COVID-19 related events to mobilise the scientific community, including an emergency scientific conference in the summer of 2020 [16]. EMBL Services, moreover, were able to contribute in extremely significant ways to global progress: for instance, EMBL Data Services at EMBL-EBI were leaders in building the European COVID-19 Data Platform, bringing together relevant COVID-19 datasets and analysis tools, while EMBL Experimental Services were able to support BioNTech in the development of more effective ways to deliver the first COVID-19 mRNA vaccine.

Concurrently, the pandemic is also providing EMBL scientists with a platform to respond externally to other related issues—such as speaking out as expert stakeholders in an international public arena (on radio, TV, etc) [17], and providing a way to analyse and debate the societal implications of the pandemic through an Infectious Disease & Society [18] seminar series.

The question we wish to examine is whether EMBL has the duty to respond to the changes in societal norms that may have been precipitated by the pandemic. In this case, one could argue that EMBL did indeed have on obligation to do so, as the nature of the crisis was relevant to EMBL's core mission, and also provided a springboard for EMBL scientists to present information within their area of expertise to the public.

However, societal movements promoting equality in underserved communities might provide a less obvious impetus for an IGO response. Over the past few years, social justice movements such as the Black Lives Matter movement or the Pride movement for LGBTQA+ rights have become increasingly important players in society and its new norms. How would external movements like these impact a life sciences institute such as EMBL? EMBL's primary mission is to promote excellence in basic life science research: to achieve this goal, EMBL depends on hiring and training outstanding research, service, and administrative personnel. Ensuring an equality and diversity balance among EMBL personnel, and fostering awareness and skills relevant to its promotion, thus directly contributes to research excellence. Therefore, EMBL is committed to creating and maintaining a culture in which diversity and equality are both valued and promoted.

Changes in EMBL's governing documents and internal policies demonstrate that its approach to social issues has changed during the course of its existence, as it aligned itself more or less closely on certain issues with the world around it.

For example, EMBL recently set up a new governance structure for equality, diversity, and inclusion (EDI) initiatives in order to deliver and implement an EMBL-wide strategy, while creating mechanisms for institutional accountability. The EDI initiative is composed of three structures: the EDI Office, which coordinates activities around equality, diversity and inclusion; the EDI Forum, which brings together staff members from all sites; and the EDI Governing Body, chaired by the Chief Operating Officer and including senior scientists and leaders.

The principle of championing societal values, while at the same time acting as a role model for research organisations, links back to the question of whether IGOs should strive to be a 'force for good'. The fact that EMBL is ultimately funded by the European taxpayer reinforces the notion that it should work both with and for society. This involves ethically driven activities, such as setting standards at a European level, or establishing models of best practice for others to follow. In this case, it may actually be argued that, rather than being a passive observer or enforcer of societal norms, IGOs have the capacity to set norms, and not just reflect them (see the concept of 'teachers of norms' [19]). This idea of IGOs embracing a position as leaders or trendsetters within Big Science will be explored in the two case studies below.

Overall, it emerges clearly here that non-state actors such as EMBL have the ability to influence the process of normative change, and therefore also have a certain duty to respond to changes in society norms in some circumstances. The conduct of IGOs cannot be considered in isolation, however, even if the duty to respond to norms can be interpreted in the limited way described above.

11.3 Case studies: Big Science at EMBL

As discussed in the previous section, IGOs such as EMBL have both the ability and, at times, the duty to interact with changing societal norms, in particular when such norms concern areas which fall within the remit or missions of the IGO itself.

In this section, we will examine two case studies which exemplify how EMBL responds to changes in societal norms within the life sciences, both as a leader and as an implementer. First, we discuss how EMBL's role in providing scientific services puts it in a position to influence society directly in impactful and far-reaching ways, including through genomic medicine and the development of better crops. Second, we will illustrate how EMBL can be part of a global movement for societal change through the example of open science, a set of values that aligns with EMBL's core missions and which EMBL is committed to advancing. These case studies will provide specific examples of how an ethical IGO can act as a 'force for good' within specific communities, and in society in general.

11.3.1 EMBL's scientific services

EMBL benefits from a multi-faceted position as a research institute, a scientific services provider, and a scientific training coordinator. The joint provision of scientific services and research is a distinctive feature of EMBL, as both activities benefit from constant co-operation and cross-talk with each other. EMBL's scientific services are crucial to life sciences research: bioinformatics services at EMBL-EBI receive millions of requests every day, while users access EMBL's beamline, cryo-EM, and light microscopy service facilities thousands of times a year.

These services are intended to provide world-class scientific support to the research communities they serve, across all EMBL member states and beyond. EMBL Service provision revolves around three key principles: accessible infrastructure, expertise, and quality assurance. Scientific services have a direct effect on society, since they enable and accelerate scientific discoveries which, in turn, have a profound impact on everyday life.

One example of the way scientific services have a huge impact on society, is the annotation and curation of a human reference genome. A reference genome is a database for genomic sequences which serves as a representative example of the genome of one idealised individual within a species. Ensembl [20], an EMBL-EBI data resource, hosts and annotates the human reference genome, in collaboration with the Human Pangenome Reference Consortium (HPRC [21]). Human reference genomes are of great importance in medical research, as they enable scientists to study how an individual's genetics can be linked to disease. Genomic medicine uses information about a patient's genome for diagnosis or to make decisions about clinical care. It is becoming increasingly important, in particular in the fields of oncology and rare diseases.

EMBL-EBI, together with the Center for Genomic Regulation (CRG), hosts and operates the European Genome-phenome Archive (EGA) [22], a service for permanent archiving and sharing of all types of genetic and phenotypic data resulting from biomedical research projects. Research enabled by the EGA is allowing researchers to understand the connections between an individual's genetics and their risk of developing cancer or heart disease, for example, and is shedding light on the mechanisms that underlie rare diseases. As a result of human genome annotation and curation, some established disease variants can now be used in making clinical decisions for healthy individuals to prevent disease, as well as informing patient treatment.

Genome annotation can also be exceptionally influential when it is used to produce better crops. The challenges of climate change and disease epidemics, together with the need to sustain a growing global population are key factors in driving the need to develop more efficient, more resistant crops. Research which aims to gain a full understanding of the biology and of the genome of crop plants is at the heart of these efforts. This work is enabled by plant reference genomes stored in databases such as Ensembl Plants [23] at EMBL-EBI.

An excellent example of the impact of annotated crop genome is the recent sequencing and annotation of the genome of bread wheat (*Triticum aestivum* L.). Over two billion people worldwide rely on wheat as a staple, and bread wheat is the most widely cultivated cereal in the world, making a reference genome for wheat hugely relevant for large parts of society. The wheat reference genome is enabling researchers to perform new wheat analyses and therefore to develop new varieties of wheat, supporting breeding programmes that aim to improve yields and increase resistance to pests and climate challenges.

While EMBL services contribute directly to society by accelerating world-changing science, they also support the development of robust experimental and data standards within the research community. The implementation of these standards ensures that data produced, managed, and accessed through EMBL services are of high-quality and consistent. Moreover, data standards shape the scientific communities which access EMBL services by informing their research and publishing practices and thus acting as *de facto* societal norms within these communities. The close connection between scientific services and data standards is made especially evident in new and emerging data types (such as biological imaging data [24] or polygenic risk scores [25]), which require both new infrastructure to collect data and support growing data deposition, and internationally recognised standards to ensure that the deposited data are impactful.

Through its scientific services, EMBL therefore drives societal norms by enabling research which shapes the future of society, and by setting data standards that regulate the research community. In the next section, we will examine how EMBL responds to wider societal norms through the case study of open science.

11.3.2 Open science: participating in a global movement

Open science is a movement which was first defined in its modern form in the late 1990s, and which has become increasingly prevalent in the scientific community. The idea behind open science is to make the methods and results of scientific research accessible to all—within and beyond academia. One may argue that open science can at this stage be considered a societal norm within the broader scientific community.

EMBL's Establishing Agreement sets out that 'results of the experimental and theoretical work of the Laboratory shall be published or otherwise made generally available'. It is therefore reasonable to suggest that EMBL has been aligned with the values of open science from its inception. Since then, EMBL has worked to implement its commitment to open science in its research and service practices, acting in unison with the broader scientific community. One self-evident example of EMBL's commitment to open science is its data resources, which are operated through EMBL Data Services at EMBL-EBI. EMBL-EBI data resources are intrinsically open and free for everyone to use. EMBL's work during the pandemic, in particular through the Covid-19 Data Platform, has further highlighted the importance not only of freely accessible data and methods, but also of the importance of international coordination in order to ensure that scientific research is truly open.

One of the key ways in which EMBL is furthering the advancement of open science, is by working to implement the FAIR data principles through its research practices and service provision. FAIR principles[3] state that data must be findable,

[3] https://www.go-fair.org/fair-principles/

accessible, interoperable, and reusable: FAIR data are not only open, but can also maximise the impact of the research that produced it by informing the work and practices of other researchers globally. The work of EMBL Scientific Services is one example of how a Big Science organisation can efficiently advance the adoption of FAIR data standards. Experimental Services at EMBL endeavour to adhere to all FAIR Data principles: this allows data generated by EMBL Services to benefit the entire research community, furthering its impact to all member states and beyond. These principles apply equally to EMBL Data Services, which are not only open, but are being constantly developed to improve the findability, accessibility, interoperability, and reusability of the data hosted.

Furthermore, EMBL is a signatory of the San Francisco Declaration on Research Assessment (DORA), which aims to improve the ways in which researchers and the outputs of scholarly research are evaluated. This includes recognising the value of a broad range of scientific outputs (including research articles, data, and software), assessing research deliverables and other research contributions based on their intrinsic merit, and discouraging the inappropriate use of proxies or metrics, such as the title or impact factor of the journal in which the work was published.

Respect for the social norms of open science and fair research assessment and efforts to drive these values forward epitomise the EMBL mission and ethos. EMBL's IGO status allows it to act freely from local constraints, demonstrating how a large life sciences research institution can conduct itself with integrity and be a powerful actor for positive change.

11.4 Conclusion

EMBL relies on Big Science concepts to deliver on its missions and goals. As a life sciences IGO, it has its own particular legal and regulatory frameworks (often adhering to established legal instruments, such as European Directives). It does not, therefore, *have* to respond to many of the external societal changes outlined above. However, in some cases, we have argued that EMBL could have an explicit duty to respond, for example, when acting in accordance with the organisational missions.

Nevertheless, there are examples where EMBL does not have a duty, *per se*, to respond to changes in societal norms, but *chooses* to do so—thereby contributing to its role as leader in the European life sciences, as well as cementing its role as a standard-setter and champion of best practice. In this way, it can be argued that IGOs such as EMBL have the power to both contribute to and to drive certain changes in societal norms. This is also true of the collaborations forged by or with EMBL, and by its commitment to outreach, which in itself helps create a more accepted culture of bilateral societal engagement among other IGOs.

The fundamental benefit of an IGO status remains its multilateralism—it is collaborative and innovative, and it democratises science. Although IGOs and the research they contribute represent just one facet of responses to, or interpretations of changing debates in the world, they provide a unique nexus of views and interpretations by nature of their global contributors and vast scientific networks (for example, the federated EGA [26], or rolling out of the COVID-19 platform, which was facilitated by the considerable groundwork laid in other research areas [27]).

It remains to be seen how future landmark changes in societal norms will impact IGOs, but it is envisaged that EMBL will continue both to respond to those challenges effectively, and at the same time to set new norms in the European life sciences—thus taking a greater role in becoming a proactive 'teacher of norms' rather than merely providing a reflection of them.

References

[1] United Nations 1973 Agreement establishing the European Molecular Biology Laboratory, Geneva, CH, 10 May 1973 (EIF 4 July 1974) *United Nations Treaty Series* **954** (353) article 2

[2] EMBL Member states https://www.embl.org/about/member-states/

[3] EMBL Council https://www.embl.org/about/info/embl-council/

[4] EMBL Scientific Advisory Committee https://www.embl.org/about/info/embl-council/scientific-advisory-committee/

[5] EMBL Collaborations https://www.embl.org/about/info/science-society/collaborations/

[6] https://intranet.ebi.ac.uk/sites/intranet.ebi.ac.uk/files/tsc/external_relations/embl-ebis_public_engagement_strategy_2020–2025.pdf

[7] EMBL 'The World of Molecular Biology' exhibition https://www.embl.org/news/lab-matters/the-world-of-molecular-biology-exhibition/

[8] About EMBLEM https://embl-em.de/company/about-emblem/

[9] EMBL Industry Partnerships https://www.ebi.ac.uk/industry/industry-programme/

[10] Technopolis group https://www.technopolis-group.com

[11] Babb R *et al* 2021 Review of the impacts of EMBL experimental services *Technopolis Group* https://www.embl.org/documents/wp-content/uploads/2021/11/technopolis-group-experimental-services-impact-report.pdf

[12] Beagrie N and Houghton N 2021 Data-driven discovery: the value and impact of EMBL-EBI managed data resources *Charles Beagrie* https://www.embl.org/documents/wp-content/uploads/2021/10/EMBL-EBI-impact-report-2021.pdf

[13] About EMBL https://www.embl.de/aboutus/science_society/index.html

[14] EMBL Science education https://www.embl.org/ells/about/

[15] Biosafety Clearing-House Peer review of the technical series on synthetic biology http://bch.cbd.int/synbio/peer_review/

[16] EMBL Events https://www.embl.de/training/events/2020/EID20–01/

[17] McKie R 2020 What is the new Covid strain—and will vaccines work against it? *The Guardian Online* 19 December https://www.theguardian.com/world/2020/dec/19/what-is-the-new-covid-strain-and-will-vaccines-work-against-it

Ansede M 2021 'We will be dying from antibiotic-resistant bacterial infections in a decade' *El País* 25 June https://english.elpais.com/usa/2021-06-25/we-will-be-dying-from-antibiotic-resistant-bacterial-infections-in-a-decade.html

Cherki M 2021 Une nouvelle mutation inquiétante détectée sur le variant britannique *Le Figaro* 3 February https://www.lefigaro.fr/sciences/une-nouvelle-mutation-inquietante-detectee-sur-le-variant-britannique-20210203

[18] About EMBL https://www.embl.de/aboutus/science_society/infectious-disease-and-society/index.html

[19] Finnemore M 1993 International organizations as teachers of norms: the United Nations educational, scientific, and cultural organisation and science policy *Int. Organ.* **47** 565–97 www.jstor.org/stable/2706939
[20] Ensembl http://www.ensembl.org/Homo_sapiens/Info/Index
[21] Human Pangenome https://humanpangenome.org/
[22] European Genome-phenome Archive https://ega-archive.org/
[23] Ensembl Plants https://plants.ensembl.org/index.html
[24] Stroe O 2021 Metadata to make microscopy images more open *EMBL-EBI Announcements* 21 May https://www.ebi.ac.uk/about/news/announcements/REMBI-launch
[25] Stroe O 2021 Getting polygenic risk scores into the clinic *EMBL-EBI Announcements* 10 March https://www.ebi.ac.uk/about/news/press-releases/PGS-Catalog
[26] Federated EGA https://ega-archive.org/federated
[27] Cantelli G *et al* The European Bioinformatics Institute: empowering cooperation in response to a global health crisis *Nucleic Acids Res.* **49** D29–D37 https://doi.org/10.1093/nar/gkaa1077

Part II

Economic and innovation impact of Big Science projects [Editor: Jason Li-Ying]

IOP Publishing

Big Science in the 21st Century
Economic and societal impacts
Panagiotis Charitos, Theodore Arabatzis, Harry Cliff, Günther Dissertori, Juliette Forneris and Jason Li-Ying

Chapter 12

Societal impact of Big Science organizations—a multifaced phenomenon

Jason Li-Ying

The impact of public research on the economy and society has long been a central concern for scholars, managers, and policymakers, and the research about the distinct effects of Big Science organizations (BSOs) has prevailed during the last three decades (Cohen *et al* 2002, Kokko *et al* 2015, Maroto *et al* 2016, Mazzucato 2013). BSOs are large research organizations established purposefully to address fundamental and complex scientific research challenges that cannot be addressed in isolation by individual universities, research institutes, or even government agencies. Unlike universities and other national research institutes, BSOs are unique scientific organizations by virtue of their sheer size, level of complexity, and uncertainty with respect to the outcomes of research and development. Several examples of BSOs, such as the Stanford Linear Accelerator (SLAC), the European Organization for Nuclear Research (CERN), European XFEL, MAX IV, the Francis Crick Institute, NASA Jet Propulsion Laboratory, Culman Centre for Fusion Energy, etc, have shown their importance in technological breakthroughs and advancement of scientific knowledge. Moreover, BSOs involve large networks of suppliers and collaborators in science, government, and business, constituting a complex system with permeable boundaries that offer opportunities for technology transfer, knowledge accumulation, and business creation (Link *et al* 2007). Hence, BSOs are influential players within complex systems of innovation, learning, and business creation. Despite their important role for national and international economies as well as society at large, our current understanding of their management and impact is underdeveloped in both theory and practice.

Over the past two decades, there has been increasing interest in the impact of BSOs on technological innovation, organizational learning, the organization of complex large-scale projects, collaborative innovation processes with suppliers, and other spillover effects (Autio *et al* 2003, Autio *et al* 2004, Vuola and Hameri 2006,

Autio 2014, Tuertscher *et al* 2014). Most recently, research has paid particular attention to the challenges and opportunities of BSOs with regard to project management (Schimel and Keller 2015), governance (Smart *et al* 2012), and data management (Bicarregui *et al* 2015) within BSOs. However, we still lack knowledge about what exactly the impacts of BSOs are, how these impacts take place following what kinds of fundamental mechanisms, and how to assess them in a context of changing economic, technological, and societal environments that arise in the broader ecosystem surrounding BSOs. Therefore, we called for submissions to part II of this book to present the multifaceted impact of BSOs and various approaches to assess the impact.

Investments into BSOs are major, long-term commitments of government funds and political attention to technologies with substantial promise but also uncertainties. Given the scale of the investments and the complex coordination tasks across countries, institutions, and supply chains, BSOs are more than just technology producers. Instead, they have substantial effects on their partners by providing new technological infrastructures, on their regions by agglomerating skilled individuals, and on their supply chains by creating demand for advanced inputs. It is important for researchers and policymakers to have an ecosystem view on the multifaceted impact of BSOs and explore interactions with all types of stakeholders to share knowledge, to create mutual stimuli, to leverage complementary assets, to define common goals and diversified means to create value, and in turn, to make positive societal impact. BSOs not only advance science and develop technologies, but also provide an organizational context for processing innovations, a risk-taking culture, market orientation, and innovation capacity for the BSO and with stakeholders in its ecosystem.

The work included in part II of this book offers some new insights regarding various aspects of impacts of BSOs. Each of these articles focuses on a specific aspect of the impact of BSOs. In chapter 13, Egleton from Sussex University contributes a work that examines the various tools for evaluating the impact of Big Science laboratories/research infrastructures by reviewing the various tools and metrics for evaluating Big Science/research infrastructures in terms of scientific impact, economic impact, and societal impact. In chapter 14, Crescenzi and Piazza provide a critical overview of the available methods to estimate the local socio-economic impact of RIs and discuss the transmission mechanisms and the channels through which RIs activities can affect local economies. The focus is mainly on procurement as this is the closest interaction between research and industry. Chapter 15, authored by Vignetti, discusses the advantages in having socio-economic impact assessment considerations built-into the design of a research infrastructure to maximize the delivery of socio-economic impact and their long-term sustainability. The author argues that not only can socio-economic impact assessment help scientific communities and governments in jointly thinking about how best to make use of scarce resources but also to reconceptualise the relationship between science and the public, fuelling a transparent and interactive process through which societal actors, researchers, and governments become mutually responsive to each other. In chapter 16, Bastianin and Florio survey models and methods used to assess the economic and societal impacts of Large Scale Research Infrastructures (LSRIs) and provide a taxonomy of existing frameworks of analysis and then review the methodologies used to quantitatively assess the impacts of LSRIs. They also survey the literature that provides quantitative and qualitative evidence on the effects of LSRIs and discuss methodological policy implications arising from such studies. Chapter 17, authored by Li-Ying and his co-authors, showcases the theoretical foundation and methodological approach used by European Spallation Source (ESS) to collect early evidence and some important aspects of socio-economic impact of ESS based on the narratives and impact pathways that are specifically relevant for the construction phase of ESS. In chapter 18, Keinz and his co-authors discuss the technological competence leveraging (TCL) method as a means to successfully master the fuzzy front-end of KT processes in Big Science projects. It investigates the short- and long-term promises of the TCL method and elaborates on its limitations. In chapter 19, Larner from STFC in the UK explores the research infrastructure system for fundamental research of the UK with specific attention on the Big Science used to study physics and related phenomena, and the methods used by the UK public sector to evaluate the impact of this investment for society. In chapter 20, Nielsen, Gustafsson, and Lauritsen introduce the development and operation of the network of European Space Agency Business Incubation Centres, ESA BICs, which has been established by the European Space Agency to promote the commercialization of space through support for early-stage start-ups. Several ESA BIC startup cases with societal impact are presented and the overall socio-economic impact of the ESA BIC network is also discussed.

We hope the chapters in part II will provide the readers with a comprehensive overview regarding multifaceted impact of BSOs and help policymakers to capture a full picture of the fundamental issues in this regard.

References

Autio E 2014 Innovation from big science: enhancing big science impact agenda *Report* UK Department of Business Innovation and Skills https://core.ac.uk/download/pdf/19888222.pdf

Autio E, Bianchi-Streit M and Hameri A-P 2003 *Technology Transfer and Technological Learning Through CERN's Procurement Activity* (Geneva: CERN Scientific Information Service)

Autio E, Hameri A-P and Vuola O 2004 A framework of industrial knowledge spillovers in big-science centers *Res. Pol.* **33** 107–26

Bicarregui J, Matthews B and Schluenzen F 2015 PaNdata: open data infrastructure for photon and neutron sources *Synchrotron Radiat. News* **28** 30–5

Cohen W M, Nelson R R and Walsh J P 2002 Links and impacts: the influence of public research on industrial R&D *Manage. Sci.* **48** 1–23

Kokko A, Tingvall P G and Videnord J 2015 The growth effects of R&D spending in the EU: a meta-analysis *Economics* **9** 1–20

Link A N, Siegel D S and Bozeman B 2007 An empirical analysis of the propensity of academics to engage in informal university technology transfer *Ind. Corp. Chang.* **16** 641–55

Maroto A, Gallego J and Rubalcaba L 2016 Publicly funded R&D for public sector performance and efficiency: evidence from Europe *R&D Manag.* **46** 564–78

Mazzucato M 2013 *The Entrepreneurial State. Debunking Public vs Private Sector Myths* (London: Anthem)

Schimel D and Keller M 2015 Big questions, big science: meeting the challenges of global ecology *Oecologia* **177** 925–34

Smart J *et al* 2012 Big Science and big administration: confronting the governance, financial and legal challenges of FutureICT *Eur. Phys. J. Spec. Top.* **214** 635–66

Tuertscher P, Garud R and Kumaraswamy A *et al* 2014 Justification and interlaced knowledge at ATLAS, CERN *Org. Sci.* **25** 1579–608

Vuola O and Hameri A-P 2006 Mutually benefiting joint innovation process between industry and big-science *Technovation* **26** 3–12

IOP Publishing

Big Science in the 21st Century
Economic and societal impacts
Panagiotis Charitos, Theodore Arabatzis, Harry Cliff, Günther Dissertori, Juliette Forneris and Jason Li-Ying

Chapter 13

Evaluating the impact of Big Science/research infrastructures

David Eggleton

This chapter examines the potential impacts of Big Science laboratories/research infrastructures. It will review the various tools for evaluating Big Science/research infrastructures in terms of economic impact, scientific impact, and societal impact. There is a shift currently in progress from measuring outputs towards achieving outcomes which is addressed as a conclusion.

13.1 Introduction

Since the evolution of the modern Big Science laboratory in the twentieth century (Heilbron *et al* 1981, Hermann *et al* 1987, Seidel 1992), they have consumed a disproportionate amount of public funds relative to their numbers (Webre 1988, Lefevre 2009). Modern day Big Science often requires significant investments from one or more national governments. It is therefore not surprising that such stakeholders will ask what the return is on these investments so that future Big Science projects can be justified to taxpayers. From the perspective of scientists, the return on investment is in terms of new understandings of fundamental phenomena; yet the traditional answer to such stakeholder queries is usually articulated in terms of economic impact (Autio *et al* 1996, 2004, Scarrà and Piccaluga 2020). This chapter seeks to outline the breadth of Big Science impacts beyond these traditional economic measurements. It begins by looking at the traditional methods for capturing the economic return on Big Science before looking at how the academic literature articulates how to measure the scientific impact and productivity of Big Science before looking at its broader societal impacts. However, it is broadly acknowledged that direct measurement of the effects of Big Science is not feasible, and most metrics capture the by-products of knowledge production (Cronin and

doi:10.1088/978-0-7503-3631-4ch13

Sugimoto 2014). As will become apparent in this chapter, there has been a transition from more traditional measures associated with trying to capture the by-products of knowledge production towards much broader metrics that seek to quantify the outcomes that accrue to society.

13.2 Economic impact

As noted above, one of the most frequently used justifications for Big Science is to emphasise the economic benefits. These include direct economic benefits as a result of concentrating spending within the geographic region. These benefits include direct spending on construction as well as the clustering effect from the demand for high technology components (Simmonds *et al* 2013). Much of the literature examining this topic focusses on the benefits of publicly funded basic research in a generalised sense (Salter and Martin 2001). However, there are still some pockets of literature specifically examining the economic benefits of Big Science which are examined below.

The most traditional method for quantifying the economic return on Big Science investment has generally been to look at patents. Using the somewhat simplistic linear model framework for understanding the return on investment from Big Science, investments in such basic research generate new knowledge that are pulled by the free market through to new products (Godin 2006). During this process the private sector patents this knowledge, so it was traditionally used by governments as a way of understanding how investments were converted to economic growth. There is some evidence that the demand created by Big Science projects can lead to new innovations, particularly in terms of new process innovations. However, while product innovations do occur, they are considered to be relatively less numerous compared to process innovations (Florio *et al* 2018). Certain authors have sought to capture the economic impact of procurement for Big Science projects by performing patent analysis (Castelnovo *et al* 2018), which unfortunately was unable to capture these process innovations. Therefore, solely using patents which primarily capture product innovations, may be inadequate for understanding the relationship between Big Science projects and economic growth.

Patent analysis is not the only way to understand the economic impact of Big Science. The laboratory itself can be a major source of demand for high technology products through its procurement policy. One of the hallmarks of Big Science is that many of the experiments can be considered instances of 'super high' technology whereby the technology is discovered in the context of application (Shenhar 1993, Shenhar and Dvir 1996). Such technologies can cause substantial changes across the entire economy. The World Wide Web (WWW), originally developed at CERN as a way of synthesising several pre-existing technologies to identify and access information held on different computers (Berners-Lee 1989), is perhaps the most famous example of such a spillover technology. Such spillovers can occur at any stage in the project and have been categorised into three components: market spillover, where enterprises are pulled into the direct supply chain for the project; network spillover, where the technology diffuses into other economic sectors; and knowledge spillover, where human capital diffuses into the wider economy (Choi *et al* 2017).

Market spillover is perhaps the most obvious economic impact that can be generated by Big Science projects. In the short-term, Big Science projects naturally create local demand for components for the apparatus. For example, in the cases of ITER and the LHC, these major projects incentivised industry to create new production lines for high tech components in the medium term. It also provided some companies with the platform to stand out in the market and create a long-term comparative advantage over competitors as their greater capacity allowed them to benefit from economies of scale. Big Science laboratories appear to be more willing to give external companies more time to experiment and refine new technologies or to even work with them compared to industrial customers (Autio *et al* 2004). Given that many Big Science projects could be using a high or 'super high' level of technological uncertainty, having a known customer base for emerging technologies may help to pull them across the gap between the laboratory bench into the market.

In the case of network spillover, the wider economy can benefit from the demand temporarily created by the construction phase of Big Science projects. Often by design, Big Science projects require significant volumes of high technology components with low tolerances for error that can incentivise industry to create new production lines to satisfy this demand (Autio 2014). This leads to two developments. First, as economies of scale mean that the unit cost of such components decreases, other industries may find lower barriers to entry for developing new products using these components (Castelnovo *et al* 2018). Second, the laboratory helps to create the specifications for the new components that become the standard for future industry (Autio 2014). This has been demonstrated to some degree through the temporary demand and specifications created for niobium-titanium wire for the Tevatron in the 1980s; this helped to make MRI machines technically and financially viable for healthcare (Hesla 2017).

Lastly, in the case of knowledge spillover, this is largely characteristic of the highly educated workforce that characterises Big Science laboratories. This has traditionally been described as the spillover of ideas created in the basic sciences whereby tools and techniques developed in the laboratory can diffuse into the wider economy through industrial participation in Big Science projects and research programmes (Choi *et al* 2017). However, this may only capture part of the reality as there will also likely be human capital exchanges as a community of practice develops and some workers move from the laboratory into industry and vice versa.

These economic impacts may be measurable at all stages of a Big Science project's lifecycle rather than being restricted to after first physics (Autio *et al* 2004, Bianco *et al* 2017). However, most of the research outlined in this chapter does not capture intangible benefits that may accrue to enterprises linked to Big Science projects. This represents something of an opportunity for future research. Comparatively modest enterprises could benefit from the reputational halo effect of being a contractor in a Big Science project. While such benefits may not materialise in either the short or medium term, they could be considered as preparatory work for the future (Shenhar *et al* 2001).

13.3 Scientific impact

This section briefly examines the scientific impacts of Big Science projects. Naturally the scientist may view the production of new knowledge or understandings of nature as the principal benefit of Big Science. Yet the scientific impacts of Big Science generally cannot capture the intangible benefit of this and tend to focus on publications. The traditional measurements for scientific productivity and impact depend on the unit of analysis, if the individual scientist is the unit of analysis metrics such as the h-index (Kumar 2009, da Silva *et al* 2018), journal impact factor (IF) (Archambault and Larivière 2009), and author-level eigenfactor can be used (West *et al* 2013); these all seek to quantify an individual scientist's productivity and centrality to the field[1]. Classically one might judge the impact of research to be best judged through the simple measure of citations on the basis that the more citations to an article, the more important the contribution to the field (Kumar 2009). But this type of metric quantifies the scientific impact of the experiments conducted with the apparatus rather than of the apparatus directly. Equally these metrics can only judge the value of the apparatus after project completion, this reality means that scientists can use the sunk cost fallacy to their advantage. Equally these metrics tend to privilege new apparatus that can quickly make new discoveries over mature apparatus that may still be producing valuable results (Bianco *et al* 2017). As an example, the Tevatron was closed relatively soon after the LHC achieved first physics, despite its experiments continuing to produce valuable data (Oddone 2011a).

While the scientific impacts of Big Science may be obvious to practitioners who work on or with the apparatus, external stakeholders such as funders or member state governments may no longer perceive those benefits quite so readily as the scientific community. It is likely the scientific community will use practical or technical parameters to judge the impact of a completed project (Hallonsten 2016). In the case of high energy physics these technical parameters will be the obvious dimensions of beam energy or luminosity or slightly less obvious technical parameters such as machine reliability or the ability to access the scarce resource of beamtime (Hallonsten 2016). There may be a tradeoff to be made between using sufficiently advanced technology to achieve new experimental possibilities and machine reliability (Hoddeson and Kolb 2003, Hoddeson *et al* 2008). This may not be the tradeoff that is expected with some scientists believing that a reliable machine may be overengineered and therefore too expensive (Hoddeson and Kolb 2003, Hoddeson *et al* 2008). Scientists may also articulate the ideal of collaborating to make new discoveries, there can also be an element of 'collaborative competition' between laboratories to be able to lay claim to the 'energy frontier' (Lederman 1983). Equally there is the intangible impact generated by making new discoveries

[1] However, there is a level of controversy over the use of research metrics at the individual level beyond simple curiosity (Else 2021). Although they may present the appearance of being useful for determining career progression rather than just as a novelty, several authors have made it clear that they are intended for comparability at the institutional or national level only in line with the Leiden Manifesto for responsible research metrics (Else 2021, Hicks *et al* 2015).

which are extremely important to advancing the scientific field (Abe *et al* 1995, Aad *et al* 2012). However, these are relatively difficult concepts to communicate to a government and the general public. Even if they could be communicated in such ways, funders may not necessarily appreciate these impacts in quite the same way as the scientific community.

Qiao *et al* (2016) classified the scientific effects of Big Science into four categories, which are how the apparatus advances the field which has already been addressed above, how the laboratory develops the organisation and trains future scientists, how the laboratory can act as a networking hub for building relationships with other scientists, and for linking with other industries to create a knowledge ecosystem.

First, in terms of the effect of Big Science projects on the development of human capital, namely future scientists, these effects have been well documented in the literature and are intrinsically linked to the exceptional intellectual environment that characterises Big Science laboratories (GSF 2014). While the traditional and usually false impression of science as the product of a single individual's toil is relatively rare, with many collaborations composed of hundreds or even thousands of scientists, it does offer the opportunity for greater networking (Blankenship 1974, D'Ippolito and Rüling 2019). The evolution of modern Big Science has had some impact on the nature of training for future scientists. Historically, experiments and accelerators were sufficiently small and on a short enough timescale that a scientist could obtain significant experience of all experimental phases from initiation to completion. The modern nature of Big Science, with very long experimental lifetimes, now means that the type of training is much more specialised but is tied to these lifecycles (GSF 2014). For example, a scientist trained during an experiment's construction phase will receive a very different training compared to a scientist trained during data analysis. Second, Big Science laboratories act as central nodes within a wide network of associated universities, private research institutions, and other private sector organisations (GSF 2014, Qiao *et al* 2016). This can create a clustering effect in the local geography as wider agglomerations develop (GSF 2014, Qiao *et al* 2016). This allows the laboratory to act as a focal point for the community as part of a local 'triple helix' system of universities, industry, and government which can support the circulation of talent within the local knowledge ecosystem (Etzkowitz and Leydesdorff 2000). Third and finally, the laboratory also acts as a networking hub for building relationships between stakeholders (Qiao *et al* 2016). While this is linked to the clustering identified above, the Big Science laboratory acts as a non-human artefact for focussing human minds from a variety of different stakeholder groups. This is particularly true during the construction of a new piece of apparatus (Qiao *et al* 2016).

13.4 Societal impact

The societal impact of Big Science is a somewhat contested concept in the literature; while some authors have linked Big Science to the development of human capital and new innovations that can benefit society (Magazinik *et al* 2019), others have said that the evidence is less clear cut, in particular when the scope of analysis is limited to the local level (Gastrow and Oppelt 2018). Some authors have examined the role of using science to address societal needs (Ciarli and Ràfols 2019). Given its fundamentally basic nature, it is not surprising that there is relatively little literature focussing on the use of Big Science laboratories for addressing societal needs. Although societal impact may not be the primary mission of the laboratory, there is evidence of the impact that Big Science laboratories have on society, this is addressed in this section.

Giudice (2012) examined the potential benefits to society of Big Science, and classified the societal benefits according to (i) the benefits created by clustering significant financial and intellectual resources together, (ii) the spinouts from technological developments during the R&D phase of the project, (iii) the close relationship between the laboratory and the private sector for component manufacturing necessitating new manufacturing techniques, (iv) the opportunity for countries to participate in scientific experiments where they lack the resources to do it alone and for the Big Science project to be used as a vehicle for promoting international links—often referred to as 'science diplomacy' (Ruffini 2017, Robinson 2019), (v) the training opportunities for early career researchers, and (vi) the intrinsic benefits of making new discoveries and uncovering new knowledge. Many of these benefits that Giudice (2012) regarded as societal have already been examined above, indicating that the societal benefits of Big Science can be conflated with economic or scientific benefits. Nonetheless, this can be considered a framework for future work.

One of the most obvious societal benefits from investment in Big Science surrounds the development of human capital. As noted above, the laboratory lies at one focus of a scientific ecosystem, with a wide variety of users at all career stages often based at academic institutions who visit to conduct experiments (Leon and Elias 2010, Qiao *et al* 2016). Each one of those users are technically competent individuals in their area of expertise trained at their home institution to be able to work on the machine and to contribute to the scientific method. As such it can be argued that the Big Science project acts as a hub for the development of social capital across the community and for the mass transfer of tacit knowledge between early career researchers and more experienced individuals (Hislop 2013, Qiao *et al* 2016).

One of the most obvious benefits that has accrued to society as a result of high energy physics has been the discovery of new radiation-based medical techniques. Such medical applications of Big Science have been explored since its beginning in the 1930s. The link between what were then the nascent fields of nuclear physics and nuclear medicine can be best exemplified by the early links between the Berkeley Radiation Laboratory and the Donner Laboratory. While Ernest Lawrence's Radiation Laboratory sought to understand the inner workings of the atom, John Lawrence's[2] Donner Laboratory aimed to direct these high energy particle beams to understand human physiology and to treat clinical conditions (Williams 1999). The first use of a radioactive isotope to treat human disease took place in 1936 where

[2] John Lawrence and Ernest Lawrence were in fact brothers, which demonstrates the synergy between Big Science and its societal benefits.

radioactive phosphorous-32 created in the Berkeley cyclotron was used to treat a patient with leukaemia (Williams 1999). This synergy between high energy physics and nuclear medicine continued such that in the 1970s, a collaboration between Fermi National Accelerator Laboratory and a private medical provider allowed patients to receive treatments using the laboratory's linear accelerator's neutron beam capabilities (Awschalom *et al* 1979). Even hadron and anti-hadron therapies were first developed using adapted linear accelerators from high energy physics (Dosanjh and Bernier 2018).

The societal impact of Big Science also includes its contribution to human development. While most Big Science laboratories tend to be constructed in more developed countries, the case of the Square Kilometre Array (SKA) offers an interesting example of using a Big Science project as a means to develop both regional infrastructure and human capital. The central mission of the SKA was to construct a radio telescope with a collecting area equivalent to one square kilometre. However, one of the challenges with radiotelescopes is that the cost to build a single very large radio telescope is much greater than an array of many small telescopes (Dewdney *et al* 2009). Two sites have been selected for construction of these radiotelescopes—the first being in Western Australia and the second being in an area not traditionally known for radio astronomy, namely the Karoo Desert in South Africa (Walker 2019). While the benefits that have been created as a result of significant investments in that region of South Africa are relatively traditional and could be covered in the economic benefits section above, a new phenomenon was identified which unusually led to a broadening of SKA's mission. This was the massive development of human capital in South Africa in both direct and indirect forms. Essentially it was the creation of a network spillover effect and its direction to develop an entire region throughout the process and entire economic sectors, creating a new dynamic in Big Science as a platform for development. Some of those who worked on SKA described it as if someone had taken a high-pressure bottle and removed the cap—all of the human potential was there, and the project served to focus that potential and provide it with an outlet (Dewdney 2019). This was also investigated by Gastrow and Oppelt (2018), who confirmed that there are benefits to the national innovation system and human development at the national level. However, the benefits at the local level in the Karoo could not be identified as there was no absorptive capacity for the local economy to engage with the advanced technologies deployed on-site. In fact, Gastrow and Oppelt (2018) argued that there was evidence that the construction of SKA in the area had a short-term detrimental impact as Internet infrastructure was constrained. The local effect of SKA was subsequently investigated by Gastrow and Oppelt (2019) who noted that there was a gap between how local communities hoped SKA would benefit them and what benefits SKA could actually deliver while staying within its project scope.

However, quantifying the scale of the benefits that accrue to society may require very different metrics drawing on interdisciplinary forms of expertise. For example, how one measures the societal benefit as a result of these new medical technologies could be captured in terms of the additional years of life as a result of treatment adjusted for the quality of that extension (Vergel and Sculpher 2008,

Devleesschauwer *et al* 2014), but obviously one does not necessarily need to quantify the benefits to the individual from a significantly longer or improved lifespan which may not be quantifiable.

The notion that the impact of Big Science cannot be captured by any single one of these dimensions or indicators is not an innovation but has existed since the 1980s (Martin and Irvine 1983, Leydesdorff 2005). Instead, the academic literature has suggested that these be used in combination as partial indicators, in line with other domains that emerged during this time period (Martin and Irvine 1983), but this itself may be inadequate for capturing the real return on investment that can be easily articulated. As will be seen below, there is a trend in progress away from indicators towards using scientific investments to achieve certain outcomes.

13.5 New dynamics in evaluation

Over the past ten years, a new dynamic in research evaluation has emerged that is still being rolled out across government portfolios. So far, the evaluation methods examined for this chapter have tended to focus on specific outputs from Big Science projects, whether in terms of scientific or economic outputs. However, governments have shifted in focus from using such outputs as the metric for judging the return on Big Science investments towards using investments to achieve certain outcomes. This shift from an output-oriented approach towards a goal-oriented approach can be seen above in the case of SKA in South Africa, where it has been used to build up human capital and develop a region (Bhogal 2018). This raises the challenging question regarding how to measure such returns on investment. However, the tools for the creation of these goals and judging whether the goals have been met are still relatively new. Some experimental proposals have been to combine traditional metrics with more narrative descriptions of the benefits from scientific funding. These proposals involve combining narrative descriptions of the beneficial outcomes with more traditional metrics that examine the collaboration's construction by mapping the diversity within the collaboration in terms of their disciplinary interrelatedness, their geographic distances, the relative similarities of their respective working environments, the social capital between collaborators, and the level to which hierarchical structure help or hinder collaboration.

A particularly tricky aspect of Big Science for policymakers is evaluating when a piece of apparatus has reached maturity whereby its productivity is at a stage where the increasing maintenance costs no longer justify the operating expenses. The decision to continue investing in a mature apparatus can be very challenging when new apparatus have been commissioned with new capabilities (Cho 2010). Compounding these challenges is the significant risk of community outcry if there is the perception that a useful facility is being closed to satisfy budgets (Oddone 2011b, Reich 2010, Samuel Reich 2011). Traditional methods for evaluation tend to also create further complications as the most eye-catching metrics are displayed early during the operating life of a Big Science project or at specific endpoints. One suggestion to try to rectify this situation is to capture data between these endpoints at designated waypoints that can capture both scientific and economic impacts

(Bianco *et al* 2017). These are described in terms of knowledge, commercialisation, or universal waypoints and broadly capture knowledge production by synthesising some of the metrics discussed above (Bianco *et al* 2017). These capture publications or patents, commercialisation grants or licensing agreements, or even whether the facility has secured additional funding to continue research (Bianco *et al* 2017).

13.6 Conclusion

As this chapter has demonstrated, the popular justification used by scientists for Big Science investments has tended to focus primarily on the economic impacts that they generate. This is not surprising as the major funders for Big Science projects tend to be national governments who must justify their investments to the citizenry and these types of impact are often the most obvious returns on investment. However, there can be very long lead times before these benefits become apparent in the real economy. But it is perfectly reasonable and often accepted by the public that Big Science investments can be substantially justified on scientific grounds alone. This has historically been the case during the early days of Big Science in the early post-war period. Since the 1970s Oil Crises in the West there has been a tendency to play down the significance of scientific impact as a justification for Big Science investments (Hoddeson *et al* 2008). Amongst the new trends that have developed over the past 20 years has been an increased focus on the societal benefits that arise from Big Science projects, whether in terms of human capital or as using the project as a way to develop a region. This was observed in the case of the Square Kilometre Array in South Africa, which is unusual amongst Big Science projects in that it is being constructed in a less economically developed country and the project is being used as a way to develop human capital to construct and service the final apparatus (Bhogal 2018). The topic of evaluation of Big Science projects will continue to develop and amongst recent developments are new efforts to combine traditional metrics with more narrative descriptions that can capture benefits from increasingly interdisciplinary research (Dave *et al* 2016).

References

Aad G *et al* 2012 Observation of a new particle in the search for the standard model Higgs boson with the ATLAS detector at the LHC *Phys. Lett.* B **716** 1–29

Abe F *et al* 1995 Observation of top quark production in collisions with the collider detector at Fermilab *Phys. Rev. Lett.* **74** 2626

Archambault É and Larivière V 2009 History of the journal impact factor: contingencies and consequences *Scientometrics* **79** 635–49

Autio E 2014 *Innovation from Big Science: Enhancing Big Science Impact Agenda* (London: Department for Business Innovation and Skills) http://data.parliament.uk/DepositedPapers/Files/DEP2014-0843/Innovation-from-big-science-enhancing-big-science-impact-agenda.pdf

Autio E, Hameri A P and Nordberg M 1996 A framework of motivations for industry Big Science collaboration: a case study *J. Eng. Tech. Manage.* **13** 301–14

Autio E, Hameri A P and Vuola O 2004 A framework of industrial knowledge spillovers in big-science centers *Res. Policy* **33** 107–26

Awschalom M, Grumboski L, Hrejsa A F, Lee G M and Rosenberg I 1979 The Fermilab cancer therapy facility: status report after 2.5 years of operation *IEEE Trans. Nucl. Sci.* **26** 3068–70
Berners-Lee T 1989 Information management: a proposal *Hypertext proposal* TBL-900620 CERN https://cds.cern.ch/record/369245/files/dd-89-001.pdf
Bhogal N 2018 The role of the square kilometre array in South Africa's economic development strategy *S. Afr. J. Sci.* **114** 7
Bianco W, Gerhart D and Nicolson-Crotty S 2017 Waypoints for evaluating Big Science *Soc. Sci. Quart.* **98** 1144–50
Blankenship L V 1974 Management, politics, and science: a nonseparable system *Res. Policy* **3** 244–57
Castelnovo P, Florio M, Forte S, Rossi L and Sirtori E 2018 The economic impact of technological procurement for large-scale research infrastructures: evidence from the large Hadron Collider at CERN *Res. Policy* **47** 1853–67
Cho A 2010 New delay of large Hadron Collider might not keep its rival on the job *Science* **327** 766
Choi W, Tho H, Kim Y, Hwang S and Kang D 2017 The economic benefits of Big Science R&D: with a focus on fusion R&D program in Korea *Fusion Eng. Design* **124** 1263–8
Ciarli T and Ràfols I 2019 The relation between research priorities and societal demands: the case of rice *Res. Policy* **48** 949–67
Cronin B and Sugimoto C R (ed) 2014 *Beyond Bibliometrics: Harnessing Multidimensional Indicators of Scholarly Impact* (Cambridge, MA: MIT Press)
da Silva J A T and Dobránszki J 2018 Multiple versions of the H-index: cautionary use for formal academic purposes *Scientometrics* **115** 1107–13
Davé A *et al* 2016 Landscape review of interdisciplinary research in the UK *Report to HEFCE and RCUK by Technopolis and the Science Policy Research Unit (SPRU), University of Sussex* http://sro.sussex.ac.uk/id/eprint/65332/1/2016HEFCE_Landscape%20review%20of%20UK%20interdisciplinary%20research.pdf
Devleesschauwer *et al* 2014 Calculating disability-adjusted life years to quantify burden of disease *Int. J. Public Health* **59** 565–9
Dewdney P E, Hall P J, Schilizzi R T and Lazio T J L W 2009 The square kilometre array *Proc. IEEE* **97** 1482–96
Dewdney P 2019 Impact of precursors and pathfinders on the SKA design *SKA History Conf. (Joddrell Bank, UK, April 4)*
D'Ippolito B and Rüling C-C 2019 Research collaboration in large scale research infrastructures: collaboration types and policy implications *Res. Policy* **48** 1282–96
Dosanjh M and Bernier J 2018 *Advances in Particle Therapy: A Multidisciplinary Approach* (London: Routledge)
Else H 2021 Row erupts over university's use of research metrics in job-cut decisions *Nature* **592** 19
Etzkowitz H and Leydesdorff L 2000 The dynamics of innovation: from national systems and 'Mode 2' to a triple helix of university–industry–government relations *Res. Policy* **29** 109–23
Florio M, Giffoni F, Giunta A and Sirtori E 2018 Big science, learning, and innovation: evidence from CERN procurement *Ind. Corp. Change* **27** 915–36
Gastrow M and Oppelt T 2018 Big science and human development—what is the connection? *S. Afr. J. Sci.* **114** 5182
Gastrow M and Oppelt T 2019 The Square Kilometre Array and local development mandates in the Karoo *J. South Afr. Stud.* **45** 711–28

Giudice G F 2012 Big Science and the large Hadron Collider *Phys. Persp.* **14** 95–112
Godin B 2006 The linear model of innovation: the historical construction of an analytical framework *Sci. Technol. Hum. Values* **31** 639–67
GSF 2014 *The Impacts of Large Research Infrastructures on Economic Innovation and on Society: Case Studies at CERN* (Paris: OECD)
Hallonsten O 2016 Use and productivity of contemporary, multidisciplinary Big Science *Res. Eval.* **25** 486–95
Heilbron J L, Seidel R W and Wheaton B R 1981 Lawrence and his laboratory: nuclear science at Berkeley, 1931–61 *California Studies in the History of Science* (Berkeley, CA: Lawrence Berkeley Laboratory and Office for History of Science and Technology, University of California)
Hermann A, Belloni L, Krige J and European Organization for Nuclear Research 1987 History of CERN volume 1—launching the european organization for nuclear research *History of CERN* (Amsterdam: North-Holland)
Hesla L 2017 IEEE honors game-changing Tevatron technology *Fermilab* 4 December https://news.fnal.gov/2017/12/ieee-honors-game-changing-tevatron-technology/
Hicks D, Wouters P, Waltman L, de Rijcke S and Rafols I 2015 Bibliometrics: the Leiden Manifesto for research metrics *Nature* **520** 429–31
Hislop D 2013 *Knowledge Management in Organizations: A Critical Introduction* (Oxford: Oxford University Press)
Hoddeson L and Kolb A 2003 Vision to reality: from Robert R Wilson's frontier to Leon M Lederman's fermilab *Phys. Persp.* **5** 67–86
Hoddeson L, Kolb A and Westfall C 2008 *Fermilab: Physics, the Frontier, and Megascience* (Chicago, IL: University of Chicago Press)
Kumar M J 2009 Editorial *IETE Tech. Rev.* **26** 165–8
Lederman L 1983 State of the laboratory *Report* Fermi National Accelerator Laboratory http://lss.fnal.gov/archive/annual/fermilab-annual-1983.pdf
Lefevre C 2009 *LHC Guide Book* (Geneva: CERN) https://cds.cern.ch/record/1164451/?ln=en
Leon G and Elias P 2010 *A Vision for Strengthening World-Class Research Infrastructures in the ERA, Report of the Expert Group for Research Infrastructures* (Brussels: EU Publications)
Leydesdorff L 2005 Evaluation of research and evolution of science indicators *Curr. Sci.* **89** 1510–7
Magazinik A, Sauza Bedolla J, Catalan Lasheras N and Mäkinen S 2019 Societal impact as cost-benefit analysis: comparative analysis of two research infrastructures *2019 IEEE Int. Conf. on Engineering, Technology and Innovation (ICE/ITMC)*
Martin B R and Irvine J 1983 Assessing basic research: some partial indicators of scientific progress in radio astronomy *Res. Policy* **12** 61–90
Oddone P 2011a Director's corner—Tevatron *Fermilab Today* http://fnal.gov/pub/today/archive/archive_2011/today11-01-11.html
Oddone P 2011b Tevatron closure: there's life in the old dog yet *New Sci.* **209** 26–7
Qiao L, Mu R and Chen K 2016 Scientific effects of large research infrastructures in China *Technol. Forecast. Soc. Change* **112** 112
Reich E S 2010 Momentum builds for tevatron extension *Nat. News* **October**
Reich E S 2011 Tevatron faces final curtain *Nat. News* **469** 141–1
Robinson M 2019 Science mega-project communities; mechanisms of effective global collaboration? *Doctoral Thesis* Durham University http://etheses.dur.ac.uk/13263/

Ruffini P 2017 Science and diplomacy: a new dimension of international relations *Science, Technology and Innovation Studies* (Cham: Springer)

Salter A J and Martin B R 2001 The economic benefits of publicly funded basic research: a critical review *Res. Policy* **30** 509–32

Scarrà D and Piccaluga A 2020 The impact of technology transfer and knowledge spillover from big science: a literature review *Technovation* **116** 102165

Seidel R 1992 The origins of the Lawrence Berkeley laboratory ed P Galison and B Hevly *Big Science: The Growth of Large Scale Research* (Stanford, CA: Stanford University Press)

Shenhar A 1993 From low-to high-tech project management *R&D Manag.* **23** 199–214

Shenhar A and Dvir D 1996 Toward a typological theory of project management *Res. Policy* **25** 607–32

Shenhar A J, Dvir D, Levy O and Maltz A C 2001 Project success: a multidimensional strategic concept *Long Range Plan.* **34** 699–725

Simmonds P, Kraemer-Mbula E, Horvath A, Stroyan J and Zuijdam F 2013 Big science and innovation *Technopolis* https://assets.publishing.service.gov.uk/government/uploads/system/uploads/attachment_data/file/249715/bis-13-861-big-science-and-innovation.pdf

Vergel Y B and Sculpher M 2008 Quality-adjusted life years *Pract. Neurol.* **8** 175–82

Walker C 2019 Cosmopolitan karoo: land, space and place in the shadow of the square kilometre array *J. South Afr. Stud.* **45** 641–62

Webre P 1988 Risks and benefits of building the superconducting super collider *Special Study* Congress of the US Congressional Budget Office, Washington, DC

West J D, Jensen M C, Dandrea R J, Gordon G J and Bergstrom C T 2013 Author-level eigenfactor metrics: evaluating the influence of authors, institutions, and countries within the social science research network community *J. Am. Soc. Inform. Sci. Technol.* **64** 787–801

Williams J E 1999 Donner laboratory: the birthplace of nuclear medicine *J. Nucl. Med.: Off. Publ., Soc. Nucl. Med.* **40** 16N, 18N, 20N

Chapter 14

The role of large research infrastructures for regional innovation*

Riccardo Crescenzi and Gabriele Piazza

14.1 Introduction

Policymakers have been attracted for a long time by the concept of innovation clusters to boost overall regional innovation, development, and employment. In this context, public research centres and large research infrastructures (RIs) have become a core part of regional innovation policies, such as smart specialization strategies (S3), with governments allocating substantial public resources to their support and promotion (Foray *et al* 2012). Beyond their scientific function, the assumption is that RIs are an important tool for capacity building, concentrating human capital and investments in the hosting region (Foray *et al* 2012). According to this logic, the geographical clustering of firms, universities, and RIs would per se support knowledge exchange and innovation.

Despite these claims, the role that RIs can play in regional innovation is not well understood. The notion that allocating public resources to basic science would automatically result in (localized) innovation came under scrutiny in recent years. The generation and diffusion of new knowledge in national, regional, and local economies depend not just on efforts and investment in research and development (R&D). Evidence suggests that this type of investment can enhance regional

* Some of the research featured in this chapter has received funding from the European Organization for Nuclear Research (CERN) under Addendum FCC-GOV-CC-0185 (KE4819/ATS), defining LSE contribution under Article 6 of the Memorandum of Understanding for the Future Circular Collider Innovation Study (FCCIS) (FCC-GOV-CC-0004, EDMS 1390795) hosted by CERN. The FCCIS project has received funding from the European Union's Horizon 2020 research and innovation programme under grant No 951754. Support from the European Research Council under the European Union Horizon 2020 Programme H2020/2014-2020 (Grant Agreement No. 639633-MASSIVE-ERC-2014-STG) is also gratefully acknowledged. All errors and omissions are our own.

doi:10.1088/978-0-7503-3631-4ch14

innovation only when coupled with a supportive endowment of human capital and other systemic conditions. This highlights the risk of the 'cathedrals in the desert' scenario, where significant R&D investments are concentrated (e.g. because of policy decisions on the location of RIs or localized incentives for private research programmes) in regions that lack the appropriate receptive environment in terms of human capital and other systemic conditions. The local mismatch between R&D and skilled labour can persistently hinder innovation and local spillovers. How can this be avoided? How can public policies facilitate the embeddedness of R&D investments and RIs into local and national innovation systems?

The relevant literature highlights the role of relations and linkages between various economic agents (domestic firms, multinationals, research centres, and universities) in promoting regional innovation. Most scholars have initially emphasized the role of geographical proximity in facilitating these collaborations. According to this view, knowledge is a spatially bound quasi-public good whose circulation is restricted to the area in which it was generated. However, more recent studies have questioned whether spatial proximity is a necessary or sufficient condition for knowledge sharing, with other types of non-spatial proximities enabling economic agents to overcome geographical distance (Crescenzi *et al* 2017).

In this context, RIs have specific mechanisms shaping the spatial footprint of their knowledge and innovation creation and diffusion impacts. More specifically, we can identify two distinct sets of mechanisms associated with the two main stages: the RI design, building, and upgrading phase, and the research phase during which scientists use experiments to produce new knowledge[1]. Regarding the first phase, most of the innovation and knowledge sharing is likely to happen through procurement. The development of cutting-edge equipment often involves very close interactions with research scientists that work in close collaboration with highly knowledge-intensive firms located in specialized clusters across the globe. These firms can then find commercial applications for the products, processes and services initially created for scientific goals and to supply the RIs. In addition, suppliers' labour force can benefit from learning-by-doing processes and develop new competences that can be applied to existing or new commercial applications. Looking at the research phase, RIs often work as platforms by hosting scientists from different countries. Their scientific output can then generate further knowledge. Depending on its commercial applicability, new scientific discoveries can be used to develop new products and services.

These mechanisms make knowledge creation and diffusion, and the derived benefits, a less spatially bounded phenomenon than often anticipated by the cluster literature and implicitly assumed by policymakers. In other words, the effects are not limited to where the RI is located. Understanding the mechanisms through which knowledge flows within and from these complex networks of firms, research centres, and universities (and their spatial footprint) is crucial for policymakers that aim to maximize local knowledge spillovers from investments in RIs. However, for this to

[1] In most RIs, particularly in large ones that host multiple experiments, these two phases can run in parallel.

be possible, local innovative actors might need to go beyond spatial proximity and clustering and leverage alternative proximities to tap into these remote knowledge pools. In these settings, cognitive proximity rather than co-location seems to play a critical role. This is because members of these networks share a common language, understanding of the scientific domain, experiments, and technology, and their ties might transcend geographical proximity.

However, these dynamics have not been yet examined in the RI context. The existing literature has highlighted the importance of the procurement channel. This is because the quest for new scientific discoveries needs more complex components ever and in larger quantities than in the past. Consequently, technology-intensive equipment takes an increasingly higher share of the budget for these scientific projects and involves a large number of specialized firms. The significant funds involved, the technology intensity of these contracts, and the required collaboration between industry and scientists to design and develop these products have made procurement the most likely and identifiable channel through which knowledge spillovers can occur.

This chapter remains aligned with this body of research and focuses on the collaboration between RIs and industrial partners during the design, building, and upgrade phase. However, the main focus of the chapter will be on the spatial footprint of these collaborations and their impacts, a perspective systematically overlooked in existing studies. The chapter starts off by reviewing the evidence on the role of collaborations in knowledge creation and diffusion and what these findings imply for RI-industry collaborations. The subsequent section looks at existing studies on RI procurement effects. While we have evidence on the benefits to suppliers, we do not understand how knowledge produced during these interactions flows beyond the boundaries of the firms involved in the process and their supply chain. The fourth section discusses some of the potential mechanisms through which knowledge spreads to firms, sectors, and regions not initially involved in procurement activities. The final section concludes by highlighting the evidence gaps and briefly outlining potential research avenues.

14.2 The role of collaboration in research and innovation

The romantic notion that a new Nikola Tesla will emerge from the lab with the next AC motor (or a death ray) increasingly belongs to a bygone era. While in the late 1970s, around 75% of EPO patent applications in the United Kingdom (UK) were filed by individual inventors, nowadays that figure is below 15%. More than 80% of all patents are registered to more than one inventor, suggesting that collaboration in research and innovation has become the norm (Crescenzi *et al* 2016). Teams within the firm or the research centre and increasingly complex networks of researchers involving different firms, often collaborating with universities, public agencies, and research centres, drive the world of invention in the early twenty-first century. As Seaborn (1979, 88) puts it, ‘big science [has] eclipsed the garage inventor [...] Edison has been superseded by a team of white-coated theoretical physicists’.

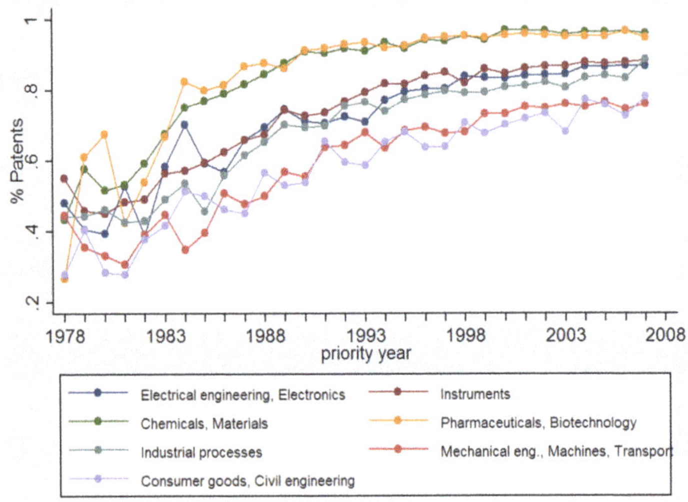

Figure 14.1. Co-invented patents by technology field in the UK, 1978–2007. (Source: Crescenzi *et al* (2016). This is an open-access article under the CC BY license (http://creativecommons.org/licenses/by/4.0/).)

This fundamental trend towards collaboration in patenting activity is documented in figure 14.1 that plots the share of co-invented patents (i.e. patents filed by two inventors or more) in the UK between 1978 and 2007.

While the trend towards the formation of ever-larger research teams and inventor networks has been well documented, we know much less about the factors that drive researchers to collaborate in the first place. For example, how vital are geographical proximity and spatial clustering for successful collaborations to happen? What role do non-spatial proximities play in facilitating and shaping these collaborations? What can be done to facilitate local collaborations and knowledge diffusion?

Crescenzi *et al* (2016) have studied empirically the behaviour of 'multiple patent' inventors—i.e. the most prolific and innovative individuals in the economy—showing that being part of the same organization plays a crucial role in forming co-patenting teams. However, social networks and cognitive proximities are key factors in shaping the selection of team members with a limited direct role of geographical proximity. The role of geographical proximity only emerges as well in interaction with other factors reinforcing their role.

Similar conclusions are reached in Crescenzi *et al* (2017), looking at university–industry collaborations (U–I). By looking at the collaborative behaviour of all Italian inventors over the 1978–2007 period, the empirical analysis shows that U–I collaborations are less likely to happen when compared to other types of collaborations (between firms or between universities) and suggests that they tend to generate patents of more general applicability in subsequent inventions—measured by forward-citations. As emphasized by the literature, geographical proximity plays an important role in facilitating all forms of collaboration. At the same time, it works as a possible substitute for institutional proximity, facilitating U–I collaborations. However, 'star inventors' on both sides of the collaboration can play an equally important role in 'bridging' universities and industry through their reputation.

This suggests that local collaborations between RIs and their local environment are possible, but other conditions need to be in place. However, the type of collaborations that RIs might give rise to, particularly when set up as global projects, might differ from the U–I ones discussed above. On the one hand, in particular when it comes to RIs involved in fundamental research, the produced knowledge might not yet have any commercial application, making it harder to predict how and where collaborations with industry partners might form, and knowledge spillovers might occur. On the other hand, RIs are often seen as platforms hosting researchers and working with firms from across the world and, consequently, can expand existing networks and create new ones. Through RI involvement, members of these scientific disciplines, from both the public and private sectors, can create epistemic communities that share a common language and understanding of a particular scientific domain, experiment or technology, developing a level of trust that facilitate communication regardless of distance (Feldman and Kogler 2010). The need to share scientific results relevant only to this group of spatially dispersed high-skilled individuals makes interaction with this community more likely than with other individuals in the spatial vicinity (Feldman and Kogler 2010). In this context, when trying to understand knowledge diffusion, this community's boundaries and their functional geography should be the unit of analysis rather than physical proximity *per se*. At the same time, for cognitive proximity to be conducive to knowledge flows between the parties involved in the collaboration, a certain level of dissimilarity in terms of know-how and know-what is needed (Huber 2012). Looking at the two stages of RIs, these differences are more likely to be found in the design, building, and upgrade phase of the project when industrial partners and RI staff, driven by different incentives, have to come together to produce the scientific equipment necessary for the experiment.

Conversely, the second phase might involve scientists working in the same disciplines, and the dissimilarities might be smaller. In addition, from an empirical perspective, it is more difficult to track how the scientific output produced during the research phase can find practical applications and lead to further innovation. And this is even more challenging when one wants to track commercial applications that build on new discoveries from basic research. This process often involves very long timeframes and a degree of serendipity[2].

Overall, these dynamics may result in complex geographies of innovation spanning multiple regions and involving various local transmission mechanisms from individuals, institutions, and firms directly participating in RIs activities and the local economies and localized networks in which they are embedded. However, this process is not yet well understood, and many research questions remain

[2] For example, it took more than a century for the articles by Albert Einstein on special relativity and general relativity to find commercial applications in GPS devices (Florio 2019).

unanswered. For example, what is the geography of these collaborations? And what is the role of different types of proximities in facilitating them? What other factors might support them? How spatially bounded are the knowledge flows from these collaborations? Do they 'leak away' to regions not initially involved in the project? Are these collaborations more valuable than others that might need a lower level of public funding?

14.3 How can RIs procurement contribute to innovation?

Before trying to identify the spatial footprint of this phenomenon and the factors that make these collaborations more likely to develop, it is crucial to review the evidence on the impact of procurement on the innovative performance of suppliers.

Most of the existing studies on RI socio-economic impact have focused on procurement (Autio *et al* 2004, Neumann 2014, Castelnovo *et al* 2018, Florio *et al* 2018, Dal Molin and Previtali 2019, Riebe *et al* 2020, Bastianin *et al* 2021, Li-Ying *et al* 2021). This is for three main reasons.

First, procurement activities involve building and operating the RI, and these contracts account for the largest share of investments. Therefore, this is where we expect to see the effects. For example, Florio *et al* (2016) showed that industrial spillovers accounted for approximately a third of the total socio-economic benefits from the Large Hadron Collider (LHC) at the European Organization for Nuclear Research (CERN).

Second, the research focus on the procurement channel can be explained by the recent attention that it has received as a policy tool to improve firms' innovation performance and potentially the local economies in which they operate (Edquist and Hommen 2000, Uyarra and Flanagan 2010, Edquist and Zabala-Iturriagagoitia 2012, Kundu *et al* 2020).

Finally, procurement is the area where we are most likely to see knowledge spillovers because of the technological composition of the scientific equipment required for these experiments. The close collaboration between suppliers and RIs can lead to the development and commercialization of new products and services. Li-Ying *et al* (2021) identify five main modes in which this relationship can facilitate innovation. First, RIs can act as early adopters of products or services that are not yet widely commercialized. Second, RIs, with the involvement of suppliers, can invent new products, processes, or methods to address specific scientific challenges. Third, to meet RI technological demand, suppliers might be the ones to come up with new solutions. Fourth, suppliers might build on the procurement experience and implement similar solutions for other scientific projects. Lastly, suppliers might use this newly acquired knowledge to develop products, services, or processes for customers outside Big Science projects.

Different quantitative and qualitative methodologies have been used to investigate the economic effects of RI procurement on the innovation output of its industrial partners. What emerged from this body of work is a positive association between this particular type of procurement and firms' productivity and innovation output.

Most of the evidence on the effects of procurement comes from CERN, which hosts the LHC, the biggest RI ever built and with an estimated cost of 13.5 billion euros[3], also one of the most expensive ones (Florio *et al* 2016, Bastianin *et al* 2021). The procurement activities for the LHC involved approximately 4200 firms between 1995 and 2015 (Bastianin *et al* 2021).

Looking at the impact on firms, a survey of CERN suppliers shows that collaborating with CERN contributed to product innovation and new R&D, starting a new business unit, and 40% of the respondents reported that they were more internationalized as a result of the collaboration (Autio *et al* 2004). Castelnovo *et al* (2018) used a simultaneous equation model to demonstrate that firms generally experienced a rise in R&D, patents, productivity, and profits after becoming CERN industrial partners. Florio *et al* (2018) showed that CERN procurement significantly affects suppliers' innovative performance when a relational type of governance is in place through cooperative relations (i.e. exchanges implying that CERN and suppliers work closely to deal with complex information that is not easily transmitted or learned). Bastianin *et al* (2021) showed that collaborating with CERN is associated with an increase in the hazard to file a patent for the first time and in the number of patent applications. Studies also found that spillover to second-tier suppliers impacts the entire supply chain (Florio *et al* 2018).

More qualitative studies also confirm the positive effect of RI procurement. For example, based on evidence from 14 Swedish firms that collaborated with CERN, Åberg and Bengtson (2015) pointed out that firms' product and process innovation occurs mainly when a development project is in place (i.e. when CERN invites firms to participate in developing products that cannot be bought off-the-shelf). Similarly, Sirtori *et al* (2019) demonstrated that thanks to the collaboration with CERN, several firms were subsequently able to develop new products for customers in other markets.

But working with RIs does not only have positive effects—for some firms, collaboration with CERN has been challenging, costly, risky, and was not necessarily financially profitable in the short term (Sirtori *et al* 2019). Despite the size of some of these contracts, the market for the most technologically intensive products developed explicitly for RIs is often small, and demand fluctuates considerably. Together with the significant initial capital investments, these patterns often discourage new players from entering the market (Peiniger *et al* 2012).

While these studies suggest that overall RI procurement can positively affect the supply chain, the effects could go beyond the first- and second-tier suppliers. At the same time—when impacts are assessed beyond firms' boundaries—new and unexpected trade-offs (such as crowding out private investment into more profitable industries) can be highlighted (and ideally incorporated into policy choices). And even when positive, it is not always possible to establish causality.

[3] This includes the operating costs of the LHC over its lifetime and this is why it is higher than the figure reported at the beginning of the chapter.

There are also distributional considerations to account for: are the gains derived from working with RI evenly spread across the population? For example, high-skilled workers employed in RI supply chain might see an increase in their wages, but we do not know whether the rest of the local workforce also experience net benefits. In addition, the development of a high-tech cluster often involves a degree of displacement, which might negatively affect the workers and firms being displaced. But these effects, both positive and negative, might not necessarily have the same geography and might be less or more spatially concentrated.

When studying these effects, one also needs to consider several factors that are likely to determine their sign and magnitude. For example, in the case of frontier technology, the possibility of commercializing these products or processes plays a crucial role. In addition, the amount and duration of these procurement contracts are also likely to affect the size of these effects.

14.4 How can knowledge from RI procurement spread to the rest of the economy?

The evidence discussed in the previous section shows the overall positive effects of RI procurement on suppliers. However, this channel needs further unpacking as we do not yet fully understand the mechanisms through which the knowledge generated in these procurement activities spreads to the rest of the economy. Borrowing from the vast literature on knowledge diffusion, we can identify three main potential mechanisms that are still relatively unexplored in RI-related studies. These are the *creation of new firms*, *publications, patents and technology licences*, and the *movement of scientists and engineers*.

14.4.1 Creation of new firms

Creating a new company can be an efficient way of transferring knowledge between different firms when the inventions are not protectable (Nilsen and Anelli 2016). New companies can exploit the founders' tacit knowledge and intellectual property acquired during the procurement activities. The evolution of the semiconductor industry is a famous example of this process (Møen 2002). If we evaluated the economic returns to US military procurement contracts awarded to pioneering firms such as Sprague Electric, Shockley, and Fairchild based on their performance alone, the returns are likely to be modest (Holbrook *et al* 2000, Møen 2002). However, historical accounts of the semiconductor industry show that key technologies later adopted by the industry leader Intel were initially developed in these pioneering firms and transferred by its employees (Holbrook *et al* 2000). And since 1976, NASA has been tracking products spin-offs, and their list of case studies includes those that have resulted from procurement (Florio 2019). But how big is this phenomenon in the context of RI procurement? How does it vary across different typologies of RIs? And what are the conditions that facilitate the creation of new firms? These are questions to which we do not yet have clear answers.

14.4.2 Publications, patents, and technology licenses

Non-commercial and commercial forms of codified knowledge are another important mechanism for knowledge diffusion. Non-commercial transfers such as publications, presentations, and informal exchanges are among the most important ways information is diffused from public research to industry (Nilsen and Anelli 2016). Agrawal and Henderson (2002) show that publications are preferred to transfer knowledge even in universities with a large patent output. However, the phenomenon of corporate publishing has received less attention until recently (Camerani *et al* 2018, Krieger *et al* 2020). Camerani *et al* (2018) argue that there are different reasons why firms decide to publish, such as accessing external knowledge and resources, attracting and retaining researchers, signalling, and reputation building.

Firms publish despite hindering the possibility of fully capturing the benefits of R&D investments. One reason for this is that firms choose what to publish. Given the level of tacit knowledge involved, publications do not ensure replicability suggesting that corporate publications might not be the most efficient mechanism to spread knowledge (Camerani *et al* 2018). Looking at the probability of publishing, the evidence shows that large firms are more likely to do so. The publication intensity—the number of publications normalized by R&D spending—varies by sector, and it is relatively low for firms in engineering (Camerani *et al* 2018). However, it is not yet clear how RI procurement activities affect corporate publishing. Do firms involved in procurement activities publish more? Which firms are most likely to publish, and why do they do it? Do suppliers co-author these publications with other organizations? And where are these co-authors located? Studying the publishing activities of suppliers might give us some further insights into how knowledge from RI gets diffused.

Patenting has been the main focus in the knowledge diffusion literature. Patents play two different main roles in spreading knowledge. First, the patent system can be a vital tool for knowledge diffusion. It represents a comprehensive repository of codified information and an important knowledge input for other inventors (Baruffaldi and Simeth 2020). Second, patent citations are often used to track how knowledge spreads: a firm citing a patent is assumed to have utilized the knowledge of the cited patent (Feldman 1999, Nelson 2009). While we know that suppliers are more likely to patent after the procurement contract (Castelnovo *et al* 2018, Bastianin *et al* 2021), we still have to understand how this knowledge gets utilized by other economic agents not initially involved in the contract. Building on the pioneering work by Jaffe *et al* (1993), patent citation studies could potentially help us track how the knowledge produced during procurement activities gets used by other firms and industries.

One challenge in doing this is that inventions that come from RI procurement activities are not often patented. This is due to the nature of the technology or because in some fields, such as high energy physics in which some RIs specialize, patenting is not part of the culture (Scarrà and Piccaluga 2020). Therefore, patent citations will only give us a partial account of how knowledge spreads, and we need to look at different mechanisms and use other types of data to track these knowledge flows.

Licensing of patents and intellectual property is the most traditional way to transfer knowledge (Nilsen and Anelli 2016). Nelson (2009) argues that patent-linked technology licences better capture knowledge spillovers as the competing interests of licensors and licensees work as a 'checks and balances' system to ensure the indicator's correctness and reliability. Thus, this could be another avenue to track knowledge flows from procurement activities. However, due to the patenting patterns discussed above, it will give us only a partial account of these knowledge flows.

14.4.3 Mobility of scientists and engineers

Few firms can internally generate all the knowledge required to stay competitive, and therefore they might rely on external sources of knowledge, but some of the knowledge is embedded in individuals (Feldman 1999). This knowledge might be the know-how on a tricky production method or the experience of failed experiments (Song *et al* 2003, Shih and Chai 2015). Businesses recognize this as 'residual knowledge' and impose non-compete provisions on their employees and target competitors when hiring (Shih and Chai 2015, Schaefer 2020). When they move, scientists and engineers take their knowledge with them and share it with their new colleagues (Miguélez and Moreno 2013). At the same time, they acquire new knowledge from their new colleagues, establish new links and social networks for future collaborations and promote new combinations of knowledge (Miguélez and Moreno 2013). In our case, this refers to the movement of scientists and engineers involved in the procurement activities, who transfer this tacit knowledge to the new employer. The movement of these skilled individuals can shape the geography of knowledge flows that derive from procurement activities. Although the literature has examined the movement of high-skilled workers between different types of organizations (Almeida and Kogut 1999, Agrawal *et al* 2006, 2017), labour mobility has not been analysed in the context of RIs and their procurement activities. As a result, we lack a clear understanding of this phenomenon. What is the number of engineers and scientists that move? What are the patterns of these movements? And which factors influence location decisions? Most importantly, we don't know whether and how they contribute to knowledge diffusion in these settings.

14.4.4 Multiple mechanisms at play

From a conceptual point of view, it is helpful to describe these mechanisms separately, but in practice, they might often be interlinked and reinforce each other. Two stylized examples might illustrate this point. For instance, scientists or engineers involved in procurement activities might leave their organization (mobility of scientists and engineers) to set up a new company (creation of new firms) and draw on the knowledge acquired during the procurement process to develop new products (publications, patents, and technology licenses). In another example, scientists involved in the procurement activities might join the R&D department of another company (mobility of scientists and engineers) and publish in the same area in which they worked at their prior organization (publications, patents, and technology licenses).

These interlinkages add two layers of difficulty for researchers that want to understand the wider socio-economic impact of RIs. First, studies attempting to understand these projects' wider socio-economic impact are likely to underestimate the effects if they focus on only one mechanism. Second, the complex interlinkages between these mechanisms make it difficult to identify the geography and timeframe in which these wider effects materialize.

14.5 Conclusion

Much attention and funding have been devoted to RIs for their scientific benefits and potential economic impacts, becoming a crucial part of local innovation strategies across the globe. This is because research and innovation are increasingly becoming a collaborative effort, and RIs, particularly through their procurement process, can forge new collaborations with industry. Industrial partners could then exploit the knowledge acquired during the production of experimental technologies for commercial purposes. However, while we have evidence on the positive effects of RI procurement activities on suppliers, further conceptual work is needed to understand how the knowledge generated during these collaborations flows beyond the supply chain. For example, scientists and engineers involved in the procurement activities might create new companies operating in more profitable industries; suppliers might patent what was developed for the scientific experiments; after the procurement collaboration, highly skilled workers might move to other organizations, taking with them the knowledge acquired during the contract. It is crucial for regional innovation strategies to identify the spatial footprint of these knowledge diffusion mechanisms, their patterns, and the role of different proximities in shaping knowledge flows. If, as we hypothesized, cognitive proximity between the agents involved plays a more critical role than other ones, what is the optimal distance that maximizes these knowledge flows? Can other types of proximities act as substitutes?

A better understanding of these aspects will allow policymakers to increase the exposure to these knowledge flows and make more targeted investments to improve the local absorptive capacity. However, these aspects have been so far overlooked in the literature and should be the priority of future research on the economic impact of RIs.

To understand these complex dynamics, empirical analyses need to rely on new methodologies and data. First, substantial progress is required to identify spatial and non-spatial networks and the geographical distribution of the benefits derived from RI-industry collaboration. While knowledge production functions have been used to estimate the innovative impact of RI procurement activities, existing applications have focused on the supply chain, excluding the effect on other firms and sectors. Second, future research should not rely solely on patent data as it does not capture all the new knowledge that is not patentable (or not patentable yet). This is a particularly relevant issue for those industries and scientific disciplines in which patenting is not standard practice. In this context, combining different data sources, including business surveys, is a step in the right direction. However, to answer some of the most complex research questions and assess the importance of the various knowledge diffusion mechanisms outlined in this chapter, more sophisticated and granular data on firms and individuals involved in these collaborations is needed. Finally, more research is also required on the public policy tools (if any) that work in practice to facilitate and support the establishment and development of these knowledge diffusion channels at a local and global level to inform policies such as smart specialization in Europe and beyond.

References

Åberg S and Bengtson A 2015 Does CERN procurement result in innovation? *Innov.: Eur. J. Soc. Sci Res.* **28** 360–83

Agrawal A, Cockburn I and McHale J 2006 Gone but not forgotten: knowledge flows, labor mobility, and enduring social relationships *J. Econ. Geogr.* **6** 571–91

Agrawal A and Henderson R 2002 Putting patents in context: exploring knowledge transfer from MIT *Manage. Sci.* **48** 44–60

Agrawal A, McHale J and Oettl A 2017 How stars matter: recruiting and peer effects in evolutionary biology *Res. Policy* **46** 853–67

Almeida P and Kogut B 1999 Localization of knowledge and the mobility of engineers in regional networks *Manage. Sci.* **45** 905–17

Autio E, Hameri A P and Vuola O 2004 A framework of industrial knowledge spillovers in big-science centers *Res. Policy* **33** 107–26

Baruffaldi S H and Simeth M 2020 Patents and knowledge diffusion: the effect of early disclosure *Res. Policy* **49** 103927

Bastianin A, Castelnovo P, Florio M and Giunta A 2021 Big science and innovation: gestation lag from procurement to patents for CERN suppliers *J. Technol. Transf.* **47** 531–5

Camerani R, Rotolo D and Grassano N 2018 Do firms publish? A multi-sectoral analysis *JRC Working Papers on Corporate R&D and Innovation* No 05/2018

Castelnovo P, Florio M, Forte S, Rossi L and Sirtori E 2018 The economic impact of technological procurement for large-scale research infrastructures: evidence from the Large Hadron Collider at CERN *Res. Policy* **47** 1853–67

Crescenzi R, Filippetti A and Iammarino S 2017 Academic inventors: collaboration and proximity with industry *J. Technol. Transf.* **42** 730–62

Crescenzi R, Nathan M and Rodríguez-Pose A 2016 Do inventors talk to strangers? On proximity and collaborative knowledge creation *Res. Policy* **45** 177–94

Dal Molin M and Previtali E 2019 Basic research public procurement: the impact on supplier companies *J. Public Procure.* **19** 224–51

Edquist C and Hommen L 2000 Public technology procurement and innovation theory ed C Edquist, L Hommen and L Tsipouri *Public Technology Procurement and Innovation* (Boston, MA: Springer)

Edquist C and Zabala-Iturriagagoitia J M 2012 Public procurement for innovation as mission-oriented innovation policy *Res. Policy* **41** 1757–69

Feldman M P 1999 The new economics of innovation, spillovers and agglomeration: a review of empirical studies *Econ. Innov. New Technol.* **8** 5–25

Feldman M P and Kogler D F 2010 Stylized facts in the geography of innovation *Handbook of the Economics of Innovation* ed B Hall and N Rosenberg vol 1 1st edn (Amsterdam: Elsevier), ch 8 pp 381–410

Florio M, Forte S and Sirtori E 2016 Forecasting the socio-economic impact of the Large Hadron Collider: a cost–benefit analysis to 2025 and beyond *Technol. Forecast. Soc. Change* **112** 38–53

Florio M, Giffoni F, Giunta A and Sirtori E 2018 Big science, learning, and innovation: Evidence from CERN procurement *Ind. Corp. Chang.* **27** 915–36

Florio M 2019 *Investing in Science: Social Cost-benefit Analysis of Research Infrastructures* (Cambridge, MA: MIT Press)

Foray D, Goddard J, Beldarrain X G, Landabaso M, McCann P, Morgan K, Nauwelaers C and Ortega-Argilés R 2012 *Guide to Research and Innovation Strategies for Smart Specialization (RIS3)* (European Commission Publications Office)

Holbrook D, Cohen W M, Hounshell D A and Klepper S 2000 The nature, sources, and consequences of firm differences in the early history of the semiconductor industry *Strateg. Manag. J.* **21** 1017–41

Huber F 2012 On the role and interrelationship of spatial, social and cognitive proximity: personal knowledge relationships of R&D workers in the Cambridge Information Technology Cluster *Reg. Stud.* **46** 1169–82

Jaffe A B, Trajtenberg M and Henderson R 1993 Geographic localization of knowledge spillovers as evidenced by patent citations *Quart. J. Econ.* **108** 577–98

Krieger B, Pellens M, Blind K and Schubert T 2020 Are firms withdrawing from basic research? An analysis of firm-level publication behaviour in Germany (2020/13; Paper in Innovation Studies) *ZEW—Centre for European Economic Research Discussion Paper* No. 20-083 https://ssrn.com/abstract=3766726

Kundu O, James A D and Rigby J 2020 Public procurement and innovation: a systematic literature review *Sci. Public Policy* **47** 490–502

Li-Ying J, Forneris J, Korsholm S B, Jensen A and Zangenberg N 2021 How European Big Science organizations and suppliers innovate through public procurement *Res.-Technol. Manag.* **64** 46–56

Miguélez E and Moreno R 2013 Do labour mobility and technological collaborations foster geographical knowledge diffusion? The case of European regions *Growth Change* **44** 321–54

Møen J 2002 Spin-offs and spillovers: tracing knowledge by following employees across firms *Discussion Paper* 5/02 Department of Finance and Management Science, Norwegian School of Economics and Business Administration

Nelson A J 2009 Measuring knowledge spillovers: what patents, licenses and publications reveal about innovation diffusion *Res. Policy* **38** 994–1005

Neumann M 2014 External effects of basic research infrastructure *European XFEL Publication Database* https://xfel.tind.io/record/141/files/?ln=en

Nilsen V and Anelli G 2016 Knowledge transfer at CERN *Technol. Forecast. Soc. Change* **112** 113–20

Peiniger M, Pekeler M and Vogel H 2012 Industrialization of superconducting accelerator technology *Rev. Accel Sci. Technol.* **5** 265–83

Riebe T, Schmid S and Reuter C 2020 Measuring spillover effects from defense to civilian sectors —a quantitative approach using LinkedIn *Defence Peace Econ.* **32** 773–85

Scarrà D and Piccaluga A 2020 The impact of technology transfer and knowledge spillover from Big Science: a literature review *Technovation* **116** 102165

Schaefer K J 2020 Catching up by hiring: the case of Huawei *J. Int. Bus. Stud.* **51** 1500–15

Seaborn T 1979 Talking about the automat. The open channel *Softw. Pat. Inst. (IEEE Comput.)* **12** 87–8

Shih W C and Chai S 2015 What to know about locating in a cluster *MIT Sloan Manag. Rev.* **57** 8–11

Sirtori E C, Florio M, Catalano G, Caputo A C, Pancotti C C and Giffoni F C 2019 *Impact of CERN Procurement Actions on Industry: 28 Illustrative Success Stories* (Geneva: CERN) https://cds.cern.ch/record/2670056?ln=it

Song J, Almeida P and Wu G 2003 Learning-by-hiring: when is mobility more likely to facilitate interfirm knowledge transfer? *Manage. Sci.* **49** 351–65

Uyarra E and Flanagan K 2010 Understanding the innovation impacts of public procurement *Eur. Plann. Stud.* **18** 123–43

IOP Publishing

Big Science in the 21st Century
Economic and societal impacts
Panagiotis Charitos, Theodore Arabatzis, Harry Cliff, Günther Dissertori, Juliette Forneris and Jason Li-Ying

Chapter 15

Rethinking how to maximise the impact of your research infrastructure

Silvia Vignetti

This chapter discusses the advantages of having socio-economic impact assessment considerations built-into the design of a research infrastructure and used as a tool to engage and exchange with key stakeholders and users, including the wider public. Socio-economic impact assessment investigates the mechanisms and interactions of the infrastructure with its stakeholders and their environment, enlightening the causality chains that are expected to lead to the generation of benefits. It also allows determining factors and conditions for impact maximisation and therefore contributing to design studies and long-term sustainability.

A more traditional approach of addressing socio-economic concerns only after implementation, upon external requests, may lead to down-scaling, re-scoping, and organisational adjustments at a later stage to better fit funders' or other stakeholders' expectations. Instead, an early commitment by scientific communities to the delivery of well-established and realistic set of socio-economic impacts, as part and not beyond the science mission, can avoid such risks at later stages. This chapter, in particular, maintains the idea that socio-economic impact assessment can help scientific communities and governments jointly think about how best to make use of scarce resources and be a tool or the reconceptualization of the relationship between science and society. If properly managed, impact assessment may fuel a transparent and structured process through which societal actors, researchers and governments become mutually responsive to each other on a set of strategic and shared objectives.

15.1 Introduction

This chapter discusses how designing and implementing a socio-economic impact assessment framework of a research infrastructure (RI) since its design and planning phases can help to steer a process of public engagement and negotiation and exchange with key stakeholders. In order to be a credible and effective tool for management and dialogue, it requires a solid conceptual framework and the full ownership of the infrastructure management.

Large RI are nowadays a constituent element of science production in all scientific domains. This goes together with increasing expectations. They are required to contribute to technological and scientific progress, but they are also expected to deliver the solution to many global societal challenges. Considering the growing research expectations, their complex and international nature, RI requires increasingly large budget commitments from different public sources and large partnerships combining a range of local, national, and global stakeholders with different constituencies and expectations. All this brings rising stakes about RI funding and management and creates competition, both nationally and internationally, for scarce public funds and expertise among scientific communities and between scientific communities and other advocacy groups, such as health and education facilities, which can show more immediate societal impacts. In this context, the demonstration of RIs contribution to well-identified social 'missions' becomes crucial for social acceptance.

This paper offers some reflections drawing from the concepts and toolbox that economists have for assessing impact of RIs. It discusses the importance of setting up, in the early phase of conception, an appropriate and stable assessment framework, going possibly beyond a long list of all existing indicators subject to ever changing external requests. In so doing, it finally offers to managers of RIs some indications about how to engage internal staff, external stakeholders and the wider public in a constructing dialogue about what to realistically expect from funded infrastructure, wisely using the tool of socio-economic impact assessment.

The paper is structured in the following way: section 15.2 sets the scene and illustrates the changed paradigm in public funding for RIs and the increasing importance of social acceptance thereof. Section 15.3 discusses the variety of actors, expected benefits, and raising expectations related to their delivery. Section 15.4 discusses how a socio-economic exercise should be understood and structured in order to support science manager to engage in a constructive dialogue with key stakeholders about credible and feasible expectations about impacts.

15.2 Public funding and social acceptance: a changing paradigm

15.2.1 Increasing budgets and public interest

Scientific knowledge creation is a key driver of economic growth (European Commission 2017, Salter and Martin 2001, Martin and Tang 2007). But cutting-edge science relies on cutting-edge instrumentations, and cutting-edge instrumentations can be extremely costly. Increasing public funding is allocated to the design, development, operation, and upgrading or expansion of RIs, both at the EU level and national/regional level. Between 2015 and 2020, the UK government committed to spend over seven billion of capital funding on research and innovation infrastructure (UKRI-UK Research and Innovation 2020). In the EU, the amount of national funding for research infrastructures in the EU member states is significant and complements the funding provided by the EU[1]. For example, the German Federal Government provides an annual budget of over 1.3 billion euros for large-scale equipment in basic research[2].

Dedicated budget envelopes are also available at supra-national level. The Cohesion Policy and the European research policy through, respectively, the European Regional Development Fund and the EU Framework Programme for Research are the two main funding instruments through which the EU supports RI as triggers for regional development and the development of supra-national RIs. In the period 2014–20 they together invested almost ten billion euros (European Commission 2019).

As a result of the interest they raise on policy and scientific grounds, RIs have been growing steadily. The Mapping of the European Research Infrastructure Landscape (MERIL) portal[3] currently stores information about 1042 openly accessible RIs in Europe, across all scientific domains. They have been constantly increasing since 2007 when they were 622 (Pancotti *et al* 2014).

In such a context, there is increasing competition for public support for the construction and operation of these often large and complex facilities. The question as to which type of research to promote and fund (e.g. in which scientific domain/basic or applied/large or small) is top on the agenda of policymakers. And this calls for a specific analysis of how to optimise the use potential and operation of these facilities, more and more based on consideration linked to the strategic evaluation of the benefits and risks of each (OECD 2020). The issue of how to improve governance, efficiency, and sustainability of RIs is more and more linked to how to increase their utilisation (and by whom) and, therefore, which types of impacts to prioritise.

15.2.2 From peer review to public scrutiny

The observed trend results from of a change in paradigm in science production from 'little science to big science' (De Solla Price 1963, Florio 2019). The increasing importance of large-scale facilities (and related big international collaborations and extensive public budgets) in science production occurred in the physics, life, and natural science domains, which traditionally require high-tech equipment and large international collaborations, but also in social sciences and humanities. The ESFRI

[1] https://op.europa.eu/en/publication-detail/-/publication/6702e82f-e4c3–11e9-9c4e-01aa75ed71a1/language-en/format-PDF/source-106123556

[2] https://www.research-in-germany.org/en/research-landscape/research-organisations/research-infrastructures.html

[3] It is a mapping resulting from a survey undertaken by the European Commission and the European Science Foundation and constantly updated. See https://portal.meril.eu/meril/

roadmap[4] of 2018 included two new RIs recognised being of pan-European interests in the field of social sciences and humanities[5].

Along with a change in the paradigm of science production, a shift was observed in the last decades in science policy and governance, particularly on the rationale for public research funding. Governments, funding agencies, and policymakers have high expectations in the context of science and innovation policies aimed at sustaining economic growth in the long term. This implies some degree of external scrutiny on scientific investments, somehow, in contrast with the traditional approach based on science's internal quality assessment. The underpinning consideration of such traditional approach was that science can be judged only by scientists, as embedded in the Haldane principle[6], so that the principal mechanism determining the quality of research and resource allocation was internal peer reviewing (Feller 2013).

A combination of a peer review process for the 'science case' and a financial model for the 'business case' is currently still part of many selection processes of RIs (Pancotti *et al* 2014). Although efficient and fair to build national roadmaps and prioritisation processes, this approach is not suitable for assessing a RI's socio-economic effects and building a solid case for broad social accountability. Increasing demand for public money goes together with increasing expectations from the broader public, and social acceptance for science is no longer taken for granted as expected benefits come together with costs. If there are good reasons to invest in RIs, there may also be arguments for not engaging in such costly investments in some circumstances. Large facilities can be characterised by excess capacity and may be expensive to maintain in the long run, often requiring more investments to remain at the frontier of science (ESFRI 2017, OECD 2017). They require specialised managerial expertise as well as ad hoc governance and legal arrangements. They may impose excessive concentration of resources at the expense of territorial cohesion or give positional rents to a small number of organisations (European Commission 2012).

In the face of increasingly pressing global socio-economic challenges, the question as to whether modern societies can afford investing in curiosity-driven basic research that does not promise precise and immediate application in society is a relevant one and need to be answered using convincing and robust argument. High scrutiny

[4] The European Strategy Forum on Research Infrastructures (ESFRI), is a strategic instrument to supporting a coherent and strategy-led approach to policy-making on research infrastructures in Europe. The EFRI Roadmap for research infrastructures (new and major upgrades, pan-European interest) was first published in 2006, with 35 projects, then updated in 2008, 2010, 2016, and 2018 bringing the number of RIs of pan-European relevance to 55. More information are available here: https://www.esfri.eu/

[5] They are the European Research Infrastructure for Heritage Science (E-RIHS) and the European Holocaust Research Infrastructure (EHRI).

[6] In British research policy: 'The Haldane principle means that decisions on individual research proposals are best taken by researchers themselves through peer review. This involves evaluating the quality, excellence and likely impact of science and research programmes. Prioritisation of an individual research council's spending within its allocation is not a decision for Ministers. The coalition Government support this principle as vital for the protection of academic independence and excellence. We all benefit from its application in the UK.' (Source: Government statement of 20 Dec 2010: Column 135WS).

requires a robust investment case. But there is indeed an inherent difficulty to promote an informed and open public dialogue around science investments due to the technical language and complexity involved. Moreover, a proper assessment of impacts requires overcoming the challenges of the long-time span for their materialisation and the possibly loose attribution link to the initial investment. This, however, brings in an accountability issue since funding agencies must justify to taxpayers the money spent in large research infrastructures instead of on schools, hospitals, or public transport that brings more immediate and easily communicable benefits to the society. In this context, RIs with high visibility can be vulnerable to political pressures and lobbying activities by the scientific communities and the wider public.

15.2.3 Mission-orientation and social responsibility in science

Since pure contribution to the advancement of scientific knowledge is no longer enough as a justification, RIs are more and more expected to serve 'grand challenges' such as climate change, energy transition, health, and population ageing. The fast deployment of relevant data, facilities, and knowledge by several RIs active in the life sciences in response to the Covid-19 emergency proved that transforming scientific knowledge into practical uses can be carried out faster than expected when large funds are available[7]. RIs are required to orient their scientific mission to well-identified societal challenges on which well-identified results can be delivered in an acceptable timeframe instead of relying on serendipity. This requires however a dedicated strategy.

Funding mechanisms are increasingly mission-oriented, requiring scientists to actively pursue socio-economic impact within politically prioritised areas. The Horizon Europe programme, the newly approved EU research programme committing around 100 billion euros to basic science and cross border research projects, earmarked about 4.5 billion euros for five top priority missions: climate change; cancer; oceans and other bodies of water; smart cities; and soil and food. Inspired by the work of Mazzucato (2018), missions are framed around ambitious goals of societal relevance,such as improving health or the living environment for European citizens. This explicit formulation is a way to engage the public, showing that science is at the service of citizens instead of being only moved by a genuine but useless interest for new knowledge.

This approach is sometimes seen as an undue request by scientific communities and indeed requires a change in the mindset of scientists. Freedom in science and social responsibility are sometimes seen as in tension or bearing a trade-off. This perception goes together with the social responsibility in science movement, which sees research and innovation as a transparent and interactive process through which societal actors, researchers, and innovators become mutually responsive to each other with a view to the (ethical) acceptability, sustainability, and societal desirability of the scientific and innovation process. Amongst the goals of the movement,

[7] See https://www.esfri.eu/covid-19

there is the engagement of scientists and researchers in public discussion to set their priorities in scientific and technological research and create venues for open and public discussion to allow a proper embedding of scientific and technological advances in society (Von Schomberg and Hankins 2019). While this reconceptualization of the relationship between science and the public builds on a call for increased outreach activities, dissemination, and scientific literacy, the key message to be conveyed in this exchange can come from showing the socio-economic benefits associated to Big Science investments.

15.3 Many have a standing

15.3.1 Which impact and for whom

Differently from the past, RIs are more and more conceived as facilities offering services to a plethora of users or players than serving the interests of a closed and self-serving scientific community. Accordingly, the voices of such players are to be heard and their expectations managed since the conceptual design of the RI (table 15.1). From the point of view of RI managers, engaging stakeholders and users as 'advocates' becomes strategic in the search for international consensus and funding so that it is not just scientists making the case around an RI. Finding the right governance structure for the partnership starts also with establishing a clear vision at the outset on the expected impacts and the way in which they are prioritised, pursued, monitored, and evaluated.

Different players are associated with different types of benefits, with may possible overlapping. The recognition and proper identification of actual and potential stakeholders and the way in which they are expected to benefit from the RI is an essential starting point of any socio-economic assessment. Regardless of the estimation technique chosen, the underlying assumptions that show that a critical mass of users exists need to be recognised and stated transparently since the beginning.

A better understanding of the RI's potential impacts and the way they are generated, may facilitate a crucial and important change in scientific practices. A new scientific governance can be developed where relations between policymakers, science, and the public can be active agents about the priorities that scientific and technological development should address (Phillips *et al* 2012). This requires an ethos of evaluation that emphasises the need to show how science outcomes may impact the public and vice versa and how to measure in a solid manner the socio-economic effect of science (Short 2013).

15.3.2 Managing expectations

RI managers are more and more required to generate and show evidence that needs to be solid, measurable, and communicated to a non-technical recipient. The report of the European expert group on cost control and management issues of global research infrastructures (European Commission 2010) noted that at the start of a RI initiative, it is important that there is a shared understanding of the foreseen scope, schedule, and cost among the policymakers, addressing inherent uncertainties and

Table 15.1. Different players, roles, and expectations. Adapted with permission from Florio *et al* (2016).

Player	Role	Expectations
Scientists	*Scientists and researchers* are science producers but also users of science results and evidence to produce further knowledge. They are internal staff and external scientists, either from the same scientific domains or others. *Young professionals, junior researchers, and students* include post-doctoral and early career researchers and students involved in scientific or training programmes. They gain knowledge and skills during the period they spend within the infrastructure.	Access to data, instrumentation, tools and research collaborations.
Businesses	*Industrial users:* it includes enterprises (both large and small) that directly enjoy the services provided by the RI in terms of access to data, instrumentations, support to proprietary experiments, open access tools (e.g. open software). It also includes spin-offs and start-ups set up to commercially exploit research results. *Industrial suppliers*: they are contractors of the RI providing high-tech, not off-the-shelves, technological components or solutions to be used for the construction or operation of the RI. They benefit from indirect spillover effects through procurement and supply chain learning effects.	Procurement contracts, access to data, instrumentation, tools and research collaborations.
Policymakers	*Funders:* These include funding agencies securing resources for the construction, operation, and upgrading of the facilities. For internationally relevant RIs, funders are typically part of international partnerships where membership contributions are agreed. The contribution can include the so-called 'host premium'. That is the agreement for the host country/institution to cover whole or partially the costs of operating the central facility. *Users:* These include public administration or agencies responsible for designing or implementing public policies in certain sectors (e.g. environment, health, education, etc) where updated and reliable data and tools provided by RIs can support the decision-making.	Scientific and industrial *juste retour.* Employment and economic effects on a specific region or country. Access to data, instrumentation, tools and research collaborations.

(*Continued*)

Table 15.1. (*Continued*)

Player	Role	Expectations
Citizens	*Users of new goods or services produced with applications of research results*: They may include for example patients of medical treatment provided at a health research infrastructure, or residents of a region in which major natural risks are better monitored/forecasted thanks to activities carried out in observatories, stations, or satellites. *Users of outreach services and activities*: These include onsite visitors to the facility and virtual visitors on websites and social networks, better aware and science supportive citizens benefitting from dissemination and other outreach activities. *Funders:* Taxpayers ultimately provide financial resources through fiscal contribution. *Non-users*: For example, most taxpayers who fund RI infrastructures without directly using or planning to use it. They may derive some utility from the mere fact that they appreciate that scientific discoveries and technological progress are possible because of the infrastructure; they create a global public good with an inherent non-use value (i.e. existence value).	Access to research results and advancements, access to better products and services, access to information and data on science, answers to the big quests of life
Research infrastructure managers	*Managers:* These include governance and management staff taking decisions on the allocation and management of internal resources (staff, financial resources, equipment, time) in order to fulfil the scientific mission of the facility but also delivering the expected socio-economic impacts.	Professional recognition, scientific reputation, long-term sustainability of the infrastructure.

Source: Based on Florio *et al* (2016).

any external constraints, and describing what must be done if deviations occur during the following phases.

Different and multidimensional stakes may imply certain conflicts or trade-offs, which requires taking critical decisions. For example, the establishment of RIs can attract investments and give opportunities for market development, contributing broadly to the socio-economic development of a given region (Cooke *et al* 1997). At the same time, RIs are typically expected to pursue excellent scientific results at the international level. Therefore, a certain balance between global excellence and territorial impacts (as part of local 'premium' for hosts) should be maintained. In the same way, balancing the provision of 'free-from-costs' access for merit-based academic use and the establishment of clear and transparent pricing policies for all potential users may not be straightforward.

Timing of impacts can be another critical aspect. The lifecycle of an RI is typically in the range of 20–50 years and spans along a set of different phases (design, construction, operation, possible upgrades, decommissioning). Impacts may materialise according to different timeframes. For example, positive technological impacts may spillover during the construction period thanks to the procurement activities (Florio *et al* 2018), while scientific impacts materialise mostly at a later stage when the RI is in full operation and education impacts may last even after the decommissioning. A socio-economic impact assessment may provide different results depending on which impacts are considered and when. However, formulating strategic objectives in terms of ambitious societal missions can help build a long-term commitment from all stakeholders to overarching goals that require joint efforts from all interested parties.

15.4 A framework for socio-economic impact assessment

As demonstrated by the results of a recent survey collecting almost 200 questionnaires from RI managers (Catalano *et al* 2018)[8], the main motivation for carrying out an impact assessment is the need to comply with formal procedures for funding applications. Since socio-economic impact assessments are produced more to face an external request than to fulfil an internal informative need, the methods and tools used to perform such assessments are typically defined on an *ad hoc* basis and fitting specific timing and resource availability, which may not be the optimal solution. On the contrary, as now widely recognised by practitioners and RI managers (Gutleber 2021, Vignetti 2021), economic and social impacts require to be identified, purposedly pursued and integrated into the design of the RI since the early stage of its conception.

15.4.1 Mixed methods and indicator based approaches

The multiple requests to provide evidence on a wide array of socio-economic impacts have led to a proliferation of indicator systems to describe and measure the possible dimensions of performance and impact in many different ways.

[8] https://ri-paths-tool.eu/en

Notable examples are the OECD-GSF framework (OECD 2019) and the ESFRI monitoring report (ESFRI 2019), but the system of tracking research performance through a battery of indicators along causal pathways has also been endorsed by the European Commission as a way to monitor and evaluate the upcoming Horizon Europe programme (see European Commission 2018).

The underpinning rationale is that a single indicator would only reveal a small part of the multidimensional picture. This argument is based on the stream of literature drawing back to the seminal contributions by Ben Martin (Martin 1996). While providing a comprehensive picture of all possible effects of RIs and being effective in supporting narratives, indicators systems also have some limitations (see Giffoni *et al* 2019). Since they rely on the most common indicators already collected by RIs, in an attempt to be simple and operational, these approaches produce a set of organised statistics describing trajectories of performance over time and on different dimensions. They are particularly helpful as a monitoring tool, but they do not provide a conclusive measure impact. Moreover, lacking a theory on how to define and measure the identified impacts consistently, but calling for the use of different possible metrics and methods for the quantification, including different approaches ranging from micro to macro-economic assessment, it has limited reliability and is open to inconsistencies and possible inaccuracies. For example, there is a risk to mix considerations on the use of resources (i.e. the cost side) with the one on the production of resources (i.e. the benefit side). This represents an element of arbitrariness and leads to possible double-counting and overlaps.

If in the evaluation of individual RIs these limitations can be accommodated at the indicator level, it is not the case when it comes to different RIs which have to be assessed in a comparative way. There is also a problem of aggregation. Rather than attempting to provide a comprehensive and synthetic judgement on the socio-economic impact of RIs projects, the approach delivers a multidimensional set of indicators, with no attempt to aggregate evidence into a synthetic and conclusive assessment. For example, the OECD-GSF framework cannot be used for comparison between RIs, rather such indicators are 'meaningful when compared with the RI's objectives' and can be used to 'evaluate trends/progresses (yearly comparison) than to interpret absolute numbers' (OECD 2019). Hence, comparative analyses need well-structured logic models to manage complexity and yield results useful for decision-making (e.g. which RI can deliver higher educational benefits for the same level of contribution? Which RI is more effective in producing technological spillover?).

15.4.2 Modelling the socio-economic impact assessment of RIs

Big Science is such a suggestive domain that it inspired many different angles of analysis (Smit and Hessels 2021). Science performance has been investigated from ethical, organisational, historical, sociological, and anthropological perspectives.

In economics, there is a long-lasting tradition to study the impact of RIs within the broader topic of public investments in research and technological development. In analysing the socio-economic impact of RIs, the economic modelling can be seen as a natural framework. Theoretical and empirical models date back to the 1950s and were developed in order to justify public and private investments in R&D, measuring the economic and social benefits they can deliver. In this respect, the most common economic frameworks are those based on impact multipliers, methodologies grounded in the knowledge production function approach and cost–benefit analysis (Giffoni *et al* 2019 for a comprehensive review). One of the key strengths of economic modelling is that they are based on solid and tested conceptual models. A conceptual model is a theoretical representation of a system specifying the logic relationship among the element of the system.

Differently from multipliers and production functions models, which are based on macro-economic modelling and thus more suitable for the assessment of aggregated impact of large expenditure programmes, cost–benefit analysis (CBA), having a microeconomic foundation, lends itself particularly well for the assessment of individual RI (Florio 2019). An important element of economic modelling is the counterfactual perspective (i.e. the comparison of the observed situation with a hypothetical alternative situation). Without a credible counterfactual, the incremental impact of an investment could be zero, positive, or negative and the heuristic power of the analysis remains limited. A model also needs a metric, which can be strong (cardinal, e.g. a number) or weak (ordinal, e.g. a colour code), but consistent throughout the analysis. Without an associated metric, a model is usually not helpful for any kind of assessment. The metric can be expressed in terms of money, as it is often the case in economic modelling and CBA in particular, but not necessarily. The key point is that it has to be unique and consistent for all the considered effects, in order for them to be aggregable. Another important aspect is that both positive and negative impacts must be considered: while a positive impact increases the utility of an agent a negative impact decreases it. Describing only positive impacts (benefits) without describing the negative impacts (costs) can lead to biased assessment. A model is also necessary to understand what must be classified as cost and what is a benefit, as well as how to add them to reach a synthetic result. Benefits for some agents are costs for others, so sometimes they cancel out. A solid model can avoid double-counting, which happens when the identification of social agents, metrics, and impacts is blurred. Finally, a model can allow benchmarking, which is very useful when competing investment venues need to be considered.

15.4.3 Ownership and evaluation strategy

Designing a solid model to assess the socio-economic impact of a RI requires the necessary professional expertise that may not be available internally. External support can be necessary to structure the assessment framework, collect the relevant data, and put them in the needed shape. The model would state which impact can be realistically delivered and how, it would describe the understanding of which pathways lead to their materialisation and indicate the tools and methods more suitable for measuring such impacts.

However, the view and full endorsement of the exercise by top management and the involvement of internal staff is necessary to provide a credible representation of how the RI interacts with its ecosystem and generates effects. This would facilitate also the systematic use of the tool for monitoring throughout the lifecycle of research infrastructures. This is a useful exercise that would allow for adjustments or adaptations to new and evolving situations. It would also allow deciding about which impacts are to be prioritised and how they should be pursued. In order to respond to expectations from the stakeholders, this exercise should be transparent, timely and credible.

15.5 Conclusion

Justified by the new roles endorsed by science in general and Big Science in particular, there is now a routine recourse to systematic approaches to evaluation of RI, in line with common practices in other public policy areas. RI funding has become part of a new social contract between science and society, which requires a common understanding of shared objectives, language, and tools for negotiating and assessing investment cases. There is a high consensus that RI needs to explain their broader impacts both to politicians and the public. This, however, requires scientific communities to move from the more traditional narrative that scientific facilities are justified by a public perspective point of view primarily by a genuine desire to advance human knowledge and cultural and ethical values related to freedom, peace, and international collaboration.

In the design of a new or expanded research infrastructure, the driving rationale of scientific communities is pushing the knowledge beyond the frontier in their scientific field. Yet, being ultimately funded by taxpayers, investment decisions related to research infrastructures shall be subject to public scrutiny and acceptance on different grounds, pointing more to expectations of prosperity, job creation, innovation booster, and start-up creation. Scientific communities should not see this as undue pressure on scientific facilities to ask more than what is reasonable. They need to think about reconciling their main scientific objectives with all additional expectations to deliver impacts beyond science.

In this context, the socio-economic impact assessment exercise can set a common ground to assess the interplay between science and technology, socio-economic and political aspects. An in-depth understanding and awareness of socio-economic impact assessment methods and tools can provide effective support to managers of RIs in the public policy arena beyond auto-regulation.

References

Catalano G, Vignetti S, Ipolyi I, DeYoung Becker E and Dostalova Z 2018 Survey of research infrastructures on impact assessment practices, RI-PATHS project, Brussels https://doi.org/10.5281/zenodo.3946309

Cooke P, Gómez Uranga M and Etxebarria G 1997 Regional innovation systems: institutional and organisational dimensions *Res. Policy* **26** 475–91

De Solla Price D J 1963 *Little Science, Big Science* (New York: Columbia University Press)

ESFRI 2017 *Sustainable European Research Infrastructures, A Call for Action* (Luxembourg: EC Publications Office)

ESFRI 2019 *Monitoring of Research Infrastructures Performance* Working Group Report

European Commission 2010 *European Expert Group on Cost Control and Management Issues of Global Research Infrastructures* (Luxembourg: EC Publications Office)

European Commission 2012 *High Level Panel on the Socio-Economic Benefits of the European Research Area, Final Report* (Luxembourg: EC Publications Office)

European Commission 2017 The economic rationale for public R&I funding and its impact *Policy Brief Series* (Luxembourg: EC Publications Office)

European Commission 2018 *A New Horizon for Europe, Impact Assessment of the 9th EU Framework Programme for Research and Innovation* (Luxembourg: EC Publications Office)

European Commission 2019 *Research Infrastructures Make Science Happen* (Luxembourg: EC Publications Office)

Feller I 2013 Peer review and expert panels as techniques for evaluating the quality of academic research ed A N Link and N S Vonortas *Handbook on the Theory and Practice of Program Evaluation* (Cheltenham: Edward Elgar)

Florio M, Forte S, Pancotti C, Sirtori E and Vignetti S 2016 Exploring cost-benefit analysis of research, development and innovation infrastructures: an evaluation framework *Working Paper* No. 1/2016 CSIL

Florio M, Giffoni F, Giunta A and Sirtori E 2018 Big science, learning, and innovation: evidence from CERN procurement *Indust. Corp. Change* 27 915–36

Florio M 2019 *Investing in Science: Social Cost-Benefits Analysis of Research Infrastructures* (Cambridge, MA: MIT Press)

Giffoni F, Schubert T, Kroll H, Zenker A, Griniece E, Gulyas O, Angelis J, Reid A and Vignetti S 2019 Literature review on socio-economic impact assessment of Research Infrastructures, RI-PATHS project, Brussels https://doi.org/10.5281/zenodo.3946298

Giunta A, Florio M, Giffoni F and Sirtori E 2017 Big science, learning, and innovation: evidence from CERN procurement *Ind. Corp. Change* **27** 915–36

Gutleber J 2021 Rethinking the socio-economic value of Big Science: lessons from the FCC study ed H P Beck and P Charitos *The Economics of Big Science. Science Policy Reports* (Cham: Springer)

Martin B 1996 The use of multiple indicators in the assessment of basic research *Scientometrics* **36** 343–62

Martin B and Tang P 2007 The benefits from publicly funded research *Electronic Working Paper Series* Paper No. 161 Science and Technology Policy Research (SPRU)

Mazzucato M 2018 Mission-oriented innovation policy: challenges and opportunities *Ind. Corp. Change* **27** 803–15

OECD 2017 Strengthening the effectiveness and sustainability of international research infrastructures *OECD Science, Technology and Industry Policy Papers* (Paris: OECD Publishing)

OECD 2019 Reference framework for assessing the scientific and socio-economic impact of research infrastructures *OECD Science, Technology and Industry Policy Papers* (Paris: OECD Publishing)

OECD 2020 Optimising the operation and use of national research infrastructures *Policy Papers* (Paris: OECD Publishing)

Pancotti C, Pellegrin J and Vignetti S 2014 Appraisal of research infrastructures: approaches, methods and practical implications *Departmental Working Papers* 2014-13 Department of Economics, Management and Quantitative Methods at Università degli Studi di Milano

Phillips L J, Carvalho A and Doyle J (ed) 2012 *Citizen Voices: Performing Public Participation in Science and Environment Communication* (Bristol: Intellect)
Salter A J and Martin B R 2001 The economic benefits of publicly funded basic research: a critical review *Res. Policy* **30** 509–32
Short D B 2013 The public understanding of science: 30 years of the Bodmer report *School Sci. Rev.* **95** 39–44
Smit J P and Hessels L K 2021 The production of scientific and societal value in research evaluation: a review of societal impact assessment methods *Res. Eval.* **30** 323–35
UKRI-UK Research and Innovation 2020 The UK's research and innovation infrastructure: opportunities to grow our capability *UKRI Infrastructure Roadmap*
Vignetti S 2021 Designing a research infrastructure with impact in mind ed H P Beck and P Charitos *The Economics of Big Science, Science Policy Reports* (Cham: Springer)
Von Schomberg R and Hankins J (ed) 2019 *International Handbook on Responsible Innovation: A Global Resource* (Cheltenham: Edward Elgar)

IOP Publishing

Big Science in the 21st Century
Economic and societal impacts
Panagiotis Charitos, Theodore Arabatzis, Harry Cliff, Günther Dissertori, Juliette Forneris and Jason Li-Ying

Chapter 16

The socioeconomic impact of large scale research infrastructures: models, methods, and data

Andrea Bastianin and Massimo Florio

Large scale research infrastructures (LSRIs) have the potential to generate a variety of socioeconomic spillovers. The aim of this chapter is to review the models and methods used to assess the economic and societal impacts of LSRIs. We first provide a taxonomy of existing frameworks of analysis and then review the methodologies used to quantitatively asses the impacts of LSRIs. Next, we review the literature that provides quantitative and qualitative evidence on the effects of LSRIs and discuss methodological policy implications arising from such studies.

16.1 Introduction

Large scale research infrastructures (LSRIs) are built to pursue specific scientific missions in different domains such as physics, astronomy, materials science, space exploration, biology, and medicine. Given the budget constraints faced by policy-makers, the decision to upgrade an existing LSRI or build a new one is taken considering two main factors. First, and most importantly, the existence of a scientific case justifying the investment of public funds in an expensive and highly uncertain project. This decision is driven by consultations between scientists and has the goal to prove that a particular LSRI offers the best chances to significantly advance the frontiers of human knowledge in a specific domain.

Second, the socioeconomic impacts of LSRIs need to be documented as a support to the scientific evidence proving the need to finance a specific project. A case in point is CERN's next generation of particle colliders. For both the High Luminosity LHC and the Future Circular Collider, the conceptual design reports of these new

doi:10.1088/978-0-7503-3631-4ch16

colliders illustrate the expected socioeconomic impacts (see Brüning and Rossi 2015, Abada *et al* 2019).

While the main socioeconomic benefits that derive from LSRIs are easily identified (these include scientific output, human capital formation, industrial spillovers, cultural benefits), there is no consensus in the literature about which methodology should be adopted for quantifying such benefits. The aim of this chapter is to review a set of widely used methodologies commonly implemented to assess the socioeconomic impacts of LSRIs. See Florio (2019) for more comprehensive coverage of the topic.

We present six main approaches to impact assessment (IA): socioeconomic assessment based on impact multipliers, methodologies applying the knowledge production function, social cost–benefit analysis, approaches based on multi-methods and multiple partial indicators, theory-based approaches, and case studies.

The rest of the chapter is organized as follows. Sections 16.2–16.7 present the six different IA methodologies commonly used in studies concerning the evaluation of the socioeconomic benefits of investments. Section 16.8 concludes this chapter.

16.2 Impact multipliers

This IA method is rooted in macroeconomic theory and aims to quantify how an investment in a given industry or sector affects and propagates to the whole economy (of one or more countries). Accurately quantifying such an effect requires a well-designed system of national and regional accounts that describes the interrelationships between different sectors of an economy.

The underlying methodology dates back to 1941 with the publication of *The Structure of American Economy, 1919–29* by Wassily Leontief. The input–output model developed by Leontief exploits national and regional statistics to quantify the mutual relationships between different sectors of an economy and build a picture of it in matrix form. An inter-industry matrix presents the monetary value of inputs to each sector in its columns, while rows collect the value of each sector's output. Such a matrix is a key ingredient for estimating the effects of an investment project on a relevant macroeconomic indicator (e.g. employment, value added, gross domestic product) over a given time span.

The final effect measured with the method of impact multipliers has three main components: direct, indirect, and induced effects. Direct impacts capture capital investment and its first order effects (e.g. changes in employment and gross value added in the sector directly affected by the new investment) as producers adjust to the increased needs created by investment stimulus. Indirect or second-round effects measure further benefits along the upstream value chain (e.g. impacts due to additional purchases of entities in sectors that directly affected by the stimulus). Lastly, induced effects capture downstream impacts on the economy due to increased household spending on goods and services due to income effects produced by direct and indirect effects. Induced impacts affect sectors and industries that are not necessarily those where the initial investment project took place.

Effects not captured by the system of national/regional accounts cannot be measured with the impact multiplier approach (i.e. non-monetary cultural effects, environmental externalities). Moreover, impact multipliers measure effects over a precise time horizon. While direct effects quickly materialize, induced effects might take time to build up and hence defining the correct time span is crucial for capturing them.

16.3 Knowledge function approach

Growth theory, in its attempt to quantify which factors—and to what extent—account for the observed economic growth of different countries, has led to the development of the so-called knowledge function approach to IA. A production function is an economic law describing how a vector of inputs (e.g. capital, labour, and energy) is transformed into a scalar output whose precise definition (e.g. gross domestic product, value added, profits, patents) depends on whether the analysis is carried out at country, industry, or firm level.

Focusing on early developments, in the 1950s Solow showed that, because of decreasing returns to scale, the accumulation of physical capital accounts only for short-term economic growth. In the long-run, shifts of the aggregate production function and hence economic growth can only be explained by increases in total factor productivity, that subsumes the idea of 'technical change'. While in the original formulation technological progress is exogenously determined, later approaches—known as endogenous growth models—showed that R&D activities are necessary to sustain long-term growth. A path breaking study by Pakes and Griliches (1984) showed how to empirically implement the knowledge function approach. In this study, the authors showed how to link and estimate measures of knowledge output—proxied by patents—to R&D and innovation expenditure at the firm level. Since then, the knowledge function approach has been used in empirical analyses at the firm, industry, and country level.

Much like IAs based on impact multipliers, analyses that rely on the knowledge function approach to estimate the socioeconomic effects of LSRI can capture only a small subset of such benefits. Moreover, while this approach can be informative for policy-makers interested in having a rough estimate of the size of impacts of investments in R&D, it does not deliver detailed information on the sources of such effects.

16.4 Social cost–benefit analysis

Social cost–benefit analysis (CBA) is used to assess whether costs and benefits generated by an investment project improve social welfare. The main output of a social CBA is thus not a financial performance indicator, but a measure of impact of the investment project on the welfare of the society.

To asses changes in social welfare CBA relies on the notion of shadow prices that reflect the social opportunity costs of goods and services. Shadow prices are used in place of market prices because these might be distorted by fiscal corrections. Moreover, shadow prices allow including into the social CBA of an investment project its intangible impacts—such as benefits of goods provided for free—that would not be reflected in market prices.

A crucial step in a social CBA is the identification of all costs and benefits; the socioeconomic impact of an investment project is the cumulative difference over time of benefits due to the project and costs for generating those benefits. In social CBA costs do not include only investment and operating costs, but also social costs such as negative environmental externalities (e.g. pollution, noise, traffic during the construction of a new LSRI). To gauge the net benefits for the society, the results of a CBA are summarized with a number of key descriptive statistics. Probably, the most important is the 'net present value' (NPV); this is the difference between discounted benefits and costs. Discounting is applied to the whole stream of costs and benefits over the lifespan of the project, that is the time period over which we expected them to be observed. An investment project passes the CBA test when its NPV is greater than zero. When more than one project is available a social CBA evaluates the incremental NPV of such alternatives.

In the computation of social costs and benefits, the analyst relies on assumptions about their time profile and a set of parameters used to compute benefits and costs (e.g. the percentage of hi-tech procurement, the salary increase due to experience in a LSRI, cost ranges). Assuming that those parameters have well-defined statistical distributions Monte Carlo methods can be used to assist social CBA in the quantification of risks associated with different projects.

Social CBA is rooted in economic theory (Florio and Pancotti 2023, Boardman *et al* 2017, Johansson and Kriström 2018) and can be easily adapted to quantitatively assess the multi-dimensional nature of costs and benefits arising from LSRI. CBA has been widely used in the evaluation of LSRIs. The 'Horizon 2020—Work Programme 2018–2020 on European Research Infrastructures' mentions that the preparatory phase of new European Strategy Forum on Research Infrastructures (ESFRI) projects should include a CBA (European Commission 2017). The European Investment Bank uses CBA methods to decide the allocation of its funds among different projects, including for research infrastructures. For examples of the application of the CBA methodology to CERN particle colliders, refer to the Bastianin *et al* (2022) and Florio *et al* (2016).

16.5 Multi-methods multiple indicators

This approach combines insights from science policy and management and was developed specifically for the evaluation of socioeconomic benefits arising from publicly funded basic research (see Martin 1996). The premise is that investments in research yield a large variety of effects and hence a wide range of indicators and methods are required to evaluate them in a comprehensive manner. Moreover, within this approach all quantitative indicators are seen—at best—only as partial indicators, because research outputs and the socioeconomic benefits deriving from them are complex and multi-faced. For instance, if the aim is to measure scientific outputs it is possible to rely on indices relating to different aspects such as activity, production, and quality (e.g. number of publications, citations).

These methods usually target four categories of benefits: scientific, educational, technological, and cultural effects. Each of these categories involves several different classes of benefits. For instance, depending on the nature of the project, technological benefits can relate to the new products, processes, and spin-offs. Partial indicators are often combined together to produce an index that summarizes a group of effects, rather than an individual impact.

The reliability of this approach hinges on a sound data collection process and ability to construct good empirical proxies of the impacts of interest. While having a large number of indicators is useful to capture the complex nature of socioeconomic impacts of research, having too many of them might not help decision-makers that are faced with too much information, possibly pointing to different conclusions. Notwithstanding these limitations, the approach based on multiple indicators might be useful for both policy-makers and managers of LSRI. Indicators are a useful tool for managers because measure progresses toward objectives and facilitate comparison over time.

16.6 Theory driven approaches

Theory driven approaches can be both qualitative and quantitative in nature and can be implemented in a variety of ways. Often theory-based approaches yield a flowchart-like representation of causal and impact pathways that are observed triggered by a policy intervention or a new investment. In the case of an LRSI, impacts are traced in terms of their distance from funded activities. The connections between causes and effects are constructed around the so-called 'theory of change'. This concept describes a set of assumptions about how an intervention an intervention achieves its goals and under which conditions.

The implementation of theory driven approaches to IA is desirable when the focus is more on the identification of causal chains than on the quantification of the impacts of an intervention. Moreover, these approaches are well suited for capturing the multi-dimensional nature of socioeconomic benefits deriving from LSRI.

Interpreting results derived from theory driven approaches might not be easy. If stakeholders disagree about which effects are important or about the way in which interventions and effects are related, it might be necessary to resort to other IA methods.

16.7 Case studies

Case studies encompass a set of widely used—mostly qualitative—methodologies on which investigators rely to provide an in-depth investigation of particular events, firms, management practices, or, more generally, interventions. Methodologies used in case studies range from surveys and in-depth interviews to data collection and analyses. The choice of such methods of analysis depends on the nature of the problem being investigated and the final goal.

There are two broad categories of case studies: within- and cross-case analyses. Within-case studies focus deeply on a single representative case from which an attempt to draw more general conclusions is made. Cross-case studies compare multiple cases to draw conclusions that are more general. Moreover, a combination of the two approaches is often used to have more information for a generalization of what has been learnt from the analysis.

The most important strength of this IA method is its ability to offer a great level of detail that reflects the uniqueness and complexity of LSRI. Moreover, an additional strength of case studies, with their narrative approach, is their ability to communicate results. It is, however, worth pointing out that selecting the right histories is key for the ability of the methodology to provide conclusions with external validity.

16.8 Discussion and conclusions

Evaluating economic and societal impacts of LSRIs is a very important task to assist decision-makers. Both policy-makers and managers of LSRIs need such an evaluation to complement the evidence showing that a scientific case does exists and a new investment is needed to advance the frontier of human knowledge.

While different IA methods do exist, there is no consensus on which should be used. This lack of consensus is due to several factors that include data availability and quality, difficulties in translating theoretical concepts into empirical proxies, identification of causal relationships, and the multi-dimensional and complex nature of economic and social costs and benefits generated by LSRI.

We have briefly reviewed six approaches to the IA of LSRIs that can ideally be grouped into three main analytical frameworks. The first group of methods, based either on microeconomic or macroeconomic theory, includes impact multipliers, the knowledge function approach, and social cost–benefit analysis. These methods—well accepted inside and outside the academic community—have solid theoretical foundations that lead to a high degree of internal consistency. Analyses based on these methods can be easily reproduced, generalized, and adapted to reflect the specificities of different LSRIs. Impact multipliers and analyses based on the knowledge function approach lead to an estimate of economic benefits accrued by an LSRI that are expressed in terms of change in one or more economic variables of interest. Results based on these methods are thus not informative concerning non-monetary impacts, such as cultural effects or environmental externalities. If the analyst is also interested in such non-monetary impacts, then social CBA should be considered.

A second set of methods includes theory driven approaches and case studies. These IA methodologies are well suited when the focus is more on the identification and explanation of causal relationships, than on the quantification of the magnitude of impacts. The strength of these methods is their ability to communicate results and describe causal nexuses in narrative form. The lack of quantitative assessment and the degree of generalizability of results might, however, represent an issue.

Lastly, we have analysed approaches based on mixed methods and multiple partial indicators. These methods yield a multiplicity of indicators that capture different socioeconomic impacts of LSRIs. Having a variety of indicators is useful because they explain in detail such impacts from different points of view. Indicators, however do not necessarily measure socioeconomic impacts. For instance, the number of patents deriving from LSRIs does not measure socioeconomic impacts directly. To have a measure of socioeconomic impact, further analysis is required to determine which patents have been commercialized and used to create profitable products.

To sum-up, none of the methodologies represents a satisfactory and comprehensive solution to be used in the analysis of socioeconomic benefits deriving from LSRIs. While each of them require specific advances for what concerns the underlying analytical framework, improvement in terms of data availability and collection is a common theme that is partly under the control of managers of LSRIs. More and better data would be useful for all the methods surveyed in this chapter and thus represent a strategic lever that can be used.

References

Abada A *et al* 2019 FCC physics opportunities—future circular collider conceptual design report volume 1 *Eur. Phys. J. C* **79** 474

Bastianin A, Del Bo C F, Florio M and Giffoni F 2022 Projecting the socio-economic impact of a big science center: the world's largest particle accelerator at CERN *Appl. Econ.* https://doi.org/10.1080/00036846.2022.2140763

Boardman A E, Greenberg D H, Vining A R and Weimer D L 2017 *Cost–Benefit Analysis: Concepts and Practice* (Cambridge: Cambridge University Press)

Brüning O and Rossi L (ed) 2015 *The High Luminosity Large Hadron Collider: The New Machine for Illuminating the Mysteries of Universe* vol 24 (Singapore: World Scientific)

European Commission 2017 Horizon 2020—work programme 2018–2020. European research infrastructures (including e-infrastructures). Annex 4 *European Commission Decision* C(2017) 7124 of 27 October 2017 https://ec.europa.eu/research/participants/data/ref/h2020/wp/2018-2020/main/h2020-wp1820-infrastructures_en.pdf

Florio M and Pancotti C 2023 *Applied Welfare Economics: Cost–Benefit Analysis of Projects and Policies* 2nd edn (London: Routledge)

Florio M 2019 *Investing in Science: Social Cost–Benefit Analysis of Research Infrastructures* (Cambridge, MA: MIT Press)

Florio M, Forte S and Sirtori E 2016 Forecasting the socio-economic impact of the Large Hadron Collider: a cost–benefit analysis to 2025 and beyond *Technol. Forecast. Soc. Change* **112** 38–53

Johansson P O and Kriström B 2018 Cost–benefit analysis *Cambridge Elements of Public Economics* (Cambridge: Cambridge University Press)

Martin B 1996 The use of multiple indicators in the assessment of basic research *Scientometrics* **36** 343–62

Pakes A and Griliches Z 1984 Estimating distributed lags in short panels with an application to the specification of depreciation patterns and capital stock constructs *Rev. Econ. Stud.* **51** 243–62

Chapter 17

Socio-economic impact for the European Spallation Source (ESS)—a narrative and pathway development approach

Jason Li-Ying, Erika Susan Dietrichson, Ute Gunsenheimer and Lenka Unge

17.1 Introduction

The European Spallation Source (ESS) ERIC is a ground-breaking scientific and technical project and a future international multidisciplinary research facility based on the world's most powerful neutron source. The ESS aims to advance the use of neutrons in scientific discovery, enable access to state-of-the-art instrumentation, and provide world-class support to the scientific community. Once fully operational, the facility's unique capabilities will both greatly exceed and complement those of today's leading neutron sources, enabling new opportunities for researchers across the spectrum of scientific discovery, including materials and life sciences, energy, environmental technology, cultural heritage, and fundamental physics. The facility is under construction in Lund, Sweden, with its Data Management and Software Centre (DMSC) located in Copenhagen, Denmark. In 2015 ESS became the first European Research Infrastructure Consortium (ERIC) established in Scandinavia.

The life cycle of the facility is divided into several phases. The construction phase started in 2013, and the ground-breaking ceremony took place in September 2014. It is currently running in parallel to the initial operations phase (2019–25). First science is planned to begin during this period in 2025.

The ESS is a partnership of 13 member countries, including the two host countries, Sweden and Denmark. The construction phase, up to August 2020, has been 72% completed. The construction is financed 53.3% by its host countries, and the remaining 46.7% by the other 11 member countries. In total, around 550 M€ of the construction costs will be supplied through in-kind contributions (IKCs). The member countries have committed to cash or in-kind towards the construction of ESS.

doi:10.1088/978-0-7503-3631-4ch17

Through the in-kind model, partner institutions—on behalf of their national governments—contribute to the ESS construction in supplying equipment, design documentation, personnel, or other services. These contributions are accounted for in the overall contribution from each member country, alongside their financial contributions. Under its in-kind model, ESS works with more than 40 European partner institutions and nearly 130 collaborating institutions worldwide. Nearly 30% of the construction cost is expected to be supplied through IKC. An in-kind model on the scale of ESS has never been attempted previously in a European Big Science project.

The construction of ESS as one of the largest research infrastructure projects to date is only made possible by dedicated public investment in Europe, paid for by tax-payer contributions. Thus it is central to the vision and missions of ESS to generate and demonstrate its return to society at large in the long run. The vision of ESS is to 'build and operate the world's most powerful neutron source, enabling scientific breakthroughs in research related to materials, energy, health and the environment, and addressing some of the most important societal challenges of our time' and its missions are reflected in its strategic objectives (see table 17.1). The vision and missions of ESS indicate clearly that ESS is not only supposed to achieve scientific breakthroughs, but also to generate a much wider range of societal impacts through, among others, research activities and dissemination of public knowledge, extension of scientific networks, societal outreach, economic growth in local regions, procurement of innovative technologies, industrial collaboration, and in-kind partnership. These impacts are the results of the *enabling science* of ESS and are expected to be evident directly on the member countries and to be extended indirectly on many countries and regions.

The assessment of the long-term impact of ESS is not easy when it has not yet begun its scientific user programme. Even though ESS is still in the middle of the construction phase, it has already been engaged in various activities and is keen on demonstrating its socio-economic impact (SEI) on its direct partners and stakeholders and indirect networks along with the pursuit of its strategic objectives. Thus, this chapter showcases the theoretical foundation and methodological approach used by ESS to collect early evidence and some important aspects of the SEI of ESS

Table 17.1. Strategic objectives based on missions, vision, and values of ESS.

Strategic objective 1: ESS produces research outputs that are best-in-class both in terms of scientific quality and in terms of socio-economic impact.
Strategic objective 2: ESS supports and develops its user community, fosters a scientific culture of excellence, and acts as an international scientific hub.
Strategic objective 3: ESS is built safely on time and on budget, operates safely, efficiently, and economically, and responds to the needs of its stakeholders, its host countries and member countries.
Strategic objective 4: ESS develops innovative ways of working, new technologies, and upgrades to capabilities needed to remain at the cutting edge.

based on the narratives and impact pathways that are specifically relevant for the period of the construction phase until 2018.

17.2 Socio-economic impact assessment: prior art and ESS methodology

17.2.1 Socio-economic impact for RIs

As recommended by ESFRI's Strategic Report on Research Infrastructures (European Strategy Forum on Research Infrastructures 2018), European research infrastructures (RIs) must set up effective means of determining the economic and wider social value of RIs, and the SEI assessment of RIs has been an increasingly important topic in recent years. SEI methodologies and challenges are regularly discussed within the European RI community, whose stakeholders have developed an interest in the effects of science on society beyond scientific breakthroughs.

Assessing the socio-economic impact of RIs is important for four distinct categories of stakeholders: (i) *RI managers* who expect value for different audiences (social economic and scientific); (ii) *funders* seeking economic and social returns from their investment; (iii) *the public* at large looking for visible change in economic conditions, business, and social activities; and (iv) *employees and users* to enhance expertise and research capabilities through excellence.

Much progress in the field of SEI in relation to large RIs has occurred during the last few years with the development of reference frameworks and assessment protocols by a variety of significant actors, including but not limited to the OECD, ESFRI, and various projects funded by the European Commission. The OECD 'Reference framework for assessing the scientific and socio-economic impact of research infrastructures' (OECD 2019) report was published in the spring of 2019. The reference framework is made up of a list of core impact indicators (CIIs) along with a comprehensive list of standard indicators. The December 2019 ESFRI Working Group report 'Monitoring of Research Infrastructures Performance' (ESFRI 2019), provides a list of 21 proposed key performance indicators (KPIs) that were developed to be used in a periodic review of ESFRI landmarks. The EU Horizon 2020 funded projects RI-PATHS (GA no. 777563) and ACCELERATE (GA no. 731112) place a stronger focus on different pathways of the impact of RIs through the use of narratives that are received as relevant for different stakeholders and supported by both quantitative and qualitative empirical evidence.

The ability to measure and evaluate socio-economic as well as scientific impact is key for an RI to make itself countable and sustainable in the long-term. Thus, during the Horizon 2020 (H2020) funded project BrightnESS (2015–18; H2020 grant agreement no. 676548), ESS had put an effort in developing a pilot framework to formulate the assessment methods for the organisation's SEI (Mangematin and Bally 2018) together with the Grenoble School of Management (GEM) and KEDGE Business School. The BrightnESS project played a critical role in laying down the foundation early in the life cycle of ESS and setting up an overall approach to systematically measure its socio-economic impact. The BrightnESS project took a first step to outline socio-economic metrics and indicators relevant for the construction and operations phases of ESS, and to collect data covering the pilot year of 2016.

To follow up with the BrightnESS project and align with the strategic priority of assessing RI's societal impact laid down by the various initiatives in Europe mentioned above, ESS has continued its efforts in further developing and operationalising an actionable assessment framework of socio-economic impact in the BrightnESS^2 project for long-term use. The BrightnESS^2 Work Package (WP) 5, task 5.3 aims to make the assessment of the socio-economic impact of ESS a systematic annual routine of ESS.

The BrightnESS^2 WP 5, chaired by the authors of this chapter, has built on the ground laid down by the BrightnESS to refine the SEI assessment framework for ESS and to collect and analyse data covering the construction phase until 2018. This has been done with the support of the management across several divisions of ESS, along with inputs from many of the industrial suppliers of ESS and in-kind partners. The original metrics suggested by the BrightnESS project were mostly kept in the current framework, adapted into clear indicators and measures. In addition, some changes have been made to regroup or rephrase the items in the current framework to better align with the new EU initiatives and programmes (RI-PATHS and ACCELERATE). Relevant data for measuring most of these indicators have been monitored and stored within the internal management system of ESS already at the time of making this report.

17.2.2 Making an impact in pursuit of ESS strategic objectives—building narratives

ESS was founded with an ambitious vision and missions because it has from its beginning put a strategic focus on making science countable and impactful for the society at large. ESS impacts its direct environment in different ways through its activities. It is expected to enable excellence in science via granted research and demonstrated by high quality scientific publications, promoting science to citizens, creating economic activities (e.g. through procurement), contributing to education and training, creating attractiveness for the local areas, and shaping scientific and economic networks (e.g. through in-kind partners). All these potential impacts on society, not only on science, must be taken into account and measured to show that ESS is not only an infrastructure focused on science for scientists, but also on science for society.

These intended impacts will take time to emerge and become evident, considering the developmental process of ESS. Currently, ESS is still under the construction phase, which runs until 2025, and moving towards the initial stage of the operations phase. This is an important context for the readers to consider.

Since the vision of ESS is to achieve scientific breakthroughs in neutron research, which in turn generates positive societal impact, it is reasonable to understand that the assessment of the long-term impact of ESS in a wide range is relatively difficult when it has not yet begun its scientific user programme. However, even though it is still in the construction phase, ESS has already been engaged in various activities and has been keen on creating and demonstrating its scientific, economic, innovation, social, and environmental impacts on its direct partners and stakeholders, and indirect networks in pursuit of its strategic objectives. Thus, the early evidence of the SEI of ESS demonstrated are collected and organised based on narratives and impact pathways that are specifically relevant for this particular period of the construction phase until 2018.

Following the principle of several recent EU forums, initiatives and programmes, such as ESFRI (ESFRI 2019), ACCELERATE (GA no. 731112), and the RI-PATHS (GA no. 777563; Griniece *et al* 2019, 2020), we propose that the pursuit of the strategic objectives of ESS, as defined by the ESS mission statement, will drive the socio-economic impact of ESS. This report therefore uses a mix of indicators relevant to the ongoing construction phase (evidence collected for the period 2013–18), and can find their references to the prior art of EU or OECD suggestions.

In pursuit of each strategic objective, an impact narrative is developed as the rationale behind the resulting impacts being demonstrated in this report. Each impact narrative is connected to relevant indicators and measures. These impact narratives are not intended to make any assessment of the state of the ESS project itself.

Strategic objective 1: *ESS produces research outputs that are best-in-class both in terms of scientific quality and in terms of socio-economic impact.*

Impact narrative pathway*: Publication–citation–collaboration–recognition*
As a world-class scientific research institution, ESS will be the centre of numerous *research activities* in the relevant scientific fields in the years to come. The *research outputs* must be excellent in a sense of scientific value, as well as creating impact to ESS research partners within the scientific community. ESS can already demonstrate considerable research outputs that reflect both its ambition to be the best-in-class regarding the technology development along with the construction of ESS and its preparedness for the future scientific breakthroughs it is going to enable. Although it is not yet an operational facility, ESS has become a leading member of the scientific community regarding accelerator, target, neutron technologies (including detectors, choppers, optics, polarisation, sample environment, control, and data infrastructure), and neutron instrumentation and method development, demonstrated by the scientific publications of ESS and collaborative research with, amongst others, many partner institutions, universities, and industry. To achieve this strategic objective, ESS generates and disseminates enabling and impactful scientific knowledge.

Several indicators are used to demonstrate the impacts achieved in pursuit of strategic objective 1. These indicators include but are not limited to the following, for example:

- *Research activities*: with and without collaboration with various partners, including several neutron source scientific networks, i.e. the League of Advanced European Neutron Sources (LENS) and the European Neutron Scattering Association (ENSA).

- *ESS publications*: including their citations, h-index, co-authorship with partner universities, other RIs, and industry, share of publications per member countries, and the relationship between ESS publications and industry R&D collaboration.

Strategic objective 2: *ESS supports and develops its user community, fosters a scientific culture of excellence, and acts as an international scientific hub.*

Impact narrative pathway 1: *Creating and shaping scientific networks and communities*
It has long been understood within the science community that the long-term sustainability of the European Spallation Source (ESS) is dependent above all on the long-term sustainability of the existing community of European neutron scientists and facilities—the future users of ESS. For its scientific success, ESS must be directed by the scientific needs of its future users. The need for ESS long-pulse neutron source and corresponding instrument suite is the result of a bottom-up neutron user driven approach.

Even though it is still at the construction phase, ESS has been actively building its future user community and broadening its network through collaborative research supported by various activities. For instance, by leading and participating in various grant projects, ESS has an opportunity to develop relationships with new partners from other research facilities, academia, and with industrial firms, within and outside of Europe, so that the status of an international scientific hub will be achieved. The role of being a scientific hub can also be demonstrated in terms of, for instance, the scientific events and user meetings organised by ESS, as well as the network benefit derived by in-kind partners from working with ESS.

Several indicators are selected to demonstrate this impact pathway in pursuit of strategic objective 2. These indicators include but are not limited to the following, for example:

- *Grant proposals*: including information about granted and non-granted proposals, involvement with in-kind partners, industry partners, and other RIs.
- *Scientific events*: including conferences, symposiums, and workshops with and without collaboration with various partners, including other neutron source scientific networks, i.e. LENS and ENSA.
- *User meetings*: including meeting in collaboration with other relevant RIs.
- *Research collaboration with industry*: including grant projects and other procured industry R&D.

Impact narrative pathway 2: *Communication, outreach, and engagement*
To foster a scientific culture around neutron source research and related fields, it is important to engage society through various public communication and outreach activities. These activities can be monitored through, for instance, site visits and online information sessions, online media channels, including social media, and

involvement in various education programmes. In the coming years, these efforts may also be demonstrated through public events.

Several indicators are selected to demonstrate this impact pathway. These indicators include but are not limited to the following, for example:

- *Informational visits*: including online sessions about the ESS construction site and research.
- *Media exposure*: including social media activities and attention received.
- *Involvement in education*: including the contribution to partner universities' relevant education and PhD and Postdoc programmes

Strategic objective 3: *ESS is built safely on time and on budget, operates safely, efficiently, and economically, and responds to the needs of its stakeholders, its host countries and member countries.*

Impact narrative pathway 1*: Safety–energy use–waste management–environmental impact*

Throughout the construction of the facility, ESS has been built with highest consideration of safety and efficiency, and environmental impact. ESS seeks to procure goods, services, and works with a reduced environmental impact throughout their life cycle. This may include, for example, the reduction of greenhouse gas emissions and air pollutants, improved energy, and water efficiency, use of renewable resources, reduced hazardous waste, and support for refuse and recycling. Meanwhile, the construction of ESS also responded to the needs of ESS member countries and its stakeholders, primarily achieved through cost-efficient procurement activities, the tendering system, and budgetary means. In later years, safety activities will be monitored by ESS.

Several indicators are selected to demonstrate this impact pathway in pursuit of strategic objective 3. For example:

- Waste volume.
- Hazardous waste.
- Energy consumption.

Impact narrative pathway 2: *Employment and procurement—rendering economic benefits to member countries through suppliers and in-kind partners*

First, public procurement is commonly used as a demand-side policy tool that renders *economic benefits* for industrial suppliers, including creating new jobs and business. These benefits are expected to be evident already during the construction phase of ESS for both host countries and member countries. Economic benefits cannot be expected to emerge without effective implementation of ESS procurement activities according to the European Spallation Source ERIC Procurement Rules. Second, the ESS construction project relies heavily on IKCs for its successful execution. The ESS IKC model was chosen by the organisation's founding partners to enable a distributed effort that would both capitalise on and enhance competencies across Europe, while expediting construction. As a result, the procurement activities and the IKC model allow ESS to leverage the collective knowledge, experience, and resources of Europe's leading research institutions and industry, support the creation of jobs, strengthen the high-tech industry, and foster collaboration between key scientific stakeholders, without which it is simply impossible for ESS to be built and keep operating successfully.

It is also important to consider these impacts of ESS during the construction phase regarding the way in which ESS interacts with suppliers and in-kind partners so that a mutually supportive relationship can foster future cooperation that leads to economic and innovation benefits. The construction and operation of ESS is not supported by an established laboratory, and therefore it is essential the project recruit skilled, mobile professionals from partner countries and other sources, and that staff are in turn able to feed back into the scientific community. Diversity goes hand in hand with recruitment goals across the ESS member countries and project stakeholders. Staff diversity and mobility (such as relocations and secondments) are also being tracked. Several indicators are selected to demonstrate this impact pathway in pursuit of strategic objective 3. For example:

- Tenders received from and contract awarded to member countries.
- Number of suppliers and their perceived economic benefits.
- ESS support to suppliers.
- Interaction and cooperation with ESS assessed by in-kind partners.
- ESS Industrial Liaison Officers Network (ILO) Meetings.
- Employment of ESS, in-kind partners, and Skanska during the construction phase.
- ESS staff diversity in terms of gender and nationalities.

Strategic objective 4: *ESS develops innovative ways of working, new technologies, and upgrades to capabilities needed to remain at the cutting edge.*

Impact narrative pathway*: Technology push through procurement and in-kind contribution and building the innovation capacity of ESS and industry in member countries*

Building a neutron facility of this magnitude in size and involvement of specialised technologies requires the world-class expertise, cutting-edge technologies, and innovation. That is why ESS works closely with suppliers and in-kind partners in the member countries to make technological and process innovations and build innovative capabilities within ESS and with partners. This is achieved through joint R&D via in-kind contributions and through specialised procurement that add innovation value to the suppliers in member countries. In doing so, knowledge transfer and positive spillover effects in a medium to long run will contribute to innovation capacity building among the member countries.

The ESS IKC model creates the opportunity to enrich knowledge and skills and to create business in each member country—thus benefitting from ESS membership already in the construction phase. The benefits are expected to exceed the direct relationship between ESS and its in-kind partners, through collaborative projects beyond the scope of the in-kind contributions, encouraging future users, and procured R&D. Similarly, by working with ESS, suppliers may benefit in terms of developing new technologies, products, and services, and potentially extending market opportunities. Thus, in pursuit of this strategic objective, ESS makes its impact by uplifting technological innovation capacity of itself and of in-kind partners and industry in the member countries.

Several indicators are selected to demonstrate these impacts in pursuit of strategic objective 4. For example:

- In-kind agreements and in-kind technical annexes.
- Cost book value of IKC by member countries.
- Innovation level of IKC.
- Innovation benefits for suppliers.
- Types of innovations supplied to ESS.
- ESS patenting.
- ESS technologies that contributed societal challenges.

17.2.3 The ESS methodology—indicators, measures, and complementary surveys

The current SEI assessment framework consists of a series of indicators and measures, built on the prior art made by the original BrightnESS project and aligned with other EU initiatives. One distinctive approach of the current ESS SEI assessment is that it maintains a clear distinction between indicators and measures, taking pains not to equate the two. The report, therefore, uses measures—that is, data—only as the means to support the assessment of a particular indicator. For instance, 'collaboration between MAX IV and ESS' is an indicator, which can be measured in different ways, among which 'publications co-authored with researchers at MAX IV' is one of the measures. Many measures are further broken down into more detailed ones, e.g. per member country.

It is a common approach recognised several European SEI initiatives and programmes and followed by the BrightnESS project as well that SEI indicators can be categorised in at least the following types:

- Scientific impact.
- Economic impact.
- Innovation impact.
- Social impact.
- Environmental impact.

The categorisation above is adopted and regrouped into concrete indicators or measures into these impact types. While most indicators are quantitatively measurable, some qualitative measures (stories and cases) are used as well. In addition, three new indicators were added into the SEI assessment of ESS, i.e. cost performance index as economic impact, technology transfer as innovation impact, and ESS inventors as innovation impact.

While a full list of indicators and measures, grouped into five types, will be too long to be included in the chapter, a summary of indicators and measures according

Table 17.2. Number of indicators and measures per ESS project phase.

Phase	Total number of indicators...	...of which are measured per member country	Total number of measures...	...of which are measured per member country
Construction only	3	(1)	6	(4)
Construction and operations	34	(8)	137	(15)
Operations only	21	(1)	62	(3)
Total	**53**	**(10)**	**205**	**(22)**

to their relevance to the construction phase and operations phase is shown in table 17.2.

To make the SEI for ESS, relevant data for the construction phase until 2018 were collected with the support from the relevant divisions across ESS who own or have access to the data. With the development of the ESS User Office, new measures have been created reflecting the User Offices' own user survey that will be much more detailed in comparison to the user survey designed by the BrightnESS project. The employee survey included in the BrightnESS framework has been replaced with new measures that are based on the biannual employee survey conducted by the HR division. The survey for funders and stakeholders from the BrightnESS framework may continue to be used, whenever necessary. The 2016 pilot test results were not re-used. In addition, the ESS annual research reports are used in this report.

The current framework includes new data sources from two additional surveys, one for suppliers and the other for in-kind partners, which complement ESS internal data concerning procurement activities and in-kind contributions. Similar approaches have been used by CERN to investigate the economic and innovation impact on its suppliers (Sirtori *et al* 2019). These two surveys provide an external perspective about the impact resulting from working with ESS on suppliers and in kind partners. The surveys focused on the type of contributions these actors have made to the ESS project, the innovation level of their products and services, their perceived relationship with ESS, and the effects on their networks. Suppliers were also asked about any economic benefits generated as a result of their work with ESS. Insight derived from these two surveys is unique to ESS and invaluable to probe the indirect SEI impacts that ESS may create. The supplier survey was sent to approximately 2600 suppliers by e-mail (approximate number after e-mail bounce backs). Eventually, 284 respondents completed the survey in full, resulting a nearly 11% response rate, which is satisfactory for a randomised industry survey. The in-kind partner survey was sent to 69 work package coordinators, across the ESS member countries. Some organisations opted to submit a single consolidated answer representing all work packages coordinated by the organisation. As such, the survey received 27 responses from 12 member countries, with Spain unavailable.

As the surveys were distributed during the COVID-19 global pandemic during March and April 2020, we presume that the response rate could have been higher if the disturbance of COVID-19 were absent. For this reason, ESS is confident that the response rate of these two surveys will be higher in the years to come, as they will be distributed and administrated on a regular basis. Thus, in this respect ESS is actively building data regarding the impact on industrial suppliers and in-kind partners.

The novelty of the ESS methodology lies in the way the indicators and measures are defined and constructed. First, the full list of indicators and measures are grouped according to the categorisation made by the BrightnESS project and well respected by the current state-of-the-art. Second, each indicator can find it roots in indicators defined by either OECD or ESFRI. Third, the indicators and measures are labelled for their relevance for different phases of the ESS life cycle. Fourth, as each indicator has multiple measures, ESS managers have the flexibility to choose the most relevant and practically feasible measures to collect and report data, depending on the needs of SEI reporting. Last but not least, the combination of ESS internal data and external surveys as data sources reduces any bias that may exist within a particular RI.

17.3 Conclusion

Europe's research infrastructures (RIs) enable major scientific breakthrough and incubate the discoveries that foster innovation. Given the overwhelming public consensus on the need to address the societal challenges facing Europe and the world, RIs must embrace the responsibility to create and sustain societal impact. Within this context, the European Strategy Forum for Research Infrastructures' (ESFRI) 'Strategic Report on Research Infrastructures' (Roadmap 2018) recommends that European RIs establish effective means to determine their economic and wider social value and prioritise the development of indicators and measures that help to establish causality between RIs and socio-economic impact (SEI). This early SEI pilot, undertaken during the ESS construction phase, marks it as a pioneering document among European RIs, which share the common responsibility to maximise their societal impact by enabling scientific discovery.

Embedded in the ESS vision and mission statement is the goal to achieve not only scientific breakthrough, but also to generate a wide range of societal and economic impacts through research activities, public dissemination of knowledge, reinforcement and extension of scientific networks, societal outreach, promotion of economic growth in local regions, procurement of innovative technologies, and research collaborations with scientific organisations and with industry. The impacts of these activities are expected to be evident directly on the member countries and to extend indirectly across many additional countries and regions in the long run. The ESS SEI assessment framework, thus, has the potential to be adopted and adapted by other European RIs according to their own scientific, economic, and political context.

References

ACCELERATE n.d. Funded by the European Union's Horizon 2020 research and innovation programme under grant agreement No. 731112 https://accelerate2020.eu/

ESFRI 2019 Monitoring of research infrastructures performance *Report of the ESFRI Working Group* https://www.esfri.eu/sites/default/files/ESFRI_WG_Monitoring_Report.pdf

European Strategy Forum on Research Infrastructures 2018 European Roadmap for Research Infrastructures http://roadmap2018.esfri.eu/media/1066/esfri-roadmap-2018.pdf

Griniece E, Kroll H, Cvijanovic V and Reid A 2019 Concept note of modular impact assessment framework *Deliverable* 4.1 Horizon 2020 Programme https://ec.europa.eu/research/participants/documents/downloadPublic?documentIds=080166e5c1e7bfac&appId=PPGMS

Griniece E, Kroll H, Zenker A, Martin C, Barberis M, Helman A and Garcia G 2020 Validated IA model, Horizon 2020 Programme https://ec.europa.eu/research/participants/documents/downloadPublic?documentIds=080166e5c1e7bfac&appId=PPGMS

Mangematin V and Bally F 2018 Report on Methodology and Criteria for Socio-Economic Impact Assessment of ESS During Construction and Operations Deliverable 1.7 Horizon 2020 Programme https://ec.europa.eu/research/participants/documents/downloadPublic?documentIds=080166e5bf93917e&appId=PPGMS

OECD 2019 Reference framework for assessing the scientific and socio-economic impact of research infrastructures *OECD Science, Technology and Industry Policy Papers* No. 65 https://oecd-ilibrary.org/docserver/3ffee43b-en.pdf?expires=1554432241&id=id&accname=guest&checksum=B5561A1B326FC5BFA51B342D89E7B97E

RI-PATHS n.d. Funded by the European Union's Horizon 2020 research and innovation programme under grant agreement No. 777563 https://ri-paths-tool.eu/en

Sirtori E C, Florio M, Catalano G, Caputo A C, Pancotti C C and Giffoni F C 2019 Impact of CERN Procurement Actions on Industry: 28 Illustrative Success Stories (Vienna: CERN) https://cds.cern.ch/record/2670056/files/CERN-BOOKLET_DIGITAL-VPC.pdf

IOP Publishing

Big Science in the 21st Century
Economic and societal impacts
Panagiotis Charitos, Theodore Arabatzis, Harry Cliff, Günther Dissertori, Juliette Forneris and Jason Li-Ying

Chapter 18

Applying a systematic technology competence leveraging approach in the knowledge transfer of Big Science

Peter Keinz, Klaus Marhold and Jan Fell

18.1 Introduction

As an answer to today's 'grand challenges', Big Science has been gaining substantial attention during the past two decades (Hand 2010, Hallonsten 2016, OECD 2016). Complex endeavors—such as building the next-generation Large Hadron Collider at CERN—cannot be undertaken by single organizations (Giudice 2012). Instead, typically large consortia consisting of different, highly specialized partners that bring in their specific expertise to collaboratively work on the project are required (van Marrewijk *et al* 2008, Woodcock *et al* 2014).

Whereas large consortia are potentially favorable to research and development (R&D) processes, technology transfer in Big Science is an intricate effort (Autio *et al* 1996). Major challenges to extracting value from any given technology are the identification and evaluation of potential use cases. These activities—frequently referred to the 'fuzzy front-end' of knowledge and technology transfer processes (Herstatt and Verworn 2004, Backman *et al* 2007, Takahashi *et al* 2018)—are particularly demanding in the case of Big Science consortia. Because of the heterogeneity of the partners involved with regard to focal technologies, domain-specific languages, networks of academic and industrial organizations, and technology transfer strategies, it is difficult to achieve the planning and managing of knowledge spillovers to create positive economic and societal impact (Autio *et al* 2004, van Marrewijk *et al* 2008). In addition, Big Science is predominantly concerned with basic research (Martino 1992), which means that potential transfer objects (i.e. scientific discoveries and/or newly developed technologies) usually display very low technology readiness levels (Vuola and Hameri 2006, Tran *et al* 2011).

doi:10.1088/978-0-7503-3631-4ch18

The scientific literature on the management of knowledge and technology transfer is vast and describes many different approaches (Bengoa *et al* 2020). One specific approach is *technological competence leveraging* (TCL). TCL refers to a systematic process entailing the search, evaluation, and exploitation of new market opportunities for technologies (Danneels 2007, Keinz and Prügl 2010). In the present chapter, we report on an approach to TCL specifically suited for usage in Big Science consortia. This approach—we refer to it as *user-community-based technological competence leveraging* (user-community-based TCL)—has been used by CERN and other international research organizations for almost ten years. We describe this method and explain how it tackles the specific challenges to knowledge and technology transfer in the context of Big Science consortia. Based on an analysis of ten community-based TCL projects in a context of European Big Science consortia, we highlight the benefits and outcomes of the method. In the concluding section, we present actionable advice for the application of the user-community-based TCL method in order to better facilitate the commercial exploitation of Big Science consortia's technological resources.

18.2 Challenges and pathways of technology transfer in Big Science

In addition to the transfer of scientific knowledge, which happens predominantly through scientific publications (Nilsen and Anelli 2016), the transfer of technologies in Big Science materializes through a variety of pathways. Focusing on planned, rather than incidental spillovers, Amaldi (1999) identified three key pathways, namely patents, procurement, and joint development[1]. In contrast to technology transfer in universities, where licensing and spinoffs dominate, the requirements of prolonged and deepened R&D collaboration and activities in the basic science context leads to a lower importance of patents and licensing as a pathway to technology transfer (Autio *et al* 2004). At CERN, which from its inception did not place an emphasis on patenting the technology it invented (Schopper 2009), other pathways have further reduced the role of licensing in recent years (Nilsen and Anelli 2016). Further, licensing is also subject to the often very specific rules for patenting and licensing the output of Big Science organizations (Nordberg 2021). Patents and technologies available for licensing are often published by the science organization, for instance on the website of the knowledge transfer office. However, this is a very passive approach to technology transfer that not only relies on an interested party approaching the Big Science organization, but also to recognize the fit between the offered technologies and commercialization opportunities.

The immense scale of Big Science projects highlights the potential for using procurement as a strategic tool to foster knowledge transfer and innovation. For example, the construction of CERN's Large Hadron Collider, currently the world's largest particle accelerator, involved more than 6300 different suppliers (Sirtori *et al* 2019). Studies on the effects of Big Science procurement have

[1] For an in-depth review of recent literature on technology transfer and knowledge spillovers from Big Science we refer to the work of Scarrà and Piccaluga (2020).

generally pointed out positive effects on suppliers in areas related to the development of new technologies, technological and market learning, and reputation (Autio *et al* 2003, Florio *et al* 2018, Sirtori *et al* 2019). A recent analysis of more than 350 CERN suppliers from 1991 to 2014 has shown that becoming a supplier for CERN increases patent filings, labor productivity, and firm revenue (Castelnovo *et al* 2018). Further it has been shown that the benefits extend to second-tier suppliers, drastically increasing the potential number of industrial firms benefitting from Big Science procurement (Florio *et al* 2018). While procurement activities of Big Science organizations undoubtedly have positive effects on the involved firms, prior research has identified key obstacles such as procurement rules (Åberg and Bengtson 2015) and indicated the long time lag between the start of a supplier relationship and measurable outcomes (Bastianin *et al* 2021). Vuola and Hameri (2006) pointed out that technologies might not always be available for procurement as some relevant companies are simply not interested in doing business with Big Science organizations. Reasons given are the specific demands coupled with a small market, i.e. the lack of a business perspective beyond a single contract. Li-Ying *et al* (2021) identified the lack of insight about applications for their technology beyond the Big Science context as a major challenge for suppliers of Big Science organizations. They recommend supporting suppliers to conduct market and competence analysis and urge firms to build up capacities to exploit alternative market opportunities.

Finally, the technological complexity of Big Science, which often requires technological developments not readily available often leads to joint R&D projects. As with other R&D collaborations, their effect on technology development and transfer depends on a variety of partner- and collaboration specific factors (Mora-Valentin *et al* 2004, Fontana *et al* 2006). While prior literature has identified a range of motivations for industrial firms to enter into R&D collaborations with Big Science institutions (Hameri 1996, 1997), a study of companies taking part in R&D collaborations with CERN and identified that 'only large companies could afford to learn without having an immediate application for the gained knowhow' (Hähnle 1998, p 21). Smaller companies, which often have to shoulder large costs in comparison to their limited resources, were in more pressing need to turn R&D results into marketable products.

The above outlined pathways to technology transfer, which constitute the major technology transfer activities of Big Science organizations, are often passive approaches that do not tackle issues at the 'fuzzy front-end' of innovation. For example, due to a lack in perspective of commercial exploitability, companies might not even get to the stage where they present themselves as suitable suppliers for Big Science organizations. In a similar vein, smaller firms might not be able to reap all the benefits of collaborating with Big Science if they fail to identify commercial opportunities for the jointly developed technologies. And while entrepreneurship from Big Science has received increasing attention in the literature (Scarrà and Piccaluga 2020), the potential of many innovations stemming from the investments in Big Science remain underutilized due to the lack of a systematic search for new applications. In this vein, various methods related to TCL present themselves as suitable methods to overcome the above-mentioned shortcomings and to complement the various pathways to technology transfer from an early stage.

18.3 User-community-based technological competence leveraging

18.3.1 What is technological competence leveraging and how does it work?

The term *technological competence leveraging* refers to a set of activities, comprising the systematic identification, evaluation, and exploitation of new market opportunities to new or existing technologies and/or related competences of an organization (Danneels 2007, Keinz and Prügl 2010). The key assumption underlying TCL is that technologies are fungible resources (Teece 1982, Danneels 2002). This means that technologies can be utilized to serve different purposes and thus potentially cater a wide range of markets beyond the initial target markets they had been developed for.

TCL is generally considered a highly important activity for technology-driven organizations (Prahalad and Hamel 1990, Barca 2017). As a means to tackle the fuzzy front-end of outbound innovation activities, it fosters technology transfer processes and increases their likelihood to generate positive effects for the organization, e.g. additional sources of revenue streams, an increase in productivity, and/or a reduction of strategic dependencies from current target markets (West *et al* 2006, Gruber *et al* 2008). Due to these potential benefits, TCL has gained growing interest from the managerial realm, and ultimately from academia as well. During the last decade, an increasing body of literature has been looking into the merits of outbound innovation and—more specifically—into the question on how to organize TCL activities successfully. For example, Danneels (2007, p 511) described a 'process of technological competence leveraging', based on the principles of de-linking a technology from its current embodiment in an existing product, and re-linking it with potential, new products. A decade later, he presented a more actionable approach to systematically identifying, evaluating, and exploiting new markets to existing technologies (Danneels and Frattini 2018). Other approaches to tackle the underutilization of technologies and provide guidance to organizations searching for alternative application fields are the 'technology-push lead user method' (Henkel and Jung 2010, p 1), 'the user-community-based approach to technological competence leveraging' (Keinz and Prügl 2010, p 269), or the 'market opportunity navigator' (Gruber and Tal 2017, p 14).

The approaches to TCL described above share some important commonalities. First, they all follow a similar underlying logic and consist of three, sometimes four subsequent steps (figure 18.1). The initial step is related to analyzing the focal technology or technological competence in its own right, i.e. decoupling it from its current use context. The goal of this step is to better understand the type of problem that can be solved by means of the focal technology. The next step builds on these insights and is dedicated to looking for fields in which analogous problems exist, and thus the technology could provide a reasonable solution. In the third step, the identified application fields are analyzed in detail to collect insights into whether or not the technology might actually deliver value to their potential users in those fields.

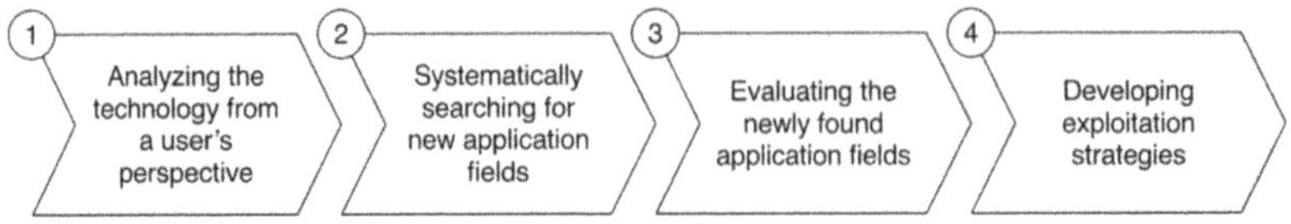

Figure 18.1. Core phases of technological competence leveraging. *Note*: Authors' drawing. Based on Danneels (2007) and Keinz and Prügl (2010).

The fourth step is usually concerned with creating an actionable go-to-market strategy, enabling the organization to actually exploit the most attractive application fields.

A second important communality of most TCL approaches is the fact that they focus on technology-driven companies in leveraging already existing technologies. Consequently, they have been shown to work quite well in this type of setting. However, in Big Science collaborations conducted by consortia consisting of various research organizations and industry partners, most of the TCL approaches described above are potentially not as effective. The reasons for this are manifold. First, the focal technologies in Big Science differ a lot from those usually leveraged by companies. In the course of Big Science projects, the consortia work on highly specific technologies that are being developed for very narrow applications, e.g. radiation-hard CMOS sensors for usage in CERN's Large Hadron Collider to detect new types of particles. Naturally, only few—if any—near-analogous applications exist to such specific technologies. This means that the focal organization has to identify far-analogous[2] applications which is very hard to do because of the phenomenon of local search bias[3]. In addition, technologies in Big Science usually have very low technology readiness levels, in some cases they do not even exist yet. This means that the consortium, in contrast to a company leveraging a well-known, existing technology—does not have a lot of knowledge on the value that can be extracted from the technology during its actual use. This lack of insight also increases the difficulties in identifying analogous, new application fields. This problem is further amplified by another decisive difference between TCL activities in companies versus Big Science consortia. In companies, there are dedicated organizational units or roles to look for, analyze, and plan the exploitation of new applications to existing technologies. These units responsible for TCL, e.g. new business developer, innovation managers, etc, usually have two important features: they know their technology, and they know—at least to some extent—how their

[2] Near-analogous application fields refer to applications that share relatively obvious surface-level attributes (e.g. similar types of users, similar types of products in which the technology is embedded, etc) with the technology's actual target application. Far-analogous application fields refer to applications that only share deep-structured relational attributes with the target application, e.g. similar types of problems to be solved, while surface-level attributes might be completely different (Franke *et al* 2013).

[3] The term local search bias refers to the phenomenon that search activities (e.g. the search for a new application field to an existing technology) tend to relate closely to prior search activities, knowledge, and experiences (e.g. existing ideas of what a technology can be used for), limiting the identification of novel solutions (e.g. radically new application ideas) (Rosenkopf and Nerkar 2001).

customers are using their technology and how they derive value from doing so. Having both of these perspectives bundled in one point makes it much easier for the company to de-link the technology from its current use and re-link it with similar applications in other fields. In Big Science consortia, the situation is different. The technology experts do have deep knowledge about the technology, but they do not have established strong ties with its users or customers, sometimes simply because they aren't any users yet. Thus, it is hard to analyze the technology on a more abstract level in order to reveal its problem-solving capability. In addition, often the scientists conducting the technological development are not really interested in working on the dissemination of their developments. Consequently, external consultants or the knowledge transfer offices of the academic consortia partners involved are asked to take care of the TCL activities (Keinz and Marhold 2021). They lack, however, a sound understanding of the technology, which might prevent them from successfully leveraging it. Another important difference between TCL activities of companies as opposed to Big Science consortia relates to the underlying intellectual property (IP) situation. Companies leveraging their own technologies usually enjoy a sound IP protection, allowing them to present their technology to potential users in the course of application discovery and validation activities, without risking their technology being stolen. In Big Science contexts, this is not so much the case. Due to the low readiness levels, the collaborative effort in the development, as well as the fact that the academic partners within the consortium have an interest in publishing scientific articles on their developments, IP rights protection often is complicated or even completely prevented. As a consequence, the technologies to be leveraged cannot be fully revealed to potential users, again making it harder to validate potential application fields.

Because of the specific situation, the promise of traditional, company-centric TCL approaches is somewhat limited in the context of Big Science. In the next section, we want to take a deeper look at the user-community-based approach *to* TCL (Keinz and Prügl 2010) that has been shown to work well in the context of three different Big Science projects (Keinz and Marhold 2021).

18.3.2 A user-community-based approach to technological competence leveraging in Big Science

As described above, the core problems of traditional approaches to TCL in Big Science projects are: (i) the technology owners' local search behavior, preventing them from identifying far-analogous applications; (ii) the fact that technology- and market-related knowledge does not reside within one role or organizational unit; as well as (iii) a lack of proper IP rights protection, complicating the process because the technology cannot easily be shown to potential users.

To address these problems and help research organizations and technology-driven companies leverage their developments, Keinz and Prügl (2010) developed the user-community-based approach to TCL over a decade ago. Their approach has been applied successfully in the course of 100+ real-life projects, many of them being conducted in the context of Big Science projects (Keinz and Marhold 2021). At its core, the user-community-based approach to TCL consists of the same four steps as the traditional, company-centric approaches. The distinct difference, however, is the idea to systematically integrate current and/or potential users in each of these steps by means of an open innovation method, i.e. crowdsourcing[4].

In the first step of the user-community-based approach to TCL, the technology is analyzed from a users' perspective. The ultimate goal is to identify the technology's two or three core use benefits, i.e. how the technology delivers value to its users and which problems it solves for them. Based on these core use benefits, alternative fields of application with similar problems and the need for the benefits delivered by the technology can systematically be searched for in the next step. To actually derive the core use benefits, current users of the technology are interviewed and asked how they benefit from the technology. In the case of a newly developed technology without current users, an easily understandable representation of the technology is presented to the future target users, who are then asked how they might benefit from using it. Focusing on a technology's core use benefits instead of its features and functionalities comes with several important advantages for TCL approaches in Big Science projects. First, users usually know much better than the inventor how and for what purposes they use a specific technology. There is ample literature on examples of misuses of technologies in ways that the inventors could not have imagined themselves[5]. Thus, learning about a technology's actual benefits might open up an unforeseeable space of far-analogous applications, based on unexpected problem-solving capabilities of the focal technology. Also, communicating a technology's benefits instead of its specifications, features, and functionalities is the most abstract way of describing it. Such a generic representation of the technology allows for a much easier identification of far-analogous application fields. The underlying argument is straightforward. Potential users from far-analogous fields might not speak the 'language' of the technology owners, i.e. they might simply not be able to understand the technology in detail. But what they can understand, irrespective of their domain, is the notion of abstract benefits a technology might deliver to them. Only if potential users from far-analogous fields are able to understand what type of problems the technology is able to solve, they might be able to recognize it as a valuable solution to them and point the technology owners to promising applications. In addition, being able to talk about a technology in terms of the benefits it delivers also reduces the necessity of revealing technological details during the search

[4] 'Crowdsourcing is a type of participative online activity in which an individual, organization, or company with enough means proposes to a group of individuals of varying knowledge, heterogeneity, and number, via a flexible open call, the voluntary undertaking of a task. [...]' (Estellés-Arolas and González-Ladrón-de-Guevara 2012).

[5] Think about the example of Viagra. Pfizer had developed this drug as a treatment for pulmonic hypertension. Only during the clinical trials and based on the feedback by actual users (the participants in the clinical studies), the company recognized the important side effect that led to a re-marketing of the product as a treatment to erectile dysfunction.

for applications, which is particularly important in Big Science settings with their weak IP protection regimes. Last but not least, focusing on benefits instead of technological details also enables external consultants or the knowledge transfer offices of consortia to conduct the search for alternative applications, even if they lack deep knowledge about the technology.

The second step of the user-community-based approach to TCL is dedicated to the systematic search for alternative fields of application. By means of broadcast[6] and pyramiding[7] search, the technology's benefits are presented to a large number of potential users from completely different domains. They are asked as to whether or not they know of problems in their professional or personal domains which might be solved by a solution providing the technology's core use benefits. The integration of potential users into the search for alternative application fields on large scale comes with some obvious advantages. Such a crowdsourcing approach is a relatively efficient way of collecting many application field ideas in relatively short time. Because of the heterogeneity of the potential users involved, the variety among the identified fields will also be relatively large, including a high-percentage of far-analogous applications.

In the subsequent third step, the application fields identified need to be assessed with regards to their commercial attractiveness. Because of the large number of crowdsourced application fields, they can usually not be analyzed in detail by the focal organizations themselves due to their limited resources. Thus, again, potential users are involved in a crowd evaluation process to identify the most promising application fields out of the initial long list of applications. After this first assessment, the focal organizations conduct detailed market and competitor analyses to gain more detailed insights into the short-listed application fields. Potential users with domain-specific knowledge are likely to be a valuable source of market-related intelligence in this step as well. When embedded in user communities, they tend to know of (the number of) other users, competitive solutions, substitutes, their prices, etc. Gaining this type of information is particularly important in Big Science consortia who depend on catering new, far-analogous application fields on which they have little to no insights.

The fourth step of the user-community-based approach to TCL comprises the development of an actionable go-to-market strategy. Questions to be answered in this phase relate to aspects such as exploring the key stakeholders and the value chain in the newly identified application field, learning about dominant business models, etc. Again, involving potential users is key in this step, since they might shed some light on the above mentioned and related aspects.

---

[6] Broadcast search refers to a search strategy which relies on publicly disclosing of a specific problem (e.g. the identification of applications to a specific technology) and inviting anyone who deems themselves qualified to solve the problem to submit solution proposals (Jeppesen and Lakhani 2010).

[7] Pyramiding refers to a search strategy consisting of a number of subsequent interviews, in course of which individuals possessing a sought characteristic provide solution-related information (e.g. ideas for applications to a specific technology) as well as referrals to other individuals with similar or even higher levels of the sought characteristic (von Hippel *et al* 2009).

18.4 An application of the user-community-based approach to technological competence leveraging

The above-described user-community-based approach to TCL has been employed in several Big Science projects, most notably in collaboration with CERN. Specifically, the TCL projects were an integral part of three Innovative Training Networks (ITNs). ITNs are multi-year programs funded by the European Union (EU) under the Marie Skłodowska-Curie Actions and are characterized as consortia of companies, research institutes (such as CERN) and universities as well as their associated researchers. In recent years, EU funded programs have become increasingly important in CERN's knowledge sharing activities as they foster collaboration and knowledge exchange with the private sector (Nilsen and Anelli 2016).

The three ITNs covered various topic of interest to CERN's key research streams including radiation detection technologies and superconductivity. TCL activities were an integral part of each ITN and represented the Vienna University of Economics and Business' key contribution to the consortia[8]. In total, over the three ITNs, ten TCL projects were conducted with both CERN and various commercial consortium members serving as the project owners. Table 18.1 provides an overview of the three ITN and ten TCL projects. Each TCL project took about 4 months and involved a project team of between five and ten people of diverse backgrounds. Direct involvement of the project owner amounted to the participation of one to three staff members, mostly scientists and engineers, during kick-off and status meetings.

The results show that for each project, between 8 and 47 different application fields were identified. Of those, a large proportion were indeed far-analogous application fields. For example, for the CVC diamond detector, the TCL project resulted in seven near-analogous fields such as the application of the detector in other Big Science settings such as neutron spallation sources, but also 12 far-analogous fields such as potential applications of the technology in oil well logging or space weather monitoring. Other examples of far-analogous application fields for technology were automated fruit sorting using CERN's superconductor technology and the use of a thermal treatment process developed for the production of superconducting magnets in scrap metal recycling (Keinz and Kretzschmar 2019) or the pharmaceutical industry (Kretzschmar *et al* 2019).

Owing to the structure of the ITNs, which bring together science, industry, and academia, the project owners in the conducted projects were either CERN itself or commercial firms ranging from small SMEs to larger multinational corporations. Most technologies on which the TCL method was applied in the collaboration with CERN were in the prototype and testing stages. A common theme encountered was that CERN developed an initial solution based on its technological needs and started looking for an industrial partner for further development and production. However, in order to attract industrial partners to share the cost and effort of further

[8] For a more detailed discussion of the TCL projects and the role played by Vienna University of Economics and Business we refer to Keinz and Marhold (2021).

Table 18.1. Overview of the conducted technology competence leveraging projects.

Innovative training network	Technology	Project owner	Technology readiness level	Identified application fields		
				Total	Near-analogous	Far-analogous
TALENT (Instrumentation for radiation detection)	CVD diamond detector	Firm (SME)	9	19	7 (37%)	12 (63%)
	Direct laser soldering	CERN	6	8	2 (25%)	6 (75%)
	HV CMOS	CERN	3	18	3 (17%)	15 (83%)
	Microscint technology	CERN	6	16	5 (31%)	11 (69%)
	Depleted CMOS sensor	CERN	6	20	5 (25%)	15 (75%)
STREAM (Advanced radiation instrumentation)	Rad-hard CMOS sensor	Firm (SME)	7	37	13 (35%)	24 (65%)
	Electron microscope	Firm (large)	9	30	22 (73%)	8 (27%)
	Near- and medium infrared sensor	Firm (large)	5	47	5 (11%)	42 (89%)
EASITRAIN (Applied superconductivity)	Superconductors	CERN	8	29	6 (21%)	23 (79%)
	Superconductor manufacturing	CERN	9	21	0 (0%)	21 (100%)
Average					27.5%	72.5%

Note: Technology readiness levels obtained from NASA (2012).

development, they need to have a perspective towards commercialization. The TCL projects supported this, as CERN was able to point out potential application fields to prospective partner companies.

In addition to the direct outcome of identifying potential application fields for the technologies, the TCL projects had learning effects for both the involved commercial firms and CERN employees. As technology transfer and the application of scientific results and technologies is an increasingly important component of grant applications, e.g. for EU grants, the involved scientists reported that they could employ the knowledge about the TCL method in other projects. The involved firms, in particular the smaller firms, predominantly staffed by scientists and researchers also reported positive benefits beyond the focus projects. For a more detailed discussion of short-term and long-term learning outcomes of the TCL projects we refer to a prior publication (Keinz and Marhold 2021), which also discusses the involvement of innovation intermediaries, in this case a university, in these projects.

18.5 Conclusion

In this chapter we set out to propose the user-community-based approach to TCL as a feasible means of mastering the fuzzy front-end of technology transfer processes in Big Science consortia. We explained the underlying logic of this methodological approach, how it differs from alternative methods, and why we consider it to be particularly applicable in Big Science endeavors. We supported our claim by reporting anecdotal evidence from ten different Big Science projects that have been conducted in the course of three European Union-funded, CERN-based Innovative Training Networks (ITNs).

The insights from these projects show that indeed, the user-community-based TCL turned out to be a highly effective approach to systematically identify and evaluate far-analogous application fields to technologies of varying, sometimes rather low technological readiness levels. Also, we found some evidence that the user-community-based TCL was successfully applied by, and yielded satisfactory results to, different types of partners usually involved in Big Science consortia. The specific challenges of highly specialized technologies, the absence of a dedicated knowledge transfer unit, and weak IP rights regimes were successfully overcome in all cases.

Our experience also indicates that the user-community-based TCL might be an important complement to other modes of technology transfer method in Big Science, i.e. licensing, spinoffs, or traditional procurement. Since these alternative modes do not specifically focus on the fuzzy front-end of knowledge transfer activities, they actually rely on the potential outcome of systematic TCL activities. For example, based on the insights generated by TCL activities, suitable suppliers and/or development partners can be identified and attracted to enter a collaboration with a Big Science consortium. In addition, due to the limited timely and monetary resources needed, the user-community-based TCL seems to be an approach particularly useful to SMEs involved in Big Science consortia.

Acknowledgments

The authors gratefully acknowledge the contribution of Linn Kretzschmar in providing key research data on the TCL projects discussed herein, as well as her helpful feedback on an earlier draft of this work.

This project has received funding from the European Union's Horizon 2020 research and innovation programme under the Marie Skłodowska-Curie grant agreement No. 764879.

References

Åberg S and Bengtson A 2015 Does CERN procurement result in innovation? *Innovation* **28** 360–83

Amaldi U 1999 Spin-offs of high energy physics to society *Int. Europhysics Conf.-High Energy Physics*

Autio E, Hameri A-P and Bianchi-Streit M 2003 Technology transfer and technological learning through CERN's procurement activity *CERN Yellow Reports: Monographs* CERN-2003-005

Autio E, Hameri A-P and Nordberg M 1996 A framework of motivations for industry–big science collaboration: a case study *J. Eng. Tech. Manage.* **13** 301–14

Autio E, Hameri A-P and Vuola O 2004 A framework of industrial knowledge spillovers in big-science centers *Res. Policy* **33** 107–26

Backman M, Börjesson S and Setterberg S 2007 Working with concepts in the fuzzy front end: exploring the context for innovation for different types of concepts at Volvo cars *R&D Manag.* **37** 17–28

Barca M 2017 *Economic Foundations of Strategic Management* 1st edn (Abingdon: Routledge)

Bastianin A, Castelnovo P, Florio M and Giunta A 2021 Big science and innovation: gestation lag from procurement to patents for CERN suppliers *J. Technol. Transf.* **47** 531–55

Bengoa A, Maseda A, Iturralde T and Aparicio G 2020 A bibliometric review of the technology transfer literature *J. Technol. Transf.* **46** 1514–50

Castelnovo P, Florio M, Forte S, Rossi L and Sirtori E 2018 The economic impact of technological procurement for large-scale research infrastructures: evidence from the Large Hadron Collider at CERN *Res. Policy* **47** 1853–67

Danneels E 2002 The dynamics of product innovation and firm competences *Strateg. Manag. J.* **23** 1095–121

Danneels E 2007 The process of technological competence leveraging *Strateg. Manag. J.* **28** 511–33

Danneels E and Frattini F 2018 Finding applications for technologies beyond the core business *MIT Sloan Manag. Rev.* **59** 73–8 https://sloanreview.mit.edu/article/finding-applications-for-technologies-beyond-the-core-business/

Estellés-Arolas E and González-Ladrón-de-Guevara F 2012 Towards an integrated crowdsourcing definition *J. Inf. Sci.* **38** 189–200

Florio M, Giffoni F, Giunta A and Sirtori E 2018 Big science, learning, and innovation: evidence from CERN procurement *Ind. Corp. Change* **27** 915–36

Fontana R, Geuna A and Matt M 2006 Factors affecting university–industry R&D projects: the importance of searching, screening and signalling *Res. Policy* **35** 309–23

Franke N, Poetz M K and Schreier M 2013 Integrating problem solvers from analogous markets in new product ideation *Manage. Sci.* **60** 1063–81

Giudice G F 2012 Big Science and the large Hadron Collider *Phys. Persp.* **14** 95–112

Gruber M, MacMillan I C and Thompson J D 2008 Look before you leap: market opportunity identification in emerging technology firms *Manage. Sci.* **54** 1652–65

Gruber M and Tal S 2017 *Where to Play: 3 Steps for Discovering your Most Valuable Market Opportunities* (Harlow: Pearson)

Hähnle M 1998 R&D collaboration between CERN and industrial companies: organizational and spatial aspects *IIR Discussion Papers* Vienna University of Economics and Business https://epub.wu.ac.at/6154/

Hallonsten O 2016 Introduction and framework *Big Science Transformed: Science, Politics and Organization in Europe and the United States* ed O Hallonsten (Cham: Springer), pp 1–42

Hameri A-P 1996 Technology transfer between basic research and industry *Technovation* **16** 51–92

Hameri A-P 1997 Innovating from big science research *J. Technol. Transfer* **22** 27–35

Hand E 2010 'Big science' spurs collaborative trend *Nature* **463** 282

Henkel J and Jung S 2010 Identifying technology applications using an adaptation of the lead user method *SSRN Electron. J.*

Herstatt C and Verworn B 2004 The 'Fuzzy Front End' of innovation *Bringing Technology and Innovation into the Boardroom: Strategy, Innovation and Competences for Business Value* (London: Palgrave Macmillan), pp 347–72

Jeppesen L B and Lakhani K R 2010 Marginality and problem-solving effectiveness in broadcast search *Organ. Sci.* **21** 1016–33

Keinz P and Kretzschmar L 2019 Impact potentials of EASITrain research on society and industry *Zenodo*

Keinz P and Marhold K 2021 Technological competence leveraging projects via intermediaries: viable means to outbound open innovation and mediated capability building? *Int. J. Proj. Manage.* **39** 196–208

Keinz P and Prügl R 2010 A user community-based approach to leveraging technological competences: an exploratory case study of a technology start-up from MIT *Creat. Innov. Manag.* **19** 269–89

Kretzschmar L, Mehner B, Hausberger M, Ldermüller F, Mayrhofer F, Schreiber D and Gutleber J 2019 Manufacturing process of superconducting magnets: analysis of manufacturing chain technologies for market-oriented industries *Zenodo*

Li-Ying J, Forneris J, Korsholm S B, Jensen A and Zangenberg N 2021 How European Big Science organizations and suppliers innovate through public procurement *Res.-Technol. Manag.* **64** 46–56

Martino J P 1992 *Science Funding: Politics and Porkbarrel* (New York: Routledge)

Mora-Valentin E M, Montoro-Sanchez A and Guerras-Martin L A 2004 Determining factors in the success of R&D cooperative agreements between firms and research organizations *Res. Policy* **33** 17–40

NASA 2012 Technology readiness level *NASA* https://web.archive.org/web/20210724213915/https://nasa.gov/directorates/heo/scan/engineering/technology/technology_readiness_level

Nilsen V and Anelli G 2016 Knowledge transfer at CERN *Technol. Forecast. Soc. Change* **112** 113–20

Nordberg A 2021 Big Science, Big Data, big innovation? ERIC policies on IP, data and technology transfer *Big Science and the Law* ed U Maunsbach and O Hallonsten (Copenhagen: Ex Tuto), pp 65–106

OECD 2016 *OECD Science, Technology and Innovation Outlook 2016* (Paris: OECD)

Prahalad C K and Hamel G 1990 The core competence of the corporation *Strategische Unternehmungsplanung—Strategische Unternehmungsführung* ed D Hahn and B Taylor (Berlin: Springer)

Rosenkopf L and Nerkar A 2001 Beyond local search: boundary-spanning, exploration, and impact in the optical disk industry *Strateg. Manag. J.* **22** 287–306

Scarrà D and Piccaluga A 2020 The impact of technology transfer and knowledge spillover from Big Science: a literature review *Technovation* **116** 102165

Schopper H 2009 *LEP—The Lord of the Collider Rings at CERN 1980–2000: The Making, Operation and Legacy of the World's Largest Scientific Instrument* (Heidelberg: Springer)

Sirtori E, Giffoni F, Pancotti C, Caputo A, Catalano G and Florio M 2019 *Impact of CERN Procurement Actions on Industry: 28 Illustrative Success Stories* (Vienna: CERN)

Takahashi M, Indulska M and Steen J 2018 Collaborative research project networks: knowledge transfer at the fuzzy front end of innovation *Proj. Manag. J.* **49** 36–52

Teece D J 1982 Towards an economic theory of the multiproduct firm *J. Econ. Behav. Organ.* **3** 39–63

Tran T, Daim T and Kocaoglu D 2011 Comparison of technology transfer from government labs in the US and Vietnam *Technol. Soc.* **33** 84–93

van Marrewijk A, Clegg S R, Pitsis T S and Veenswijk M 2008 Managing public–private megaprojects: paradoxes, complexity, and project design *Int. J. Proj. Manage.* **26** 591–600

von Hippel E, Franke N and Prügl R 2009 Pyramiding: efficient search for rare subjects *Res. Policy* **38** 1397–406

Vuola O and Hameri A-P 2006 Mutually benefiting joint innovation process between industry and big-science *Technovation* **26** 3–12

West J, Vanhaverbeke W and Chesbrough H 2006 Open innovation: a research agenda *Open Innovation: Researching a New Paradigm* ed H Chesbrough, W Vanhaverbeke and J West (Oxford: Oxford University Press), pp 285–308

Woodcock J, Brumfield M, Gill D and Zerhouni E 2014 The driving role of consortia on the critical path to innovative therapies *Nat. Rev. Drug Discov.* **13** 781

Chapter 19

Research infrastructure in the UK: evaluating large-scale fundamental physics

Fiona Larner

19.1 Introduction

Research infrastructure is both the foundations and scaffolding that we build our world on. An essential bridge between the local laboratories of universities to industry, it allows great science to be explored in a collaborative manner, within facilities that are out of reach for a single university and unfeasible for industry. Research infrastructure has proved itself time and again to be the vital underpinning of an agile, adaptable, and responsive world, not least demonstrated during the Covid-19 pandemic [1].

The leap from fundamental science to positive daily impacts for society can seem to some as a jump too far. When quarks, bosons, and black holes are mentioned, rarely does it make one think of solving world hunger [2] or diagnosing cancer [3]. However, the infrastructure that has been created to enable the investigation of fundamental physics and related phenomena not only enables the development essential for real-life applications to come to fruition [4] and provides a network that allows rapid response to emerging threats [5], but also provides the opportunity for unpredictable, serendipitous discoveries to be made, which bring significant development to our world [6].

Evaluation is the key mode by which the outcomes and impacts of research can be identified and assessed, enabling scientists to demonstrate the benefits of their work both within their specialty and further afield. Fundamental science has the added challenge of the time-horizon for benefits realisation, with socio-economic impacts (where they occur) generally not detectable immediately, and taking years or decades to emerge after the initial investment was made. Robust methods for evaluating the positive impacts of fundamental science, both short and long term, still require some development to allow a fair and accurate economisation of benefits

doi:10.1088/978-0-7503-3631-4ch19

that elements of this research brings, and also to establish a widely accepted framework that adequately recognises the value of the societal benefits that cannot be converted into pounds and pence.

In this chapter, the research infrastructure system for fundamental research of the UK is explored, with specific attention on the 'Big Science' used to study physics and related phenomena, the multidisciplinary facilities underpinned by physics-based technologies, and the methods used by the UK public sector to evaluate the impact of this investment for society.

19.2 Research infrastructure and Big Science in the UK

19.2.1 The UK research infrastructure landscape

The UK has a vast range of publicly funded research infrastructures that provide an internationally competitive, yet accessible base on which to perform high-quality investigations and developments, as outlined in UK Research and Innovation's (UKRI) Landscape Analysis [7]. The analysis identified more than 750 infrastructures of significance, which employ just under twenty-five thousand full-time-equivalent (FTE) staff in the UK, 33% in research roles and 38% in technical roles. The collaborative nature of research infrastructures is demonstrated through the more than 75% that work with UK businesses, the 92% working across more than one research sector, and the 92% with international partnerships. International users of UK research infrastructures make up 39% of all users, typically due to the unique capability these research infrastructures offer. The research and innovation infrastructure covered by the analysis cuts across six sectors and a breadth of different 'types' of infrastructure including: large-scale physical multidisciplinary research facilities, research ships, satellites, imaging technology networks, knowledge collections, people, data, computing systems. Of this, a significant proportion is administered by the Science and Technology Facilities Council (STFC), one of UKRI's nine constituent organisations, which is ultimately financed by the UK taxpayer, through the Department for Science, Innovation and Technology.

STFC is a world-leading multidisciplinary science organisation within UKRI that supports researchers investigating fundamental sciences such as particle physics, nuclear physics, astronomy, and space science, as well as many other disciplines such as chemistry, Earth science, engineering, environmental science, and life sciences. The tools required to interrogate phenomena in these fields include particle accelerators, satellites, telescopes, and high performance computing. As such, large-scale infrastructure is a key component of STFC's offer, both nationally and internationally (figure 19.1). These facilities do not come pre-packaged; a large amount of STFC-funded research is dedicated to pushing the boundaries of science and engineering through the design and construction of the facilities and constituent instrumentation to allow scientific concepts to be explored. Therefore, comparatively to other research organisations, a large proportion of STFC's budget is allocated to a suite of UK-based research infrastructures and select international facilities (figure 19.2).

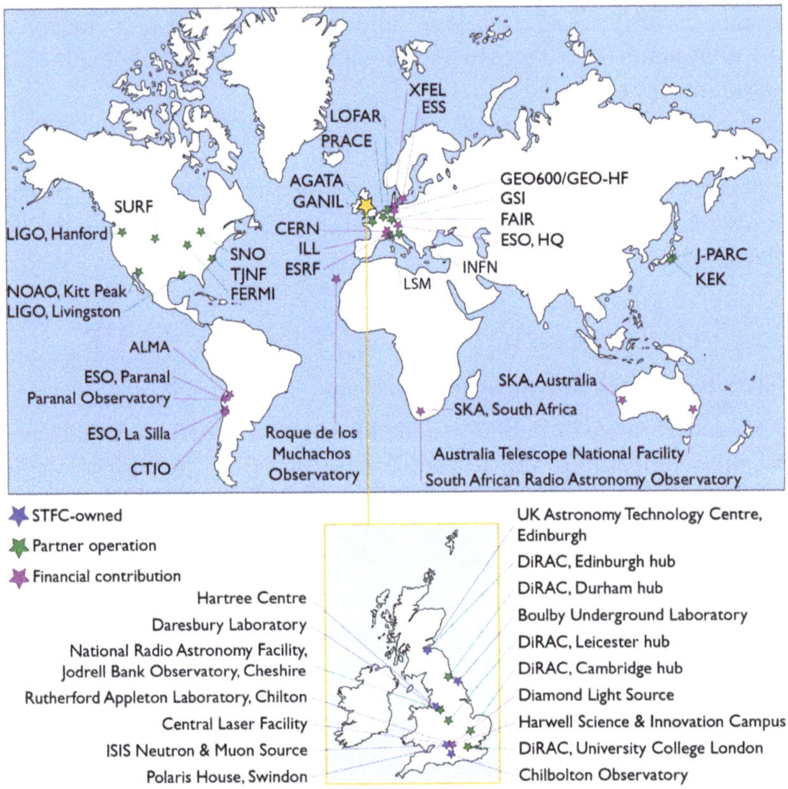

Figure 19.1. Ground-based research infrastructures funded wholly or in-part by the Science and Technology Facilities Council (STFC; image credit: Fiona Larner, Karen Clifford): Advanced Gamma Tracking Array (AGATA); Atacama Large Millimeter Array (ALMA); European Organisation for Nuclear Research (CERN); Cerro Tolo Inter-American Observatory (CTIO); Distributed Research Utilising Advanced Computing (DiRAC; computing resource hubs also in Cambridge, Edinburgh, and Leicester); European Southern Observatory (ESO); European Synchrotron Radiation Facility (ESRF); European Spallation Source (ESS); Facility for Anti-Protons and Ion Research (FAIR); FERMI National Accelerator Laboratory; Grand Accelerateur National d'Ions Lourds (GANIL); Helmholtz Centre for Heavy Ion Research (GSI); ASTRON Netherlands Institute for Radio Astronomy Low Frequency Array (LOFAR); Institut Laue-Langevin (ILL); Laboratori Nazionali di Legnaro (INFN); Japan Proton Accelerator Research Centre (J-PARC); High Energy Accelerator Research Organisation (KEK); Laser Interferometer Gravitational-Wave Observatory (LIGO); Laboratoire Souterrain de Modane (LSM); National Optical Astronomy Observatory (NOAO)—Kitt Peak National Observatory; Partnership for Advanced Computing in Europe (PRACE); Square Kilometre Array (SKA); Sudbury Neutrino Observatory (SNO); Science and Technology Facilities Council (STFC); Sanford Underground Research Facility (SURF); Thomas Jefferson National Laboratory (TJNF); European X-ray Free-Electron Laser (XFEL). Image credit: Fiona Larner, Karen Clifford.

Fundamental research infrastructure funded by the UK taxpayer is not limited by national borders, as exhibited by the distribution of those administered by STFC (figures 19.1 and 19.2). There are a number of factors that can influence decisions regarding the best host for differing facilities, such as size, location, cost, heritage of a site/country, politics, and funding available. The UK pays subscriptions to be part

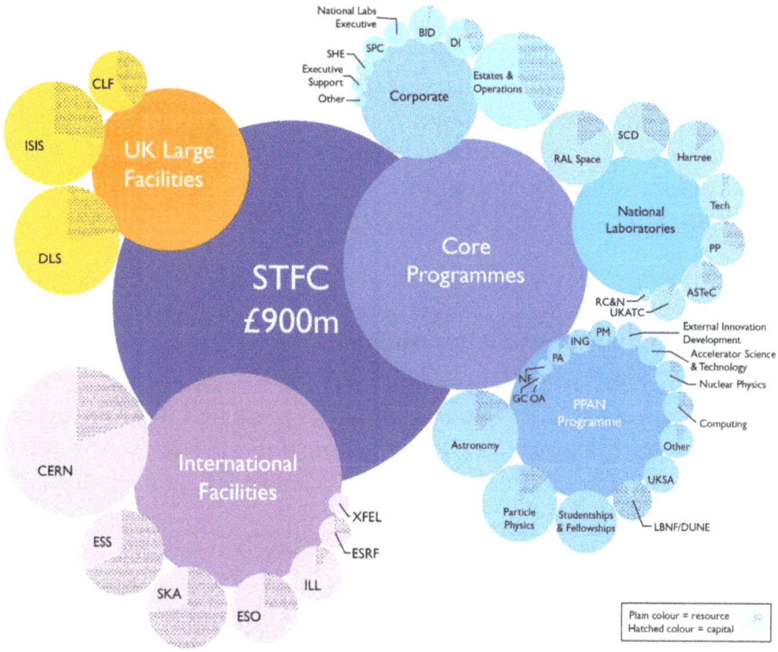

Figure 19.2. STFC's budget allocation 2021/22 (image credit: Ian Fuller, Sarah Hanrahan, Fiona Larner): Accelerator Science and Technology Centre (ASTeC); Business and Innovation Directorate (BID); European Organisation for Nuclear Research (CERN); Central Laser Facility (CLF); Digital Infrastructure (DI); Diamond Light Source (DLS); European Synchrotron Radiation Facility (ESRF); European Spallation Source (ESS); European Southern Observatory (ESO); Global Challenges Programme (GC); Hartree National Centre for Digital Innovation; Institut Laue-Langevin (ILL); Isaac Newton Group of Telescopes, La Palma (ING); ISIS Neutron and Muon Source; Long-Baseline Neutrino Facility/Deep Underground Neutrino Experiment (LBNF/DUNE); Newton Fund (NF); open access (OA); Particle Astrophysics (PA); Programme Management (PM); Particle Physics (PP); Particle Physics, Astronomy and Nuclear Physics (PPAN); Rutherford Appleton Laboratory (RAL); Research Complex and Networks (RC&N); Scientific Computing Department (SCD); Safety, Health, and Environment (SHE); Square Kilometre Array (SKA); Strategy, Planning, and Communications (SPC); Science and Facilities Technology Council (STFC); Technology (Tech); UK Astronomy Technology Centre (UKATC); UK Space Agency (UKSA); European X-Ray Free-Electron Laser Facility (XFEL). Image credit: Ian Fuller, Sarah Hanrahan, Fiona Larner.

of large international bodies that undertake projects which could not happen through the actions of one country alone. Sites needed to investigate fundamental physics often require large spaces, dark skies, extensive networks of high performance computing, or a combination of these. The scientific aims of these facilities are underpinned by ground-breaking, cutting-edge bespoke technology that is not available 'off-the-shelf'. It is important to note that aside from the necessity of collaboration to share the burden of costs, science is an international endeavour—collaboration is a fundamental part of discovery, and working across national borders to engage with experts is a standard part of scientific exploration.

The Landscape Analysis [7] highlights the dichotomy of funding and directional uncertainty that research infrastructures face in the UK. More than half of research infrastructures have operational lifespans of more than twenty years, underpinning development, growth, and innovation in all sectors, and a recent review recommended funding to be committed in at least five-year tranches [8]. However, three- to four-year funding horizons, single-year funding commitments and changes in priority taken by the Government (notwithstanding changes in the UK Government itself, which operates on a $\leqslant$5 year time scale) mean that many research infrastructures find it impossible to plan more than three years into the future [7].

19.2.2 The future of UK research infrastructure

UKRI was created in 2018 to bring the seven UK research councils, Research England, and Innovate UK together as a cohesive entity. UKRI is responsible for the overall research strategy of the UK, delivering it to the UK Government with the strengthened voice of a collective research community, advancing engagement between policymakers and research funders. UKRI strives to reduce bureaucracy and administrative burden from researchers and improve funding streams, bringing agility to the science budget including enhanced support and management of multi- and interdisciplinary research and innovation. This agility in research funding extends to the investment in research infrastructures, and will enhance sustainability of the cutting-edge research, development, and innovation that these world-class facilities enable [9]. A new UKRI Infrastructure Fund was conceived in 2020, addressing the prioritisation of potential new research and innovation infrastructure concepts and taking a long-term strategic view towards an infrastructure portfolio, with a view to de-risking future investments. To underpin this fund, an Infrastructure Advisory Committee has been established whose objectives include providing advice and recommendations on prioritisation of projects for UKRI's infrastructure funding, utilising a portfolio approach.

A strategic guide for the future development of the UK's research and innovation infrastructure was presented most recently in UKRI's *Opportunities to Grow Our Capacity* [10]. The multidisciplinary and multi-sector nature of large-scale research infrastructure was showcased by the presence of STFC-funded facilities in all six sectors of UK research infrastructures: physical sciences and engineering, social sciences, arts and humanities, environment, energy, and computational and e-infrastructure. *Opportunities to Grow Our Capacity* highlights that challenges in skills, data, and sustainability are significant obstructions to maximising the impact of the UK's research infrastructure investments.

The UK has a shortage of expertise in STEM subjects (science, technology, engineering, and maths), which has been reported to cost the UK business sector £1.5 billion a year through the expense of training and staff turnover [11]. The data skills gap is of specific concern, not only for large-scale research infrastructure but for the UK as a whole, particularly in the short-term. A 2021 survey by the Department for Digital, Culture, Media and Sport reported around 200 000 empty data-related roles in UK companies in comparison with a predicted data-science graduating cohort of approximately 10 000 a year [12]. The opportunities to improve the pipeline of STEM expertise in the UK are numerous, including focuses on education, communication, inspiration, funding, and sustained and embedded cultural change to open pathways for discriminated groups. Additionally, the presence of accessible research infrastructure in fundamental physics and related disciplines is essential to maintain the STEM expertise pipeline in the UK. STFC is an innovative and supportive employer, committed to continuously improving its approach and processes, embedding a culture where a diverse workforce is promoted and supported. Three programmes that reflect this, and work actively to create new pipelines and maintain expertise in STEM, are STFC's Technician Commitment [13], excellent [14] graduate recruitment [15] and award-winning [16] apprenticeship [17] schemes. The combination of initiatives such as these with sustained commitment by governments to invest in the UK's research infrastructure can contribute significantly to addressing the shortage in UK STEM expertise.

Robust e-infrastructure is symbiotic with physics research. Whilst current researchers of physics phenomena rely on digital infrastructure for data communication, storage, and analysis, future physics research infrastructure is also a driver for technological development in this area. A key example of this is the Square Kilometer Array (SKA), which on completion will be the largest radio telescope and scientific instrument on Earth. SKA was conceptualised in 1991, however, the daily data needs of this infrastructure once online still outweighed the total global Internet traffic *per diem* in 2022. The increase in capability needed for Big Data projects such as SKA motivate innovation in the digital space, which importantly, do not sit siloed within the physics community, but permeate across the research landscape, as well as to industry and public services [18].

High energy, emissions, and resource costs during construction and operation renders sustainability an important criterion to consider when commissioning new research infrastructures. Eventual research infrastructure decommissioning has an additional environmental cost but may also provide an opportunity for repurposing and recovering previous expenditures, making a full life-cycle analysis important. The ability to adapt and evolve in response to a changing research and innovation environment is key to ensuring initial investments are well placed. For example, STFC's ISIS Neutron and Muon Source facility, commissioned in 1984, had an original operational lifespan of 20 years, however, refurbishment and extension has enabled this infrastructure to remain relevant to 2030 and beyond [11].

Fundamental physics research at the edge of knowledge is by nature exciting, yet somewhat unpredictable. However, this is counteracted by the responsiveness of its innovative environment, meaning that with proper sustainability assessment in the first phase of development, continued monitoring, and effective measures placed to de-risk skills and data capabilities, Big Science research infrastructures are in good stead to deliver on their investment potential.

19.2.3 Research infrastructure design

The design process involved in research infrastructure creation drives development in many ways (e.g. [4, 19]). This is not only in the research outputs, but in the supporting technology and applications, which can subsequently find their way into public goods and services. The investment in these next-generation systems on behalf of the whole UK community is managed through strategic decisions at the national level. A decade-or-so-long ideation stage of a new (phase of a) research infrastructure occurs before financial support is sought, which can include early design studies and conceptual and technical design reviews. This ensures that the scientific research taken forward is a robust choice for funding and in line with the organisational strategy. A five-case business case [20] is then prepared to outline the strategic, economic, commercial, financial, and management dimensions of the proposed research infrastructure, and adheres to the guidelines given by *The Green Book* from HM Treasury [21]. While it is hard to predict the full benefits of a research infrastructure before its operation (and perhaps a decade or so after), cost–benefit analysis (CBA) is performed as part of the economic appraisal of the research infrastructure to demonstrate value for money of the programme to the public purse. A decision of whether to proceed with the research infrastructure proposed is made based on the information provided in the business case, where the scientific benefits take prime focus. Following approval, monitoring, and evaluation of the project at a series of predefined time-points takes place across programmes to ensure these CBAs remain realistic and representative, and that impacts are captured. STFC is involved in research infrastructures across the world (figure 19.1), and funds them in several different ways including UK campus-based infrastructure, UK distributed infrastructure and subscriptions to international infrastructure. Examples of each are given below.

19.2.3.1 STFC UK campus-based infrastructure: ISIS, Harwell

A large proportion of STFC-funded infrastructure is hosted on campuses, co-locating major science facilities with STFC's departments, other multidisciplinary research organisations, innovation-led companies, and business incubator centres. One such facility is the ISIS Neutron and Muon Source. ISIS is a world-leading multidisciplinary centre for research in the physical and life sciences based at the STFC Rutherford Appleton Laboratory, part of the Harwell Campus in Oxfordshire. The facility, owned and operated by STFC, is centred on a particle accelerator that is utilised to enable scientists to study materials at the atomic level using a suite of over 30 instruments. Directly through STFC core funding, the facility employs over 500 scientists, technicians, engineers, and administrators to support, maintain, and evolve the instrumentation used by researchers. Facility users are composed of members of the academic community and industry, from both the UK and internationally, as well as internal facility staff. Users have a wide range of scientific backgrounds, with applications seen in pharmaceuticals, clean energy, and nanotechnology to name a few. Facility use is administered through applications for beam time and after a peer-review process, those with successful proposals are provided access and support free-of-charge, on the condition that results are publicly available. Confidential use of instruments is also available for a fee, providing a service for industry-led innovation (figure 19.3).

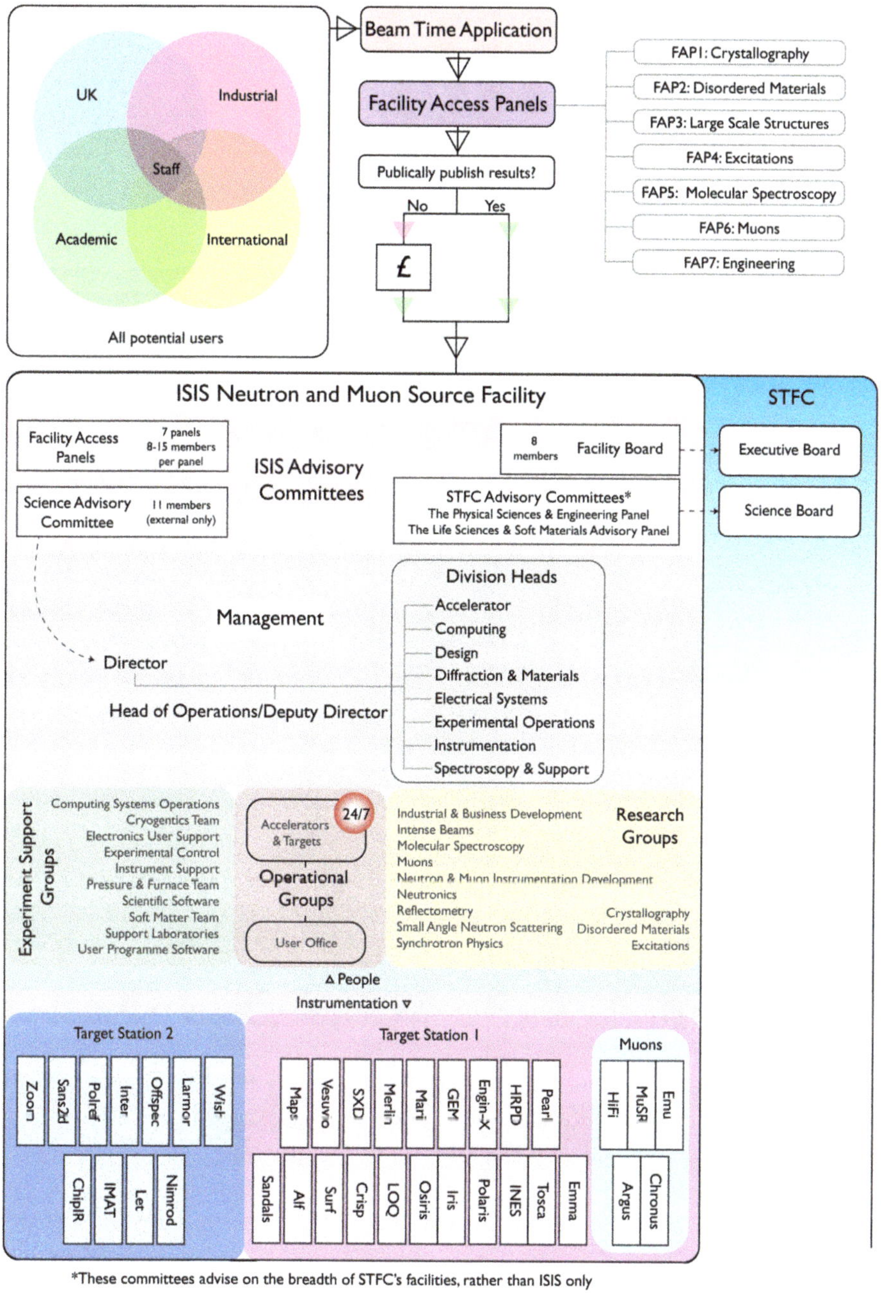

Figure 19.3. ISIS Neutron and Muon Source Facility governance structure.

19.2.3.2 STFC UK distributed infrastructure: DiRAC

A single, distributed, UK-wide supercomputing facility funded by STFC, DiRAC deploys supercomputers at four universities: Cambridge, Durham, Edinburgh, and

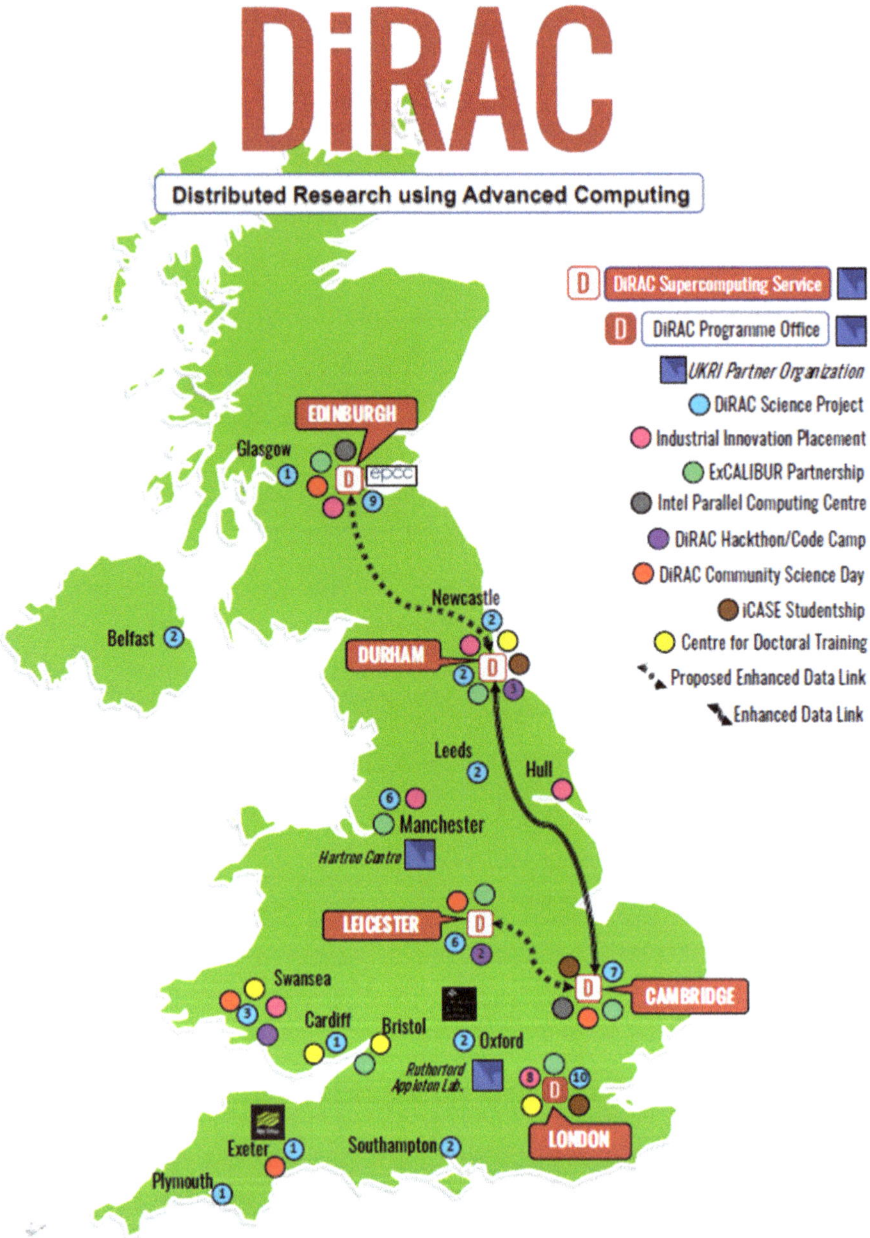

Figure 19.4. A map of the DiRAC programme. (Image credit: Clare Jenner.)

Leicester; the Project Office is at University College London (figure 19.4). Technical teams at the four sites provide expert support to DiRAC users and work with industry partners on innovation projects. DiRAC was established to provide distributed high performance computing services to the whole STFC particle physics theory community and training this community to efficiently and effectively use

DiRAC resources is a significant part of its offer. The distribution of the single facility over a few sites enables the DiRAC programme to offer efficient, agile services with a range of architectures, providing data intensive, memory intensive and extreme scaling services.

DiRAC is funded through the STFC business case process, and the programme is both academic-led and academic-supervised—the team that administer the facility are university-based staff, rather than directly employed by STFC. The call for proposals to access DiRAC is opened annually, and proposals are assessed by the Resource Allocation Committee, who award computation time, storage, and research engineering effort. The committee itself is composed of two sub-panels, Astronomy/Cosmology and Particle Physics/Nuclear Theory, both of which are comprised of university-based academics. In addition to the annual standard call, a Director's Discretionary Call is open for large, ambitious projects of outstanding scientific interest, and an anytime Seedcorn Time programme is available for small projects, typically for code testing and orientation before making a full application.

19.2.3.3 STFC international subscription: European Southern Observatory

The European Southern Observatory (ESO) is an intergovernmental astronomy organisation with the purpose of designing and operating state-of-the-art telescope observatories for its member state astronomers and astrophysicists. The administrative, scientific, and technical headquarters of ESO are based in Germany, supporting the operation of observational sites in Chile (figure 19.1). Broadly speaking, the scientific capabilities of ESO are composed of the unique combination of optical telescopes and the instrumentation associated with them to detect celestial events. The UK became the tenth member in 2002 through STFC-administrated funding, and following a one-off accession fee, pays an annual membership proportional to national income. This investment (i) provides UK scientists with preferred access to ESO facilities, (ii) allows UK companies to bid on contracts to develop new technology, and provide supplies and services, and (iii) join consortia to develop new ESO instrumentation. Application windows for observation time are opened twice annually, stratified by the instrumentation available, and proposals are assessed by the Observing Programmes Committee, which is composed of a dynamic collection of 17 international astronomy experts. In addition, those involved in instrument development are awarded guaranteed observational time on ESO telescopes. The facilities are available for non-member state access; however, proposals are subject to specific stringent assessment [22].

19.3 Why, how, what, and when to evaluate

Evaluation, for the purposes of UK government investments, is defined as 'a systematic assessment of the design, implementation and outcomes of an intervention' [21, 23]. In this case, the intervention is the research infrastructure that has been funded by public money. Evaluations seek to understand (i) how the research infrastructure and activities within are being implemented, (ii) the effects these have, (iii) any improvements that can be made to on-going or future investments, (iv) the

Table 19.1. STFC evaluation options for research infrastructure.

Evaluation	Time point	Description	Current utilisation by STFC
Baseline	On approval of funding	Determines the status of impact metrics prior to investment. Defines the starting point against which the impact of the investment is measured.	New funded research infrastructure business cases. Previous evaluations of preceding or related infrastructure can form at least a significant part of these evaluations.
Phase-end	Construction-operation phase interface	Evaluation of all aspects related to the (UK contribution of the) construction of the research infrastructure. Impacts tend to be relatively limited as this is prior to the research potential of the research infrastructure being realised.	National and international research infrastructure that has recently completed construction, and as standard for current research infrastructure under construction and future investments. Includes baseline for the operational phase.
Periodic	Every 5–10 years	Process and/or impact evaluation at strategic points, looking at the previous 10 years investment, or to the time point last evaluated.	International subscriptions and established UK-based research infrastructure. Five years after operational phase begins for current and future investments.
Project-end	Decommissioning	Lifetime evaluation of a research infrastructure.	Rarely, due to long time-horizon of infrastructure.
Post-project	5–10 years after project end	Evaluation to capture persisting impacts or impacts with long realisation times.	Not currently utilised. Relationship of current research infrastructure to previous (decommissioned) programmes means that impacts are often captured through live monitoring and evaluation.

overall impacts, and (v) cost-effectiveness [21, 23]. STFC undertakes both process and impact evaluations. Process evaluations assess the overall operation of the research infrastructure, including progress relative to objectives, quality and effectiveness, decision-making, and financial processes. Impact evaluations broadly focus on capturing knowledge, economic and societal outcomes, and impact. Depending on the stage of the research infrastructure, different types of evaluation are undertaken (table 19.1).

19.3.1 Evaluation purposes

The principle reasons for evaluating research infrastructure funded by UK public money are learning and accountability. Evaluations teach us (i) what works and what does not, (ii) how to manage risk and uncertainty for innovative programmes, and (iii) how to maximise efficiency and effectiveness of delivery. Maximising public value and outcomes delivered for taxpayers money is a key responsibility of the UK Government. Evaluations of research infrastructures helps inform the public of the value and outcomes of these investments made on their behalf and brings transparency to public investment decisions [23].

The impacts captured through evaluation create STFC's impact evidence base, which is then used in a wealth of material and environments to demonstrate the incredible socio-economic benefits investing in long-term fundamental science brings. The evidence base is utilised in business cases to support future investment, to inform and influence governmental spending reviews (e.g. [24]), shape government research and innovation strategy (e.g. [25]), improve internal processes, engage industrial stakeholders, and communicate to the general public the contribution fundamental science research has in daily life [1–6].

Alongside providing scientists with robust material to use in future funding bids and evidence of the value for money the investment has for the contemporary funding body, engagement in evaluation can stimulate and inspire scientists to think of the wider application of their work. This not only includes exploring unchartered scientific territory, but to consider potential translations of their discovery process to resources that can serve the public in their everyday lives. Evaluations can bring to the fore both processes that work and those that could be improved, allowing both the positive and negative lessons learned to be used to develop research infrastructures in the future.

High-risk discovery science comes with high-reward opportunities; stretching the outer extremities of our knowledge comes with a considerable amount of unpredictability regarding the positive impacts that may arise. This is one important reason to evaluate—by trying to capture and measure unpredictable positive impacts, we ultimately demonstrate that even though we may not be able to predict *what* additional benefit may come from pursuing a particular scientific endeavour, we can at least confidently predict that *something* will, and all that remains is the scale of the benefit. By establishing an efficient evaluation for research infrastructure, both anticipated and unpredictable positive impacts can be captured, and used to exhibit the extraordinary benefit Big Science brings to everyday lives. Indeed, with relative regularity, unexpected mass societal change occurs from a development in fundamental physics, or the science and engineering required to reach the goal; examples include touchscreens, satellites, the Internet, and MRI scanners.

19.3.2 Evaluation procedure

Evaluation of STFC's investments is a collaborative activity between the central-STFC evaluation team and the various facility and departmental programme teams, with support from central-UKRI evaluation specialists. This sits on a network of evaluation support and best practice development comprised of evaluation and evidence base experts from all of the constituent organisations across UKRI and a comprehensive collection of guidance [21, 23, 26] and frameworks [27] designed to optimise evaluation for different types of portfolio.

A key component of evaluation is contemporary collection of data (termed monitoring) which form the basis of analysis in future evaluations. This not only ensures the accuracy of data assessed across the lifetime of the programme, but also increases the likelihood of capturing impact whilst significantly reducing the burden of data collection at the point of evaluation. The monitoring of activities to allow evaluation can be regarded as an unnecessary burden by project managers and scientists—an extra layer of bureaucracy; another means to hold scientists to account; an additional channel for money to be diverted away from the science itself. Whilst these may have been unintended consequences of evaluation in the past, the current approach in the UK is focussed on supporting science teams to demonstrate the impact of their research in an objective way. STFC's position on this is 'maximum impact, minimum burden': the evaluation team work closely with research infrastructure teams to design a *monitoring and evaluation plan* to identify a collection of metrics deemed to be the most powerful and efficient way to capture both intended and additional benefits. The first step in forming a monitoring and evaluation plan is the creation of a logic model which summarises the inputs, activities, outputs, outcomes, and impacts for the investment (figure 19.5). The logic

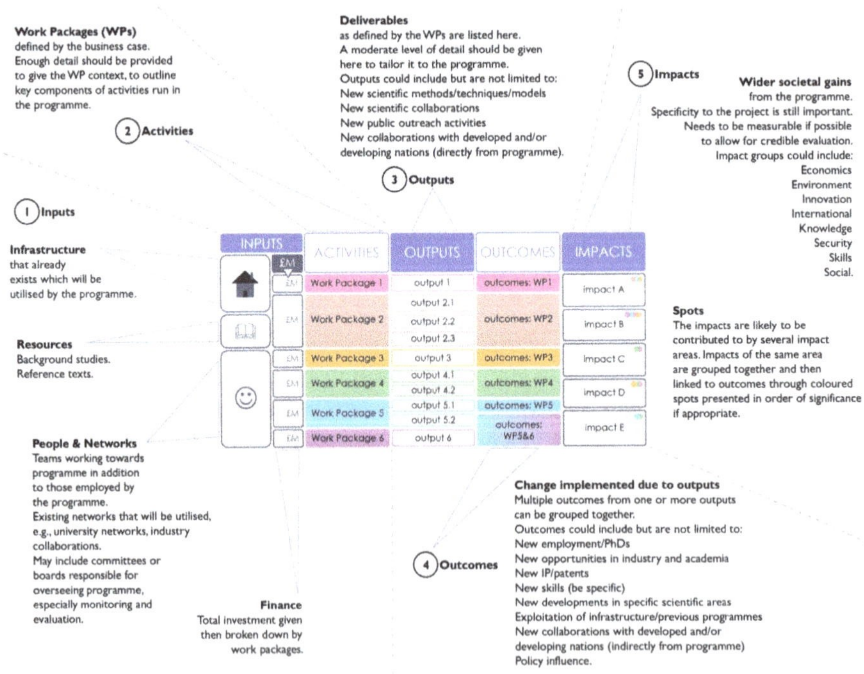

Figure 19.5. A template logic model used by STFC to initiate monitoring and evaluation plans.

model is then used to define the set of metrics best suited to capture the outcomes and impacts of the investment, and sets the framework for evaluation. At this stage, special care is taken to identify metrics which are already captured through present reporting, such that duplication of reporting (and counting) is avoided and subsequently burden of reporting minimised. Required data not already captured elsewhere is then assessed to determine if internal collection is possible (e.g. new skills learned by staff) or if external parties are required to maintain integrity of data (e.g. to perform interviews to canvass opinions of external contractors). The appropriate time point for each reporting window is defined, with as much unification between metrics as possible. Following this, depending on the timeline, the size of investment and previous related evaluations that have taken place, the schedule for evaluation is determined, and budgets stratified to support this.

The critical need to demonstrate objective assessment of research infrastructure investments means that while data are collected in real-time by STFC, external consultants are commissioned to run independent evaluations of the evidence provided, in addition to collecting or interrogating any data that is optimally performed by a third-party intermediary (e.g. stakeholder interviews, economic analysis, assessment of internal processes). Evaluations of STFC-funded research infrastructures are designed relative to STFC's Strategic Goals:

- *World-class research*: to champion UK global leadership in research to understand the Universe, its fundamental constituents, and their interactions.
- *World-class multidisciplinary facilities*: to ensure STFC's national laboratories and international facilities deliver world-leading science.
- *World-class innovation*: to be innovative across all STFC's activities, developing advanced technologies and creating new business opportunities
- *World-class skills*: to utilise STFC's national laboratories and science programme to develop a pipeline of skilled engineers, technicians and scientists
- *Operational excellence*: to ensure that STFC is a modern and effective organisation that values its staff and delivers for its stakeholders.

A further area of assessment—science diplomacy—is often added into evaluations of STFC's international subscriptions. Science diplomacy in this case is the term used to describe the (perceived) influence the UK has on decision-making in fundamental science, in terms of specific facilities, individual programmes of work, and the future direction of research in fields which rely on international collaboration.

Each evaluation is accompanied by several recommendations, both in terms of how to improve greater realisation of impacts and how to better capture those that have been realised. These recommendations are converted into a management plan, with responsibilities assigned to either the facility or the central evaluation team to find solutions. Associated Advisory Committees and/or Boards hold the teams to account with regards to the management plan. Additionally, the teams are encouraged to disseminate improvements that could be applied in other research infrastructures, leading to continuous improvement across the organisation. In particular, these recommendations are a key factor with regards to how evaluation methodology evolves over time, be that in terms of data collected or how it is analysed.

19.3.3 Evaluation challenges

19.3.3.1 Timeline and attribution

The time for the true extent of both the scientific and socio-economic impacts from Big Science infrastructure to become apparent can vary from a few years to many decades, however, when stretching themselves to understand the secrets of the Universe, be that at a galactic or sub-atomic size, physical scientists and engineers push the boundaries of technology. This impetus is the underlying driving force behind the technological breakthroughs that bring about advances in household devices that we use day-to-day. If these benefits can be captured effectively through evaluation, there can be no doubt that investment in fundamental science is not only well justified but essential for our growth as a nation. The cost–benefit time-horizon of an investment is advised to match the lifetime of a project. Ten years is accepted as a typical project duration for forward-looking cost–benefit analyses, and therefore appropriate for average timeframes in evaluation, even if the lifetime of the investment is longer. It is recognised a programme can extend to 60 years for infrastructure, therefore multiple ten-year evaluations would be expected to take place over the lifetime of these investments. Decade-long time spans naturally encompass changes to staff and evolution in data storage systems, both of which can contribute to loss of organisational memory for the research infrastructure under evaluation. Data used in evaluation is not restricted to the infrastructure itself, and involves external sources, such as industrial partners, research institutions, and academic bodies. These parties may not collect the data of interest to future research infrastructure evaluators without request, and/or may not have the capacity to maintain all data over the lifetime of the investment. These factors mean that contemporary data monitoring through robustly established processes into a dedicated repository is key to capturing impact data over long time-horizons.

The long time-horizon of many impacts in fundamental science means that whilst the value of an impact as a whole may be reasonably estimated, the attribution a research infrastructure can claim has a large associated uncertainty. Attribution can be diluted over time as further events not related to the research infrastructure contribute to development. Uniqueness of the environment leading to impact generation also influences attribution. This is particularly poignant for STFC subscriptions to international research infrastructures. An impact may arise that was facilitated through UK investment or provision of scientific expertise; however, it could be argued that the impact could have arisen without the UK being involved. Assigning full or partial value of the impact as directly derived from the UK's involvement is therefore difficult when there is potential that the UK could have enjoyed the benefits without contributing to their development. The large uncertainty means a conservative percentage of typically 1% attribution is given to research infrastructures relative to a long-term impact. The challenge of attribution to a single research infrastructure for life- and world-changing developments means the UK only considers the micro-economic effects these investments have and therefore macro-economic methods, such as multipliers are not considered a valid approach in their economic evaluation. In contrast, OECD guidance [28] allows the use of multipliers in the same circumstances if 'restricted to well-known indicators within well-defined context' to allow appropriate attribution. As such, research infrastructure evaluations by the UK may not always be directly comparable to evaluations by equivalent international bodies.

The impact that UK public investment in research and innovation has made is represented relative to a counterfactual. A counterfactual is the scenario that would have happened if the investment had not taken place and can be characterised by scenarios where no investment or an alternative investment was made. The projection of the counterfactual relies on indicator trends and comparative data from previous interventions (not necessarily by the same funding body). However, due to the uniqueness of Big Science research infrastructure, counterfactuals are more challenging to define. The afore-mentioned international collaboration and investment required for the vast majority of Big Science infrastructure precludes the presence of an infrastructure on which to draw comparisons. Despite this challenge, counterfactuals are still developed, with all assumptions explained to allow transparent assessment. In particular, external expert peer review can play an important role in defining the additional benefits a particular programme brings.

19.3.3.2 Monetisation

It is accepted that research and development drive economic growth, which underpins the UK Government's target to increase total research and development investment by 2.4% of GDP by 2027. UKRI economic modelling estimates that the net-economic benefit of research and development investment is £7 for every £1 invested [7]. Monetisation of impacts is challenging in fundamental science, due to the long-term nature of impact generation and determining a reasonable level of attribution as outlined above, as well as making robust choices of non-market valuation with respect to knowledge-based entities. However, when appropriate monetisation of impacts is possible, it is worth pursuing as it provides a tangible benchmark to demonstrate and compare value.

As a government funded entity, the methods used in economic analysis are subject to compliance with HMT guidance [21, 23, 26]. Approved methods for monetisation of impact first state to use market prices where possible, such as opportunity cost, employment, and productivity. However, due to the nature of the research, non-market valuation plays a larger role in the UK's evaluation of fundamental science. A hierarchy for non-market valuation is given by HMT; the highest-ranking most typically applied methods by STFC are 'stated preference willingness to pay' and 'stated preference willingness to accept'. Both methods are underpinned by externally designed and directed questionnaires of key stakeholder groups and the robustness of their use in calculations is only as good as the parameters defining them. For example, asking a group of fundamental scientists how much they would be willing to pay to continue to access the facility they use for their daily work may illicit responses with high monetary value, however, the bias of this relatively small community is not suitable for extrapolation to the whole population of UK taxpayers. In this case a representative group from the general public may be more suitable in terms of extrapolative power, however, the value assigned would also be compounded by the (lack of) understanding or awareness of the group with regards to the facility in question. Contrarily, asking a small group of suppliers how much they would be willing to accept to forego the business and connections that working with a particular infrastructure brings them provides a robust contribution to the economic analysis. Whilst this group is still small, the value does not need to be extrapolated to or averaged over a larger group and can be directly added to the total monetisable value a research infrastructure contributes.

There are impact areas that, whilst suitable for monetisation, are not included in the final published figure. Fundamental science research infrastructures in the UK are closely connected with numerous industrial users, from large corporations to small enterprises, and spread across a diverse set of sectors. The advantage that working with STFC-funded facilities brings to these companies renders them unwilling to share the (value of the) collaboration in a public domain. This competitive edge in a purely business-driven sector demonstrates the vital support STFC's research infrastructures have in innovation and their significant role in the UK economy. This scenario illustrates why long-term evaluation is key—these impacts can be captured, but only when commercial sensitivity ceases to be a factor. It is exciting to know that some of the key day-to-day advances experienced in life are driven by STFC's facilities, and it is important for government and society to understand that whilst they cannot yet be made public, real-life impacts are being made.

Economic benefits are only a proportion of the tangible positive impacts that science brings to the UK, and the ability to appropriately evaluate the impact fundamental physics and related infrastructure make on society is pivotal to justifying continued investment in this area.

19.3.4 Evaluation evolution

STFC began evaluating research infrastructures in 2007 and over the past 14 years, the methodologies employed have evolved and developed significantly. Working with external consultants and internal colleagues, new innovative methods are tested regularly with the aim of comprehensively capturing the diverse impacts of STFC's investments. STFC's evaluation history started with the world's first lifetime social and economic study of a large science facility [29]—The Synchrotron Radiation Source (SRS), operated at the Daresbury Laboratory from 1981 to 2008. This study was in the wake of recommendations [30] to the UK Government that Research Councils should measure outcomes and report on impact, with a view to increasing the recognised economic impact UK public investment was creating. At this time, there was no evaluation culture, and no consensus on the definition of research impact. The challenge of retrospective evaluation was evident throughout this study, as there was no requirement during the SRS lifetime to collect certain data, and contemporary tools we take for granted, such as the wide use of desktop computers, were not available. Despite these challenges, this benchmark study created a new framework for capturing research impacts, providing a platform for STFC to develop methodologies in this area. Following this study, the next approach was to build a collection of case studies, placing monetary value against as much as possible, and then extrapolating this data to provide a representation of impacts across a whole facility through its lifetime. It was recognised over time that this approach had too many assumptions, and therefore uncertainties, to capture impact effectively. The present best practice for STFC's evaluations is a mixed method approach, which outlines a range of impacts. Bibliometric analysis, appropriate economic estimates and both quantitative and qualitative capture of societal impacts are used in evaluation, with a suite of case studies to illustrate the context these analyses are performed within.

Alongside continuous efforts of the STFC evaluation team to appropriately capture research impacts and the evolution in approaches that this brings, the methods used are also subject to changes in Government guidance. This ultimately means that economic impact valuation that was acceptable in a previous evaluation, is not automatically transferable to subsequent studies, and therefore caution should be taken in comparing the impacts of research infrastructures that have been assessed under different frameworks.

It will never be possible to assign a monetary value to all of the impacts arising from STFC's investment in research infrastructures. The principal reason for fundamental science research infrastructures is exactly that—to perform scientific research. Facilities are not set up as businesses, however, the small proportion of impacts that can be monetised often outweigh the investments made within a ten-year period [11, 22, 31], and when presented alongside cutting-edge science, innovation, and societal impacts of the research infrastructures, the justification for the use of public funds is particularly compelling.

19.4 The future of evaluation

An increased importance of the evaluation of research to the UK Government has meant that a greater amount of funding has been directed towards it. This has resulted in reduced conflict for scientists—budgets are not top-sliced retrospectively for monitoring and evaluation activities, and instead ring-fenced funding for evaluation is added once the science budget has been defined. This allows research infrastructures to sufficiently resource both research and monitoring and evaluation activities. Detailed monitoring and evaluation plans are now also a requisite part of business cases. This establishes an evaluation-oriented culture at the beginning of the project, which is then more easily embedded during the programme, significantly improving the quality of data available at evaluation points.

External consultants are traditionally used to perform evaluations due to the expertise they offer and the objective perspective their assessment brings. The growth of evaluation expertise within both the administrative centre of STFC and each research infrastructure, however, allows new approaches to be piloted. In efforts to reduce the cost of evaluation to the public purse, only activities that require objectivity are outsourced to a third-party for baseline evaluations. Surveying collaborators and beneficiaries is one instance where maintaining objectivity is essential. Commissioning market researchers to perform this activity on behalf of the organisation, coupled with internal analysis of monitoring data, provides a cost-effective approach to perform the necessary assessment. Additionally, with increased engagement of research infrastructure teams with monitoring, contemporary analysis of data can be provided on-demand when an evidence base is needed. Real-time data can be used to build on existing evaluation evidence and smaller, targeted studies can be commissioned to provide specific insight. Major evaluations continue to be outsourced to independent consultants, who benefit from a more comprehensive internal data set on which to base their objective assessment. All evaluations require a level of overarching objectivity, which is provided by expert external peer review.

In recent years, the visibility, interest, and prioritisation of two topics has increased in the UK: (i) environmental sustainability and (ii) equity, diversity and inclusion (EDI). The UKRI Environmental Sustainability Strategy [32], published in 2019, outlines the organisation's key objectives and commitment in this area, categorised by (i) decision-making and engagement, (ii) net zero carbon, (iii) procurement, (iv) efficient use of resources, (v) fostering a healthy living environment, and (vi) interactions with the physical environment. STFC, along with UKRI's eight other constituent organisations, are in the process of establishing Sustainability Action Plans to drive change at a local level in line with the strategy. Evaluation of environmental sustainability at STFC-funded research infrastructure will be assessed in both process and impact evaluations going forward. Process evaluations will capture the interventions made at an operational level, for example in the choices made during construction, the suppliers and sources of energy used to run the facilities, and handling of waste. Impact evaluations will focus on the technological developments and fundamental discoveries that enable the environmental crisis to be addressed, such as new materials which increase efficiency of energy sources [33] or remedial action that can prolong the use of already established infrastructure [11].

Equity, diversity, and inclusion is a fundamental building block in UKRI's development of a research and innovation system that benefits all parts of society [34]. Evaluation teams are in a privileged position to drive positive societal change. Introducing EDI metrics into monitoring plans can act as a catalyst to inspire ambitious inclusive initiatives and monitoring and evaluation holds people to account in terms of delivery and impact. At STFC, EDI baselining will be paired with an exploration of other metrics to place EDI characteristics in context to identify both barriers to address and positive practices that should be embedded. Through establishing EDI baselines, evaluators can contribute to the design of interventions specific to each research infrastructure, and the appropriate metrics to monitor and evaluate the effectiveness of these in addition to broad interventions at a parent-organisation level.

The importance and prominence of these two fields are only set to increase in the future, such that new sections in both process and impact evaluations will become standard in STFC research infrastructure evaluations, and smaller, focused evaluations in these areas are on course to be added to the suite of evaluation tools at STFC.

Fundamental science research infrastructure takes decades to be planned, funded, constructed, and then placed into operation. The combination of these long time-horizons coupled with the abstract nature of fundamental physics at first sight can render justification for public investment challenging. Evaluation will remain a cornerstone of these processes in the UK—helping to guide decision-making, refine processes and capturing the extraordinary impacts Big Science research infrastructure ecosystems make across the breadth of scientific research and for society.

References

[1] COVID-19 Response *UKRI* https://www.infraportal.org.uk/browse-all-case-studies?f%5B0%5D=case_study_filter_categories%3Acovid (Accessed: 23 February 2023)

[2] *STFC Food Network+* https://www.stfcfoodnetwork.org/ (Accessed: 5 August 2021)

[3] Accelerator Science and Technology Centre Science Highlights 2018–19 *Science and Technology Facilities Council* https://www.astec.stfc.ac.uk/Pages/ASTeC%20Science%20Highlights%202019.pdf (Accessed: 23 February 2023)

[4] Brown N, Simmonds P, Rosemberg C, Rentel M, Glass C and De Santo A 2020 Evaluation of the Benefits that the UK Has Derived from CERN *Evidence Document* Technopolis https://www.ukri.org/wp-content/uploads/2020/07/STFC-200522-CERNEvaluation2019EvidenceDocument.pdf (Accessed: 23 February 2023)

[5] https://www.scd.stfc.ac.uk/Pages/CCP_Partners_offer_time_resources_and_expertise_for-research_into_Covid_19.aspx (Accessed: 23 February 2023)

[6] The birth of the web *CERN* (Accessed: 23 February 2022) https://home.cern/science/computing/birth-web

[7] UKRI 2020 The UK's research and innovation infrastructure: landscape analysis *UKRI* https://www.ukri.org/wp-content/uploads/2020/10/UKRI-201020-LandscapeAnalysis-FINAL.pdf (Accessed: 23 February 2023)

[8] Elsy D 2017 *Independent Review of National Facilities at Harwell*

[9] Nurse review of research councils *GOV.UK* https://www.gov.uk/government/collections/nurse-review-of-research-councils (Accessed: 5 August 2021)

[10] UKRI 2020 The UK's research and innovation infrastructure: opportunities to grow our capacity *UKRI* https://www.ukri.org/wp-content/uploads/2020/10/UKRI-201020-UKinfrastructure-opportunities-to-grow-our-capacity-FINAL.pdf Accessed: 23 February 2023

[11] Simmonds P, Brown N, Rosemberg C, Varnai P, Rentel M, Koc Y, Angelis J and Kosk K 2016 ISIS neutron and muon source lifetime impact report *Technopolis* https://www.ukri.org/wp-content/uploads/2022/07/STFC-040722-ISISLifetimeImpactStudy.pdf (Accessed 23 February 2023)

[12] UK Government 2021 Quantifying the UK data skills gap—full report *GOV.UK* https://www.gov.uk/government/publications/quantifying-the-uk-data-skills-gap/quantifying-the-uk-data-skills-gap-full-report#conclusionsobservations (Accessed: 10 February 2022)

[13] Technician commitment *UKRI* https://stfc.ukri.org/skills/technician-commitment/ (Accessed: 5 August 2021)

[14] National Graduate Recruitment Awards 2022 https://targetjobsawards.co.uk/winners/2022 (Accessed: 23 February 2023)

[15] Graduates *Science and Technology Facilities Council* https://stfccareers.co.uk/graduates/ (Accessed: 5 August 2021)

[16] Science and Technology Facilities Council—large employer (250 to 4,999 employees) *Institute of Physics* https://iop.org/about/awards/apprentice-employer-award/science-and-technology-facilities-council#gref (Accessed: 4 January 2021)

[17] Apprenticeships *Science and Technology Facilities Council* https://stfccareers.co.uk/apprenticeships/ (Accessed: 5 August 2021)

[18] Innovation placements DiRAC https://dirac.ac.uk/innovation-placements/ (Accessed: 23 February 2022)

[19] Magnet technology in MRI scanners leads to better care at lower cost *UKRI* https://www.ukri.org/about-us/how-we-are-doing/research-outcomes-and-impact/stfc/magnet-technology-in-mri-scanners-leads-to-better-care-at-lower-cost/ (Accessed: 23 February 2022)

[20] HM Treasury 2018 HM Treasury guide to developing the project business case. Better business cases: for better outcomes https://assets.publishing.service.gov.uk/government/uploads/system/uploads/attachment_data/file/749086/Project_Business_Case_2018.pdf (Accessed: 5 August 2021)

[21] HM Treasury 2022 The green book: central government guidance on appraisal and evaluation https://assets.publishing.service.gov.uk/government/uploads/system/uploads/attachment_data/file/938046/The_Green_Book_2020.pdf (Accessed: 4 August 2021)

[22] Brown N, Glass C, Simmonds P and Vinnik A 2022 Socio-economic impact evaluation study of the UK subscription to ESO *Technopolis* https://www.ukri.org/wp-content/uploads/2022/07/STFC-040722-SocioEconomicImpactEvaluationUKSubscriptionESO-FinalReport.pdf (Accessed 23 February 2023)

[23] HM Treasury 2020 Magenta book. central government guidance on evaluation https://assets.publishing.service.gov.uk/government/uploads/system/uploads/attachment_data/file/879438/HMT_Magenta_Book.pdf (Accessed: 5 August 2021)

[24] Autumn Budget and Spending Review 2021 *UK.GOV* https://gov.uk/government/topical-events/autumn-budget-and-spending-review-2021 (Accessed: 23 February 2022)

[25] UK innovation strategy: leading the future by creating it *Department for Business, Energy and Industrial Strategy* https://assets.publishing.service.gov.uk/government/uploads/system/uploads/attachment_data/file/1005000/uk-innovation-strategy.pdf (Accessed: 5 August 2021)

[26] HM Treasury 2015 The aqua book: guidance on producing quality analysis for government https://assets.publishing.service.gov.uk/government/uploads/system/uploads/attachment_data/file/416478/aqua_book_final_web.pdf (Accessed: 5 August 2021)

[27] Department for Business, Energy and Industrial Strategy BEIS monitoring and evaluation framework *BEIS Research Paper* Number 2020/016 https://assets.publishing.service.gov.uk/government/uploads/system/uploads/attachment_data/file/947722/beis-monitoring-evaluation-framework.pdf (Accessed: 5 August 2021)

[28] OECD 2019 Reference framework for assessing the scientific and socio-economic impact of research infrastructures *OECD Science, Technology and Industry Policy Papers* No. 65 https://www.oecd-ilibrary.org/science-and-technology/reference-framework-for-assessing-the-scientific-and-socio-economic-impact-of-research-infrastructures_3ffee43b-en (Accessed: 23 February 2023)

[29] New light on science—the social and economic impact of the Daresbury Synchrotron Radiation Source, 1981–2008 *UKRI* https://www.ukri.org/wp-content/uploads/2022/07/STFC-210722-NewLightScienceSocioeconomicImpactStudyDaresburySRS.pdf Accessed 23 February 2023

[30] Brown N, Glass C, Simmonds P, Rosemberg C, Seth V, Boscaro-Clarke I, Cesmat F and Coffer J 2006 Increasing the economic impact of the research councils *Economic Impact Group Report* https://webarchive.nationalarchives.gov.uk/ukgwa/+/http:/www.berr.gov.uk/dius/science/page32834.html (Accessed: 12 February 2022)

[31] 2021 Socio-economic impact study report, Diamond Light Source *Technopolis* https://www.ukri.org/wp-content/uploads/2022/11/STFC-111122-Socio-Economic-Study-Report-Diamond-Light-Source.pdf (Accessed: 23 February 2023)

[32] UKRI environmental sustainability strategy, 2020 *UKRI* https://www.ukri.org/wp-content/uploads/2020/10/UKRI-050920-SustainabilityStrategy.pdf (Accessed: 5 August 2021)

[33] de Laune R 2021 2021 roadmap to sodium ion batteries *ISIS Neutron and Muon Source* https://isis.stfc.ac.uk/Pages/SH21_SodiumIonBatteries.aspx (Accessed: 23 February 2022)

[34] Equality, diversity and inclusion *UKRI* https://www.ukri.org/what-we-offer/supporting-healthy-research-and-innovation-culture/equality-diversity-and-inclusion/ (Accessed: 2 February 2021)

IOP Publishing

Big Science in the 21st Century
Economic and societal impacts
Panagiotis Charitos, Theodore Arabatzis, Harry Cliff, Günther Dissertori, Juliette Forneris and Jason Li-Ying

Chapter 20

Space technology entrepreneurship drives societal and economic benefits

Poul Zimmermann Nielsen, Stefan Gustafsson and Sune Nordentoft Lauritsen

The European Space Agency (ESA) is Europe's gateway to space. Its mission is to shape the development of Europe's space capability and ensure that investment in space continues to deliver benefits to the citizens of Europe and the world.

To drive the societal and economic benefits of Europe's investment in space more broadly, 20 years ago the agency launched its ESA Business Incubation Centre initiative. Today, the initiative has grown to a pan-European network of 24 national centres in 21 countries, with several more on the drawing board. It has turned out to be an international powerhouse for innovative space entrepreneurship and the fostering of new companies.

The ESA Business Incubation Centres (ESA BICs) are part of ESA Space Solutions, the go-to place for great business ideas involving space in all aspects and areas of society and our economy. The mission is to support entrepreneurs in Europe in the development of business using satellite applications and space technology.

The ESA BICs offer business incubation support at over 70 sites throughout Europe and yearly foster about 200 new start-ups. At the beginning of 2022, more than 1200 new companies had been rolled out of this 'space business start-up-fostering initiative' thanks to tapping into the ever-growing number of innovative space technologies and satellite services, as well as introducing leading-edge solutions for future space programmes and space-based services.

With the overall vision of improving life on Earth, underpinning sustainable growth, and at the same time strengthening Europe's economy by fostering new, creative start-ups based on space solutions, the ESA BIC network has produced numerous prosperous and innovative high-tech businesses throughout Europe.

High-value forward-thinking start-ups in Europe have been created developing smart B2B, B2C, and in some cases also B2G solutions. Many traditional non-space

doi:10.1088/978-0-7503-3631-4ch20

industries have gained from innovative solutions, tools, and services helping them to improve their competitiveness and keeping up with today's global economy.

Driven by visionary inventors and high-tech entrepreneurs born out of the IT age, the ESA BIC start-ups combine the latest technologies and business concepts to address all segments in our society.

The newest technologies, such as block chain, data in the sky, 3D-augmented visualization, robotics, drones, artificial intelligence, and machine learning—just to mention a few—are combined with advanced space technologies and the growing wealth of satellite data and have formalized space assets as an advanced platform technology upon which all kinds of services for many parts of society can be built.

More and more forefront solutions from this ESA BIC network powerhouse are hitting the market every day. In addition to boosting Europe's economy, it has resulted in improving the quality of life here on Earth. Also, beyond Europe's borders many of these technological based solutions can fit and improve society based on satellites providing their global coverage.

In Europe, the start-ups coming out of ESA BICs have also been in line with a more general demographic trend. As it is not important where IT-based companies settle, their solutions can be offered and operated from anywhere, serving users everywhere, globally. This disrupting industrial concept can almost be seen as a new industrial revolution, overthrowing many fundamental demographics conditions and leading to new high-tech jobs being created in less populated regions, also places with a high level of unemployment. Many of the ESA BIC start-ups have chosen to settle in smaller towns and rural areas, but still offering their services globally, and thus also helping to reduce well-known demographic problems of larger city areas.

20.1 From technology transfer to space solutions

It all started in the late 1990s when ESA decided to proactively spin off its cutting-edge technology research and development (R&D) activities to the wider society by forming a dedicated an ESA Technology Transfer Office. The objective was to make the wealth of advanced space high-tech available to develop innovative and smarter solutions, also in non-space sectors.

Initially, the focus was on the re-use of the technologies which ESA had developed for its space programmes, such as for satellites to cope with the very special operations and harsh conditions in space. As it is not possible to send a repair man to fix a satellite, smart state-of-the-art solutions were developed to avoid single point of failures and enabling successfully operating missions in space over many years with built-in work-around solutions in case of failures. Exceptional materials were developed to withstand the extreme temperatures in space, and special highly advanced control systems were designed to manage remote satellites far away from Earth. It was realized that many of these cutting-edge technologies also have great potential outside space missions, and could help provide smarter and valuable solutions in non-space sectors.

Later, the ever-growing number of satellites provided a wealth of data that provided the ground for developing smarter and more efficient systems for applications on Earth. Dedicated satellites were launched to provide more precise weather forecasts and prediction of extreme conditions such as hurricanes; others were put into space to provide global positioning services that improve navigation wherever you are, and yet others were launched to broadcast television worldwide and to providing today's global communications services.

Also more and more companies have been developing innovative solutions—based on the data from the growing number of Earth observation satellites—to improve all parts of our society. Farmers have access to better information which help them grow more and better crops, captains at sea can route their ships better and safer, and city administrations have means to develop better initiatives to reduce pollution and heat loss, just to mention a few. At the same time as improving production and making activities on Earth better and safer, these satellite data-based opportunities also result in the application of more environmentally sustainable solutions in our society.

As part of the ESA Technology Transfer Programme, this space spin-off activity was backed by the set-up of a pan-European national ESA Technology Brokers network which also today helps industries tap into space R&D and technologies, both from ESA programmes, but also from other space programmes.

In 2004 the next proactive outreach initiative was launched to support inventors and entrepreneurs to start new companies based on the use of space technologies. To this end, ESA opened their first European Space Agency Business Incubation Centre (ESA BIC) in Noordwijk in the Netherlands, near the Agency's space technology development centre ESTEC. Since this first opening, more centres have been added in Germany, Italy, the UK, Belgium, France, Spain, Portugal, Sweden, the Czech Republic, Austria, Ireland, Switzerland, Estonia, Finland, Hungary, Norway, Denmark, Greece, and Poland, all in close collaboration with national partners. In addition, Luxembourg has set-up the European Space Resources Innovation Centre (ESRIC) which is also part of the network.

Later around 2009, ESA initiated its Business Application initiative to support European companies to use space technologies and satellite data to boost their products, services, and operation, and to develop new solutions for non-space sectors. This was backed by a network of national ESA Business Application Ambassadors to promote the potential in using satellite services more widely in society.

All these initiatives were in 2019 combined under ESA Space Solutions (figure 20.1).

20.2 What is an ESA BIC start-up?

In short, an ESA BIC start-up can be defined as, 'An innovative entrepreneurial idea having a space link with the potential to improve or even disrupt current practices and techniques in a specific application area, space or non-space, and being developed by a registered company at an ESA BIC into a viable business with potential'.

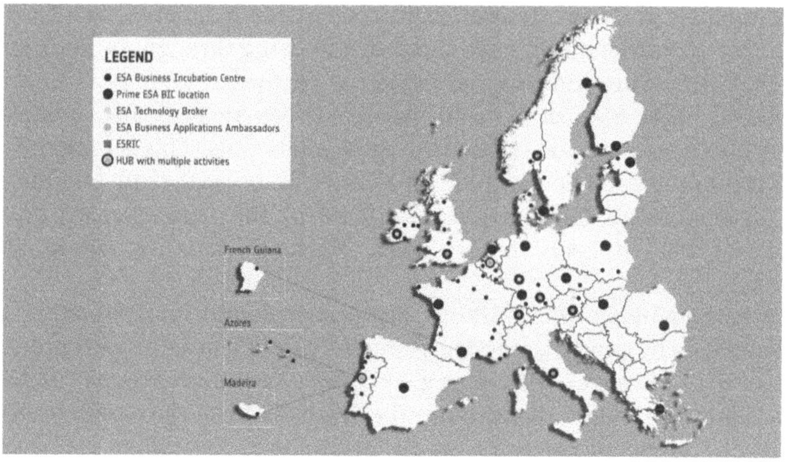

Figure 20.1. ESA Space Solutions network 2021. Credit: ESA.

To be accepted into an ESA BIC start-ups must have less than five years on their back. Some of the start-ups are actually first formed as registered companies after the entrepreneurs' business idea has been accepted by an ESA BIC. Others are young companies that need business incubation support to move forward faster, and to get their company really going and off the ground. Reasons can be the need for special space-related technical assistance which the ESA BICs can facilitate, access to space and industrial networks through international partnerships, business support often needed by engineering centred high-tech start-ups, or just economic incentives.

There are basically two types of ESA BIC start-ups. The first are called 'downstream' where the space assets, satellite technology, and satellite data services are used for the development of a new application down here on Earth. The second type is called 'upstream', where a technology from Earth is transferred to an application for future use in space, a new development for future space missions, or even the development of a full satellite constellation to fulfil a specific purpose, most often tailored to the needs of a specific user segment on Earth. A few ESA BIC start-ups develop business solutions targeting both downstream and upstream areas. A recent internal assessment revealed that roughly 89% of all ESA BIC start-ups belong to the downstream group while the rest are upstream.

20.3 Biggest space business incubation network

Together with the other outreach activities in ESA Space Solutions, the ESA BIC network forms the world's largest business incubation network. The main objective is to support entrepreneurs presenting a space-related business idea, thereby creating and growing start-ups in Europe to develop new upstream and downstream solutions in our society.

It is by no means a straightforward task to create a business from a space asset, to turn the use of satellite data into viable economic terrestrial ventures, to use Mars robot technology to create smarter industrial productions, etc. But this is what a great number of entrepreneurs have succeeded in doing throughout Europe with the help of ESA BIC support teams and industrial partners.

Statistically, over 85% of start-ups accepted into an ESA BIC complete their business incubation cycle and continue to grow afterwards. This is primarily because of two important elements: the in-depth evaluation of applicants and the comprehensive business incubation curriculum offered by the centres, based on years of experience of what it takes to turn space business ideas into sound, successful businesses.

Before entrepreneurs and start-ups are accepted into an ESA BIC, they are assessed carefully by expert teams consisting of ESA officials, business professionals, entrepreneurial advisors, and, where relevant, scientific experts. Based on a submitted business plan application, the start-ups will be scrutinized on the feasibility of turning the foreseen space technologies into what is proposed, the overall business potential of the proposed idea, and whether the combined entrepreneurial team possesses all the different qualifications needed to develop a sound company with scale potential. With the objective of fostering new innovative space-based solutions in all parts of society, candidates with business growth potential have a chance to be admitted to ESA BIC incubation support, within the capacity of each centre.

The second important element is the extensive business incubation curriculum put together by the ESA BICs to support the early start-ups on their business journey to become fully fledged successful companies. Over time, this business incubation scheme has been refined based upon experience dealing with the challenges for developing space-related ventures. It includes general aspects relevant for any early start-up but also specific elements related to the space assets. Support is always specifically tailored to the needs of each start-up, and complemented by expertise from the ESA BIC national partners.

All the ESA BICs operate as a combined network sharing experience and good practices to improve business incubation continuously, but are at the same time well connected to their local industry, universities, research organizations, government, and investor communities. Being part of a pan-European space entrepreneurship network while at the same time being well embedded in local communities reinforce the operation of each ESA BIC with its hosted start-ups as well as becoming the place of choice for space-related innovation and business.

The combined excellence of the individual ESA BIC, the ESA Space Solutions network, the local scientific and industrial partners, and support from ESA specialists provide a rich umbrella of opportunities to support the start-ups on their way to become independent, viable, and financially sound companies. All this and working under the ESA brand, speed up the development of their products and services, their market entry, finding the best business partners, as well as open up for investors, business angles and VCs.

20.4 Unique support package turns ideas into viable businesses

The ESA BICs inspire and support entrepreneurs in turning space-related business ideas into commercially successful companies. Inventors, entrepreneurs, and young start-ups are invited to offer new ideas on how a problem can be solved by use of space assets, how satellite data can improve a certain case in a specific area, or maybe how the introduction of a new satellite-based services can revolutionize and make a critical functionality significantly better than how it is done today.

Entering an ESA BIC, the start-ups are offered a comprehensive business support package (figure 20.2) consisting of:

- Up to two years of business incubation at an ESA BIC of choice. The start-up must be registered in the country of the selected ESA BIC.
- Up to 80 h of technical support from leading experts in the regions and from ESA on a case-by-case basis.
- Business coaching including legal and IPR advice, and participating to workshops and training on aspects related to develop a sound company.
- Access to international networks helping the start-ups to grow outside of their region and country.
- The use of ESA BIC and ESA Space Solutions branding which have been demonstrated to be very valuable for customers, industrial partners, and investor communities.
- Receiving 50 000 euros as seed money to fund the development of products, services, and IPRs.
- Guidance to fundraising and access to ESA BIC initiated fundraising events with international business angles and VCs.
- Access to office space and logistics support.
- Support to strengthen the integration with local networks, industry, and relevant business partners.

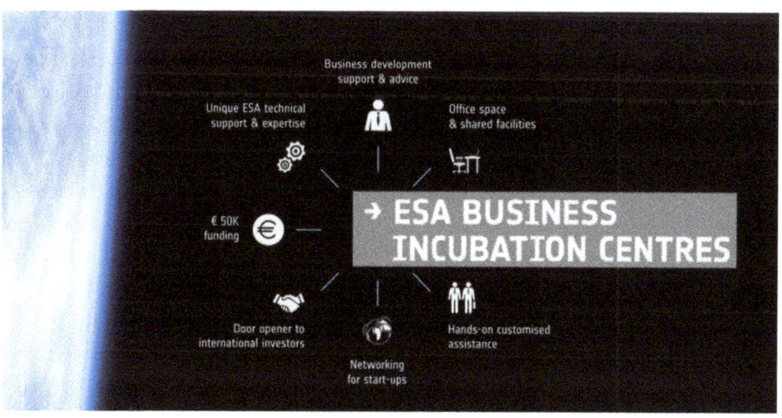

Figure 20.2. ESA Business Incubation Centre start-up support package. Credit: ESA.

The conditions can vary dependent on national conditions with some centres even offering additional support, e.g. for access to laboratories for test and development, larger economical incentives, and arranged loan opportunities.

Each ESA BIC teams up with national research centres, local businesses, technical support professionals, and national business networks to provide the best support for each start-up.

The unique support package which the start-ups receive at their hosting ESA BIC is tailored to the exact needs of space entrepreneurs, including the technical expertise and networking of the ESA Space Solutions network. This makes the business incubation at an ESA BIC particularly attractive to start-ups focusing on using space assets to develop innovative solutions for non-space customers, both for B2B and B2C. Also for upstream start-ups, the support from ESA provides access to a wealth of space expertise from ESA engineers, which is a unique business incubation support opportunity.

After the ESA BIC hosting period, the start-ups can also get support in their further business development journey through the ESA Space Solutions' Business Applications programme. On equal level with other European companies, the start-ups can apply for ESA co-financing to design and develop new space business solutions through feasibility studies and demonstration projects.

20.5 Space connections

ESA BIC start-ups have developed businesses based on assets from space programmes, space technologies, and satellite data services. Space assets can be from national European and ESA space programmes, as well from other international players such as NASA, NOAA satellites, and GPS services.

Initially, the majority developed their businesses using technologies, materials, methodologies, hardware, and software from space developments. Over the past ten years, more and more have developed solutions where satellite data are the prime source for systems and services targeting non-space sectors.

Lately more and more are also developing upstream solutions targeting the fast-growing new space sector. This is a growing area, and a recent internal assessment of ESA BIC start-ups shows that about one in nine now primarily works in developing upstream solutions, while the rest operate in downstream applications.

Upstream applications can address all aspects of the value chain of building spacecraft, launching satellites, and the control of operation when in orbit. The solutions offered can range from developing innovative booster launching technologies, special onboard instrumentation to observe specific situations and objects from space, smarter onboard handling electronics, or even establishing a full constellation of satellites to fulfil a specific business purpose.

However, downstream applications are the business area for the majority of ESA BIC start-ups. With the network taking in over 200 new start-ups each year, developing solutions during their business incubation, the space assets used include all possibilities. However, from our internal assessment, the general trend is that more and more start-ups, today around 60%, use satellite data to develop solutions for non-space customers. The remaining downstream start-ups are spinning off a variety of specific spacecraft related developments, which include the use of innovative space technologies, IPs, methodologies developed for spacecraft, as well as processing applications developed for e.g. satellite operations.

From our internal survey of the larger group using satellite data, figure 20.3 illustrates that the solutions provided by ESA BIC start-ups target a wide range of sectors in society.

Satellite data for downstream solutions can be divided into three groups, each provided by different types of satellite systems:

- Navigation and positioning data from GPS, Europe's Galileo system, and others in the Global Navigation Satellite System (GNSS).
- Earth observation data providing satellite imagery of the Earth, such as from Europe's Copernicus programme but also from US and other nations' satellite services.
- Communication satellites providing connectivity worldwide for TV, radio, phone, and data transmissions.

There are also dedicated satellite systems to provide information on the weather on the Earth as well on the space weather.

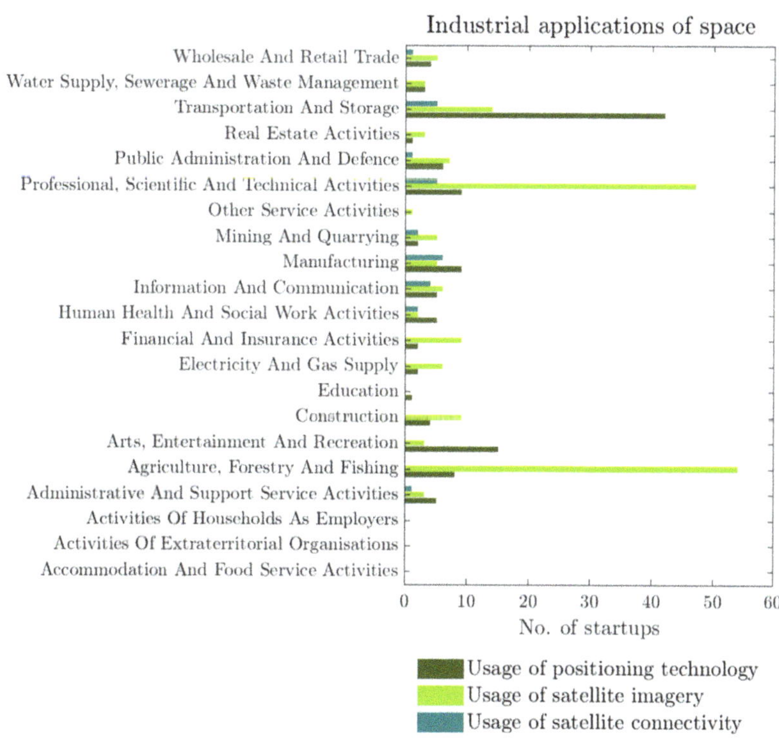

Figure 20.3. ESA BIC satellite data start-ups' business areas (ESA BIC Denmark internal survey).

20.6 Sectors targeted by the start-ups

The ESA BIC start-ups address many sectors and thus have a wide impact in our society. Our internal assessment reveals that around eight in nine develop solutions for customers in non-space sectors. The following list the seven largest business areas of these downstream start-ups:

- 22%: Professional, scientific, and technical activities
- 16%: Transportation and logistics
- 15%: Manufacturing and production techniques
- 8%: Information and communication services
- 8%: Agriculture, forestry, and fishing
- 6%: Human health
- 5%: Arts, entertainment, and recreation

Specifically, the three areas 'Professional, scientific, and technical activities', 'Transportation and logistics', and 'Agriculture, forestry, and fishing' can be broken further down as shown below (figure 20.4).

The two areas 'Manufacturing and production techniques' and 'Information and communication services' are difficult to explicitly break down into different societal areas, as many of the same solutions are used in different industrial sectors. 'Human health' and 'Arts, entertainment, and recreation' have not been broken down in our assessment.

From this survey it can be concluded that start-ups from the ESA BIC network address situations in more or less all sectors of society. One of the primary reasons is that today's space programmes and in particular satellites provide an infinity of data relevant to all kinds of situations and areas here on Earth. This enables innovative entrepreneurs to develop smarter ways of doing almost anything here on Earth by using these data and the knowledge they provide, often adding additional innovative data processing methodologies such as e.g. machine learning and artificial intelligence.

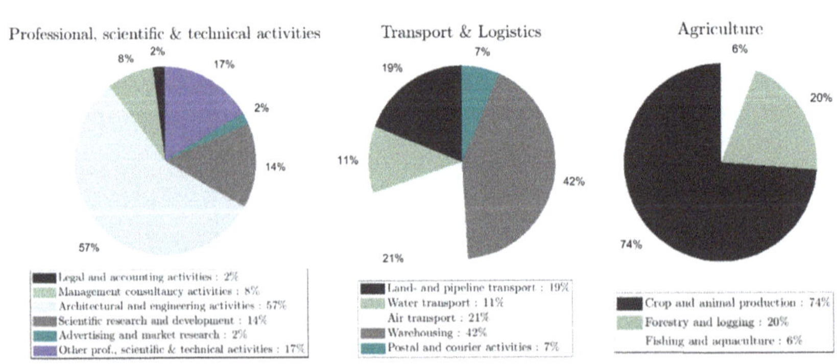

Figure 20.4. ESA BIC start-ups' key business areas (ESA BIC Denmark internal survey).

20.7 ESA BIC start-up business cases

The following examples illustrate the success cases of some of the over 1200 start-ups that have graduated from or currently are under business incubation at the ESA BIC network:

- Vultus from ESA BIC Sweden helps farmers optimize fertilization and grow healthier plants.
- Sternula from ESA BIC Denmark is launching a low-cost data efficient communication network to link shipping operators to their vessels globally.
- Thruvision from ESA BIC UK has developed people-screening safety and security systems revealing hidden weapons and explosives.
- Active Space Technologies from ESA BIC Noordwijk has developed an autonomous rover, automating car assembly.
- Arctic Space Technologies from ESA BIC Sweden is offering complete satellite ground station infrastructures.
- Vake from ESA BIC Norway helps identifying illegal fishing and environmental crimes at sea.

All these start-ups have developed solutions, either using space assets or new solutions for space missions.

20.8 Optimize fertilization, get healthier plants, and boost yield

About 8% of all ESA BIC start-ups offer business services to agriculture, forestry, and fishing customers, to improve and optimize production by innovative solutions, many of them based on satellite data from Europe's Copernicus programme.

From ESA BIC Sweden, Vultus illustrates this trend well. Their service helps farmers increase yields by cutting off up to 30% of fertilizers, resulting in healthier plants at lower costs. In addition, their innovative system also leads to a reduction in the environmental impact from agricultural production.

Inspired by the wish to reduce waste in the sector, Vultus developed their system in close collaboration with farmers to modernize operations. The aim was to combat the over-fertilization of the soils by analysing the specific nutrient needs of the fields and providing farmers while on the tractor with detailed near-real-time nitrogen recommendations based on satellite imaging (figure 20.5). As the exact location of tractors are known using satellite navigation data, farmers can easily apply the exact guidance while working the land.

Knowing that each crop type has a specific spectral signature by analysing satellite data, Vultus can estimate the health of the plants as well as the fertilizer and irrigation needs, providing specific recommendations to the farmers. By using synthetic aperture radar satellite imaging to penetrate clouds, Vultus can guarantee results at least twice a week, regardless of cloud cover.

Thanks to the simple integration in existing farm management, Vultus' system is now used by farmers in over 12 countries. Farmers have in average cut fertilization costs by 30%, while increasing their production by at least 3%–5%.

Figure 20.5. ESA BIC Vultus guides farmers to reduce use of fertilizers. Image courtesy of Vultus AB.

20.9 Better safety and communication at sea

Many ESA BIC start-ups have developed innovative solutions through the use of satellite services to support the shipping industry. To manoeuvre well and safely at sea, ships rely first of all on the experience of the captain and his crew, but smarter tools can help in doing this better and safer, also saving time and fuel.

With today's global satellite services, ships can receive accurate local weather forecasts, exact positioning information, as well as data on sea currents and other relevant information for route planning. However, communication with ships does not always provide optimal means for the latest and most detailed guidance to arrive often due to a low data rate between land and ships at sea. They can use low data rate satellite communication networks that are optimized for voice services, or expensive high data rate solutions which require stabilized satellite dishes to be mounted on board the ship.

The Danish ESA BIC start-up Sternula is changing this by setting up a global satellite communication constellation to meet the needs of the maritime industry today, and provide it at a highly cost-competitive price (figure 20.6). Their system utilizes the new VHF Data Exchange System (VDES) technology currently being rolled out to provide better maritime connectivity for safety and navigation as well as helping shipping companies reduce costs and environmental pollution.

From 2023, Sternula will start offering global VDE-SAT connectivity for maritime authorities and shipping industry through their own fleet of advanced micro-satellites in low-Earth orbit (LEO). This will provide a 30 times better data rate than today's commonly used AIS, together with two-way connectivity over the satellites. It is being designed to re-use the existing onboard VF equipment so ship owners do not need new installations, but just a simple update to get this improved service. Sternula's solution opens up stronger satellite communication to a shipping market previously out of reach, due to prohibitive costs.

Figure 20.6. ESA BIC Sternula will help ships obtain better communication facilities. Image credit: GateHouse SatCom.

VDES will enable a range of new digital services such as distribution of maritime safety information by local maritime authorities including weather and ice warnings, collection of data from critical onboard systems like engines, and environmental monitoring from sensors on smart buoys.

Sternula from ESA BIC Denmark is the world's first commercial provider of the advanced satellite-based VDES services. Launching the first satellites in 2023, the system will be fully operational with 50 satellites by 2028.

20.10 Boost security using space technology

Thousands of daily passengers on the Los Angeles Metro ride more securely with the deployment of cutting-edge ESA-patented screening technology to detect concealed weapons or explosives.

The new screening cameras from the English company Thruvision, a start-up from ESA BIC UK, are placed at various locations in the LA Metro system and can detect both metallic and non-metallic objects on a person's body (figure 20.7). They are capable of screening up to 2000 passengers per hour, and provide early visual warning of hidden person-borne improvised explosive devices or weapons that could result in mass casualties.

The system is based on a space R&D effort carried out in the early 2000s by an ESA-led team working out of the UK Rutherford Appleton Laboratory, aimed at future space missions. The work resulted in an innovative sensor to capture terahertz radiation which can be used to observe galaxies that could help to better understand how they were formed in the early Universe, and how stars have been formed throughout history.

Using this special ESA-patented technology, Thruvision developed body scanners that are completely non-invasive, passively screening individuals for suspicious objects. They detect the absence of the naturally occurring, extremely high-frequency 'terahertz' waves, emitted by anything warm, including the human

Figure 20.7. ESA BIC Thruvision boosts people's safety by ESA space technology. Image credit: Thruvision Ltd.

body. When an object is hidden in clothing or strapped to a person, these waves are blocked and their absence is detected by the system.

Thruvision now has over 300 people-screening cameras deployed in 18 countries around the world. The cameras put in place at escalators in the LA Metro each screen up to 2000 people passing by every hour and spot hidden objects at 10 metres' distance without slowing down their daily commute.

20.11 Mars rover modernizes auto assembly

Small automated self-guided vehicles are now used to quickly transport parts around the Volkswagen car assembly plant in Palmela near Lisbon, autonomously operating thanks to spin-offs from the ESA Mars rover.

Developed by the Portuguese start-up Active Space Technologies from ESA BIC Noordwijk, its automatically guided vehicle (AGV) ActiveONE is based on the technology and expertise gained from work with ESA. Engineers at this start-up took part in the design of several mechanical space mission structures and specifically on the development of systems to test the locomotion sub-systems and motors for a Mars rover to be used in surface and underground exploration on ESA's ExoMars mission.

What earlier was done by factory workers on trolleys and forklifts is now done autonomously by these computer-controlled ActiveOne AGVs (figure 20.8).

When presented to Volkswagen Autoeuropa, the company saw the opportunity immediately and introduced thirty ActiveONE AGVs in their demanding Palmela logistics production chain.

What differentiates ActiveONE from other AGVs already on the market is its innovative architecture and control system, a direct result of ESA's R&D on the

Figure 20.8. ESA BIC Active Space Technologies Mars robot technology makes car assembly more efficient. Image credit: Active Space Technology.

space technology for the advanced Mars rover. The spin-off from working on space structures and robots made the difference, in particular in its central differential traction system.

Now also employed in other factories, this space spin-off has opened up a very successful entry for the ESA BIC start-up to the fast-growing industry 4.0 market.

20.12 Automate and outsource satellite operation

Among the around one in nine upstream ESA BIC start-ups, Swedish Arctic Space Technologies is a good example on how to automate and outsource the handling of communication with a satellite or a constellation of satellites (figure 20.9). The company offers solutions with ground segment infrastructures fulfilling the requirements to operate satellite, for operators who prefer outsource this activity.

The solution includes everything components to communicate with satellites, including ground station antennas and other infrastructures, and facilitates operation of low-Earth orbiting (LEO) and medium-Earth orbiting (MEO) satellites in the VHF, UHF, S-, and X-band frequencies. Polar Earth orbits and Sun synchronous orbits are the most common orbits for Earth observation and satellite communication missions and with its ground stations located in Northern Sweden, the company can easily service all of those.

Arctic Space Technologies allows operators to connect directly to a satellite Earth station antenna and its secure data processing hardware and software inside data centres. Real-time processing of satellite data is offered without additional latency and fulfilling all necessary bandwidth requirements.

Figure 20.9. ESA BIC Arctic Space Technologies simplify the operation of satellite constellations. Copyright picture: Arctic Space Technologies.

20.13 Spot illegal fishing and environmental crimes at sea

Identifying illegal fishing, environmental crimes and protection of assets at sea are maritime challenges faced by governments and businesses around the world. The vastness of the sea and the limited traceability of ships create large information gaps, adding to the complexity of these challenges.

Vake from ESA BIC Norway now provides an innovative identification and tracking system of vessels at sea, even if they try to hide by switching off their AIS transmitters (figure 20.10). This is done by combining machine learning and satellite imagery to find, classify, and track ships and in particular those that are not voluntarily reporting and disclosing their location. Vake looks at multispectral satellite images made available through the ESA's Sentinel programme. These images contain more information than what is visible to the naked eye, and make it possible to track vessels even through clouds.

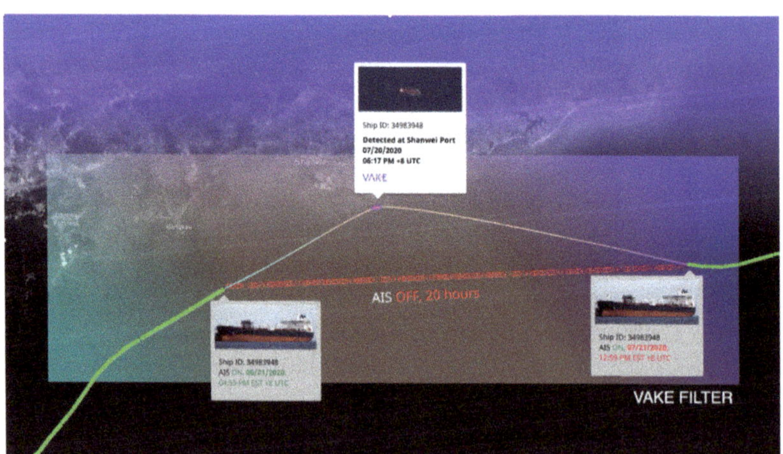

Figure 20.10. ESA BIC Vake tracks ships at sea, even when they try to hide. Image credit: VAKE.

This service is in particular very useful for coast guards, national administrations, offshore assets owners, and NGOs.

20.14 ESA BIC start-up opportunities in the New Space Era

Over the next decade, hundreds of conventional large satellites are expected to be launched into Earth orbit as well as thousands of micro- and nano-satellites. They will boost a variety of global services in telecommunication, navigation, weather forecasting, as well as Earth observation that will play a vital role in how our society will evolve.

While space programmes until some years ago were driven by national and international publicly funded entities such as NASA and ESA, private enterprises are now taking the lead within space systems.

Morgan Stanley estimates that the global space industry could more than triple over the coming 20 years, up from today's 350 billion US dollars to over one trillion US dollars by 2040 [1].

This is driven by several factors. One of the most important is the substantially lower cost of putting satellites into orbit, first of all thanks to the development of commercial launch systems. In the 1990s, NASA's space shuttle could launch satellites into LEO at typically USD 54 500/kg while today's SpaceX's Falcon 9 can do the same for around USD 2720/kg, a cost-reduction factor of 20 that creates many new opportunities for using space and satellites to do much more [2].

Another key factor on the space scene is that climate change is now commonly recognized as an important threat to our environment, and thus to our society. Satellite technologies can provide essential information for monitoring the Earth and climate change globally. This can help validate models and interventions to improve initiatives to move towards a more sustainable balance. Satellite have already proven to be effective to locate large emission sources previously unknown.

Also, technological miniaturizing of functionality is an important factor and has opened up for a whole new business market based on developing micro- and nano-satellites that can offer innovative and smarter services to selected customers. The possibility of developing increasingly smaller satellites has redefined the opportunities of using space.

Furthermore, the trend of developing autonomous transportation means, on land, in the air, and at sea, requires more advanced and accurate satellites services, but also more entrepreneurs to develop smart systems to 'make it happen'.

Even with high-accuracy navigation satellites, detailed Earth observation satellite data, and fast communication satellites, there is always a need for innovative companies to develop intelligent systems to bring it all together to create operational scenarios for us on Earth—and in the end to help develop a sustainable society for us all and for the close to 11 billion people estimated to constitute the world population at the end of the twenty-first century.

With recent technology advances, satellites are becoming a type of platform technology in the sense that they are used as infrastructure upon which digital applications can be developed for a multitude of end-users. It will assist in solving numerous problems on Earth, and the data will just be seen as a platform to build innovative services upon, somewhat like apps on a smart phone.

The ESA BIC network has been a significant contributor to this transformation with its over 1200 space start-ups throughout Europe, growing the base of space-related companies and speed up innovation related to making use of space in our society. They also complement existing smaller and larger companies in the steadily growing business opportunities from adding space services, as a value chain of innovation and progress that can help other companies step up their performance, competitiveness, and overall business potential.

This has demonstrated that space has potential everywhere and that start-ups with creative entrepreneurs are a great way to realize this potential. The outstanding impact has already resulted in a total revenue generated of nearly 200 million euros in 2017 according to internal estimates, more than several European countries contribute to space programmes and ESA.

Based on the results achieved over the first almost 20 years, the ESA Business Incubation Centre network has proven to be an excellent driver for supplying solutions to use space to address common needs and very often in an environmentally friendly way, for supporting creative entrepreneurs to turn forward-thinking ideas into viable successful businesses, and for bringing value to our society here on Earth by fostering innovative space start-ups.

The commercialization of space is the next big frontier, also here on Earth.

ESA Business Incubation Centres are making it a successful journey.

References

[1] Morgan S 2020 Space: investing in the final frontier https://www.morganstanley.com/ideas/investing-in-space

[2] Jones H W 2018 The recent large reduction in space launch cost 48th Int. Conf. on Environmental Systems https://ttu-ir.tdl.org/bitstream/handle/2346/74082/ICES_2018_81.pdf

Part III

Historical perspectives on Big Science [Editor: Theodore Arabatzis]

IOP Publishing

Big Science in the 21st Century
Economic and societal impacts
Panagiotis Charitos, Theodore Arabatzis, Harry Cliff, Günther Dissertori, Juliette Forneris and Jason Li-Ying

Chapter 21

Introduction to part III: Big Science in a historical perspective

Theodore Arabatzis

Big Science has a long history. In some form or another, it can be traced back to the days of Tycho Brahe, the great Danish astronomer of the sixteenth century (chapter 22), or even to the early days of meteorology in antiquity (chapter 29). Its more recent origins, however, can be found in early twentieth century American physics and the big technoscientific projects of World War II, such as the Manhattan Project for the construction of the atomic bomb (chapters 23 and 24). Since then, different forms of Big Science have developed, from gigantic laboratories with a centralized, 'vertical' organization, such as CERN, to decentralized, 'horizontally' organized enterprises involving huge networks of researchers, such as the Human Genome Project (chapters 25 and 30).

Part III of this volume treats the historical development of Big Science, with a chronological focus on post-World War II episodes. Its aim is to put Big Science into a historical perspective and, thereby, to understand historically the ways and whys of large-scale forms of scientific life. An additional, albeit indirect, goal is to shed new light on contemporary discussions about the merits, drawbacks, and challenges of knowledge production on a big scale.

The contributors to this part of the volume represent a wide range of historical, sociological, and philosophical expertise on Big Science, that extends across geographical and thematic boundaries, covers different parts of the globe (Europe, US, Russia), and a variety of disciplines from the physical, biological, and engineering sciences. Taken as a whole, they indicate various ways in which a science can be 'Big': It may involve big centralized infrastructures (chapters 23, 25, and 28), big networks of collaboration (chapters 28 and 30), big geographical scope, big state intervention (chapter 27), a big object of study and big data (chapters 28 and 29). Furthermore, it may require big funding (chapter 33) and have big economic and social impact (chapter 26).

doi:10.1088/978-0-7503-3631-4ch21

Thus, the history of Big Science has several dimensions and can be told from various points of view. Historical studies of Big Science have focused, among other things, on the big instruments and infrastructures associated with it; on the collaborative practices among scientists and engineers with different kinds of expertise that enable collective knowledge production; on the embrace of scientists, policy makers, and politicians involved in planning, designing, and implementing large-scale projects; and on the public outreach of its activities and results that secures its continuous flourishing. The following chapters touch upon all these different dimensions. Furthermore, they indicate that the history of Big Science has not been cloudless. Rather it has been characterized by significant opposition. Its critics have complained about its often reductionist approach, its uncertain epistemic and technological payoff, its factory-like style of knowledge production that presumably stifles creativity, and the massive resources that it commands at the expense of smaller scale scientific endeavors (chapters 24, 28, 30, and 31).

Part III starts with a wide-ranging historical overview in chapter 22 by Helge Kragh, who sketches the long-term development of Big Astronomy and Big Physics, from Tycho Brahe's Observatory in the sixteenth century to CERN in the twentieth. The next two chapters, 23 and 24, by Robert Seidel and Joseph Martin, respectively, focus on Big Physics in America. Seidel discusses the 'industrialization' of nuclear and particle physics, the technological infrastructures associated with it (accelerators, detectors, nuclear reactors), its institutions and sources of funding, and its wider historical context from the pre-World War II to the post-Cold War period. Martin's chapter focuses on the critics of Big Physics, inside and outside the particle physics community, who have complained, among other things, of the bureaucratization of Big Physics, its constant need for huge financial support, its limited social relevance, and its reductionist approach. Martin then draws upon his historical analysis to throw into perspective contemporary critiques of Big Physics.

The next chapter, by Grigoris Panoutsopoulos and Kostas Gavroglu (chapter 25), shifts the focus of the volume to the European scene. They examine the birth of the Large Hadron Collider at CERN, one of the most emblematic Big Science organizations in the world. They highlight the importance of financial, political, and diplomatic factors for understanding how Big Science projects get off the ground and develop. CERN was also the birthplace of the World Wide Web (WWW), perhaps the most impactful by-product of Big Science. The WWW is the subject of chapter 26 by Michael Riordan, which traces its origins to the communication and coordination needs of big projects at CERN and other high-energy-physics (HEP) laboratories in the US. Riordan discusses the momentous growth of the WWW, its global economic and political impact, its positive feedback on HEP, as well as the dangers that it poses to democracy by facilitating the spread of misinformation.

Chapter 27, by Paul Josephson, provides a much-needed comparative perspective on Big Science and technology, by focusing on the Soviet Union and Russia. He argues that big forms of organization were a feature of Soviet science well before World War II, and offers a detailed and wide-ranging overview of big scientific and technological projects since the early days of the Soviet Union. He closes his essay with some reflections on the persistence of Big Science in contemporary Russia and the adverse impact of the war in Ukraine.

The next four chapters have a thematic focus. In chapter 28 David Baneke discusses the transformation of astronomy from a Big to a 'mega' science. Astronomy changed from being big, in terms of relying on big telescopes, to being big in terms of involving huge networks on a global scale and producing publications signed by hundreds of authors. He also discusses the legitimization of Big Astronomy through a 'double' narrative that emphasizes either the cultural significance or the instrumental utility of big astronomical projects.

In chapter 29 James Fleming provides a *longue durée* history of meteorology, arguing that it has always been a Big Science because of the size of its object of study, the atmosphere. Among the things that make meteorology a big enterprise is the scale of its practices of data gathering and data processing. Those practices extend across the globe and involve big initiatives, big money, and big computing power.

In chapter 30 Niki Vermeulen, Rosalind Attenborough, and Rodrigo Liscovsky discuss the history and contemporary state of Big Biology, and contrast it with Big Physics. The emergence of the former went hand in hand with the formation of large decentralized international networks and was shaped by a social ideology that emphasized the direct applicability of research; whereas the early development of the latter involved huge centralized infrastructures in an environment that placed a premium on the acquisition of fundamental knowledge with no requirement for immediate applications.

In chapter 31 Thomas Turnbull and Cyrus Mody discuss a shift in American science towards 'mezzo' projects in the late 1960s and 1970s. They concentrate on another emblematic Big Science organization, NASA, during that period and show how it turned its attention to medium scale projects involving alternative energy sources. They also examine how, during the same period, the National Science Foundation, an American institution that had funded small-scale research, reoriented itself towards bigger science by sponsoring likewise alternative energy projects.

The final three essays, by Robert Crease, Olaf Hallonsten, and Michela Massimi, bring the discussion to contemporary themes and concerns. In chapter 32 Crease documents the emergence of a 'new Big Science', associated with a shift in the US from HEP to materials science. That shift went hand in hand with the growth of the number of users of big infrastructures, the rise of interdisciplinary projects, and the development of decentralized networks of small teams. It also generated new management challenges. Hallonsten (chapter 33) argues that the change in the historical context of Big Science, from the Cold War to contemporary innovation-driven economies, has transformed its character and raised anew the issue of its legitimization. According to Hallonsten, the promotion of Big Science projects, as a result of their uncertain epistemic and technological payoff, often entails 'overselling' marketing strategies and exaggerated promises. Finally, Massimi's essay (chapter 34) addresses issues of justice in relation to Big Science. For that purpose, she outlines a characterization of scientific progress that integrates its epistemic features with societal progress. In this all-encompassing conceptualization of scientific progress, scientific advancements are treated as public goods, whose fair distribution on a global scale requires the development of transnational institutions.

In all, the chapters of part III give us a broad and up-to-date historical overview of Big Science in Europe, the US, and Russia, and address some of the challenges faced by large-scale scientific and technological research. We hope that these chapters will stimulate further historical, sociological, and epistemological studies of Big Science in these and other parts of the World, including China, Japan, Africa, and Latin America.

IOP Publishing

Big Science in the 21st Century
Economic and societal impacts
Panagiotis Charitos, Theodore Arabatzis, Harry Cliff, Günther Dissertori, Juliette Forneris and Jason Li-Ying

Chapter 22

Origins and developments of Big Science

Helge Kragh

22.1 What is Big Science?

Although the term 'Big Science' can be found earlier, it was effectively coined and first discussed in an influential paper of 1961 written by the American nuclear physicist Alvin Weinberg. As director of the Oak Ridge National Laboratory and newly appointed member of the President's Science Advisory Committee, Weinberg was centrally placed and well prepared to reflect on the growing trend in US large-scale research. Conceiving Big Science as essentially government-funded, very expensive large-scale science, Weinberg likened the towering rockets and mighty high-energy accelerators to earlier cultural monuments such as the great Egyptian pyramids and the medieval European cathedrals. Moreover, he critically discussed what he called the triple diseases of the new form of science, namely 'journalitis, moneyitis, administratitis' (Weinberg 1961). Two years later, the historian and sociologist of science Derek J de Solla Price made the concept more widely known with his book *Little Science, Big Science*. Focusing on bibliometric and other quantitative data, Solla Price pictured Big Science as an inevitable late phase in the modern growth of science. Although the notion of Big Science was well known in the 1960s, a search in Google Scholar suggests that the term was used only sparingly in the period.

Paradigmatic but rather extreme examples of multibillion twenty-first-century Big Science include CERN's Large Hadron Collider (LHC) and the International Thermonuclear Experimental Reactor (ITER), both of which projects are truly international and requiring large teams of managers and engineers. They are sometimes described as 'megascience' projects to emphasize that they are even bigger and more complex than traditional Big Science projects. Whether mega-science or just Big Science, this kind of scientific-technological research is often defined in terms of certain characteristics such as the following. (i) The project is so expensive that it must rely, wholly or in part, on funding from one or more

doi:10.1088/978-0-7503-3631-4ch22

governments. (ii) It includes a big staff, many of whom are scientists but others and no less important are managers, administrators, and media and technical personnel. (iii) It is in many but not all cases centered around a big and technologically sophisticated machine. (iv) It is—again often but not always—based in a centralized system of big laboratories.

However, it is important to realize that Big Science comes in many versions and that none of the criteria listed above are strictly necessary for a research project to qualify as 'Big' (Capshew and Rader 1992, Hevley 1992). Big Science is not only a matter of scale, science carried out with expensive instruments and funded accordingly by governments and private donors. It also involves a new form of organization, including new ways of conducting scientific work and a correspondingly new role for the individual scientist. While the classical Big Science projects during the Cold War period were hierarchically and 'vertically' organized around a large machine, such as a high-energy accelerator, another kind of Big Science activity is 'horizontally' organized: it is spread out among many geographically separate and not necessarily very large research centers. CERN-type Big Science is of the vertical type, whereas the very differently organized Human Genome Project (HGP) belonged to the horizontal type. At an approximate total cost of $3 billion, the HGP began in 1990 and was completed in 2003. The US-initiated project mainly consisted of 20 American, European, and Asian institutions, and it did not rely on a single big machine of any kind.

What in the present context is more to the point, 'Big' is a relative term which is only meaningful in a historical context and needs not be restricted to modern standards of large-scale and advanced technology. As Solla Price (1963, p 4) pointed out, Big Science predecessors can be found as early as in 'the great observatories of Ulugh Beg in Samarkand in the fifteenth century [and] of Tycho Brahe on his island of Hven in the sixteenth century'. Indeed, although it is customary to start the historiography of Big Science with the cyclotron adventure of Ernest Lawrence in the 1930s, there are many earlier examples which share at least some of the characteristics which we associate with the emergence of modern Big Science. In what follows I shall briefly refer to a few of these isolated examples without suggesting that the era of Big Science began centuries ago. They are instructive in their own right but exceptional as they were they did not indicate a general change in how to obtain knowledge about nature, a transition from Little Science to Big Science.

22.2 Pre-twentieth-century Big Science

Tycho Brahe's fabled observatory on Hven has often and with some justification been referred to as an early Big Science installation. Not only was it lavishly funded by the Danish Crown with a little more than 1% of the nation's total budget, it also included a large number of assistants, mechanics and calculators, and to disseminate his results Tycho established his own printing press. As noted by historian of science John Heilbron (1992, p 45), Tycho 'ruled over his staff with an authority more feudal than project leaders enjoy at CERN'. The Uraniborg institution was aimed at pure astronomical research with only little concern for spinoffs of social or economic relevance. The same was the case with the later projects during the enlightenment era to study in detail the rare passages of Venus over the disc of the Sun (Sheehan and Westfall 2004). Both were Big Science, Tycho's of the vertical kind and the Venus passage or transit projects of the horizontal kind.

The goal of the two transit projects of 1761 and 1769 was to measure the size of the Solar System by means of a precise value of the solar parallax from which follows the Earth–Sun distance (the astronomical unit). To achieve it multiple observations across the globe were necessary, which again necessitated a collaborative effort among the leading scientific nations, at the time primarily England and France and secondarily Russia, Austria, and Sweden. Computations based on the many worldwide observations resulted in a mean radius for the orbiting Earth between 148 and 155 million km, in fair agreement with the currently known value. The Venus transit projects were impressively big in a spatial, social, and organizational sense, but not in a technological sense as the telescopes on which they relied were neither big nor particularly advanced instruments. The 1761 project involved about 120 astronomers (some professional, others amateurs) observing from 62 different locations, and in the bigger and better organized 1769 project participated at least 138 astronomers of a dozen nationalities. According to one scholarly study (Raj 2000, p 79), the transit expeditions were 'more akin to today's "big science" than to an activity of free-minded, autonomous gentlemen'.

In France, the 1769 transit project was coordinated by a committee under the Paris Academy of Science and funded by the absolutist monarch Louis XV. Likewise, in England the project was planned and executed by a committee under the Royal Society and financed by a £4000 donation from the king. To obtain the necessary funding the Royal Society suggested that the project would benefit navigation and overseas trade, but this was merely a rhetorical move to hide that in reality it was a pure science project. The best known of the British transit expeditions is undoubtedly James Cook's epic voyage with HMS *Endeavour* to Tahiti from where he observed Venus passing the Sun. Continuing from Tahiti, Cook famously explored (or 'discovered') New Zealand and Australia, but originally it was astronomical observations and not geographical discoveries that were the primary purpose of his first expedition to the Pacific Ocean.

Astronomy was not the only science in the era of the enlightenment that provided examples of pre-twentieth-century Big Science. Another example, more in the machine-focused CERN style than in the style of the horizontal Human Genome Project, was the large electrostatic generator displayed since 1785 at the Teyler Museum in Haarlem (figure 22.1) (Dibner 1963). The monster machine consisting of two glass plates 165 cm in diameter was designed by the director of the museum, the Dutch physician and natural philosopher Martinus van Marum, but its actual construction was due to an English instrument maker. The electrostatic generator was able to produce 60 cm long sparks corresponding to a voltage of approximately 120 000 volts. When connected to a series of 135 Leyden jars (capacitors) it charged them with an energy of about 3000 J. Of course, these values are anachronistic

Figure 22.1. Painting of van Marum's electrostatic generator at the Teyler Museum. (Source: this [Oval-room.JPG] image has been obtained by the author from the Wikimedia website, where it is stated to have been released into the public domain. It is included within this article on that basis https://en.wikipedia.org/wiki/File:Oval-room.JPG.)

insofar that the units volt and joule, not to mention the concepts of voltage and energy, were unknown at the time of van Marum's experiments.

van Marum's machine was big and spectacular, but of limited scientific value. Although van Marum and his assistants conducted a long series of experiments with the machine, including chemical and physiological studies, the results were disappointing and of little significance except that they gave some support to Benjamin Franklin's theory of a single electrical fluid. With the discovery of the voltaic pile in 1800 followed a new and exciting form of electrical apparatus, this time appealing more to chemists than to physicists. In 1813 Humphry Davy constructed at the Royal Institution the world's largest battery with 2000 double plates of copper and zinc. For funding, he relied on enlightened subscribers. Davy used the chemical Big Science machine, which was actually built by his colleague William Wollaston, to decompose earth alkalis and isolate for the first time the metals magnesium, calcium, strontium, and barium. In the hands of Davy, big electrical science proved its worth.

Yet another Big Science monster saw the light of day in 1845 with the completion of the mirror telescope sometimes known as the 'Leviathan of Parsonstown' (figure 22.2) (Hoskin 2002). Constructed by William Parsons, Third Earl of Rosse, on his home estate in Ireland, the refractor was the largest in the world and the result of a 12-year long building process based on advanced engineering technology. To grind and polish the two mirrors made of a new copper–tin alloy Parsons invented a special machine driven by a steam engine of his own design. The 72 inches (1.83 m) mirror alone weighed about 4 tons and with the enclosing 15 m tube the weight increased to 12 tons.

Figure 22.2. The Leviathan of Parsonstown. (Source: this [BirrCastle 72in.jpg] image has been obtained by the authors from the Wikimedia website, where it is stated to have been released into the public domain. It is included within this article on that basis https://upload.wikimedia.org/wikipedia/commons/6/68/BirrCastle_72in.jpg.)

While some large machines are of limited scientific worth (such as was van Marum's generator), Parsons' telescope revolutionized astronomy. It was with this instrument that Parson discovered the first spiral nebulae which would soon occupy center stage in astronomy and cosmology. Although his giant reflector was a kind of Victorian Big Science, it involved neither government money nor a large team of researchers and other personnel. The entire telescope project was a private adventure funded by the wealthy Parsons, who used between £20 000 and £30 000 on it. As far as size is concerned, it was only surpassed in 1917, when the 100 inch (2.5 m) Hooker Telescope was installed at the Mount Wilson Observatory in California, funded by industrial philanthropists. Another early Big Science instrument, it was used by Edwin Hubble in the 1920s to gain knowledge about the universe even more revolutionizing than what flowed from Parson's telescope.

22.3 Early low-temperature physics

To further illustrate the diversity of the Big Science concept, consider the highly successful low-temperature laboratory in Leiden directed by Heike Kamerlingh Onnes in the early part of the twentieth century (Kragh 1999, pp 76–86). When Kamerlingh Onnes succeeded in liquefying helium in 1908—an achievement which secured him a Nobel Prize five years later—he had turned the laboratory into an efficient large-scale physics institution with some of the characteristics of Big Science. Although it was not built around a particular large machine, the laboratory was run according to a Big Science factory system somewhat similar to the later high-energy accelerator facilities (van Delft 2007). Kamerlingh Onnes paid as much attention to technical, managerial, and organizational matters as to purely scientific matters. Technicians and instrument makers, not to mention industrial and political connections, were of no less importance than the physics-trained research staff. The laboratory, which was generously funded by government money and private donations, was a favorite destination for a large number of foreign scientists.

As the undisputed leader of the Leiden cryogenic laboratory, Kamerlingh Onnes ran it in an almost feudal manner, much like Tycho Brahe had run his astronomical empire at Hven more than 300 years earlier. As Tycho installed a printing press to disseminate his findings, so Kamerlingh Onnes established an in-house journal to secure wide and rapid circulation of the results obtained in Leiden. The title of the journal was *Communications of the Physical Laboratory at Leiden*. Moreover, Kamerlingh Onnes stimulated in his scientific and technical staff a spirit of commitment and cooperation, an ethos according to which the individual was subordinated to the institution he worked in (the exception to the ethos was, not unexpectedly, the highly visible director of the laboratory).

For more than a decade after 1908 the Leiden laboratory had a worldwide monopoly on liquid helium, which gave it a decisive advantage over other laboratories specializing in low-temperature physics. Without the facilities producing liquid helium the unexpected and hugely important 1911 discovery of superconductivity—and later of superfluidity—would not have taken place. The large-scale Leiden research program may have been the first case in the history of physics in which Big Science led to major discoveries in fundamental physics.

22.4 The Manhattan Project

More than 140 years after van Marum had presented his electrostatic machine, much larger and more powerful generators of electrical energy appeared as instruments of nuclear physics. In 1931 the American engineer Robert Van de Graaf constructed an electrostatic accelerator with a maximum voltage of 1.5 million volts, and Lawrence and his student David Sloan described in *Physical Review* a machine which accelerated mercury ions up to 1.3 MeV of energy. However, it was another of Lawrence's machines, the cyclotron, that ushered in the Big Science era in nuclear and particle physics. The development of ever bigger, more expensive, and more powerful cyclotrons at Berkeley's Radiation Laboratory has been described in detail by historians of science (Heilbron and Seidel 1989). What matters is that by the late 1930s cyclotron physics had become Big Science and Lawrence's laboratory a new type of large-scale physics institution that transcended earlier ones such as Kamerlingh Onnes' in Leiden. The cyclotron-based Radiation Laboratory was based on a new mode of organization that increased the demands of external funding and inspired work in groups larger than seen previously. All of the groups related to the center of the laboratory, namely the cyclotron. The Radiation Laboratory would set the standards for experimental nuclear and particle physics after World War II.

The expertise that was acquired in Lawrence's laboratory turned out to be crucial for the huge Manhattan Project, the specific aim of which was to develop a nuclear bomb. For this purpose, it was necessary to integrate scientific, technological, and industrial research efforts on an unprecedented large scale under the close supervision of the US military authorities (Hughes 2003). The project was enormous by all standards: at its peak it employed more than 130 000 staff members, and the total cost amounted to $2.2 billion or what in current money is about $29 billion. This massive and resource-intensive collaboration between science, industry, and government involved the mobilization of thousands of predominantly American scientists in an engineering project of unprecedented scale. It is often stated that the Manhattan Project was history's first Big Science project or at least that it, as a kind of side effect, ushered in the era of Big Science. The last statement comes probably closer to the truth than the first one. According to Gian Giudice (2012, p 96), a CERN theoretical physicist, 'the Manhattan Project was only an episode that accelerated an already inevitable evolutionary process'.

There are reasons to question whether the Manhattan Project itself qualifies as Big Science, not least because its aim was not scientific, but military and political. The alpha and omega of the project was secrecy, which alone would seem to disqualify it from being scientific in the sense of earlier and later Big Science projects. In addition, the Manhattan Project had a well-defined goal, namely to produce the bomb, and was in this respect quite different from most other projects of the Big Science type, which are oriented towards increased knowledge about nature. Finally, the Manhattan project consisted of hundreds of very different sites spread over the entire country, some small and others large, which makes it an example of horizontal Big Science and in this respect somewhat similar to the earlier Venus transit projects.

Rather than characterizing the Manhattan Project as Big Science it may better be seen as a huge military and multi-sited project which incorporated, coordinated, and greatly expanded earlier Big Science efforts. It also needs to be emphasized that the Manhattan Project was more a feat of complex engineering than of physics proper. Indeed, engineering and management were the key elements in the project (Hughes 1989, pp 381–87). Although the contributions of the nuclear physicists were crucial for the success of the project, in the big picture they filled relatively little and were hardly more important than the extensive works done by chemists, metallurgists, and engineers. The much publicized and to some extent romanticized work done at Los Alamos under the charismatic leadership of J Robert Oppenheimer counted for only 4% of the project's total cost. The figure should be compared to the 84% of the mega factories at Oak Ridge and Hanford, which produced the fissile material for the bombs, uranium-235 at Oak Ridge (63%) and plutonium at Hanford (21%). The two factory systems depended on engineering and chemical technology, whereas research in nuclear physics was of less significance. On the other hand, physics was crucially important for the design of nuclear reactors and the electromagnetic separation of uranium isotopes.

22.5 Aspects of high-energy physics

Although postwar Big Science was not limited to physics, it developed most spectacularly in connection with the accelerators, detectors, and other instruments used in the new field of high-energy physics (Brown *et al* 1989; see also

Kragh 1999, pp 302–11 for a summary account). The general trend was towards bigger and more complex machines with still larger beam energies which now replaced the cosmic rays as the physicists' favored area for studying processes at very high energy. The early development was dominated by American laboratories benefitting from the experiences of the Manhattan Project and the government's willingness to spend large sums of money on national accelerator research facilities. For example, in 1961 Congress appropriated $114 million to construct a 2 miles long linear accelerator at Stanford, at the time the largest and most expensive scientific instrument in history. SLAC, an acronym for the Stanford Linear Accelerator Center, began operations in 1966 and soon proved to be of crucial importance to high-energy physics.

As an earlier, more modest but still very expensive example of Big Science physics, the 10 000 tons and $10 billion Bevatron funded by the Atomic Energy Commission merits attention. The machine was completed in 1954 at the Berkeley Radiation Laboratory, which after Lawrence's death in 1958 was renamed the Lawrence Radiation Laboratory and today, with a budget of $1.1 billion, is known as the Lawrence Berkeley National Laboratory (LBNL) (Heilbron and Seidel 1989). The Bevatron was able to accelerate protons up to an energy of 6.2 GeV, which made it ideal to produce the antiprotons that Paul Dirac had predicted as early as 1933 but still remained undetected. Expectedly, in 1955 the machine revealed the existence of the antiproton, a discovery which four years later was rewarded with a Nobel Prize to the Berkeley physicists Emilio Segré and Owen Chamberlain.

Despite the many exciting discoveries flowing from the new Big Science accelerators, not all physicists were happy with the development. In a paper of 1971, the condensed-matter physicist Phillip Anderson, who six years later would become a Nobel laureate, questioned the necessity of the big machines. Without referring specifically to Big Science, he objected that 'the tremendous effort which has gone into high-energy theory has not, in the final analysis paid off proportionally' (Anderson 1971). He further doubted the claimed spinoff effects of high-energy experiments and, alluding to the Bevatron discovery of the antiproton (which he referred to as a 'discovery' possibly because it was manufactured), he opined that 'To do an experiment just because it is feasible is all right, but not if you are effectively closing down tens of hundreds of equally valid investigations'.

Anderson had no confidence in the reductionist epistemology which, to his mind, governed high-energy physics but was useless in all other parts of science. Basing his arguments on Thomas Kuhn's philosophy of science as expounded in Kuhn's *Structure of Scientific Revolution* from 1962, Anderson denied the hierarchic picture of science with fundamental high-energy physics at the top of the pyramid. Incidentally, at Harvard Graduate School Anderson and Kuhn were classmates and had the same thesis supervisor, John van Vleck (Zangwill 2021, pp 29–32). Twenty years later, Anderson reinforced his criticism in connection with the failed proposal of building a huge 40 000 GeV proton accelerator at a cost of about $10 billion, the so-called Superconducting Supercollider or SSC. Whether in the early

1970s or 1990s, Anderson's attack on Big Science accelerator physics was not kindly received by the high-energy physics community (Cat 1998).

The new kind of accelerator-based Big Science led to a marked shift of the work ethos and responsibility of the high-energy physicist. From the classical role as an individual researcher he or more rarely she turned into a small wheel in a collective factory-like research effort. Cooperativeness became higher rated than creativity. In Europe, the trend was seen as an Americanization of physics. Not only was there little room for independent work and individuality in the large-scale experiments at high energies, it also seemed as if the expensive instruments lived their own lives. The changed relationship between physicists and machines was unintentionally spelled out in a memorandum of 1956 that Samuel Goudsmit, head of physics at the Brookhaven National Laboratory, circulated to his staff: 'In this new type of work experimental skill must be supplemented by personality traits which enhance and encourage the much needed cooperative loyalty. … I shall reserve the right to refuse experimental work in high energy to any member of my staff whom I deem unfit for group collaboration. I must remind you that it is, after all, not you but the machine that creates the particles and events which you are now investigating with such great zeal' (Heilbron 1992, p 44).

The growth in accelerator size and energy was followed by new and more complex methods of detection. After all, accelerators are of no use without detectors. By the mid-1950s traditional methods of visualizing particle tracks, such as cloud chambers and photographic emulsions, were increasingly seen as inadequate and outdated. In close association with the new generation of large accelerators they were replaced by the bubble chamber method invented by Donald Glaser in 1952. The large bubble chambers introduced by Luis Alvarez and his group in Berkeley were 'undemocratic' in the sense that they could only be afforded and operated by Big Science research facilities. Alvarez' 72 inch hydrogen bubble chamber of 1959 cost $2 million and required a $1 million computer to analyse the data, not to mention a small army of assistants and operators. When Alvarez was awarded the 1968 Nobel Prize, it was primarily for his development of the bubble chamber as an instrument of particle physics.

While in the earlier experiment tradition physicists often built and operated the detectors themselves, now they were forced to pass on the responsibility to a new kind of workforce. As argued by Peter Galison (1997, p 318) in a detailed study of the development of early bubble chambers, 'The removal of the physicists from the apparatus, the specialization of tasks, the increased role of computation, and the establishment of hierarchical collaborations have become hallmarks of high energy physics experiments.' The large-scale bubble chambers were supplemented with computerized 'reading machines' which more or less automatically could follow and analyse the enormous amount of photographs. The first reading machine installed in Berkeley was a technological monster appropriately nicknamed 'Franckenstein', a name alluding not only to Mary Shelley's classic horror novel of 1818 but also to the engineer Jack Franck who developed the machine.

22.6 Towards hyperauthorship

Instead of a conclusion, let me briefly point to yet another of the consequences of Big Science, or even one of the defining features of the concept, namely the drastic change in the authorship of scientific papers. The 1955 paper in *Physical Review* announcing the discovery of the antiproton was authored by four physicists acknowledging the operating crew of the Bevatron. Nine years later the theoretically predicted Ω^- hyperon was observed in bubble chamber experiments at Brookhaven National Laboratory. The discovery, of crucial significance to the quark model of hadrons, appeared in a paper with 37 authors. At the end of the paper they acknowledged not only 'the staff at the AGS [Alternating Gradient Synchrotron]' but also 'the untiring efforts of the 80-in. bubble chamber and scanning and programming staffs' (Barnes *et al* 1964). While 37 authors were many by the standards of physics at the time, in the early twenty-first century it was a very small number.

Large-scale multiple authorship grew into so-called 'hyperauthorship' characteristic of very Big Science or megascience projects (for the term and its meaning, see Cronin 2001; see also Galison 2003). The much publicized observation of the Higgs boson in CERN-LHC experiments of 2012 was communicated in a paper from the ATLAS detector collaboration with 2932 nominal authors. A hysterically large number, some might say, but three years later it was dwarfed by another CERN paper narrowing down the mass of the Higgs boson to 125.09 ± 0.21 GeV. The 33-page article carried the names, alphabetically ordered, of no less than 5154 'authors' coming from a large number of institutions and nations (Aad *et al* 2015). This kind of hyperauthorship is exceptional of course, but in modern Big Science there is a growing number of papers with more than 1000 authors, not only in high-energy physics and space science but also in genetics and allied sciences. According to a study of the Institute for Scientific Information, in the period 2009–13 there were 573 papers in this category, which in the next five-year period had grown to 1315 papers (Chawla 2019).

Let me end by referring to three pioneering discovery papers in nuclear and particle physics dating from the early pre-Big Science period. Ernest Rutherford announced his discovery of the atomic nucleus in a 1911 paper in *Philosophical Magazine*; in 1932, *Nature* included James Chadwick's communication on the existence of the neutron; and the same year Carl Anderson reported in *Physical Review* that he had found positive electrons or what he called positrons in cosmic rays. All three of these very important discovery papers were communicated by a single author (they are reprinted in Beyer 1949 and elsewhere).

The history of Big Science as discussed cursorily in this essay is scarcely of direct relevance to contemporary Big Science of the CERN type and the debates relating to this kind of science. The principal object of history is the past, not the present. Nonetheless, a look at the past tells us that the Big Science phenomenon has a rich history and that it exists and has existed in a variety of forms. It also suggests that the development toward still bigger machines and more complex forms of Big Science is neither inevitable nor irreversible. It is possible that in the future the Big

Science era will come to an end or evolve in more complicated forms than we can presently imagine. A more comprehensive history would not only include more examples from the history of physics, it would also have to refer to Big Science projects outside the physical sciences, as in chemistry, biology, geology, oceanography, and even archeology. I believe that such a comparative study would be valuable in many ways and throw a new light on how Big Science has evolved and what it currently is.

References

Aad G *et al* 2015 Combined measurement of the Higgs Boson Mass in pp collisions at $s = 7$ and 8 TeV with the ATLAS and CMS experiments *Phys. Rev. Lett.* **114** 191803

Anderson P 1971 Are the Big Machines necessary? *New Sci.* **82** 510–4

Barnes V E *et al* 1964 Observation of a hyperon with strangeness minus three phys *Rev. Lett.* **12** 204–6

Beyer R T 1949 *Foundations of Nuclear Physics* (New York: Dover)

Brown L M *et al* 1989 *Pions to Quarks* (Cambridge: Cambridge University Press)

Capshew J H and Rader K A 1992 Big Science: Price to the present *Osiris* **7** 2–25

Cat J 1998 The physicists' debates on unification in physics at the end of the 20th century *Hist. Stud. Phys. Biol. Sci.* **28** 253–99

Chawla D S 2019 Hyperauthorship: global projects spark surge in thousand-author papers *Nature News* 13 December www.nature.com/articles/d41586-019-03862-0

Cronin B 2001 Hyperauthorship: a postmodern perversion or evidence of a structural shift in scholarly communication practices? *J. Am. Soc. Inf. Sci. Tech* **52** 558–69

Dibner B 1963 The Great Van Marum electrical machine *Nat. Philosopher* **2** 87–104

Galison P 1997 *Image and Logic: A Material Culture of Microphysics* (Chicago, IL: University of Chicago Press)

Galison P 2003 The collective author *Scientific Authorship: Credit and Intellectual Property in Science* ed P Galison and M Biagioli (New York: Routledge) pp 325–53

Giudice G F 2012 Big Science and the Hadron Collider *Phys. Persp* **14** 95–112

Heilbron J L 1992 Creativity and Big Science *Phys. Today* **45** 42–7

Heilbron J L and Seidel R W 1989 *Lawrence and His Laboratory: A History of the Lawrence Berkeley Laboratory* (Berkeley, CA: University of California Press)

Hevley B 1992 Reflections on Big Science and big history *Big Science: The Growth of Large-Scale Research* ed P Galison and B Hevley (Stanford, CA: Stanford University Press) pp 355–66

Hoskin M 2002 The Leviathan of Parsonstown: ambitions and achievements *J. Hist. Astr.* **33** 57–70

Hughes J 2003 *The Manhattan Project: Big Science and the Atom Bomb* (New York: Columbia University Press)

Hughes T P 1989 *American Genesis: A Century of Invention and Technological Enthusiasm 1870–1970* (New York: Viking)

Kragh H 1999 *Quantum Generations: A History of Physics in the Twentieth Century* (Princeton, NJ: Princeton University Press)

Raj K 2000 18th-century pacific voyages of discovery, 'Big Science', and the shaping of a European Scientific and Technological Culture *Hist. Tech* **17** 79–98

Sheehan W and Westfall J E 2004 *The Transits of Venus* (Amherst: Prometheus)

Solla Price D J de 1963 *Little Science, Big Science* (New York: Columbia University Press)

van Delft D 2007 *Freezing Physics: Heike Kamerlingh Onnes and the Quest for Cold* (Amsterdam: Royal Netherlands Academy of Arts and Sciences)
Weinberg A 1961 Impact of large-scale science on the United States *Science* **134** 161–4
Zangwill A 2021 *A Mind Over Matter: Philip Anderson and the Physics of the Very Many* (Oxford: Oxford University Press)

IOP Publishing

Big Science in the 21st Century
Economic and societal impacts
Panagiotis Charitos, Theodore Arabatzis, Harry Cliff, Günther Dissertori, Juliette Forneris and Jason Li-Ying

Chapter 23

Physics enters the factory age

Robert Seidel

The origins of CERN, FERMILAB, and other factories of Big Science predate World War II, which fertilized the application of accelerator physics to problems of interest to the military and made large-scale state support possible. I will reflect upon the persistence of such institutions into the new millennium and the stimulus given to science, technology, and the academic–industrial–military complex represented by national and international high-energy physics laboratories.

> An Institution is the lengthened shadow of one man.
>
> —Ralph Waldo Emerson *Self-Reliance*

Institutions also cast their shadows a long way. Although most of the Big Science laboratories are over half a century old, they persist for reasons that are of more than scientific interest. As the centenary of Big Science laboratories approaches, I offer some of my own historical reflections on persistence of these institutions of Big Science. My focus is on those factories of science nuclear and particle physics laboratories. I have reflected elsewhere upon semi-centennials of these institutions [1].

23.1 Origins

Big Science and Engineering are reflected in large accelerators and detectors built in the past century [2]. By the end of World War I physicists were ready to exploit factory-sized laboratories to mass produce nuclear reactions. During the nineteenth century, astronomers had used large sophisticated telescopes to produce astrophysical data and chemical engineers had applied unit operations to industrial processes of chemical manufacture, foreshadowing the scaling up of physics to penetrate the heart of matter in particle factories [3]. The discoveries of x-rays and radioactivity also stimulated the medical equipment industry to develop x-ray tubes and radium

doi:10.1088/978-0-7503-3631-4ch23

sources of increasing size and power. By World War II the cyclotron and Van de Graaff had become analogous to the steam engine in particle factories in Europe and the United States.

The high energies required to transform nuclei produced using natural and artificial sources proved inadequate for most elements and lightning, cosmic rays, and radium sources proved either too expensive, too dangerous, or too unreliable for particle production on a factory scale. Nuclear physics provided a new basis for the industrialization of physics. Nuclear physicists converted Tesla coils, voltage multipliers, and high-voltage x-ray tubes to proton and alpha-particle accelerators capable of transforming lighter elements in several university laboratories. Radio technology enabled the invention of more powerful particle accelerators in the 1920s [4]. For example, the Poulsen arc generator and the triodes used by the Navy for long-distance voice transmission made possible radio-frequency oscillators that enabled resonant acceleration of ions in a cyclic particle and linear accelerators [5]. Although the relativistic increase of mass of particles accelerated to near light speed limited the technique at first, the later development of frequency-modulated accelerators in which the applied potential was synchronized with the slowing ions extended the available energies exponentially. Radio-frequency and frequency modulation remain the principal means of particle acceleration in modern devices, such as those at CERN [6].

23.1.1 Scaling up

The first accelerators large enough to demand factory scale settings were cyclotrons and Van de Graaff accelerators. The cyclotron was invented by Ernest Lawrence and constructed by M Stanley Livingston, his graduate student in the physics department at the University of California in Berkeley (UC). UC president Robert Gordon Sproul made an old Civil Engineering Laboratory available to house Lawrence's Radiation Laboratory (LRL)[1] (figure 23.1).

The Van de Graaff accelerator was invented by Robert J Van de Graaff, who established his own factory at MIT's Round Hill Laboratory in 1933 (figure 23.2). It had been developed for radio broadcasting and dirigible research and included two Zeppelin hangers. These housed nuclear experimentation with two Van de Graaff 40 foot accelerators producing five million-volt potentials.

After Van de Graaff built his Round Hill generator, Westinghouse developed its own Van de Graaff accelerator, the 'Atom Smasher', for research into the medical applications of neutrons and radioisotopes in 1937. A special laboratory was built in Forest Hill, Pennsylvania to house the 5 MeV machine, the first industrial Van de Graaff in the world, surrounded by a 60-foot-high pressurized tank to suppress electrostatic breakdown [7].

These installations rivaled in size those of many academic physics laboratories built in the first part of the twentieth century [8]. Those collections of workshops of

[1] The University of California Radiation Laboratory was named Lawrence Radiation Laboratory (LRL) after his death. I use that acronym throughout.

Figure 23.1. Old Radiation Laboratory 1931. Copyright LBL.

Figure 23.2. Round Hill Van de Graaff laboratory. Copyright MIT Institute Archives & Special Collections.

individual physicists were scaled up to factory size plants centered on specialized accelerators and ancillary scientific equipment such as cloud chambers, nuclear emulsions, radiation counters, and lead and/or water tanks that protected operators from the particles manufactured in them and the 'artificial radioactivity' and neutrons produced in nuclear disintegration. New subdisciplines such as health physics were eventually created to detect and protect workers from the damage caused by these atomic bullets.

Once 'artificial radium' and neutron beams were made by these facilities, it became commonplace to site the accelerator in purpose-built medical facilities, such as the Crocker Medical Cyclotron Laboratory at the University of California.[2] Artificial radioactivity—the production of unstable isotopes—became the staple byproduct of this and other laboratories. Lawrence divided his staff into shifts to facilitate their mass production. His brother John provided inhouse quality control. Isotopes made in the cyclotron could be shipped to hospitals at other universities, which might also build their own machines.

These physics factories merged engineering, mass production, quality control, teams of operators, production schedules, shipping and receiving, and other traditional features of industrial production with academic research and instruction: Postdoctoral and PhD students and research assistants in research teams, research administrators who oversaw production of radioisotopes and neutron beams for medical applications comprised the staff, who found spare time for research in fundamental physics in their spare time. Lawrence's acolytes departed to start new projects elsewhere that operated on a smaller scale, and some aspired to create even bigger factories of physics. For this, they would need more funding than was available before World War II from philanthropic venture capitalists in physics.

23.1.2 Venture capital

The capital requirements for these early machines came from the Research Corporation of New York (RC), founded by UC engineer Frederick G Cottrell, using the proceeds of his patent on an electrostatic precipitator, and the Chemical Foundation, using the royalties of German patents seized by the Alien Property Custodian in World War I. The Crocker Medical Cyclotron was also financed by California banker W H Crocker, who was interested in cancer therapy. The discovery of the deuteron provided a way to bombard tumors with neutrons, van de Graaff generators, cyclotrons and other accelerators won well-funded laboratories specializing in cancer research. Although the great philanthropies founded by Carnegie and Rockefeller avoided cancer research at first, once the accelerators were shown to be more promising than radium or high-voltage x-rays in producing both neutrons and radioisotopes, medical research grants from both philanthropies as well as smaller medical funds such as the Merkle and Macy Foundations became

[2] Built at the Berkeley between 1936 and 1939, its cyclotron chamber was 60 inches in diameter, the custom-designed magnet was 220 tons in weight, and the radio-frequency system as capable of producing 200 kilovolts of potential on the dees the cyclotron electrodes, so-called because of their 'D' shape. It was much larger than theorists thought feasible because Lawrence could get it funded. Despite pessimistic theoretical predictions, it operated successfully on the first trial. See [9].

available. By World War II, this support led to the erection not only of the Crocker Medical Cyclotron, but also the Donner Radiation Laboratory led by John Lawrence. The New Deal provided funding through the National Cancer Advisory Council of the National Cancer Institute, which was created in 1937. Together with state government support, this federal grant provided up to a quarter of UCRL's funding before World War II.

The RC made patent agreements covering various applications of the new Big Science. Since it depended upon royalties from patents of inventions resulting from its sponsorship of scientific research, it had its own patent lawyers who worked with MIT and UC formulating their patent policies. Although patents were unusual in university research in the inter-war period, they became central features of accelerator R&D. The Chemical Foundation patents expired in 1937, and the philanthropies that succeeded them as patrons of particle physics were reluctant to patent products of scientific research [10].

The ultimate pre-war factory of physics was the 184 inch cyclotron erected on Charter Hill behind the UC campus (figure 23.3). Following Lawrence's Nobel Prize for Physics in 1939, UC agreed to grant land and the Rockefeller Foundation (RF), RC, and gave $1 150 000 to build the giant cyclotron [11]. Its 400 ton magnet diameter, 184 inches, was limited only by the dimensions of the largest commercially available steel plate. The 90 foot high dome that covered the accelerator and its appendages spanned a circular laboratory 160 feet in diameter. It produced particles used for nuclear physics, nuclear medicine, and biophysics as well as medical experiments. It would produce a new generation of explosives, far more powerful than Nobel's dynamite (figure 23.4).

Figure 23.3. 184 inch cyclotron laboratory under construction. Copyright LBL.

Figure 23.4. Wartime staff in the 184 inch magnet, 1945. Copyright LBL.

23.1.3 War research propagates factory physics

World War II expanded existing labs and generated new ones. The National Defense Research Committee (NDRC) and its successor, the Office of Research and Scientific Development (OSRD) both headed by former MIT vice-president Vannevar Bush, established the 'Radiation Laboratory' not to build accelerators, as its name misleadingly suggested, but to develop microwave radar. MIT Trustee Alfred Lee Loomis, who headed the NDRC Radar section, was a wealthy physics 'angel', who had underwritten Lawrence's 184′ cyclotron and the *Physical Review* as well as equipping a lab of his own in Tuxedo Park [12].

The NDRC and OSRD accounted for over $500 million dollars in support of research, including the early research into the possibilities of an atomic bomb by Lawrence, J Robert Oppenheimer, and their Berkeley students. Early in 1941, LRL chemist Glenn Seaborg manufactured a new nuclear explosive, plutonium (Pu), in the Crocker Cyclotron. After Pearl Harbor, the so-called 'Metallurgical Laboratory' (Met Lab) at the University of Chicago consolidated plutonium research with Enrico Fermi's reactor research previously pursued at Pupin Laboratory at Columbia University. Frank Spedding provided purified uranium at the Ames Project at Iowa State University. Fermi constructed the first successful nuclear reactor under the stands of Stagg field [13], followed by a pilot production reactor at Oak Ridge [14].

The Clinton Reactor was part of the Manhattan Engineer District (MED)'s Clinton Engineer Works (CEW) in Oak Ridge Tennessee established by General Leslie R Groves, who had built the world's largest office building at the time, the Pentagon [15]. CEW also contained the Y-12 plant, where uranium-235, separated from uranium-238 electromagnetically by Lawrence's calutrons (California University Cyclotrons). A half-dozen 'race-tracks' built for $427 million, not including the cost of silver borrowed for magnet coils from the Fort Knox United States Bullion Depository, employed 22 000 workers. Twelve thousand workers operated the gaseous diffusion plant, K-25, which covered more area than any building ever built, cost $500 million. Groves added a supplemental thermal diffusion plant, S-50, at a cost of $3.5 million in the last year of the war to supplement the uranium feed to the other plants [16]. These facilities encouraged physicists to enlarge their vision of possibilities for postwar laboratories as elaborated in the Jeffries Report from the Met Lab, and the Tolman Report solicited by Groves [17]. In May 1945, President Truman created an Interim Committee to set policies for use of atomic bombs and for the postwar MED laboratories and production facilities.

23.2 Creating the nuclear industry

The MED weapons complex was a vast industrial empire, with scientists in laboratories doubling as consultants to private firms that supplied the production facilities. Indeed, development became an essential function of most wartime scientists, who sacrificed basic research to scaling up laboratory equipment into the largest factories then existing. Leaving university laboratories at Berkeley, Chicago, Columbia, Harvard, MIT, and throughout the nation, physicists learned to take orders, work in teams, maintain secrecy, obey security regulations, and use aliases when traveling. At Clinton, Los Alamos, Hanford, Y-12, and other secret sites, they labored invisibly to build the weapons of war.

The wartime effort also included many firms that developed the results of radar research. OSRD contractors included 69 different academic institutions and hundreds of private firms including General Motors, Chrysler, Studebaker, Western Electric & Bell Labs, RCA, Sperry, Westinghouse, Philco, Sylvania, Standard Oil Co, and Dupont. In addition to the MIT Radiation Laboratory, which produced over $3 billion worth of radar equipment of over 150 varieties, the OSRD established the Jet Propulsion laboratory in Pasadena operated by Caltech. The JPL produced rockets for the Navy and Army Air Forces at the rate of $1 billion per year and spawned the Navy Ordnance Test Station at Inyokern, which became the Michelson Laboratory after the war [18]. Industrial laboratories and non-profit research institutions created or expanded during the war often perpetuated their work afterwards.

The crucible of World War II melted barriers between the federal government, academia and private industry. Entrepreneurs like Ernest Lawrence quickly responded to opportunities to remold the relationship to insure postwar support for the Radiation Laboratory. The result was funding for the completion of the 184 inch cyclotron (redesigned as a synchrocylotron) (figure 23.4), a 40 foot long linear accelerator, and an electron synchrotron. These were the prototypes of postwar accelerators built throughout the world, just as his laboratory became the prototype of the institutions that housed them.

23.2.1 The diaspora

In the bomb's early light, many saw accelerators and reactors as crucial for postwar nuclear research. As physicists returned to universities where they were now in unprecedented demand, they sought to build laboratories to house these instruments. This required an investment beyond the resources traditionally available to academic and industrial researchers. The Atomic Energy Commission [19] (AEC) and the Office of Naval Research [20] (ONR) shoveled cash into new physics laboratories, often run by consortia of universities. Before the AEC was appointed, the MED established 'National Laboratories' at Argonne (ANL), Brookhaven (BNL), and Oak Ridge (ORNL) at the end of WWII [21].

By 1950 it was clear that 'the AEC needs the National Laboratories because it needs secrecy and big groups of scientists who will take orders, and big equipment' [22]. The nuclear weapons complex, which handled every aspect of nuclear weapons production from uranium mining to weapons operations, relied on the national laboratories for research and development and trained manpower. The High-Energy Physics program produced many more graduates than could be employed in academic institutions providing a labor pool for the complex [23].

The economic infrastructure of Big Science was transformed by the socialization of nuclear science. Eventually, $5.8 trillion was invested in the complex, of which the national laboratories were only the most visible manifestation [24]. The Department of Defense (DOD) and the AEC were the primary sources of support. To cut the red tape associated with government laboratories and to control security within the AEC, Congress created the Joint Committee on Atomic Energy (JCAE) and the Military Liaison Committee. Between them, they approved budgets for research, development, testing, and evaluation of nuclear weapons, which were the primary activities of the AEC. The AEC General Advisory Committee and its Division of Research determined the role of fundamental research in AEC facilities [25]. The laboratories had their own networks of directors, divisions, and programs, reacting to the bureaucracy of big government.

These labs housed the big equipment for research as well. This included nuclear reactors, which often were used in classified research in the early years, and accelerators, not originally included in their remit. Lawrence made sure that they were at a meeting at the Bohemian Grove in California of AEC and GAC staff as well as laboratory directors. As a result of this informal and unrecorded meet-and-greet, he was able to negotiate funding for the Bevatron and well as the Cosmotron. This led, in turn, to a series of high-energy physics accelerators built at the two laboratories and, later, the Los Alamos Meson Physics Facility, Stanford Linear Accelerator Center (SLAC), and Fermilab (FNL) (figure 23.5) [26].

Fermilab was the largest of the single purpose accelerator laboratories constructed by the AEC. The 200 GeV proton synchrotron designed by LBL to follow

Figure 23.5. Fermi National Laboratory Super Proton Synchrotron and Storage Ring for Collider Experiments. Copyright Fermilab.

the BNL AGS became the object of intense political conflict, a national site search, and extensive redesign by Robert R Wilson which produced a cheaper 400 Gev machine subsequently enlarged using superconducting magnet technology into the Tevatron [27].

23.3 Cold warriors

UC, which ran Lawrence's RL as well as the Los Alamos Scientific Laboratory, maintained a degree of autonomy, thanks to Lawrence, who could and did go over the head of the AEC to the JCAE to accomplish his ends. These included the Livermore branch of the RL, set up to use huge linear accelerators to manufacture fissionable material and converted to a second weapons research laboratory to manufacture Edward Teller's novel weapons designs [28].

The Soviet Union pursued accelerators as part of their nuclear weapons program, which produced their first fission weapon in 1949. Beginning with small unfinished prewar cyclotrons or 'Lawrences', it developed accelerators rivaling those of the west in size, if not in output [29]. The synchrophasotron completed at Dubna in 1957 was the largest accelerator in the world prior to the BNL Alternating Gradient Synchrotron (AGS) in 1960. Its revelation at an international particle physics conference led the AEC, under the direction of the JCAE, to commission the Zero Gradient Synchrotron (ZGS) at ANL in 1956 in hopes of catching up. As one of the machine's builders wrote, 'It [had] an energy of 10 to 15 GeV to just beat the 10 GeV Dubna synchrophasotron which was under construction at the time. ...designed using only well-established principles and conventional technology, [with] a 'crash' construction schedule to beat the competition of the Dubna machine'. Unfortunately for the US, it was not completed until 1964 [30].

Accelerators were a scientific ‘macht-ersatz’. As another ZGS builder lamented ‘the AGS and the CERN machines were being built, and our task, clearly given to us by Congress, was to build a machine a little bit bigger than the Russians, and a little bit faster and with more intensity. The general idea of Congress was to move the United States far ahead of the Soviet Union. It’s difficult now to understand why that was, and I was always absolutely convinced that Congress really did not understand the difference between high-energy physics and nuclear weapons. And of course nobody bothered to explain the difference. I believe that to be a true statement, and I feel rather ashamed of participating in what I believe was a deliberate delusion’ [31].

Despite the JCAE, the ZGS eventually became a useful machine for high-energy physicists in the Midwest. The Midwest Universities Research Association (MURA) had lobbied for a high-energy accelerator for decades, incorporating novel concepts, e.g. the Fixed Frequency Alternating Gradient (FFAG) accelerator and colliding beams, but had failed to win support because it preferred to be independent of ANL. The ZGS and the Fermilab Proton Synchrotron eventually provided them with local access to machines, not least because they were determined to secure a place in high-energy physics in a bicoastal AEC accelerator establishment, where LRL and BNL alternately won support for the largest accelerators from the government [32].

23.4 CERN: an international laboratory

CERN distilled the efforts of European accelerator builders in the 1950s. World War II had convinced European nations that the achievements of the United States symbolized by the atomic bomb were sufficient reason to underwrite large-scale accelerator research and development [33]. Inspired by I I Rabi, one of the founders of BNL, representatives from eleven nations signed the document establishing CERN (Conseil Européen pour la Recherche Nucleaire) on February 15, 1952 [34]. Motivated by the achievements of LRL and BNL accelerator builders, European physicists aimed for a 600 MeV synchrocyclotron and a ‘Bevatron’. The first machine permitted them to cut their teeth on accelerator construction and pion and muon physics. It also provided a training ground for CERN experimental teams, most of whom had little experience in high-energy physics, in anticipation of a larger proton synchrotron [35]. By the end of the decade, CERN had developed a 28 GeV proton synchrotron which eclipsed the USSR’s Synchrophasotron’s world-record energy of 10 GeV.

23.5 Factories of new particles

The discoveries made in these accelerator laboratories prompted physicists to seek higher energies in order to created new particles that could resolve theoretical anomalies that had arisen first in the meson problem of the late 1940s, spread with the discovery of strange particles in the early 1950s, and resulted in the discovery of parity non-conservation in strange-particle decay (the τ–θ puzzle) and weak interactions in 1957. A variety of theoretical attempts to organize the 'particle zoo' spawned by accelerators led them down the 'eight-fold way' to the quark model of hadrons as well as to the Standard Model that incorporated electromagnetic, weak, and strong forces. The search for a grand unified theory extended to gravity and the postulation of a particle that could 'carry' mass: the Higgs boson. Like the photon, pion and gluon in the forces of the Standard Model, it was an exchange particle carrying the force that produced mass in particles, e.g. the W and Z particles discovered at the end of the century, which the Standard Model predicted would be massless. Experiments showed they had mass and that the Higgs force could be excited by exceedingly high energies, beyond those of existing accelerators.

Simultaneously, the development of massive detectors and minicomputers enabled physicists to expand their laboratories with user facilities that could be remotely accessed, as well as visited onsite for extended periods of time. The detectors added another dimension to the search for particles. Imaging devices such as the bubble chamber, spark chamber, and time projection chamber spatial reconstructions of particle interactions. The logic tradition presented computer analyses of data from electronic counters [36]. There was a consequent loss of empirical data when data analysers were narrowly programmed in order to exclude events unanticipated by theory as well as unwanted data, and the realism of the results experiments mediated by these processors has been a subject of ongoing controversy [37].

Notwithstanding the erosion of ontological validation, the quest for the Higgs boson became the holy grail of particle physics. It justified the Superconducting Supercollider (SSC) in the United States and the Large Hadron Collider (LHC) at CERN. After the SSC was designed by a multi-laboratory team at LBL and a site was selected by a complex process involving several new scientific and federal agencies, the AEC prioritized its construction over smaller colliding beam machines—Isabelle under development at BNL and the Continuous Beam Electron Accelerator Facility (CEBAF) in Newport News, Virginia. After ten years, the SSC was canceled by the US Congress in 1993, as costs, not protons, spiraled out of control. The original estimate for completion of the laboratory rose from \$2–3 billion to well over \$10 billion, and finally government support for the project collapsed. The already-excavated tunnel at Waxahachie, Texas was filled in, and the laboratory scientists and staff left to fend for themselves. The effect was slightly mitigated by the exhumations of the BNL Isabelle accelerator as a Relativistic Heavy Ion Collider (RHIC) and of the CEBAF as Thomas Jefferson National Accelerator Facility [38].

23.6 A peace dividend

The end of the Cold War exhausted the national security impulse behind the giant American and Russian accelerators. The inertia of accelerator building continued into the new millennium, however, as Fermilab persisted in its efforts of produce the Higgs boson right up until the Large Hadron Collider at CERN succeeded in doing so. Many of Fermilab's high-energy physicists had joined the refugees from the SSC debacle at CERN, but those who remained sought to modify the Tevatron to produce energies theoretically sufficient to generate a particle of the predicted mass by using colliding proton–antiproton beams. Fermilab failed. Their colleagues at CERN succeeded in manufacturing the Higgs boson in 2012. This verified the theoretical models [39] and established that laboratory as pre-eminent in high-energy physics. The political dimensions of accelerator construction in US national laboratories frustrated their pursuit of the 'God' particle [40].

The early external justifications for the construction of physics factories have been attenuated. Cancer therapy with accelerators, radioisotope production, nuclear weapons research, high-voltage x-rays, reserve armies of physicists, cold war competition, technology transfer, and other rationales transparent to the public, philanthropists, the press, and politicians have become stale substitutes for twenty-first century concerns about global warming, energy conservation, and economic stimulation of postmodern economies.

More profoundly, there has been a loss of faith in modern science as a result of 'science wars', corruption of science by those who seek to perpetuate corporate interests in profit by hiring scientists to provide conflicting points of view, and the perception that arrogance is not the same as understanding when used to defend established social and political systems [41].

23.7 Conclusion

'This is the way the world ends, not with a bang, but with a whimper.' T S Eliot, 'The Hollow Men'.

One by one, the factories of physics were repurposed at the turn of the millennium. LBL converted the Bevatron to the Bevalac by coupling it with the SuperHILAC Heavy-ion Linear Accelerator, itself a rebuild of one of the first heavy-ion accelerators, but was unable to preserve the machine beyond 1992 when the last of the 'Lawrence style' accelerators was denied further redemption [42]. BNL's cosmotron survived until 1969, after its successor, the AGS came into operation. After the demise of the SSC, the AGS became an injector for RHIC, ceding high-energy physics to CERN Fermilab's Tevatron, having lost the Higgs boson to CERN, was shut down in 2011. CERN has designs on the High-Energy Large Hadron Collider [43].

When I visited the Bevalac control room in the mid-1990s to be filmed for a NOVA episode that was never aired, I sat in front of an instrument panel which had been cannibalized of most of its instruments, at the edge of the vast structure specifically built to house the machine and remodeled to accommodate its various transformations, and pondered the meaning of the massive structures that will surely seem the twentieth century equivalent of medieval cathedrals, just as Henry Adams did the steam turbine at the Columbian Exposition in 1892, and Robert Wilson did with particle accelerators in 1968. Although the objectives of all were met, the faith which sustained them seems to have faltered with the turn of the third millennium. A synchrotron light source was installed in the domed 184 inch cyclotron laboratory in 1993, but the Bevatron building was demolished shortly after my visit [44].

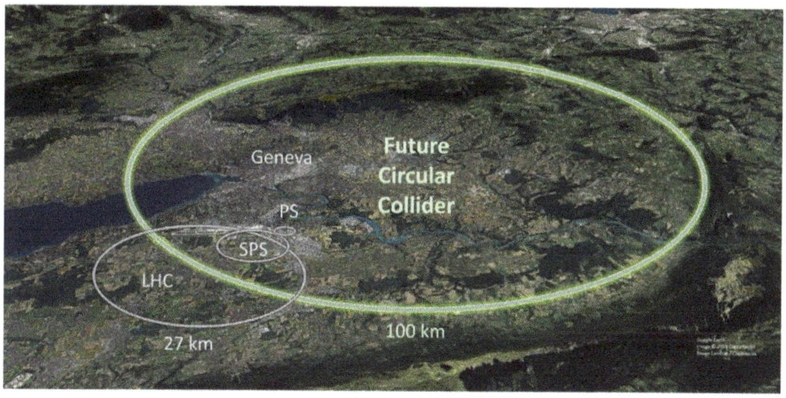

Figure 23.6. Aerial view of CERN with proposed Future Circular Collider Image Credit: Google Earth Image, copyright 2016, DigitalGlobe Image Landsat/Copernicus.

Although most of the factories of physics have been re-equipped with machines of varying functions, they continue to provide research employment for physicists as well as physicians, geneticists as well as engineers, and chemists as well as computer programmers. The regimentation of high-energy physics has remained, but new scientific styles characterize the 'job shops' within the laboratories, reflecting the 'small science' traditions eclipsed by Big Science in the twentieth century [45]. Factory physics has, in job shops, given way to guild physics, with masters (principal investigators), journeymen (postdocs), and apprentices (graduate students).

At CERN, the engrossment of experiments persists. 1300 contributors from 150 universities, research institutes and industrial partners actively participated in the design for a possible future circular collider, the High-Luminosity LHC (figure 23.6). They also foresaw expanding the machine to gargantuan proportions. This machine would transcend the physics factories of the past in size, energy, manpower and costs:

> A 90-to-365-GeV electron–positron machine with high luminosity could be a first step. Such a collider would be a very powerful 'Higgs factory', making it possible to detect new, rare processes and measure the known particles with precisions never achieved before. These precise measurements would provide great sensitivity to possible tiny deviations from the Standard Model expectations, which would be a sign of new physics. The cost of a large circular electron–positron collider would be in the 9-billion-euro range, including 5 billion euros for the civil engineering work for a 100-kilometre tunnel. This collider would serve the worldwide physics community for 15 to 20 years. The physics programme could start by 2040 at the end of the High-Luminosity LHC. The cost estimate for a superconducting proton machine that would afterwards use the same tunnel is around 15 billion euros. This machine could start operation in the late 2050s [46].

Like the dreams of particle physicists from Lawrence to Weinberg, the visions of future physics factories are grandiose and limited only by the budgets of nation-states, whose concern for the practical military and social applications of physics exceeds their grasp of the science involved. For them, it suffices to say:

> The complex instruments required for particle physics inspire new concepts, innovation and groundbreaking technologies, which benefit other research disciplines and eventually find their way into many applications that have a significant impact on the knowledge economy and society. A future circular collider would offer extraordinary opportunities for industry, helping to push the limits of technology further. It would also provide exceptional training for a new generation of researchers and engineers [47].

Robert Wilson's hope as he embarked on the adventure of building Fermilab may prove to be prophetic:

> Cathedrals were technological constructions of great beauty and significance that expressed the aspirations and spirituality of their age. Who of that period, if he lived now, would not be proud of what had been done? I like to think, that six centuries hence, we will be as proud of our accelerators. I like to think that, even though they will only endure then as the Stonehenges of the future, the discoveries we make today will be part and parcel of the culture of that future [48].

Like Stonehenge, the physics factories of the twentieth and twenty-first centuries represent a quest for meaning in a mysterious world. Whatever the meaning assigned to the results of that quest by theorists, their historical meaning is clearly a function of an elite priesthood reading increasing masses of data extrapolated from a micro-universe inaccessible to human senses via Big Science.

References

[1] Seidel R W 1999 The Golden Jubilees of Lawrence Berkeley and Los Alamos National Laboratories *Osiris* **14** 187–202

[2] Landes D 1969 *The Unbound Prometheus: Technological Change and Industrial Development in Western Europe from 1750 to the Present* (Cambridge: Cambridge University Press)
Musson E and Robinson E 1969 *Science and Technology in the Industrial Revolution* (Manchester: Manchester University Press)

[3] Miller H S 1970 *Dollars for Research: Science and its Patrons in Nineteenth-century America* (Seattle, WA: University of Washington Press)

[4] Aitken H G J 1985 *Syntony and Spark: The Origins of Radio* (Princeton, NJ: Princeton University Press)

[5] Livingston M S (ed) 1966 *The Development of High-Voltage Accelerators* (New York: Dover)

[6] Pralavorio C 2017 A new acceleration system for the PS booster *CERN News* 18 July https://home.cern/news/news/accelerators/new-acceleration-system-ps-booster

[7] Shoupp W E 1948 The electrostatic generator used in nuclear studies *Am. Inst. Elect. Eng. Elect. Eng.* **67** 668–71

[8] Forman P, Heilbron J L and Weart S 1975 *Physics Circa 1900, Historical Studies in the Physical Sciences Annual* ed R McCormmach 5th edn (Princeton, NJ: Princeton University Press) pp 90–114

[9] Heilbron J L and Seidel R W 1989 *Lawrence and His Laboratory: A History of the Lawrence Berkeley Laboratory* (Oakland, CA: University of California)

[10] Cottrell F G 1912 The research corporation: an experiment in public administration of patent rights *Sci. Am. Suppl.* **74** 191–2

Cottrell F G 1934 The protection by patents of scientific discoveries *Occas. Publ. AAAS 1; Sci. Suppl.* **77**

Turchetti S 2009 A contentious business: industrial patents and the production of isotopes *Dynamis* **29** 191–217

[11] Heilbron J L and Seidel R W 1989 *Lawrence and His Laboratory: A History of the Lawrence Berkeley Laboratory* (Oakland, CA: University of California) pp 466–84 (note 10)

[12] Conant, J 2002 *Tuxedo Park: A Wall Street Tycoon and the Secret Palace of Science that Changed the Course of World War II* (New York: Simon and Schuster)

[13] Holl J M 1997 *Argonne National Laboratory, 1946–96* (Urbana, IL: University of Illinois Press)

[14] Johnson S and Schaffer D 1994 *Oak Ridge National Laboratory: The First Fifty Years* (Knoxville, TN: University of Tennessee Press)

[15] Groves L 1962 *Now It Can Be Told* (New York: Harper)

[16] Johnson S and Schaffer D 1994 *Oak Ridge National Laboratory: The First Fifty Years* (Knoxville, TN: University of Tennessee Press) (note 15)

[17] Fermi E, Franck J, Hogness T R, Jeffries Z, Mulliken R S, Stone R S and Thomas C A 1944 Prospectus on nucleonicsMUC-RSM-234 https://atomicinsights.com/wp-content/uploads/Prospectus-on-Nucleonics.pdf (Accessed: 17 May 2021)

Tolman R 1944 Piles of the future review *Oak Ridge National Laboratory* MUC-SKA-811

These and other reports were consolidated with individual laboratory prospectuses from the MED laboratories at Ames, Berkeley, Columbia, Los Alamos and Oak Ridge.

[18] Baxter J P 1968 *Scientists Against Time* (Cambridge, MA: MIT Press)

[19] Hewlett R G and Anderson O E 1962 *A History of the United States Atomic Energy Commission.* (University Park, PA: Pennsylvania State University Press)

[20] Sapolsky H M 1990 *Science and the Navy: The History of the Office of Naval Research* (Princeton, NJ: Princeton University Press)

[21] Westwick P J 2003 *The National Labs: Science in an American System, 1947–74* (Cambridge, MA: Harvard University Press)

Crease R P 1999 *Making Physics: A Biography of Brookhaven National Laboratory, 1946–1972* (Chicago, IL: University of Chicago Press)

[22] Smyth H D 1950 The role of the national laboratories in atomic energy development *Bull. At. Sci.* **6** 5–8

[23] US Atomic Energy Commission 1969 High Energy Physics Advisory Panel *High Energy Physics Manpower Survey* (Washington, DC: US GPO)

[24] Schwartz S I (ed) 1998 *Atomic Audit: The Costs and Consequences of US Nuclear Weapons Since 1940* (Washington, DC: Brookings Institution Press) p 3

[25] Seidel R W 1986 A home for Big Science: the atomic energy commission's laboratory system *Historical Stud. Physical Sci.* **16** 135–75

Sylves R T 1987 *he Nuclear Oracles: A Political History of the General Advisory Committee of the Atomic Energy Commission, 1947–77* (Ames, IA: Iowa State University Press)

[26] Greenberg D S 1967 *The Politics of Pure Science* (New York: New American Library)

Greenberg D 2001 *Science, Money and Politics: Political Triumph and Ethical Erosion* (Chicago, IL: University of Chicago Press)

[27] Hoddeson L, Kolb A W and Westfall C 2008 *Fermilab Physics, the Frontier, and Megascience* (Chicago. IL: University of Chicago Press)

Westfall C 2017 Between the lines: a first-person account of Berkeley's loss of Fermilab *Phys. Persp.* **19** 91–104

Lebedev V and Shiltsev V (ed) 2014 *Accelerator Physics at the Tevatron Collider* (New York: Springer)

[28] Seidel R W 2001 The national laboratories of the Atomic Energy Commission in the early Cold War *Hist. Stud. Phys. Biol. Sci.* **32** 145–62

Seidel R W 1983 Accelerating science: the postwar transformation of the Lawrence Radiation Laboratory *Hist. Stud. Phys. Sci.* **13** 375–400

[29] Jungk R 1968 *The Big Machine* (New York: Scribner's)

Fraser G 1997 *The Quark Machines: How Europe Fought the Particle Physics War* (Philadelphia, PA: Institute of Physics)

[30] Teng L C 1980 The ZGS conception to turn-on *History of the ZGS (Argonne, 1979)* ed J S Day *et al* (New York: American Institute of Physics) p 11

[31] Teng L C 1980 The ZGS conception to turn-on *History of the ZGS (Argonne, 1979)* ed J S Day *et al* (New York: American Institute of Physics) pp 19–20

[32] Jones L *et al* 2010 *Innovation Was Not Enough: A History of the Midwestern Universities Research Association (MURA)* (Singapore: World Scientific)

[33] Samios N P 1998 First accelerator fruits *The Particle Century* ed G Fraser (Philadelphia, PA: Institute of Physics) pp 24–5

Pickering A 1989 Pragmatism in particle physics: scientific and military interests in the post-war United States *The Development of the Laboratory: Essays on the Place of Experiment in Industrial Civilization* ed F A J L James (New York: American Institute of Physics)

[34] Rigden J S 1987 *Rabi, Citizen and Scientist* (New York: Basic Books) pp 235–9

[35] Mersits U 1984 Construction of the CERN synchro-cyclotron (1952–1957) *Studies in CERN History* CHS-13 https://cds.cern.ch/record/154789 (Accessed: 29 May 2021)

[36] Galison P L 1997 *Image and Logic: A Material Culture of Microphysics* (Chicago, IL: University of Chicago Press)

[37] Richard D 2019 Scientific realism and high energy physics *PhilSci Archive* http://philsci-archive.pitt.edu/id/eprint/12900 14 March 2017 (Accessed: 31 May 2022)

Dardashti R, Dawid R and Thebault K (ed) 2019 *Why Trust a Theory? Epistemology of Fundamental Physics* (Cambridge: Cambridge University Press)

Cartwright N 1983 *How the Laws of Physics Lie* (Oxford: Oxford University Press)

[38] Riordan M, Hoddeson L K and Adrienne W 2015 *Tunnel Visions: The Rise and Fall of the Superconducting Super Collider* (Chicago, IL: University of Chicago Press)

[39] Englert F 2015 The BEH mechanism and its scalar boson *Int. J. Modern Phys.* A **30** 1

[40] Castillo L R F 2015 *The Search and Discovery of the Higgs Boson: A Brief Introduction to Particle Physics* (San Rafael, CA: Morgan and Claypool)

Lederman L M and Teresi D 1993 *The God Particle: If the Universe Is the Answer, What Is the Question?* (Boston, MA: Houghton Mifflin)
Weinberg S 1992 *Dreams of a Final Theory* (New York: Pantheon)
[41] Oreskes N and Conway E M 2010 *Merchants of Doubt* (*New* York: Bloomsbury)
[42] Goldhaber J 1992 Bevalac had 40-year record of historic discoveries *Lawrence Berkeley National Laboratory* https://www2.lbl.gov/Science-Articles/Archive/Bevalac-nine-lives.html
Bodvarsson M 1993 Bevatron shutdown in nostalgic ceremony *Lawrence Berkeley National Laboratory* https://www2.lbl.gov/Science-Articles/Archive/Bevalac-shutdown.html (Accessed: 24 May 2021)
[43] Abada A *et al* 2019 HE-LHC: the high-energy Large Hadron Collider *Eur. Phys. J. Spec. Top.* **228** 1109–382
[44] Adams H 1959 *Mont-Saint-Michel and Chartres* (New York: Doubleday Anchor)
Adams H 1918 *The Education of Henry Adams* ed H C Lodge (Boston, MA: The Massachusetts Historical Society) chs 23–25
Wilson R R 1968 The Richtmyer Memorial Lecture—particles, accelerators, and society *Am. J. Phys.* **36** 490–5
[45] Price D S 1963 *Little Science, Big Science* (New York: Columbia University Press)
Weinberg A 1967 *A Reflections on Big Science* (Cambridge, MA: MIT Press)
Fortin J-M and Currie D J 2013 Big Science vs little science: how scientific impact scales with funding *PloS One* **8** E65263
Kleiner K 1994 Every lab for itself—America's big nine national laboratories flourished during the Cold War. But with communism no longer perceived as a threat, they are scrambling to find new roles *New Sci.* 1930 18 June https://newscientist.com/article/mg14219302-400 (Accessed: 15 May 2021)
[46] CERN 2019 International collaboration publishes concept design for a post-LHC future circular collider at CERN *CERN Press Release* 15 January
[47] CERN 2019 International collaboration publishes concept design for a post-LHC future circular collider at CERN *CERN Press Release* 15 January (note 47)
[48] Wilson R R 1968 The Richtmyer Memorial Lecture—particles, accelerators, and society *Am. J. Phys.* **36** 490–5 (note 45)

IOP Publishing

Big Science in the 21st Century
Economic and societal impacts
Panagiotis Charitos, Theodore Arabatzis, Harry Cliff, Günther Dissertori, Juliette Forneris and Jason Li-Ying

Chapter 24

Critiques of Big Science

Joseph D Martin[1]

The name 'Big Science', like 'Big Bang' or 'Big Pharma', began as a slur. Big Science, that is, has been the subject of critique for at least as long as we have had a name for it. This chapter explores that legacy. It demonstrates how Big Science became a pivot point in disagreements over the social and cultural value of scientific inquiry, how those debates exposed disciplinary rivalries and jealousies, and how debates about the merits of Big Science within the science community translated into the world of policymaking, in particular as Big Science became megascience toward the end of the twentieth century.

24.1 Introduction

Monuments are juicy targets for skeptics of the ideals they instantiate. In Imperial Rome, the ousting of a particularly reviled emperor would be accompanied by the *damnatio memoriae*, the erasure of the public image of a hated despot, including the destruction of statuary (Stewart 1999)[2]. Within days of the United States' Declaration of Independence in July 1776, a gilded statue of King George III in New York had been torn down, its head and body disfigured (Hooke 2010, pp 50–51). And the Black Lives Matter protests that erupted in response to the May 2020 murder of George Floyd in Minneapolis, Minnesota, felled many monuments, among them a statue of the slave trader Edward Colton in Bristol. Such acts are symbolic, but they also aim to create space into which new values,

[1] Joseph D Martin is an associate professor of the History of Science and Technology at Durham University, and the author of *Solid State Insurrection: How the Science of Substance Made American Physics Matter* (2018; Pittsburgh, PA: University of Pittsburgh Press). An abridged version of this essay was published in *History in Politics* (Martin 2021).

[2] *Damnatio memoriae*, meaning 'condemnation of memory', is a nineteenth-century coinage.

doi:10.1088/978-0-7503-3631-4ch24

ideals, and practices can move. When we regard attempts to topple a monument, we should of course inquire what that monument embodied that inspired such attacks, but we should also investigate what the people doing the toppling would like to erect in its place.

This chapter applies this way of thinking to attacks on *scientific* monuments—that is, the monuments of Big Science, particularly as they appeared in twentieth-century American physics. Both the growth of Big Physics in the early Cold War and the backlash against it in the late Cold War can be understood as stemming from a similar kind of opposition to values encoded in monumental projects—high-energy physics initially cast itself as the foil to militarized nuclear research; subsequently, skeptics of Big Physics took aim at particle accelerators. I begin with the revolution in high-energy physics that led to what is known as the Standard Model, which describes elementary particles and forces governing them. High-energy physics underwent a conceptual transformation after World War II. It proposed a radically revised understanding of microphysics; but it simultaneously sought to distance itself from what some saw as the pernicious effects of nuclear weapons, which cast a monumental shadow over the American physics community and tightly tied nuclear physics—previously regarded as a rather arcane branch of inquiry—to Cold War politics. The research program that resulted from efforts to distance high-energy physics from those entanglements was carried out using the iconic machines of Big Physics, particle accelerators, which grew in size and expense as the century wore on.

But the massive machines required to probe the smallest scales of matter and energy generated a backlash. Some physicists in the late-twentieth century sought to topple the monuments of Big Physics themselves, arguing that they consumed too many resources for too little payoff, and that the reductionist worldview they encoded devalued other forms of knowledge. By seeking to distance physics from politics, these critics suggested, high-energy physics had become political in another sense, forcing a confrontation with questions about why, how, and in what measure societies do and should support scientific research. Interrogating these late-century critiques, and the ways they have carried over into twenty-first century debates, illuminates the distinctive place Big Physics held as a politically important development, as well as a scientifically important one.

24.2 Making physics remote

Many historians would identify the Manhattan Project as the archetype of Big Science—the first scientific undertaking that combined large facilities and sprawling research groups with lavish funding, all against a background of high stakes (Hughes 2003). The monument the Manhattan Project produced was the bomb itself, which cast a grim shadow over the post–World War II world. On 6 August 1945, a uranium bomb code-named 'Little Boy', dropped from an American B-29 Superfortress, detonated over the Japanese city of Hiroshima. Tens of thousands of people, mostly civilians, were killed almost instantly. Tens of thousands more later suffered from or succumbed to injuries sustained in the blast, including the effects of the bomb's radiation. Three days later, a plutonium bomb code-named 'Fat Man' wrought similar destruction on Nagasaki. Japan surrendered[3].

The United States had announced nuclear weapons to the world, bringing physicists new and unfamiliar notoriety. Through the 1930s, physicists had painstakingly investigated the structure and behavior of the atomic nucleus, laboring mostly in popular obscurity. Even in 1944, the president of the American Physical Society could remark, 'It is a rare occurrence that a census taker has ever heard of a physicist, and the task of explaining is such that one is often tempted to register as a chemist' (Hull 1944, p 66). The bomb, however, positioned physicists as custodians of an awesome power with the potential to reshape the global order.

Some embraced the bomb. Most notoriously, the Hungarian émigré Edward Teller championed an arsenal of bigger, more powerful thermonuclear weapons, which he hoped could be used for applications ranging from nuclear deterrence to carving out a new harbor in Alaska (O'Neill 1998). Others took a less hawkish approach, seeking to subject nuclear weapons to sober, civic-minded custodianship. J Robert Oppenheimer, the head of scientific operations at the Los Alamos laboratory where the bombs were designed and assembled, advocated for international control of nuclear weapons and, when that failed, sought to exert his influence to rein in the excesses of Teller and his allies. Oppenheimer's skepticism of an expanded nuclear arsenal contributed to the Atomic Energy Commission's (AEC) revocation of his security clearance, over the strenuous objections of many of his colleagues (Bird and Sherwin 2007). Nuclear physics, which had made huge strides in the 1930s through table-top experiments conceived by researchers driven mostly by curiosity about the structure of the atom, had become both Big Science and big politics.

Still other physicists refused what they saw as the Faustian bargain of Cold War nuclear research. Robert R Wilson provides an illustrative example. Wilson had led the Research Division at Los Alamos for the Manhattan Project. But the bombings of Hiroshima and Nagasaki disturbed him profoundly. 'I remember being just… ill. Sick', he told the documentarian Jon Else in 1981, 'to the point I thought I would … vomit. I was so overwhelmed when it happened, that that thing had happened. Still am.' His reaction soured him on nuclear research: 'At the end of the war then, I gave up my clearance, and have not worked on … nuclear energy in any of its aspect[s] … or on bombs' (Else 1981)[4].

An influential segment of the physics community shared Wilson's distaste for applications that might be misused. High-energy physicists, many of them Manhattan Project veterans, prided their field for its intellectual purity, and even boasted of its remoteness from earthly affairs. Upon being awarded a share of the 1976 Nobel Prize for the experimental discovery of the J/Ψ meson, Burton Richter

[3] The bomb prompted Japan to accept the United States' unusual demand of unconditional surrender. The necessity of using the bomb to end the war in the Pacific has, however, been forcibly challenged. See e.g. Alperovitz (1995).

[4] The quoted passages can be found at 1:03:08–1:03:29 and 1:07:58–1:08:11 of *The Day After Trinity* (Else 1981).

bluntly informed a reporter who asked after his discovery's uses: 'The significance is that we have learned something more about the structure of the Universe. In terms of practical application right now, it's got none' (Chicago Tribune 1976)[5]. Wilson himself was one of the first to liken particle accelerators to gothic cathedrals, suggesting that modern accelerator designers experience a tingle of excitement similar to what medieval cathedral builders must have felt when gazing on their creations (Wilson 1958). The analogy reflected his 'desire to combine aesthetic values and ethical commitments with the political and institutional circumstances of big science', as Jessica Wang (2017) suggests.

Wilson generated the most iconic expression of this outlook during a 1969 congressional hearing on support for the National Accelerator Laboratory (NAL)—better known as Fermilab (figure 24.1)—of which Wilson would become the founding director (Hoddeson *et al* 2008). When asked by Rhode Island Senator Joe Pastore whether the facility would contribute to national security, or competition with the Soviet Union, Wilson insisted that it would not. Instead, he insisted, 'it has to do with: Are we good painters, good sculptors, great poets? I mean all these things that we really venerate and honor in our country and are patriotic about. In that sense, this new knowledge has all to do with honor and country, but nothing to do directly with defending our country except to make it worth defending' (US Congress 1969, p 113).

This eloquent defense of the cultural value of physics has been widely quoted, but its context widely misrepresented. Wilson is often cast as offering his noble resistance

Figure 24.1. Robert Rathbun Wilson Hall at Fermi National Accelerator Laboratory made the metaphorical link between particle accelerators and cathedrals explicit with a design inspired by the Cathédrale Saint-Pierre de Beauvais. Photo by Steven Keys, KeysPhotography.com, distributed under CC BY 4.0.

[5] For further discussion of this attitude, see Martin (2017) and Stevens (2003).

in the face of pressure to either kowtow to national security demands or jeopardize federal funding. But Pastore, a supporter of the NAL, had different concerns in mind. As the Senator charged with selling the project to the appropriations committee, he wanted to be armed with a clear justification for the project. Vague promises that the facility would furnish 'a better understanding of the subnuclear Universe', but that, specifically, physicists 'do not know what we will find', as the AEC Research Director Paul W McDaniel suggested, left Pastore cold. He lamented: 'We have these Senators … going all over the country showing how many people are starving, how many people are hungry, how many people live in rat-ridden houses. Here we are asking for $250 million to build a machine that is an experimental machine, in fundamental high-energy physics, and we cannot be told exactly what we are trying to find out through that machine' (US Congress 1969, pp 111–12).

We should understand Wilson's oft-quoted defence of science as a cultural good within that context. Before the exchange leading to Wilson's famous paean to pure physics, Pastore entreated the scientists testifying at the hearing: 'I would like to have some definitive answers here as to priorities, because that is going to be thrown right into my face' (US Congress 1969, p 112). In that context, Wilson's response was singularly unhelpful. His reluctance to engage in questions about priorities—to measure high-energy physics against pragmatic goals at all—even in the face of repeated, leading question from an ally seeking to build the best case possible for supporting a new facility—shows the depths of Wilson's commitment to maintaining a firewall between high-energy research and anything but the most remote and notional practical justifications.

The NAL would get its funding, but Pastore's frustration reflected growing disquiet about the effects of large-scale science, particularly in the context of other major government expenditures. In 1969 the United States was embroiled in a costly (in several senses) war in Vietnam. The Apollo Program was in full swing, and would send astronauts to the Moon in July, but would also consume almost $30 billion. With funding for the NAL recently approved, solid state physicist Benjamin Lax complained that 'millions are being spent on redundant facilities' while his own magnet laboratory at the Massachusetts Institute of Technology languished (Lax 1967). The poet Gil Scott-Heron was equally sharp-penned in 1970, skewering spending on the Moonshot in the face of poverty, racial inequality, and infrastructural decay, in his poem 'Whitey on the Moon': 'With all that money I made last year / For whitey on the Moon / How come I ain't got no money here? / Hmm, whitey's on the Moon' (quoted in Russell and Vinsel 2017). Pastore, Lax, and Scott-Heron, though they represented different perspectives, shared some motivating questions: What cultural priorities did Big Science projects reflect? Whose interests did they entrench? At what cost? Those same questions motivated the ongoing critiques of Big Physics in the second half of the twentieth century.

24.3 The critics of monumental physics

American high-energy physics was in the monument business by the 1960s. Ernest O Lawrence and M Stanley Livingstone had invented and developed the cyclotron in the early 1930s. This instrument for accelerating subatomic particles facilitated physicists' efforts to create interactions in the laboratory that revealed the workings of nature at its smallest scales. But the first cyclotron was small enough to fit in one hand. Later iterations grew larger as physicists grew hungry to accelerate particles, and smash them together, at higher velocities, which required bigger magnets (see Pais 1988). By 1960 this process had led to machines such as the Alternating Gradient Synchrotron (AGS) at Brookhaven National Laboratory on Long Island, 843 feet in diameter (figure 24.2) (see Heilbron and Seidel 1989, Crease 1999). Such monumental edifices revolutionized physical understanding of the subnuclear world. But they also stoked revolution of a different type: a revolt against the values mixed into their mortar.

In 1961 Alvin Weinberg published an article in *Science* entitled 'Impact of Large-Scale Science on the United States', which lamented the influence monumental facilities like the AGS exerted on the American scientific community (Weinberg 1961). Weinberg was an unlikely revolutionary. As the director of Oak Ridge National Laboratory, a sprawling national research facility that had its roots in the Manhattan Project, he was an established member of the scientific–managerial class that guided American science during the Cold War. He was a physicist who had worked on the Manhattan Project, often regarded as the first true example of Big Science (de Solla Price 1963). But he nevertheless emerged as one of the most vocal opponents of the edifice complex that gripped the US scientific establishment in the 1960s. He would credit the comprehensive, humanistic undergraduate education he received at the University of Chicago in the 1930s with conditioning him to

Figure 24.2. Aerial view of the AGS at Brookhaven National Laboratory during its construction. Credit: Brookhaven National Laboratory, distributed under CC BY-NC-ND 2.0.

understand science in terms of how it intertwined with human affairs (Weinberg 1994, p 3).

Weinberg examined the intertwining of Big Science with human affairs, and he did not like what he saw. Like Wilson, he compared the monuments of Big Science to monuments like the Pyramids and medieval cathedrals. Whereas Wilson drew that comparison to highlight the cultural importance of particle accelerators, however, Weinberg drew it instead to question their value. It might be that Big Science was 'here to stay', but 'those cultures which have devoted too much of their talent to monuments which had nothing to do with the real issues of human well-being have usually fallen upon bad days' (Weinberg 1961, pp 162, 164). Without concerted efforts to counter the ill-effects of Big Science, Weinberg warned, such bad days might well be in store for the United States, particularly if it permitted questions of scientific merit to be arbitrated 'in the congressional committee room rather than in the technical-society lecture hall', which would ensure that 'the spectacular rather than the perceptive becomes the scientific standard' (Weinberg 1961, p 161). Spectacle, in other words, should not distort our sober evaluation of a scientific undertaking's worth in terms of benefits to society[6].

This critique focused attention on the values that underwrote Big Physics. Weinberg himself, as well as critics who followed his lead, focused in particular on *reductionism*. This term has been used to name all manner of sin, but in the case of high-energy physics in the second half of the twentieth century, it typically refers to the conviction that the whole natural world can in principle be described in terms of a single set of laws governing the workings of the Universe at its smallest scales. Such a view was attractive to particle physicists in the twentieth century. Deciphering the workings of the elements of matter and energy held the promise of answering *the* most fundamental questions about the Universe: an intellectual spectacle worthy of spectacular research facilities. And a reductionist approach led physicists of Wilson's stripe away from the uncomfortable prospect of generating applications they could not control.

But many physicists became concerned when reductionism placed disproportionate value on the kind of science that was disproportionately distant from human-scale affairs. Weinberg worried that by equating fundamental physics with reductionist physics, high-energy physicists were neglecting more important dimensions of fundamentality. He voiced this worry as a preamble to his articulation of what has become known as the 'Weinberg criterion': 'other things being equal, *that field has the most scientific merit which contributes most heavily to and illuminates most brightly its neighboring scientific disciplines*' (Weinberg 1964, p 45). The type of Big Physics that was emerging in the 1960s, from Weinberg's perspective, rated quite low by this measure of scientific merit; its reductionist thrust had focused it away from areas that could be the source of rich connections with other subjects.

Weinberg's articulation of this critique in 1964 was precocious; similar worries blossomed through the 1970s and 1980s. Philip W Anderson, who conducted

[6] This view is expounded further in Weinberg (1964). For further discussion, see Martin (2018, ch 7).

pioneering and fundamental work in solid state and condensed matter physics that clarified the mechanisms that underlie magnetism, among other phenomena, wrote a famous broadside against this form of reductionism, titled 'More Is Different', in 1972. Although we often learn about systems by breaking them down into their components, Anderson argued, we cannot then assume that knowledge of those components permits us to reconstruct or predict the properties of the higher-level systems they might constitute—to presume otherwise was simply 'the arrogance of the particle physicist' (Anderson 1972, p 396; see also Martin 2015). In 1972, Anderson limited himself to a philosophical critique of the brand of reductionism he perceived in high-energy physics, but would later expand it to include the institutional effects of Big Physics on his own community of condensed matter physicists—such as a dearth of funding for curiosity-driven research, less influence in prestigious organizations, and difficulty recruiting talented graduate students, a situation he thought was maintained by the narrowness and insularity of the influential high-energy physics community (Anderson 2001).

Another condensed matter physicist, Leo Kadanoff, argued along similar lines. Like Anderson, he accepted that a reductionist approach, which often led to the unification of physical laws once thought to be separate, had been a great driver of scientific progress. But, he qualified, 'another part of the grand structure is the opposite view—that natural laws are diverse. Different laws may apply at different levels of organization' (Kadanoff 1986, p 7). Kadanoff echoed Weinberg in suggesting that physicists had neglected the contributions science could make beyond narrow communities of savants when pushing for big projects:

> We scientists have chosen to follow fashion and have not continued research in less fashionable areas. We have thoughtlessly pushed our own subjects of research without concern for whether these areas have a value commensurate with their cost. And we have indeed pursued big science. While giving lip service to statements that a balanced and solid physics program requires a major investment in small science, physicists have pushed for each of the big projects. It is easier to argue for another synchrotron light source or for a supercomputer or supercollider than for the broad programs that will provide the real intellectual base for understanding our natural world.

Insufficient appreciation of small science had encouraged US science budgets that were 'increasingly misdirected toward grandiose projects' by the late 1980 (Kadanoff 1988, p 11).

The opposition to monumental physics that grew from the 1960s through the 1980s was multifaceted. It hinged on evaluating the connections among scientific fields and between those fields and technical and social concerns, on philosophical commitments about the merits of different types of knowledge, and on internecine squabbles and personal politics. But at core, it was a question about priorities. By casting fundamental knowledge of the very small as valuable principally as a form of culture, high-energy physicists distanced themselves from practical justifications for their work. This was in some measure intentional. Some were scarred by what they saw as the misuse of physics in wartime, and others perceived the necessity for a fierce defense of abstract physical knowledge as an end in itself within a notoriously pragmatic American context. But by pursuing this type of argument, and tying Big Physics into it, they also left a question open for the asking: if we value physicists for the same reasons we value sculptors, artists, and poets, why are physicists funded so much more richly?

Even in the 1960s, as Big Physics was just beginning to establish its toehold on the research landscape, skeptics were asking that question. They believed, in line with Alvin Weinberg, that 'society does not *a priori* owe the scientist, even the good scientist, support any more than it owes the artist or the writer or the musician support' (Weinberg 1964, p 44). And they were driven by an understanding that science is political. The choices we make as a society about what science to pursue and how to pursue it both speak to and reinforce societal values. In that sense, high-energy physicists miscalculated if they hoped to distance themselves from political concerns by retreating to the distant world of quarks, bosons, and neutrinos. By aiming to distance themselves from one political realm, physicists had made their science big, but the bigger it became, the more central another political realm—the politics of scientific choice—became to its fate.

24.4 Contemporary critiques

At a standing-room-only seminar on 4 July 2012, physicists from the Large Hadron Collider (LHC) at CERN announced that data from the machine's recent run revealed decisive evidence for a Higgs-like scalar boson[7]. The Higgs was the last elementary particle predicted by the Standard Model of particle physics to be experimentally detected, and its detection sent a frisson of excitement through the high-energy physics community. But with further data analysis, that excitement began to turn to concern. Many theorists had hoped that the LHC would turn up a Higgs with unexpected properties, which would be the key to unlocking deeper understanding and give experimental traction to speculative theories such as supersymmetry and string theory. But the Higgs that showed itself at the LHC came with no strings attached. It appeared fully consistent with the Standard Model. This became known as the 'nightmare scenario' in high-energy physics circles, since it was clear that it jeopardized ongoing support for expensive experimental programs. Indeed, the lack of support the Higgs gave to the speculative theoretical programs that played an important role in justifying future Big Physics experiments bolstered contemporary arguments against Big Science.

The early 2000s brought a backlash against string theory. The speculative theoretical framework aspired to pave the way for the unification of relativity and quantum mechanics, which otherwise remained confined to their respective domains of applicability. But after several decades in which it offered no chance of

[7] Decisive evidence meaning a signal in the data with a significance over the background of five standard deviations. On the history of the five-sigma standard in particle physics, see Franklin (2013).

experimental test and an almost inexhaustible theoretical malleability, some physicists grew frustrated, in particular as they watched string theorists dominate prestigious institutions and drive the public conversation about the future of physics. Peter Woit argued forcibly that string theory was 'not even wrong', and was consuming both intellectual and financial resources that could do more good elsewhere (Woit 2006). Lee Smolin was similarly skeptical, suggesting that the apparently inexhaustible malleability of string theory meant that it shouldn't qualify as a theory at all (Smolin 2006).

In 2006, however, when both Woit and Smolin published their broadsides against string theory, the future of experimental high-energy physics was secure. The LHC was under construction, its experimental goals were clear, and so concern about the influence of speculative theoretical programs was not directly pertinent to Big Physics as it was then being practiced. That would change after the Higgs was discovered and its characteristics failed to arbitrate key questions about the viability of various forms of string theory, supersymmetry, dark matter, cosmic inflation, and other longstanding questions in fundamental physics.

Sabine Hossenfelder, who describes herself as a recovering high-energy theorist, has emerged as this century's sharpest critic of these programs. Like Woit and Smolin, she began with concerns about the intellectual viability of speculative theoretical programs such as string theory and supersymmetry, which, she contends, suffer from a misguided commitment to aesthetic criteria over empirical ones (Hossenfelder 2018). Her critique has also extended to CERN's plans for its Future Circular Collider (FCC), the planned successor to the LHC, which she criticizes both for its cost and for its uncertain theoretical payoff. 'It's time for particle physicists … to ask why none of the exciting predictions they promised have resulted in discoveries', she wrote in the *New York Times*. 'Money will not solve this problem. And neither will a larger particle collider' (Hossenfelder 2019).

Some of the arguments against the FCC are redolent of worries from the 1960s–1980s. Hossenfelder points out that the world as yet lacks an international center for climate predictions or epidemic modeling, echoing Weinberg's concern that high-energy physics competes poorly when we rate it in terms of its social benefits. Like Anderson, Hossenfelder critiques the world of Big Physics for its insularity, which means that Big Physics gets bigger because of inertia, rather than necessity (Hossenfelder 2020).

But twenty-first century critiques of Big Physics nevertheless spring from a different well. Whereas Weinberg had focused on Big Science as it interlocked with national priorities, more recent critics have encountered an increasingly international high-energy physics community. Weinberg, Anderson, Kadanoff, and other twentieth century critics, for all their concern over the comparative value of large accelerators, were ready to concede that the knowledge they produced would be of value, however narrowly. Twenty-first century critics are less sanguine about the intellectual prospects of higher energy machines. They consider them profligate *and* question their payoff for fundamental physics.

Despite this contextual difference, however, the spirit of the opposition is the same. It derives from the recognition that the scale of these projects involves the entrenchment of certain ideals. And it proceeds with the conviction that derailing those projects is therefore necessary to create space where alternative ideals might flourish. Although I have not discussed it here, this pattern also holds for critiques of Big Biology, which point out that the scale of modern biotech entrenches certain commercial interests and economic ideals with which we might have reason to quarrel (Kozubek 2017). Critics of Big Science often object to size itself, for instance by pointing out the vast price tags some projects come with, but they are more often animated by the fact that scale bespeaks momentum. That scale solidifies one group's idea about what science is valuable, where it should be going, and who should benefit, and it thus makes such projects irreducibly political.

24.5 Epilogue

In 1993 a monument fell. The Superconducting Super Collider (SSC) was to have been a massive particle accelerator, 54.1 miles in circumference. As a research facility it was, like the NAL before it, an effort to generate data that would produce novel, unpredictable insights into particle physics. As a monument, it was an effort to ossify the reductionist worldview, and the system that had patronized it for the preceding decades, by entrenching it in a colossal facility. But by the early 1990s, the US Congress could not be convinced to support such an effort, and the monumental SSC, already in the process of being erected, lost its funding (Kevles 1997, Riordan *et al* 2015, Martin 2018, ch 9).

The SSC was unusual in facing vocal and determined opposition from within the physics community. Even Weinberg, when castigating Big Science, restrained himself from actively opposing any specific project. The likes of Philip Anderson and Leo Kadanoff, however, were not content to follow the disciplinary tradition of keeping criticism of particular high-profile projects to themselves. They were not solely motivated, however, by a yen to tear the SSC from its pedestal; they also held a conviction that something intellectually just as worthwhile, and at the same time more attuned to contemporary social, economic, and political priorities—namely, more generous support of basic research in small science—could go in its place.

Monuments are erected with the goal of setting certain ideals in stone. Tearing them down is therefore not only a symbolic rejection of those ideals, but also a meaningful attack on the material conditions that perpetuate them. Anderson, Kadanoff, Weinberg, Hossenfelder, and other critics of Big Science worried that the conditions that perpetuated Big Physics were unhealthy for other areas of science. By opposing it, they sought to usher in a new age, once in which the basic pursuit of science worked hand-in-glove with its practical employment.

The critiques of Big Science that contributed to the demise of the SSC exhibit another important feature: they embraced, rather than rejected, the political entanglements of science. The idea of science as apolitical became a powerful one during the Cold War. For the State Department and the clandestine services, the very idea of apolitical science became a tool in deeply political American efforts to exert soft power abroad (Wolfe 2018). For the nuclear physicists who turned toward the high-energy frontier, it soothed a wounded conscience and justified ongoing expenditures in support of increasingly remote research. But for practitioners of small science, whose work often exhibited closer connections to technology, the political consequences of scientific choice were easier to discern. The revolution they sought to enact by attacking monumental physics was a turn, with a look of recognition, toward the political processes that determine what sort of science we value, and why.

References

Alperovitz G 1995 *The Decision to Use the Atomic Bomb* (New York: Alfred A Knopf)

Anderson P W 1972 More is different: broken symmetry and the nature of the hierarchical structure of science *Science* **177** 393–6

Anderson P W 2001 More is different—one more time *More Is Different: Fifty Years of Condensed Matter Physics* ed N Phuan Ong and R N Bhatt (Princeton, NJ: Princeton University Press) pp 1–9

Bird K and Sherwin M J 2007 *American Prometheus: The Triumph and Tragedy of J Robert Oppenheimer* (New York: Knopf)

Chicago Tribune 1976 Yanks sweep science field for this year's Nobel Prizes, 19 October

Crease R P 1999 *Making Physics: A Biography of Brookhaven National Laboratory, 1946–1972* (Chicago, IL: University of Chicago Press)

de Solla Price D J 1963 *Little Science, Big Science* (New York: Oxford University Press)

Else J 1981 *The Day After Trinity* KETH, San Jose, CA (documentary film) https://youtube.com/watch?v=Vm5fCxXnK7Y

Franklin A 2013 *Shifting Standards: Experiments in Particle Physics in the Twentieth Century* (Pittsburgh, PA: University of Pittsburgh Press)

Heilbron J L and Seidel R W 1989 *Lawrence and His Laboratory: A History of the Lawrence Berkeley Laboratory* vol 1 (Berkeley, CA: University of California Press)

Hoddeson L, Kolb A W and Westfall C 2008 *Fermilab: Physics, the Frontier, and Megascience* (Chicago, IL: University of Chicago Press)

Hooke H 2010 *Empires of the Imagination: Politics, War and the Arts in the British World, 1750–1850* (London: Profile)

Hossenfelder S 2018 *Lost in Math: How Beauty Leads Physics Astray* (New York: Basic)

Hossenfelder S 2019 The uncertain future of particle physics *New York Times* 23 January https://www.nytimes.com/2019/01/23/opinion/particle-physics-large-hadron-collider.html

Hossenfelder S 2020 The world doesn't need a new gigantic particle collider *Sci. Am.* 19 June https://www.scientificamerican.com/article/the-world-doesnt-need-a-new-gigantic-particle-collider/

Hughes J A 2003 *The Manhattan Project: Big Science and the Atom Bomb* (New York: Columbia University Press)

Hull A W 1944 The outlook for the physicist and prospective physicist in industry *Am. J. Phys.* **12** 62–70

Kadanoff L P 1986 Cathedrals and other edifices *Phys. Today* **39** 7–9

Kadanoff L P 1988 The big, the bad and the beautiful *Phys. Today* **41** 9–11

Kevles D J 1997 Big Science and Big Politics in the United States: reflections on the death of the SSC and the life of the human genome project *Hist. Stud. Phys. Biol. Sci.* **27** 269–91

Kozubek J 2017 The trouble with Big Science *Los Angeles Review of Books* 25 November https://lareviewofbooks.org/article/trouble-big-science/

Lax B 1967 Letter to Leland Haworth, 10 May *NMLR, Francis Bitter National Magnet Laboratory* MIT Archives and Special Collections, Cambridge, MA
Martin J D 2015 Fundamental disputations: the philosophical debates that governed American physics, 1939–1993 *Hist. Stud. Nat. Sci.* **45** 703–57
Martin J D 2017 Prestige asymmetry in American physics: aspirations, applications, and the purloined letter effect *Sci. Context* **30** 475–506
Martin J D 2018 *Solid State Insurrection: How the Science of Substance Made American Physics Matter* (Pittsburgh, PA: University of Pittsburgh Press)
Martin J D 2021 On monuments and scientific revolutions *History in Politics* **1** 68–73
O'Neill D 1998 Alaska and the firecracker boys: the story of Project Chariot *The Atomic West* ed B W Hevly and J M Findlay (Seattle, WA: University of Washington Press) pp 179–99
Pais A 1988 *Inward Bound: Of Matter and Forces in the Physical World* (Oxford: Clarendon)
Riordan M, Hoddeson L and Kolb A 2015 *Tunnel Visions: The Rise and Fall of the Superconducting Super Collider* (Chicago, IL: University of Chicago Press)
Russell A and Vinsel L 2017 Whitey on Mars: Elon Musk and the rise of Silicon Valley's strange trickle-down science *Aeon* 1 February https://aeon.co/essays/is-a-mission-to-Mars-morally-defensible-given-todays-real-needs
Smolin L 2006 *The Trouble with Physics: The Rise of String Theory, the Fall of a Science and What Comes Next* (New York: Penguin)
Stevens H 2003 Fundamental physics and its justifications, 1945–1993 *Hist. Stud. Phys. Biol. Sci.* **34** 151–97
Stewart P 1999 The destruction of statues in late antiquity *Constructing Identities in Late Antiquity* ed R Miles (London: Routledge) pp 159–89
US Congress Joint Committee on Atomic Energy 1969 AEC Authorizing Legislation, Fiscal Year 1970 *91st Congress* (17 and 18 April)
Wang J 2017 'Broken symmetry': physics, aesthetics, and moral virtue in nuclear age America *Epistemic Virtues in the Sciences and the HumanitiesBoston Studies in the Philosophy and History of Science* vol 321 ed J van Dongen and H Paul (Cham: Springer) pp 27–48
Weinberg A M 1961 Impact of large-scale science on the United States *Science* **134** 161–4
Weinberg A M 1964 Criteria for scientific choice *Phys. Today* **17** 42–8
Weinberg A M 1994 *The First Nuclear Era: The Life and Times of a Technological Fixer* (New York: American Institute of Physics)
Wilson R R 1958 Particle accelerators *Sci. Am.* **198** 64–80
Woit P 2006 *Not Even Wrong: The Failure of String Theory and the Search for Unity in Physical Law* (New York: Basic)
Wolfe A 2018 *Freedom's Laboratory: The Cold War Struggle for the Soul of Science* (Baltimore, MD: Johns Hopkins University Press)

IOP Publishing

Big Science in the 21st Century
Economic and societal impacts
Panagiotis Charitos, Theodore Arabatzis, Harry Cliff, Günther Dissertori, Juliette Forneris and Jason Li-Ying

Chapter 25

From Big Science to Bigger Science: preparing for the LHC

Grigoris Panoutsopoulos and Kostas Gavroglu[1]

25.1 Introduction

The Large Hadron Collider (LHC) at CERN may very well be the most heavily publicized scientific machine of all time. Hundreds of articles, documentaries, and books have praised it throughout the years. Yet, the majority of them were content with presenting impressive records regarding its energy level, size, and cost, the number of detected events or discussing the legendary discovery of the Higgs boson. However, the almost twenty-year long period of planning and construction, that led up to the LHC's commencement of operations in 2009[2], has not been studied systematically. It is crucial to examine what took place throughout the development of the LHC, from its inception in 1984 to its firm establishment in 1996, as a Big Science infrastructure, while witnessing contingency at work at almost every step. The LHC was not simply an enormous machine constructed for scientific research. Its planning was affected by the fall of the Berlin Wall, the collapse of the USSR in 1991 and by the demise of its American competitor, the Superconducting Super Collider (SSC) in 1993. Furthermore, it developed as a new model for Big Science and it became a decisive factor for the world hegemony of CERN and Europe in the field of experimental high-energy physics (HEP).

There are two main issues that we will discuss in this chapter. First, we will argue that, as the material culture of HEP had become more complex, more costly, and the

[1] Grigoris Panoutsopoulos is a PhD Candidate in the Department of History and Philosophy of Science at the National and Kapodistrian University of Athens. His research has focused on the history of CERN and Big Science. Kostas Gavroglu is Emeritus Professor of the History of Science at the National and Kapodistrian University of Athens.

[2] While construction of the LHC ended in 2008, the collider began operations in 2009, since a magnet quench incident caused extensive damage to over 50 superconducting magnets and instigated a one year delay.

existence of the field had been tellingly dependent on colliders, what critically shaped the practices of the HEP community can be further perceived through the details of the long-term planning of these machines. In this context, these infrastructures are much more than instruments to test a theory, to hunt for new particles, or even to do exploratory experimentation. The character and the dynamics of HEP were co-created not only by the complex scientific and technological structure of these machines, but also through the intricate political, diplomatic, industrial, and financial relations, which took center stage in their planning period. The actual construction and operation of an 'ultimate' machine, such as the LHC, became an end in itself. By examining the shifts in strategies, the structure of the debates and the rhetoric revolving around the LHC, since the Lausanne workshop in 1984, when the idea of the collider was first discussed officially, we will attempt to gain an understanding of how the initial design period of the LHC delineated its trajectory rather independently of the theoretical predictions[3]. Second, we will argue, that in the aftermath of the momentous political developments, which heralded the end of the Cold War, the whole process of the realization of the LHC played a pivotal role in the integration of the 'new Europe'. The transformation of the LHC from a big collider in Geneva to a 'world machine' was indicative of the role that Europe would play in the new era of globalization (Mobach and Felt (2022), Cramer (2020)). CERN's management, having a clear and well-thought-out vision, exploited the new political realities, enhanced the potential capacity of CERN and also contributed towards the emerging new role of Europe.

25.2 Conceptualizing the LHC

The postwar decades saw the construction of impressive machines, such as the Tevatron in the United States, the Large Electron-Positron (LEP) collider in Europe, and the U-70 Proton Synchrotron in the Soviet Union. The long-term careful planning and the intensive work they demanded, the technical complexity, the cost, and the overall symbolism of these infrastructures further emphasized the high profile of HEP instrumentation. It was around these machines that the multinational HEP community developed progressively stronger links with different governments and industries, acquiring at the same time indispensable diplomatic skills, while the search for funding, the management and the division of labor became inextricable aspects of the HEP research. We will try to bring out these features, by focusing on the period of the 'becoming' of the emblematic LHC.

LHC's design started with three workshops which were initiated jointly by CERN and the ECFA (European Committee for Future Accelerators)[4]. These workshops,

[3] Nevertheless, we have to mention that theory and theoreticians played a significant role in the design of the specific experiments, such as the ATLAS, CMS, and ALICE, in the later stages of the planning of the LHC project.

[4] The three workshops were: 'ECFA-CERN Workshop on Large Hadron Collider in the LEP Tunnel', in Lausanne (CERN 1984); 'Workshop on Physics at Future Accelerators', in La Thuile (CERN 1987); and 'Large Hadron Collider Workshop' in Aachen (CERN 1990c).

attended by hundreds of experimentalists, machine builders, and theoreticians, set the foundations for an initial consensus around the machine, within and outside of CERN, and played a decisive role in legitimizing the idea of the LHC. The first workshop took place in 1984, in Lausanne, and it marked the official start of the project (figure 25.1). There, 150 participants discussed the first ideas for the construction of a proton–proton collider by using the LEP tunnel, even before its excavation had begun[5]. Most of the discussion was focused on the technical aspects of the prospective collider, which included machine parameters, superconducting

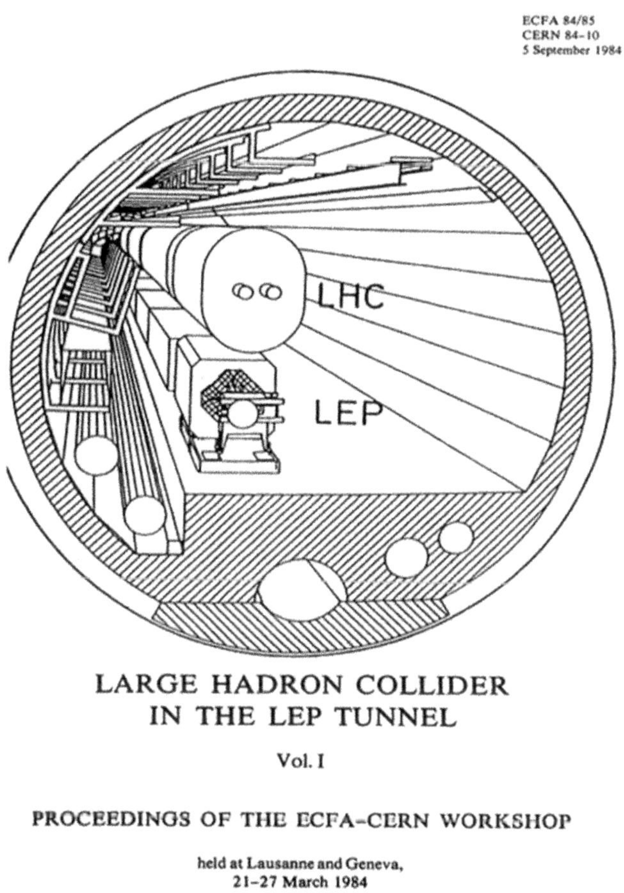

Figure 25.1. The front cover of the proceedings of the Lausanne workshop in 1984. Note the depiction of the early design, which foresaw the coexistence of the LHC and LEP in the same tunnel simultaneously. (Credit: CERN 1984.)

[5] The conviction that a newer, bigger collider would succeed LEP was such that it drove CERN to bore a larger tunnel in order to accommodate the LHC as well. In 1979, when the ECFA issued a recommendation in favor of the construction of the LEP collider, it included a request 'to preserve the full potential of the tunnel for further use' (CERN 1991c, p 5).

magnets, and suitable materials. What is particularly worth noting is that the section devoted to the 'physics case' was rather vague. In essence, it consisted of the presentation of a series of theoretical claims (technicolor theories, supersymmetry, the Higgs boson, compositeness), without clarifying whether the LHC was indispensable for their experimental exploration. As the Summary Report of the Organising Committee characteristically noted:

> [o]ne may say that there is at present a theoretical consensus that the once fashionable desert will actually bloom, but there is no consensus on what flowers exist there (CERN 1984, p 12).

Christopher Llewellyn Smith, the theoretical convener of the workshop and later Director-General of CERN, left no doubt as to whether the theoretical predictions would be decisive in the construction of the LHC:

> A large hadron collider has always seemed an obvious option to follow LEP and it is clearly becoming time to start R&D on suitable magnets. It is less clear that it is sensible to discuss the physics that might be studied with such a machine [...] A theoretical consensus is emerging that new phenomena will be discovered at or below 1 TeV. There is no consensus about the nature of these phenomena but it is interesting that many of the ideas which have been suggested can be tested in experiments at an LHC. Although many, if not all, of these ideas will doubtless have been discarded, disproved, or established by the time an LHC is built, this demonstrates the potential virtues of such a machine (CERN 1984, pp 27 and 47).

Three years later, in 1987, in La Thuile, under the auspices of the Long-Range Planning Committee[6], chaired by Carlo Rubbia, the discussions regarding CERN's new machine became more focused. Three types of particle colliders with energies in the TeV region were discussed by almost 150 physicists, along with the feasibility of experimenting with them. The candidate machines were: a Large Hadron Collider (LHC) placed in the LEP tunnel, an electron–proton collider, which would utilize protons from the LHC and electrons from LEP, and an electron–positron linear collider (CERN 1987, p III). But there were hardly any new ideas concerning the experimental testing of theoretical predictions and there was still 'no consensus for the flowers of the desert'. The workshop involved the participation of two distinct groups of theoreticians: The 'standard theory group', focusing on physics within the framework of the Standard Model (SM), with the main emphasis on the search for the Higgs particle, and the 'beyond the Standard Model group', concentrating on ideas such as supersymmetry, Z' Physics, leptoquarks, and compositeness. It is worth noting that there was still a great deal of ambiguity in the discussions on

[6] It was a committee set up by the CERN Council in 1985 to explore various options for the long-term future of the laboratory.

whether the LHC would be in a position to clarify these theoretical claims. Interestingly, some theoreticians believed that even the least controversial theoretical prediction, that of the Higgs boson, could be validated with LEP, even before construction of the LHC had begun[7].

In the Aachen workshop in 1990, the LHC was the undisputed protagonist (figure 25.2). The key figure of this meeting, which involved 500 participants, was once again Carlo Rubbia, a staunch supporter of the LHC, who in the meanwhile had been appointed Director-General of CERN. The workshop represented a milestone for the strategy that CERN's management would follow for the construction of the LHC, laying out its preliminary physical, technical, as well as political aspects. Rubbia, in his opening talk, promoted the LHC as a multipurpose exploratory infrastructure, highlighting 'the need for several diversified and complementary experimental paths', as 'the very nature of experimental research does not allow one to predict beforehand the outcome of any single approach' (CERN 1990c, p 9).

The main focus of the discussions were the virtues of the LHC compared to existing colliders, the Tevatron and LEP, but also to its upcoming competitor, the SSC. Hence, the center of interest lay in the technical parameters of the LHC, which posed numerous challenges. In contrast, the theoretical considerations for building the machine do not appear to have become a decisive factor. Except for the mass of the Higgs particle which was rather uncertain, the top quark was thought most likely to be detected by the Tevatron—which, in fact, is what happened[8]—, while the testing of the various exotic ideas regarding 'beyond SM' physics remained doubtful. In the workshop's conclusions it was explicitly stated:

> In this area of physics a very large number of possibilities has been examined. Since no physics beyond the standard model has so far been experimentally confirmed it is difficult to judge the relative importance of the various possibilities (CERN 1990c, p 587).

The workshop would end with the prophetic remarks by George Kalmus that during the meeting, 'the LHC project graduated from being still speculative to being the way forward for European particle physics' (CERN 1990c, p 581).

By studying the initial technical surveys, the workshop proceedings and the minutes of the various councils and committees regarding the LHC's design, during the 1980s and 1990s, it can be argued that the collider was not conceived as a machine for the validation of a theory, or for the discovery of a single particle such as the Higgs boson, but as a multipurpose exploratory machine, a 'logical' next step in the chain of HEP accelerators and colliders that were reaching increasingly higher

[7] The mass of the Higgs could not be accurately determined at the time. As is mentioned in La Thuile's proceedings: 'It is quite plausible that the Higgs boson will be relatively light, say $m_H < m_{w/z}$, and that it will be discovered at LEP' (CERN 1987, p 37).

[8] The top quark was discovered at the Fermilab Tevatron Collider, by the CDF and DØ experiments in 1995.

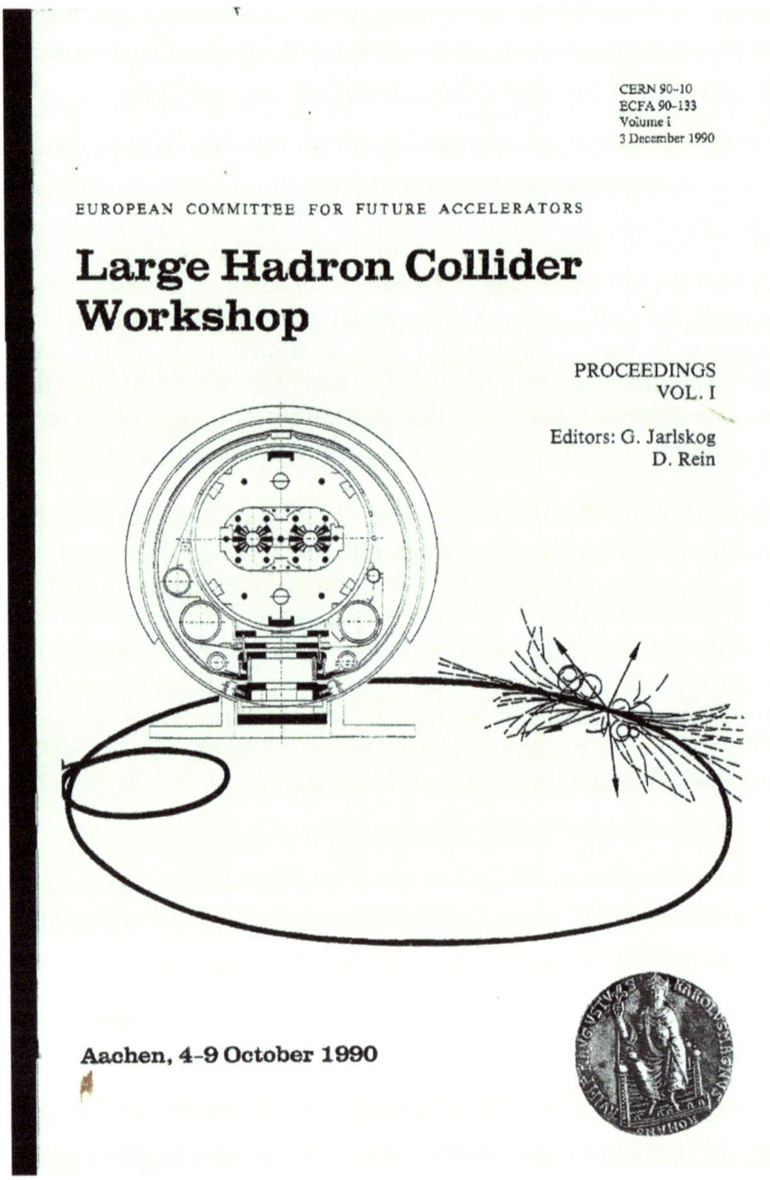

Figure 25.2. The front cover of the proceedings of the Aachen workshop in 1990. (Credit: CERN 1990c.)

energy frontiers. Of course, the role and function of an exploratory machine transcend the testing of specific predictions, which was a characteristic that further legitimized the LHC.

The fact that the design of the LHC began well before LEP had even started to operate is quite indicative. LEP was not an ordinary machine. It was designed and built as a unique engineering undertaking, becoming the pride and joy of CERN and of European HEP and holding the title, between 1983 and 1988, of the largest civil engineering project on European soil[9]. It was officially inaugurated on 13 November 1989 in the presence of some 1500 guests, including heads of state and ministers from all of CERN's 14 member states. The same day CERN issued a press release with the striking title 'LEP: the product of some 4000 man years contributed by over 500 European firms'. It was a majestic tribute for LEP, projecting it as 'the largest scientific instrument ever built' and 'the world's most powerful particle accelerator' (CERN 1989). Nonetheless, before the experimentalists had a chance to fully get the feel of the machine and certainly before they familiarized themselves with its potentialities, CERN, in December 1990, issued a press release for the LHC, which, in a way, invalidated the prospects of LEP that had been so panegyrically announced a year earlier:

> [T]he physics goals for a next generation of accelerator facility can be fully met by the LHC. In particular it is expected that the machine will reveal the clues to one of the most fundamental questions in our present understanding of the structure of matter. [...] there is a large community eager to prepare for and later to exploit the experimental opportunities of the LHC (CERN 1990b).

It is thought-provoking that the same grandiose, yet abstract, goals were stated, at the same time, as the main aspirations of LEP. More specifically, as we can read in the press release launched only a year earlier, LEP 'was conceived as the ideal scientific instrument to further knowledge of the nature of matter' (CERN 1989). It, thus, becomes clear that these basic and fundamental questions about nature, constitute grounds for the legitimization for all upcoming 'next generations' of machines, that will be aiming towards ever higher energy levels. Besides, the philosophical backbone of the HEP rhetoric is strongly reductionistic, which places the field in a privileged position as ideally suited for revealing the deepest truths of the Universe. As one of the most influential HEP figures, Steven Weinberg, claimed:

> We live in one world, and in trying to understand it we discover chains of explanation that cannot be followed farther without looking deeper into the physics of elementary particles (Weinberg 1988, 476).

Hence, it appears that the exigencies of a new collider were such that the smooth operation, the production of important experimental results[10] and the significant potentialities of existing scientific infrastructures were overshadowed by what the planned 'new' generation of machines promised. Necessarily any new generation machine has to have a long planning stage[11]. However, in the case of the LHC,

[9] For more details see Myers (1991).

[10] It is to be noted that CERN in September of 1990 issued a press release with the title 'LEP Successfully Concludes its First Year of Physics Experiments', where the richness of the first experimental results from LEP was described (CERN 1990a).

[11] The overlap between LEP and the LHC wasn't unprecedented in the history of CERN. For instance, it was only a few years earlier, in 1976, when studies for the design of LEP began the same year that the promising SPS (Super Proton Synchrotron) started its operation (Myers 1991, p 1)

local and international conditions were exerting so much pressure for its completion that LEP's demise seemed almost inevitable, irrespective of the results it would provide.

Therefore, despite the fact that the theoretical motivations regarding the LHC remained rather vague for quite a number of years, the construction of the LHC became a pressing necessity for CERN. For this reason, exceptionally delicate diplomatic maneuvers and an entire strategy of legitimization were articulated by CERN's management, with the purpose of keeping the project alive during some serious and complicated events. One of them was when it had to compete with the much more powerful SSC, which had been approved in 1987. The US machine became such a focus of attention for a few years that the dilemma of whether the European HEP community should remain attached to the LHC or join the American megaproject was intensely debated[12]. During this period, CERN's management attempted to promote the LHC as a complementary machine to the SSC. This managerial strategy stemmed from the realization that they were not in a position to compete with the American collider with respect to the attainable energy, since the SSC was planned to have an energy level of about 40 TeV, almost three times what the LHC could reach. In contrast, the LHC could prove competitive in another crucial experimental parameter, that of luminosity[13]. The complementarity of the machines would be a strategy for convincing the member states, while the creation of a vital space for HEP in Europe would remain on the agenda. Furthermore, the fact that the LHC could utilize the existing tunnel of LEP and its injector system made it a cheaper and more sustainable machine, though it lacked prestige when compared to its gigantic opponent. Indeed, during a special Council session on 19 December 1991 at CERN (figure 25.3), which was dedicated exclusively to securing a broader consensus regarding the LHC, the Director General Rubbia stated:

> [t]he press had over-exaggerated the competitive aspect of the LHC and SSC. That was a very superficial view, whereas a closer look at the scope of the two machines revealed a lot of complementarity. There was room for a machine with highest energy and for another that would go for the highest luminosity and which in addition would have other avenues (e-p collisions, heavy-ion collisions) to explore once SSC had started operation (CERN 1991c, p 5).

Even from the very early days, those involved in the planning of the LHC were worried about the SSC. For instance, in the workshop of La Thuile, the 'standard theory group' stated: 'Certainly the LHC cannot compete with the SSC. If the

[12] It is worth noting that Herwig Schopper, the DG of CERN at that time, traveled in 1987 to Washington to discuss a possible European participation in the SSC (Schopper 2009, p 188).

[13] This strategy was called the *energy vs luminosity trade*. As Luciano Maiani, the former CERN DG, arguued, the energy vs luminosity trade argument was not entirely convincing. The SSC could have increased its luminosity in a second time, while the LHC could not easily increase its energy, but it kept most of European physicists behind the LHC, until finally the SSC was cancelled' (Maiani and Bonolis 2017, p 482).

Figure 25.3. The first complete presentation of the LHC to the CERN Council at a special open session in December 1991. Sitting from left to right: Jean-Eudes Augustin, Chairman of the European Committee for Future Accelerators (ECFA), Carlo Rubbia, Director-General of CERN, Sir William Mitchell, President of CERN Council, Christopher Llewellyn Smith, Chairman of the Scientific Policy Committee of CERN (Credit: CERN PhotoLab.)

SSC is built, then the LHC must necessarily be operational from three to five years before the SSC' (CERN 1987, p 58). Nonetheless, CERN management achieved to remain in the race and consolidated the support of the European HEP scientists. The importance of this endeavor would become clear a couple of years later, when the SSC project would be canceled and the discussions and planning of the LHC would proceed within a totally new historical context. The well-articulated vision of CERN's management, the grasp of the political shifts that were underway, and an adherence to a strategy of conceptualizing and designing the machines with relative independence from theoretical and experimental developments were all factors that would further strengthen CERN's hegemonic position.

25.3 A new Europe and the LHC as a 'world machine'

The demise of the SSC in 1993 was a major drawback to the ambitions of the HEP community and a sudden realization that the historical context, within which Big Science grew so much both in scope and ambition, was rapidly going through a transformation. At the same time, a number of countries, part of the Socialist Bloc until 1990, were yearning to join the scientific, technological, political, cultural, financial, and industrial agenda of Europe—whatever content this term acquired by the different stakeholders.

The political situation accelerated the procedures for the realization of the LHC. The first big step in this direction was the approval of the project by the CERN

Council, the highest authority of the organization. What is particularly interesting with regard to this specific council is the fact that it is composed by both physicists and diplomats[14]. In this way, the scientists' agenda is intertwined with that of the diplomats', in a very particular mix of scientific and political goals.

The meetings of the CERN Council, during the following period, dealt with the details pertaining to the approval of the LHC and took place while the global HEP community was still in a state of reorienting its future plans. The LHC was by now without a competitor and the International Committee for Future Accelerators (ICFA) as well as the High Energy Physics Advisory Panel (HEPAP) in the US would end up becoming its ardent supporters. Canada, Japan, and Russia would soon follow, having started to change their national scientific policies in order to better align them with the prospective collider. A characteristic example was Canada, which 'announced in March [1994] that it was abandoning its Kaon project' to strengthen its 'role in particle physics through collaboration with CERN on the LHC' (CERN 1994b).

In 1993, Germany and the United Kingdom expressed their reservations to approve the requested increase in the budget that was essential for the construction of the LHC. The difficult and complicated negotiations with the representatives of the various countries would be handled with diplomatic mastery by Christopher Llewellyn Smith, the Director-General of CERN from 1994 to 1998. Thus, on 16 December 1994, the CERN Council reached a consensus and approved the construction of the LHC, on the basis of a two-stage construction plan.

A rather decisive aspect of the strategy for the realization of the LHC was the decision to transform the collider from a European to a world machine. The LHC became a symbol of the first large international collaborative scientific project following the Cold War, that would unite the HEP communities all over the world, from the US to Russia and from Canada to Pakistan. As Sir Mark Richmond, who represented the UK at the CERN Council, pointed out: 'in the present changing political and economic environment the Organization was inexorably moving from its present position as a European Laboratory located in Europe to a position where it would be a world laboratory located in Europe' (CERN 1993a, p 10). And furthermore:

> The LHC may become the first ever world-wide science initiative [...] The foundation of CERN in the postwar years set a precedent in uniting the nations of Europe to carry out high quality 'Big Science' research. The LHC now offers the exciting opportunity of establishing a model for future world-wide collaboration in 'Big Science' (CERN 1994b).

CERN's management was able to formulate a masterful roadmap: newcomers were welcome to use the machine and get involved in the relevant decision-making

[14] For more details see CERN (2021).

process, as long as they contributed financially towards its construction. As Llewellyn Smith underlined, 'CERN would welcome the valuable intellectual input from non-Member State physicists to the LHC programme, but that participation should be linked with a financial contribution to the project' (CERN 1993b). The press release, characteristically titled 'The LHC, A world project', was rather clear:

> CERN's governing body, the Council, is receptive to the idea of offering a voice in LHC decision-making to countries which provide substantial contributions. Such a 'globalization' of the LHC project would establish a precedent for future megascience projects, not only in particle physics but also in other fields (CERN 1994b).

Japan was the first to make a generous contribution towards the new machine (figure 25.4). Soon, other nations, such as India, Russia, and Canada, would follow. Last but not least would be the contribution on behalf of the US, which came in at a whopping $531 million (CERN 1997). This particular decision on the part of the US would also signal the end of an era, since, following decades of dominance in the field, this was the first time that it would contribute significantly towards the construction of an accelerator outside its territory.

Figure 25.4. The Japanese Minister for Education, Science and Culture, Kaoru Yosano, and CERN's Director-General, Christopher Llewellyn Smith, at the ceremony marking the start of collaboration between CERN and Japan in June 1995. At this ceremony, Minister Yosano presented Professor Llewellyn Smith with a wooden Daruma doll. In line with Japanese tradition, the doll was painted with one eye to mark the start of the LHC project and the second eye would be added upon the project's completion. (Credit: Bulletin Issue: 08/2005, Mon 21 Feb 2005, Photograph: Laurent Guiraud.)

Another central feature of the LHC was its role as a catalyst for the integration of the 'new Europe', which was emerging in the post-1989 developments. Europe, being in constant search of its identity, with its new borders being a 'grey area', as the Chairman of Council Sir William Mitchell described them, was looking for ways to establish itself in the global map[15]. CERN had the unique potential to contribute towards such a goal. For Poland, the Czech Republic, Hungary, and Slovakia, which were seeking to re-negotiate their relations with the Western world, CERN became the accessible and obliging antechamber for such an integration in the reborn European Union[16]. The LHC provided a great opportunity for these states for a series of reasons. First, by constructing parts of the new collider, they were strengthening appreciably their technological and industrial sectors. Second, their scientific communities had the chance to participate, *en masse*, in the new international experimental collaborations, by working on a cutting-edge experiment alongside their western colleagues. Scientists from those countries, through their network of contacts, communicated in a common language and shared similar aspirations. Quite a few had some earlier collaborations with CERN as well. Third, they became players in European science diplomacy, by participating in an experiment which had important geopolitical extensions. Therefore, the LHC was not only CERN's next collider, but also an enterprise that could help to splice together the fragmented scientific, diplomatic, industrial, and financial European space in the aftermath of USSR's dissolution and the disintegration of the Eastern Bloc, and become a carrier for the circulation of knowledge, technology, and workforce. In other words, the LHC became an active, dynamic part in what Thomas Misa and Johan Schot called the 'hidden integration of Europe'[17]. Such a prospect was greatly facilitated by the European Community, the forerunner of the EU, which had begun since 1985 to involve itself in a more direct and official way with CERN. As we read in the press release in February 1991 with the title 'Stronger Links Developing Between the European Community and CERN':

> Collaboration between CERN and the EC is longstanding [...] This cooperation grew stronger during the early 1980s, and was given a new impetus in 1985. Since then the Commission regularly attends meetings of the CERN Council, the governing body of the Organization, as an Observer and the Commission has taken an active interest in the

[15] The complete quote by Sir William Mitchell was: '[...]CERN was a European Laboratory although the borders of Europe was a grey area. At the same time one should be conscious that high-energy physics was moving towards world-scale machines and that CERN should evolve in that direction.' (CERN 1993a, p 12).

[16] The European Union was formally established when the Maastricht Treaty came into force in November 1993.

[17] Misa and Schot, using the term 'hidden integration of Europe', attempted to describe European integration not only as a political process limited to treaties, but also as an emergent outcome of a process of linking and delinking of infrastructures, as well as a phenomenon that is shaped by material networks, technical systems, and the circulation of knowledge and artifacts. CERN, from our perspective, has been a major part of this mechanism. For more details, see Misa and Schot (2005).

> development of CERN's scientific and administrative programmes and in strengthening links between the Organizations (CERN 1991b).

Among the many proclamations expressed by the new states joining CERN, Poland's Vice Minister for Foreign Affairs Antoni Kuklinski's address was rather characteristic:

> The admission of Poland to CERN has a triple importance for our country, scientific, political and industrial [...] there is a process of integration in Eastern Europe and our country's membership of CERN is not only a Polish occasion but also has an important European dimension (CERN 1990b).

Likewise, the Hungarian representative, György Vesztergombi, stressed 'the importance which the European character of the Laboratory had for Hungary which, apart for [sic] its scientific merit, valued the Organization as a step towards the European community' (CERN 1993a, p 11)[18]. All in all, Europe was yearning for a dramatic symbol through which it could proudly present its new dynamic to the rest of the world, and the LHC fitted that role perfectly.

Finally, in December 1996, a single stage construction of the LHC was approved by the CERN Council, on the basis of the contributions offered by non-Member States and a large loan that the organization would take out (Smith 2015, p 9). Yet, this was not to be the final act in the LHC saga, as a series of troublesome issues would arise before it finally became operational: intense negotiations between member and non-member states, a loan from the European Investment Bank, tensions between the experimenters and CERN's management regarding the indications for the existence of the Higgs particle at LEP, and a magnet quench incident that caused extensive damage to the accelerator's superconducting magnets[19].

25.4 Conclusions

The history of the LHC has become a paradigm of the ways in which the world of HEP is entangled with those of geopolitics, industry, finance, and diplomacy. As the former CERN Director-General and a key figure behind the establishment of the LHC, Llewellyn Smith, put it: 'Approval of a project the size of the Large Hadron Collider is an exercise in politics and high finance' (Smith 2007, p 281). In order to grasp the full extent of the collider's significance—of a much larger scope than that of a scientific instrument—, we have attempted to place it within a wider historical context. The characteristics of this context, which continually re-configured the role of the LHC, are the following.

[18] Similar views were also expressed by the Russian representative. See (CERN 1991a).

[19] See also Maiani and Bonolis (2017).

First, following the demise of the SSC, the LHC represented the future prospects of the entirety of the global HEP community. From this perspective, the construction of the LHC was intertwined with the continuation of the community into the twenty-first century, given the symbiotic relationship between the field and the constantly growing colliders. Even CERN's very existence was on a shaky ground without the prospect of the LHC. During a critical meeting of the Council in June 1994, which approved the LHC project, the physicist Walter Majerotto, representing Austria, stated 'that a decision on the LHC appeared to be crucial to CERN's survival', while the Swiss delegate, Jacques Vernet, argued that 'the Council would be invited to take a decision that went far beyond the LHC machine since the Council would either choose a future for CERN or it would programme its gradual closure' (CERN 1994a, pp 10 and 18).

Second, the LHC was one of the first scientific projects of the post-Cold War period, and would function as a prototype for world-wide collaborations and the transformed Big Science of the new era. Its accomplishment would also express new kinds of relationships between the financial institutions, industry, and state bureaucracies.

Third, in the emergence of the post-1989 'new Europe', the LHC would play a crucial role in the European integration. The new states would be given the opportunity, through the LHC, to participate in the construction of an astonishing scientific machine, based on cutting-edge technology. In this way they could construct scientific, industrial, and diplomatic networks and strengthen their relationship, both symbolically as well as substantially, with western Europe. The pressure and impetus to bring the LHC to fruition were such, that in order to circumvent the various financial problems that arose, mainly owing to the continuous demands raised by Germany and the UK to reduce the budget, CERN's management decided to shut down a series of ongoing noteworthy experimental programs, such as the LEAR (Low Energy Antiproton Ring)[20]. Llewellyn Smith described the closure of these facilities as 'sad news, but the price we must pay for the LHC' (CERN 1996).

Fourth, the demise of the SSC was undoubtedly an appreciable boost for the LHC. Yet, it brought to the surface two different approaches to the issue of world hegemony by the USA and Europe. By 1993 the former was confident that it won the Cold War and was the incontestable leader of the 'free world'[21]. In contrast, almost all the European leaders believed that there cannot be any European hegemony unless the countries formed after the events of 1989 will be assimilated in the (West) European family. It is this difference from the USA, that allowed CERN management to maximize the advantages of the LHC after the scrapping of the SSC.

---

[20] It is characteristic that during the Council meeting on 22 June 1995, the DG, Llewellyn Smith, responding to the pressure from the scientific community to run LEAR for a further year in 1997, informed the Committee that the Research Board 'had reluctantly concluded that the physics case, though excellent, was not commensurate with the considerable resources required' (CERN 1995, p 3).

[21] Cf. Kevles (1997) and Riordan (2000).

CERN's management did not simply read the political developments as events which would greatly facilitate the planning of an uninterrupted path for the final construction of the LHC. It also understood that, through the LHC, it had the means of boosting the new role of Europe in the era of globalization, while simultaneously becoming the rallying point of the HEP community all over the world and the uncontestable point of reference for the future of the discipline. It was in this context that the LHC was transformed from a machine at CERN to a 'world project' (CERN 1994b).

Acknowledgements

We are grateful to the editors of this volume, Theodore Arabatzis and Panagiotis Charitos, for their insightful comments and their suggestions on developing the paper into its present form. We are also thankful to Stathis Arapostathis, Vasiliki Christopoulou, Stelios Kampouridis, Iraklis Katsaloulis, Telly Tympas, Emmanuel Floratos, Vally Athanasopoulou and Themistocles Chalikias for helpful discussions. Furthermore, we are indebted to the CERN Director for International Relations, Charlotte Lindberg Warakaulle, for granting us access to the CERN archives, and to the CERN archivists, Anita Hollier, Jens Vigen, and Sandrine Reyes for their substantial help in navigating those archives. GP's research work was supported by the Hellenic Foundation for Research and Innovation (HFRI) under the 'First Call for HFRI Research Projects to support Faculty members and Researchers and the procurement of high-cost research equipment grant' (Project Number: 875).

References

CERN 1984 *Proc. ECFA-CERN Workshop on Large Hadron Collider in the LEP Tunnel (Lausanne, Switzerland, 21–27 March)* CERN-84-10-V-1, CERN-84-10-V-2

CERN 1987 *Proc. of the Workshop on Physics at Future Accelerators (La Thuile, Italy, 7–10 January)* CERN87–07-V-1

CERN 1989 LEP: the product of some 4000 man years contributed by over 500 European firms *CERN Press Release* PR 23.89

CERN 1990a 91st Session of the CERN Council *CERN Press Release* PR 13.90

CERN 1990b LEP successfully concludes its first year of physics experiments *CERN Press Release* PR 07.90

CERN 1990c *Proc. of the Large Hadron Collider Workshop (Aachen, Germany, 4–9 October)* CERN90–10-V-3

CERN 1991a Meeting of CERN governing bodies *CERN Press Release* PR 11.91

CERN 1991b Stronger links developing between the European community and CERN *CERN Press Release* PR 02.91

CERN 1991c 93rd Session of CERN Council *Special LHC Session* CERN/1914/

CERN 1993a 210th Committee of CERN Council *Minutes* CERN/CC/1990

CERN 1993b Large Hadron Collider is presented to CERN Council *CERN Press Release* https://home.cern/news/press-release/cern/large-hadron-collider-presented-cern-council (Accessed: 13 January 2021)

CERN 1994a 215th Committee of CERN Council *Minutes* CERN/CC/2051

CERN 1994b The LHC, a world project *CERN Press Release* https://home.cern/news/press-release/cern/lhc-world-project (Accessed: 13 January 2021)
CERN 1995 219th Committee of Council *Minutes* CERN/CC/2100
CERN 1996 LHC to be completed in 2005 within reduced CERN budget envelope *CERN Press Release* https://home.cern/news/press-release/cern/lhc-be-completed-2005-within-reduced-cern-budget-envelope (Accessed: 13 January 2021)
CERN 1997 US to contribute $531 million to CERN's Large Hadron Collider project *CERN Press Release* https://home.cern/news/press-release/cern/us-contribute-531-million-cerns-large-hadron-collider-project (Accessed: 13 January 2021)
CERN 2021 Our governance *CERN* https://home.cern/about/who-we-are/our-governance (Accessed: 30 September 2021)
Cramer K 2020 *A Political History of Big Science: The Other Europe* (Palgrave Macmillan)
Kevles D 1997 Big Science and big politics in the United States: reflections on the death of the SSC and the life of the human genome project *Hist. Stud. Phys. Biol. Sci.* **27** 269–97
Maiani L and Bonolis L 2017 The LHC timeline: a personal recollection (1980–2012) *Eur. Phys. J.* H **42** 475–505
Misa T and Schot J 2005 Inventing Europe: technology and the hidden integration of Europe *Hist. Technol.* **21** 1–19
Mobach K and Felt U 2022 On the entanglement of science and Europe at CERN: the temporal dynamics of a coproductive relationship *Sci. Cult.* **31** 382–407
Myers S 1991 The LEP collider, from design to approval and commissioning *Report* CERN-91-08
Riordan M 2000 The demise of the superconducting super collider *Phys. Persp.* **2** 411–25
Schopper H 2009 *LEP—The Lord of the Collider Rings at CERN 1980–2000* (Berlin: Springer)
Smith L C 2007 How the LHC came to be *Nature* **448** 281–4
Smith L C 2015 Genesis of the Large Hadron Collider *Philos. Trans. Royal Soc.* A **373** 20140037
Weinberg S 1988 The limits of reductionism *Nature* **331** 475–6

IOP Publishing

Big Science in the 21st Century
Economic and societal impacts
Panagiotis Charitos, Theodore Arabatzis, Harry Cliff, Günther Dissertori, Juliette Forneris and Jason Li-Ying

Chapter 26

Digital fire: the origins and global impacts of the World Wide Web

Michael Riordan

The World Wide Web is by far the most impactful contribution of high-energy physics to the global economy and culture. Developed at the European Center for Particle Physics (CERN) in 1989–91 by Tim Berners-Lee as a universal system for computers to exchange information via the Internet, it was soon adopted at other high-energy physics laboratories in Europe and the United States. At the Stanford Linear Accelerator Center (SLAC), Paul Kunz established the first US Web site in late 1991, and Tony Johnson programmed an early Web browser. The Internet and Web together provided a flexible, powerful information-sharing system onto which new, innovative contributions could be added easily by other scientists and engineers. With the advent of the Mosaic browser in 1993, the Web became fully globalized, truly deserving of the name World Wide Web.

I survey the early history of the Web, particularly its development at CERN and extensions at other high-energy physics labs, in particular SLAC. Then I recount how it became commercialized during the mid-1990s after Mark Andreessen developed the graphics-enhanced Mosaic browser, soon followed by Netscape Navigator. Other major applications such as the Yahoo! portal and Google search engine are briefly reviewed.

The economic and cultural impacts of the Internet and Web are vast and probably incalculable, likely worth tens of trillions of dollars. They stand at the pivot of the current wave of globalization, which began in the wake of the 1991 collapse of the Soviet Union and end of the Cold War. Modern life has since been radically transformed, with diverse information exchanges moving from paper to electronic forms of communication. Direct interactions have begun replacing the previously dominant newspaper, magazine, and broadcast information sources. The cultural and political ramifications of this process of 'disintermediation' are both positive and negative, contributing to the polarization, populism, and nationalism currently gripping the globe.

doi:10.1088/978-0-7503-3631-4ch26

26.1 Introduction

From 1989 to 1991 one world was dying, and another was quietly being born at high-energy physics laboratories in Europe and the United States. Despite the brutal suppression of Chinese student demonstrations in Tiananmen Square on 4 June 1989, pressure for democracy continued unabated in Central and Eastern Europe. The landslide electoral victory of Solidarity in Poland that very same day was followed in August by Hungary's opening of its Austrian border and by the dismantling of the Berlin Wall in November 1989. The disintegrating bipolar world order of the Cold War finally came to an end officially on 26 December 1991, the day after Mikhail Gorbachev resigned as its president, when the Supreme Soviet met in Moscow to dissolve the Union of Soviet Socialist Republics (for a behind-the-scenes account of the geopolitics of this transformation, see [1]).

During the same period, scientists at CERN and in other high-energy physics laboratories were pioneering the computer communications system that would eventually connect the farthest reaches of the globalized world, helping to replace the bifurcated one in existence since the end of World War II. Based on the Internet, which had reached CERN the previous January, British computer programmer Tim Berners-Lee outlined what became known as the World Wide Web in a March 1989 memo titled 'Information management: a proposal' [2]. Working at CERN with Belgian computer engineer Robert Cailliau the following year, he developed the core software for such a system. And in December 1991, two weeks before the Soviet Union ended, US physicist/programmer Paul Kunz established the first Web server in North America—in fact, the first one beyond Europe—at the Stanford Linear Accelerator Center in Menlo Park, California [3]. As this server allowed easy access to SLAC's popular SPIRES database of high-energy physics preprints and other publications, one can reasonably assert that Berners-Lee's innovative computer communications system actually became world-wide in December 1991, just as the Soviet Union was disintegrating.

In this article I survey the origins of the World Wide Web from 1989 to 1992, when it was largely in the hands of computer programmers and physicists at high-energy physics labs in Europe and the United States. I then cover the initial commercialization of the Web from 1993 to 1995, when easy-to-install Web 'browsers' Mosaic and Netscape Navigator led to explosive growth in its servers and users. The Web had by then become a supple, flexible system for rapid software innovation. I conclude with a discussion of the tremendous impact the Internet and Web have had on modern culture, economics and politics—in particular their pivotal role in enabling the current wave of globalization.

26.2 Weaving the Web at high-energy physics labs, 1989–92

During the 1980s, the growing diversity of computer networks became consolidated into a single, all-encompassing 'network of networks' called the Internet. BITNET, Cyclades, DECnet, NSFNET, and other early networks built on the pioneering work of Robert Taylor, Paul Baran, Donald Davies, Vinton Cerf, Robert Kahn, and others on what came to be called the ARPANET (James Gillies and

Robert Cailliau provide a good overview of the ARPANET in their 2000 book [4, pp 11–33]). Emerging during the depths of the Cold War, this was a project of the US Defense Department's Advanced Projects Research Agency (ARPA and, later, DARPA) to construct a decentralized, nationwide communications network that could survive major disruptions (such as would occur in a nuclear war) and continue to function. All these networks used the technology of 'packet-switching', in which digital messages between remote computers are chopped up into separate packets that are then routed through diverse communications channels (including telephone lines) and reassembled at their eventual destination.

During the 1980s, following the lead of ARPANET, networks increasingly adopted its Transmission Control Protocol and Internet Protocol (TCP/IP); these two protocols permitted computer communications across other networks and therefore allowed the Internet to spread widely [4, pp 42–3]. It fostered nationwide and eventually world-wide email, which had previously been available to a lesser extent on more limited networks such as BITNET. Europe was a latecomer to the Internet, remaining isolated from it until 1988 (although the French Cyclades network was based on these ARPANET protocols). CERN, for example, used TCP/IP for its internal computer networks but opened external connections to the Internet only in January 1989 [5].

In retrospect, CERN was the ideal site for the germination and birth of the Web, as well as the nursery for its earliest growth. During the 1980s physicists came there by the thousands (as they still do) from all over Europe—and from other nations including India, Japan, and the United States. They increasingly brought with them their personal computers, which communicated with one another through a great variety of systems and networks set up by the various experimental collaborations, some then numbering in the hundreds of members. Most would return to their home institutions, but they still needed computerized communications with their colleagues at the lab and in other institutions.

'CERN is a model in miniature of the rest of the world in a few years time', observed Berners-Lee in his proposal (see figure 26.1). The lab 'meets now some problems the rest of the world will have to face soon. In ten years there may be many commercial solutions to the problems, while today we need something to allow us to continue' [2]. His solution was to marry the Internet with 'hypertext', an emerging software technology that linked text on computer screens to related information and data, which are typically accessed by point-and-click techniques. 'Imagine, then, the references in this document, all being associated with the network address of the thing to which they referred', he envisioned, 'so that while reading this document you could skip to them with the click of the mouse' [2].

But hardly anyone at CERN was listening in 1989–90. Thus Berners-Lee submitted a revised proposal in May 1990, again with few reactions. One person did pay attention, however—Robert Cailliau, who had been working at CERN since the 1970s. Then the head of the Office Computing Systems in the Data Handling Division, he was also very interested in hypertext. And he had much more experience in negotiating for CERN resources than Berners-Lee, who had been working only as a consultant or on a fellowship until the mid-1980s. They came up

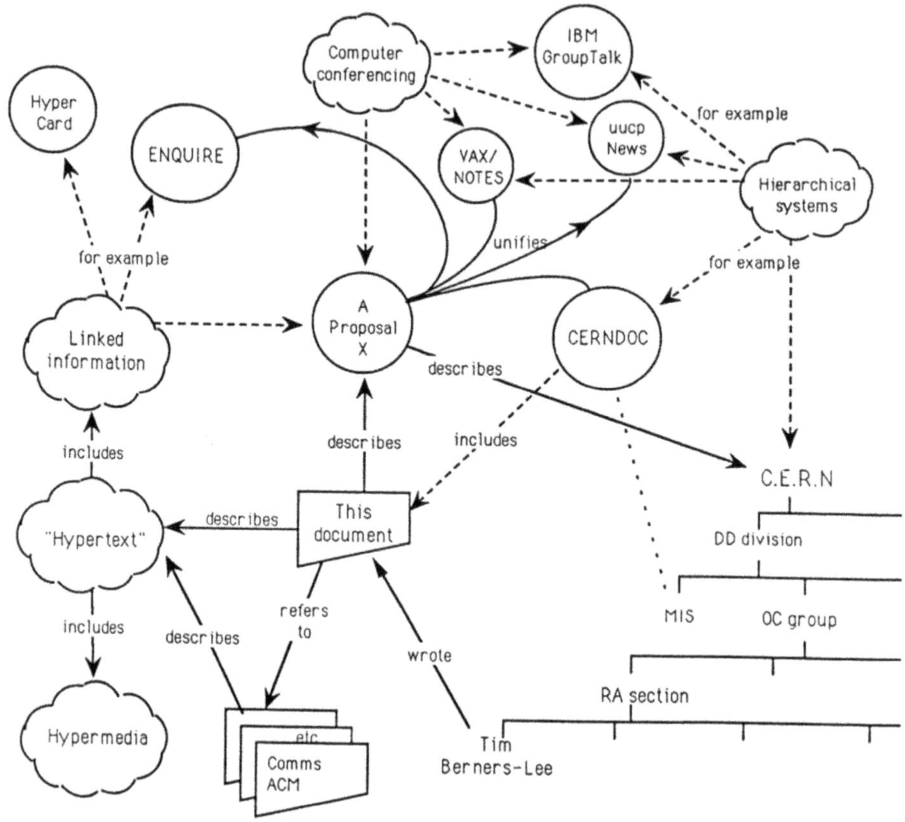

Figure 26.1. Diagram of data and information flows presented in Berners-Lee's March 1989 memorandum, 'Information management: a proposal'. (Image credit: CERN.)

with an audacious name for their project, dubbing it the 'World Wide Web', and in October 1990 they issued a third, much more detailed proposal, 'World Wide Web: proposal for a hypertext project' [6]. As Berners-Lee wrote in his book, *Weaving the Web*, 'In the marriage of hypertext and the Internet, Robert was best man' [7, p 26].

Given the green light by his boss, he began developing the Web that month on a NeXT computer (an innovative workstation developed by the firm founded by Steve Jobs after leaving Apple Computer Company in 1985). In an amazing burst of creativity, Berners-Lee programmed the hypertext transfer protocol (HTTP), hypertext mark-up language (HTML), and universal document identifier (UDI, changed later to URL for uniform resource locator) in only a few months. HTTP is the language that computers use to communicate hypertext over the Internet and the UDI (currently URL) represents the individual Internet addresses of information resources. HTML determines the on-screen format of pages that include hypertext links. One computer acts as the 'server', providing the required digital information, while other computers act as 'clients' or 'browsers' that seek this information in this 'client–server' approach. By Christmas 1990, Berners-Lee and Cailliau had the system up and running on their NeXT computers, communicating over the Internet with the server info.cern.ch—an alias for the world's first Web server nxoc01.cern.ch, which resided on Tim's machine (see figure 26.2) [7, p 30].

But NeXT computers were extremely rare at CERN in the early 1990s. At an initial purchase price of $6500 (equivalent to about $15 000 today), they were beyond the financial capabilities of nearly all individual scientists. And CERN's computing division had not officially approved them, which made it very difficult to impossible for groups to purchase them. So Berners-Lee and Cailliau encouraged a young CERN intern, Nicola Pellow, to program what they called a 'line-mode browser'—a much simpler, text-based interface to the Web than the elaborate point-and-click 'graphical user interface' (or GUI) that came bundled with NeXT workstations. This browser could be easily loaded onto many of the computers common at CERN', according to Berners-Lee, so that 'no matter what machine someone was on, he would have access to the Web' [7, p 32].

He also succeeded in getting the CERN phone book uploaded onto a Web site, which provided easy access to an information resource that nearly everyone at the lab needed almost daily. 'Mundane as it was', recalled Berners-Lee, 'this first presentation of the Web was, in a curious way, a killer application' [7, p 33]. Slowly but surely, the people who worked at CERN—including the resident staff members and thousands of visiting scientists from Europe and around the world—were becoming familiar with the new communications technology via the then-rapidly spreading Internet.

In September 1991 SLAC physicist Paul Kunz made a fateful visit to CERN on his way back from Sweden. A NeXT user himself, he was also an early advocate of employing the Internet for communications in high-energy physics. Berners-Lee

Figure 26.2. Robert Cailliau and Tim Berners-Lee (seated in foreground). Just behind Berners-Lee sits Nicole Pellow, who programmed the Web's first line-mode browser. (Image credit: CERN.)

sought him out, inviting him to his office on September 13 to view a demonstration of the new software. Kunz was impressed with how easily and rapidly they could access his NeXT computer at SLAC. 'I was really excited', he recalled. 'I told Tim I was going to put SLAC's SPIRES database on the Web as soon as I got home' [4, p 219].

The Stanford Public Information Retrieval System had a huge database that consisted largely of high-energy physics 'preprints'—advance copies of (not yet peer-reviewed) publications that had usually been submitted to physics journals or conference proceedings. Maintained by the SLAC library on the lab's mainframe computer, it had become 'the most valuable information resource in the world of particle physics', wrote Gillies and Cailliau [4, p 218]. But accessing it from beyond SLAC had been difficult. Kunz (with help from George Crane) had tried to solve this problem by developing a computer interface called QSPIRES that functioned on BITNET, but that had experienced only halting success. So he was attuned to the need for a better solution, especially because the Internet was then making that earlier network obsolete.

Soon after he returned, Kunz visited SLAC librarian Louise Addis. 'I've just been at CERN and found this wonderful thing a guy named Tim Berners-Lee is developing', he told her. 'It's called the World Wide Web, and it's just the ticket for what you guys need for your database' [4, p 219] (see figure 26.3). After demonstrating the Web to her on his NeXT workstation, he asked whether she'd like him to set up a server on the SLAC mainframe computer. 'Yes!' Addis eagerly replied. 'By all means' [4, p 219].

Figure 26.3. Several of the leading WWW Wizards who promoted the Web at SLAC: (left to right) Louise Addis, George Crane, Tony Johnson, Joan Winters (standing), and Paul Kunz (seated). (Credit: SLAC.)

But Kunz somehow was not interested enough to do this work himself; he passed the job off to a SLAC programmer, Terry Hung. By early December the promised Web site wasn't up on the SLAC computer yet, so Addis and Berners-Lee pressed Kunz back into service to finish the task. On Thursday, 12 December 1991, the first Web server outside of Europe slacvm.slac.stanford.edu went online, making SPIRES available throughout the world via the Internet. Less than 24 h later, Berners-Lee posted a note about the new SLAC server to the WWW interest group on the Internet: 'There is an experimental W3 server for the SPIRES High Energy Physics preprint database, thanks to Terry Hung, Paul Kunz, and Louise Addis at SLAC. It's only just been put up, so don't expect perfection' [8].

Exactly two weeks later, the Soviet Union was officially liquidated.

By year's end, the Web was truly becoming a world-wide phenomenon, as had been advertised in its name. There were about ten servers in existence in Europe and the United States, almost all at high-energy and nuclear physics labs, and SPIRES was becoming the Web's killer application. 'A world of information is now available online from any computer platform', announced Berners-Lee in the October–December issue of CERN's computing newsletter. 'Information sources at CERN and across the world span subjects from poetry to biochemistry and supercomputing' [9], p 5.

— — —

Interest in the Web exploded in 1992. In the first two months, for example, the number of documents downloaded from info.cern.ch rose from a flat level of about 4000 in November and December 1991 to more than 12 000 in February 1992 [4, graph on p 224]. Part of the reason for the surge was that Berners-Lee and Cailliau hit the lecture circuit to promote their globe-spanning information system, beginning with a Web demonstration at the San Antonio Hypertext '91 conference in mid-December. At that meeting, Berners-Lee was able to feature links to SPIRES in his demonstration.

Another pivotal Web presentation featuring SPIRES came that February at a workshop in southern France on artificial intelligence in high-energy physics. Computer scientist Jonathan Streets returned to the Fermi National Accelerator Laboratory (known widely as Fermilab) west of Chicago, IL, telling people that the Web was 'the best thing around' in online data-acquisition systems [10]. When Berners-Lee visited Fermilab that June at the invitation of Ruth Pordes, head of the Online Systems Department, he and Streets set up a Web server—the second site in the United States—and linked it to the CERN site [11].

Berners-Lee was on leave from CERN and in the United States for three months beginning that June, first at the MIT Laboratory for Computer Science and later at the Xerox Palo Alto Research Center (PARC) [7, pp 60–4]. MIT set up a Web site that month, at about the same time as did Fermilab. While at MIT, he attended a gathering of the International Engineering Task Force in Boston, at which members decided (against his preferences) to rename the Web's 'universal document identifier' the 'uniform resource locator' (URL)—as it's recognized today.

Xerox PARC was the institution where the first graphical user interface had been developed. Witnessed there by Steve Jobs, it was adapted by Apple Computer Company in 1983–84 for its innovative MacIntosh personal computer. While he was in the San Francisco Bay Area, Berners-Lee would occasionally drop by SLAC and interact with Addis, Kunz and other SLAC Web pioneers, who had dubbed themselves the 'SLAC WWW Wizards' [12] (see figure 26.3). One of the wizards he encountered there, high-energy physicist Tony Johnson, was developing a flexible browser for Unix-capable computers that he called Midas (or MidasWWW).[1] Like the browser Berners-Lee had developed for his NeXT workstation, it could import and display images in a separate window. It was becoming popular at SLAC, but Johnson was initially reluctant to make it generally available, as 'he didn't like the idea of supporting Midas for a group any wider than that of his colleagues' [7, p 64].

Berners-Lee returned to CERN that September, just in time to attend the 1992 Computing in High-Energy Physics Conference in Annecy, France. After that meeting, the World Wide Web became increasingly well known and popular among members of this global scientific community [10]. He and Cailliau gave Web demonstrations there and a prominent lecture titled 'World-Wide Web'. In that lecture they showed a map of the rapidly growing number of Web sites existing in Europe and North America, plus others planned in Asia and Australia—about 20 in all [13]. Their presentation was the principal highlight for US physicist Terry Schalk, who in summarizing the meeting remarked that 'if there is one thing everyone should carry away with them from the conference, it is the World Wide Web' [4, p 223, 10, 14].

Schalk urged the 200-odd conference participants to install servers at their home institutions. And many of them did. As of November 1992, there were at least 26 Web servers in existence—most of them at high-energy and nuclear physics laboratories in Europe and the United States [4, p 233]. The Web was then almost two years old, still very much a toddler.

26.3 Commercializing the Web, 1993–95

As 1993 began and Bill Clinton moved into the White House, the World Wide Web community was confined largely to academic institutions and government or industrial laboratories. Users often employed browsers based on the Unix operating system to access local and remote information. But this was about to change abruptly that year. Together with Clinton came Vice President Al Gore, an early political champion of the Internet [15]. As a Tennessee Congressman and Senator, he had foreseen the economic, social, and cultural impacts of what he (and others) began calling the 'information superhighway' during the 1992 elections. Before that, Gore had sponsored the High Performance Computing Act of 1991 [16], which was passed by Congress and signed by George H W Bush on 9 December 1991— just three days before Kunz set up the first US Web site at SLAC.

In part, this Act helped fund the National Science Foundation's National Center for Superconducting Applications (NCSA) at the University of Illinois in Urbana-Champaign. There a young computer-science undergraduate student named Marc Andreessen became enamored of the Web in November 1992 while he was working

[1] Unix is a family of computer operating systems developed by Bell Labs during the 1970s that became popular in academic and government circles during the 1980s. It featured a powerful graphical user interface called X-Windows.

on computer visualization in NCSA's software development group. On November 16, Tony Johnson released his MidasWWW browser on the FREEHEP site, announcing it to the www-talk newsgroup that Berners-Lee had organized in 1991 [17].

Among the first to download the software from that site and try it out, Andreessen emailed Johnson the very next day. 'MidasWWW is superb!' he exclaimed. 'Fantastic! Stunning! Impressive as hell!' (Quoted in [4, p 239]. The ensuing email conversation between Andreessen and Johnson is quoted in detail on that page.) In subsequent emails, he suggested they collaborate on developing a much-improved version incorporating graphic files, animations, and other advanced features while fixing a number of software bugs that he had found. Intrigued at first, Johnson thought twice about how such an intense programming activity could conflict with his regular job as a SLAC physicist, and turned down the offer—a decision he would later regret [4, p 239].

Andreessen instead teamed with NCSA staff member Eric Bina to develop their own X-Windows browser for Unix-based computers. After a feverish two months of round-the-clock programming, he released what he called NCSA X Mosaic 0.5 on 23 January 1993, three days after the Clinton–Gore inauguration [18]. 'Brilliant!' Berners-Lee posted on the www-talk newsgroup two days later [19]. 'Having the thing self-contained in one file makes life a lot easier .... X Mosaic has a lot of practical things which real users want—like mail and print and bookmarks and stuff. Well done!'

Over the next two months, Andreessen and Bina produced six additional versions of X Mosaic in rapid succession. By early March, their software could include in-line, embedded graphics (from JPEG and GIF files) on computer screens for the first time—rather than requiring a separate window to be opened to view them. This was a pivotal salient into commercial territory. Whereas earlier Web pages had been dominated by text (with underlined text used for hypertext links), they could now begin to look like glossy magazine pages. Soon graphics could serve as links, too. According to Cailliau, graphics features fulfilled the NCSA criterion of 'being somewhere in between academic and commercial software' [4, p 240]. That April, NCSA officially released X Mosaic version 1.0.

But Unix-based X-Windows operating systems were largely the domain of the academic community, plus government and industrial laboratories. To commercialize the Web, MacIntosh and IBM-PC browsers were required, too. With Nicola Pellow, Cailliau had been working on a Mac browser on and off since late 1991, and a group of students at CERN was developing an IBM-PC browser [4, p 240]. These browsers were finally released in early 1993. But their efforts had suffered from lack of high-level support—getting the CERN management to take the World Wide Web project seriously enough to devote the needed resources.[2]

In September 1993 (just weeks before the United States Congress terminated the Superconducting Super Collider), NCSA released Mosaic browsers for Mac and

[2] CERN was (and is) focused on doing world-class physics, not computer science (which also happens there), and in the early 1990s the expensive, multibillion-euro Large Hadron Collider project loomed large in management thinking and planning.

IBM-PC computers to great fanfare (see figure 26.4). These easy-to-install (and to use) browsers, with their superior graphics capabilities, were soon being downloaded by thousands of personal computer users every month. 'We released our Mac browser first', Cailliau recalled somewhat ruefully, 'but when they came out with theirs it was obvious we didn't stand a chance' [4, p 242].

New York Times technology writer John Markoff was among the first to pick up on the news among mainstream reporters. 'A new software program available free to companies and individuals is helping even novice computer users find their way around the Internet', he declared on page 1 of the Business Day section. 'Since its introduction earlier this year, the program, called Mosaic, has grown so popular that its use is causing data traffic jams on the Internet' [20, p D1].

Marc Andreessen is not mentioned anywhere in the article, however, and Berners-Lee doesn't appear until the fourteenth paragraph, on an inside page: 'The Web is an international string of computer data bases that uses an information-retrieval architecture developed by Tim Berners-Lee, a British computer specialist at the CERN physics laboratory in Geneva' [20, p D5]. Markoff portrays Mosaic as the people's gateway to the Internet.

In reality, the Mosaic browsers and Web together provided easy on-ramps to the much-touted information superhighway for the rapidly expanding legions of personal computer users. With such a burgeoning audience now available, the number of Web servers exploded. By year's end, even the White House had one.

— — —

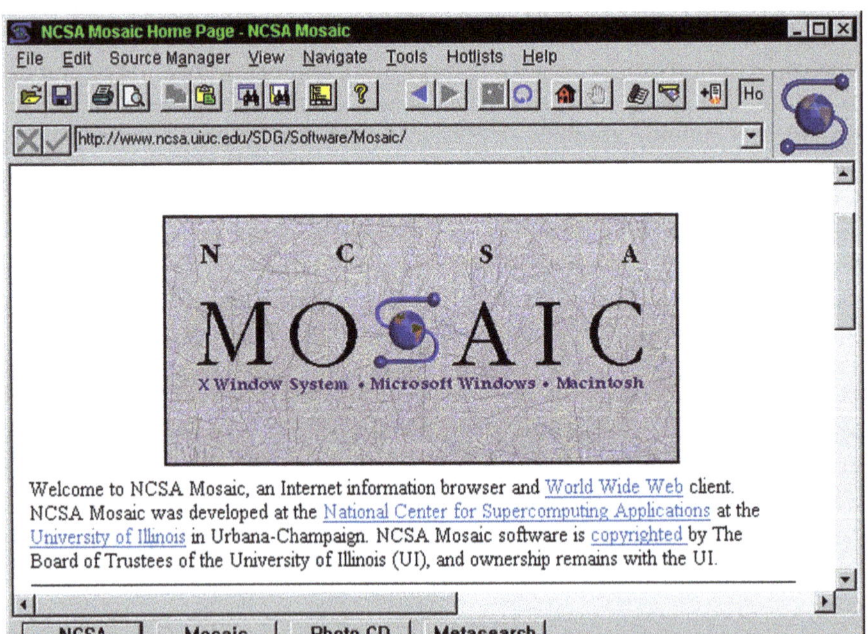

Figure 26.4. Home page of the NCSA Mosaic browser, as viewed on an IBM-PC computer (courtesy of the National Center for Supercomputing Applications (NCSA) and the University of Illinois Board of Trustees.

The inclusion of graphics—and soon audio and video—files on servers and the the Mosaic browsers' ability to read and present them embedded on Web pages was crucial to its commercial exploitation. That new feature fostered advertising applications, which require colorful visual (and audiovisual) presentations to be really effective. 'I'm convinced that very quickly we'll have a new Madison Avenue kind of industry devoted to this style of advertising', Markoff quoted a Sprint executive as saying [20, p D5].

Curiously, the high-energy physics community did not foresee this revolution, and in several cases resisted commercialization of the Web.[3] Berners-Lee, for example, has been variously mentioned as opposed to including the embedded graphics. According to Andreessen, 'Tim bawled me out in the summer of '93 for adding graphics to the thing' during a visit to NCSA [4, p 242]. Berners-Lee had a different interpretation of the matter. He had long been promoting the idea that browsers should be able to edit, too—rather than just serving as a powerful new way to consume information passively. 'With Mosaic on the loose, that idea would have to go on the back burner', recalled Cailliau, 'as the Web became another consumer medium with many readers but relatively few publishers' [4, p 243].

Marc Andreessen, by contrast, appreciated the commercial potential of the Web almost from the start. Soon after graduating from Illinois in December 1993, he headed west to Silicon Valley to seek his fortune. In April 1994 he met with venture capitalist Jim Clark, the principal of Silicon Graphics, Inc., in downtown Palo Alto and hatched a plan to lure Bina and other key members of the NCSA software development team to the Valley. Forming the core of Mosaic Communications Corporation, they released test versions of browsers called Mosaic Netscape that October. But when NCSA leaders objected the following month to the use of this name, in part because rights to use it had already been granted other companies, they changed it to Netscape Navigator [4, p 257].

This was the browser that truly brought the Web to the masses. Within four months, it had captured three quarters of the market for browsers, and it eventually reached 95 percent [21, p 104]. Netscape Navigator had many additional features beyond the ones available on NCSA Mosaic—including email on later versions. When Netscape went public on 9 August 1995, its stock value quickly tripled to $58, giving the company a market value of nearly $3 billion. That effectively made Andreessen a multi-millionaire, illustrating the tremendous potential value of the Web.

— — —

As the number of Web sites proliferated in 1993–94, keeping track of them and navigating among them became increasingly difficult. That problem inevitably fostered the development of Web libraries and search engines to help computer users locate and focus upon information sources most relevant to their work. Here CERN again took the lead with its WWW Virtual Library [4, p 245], set up in mid-1993 by another of the temporary staffers who worked with Berners-Lee and

[3] The SLAC Website, for example, was dominated by text for a few more years, as was the Fermilab site.

Cailliau. Web sites were listed under 14 categories, from 'architecture' to 'society', and a 'Search' box allowed users to enter keywords and find appropriate sites.

But back in Silicon Valley, a similar effort was under way in 1994 at Stanford University—only a few miles away from where Andreessen, Clark and their team were developing the Netscape browser software. That February David Filo and Jerry Yang, two electrical engineering graduate students, began compiling a list of their favorite Web servers on Yang's workstation, complete with hypertext links to these sites. At first they dubbed it 'Jerry and David's Guide to the World Wide Web' but changed it to Yahoo! [22]. Soon hundreds of users were accessing the site regularly.

By that fall the site experienced a million hits a day, from over 100 000 individual visitors. Recognizing its commercial potential, Filo and Yang sought input from the local venture-capital community, conveniently headquartered just across from SLAC on Sand Hill Road. In April 1995, Sequoia Capital provided an initial $2 million investment and recommended professional managers to run the business. Yahoo! went public a year later, by which time it had become a leading Internet 'portal' offering many other services such as email, news, and financial information as well as an early search engine—and of course plenty of advertising.

In addition to the Yahoo! engine, the mid-1990s was characterized by other early search engines such as AltaVista, Excite, and Lycos, most of which used 'web-crawler' software to compile relevant lists and indexes of Web servers associated with key words and phrases. As each had its strengths and weaknesses, the results of searches were often unsatisfactory and therefore frustrating. But in 1995–96 two Stanford computer-science graduate students, Larry Page and Sergey Brin, hit upon the novel idea of ranking a Web site according to the other sites linked to it—much as the number and nature of citations to a scholarly article are used to assess its value for an academic community [23]. Their 'Page Rank' algorithm formed the basis of the Google search engine released on the Stanford Web site in August 1996. Incorporated in 1998, Google, Inc. (now known as Alphabet, Inc.) has since come to dominate searching and other aspects of the Web. The corporation they founded finally went public in 2004, at a market value of $23 billion.

A common element in this explosion of commercial Web activity was the close proximity of Silicon Valley's venture-capital industry headquartered just across Sand Hill Road from SLAC in the Sharon Heights district of Menlo Park. Technology financiers at these firms began viewing computer software as a high-growth, high-tech industry after the 1980s successes of Apple and Microsoft. With innovative Web-based software firms bursting out literally at their feet, they were quick to fund them with the necessary capital and management expertise required to guide rapid growth.

— — —

'By 1994, the centre of gravity of the World Wide Web had crossed the Atlantic to a place where the entrepreneurial heart beats stronger', observed Gillies and Cailliau at the turn of the millennium [4, p 246]. And that year Tim Berners-Lee was contemplating his own move across the Atlantic to follow his brainchild, now growing into a potentially unruly adolescence. With the commercialization of the Web also came the probability of its fragmentation, as powerful business interests tried to pull it in specialized directions more amenable to their profit motives. To counter this balkanization of the Web, Berners-Lee and Cailliau began stumping for an international organization of Web users and software purveyors to establish and maintain universal standards and practices for all to follow [7, pp 75–89].

Leaders of the MIT Laboratory for Computer Science had been courting Berners-Lee since late 1993, discussing ways they could work together on such an organization [7, pp 87–89]. He finally assumed an MIT position in September 1994 as the director of the World Wide Web Consortium—to be headquartered at MIT with another, intentionally co-equal European branch at CERN. It would involve both academic Web users and commercial purveyors of Web software and services, with an industrial advisory group representing their interests [4, pp 278, 293]. But on 16 December, the CERN Council decided to proceed with the planned construction of its multibillion-euro Large Hadron Collider, which meant CERN could not seriously embrace non-physics activities such as this consortium due to inevitable belt-tightening. Instead, the French Institut National de Recherche en Informatique et en Automatique (INRIA) stepped up to assume European leadership in the new international organization. CERN gracefully stepped aside, having in April 1993 put the Web software in the public domain without insisting on being paid royalties, asking only that the Center be recognized as the institution where the Web had been created [4, p 261].

High-energy physics labs had one more important contribution to make that year before exiting center stage in the development of the Internet and Web: extending them into China. Before 1994, that nation had had international telephone connections with the outside world but had no Internet linkage. SLAC physicist Les Cottrell had been trying to remedy the problem since 1991, working with the computing staff at the Institute of High Energy Physics (IHEP) in Beijing [24]. With approval of the US Departments of Commerce, Defense and Energy, they set up an AT&T SkyNet satellite link from SLAC to Beijing Airport. Fiber-optic and copper landlines relayed signals from the airport to an IHEP mini-computer. On 17 May 1994, the first Internet link to China began operating. Soon hundreds of leading Chinese scientists could access the Web via electronic links to IHEP.

'China is a huge country with enormous potential and ambitious plans to open up electronic communications', observed Cottrell and his coauthors in a paper presented at the 1994 Computing in High-Energy Physics Conference, held in San Francisco [24]. 'The HEP [high-energy physics] community should be proud to have pioneered one of the first direct electronic links from the outside world into China, helping to open it up to the global Internet community.'

The Web also proved crucial to implementation and construction of the Large Hadron Collider, which began in 1995 and continued through September 2008. Whereas the SSC Project in Texas had had to rely largely on paper and email communications among its thousands of participants and contributors, the LHC Project could put much of the design, manufacturing, and construction information for the accelerator and its four enormous detectors up on Web sites readily accessible to those who required it. Some individuals could also input changes as needed, enabling members of the international detector collaborations—two of which,

ATLAS and CMS, numbered in the thousands—to remain abreast of new information relevant to their ongoing work. They could also remotely monitor the research and development efforts in progress at other institutions and the many manufacturers of highly sophisticated detector components [25].

And LHC project manager Lyn Evans kept informed of important developments on a large tabletop computer display of the collider project that he demonstrated for me in a June 2000 visit. Using it, he was able to monitor and coordinate many elements of this gargantuan international project (in particular the superconducting-magnet cores then being manufactured by companies in France, Germany, and Italy) at a finger's touch, in a way that was unimaginable a decade earlier. The Web permitted Big Science to successfully attain the scale of multibillion-dollar (or -euro) projects—at which level the SSC Project had foundered, in part due to inadequate project management [26]. This was an instructive example of how the Web could serve as a crucial tool in managing global supply chains.

26.4 A world transformed

In the three decades since its creation, the cultural and economic impacts of the World Wide Web have been little short of immense. Where we once visited libraries or consulted paper encyclopedias for information, many of us now employ Google or Wikipedia daily. Money has largely become digital and can be sent across a nation or continent—and in many cases around the world—at light speed with the click of a mouse. We pick up our cell phones or iPads and access the news from a cacophony of sources, not just in newspapers and magazines. Electrons and photons have largely replaced paper as the principal medium of financial and information exchanges.

Of course, other technological factors and innovations have been at work here, too—in particular the availability of cheap, highly sophisticated personal computers and cell phones, and the extension of the Internet into nearly every corner of the planet, including in China and Russia. In fact, the expansion of the information superhighway continues today, as fiber-optic cables are reaching out to rural areas and increasingly remote users. But these would be useless hardware without software innovations such as email and the Web to encode, send, receive, decode, and display messages and images on computer screens. The Internet and Web have coevolved into the dominant information-sharing platform universally in use today, putting the world on our screens and at our fingertips.

Putting a monetary figure on the value of Web activity is a speculative endeavor, but it certainly falls in the trillions of dollars (or euros) annually. In their 2000 book, Gillies and Cailliau put it at 'some three times the total amount of money invested in CERN' since its inception, or on the order of $100 billion per year [4, p 309, 27]. Alphabet, Inc.—the parent company of Google and YouTube—now enjoys annual revenues approaching $250 billion and has a market capitalization that exceeds $1.1 trillion. With a business model depending heavily on Web usage, Amazon, Inc. has over $500 billion in annual revenues at this writing. And these figures ignore the value of the myriads of other ways in which individuals and institutions now employ the Web for information gathering and sharing, purchasing, financing, manufacturing, and entertainment. The World Wide Web has thus far contributed tens of trillions of dollars worth of value to the global economy.

These gains, however, have not come without losses—particularly in the news profession. While a few national publications such as the *New York Times* and the *Wall Street Journal* have growing online readerships in the millions despite their declining print circulations, the bulk of smaller US newspapers are failing or have already shut down. According to a 2020 report from University of North Carolina's Hussman School of Journalism and Media, over 2100 local or regional US newspapers have been closed since 2005 due to the loss of advertising revenues to Web-based publications, leaving more than 1800 US communities without local news sources [28]. And more than half the nation's journalists have left the news profession, too. 'In the absence of a local news organization', observes this report, 'social media and Internet sites have often become the default media for reading, viewing and sharing information—as well as rumor and gossip—exacerbating political, social and economic divisions in a polarized nation.'

This is a troublesome example of a more general phenomenon that is known as 'disintermediation', whereby the Web lessens the need for middlemen (and women) to facilitate exchanges, in particular economic ones [29]. In the above example of the news, the intermediaries being eliminated are the media outlets, editors, and reporters that had previously attempted to establish credible common grounds for public discourse. In their stead has arisen, in particular during the past decade, a vast, fragmented array of competing sources of information with few agreed-upon standards for what is true and what is false. It reached its apotheosis during the recent Trump presidency, when 'alternative facts' and 'fake news' became the norm, believed by tens of millions of credulous citizens and voters [30]. And Web-based social-media platforms such as Facebook and Twitter have unfortunately made it far too easy to spread misinformation, disinformation, and lies.

Another good example of disintermediation enabled by the Web is the continuing transfer of retail commercial activity from so-called brick-and-mortar stores in downtown locations to online outlets such as Amazon (and its many warehouses). While this process has greatly benefitted that corporate colossus and delivery services such as FedEx and UPS, it (along with Walmart) has helped to hollow out these districts in cities and towns across America (and elsewhere). What were once thriving centers of local commerce—and citizen involvement in public affairs—have often become just withered crossroads with many shuttered or boarded-up shops, especially in the US Midwest.

The Web has also become one of the principal engines of the recent wave of globalization, which began during the 1990s after the Cold War ended. By allowing design information to be shared among manufacturers and by enabling related data and financial exchanges, it has permitted global supply-chain management to occur across continents. Apple iPhones, for example, are assembled in China by Foxconn, Inc. from semiconductor and other miniature components manufactured in Asia, Europe, and North America. While these far-flung exchanges have benefitted technological elites and the corporations that employ them, they take meaningful low-skilled labor away from blue-collar workers in developed nations, exacerbating existing inequalities. Other factors are involved in this work emigration, of course—for example, email, fiber optics, and the growth of containerized shipping and trucking. But it is difficult to imagine this vast global commercial transformation ever happening without the ultra-dense, fine-grained information sharing that occurs at light speed every second over the World Wide Web.

Thus the obvious benefits that have accrued from the Web have not come without tremendous dislocations in the economies and societies that existed before its early-1990s emergence from the high-energy physics community. Those dislocations have damaged important sectors of the US and other national economies, leading to angry 'populist' reactions in Europe and America [31]. Brexit and the electoral victory of Donald J Trump were two results of this surge of anti-globalization furor on opposite sides of the Atlantic. The world has indeed been transformed by the Web, but not entirely for the better.

The Internet and Web have proved instrumental in keeping people connected during the coronavirus pandemic. Amazon boomed as customers moved ever more of their purchases to its Web site, while online companies such as Zoom and Door Dash filled gaps that suddenly opened in video meetings and grocery purchasing. Few figures exist to support this speculation, but it is easy to imagine that the pandemic could have become much more onerous had the Web not existed and people instead had to leave their homes to engage in the activities of what previously was considered normal life.

But this withdrawal from community intercourse to the Web comes with the danger that people retreat from public life into so-called information silos or 'filter bubbles' [32]—where the random encounters that promote and enhance diversity occur much less frequently [33]. Thus the pandemic has probably hastened the fragmentation of Western societies into mutually exclusive 'cultural tribes', each with its own set of incommensurable beliefs and purported 'facts' to support them. Through the social-media that it enables, and the decline of the paper-based news media that it has caused, the Web has contributed significantly if indirectly to the political polarizations that currently afflict these societies.

Thus the commercialization of the Web unfortunately came with an unforeseen cost: it has become burdened with disinformation, falsehoods and trash. And it has become difficult for ordinary folks to distinguish good information from bad. Writing about the aftermath of the 2020 US presidential elections, author George Packer asked, 'Why did exhausted election officials across the country have to stay up late night after night to kill off the thousands of lies spreading through the sophisticated digital pathways that were invented by the country's most brilliant and successful entrepreneurs and channeled into millions of minds that have grown strangely vulnerable to contagion?' [34].

We all know how that farce ended early the following year.

— — —

Berners-Lee had begun to recognize these problems during 2019 celebrations of the Web's thirtieth anniversary. 'We suddenly realized [that] we have to make sure that the Web is serving humanity', he admitted in an interview published in *Time* magazine. 'Not just by keeping it free and open, but [by] making sure that the things people build in this permissive space are actually helping democracy' [35].

Almost any new technology is a double-edged sword that can cut both ways, for good or for evil. How that choice comes about depends upon how it is embraced by and becomes embedded in a given society. Consider firearms, for example. There are great differences in how they were embraced—and in their social impacts nationally—in Australia, Britain, and Canada versus the United States. Or take fire, one of the earliest technologies. It can be used to illuminate, cook food, and forge useful new tools, but it can also burn and destroy.

We are now facing a similar choice in much of the Western world. The Web is being used increasingly for authoritarian, harmful, and illegal purposes, undermining the hopeful, democratic dreams of its idealistic founders, who perceived open information exchange as an unalloyed public good. But not all information is necessarily good information. We are thus entering a period of serious reassessment, when individuals, governments and societies attempt to come to grips with the polarized, information-drenched world that the Web hath wrought.

Acknowledgments

I thank my frequent coauthor Lillian Hoddeson for her early input on this manuscript, and Louise Addis, Bebo White, and in particular James Gillies for their thoughtful comments after reviewing parts or all of it. Any remaining errors are my own.

References

[1] Bush G W and Scowcroft B 1998 *A World Transformed* (New York: Alfred Knopf)

[2] Berners-Lee T 1989 *Information Management: A Proposal* March 1989 https://w3.org/History/1989/proposal.html

[3] Levy D 2001 Tech pioneer recalls how he brought the World Wide Web, now 10, to America *Stanford Report* 12 December

[4] Gillies J and Cailliau R 2000 *How the Web Was Born* (Oxford: Oxford University Press)

[5] Segal B 1995 A short history of Internet protocols at CERN *CERN* http://ben.home.cern.ch/ben/TCPHIST.html

[6] Berners-Lee T and Cailliau R 1990 *World-Wide Web: Proposal for a HyperText Project* 12 September 1990 https://w3.org/Proposal.html

[7] Berners-Lee T and Fischetti M 1999 *Weaving the Web: The Original Design and Ultimate Destiny of the World Wide Web* (New York: Harper Collins)

[8] Berners-Lee T 1991 Email to www-interest@cernvax.cern.ch 13 December https://lists.w3.org/Archives/Public/www-talk/1991NovDec/0022.html

[9] Berners-Lee T 1991 World Wide Web: online information for everyone *CERN Computer Newsletter* October–December https://cds.cern.ch/record/1292517/files/n-204.pdf

[10] White B 1998 The World Wide Web and high-energy physics *Phys. Today* **51** 34

[11] Berger E 1996 Birthplace of the web *FermiNews* **9** 1 https://fnal.gov/pub/ferminews/FermiNews96-08-16.pdf

[12] SLAC Archives, History and Records Office *SLAC WWW Wizards* https://slac.stanford.edu/history/earlyweb/wizards.shtml
[13] Berners-Lee T and Cailliau R 1992 World-wide web *Proc. Int. Conf. on Computing in High Energy Physics (Annecy, France, 21–25 September)*C Verkerk and W Wojcik (Geneva: CERN), pp 69–74 Report No. 92-07
[14] Terry Schalk, private communication
[15] Kahn R and Cerf V 2000 Al Gore and the Internet *The Register* 2 October https://theregister.co.uk/2000/10/02/net_builders_kahn_cerf_recognise/
[16] Gore S A Jr 1991 S.272—High-Performance Computing Act of 1991 *Congress.gov* https://congress.gov/bill/102nd-congress/senate-bill/272
[17] Deken J M The Early World Wide Web at SLAC: Early Chronology and Documents https://www.slac.stanford.edu/history/earlyweb/history.shtml
[18] Andreessen M 1992 NCSA X Mosaic 0.5 Released. Email to www-talk@nx0c01.cern.ch http://1997.webhistory.org/www.lists/www-talk.1993q1/0099.html
[19] Berners-Lee T 1992 Re: NCSA X Mosaic 0.5 Released. Email to www-talk@nx0c01.cern.ch http://1997.webhistory.org/www.lists/www-talk.1993q1/0103.html
[20] Markoff J 1993 A free and simple computer link *New York Times* 9 December D1 https://www.nytimes.com/1993/12/08/business/business-technology-a-free-and-simple-computer-link.html
[21] Zook M A 2005 *The Geography of the Internet Industry: Venture Capital, Dot-Coms, and Local Knowledge* (Oxford: Blackwell Publishing)
[22] History Computer n.d. *Yahoo! History—Complete History of the Yahoo Search Engine* https://history-computer.com/yahoo-history-complete-history-of-the-yahoo-search-engine/
[23] Battelle J 2005 The Birth of Google *Wired* August https://wired.com/2005/08/battelle/
[24] Cottrell R L A *et al* 1994 Networking with China *1994 Conference on Computing in High-Energy Physics* Report No. SLAC-PUB-6478 (April) Stanford Linear Accelerator Center, Menlo Park, San Francisco, CA https://www.slac.stanford.edu/pubs/slacpubs/6250/slac-pub-6478.pdf
[25] Della-Negra M and Tonelli G private communications
[26] Riordan M, Hoddeson L and Kolb A W 2015 *Tunnel Visions: The Rise and Fall of the Superconducting Super Collider* (Chicago, IL: University of Chicago Press), pp 264–74 https://press.uchicago.edu/ucp/books/book/chicago/T/bo21803804.html
Evans, Lyn, and James Gillies, private communications
[27] James Gillies, private communication
[28] Abernathy P M 2020 *News Deserts and Ghost Newspapers: Will Local Journalism Survive?* (Chapel Hill, NC: UNC Huffman School of Journalism and Media) https://usnewsdeserts.com/reports/news-deserts-and-ghost-newspapers-will-local-news-survive/
[29] Hawken P 1981 Disintermediation: an economics buzzword that neatly explains a lot of what is going on *CoEvol. Quart.* Spring 6
Gellman R 1996 Disintermediation and the Internet *Gov. Inf. Quart.* **13** 1
[30] Tufekci Z 2018 How social media took us from Tahrir Square to Donald Trump *Technol. Rev.* **121** 10 https://www.technologyreview.com/2018/08/14/240325/how-%20social-media-took-us-from-tahrir-square-to-donald-trump/
[31] Judis J B 2016 *The Populist Explosion: How the Great Recession Transformed American and European Politics* (New York: Columbia Global Reports)
Eichengreen B 2018 *The Populist Temptation: Economic Grievance and Political Reaction in the Modern Era* (Oxford: Oxford University Press)

[32] Pariser E 2011 *The Filter Bubble: How the New Personalized Web Is Changing What We Read and How We Think* (New York: Penguin)

[33] Riordan M 2020 Coronavirus and community *Orcas Currents* 26 October https://orcascurrents.com/coronavirus-and-community/

[34] Packer G 2021 *Last Best Hope: American in Crisis and Renewal* (New York: Farrar, Straus and Giroux), pp 9–10

[35] Perrigo B 2019 The World Wide Web turns 30 today. Here's how its inventor thinks we can fix it *Time* 12 March https://www.yahoo.com/now/world-wide-turns-30-today-123428470.html See also ed Berners-Lee T 2019 *30 Years on, What's Next For the Web?* (Washington, DC: World Wide Web Foundation) https://webfoundation.org/2019/03/web-birthday-30/

Chapter 27

Big Science and technology in the former Soviet Union

Paul Josephson

On the eve of the its collapse, the Soviet Union had one-third of the world's engineers and one-quarter of its physicists, and according to official data there were 1.5 million specialists in all. It is difficult to compare these data with those from other countries because of differences in definitions of specialists and their level of education, but the numbers still reveal 'Big Science': a massive, state-run scientific enterprise. In general outline, institutes of the USSR Academy of Sciences were the locus of basic research in this enterprise, branch industrial and defense ministries were the site of applied research and development (R&D), and universities were a site of training, not extensive research. The state-supported nature of R&D, its centrality to the nation's political, economic, and cultural programs, and its massive scale and its locus in prestigious research institutes, some of them with thousands of employees, developed from the first years of the Russian Revolution in 1917 [1].

Soviet science was Big Science and Technology par excellence. Many people consider the US Manhattan atomic bomb project to be the first major Big Science project and the paradigm for future Big Science programs. Yet an examination of the history of the Soviet R&D system indicates that the features of Big Science and Technology (S&T) were already present in the USSR before World War II. These include not only scale, but the organization, focus, and resource intensity of the scientific enterprise. The roots of Big Soviet S&T were its initial organization in the 1920s under the Bolsheviks, the development of the command economy with its high degree of central control in the Stalin era, and the Cold War impetus to scientific achievements in competition with the US and to some extent with European science. Generally speaking, Big S&T moved ahead because of the unbridled interests of state officials, the hubris of engineers, the institutional momentum of engineering institutions, and the coalescence of the masses under a national ideology of glorious

doi:10.1088/978-0-7503-3631-4ch27

achievement and military necessity. It is not the focus of this essay why, given its importance to the state, Soviet Big Science did not do better at achieving plan targets and competing with capitalist science than it did [2]. Rather we should remember that an agrarian nation in 1917 by the 1950s and 1960s had built nuclear reactors, particle accelerators, tokamak fusion facilities, rockets, and satellites, in many cases being first in the world.

27.1 Bolshevik science policy

No sooner had the Bolsheviks seized power in 1917 than scientists set out to establish new research institutes. They recognized that the Bolsheviks would support science in a way that the Tsarist government had not. The number of research institutes and scientists in most fields doubled, tripled, and even quadrupled in the 1920s and 1930s. Such leading centers of research came into existence as the Leningrad Physical Technical Institute (LFTI), the Lebedev Physics Institute, and the Institute of Experimental Biology. Others newly established centers represented branches of industry or received funding from branch industry. In creating institutions to promote science and incorporate its achievements into the economy, the Bolsheviks were ahead of most countries, in particular the Americans. Granted, the American system of universities and such industrial laboratories as Westinghouse, General Electric, and Dupont was effective, but federal support for research and development took off much later, during World War II. Of course, the Germans had already established major industrial laboratories too, and the institutes of its multi-profile Kaiser Wilhelm Gesellschaft (Society), founded in the early 1900s, ensured a dynamic relationship between science, engineering, and the economy. However, astronomical inflation and the inattention of policy makers hurt those laboratories, the society, and universities during the Weimar regime into the mid-1920s. (The initial creation of Soviet R&D centers, with the Academy of Sciences becoming the central, prestigious location for basic research, has been characterized as 'a combination of revolutionary innovation and international borrowing'. See [3].)

Bolshevik officials, pushed by scientists, formed two organizations that explicitly encouraged the expansion of the scientific enterprise, Glavnauka and NTO. Glavnauka—the main scientific administration—supported the formation of institutes and gave scientists relative autonomy. One of the leading sites of physics research that set the stage for expansion of the entire scientific enterprise, LFTI, was founded in 1918 under its jurisdiction. Working with communist party officials the LFTI leadership pushed the creation of a series of other major physics institutes throughout the USSR including in Kharkiv (UFTI), Sverdlovsk (UralFTI), Dnepropetrovsk (DFTI), and Tomsk in Siberia (SFTI). The FTIs would focus on concrete economic problems, for example, new energy-industrial complexes in the Don coal basin (DFTI) [4].

NTO, the Scientific Technical Department of the Supreme Economic Council, was set up to contribute to the creation of a network of factory laboratories to advance modern production processes, foster innovation and invention, and include scientific experts in economic planning. In its first years, NTO set up such new research centers as the Automobile Laboratory, the Central Aerodynamic Institute, the Central Scientific Technical Laboratory, and the State Pulp and Paper Experimental Station. Such leading scientists and engineers as V N Ipatieff, a chemist of international reputation who eventually emigrated to the United States, and Peter Palchinsky, whom Lenin directly suggested serve as the head of NTO, but was later murdered by the secret police as an enemy of the state, were involved in its activities. NTO eventually was incorporated into the Commissariat of Heavy Industry, an act in itself that showed the determination of Party officials to link Big Science and industry [5].

Big Science and Engineering projects in the former Soviet Union persisted throughout all eras, largely because of continuities in the nature of one-party political rule. Communist Party dictates required that institutes, agencies and laboratories fulfill plan targets. Those institutions became adept at pursuing new projects. Because there was no unemployment in the USSR, institutions found new projects to maintain employment levels, and this went for engineering and scientific institutions as well. The result was a kind of technological momentum where research, development, geophysical, charting, and defense programs grew in service to industrial ministries in Moscow, each with its own branch scientific research institutes [6].

27.2 Stalin and Big Science

Under Joseph Stalin the Communist Party introduced a series of policies and organizational steps that created the foundation for Big Soviet Science. First, it centralized control of economic planning including that for science in Moscow under such bodies as the State Planning Commission (Gosplan). It subjugated the Soviet Academy of Sciences to these agencies to ensure that ‘pure’ or basic science also was planned and served state industrialization plans. Officials believed central planning would avoid duplication of research in a variety of industrial settings that they argued plagued Western science. Central planning enabled marshalling of resources for campaigns for important projects: the dams and canals and railroads of the 1930s, the atomic bombs and rockets of the 1940s and 1950s, and so on. These changes in the scientific enterprise accompanied the so-called Great Break of Stalin with the past Soviet reality: rapid industrialization, forced and murderous collectivization of agriculture, and cultural revolution that had great impact on education and science, all of which led to its Big Science and industrial profile.

The planning of R&D activities in one- and five-year plans was intended to tie the ‘achievements’ of science into the production process. The Party also sought autarky in the economy to create a powerful Soviet state that was not reliant on Western S&T or industry. This foreshadowed the status of Cold War Soviet science as a crucial tool in the struggle with the west for military parity. The creation of this system enabled ideological control over scientists, many of whom had been trained in the Tsarist period and thus were politically suspects as members of the bourgeoisie [7]. The loss of autonomy for scientists, the emphasis on applied science at the expense of fundamental research, and the tying of science and engineering to rapid industrialization plans were crucial features of Cold War Big Science. Yet in connection with rapid industrialization, the number of research centers outside of Moscow and Leningrad increased, and state funding expanded, all of which gave scientists greater flexibility in pursuing new directions of research. Scientists managed to embark on research in such new areas as nuclear physics and low temperature physics in the early 1930s before there was any promise of application and before state support reached levels that physicists believed they required [8]. Overall, there was a 50% increase in the number of research institutes from 1928 to 1932. The authorities funded the establishment of new academies in several republics, a process that accelerated after World War II, and also the All Union Academy of Agricultural Sciences (VASKhNiL) among many others [9].

Another step in the establishment of Big Science was the transfer of the Academy of Sciences presidium to Moscow from Leningrad in 1934 to bring it closer to Party direction and scrutiny. Soon enough the Soviet Academy was incorporated into the Stalinist industrialization effort in a variety of 'hero projects'—a new Trans-Siberian railroad anointed the Baikal–Amur Magistral (BAM, begun in the mid-1930s but finished only in the early 2000s), a series of dams on the Volga and other rivers, and a variety of canals. Newly created industrial laboratories and research institutes were joined into the efforts with geological and geographical surveys, hydrological studies and charting of valuable resources. Various laboratories of academy institutes eventually became special institutes of the atomic bomb project and the foundation of the nuclear enterprise in the Ministry of Middle Machine Building.

27.3 The Soviet atomic bomb project

Like in the US, so in the USSR the Soviet atomic bomb project and the Cold War provided further impetus to the rise of Big Science [10]. Stalin and the head of the secret police (NKVD), Lavrenty Beria, spared few resources in the race to build nuclear weapons, in particular after the US nuclear attacks on Hiroshima and Nagasaki, Japan. The state established priority on machinery and equipment, and it designated not only new university graduates to fill expanding programs, but Gulag prisoners and German prisoners of war in construction projects. Mining, enrichment, and production facilities expanded rapidly. Later other laboratories and weapons tests sites paradigmatic of nuclear Big Science were added. The major institutional homes of the Soviet bomb effort were laboratory no. 1 (UFTI) and laboratory 2 (the new institute of the head of the bomb project, Igor Kurchatov, later called the Kurchatov Institute for Atomic Energy, KIAE, to this day a leading center of Russian science). In 1945 the Soviets opened another institute to have competing research teams and calculations and to ensure program continuity should KIAE fall under attack, Arzamas-16 (originally KB-11, now the All-Russian Scientific Research Institute of Experimental Physics [VNIIEF]), often jokingly called Los Arzamas by Soviet physicists. VNIIEF had 20 000 employees in 2022.

The extensive nuclear program involved the building of experimental and then plutonium production reactors; massive ore processing and enrichment factories;

and bomb test sites. The Soviets built three major testing grounds: at Semipalatinsk (18 000 km^2) in Kazakhstan where 456 tests were carried out (340 underground and 116 atmospheric tests suspended from towers or dropped from aircraft), Novaia Zemlia in the Arctic Ocean with 130 tests, and a site in the Ural Mountains. Semipalatinsk alone had four major testing areas at the site, along with two research reactors [11]. A total of three sites were deployed on Novaya Zemlya: Sukhoi Nos, Chernaia Guba, and Matochkin Shar [12].

Within the limits of this essay it is not possible to devote as much attention to space R and D as to nuclear [13]. Suffice it to say that the USSR won the space race on many levels.[1] The USSR had the greatest number of launches. It was first into space with a satellite, it developed expendable launch vehicles, and it had an extensive human space flight program that put the first man and woman into space, conducted the first space walks, and so on and on. The Soviets first achieved a soft lunar landing, launched the first space station (Salyiut 1), and some of the first interplanetary probes. The US put significantly more funding into its program than USSR, for example, in manned programs from 1961 to 1969 spending some $23 billion versus about $10 billion for the USSR. But considering the size of the Soviet economy this was a roughly equal amount. The Big Science of space covered the entire Soviet landmass from its Baikonur launch complex on desert steppe east of the Aral Sea in Kazakhstan, to its Kapustin Yar (1946–present) and Plesetsk bases. (The cost of maintaining the base in another country means that Russia is building a new space center in Konechny, which is well behind construction targets and has experienced significant cost overruns [15].) As was typical for Soviet Big Science, the construction of Baikonur involved a campaign that targeted national resources including thousands of soldiers and construction workers to open it within three years.

In addition to its prevailing military components, the space program consisted of a major Academy of Sciences institute, the Institute of Cosmic Research, ten ground tracking stations within the USSR and ten tracking ships around the globe operated by the Soviet Navy and Strategic Rocket Force. Major rocket production facilities were built in Moscow, Samara, Miass, Omsk, and Dniepropetrovsk (Ukraine). At least nine military industrial ministries were tied to the space program. The program had vital successes in the Cold War competition with the US that were ideological ('First in Space!') and military-economy, that is, in achieving parity in the arms race. It also experienced significant crises (the loss of life of a number of cosmonauts in accidents, the Plesetsk 1980 launch pad disaster that killed at least 44 people, the loss satellites, and the failure to beat the US to the Moon). But the prevailing message of the Soviet space program was that the USSR was first into space, had beaten the Americans, and the Soviet system was better than capitalism.

[1] The space program was hatched in the 1930s and carried out within a special gulag prison in which the leaders of Soviet aero-astronautics served terms. The head of the space program, Sergei Korolev, who developed the Soviet equivalent of the V2 rocket, the first intercontinental ballistic missile, the R7, first tested in August 1957, served in this aero-astro gulag, see [14].

Cold War Big Science competition extended to powerful particle accelerators in which Soviet specialists were pioneers. The devices include the Joint Institute of Nuclear Research (Dubna) synchrophasotron at 10 Gev (1957–2003), the ideas for which were developed by Vladimir Veksler at the same time as Edwin McMillan (a Nobel laureate; Soviet specialists contended that Veksler ought to have shared the prize); the Institute of High Energy Physics U-70 GeV accelerator at Protvino (Serpukhov) 1967–present (in the financial and political crises of the Gorbachev and Yeltsin years Protvino had to abandon its 3 TeV large superconductor—with the U-70 to be its injector—after 11 years of construction with its 21 kilometer tunnel already completed); at the Budker Institute of Nuclear Physics (Akademgorodok, near Novosibirsk), a series of VEPP electron–positron colliders from 1964 to the VEPP-4M 6.0 GeV from 1994; and Gatchina's Institute of Nuclear Physics, a synchrocyclotron at 1 GeV (1967–) [16].

Soviet Big Science was carried out in a series of closed military research cities, that is, in the central loci of the military–industrial complex, the Closed Administrative-Territorial Formation (ZATO in its Russian acronym). Many of these cities never appeared on maps and were known only as a post office box number. The people who lived and worked in these cities needed special passes to enter and leave. Even today there are over forty ZATOs in Russia with one-million-plus residents that house Russia's military–industrial complex including its weapons production and waste facilities: Cheliabinsk-65 (now Ozersk, home of the major Soviet plutonium production facility); Tomsk-7 (now Seversk, home of a fuel reprocessing factory and of a major nuclear accident in 1993); Krasnoiarsk-24 (now Zheleznogorsk, home of another plutonium processing facility); Sarov (Arzamas-16, where bombs were designed); and the nuclear shipbuilding center Molotovsk (today Severodvinsk, on the North Sea) [17].

27.4 Khrushchev era reforms and Big Science

Nikita Khrushchev (1953–64), who followed Stalin, initiated a series of reforms in Soviet society, abandoning some aspects of Stalinism, although maintaining one-party rule. The reforms had an impact on science policy as well. The most significant impact was the growth of the scientific enterprise. The total number of scientists increased from 162 500 in 1950 to 665 000 in 1965, including an increase in the number of senior and junior specialists from 62 000 to 140 000. A second aspect of reform was decentralization of the scientific enterprise, in part because of the growth of the nuclear establishment. The goal of scientific reforms was to rejuvenate the Soviet economy and invigorate science. Leading specialists and party officials alike recognized that for all of the quantitative indicators of scientific excellence, the quality of research results and the lag of innovation in industry left much to be desired. In spite of official pronouncements to the contrary—and in spite of such evidence as Sputnik (1957) and the tokomak fusion reactor—Soviet science lagged behind European and American science.

One solution was the establishment of new science cities outside of Moscow and Leningrad. The growth of the nuclear enterprise in the Cold War made billions of rubles available to establish the Troitsk center of nuclear research, 20 km outside of Moscow; the Joint Institute of Nuclear Research in Dubna on the Volga River; the Serpukhov-based Institute of High Energy Physics; and a multitude of ZATOs. Siberia occupied the first position in the expansion of the scientific enterprise, with new branches of the Siberian division of the Academy established in Yakutia, Irkutsk, Krasnoiarsk, Buriatia, Tomsk, and Akademgorodok (home of division headquarters), and research centers in Chita, Kemerovo, Tiumen, Barnaul, Kysyl, and Omsk [18]. Akademgorodok was the most important of these centers from all points of view: size, breadth, and vitality of research community; political and cultural significance; and international and national reputation. Others had narrow profile which reflected both local concerns and the difficulty in attracting personnel to out-of-the-way places. For example, the Yakut Center boasted a Institute of Ice Science and an Institute of Problems of Operating Machines in the North; the Irkutsk center was of broader profile, but specialized in study of Lake Baikal through its Limnological Institute; the Ural Scientific Center grew to massive size of twelve Academy institutes and 35 000 personnel in all, but its ecological, machine building, and metallurgical research all reflected problems associated with the nuclear weapons enterprise of the Cheliabinsk region; and so on.

Akademgorodok, a city of science built in the early 1960s near the major Trans-Siberian Railway junction of Novosibirsk in Siberia, grew to 21 institutes covering all fields of research, including computer science, sociology,and economics, and included Novosibirsk State University. From modest beginnings, the number of scientific workers in Akademgorodok grew from 654 in 1957 to 13 465 in 1964, of whom nearly 2700 were specialists. Fairly rapid growth continued throughout the early 1970s. Akademgorodok reached its apogee in 1991 when among 100 000 residents there were 52 736 workers in the city, of whom almost 12 000 were scientists, including nearly 6700 with advanced degrees [19].

Under Brezhnev a number of the Khrushchev era reforms were abandoned. Rather than experiment with new forms of organization or new directions of research, the Brezhnev administration further centralized policy making in major bureaucracies, raised the level of ideological control, and re-established vigilance toward contact with Western scientists.

27.5 Soviet Big Engineering

Engineering had these same features of centralized control in service of economic development and military strength, and of being large-scale science-based efforts at resource management. Each Soviet leader pursued Big Engineering projects. Lenin embraced electrification projects as a panacea to overcome the technological backwardness of the nation. Stalin set in motion the rapid industrialization of the USSR through the five-year plans to secure the nation against military threat from without and enemies within. The Stalinist development model of emphasis on heavy industry and resource development persisted over the decades. Under Nikita Khrushchev and in the era of Leonid Brezhnev (1964–85) the effort to control resources in service of the Kremlin expanded into the Arctic and Siberia. Armed with their own applied research branch institutes, the nation's ministries built infrastructure of oil and gas, roads, railroads, bridges, canals, and hydroelectric power stations from one end of the empire to the other. All these projects gave the Kremlin control over resources and landmass. In contemporary Russia the Putin administration continues to find value for the Kremlin in megaprojects that in many ways reflect Soviet approaches, plans and dreams—in oil and gas, dams, railroads, and reactors.

Several examples indicate this general pattern. A first example: Stalin intended to remake nature through waterworks. He saw natural barriers of climate, season, geography, and topography as obstacles to state priorities no less than people he believed were enemies of the state. From the commencement of rapid industrialization at the end of the 1920s, the Communist Party pursued geoengineering projects that grew in scale over the decades and extended the reach of the state geographically to the periphery of the empire, all with a goal of creating a unitary, centrally controlled economy to serve the Kremlin. Soviet waterworks projects originated in gulag labor camps in the early 1930s; many of them grew precisely out of the operations of the White Sea–Baltic Canal (1931–33) and the Moscow–Volga Canal (1932–37), dug by prisoners and designed by geologists, hydrologists, and civil engineers, many of whom were also gulag prisoners. For its waterworks projects the secret police ultimately established a special division, as it did for railway work and other 'branches' of prison industry, in this case the Main Administration of Prisons of Hydrotechnical Construction of the NKVD that by 1951 held almost 200 000 laborers [20].

When Stalin died this giant gulag hydroelectric and melioration organization became the Zhuk Gidroproekt Design Institute in Moscow (and subsequently Rusgidro under President Putin; many contemporary Russian engineering projects were conceived in the Brezhnev period). Gidroproekt celebrated not its gulag origins, nor the extensive, unresolved environmental concerns connected with their projects, but its institutional growth. Gidroproekt engineers designed most of the hydroelectric dams on the Volga and Kama Rivers, a large number of those on the Angara and Enisei Rivers (others were designed by engineering institutes that were hived off from Gidroproekt); hydro stations in the union republics; and all of the major Soviet and Central Asian canals. In 1990 in one of Gidroproekt's 'hived off' institutes, Sredazgidroekt, there were 1500 employees, most of whom were engineers. Gidroproekt, Lengidroproekt, and Lengiprovodkhoz contributed to a canal, dam, and irrigation frenzy.

The most audacious engineering project involved a multi-billion dollar effort to transfer water from Siberian rivers to Europe and Central Asia through massive transfer canals. The project eventually involved over 250 organizations including military ones. It centered on a massive feeder canal longer than 1000 km, up to 300 m in width and 15 m deep, with other canals perhaps totaling another 1500 km, with pumping stations, dams, and holding ponds for industrial and agricultural purposes, and with excavation in part through up to 100 peaceful nuclear explosions to accelerate the work and lower the costs. During perestroika, in 1986, the Communist Party eventually killed the project as too expensive and perhaps foolhardy, but contemporary policy makers have begun to consider Siberian transfer canals again [21].

Soviet planners and engineers focused special attention on Arctic regions that were rich in a variety of resources. Stalin established a special Main Administration for the Northern Sea Route to chart and develop Arctic resources of oil, gas, coal, nickel and copper, and platinum, relying at first on gulag prisoners, and later scientists and engineers [22]. The Academy and branch institutes focused on transport, forestry, pulp and paper, geology, and mining. A major Big Science institution of Arctic development was the Kola Scientific Center of the Academy of Sciences that dates to 1930 and which evolved from a small research building into a major institution with roughly with 4600 workers including 1500 engineers and 860 scientists at its peak. The geological and mining profile of KNTs was not surprising given the rich deposits nearby of minerals (including apatite) that can be used to produce aluminum and agricultural products [23].

27.6 Conclusion

When the USSR collapsed in 1991, and R&D funding with it, a real crisis enveloped the scientific enterprise that threatened to destroy the legacy of achievements in space, physics, computer science, and other fields. The number of researchers dropped precipitously and institutes closed as directors could barely pay for electricity and heating. Even such centers of excellence as Akademgorodok—not to mention Academy institutes located in Moscow and St Petersburg—suffered irreparable losses of research programs and personnel. Since more than 90% of scientific institutions relied on government financing, when it dried up many institutes were forced to meet the challenge by cutting staff. From 1990 to 1992 the amount of government funding fell roughly twofold, and the percentage of the national product spent on science fell to 0.57% of GDP, while in the other leading scientific powers (the US, Japan, and Germany) it was six times higher. In the absence of regular funding, the number of scientific researchers plummeted, from 992 000 in 1990 to 484 400 people in 1996 [24].

The administration of Vladimir Putin has decided to focus on Big S&T again, particularly on space, military, and nuclear power programs. The administration also hopes two new targeted applied science projects will bear results: nanotechnological innovations through a state corporation, Rosnano, and the construction of the Skolkovo Science Park and university (Skoltekh) [25]. But Skolkovo, nanotechnology, and other state-sponsored endeavors are not panaceas for the underperformance of science precisely because of political interference in research, including xenophobic attitudes toward international funding, weakly developed institutions of scientific mobility, insufficient clarity in determining which projects and institutions to fund, and the diversion of resources to the military–industrial complex. To this day Russian puts great emphasis on military R&D, although striving to pick winners in computers, biotechnology, and nanotechnology. In 2021 overall, the military R&D infrastructure consisted of approximately 600 organizations, of which there were roughly 300 research institutes, 130 design bureaus, and 170 scientific production organizations. They represented sectors crucial to the remilitarization of the nation: aviation; space technology; shipbuilding;

conventional arms; radio, electronics, and communication systems; ammunition and special chemicals; and nuclear energy and weapons [26].

Putin, like past Soviet leaders, prefers big, symbolic technology—rockets, reactors, military devices—not high tech, cutting edge fantasies. Paradoxically, since returning to the presidency in 2012, he has undermined the atmosphere in which such a center as Skolkovo might thrive [27]. Putin has discouraged foreign involvement, in part by ordering a government crackdown on international NGOs and foundations as 'foreign agents', including a number that have generously supported Russian scientists [28]. The Russian state has long been the major mover in the economic sphere and in supporting science for the purposes of national security. The Russian approach to industrialization and innovation, honed to a skill under the Soviets, remains precisely emphasis on big projects, unjustified faith in planning or state sponsorship that hides the irrationality of projects, and pressure on engineers and scientists to move discoveries into production. Russian leaders tend to rely on a small circle of technocratic advisors with a production ethos. The Russian government continues to stifle social mobility needed for innovation through residency permits, work 'passports', and other legal tactics [29]. It would seem that the Big Science and Technology of contemporary Russia will repeat Soviet tactics, and also have fewer grand successes than the leadership wishes. The war on Ukraine has further isolated Russian R&D and led to attacks on the autonomy of Russian specialists reminiscent of the Stalin era.

References

[1] Graham L R 1992 Big Science in the last years of the Big Soviet Union *Osiris* **7** 49–71
On what makes science big (extensive, regular funding and importance to national economic and military programs), see ed P Galison and B Hevly 1992 *Big Science. The Growth of Large-Scale Research* (Stanford, CA: Stanford University Press)
On the US experience, see ed Leslie S 1993 *The Cold War and American Science: The Military-Industrial-Academic Complex at MIT and Stanford* (New York: Columbia University Press)

[2] Gustafson T 1980 Why doesn't Soviet science do better than it does? *The Social Context of Soviet Science* ed L Lubrano and S Solomon (Boulder, CO: Westview) pp 31–68

[3] Graham L 1975 The formation of Soviet research institutes: a combination of revolutionary innovation and international borrowing *Soc. Stud. Sci.* **5** 303–29

[4] Anokhina I N *et al* 1998 *Fiziki o Fizike i Fizikakh* (Tomsk: Izdatel'stvo NTL)
Fominykh S F (ed) 2006 *Pism'a Fizika iz Tomska* (Moscow: Znanie-Inform)
Josephson P 1988 Science policy in the Soviet Union, 1917–1927 *Minerva* **26** 342–69

[5] On Glavnauka and NTO, see; Kosnichenko V and Shoshkov E 1988 Shtab Nauki: K 70-letiiu NTO VSNkh *Leningradskii rabochiii* 9 December, p 10
Ipatieff V N 1946 *The Life of a Chemist* (Stanford, CA: Stanford University Press)
On Palchinsky and his fate, see ed Graham L 1993 *Life of the Executed Engineer* (Cambridge, MA: Harvard University Press)

[6] On large-scale systems, see ed Hughes T 1987 The evolution of large technological systems *The Social Construction of Technological Systems: New Directions in the Sociology and History of Technology* ed W E Bijker, T P Hughes and T J Pinch (Cambridge, MA: MIT Press) pp 51–82

[7] On the pressures of working at UFTI in this environment, see ed Ranyuk I 1990 *Delo UFTI A I Leipunskii: Izhrannye Trudy i Vospominaniia* ed B F Gromov (Kiev: Naukova Dumka)

[8] Tuchkevich V M and Frenkel V 1980 *Vklad Akademika A F Ioffe v stanovlenie iadernoi fiziki v SSSR* (Leningrad: Nauka)

[9] Josephson P 1988 Physics, Stalinist politics of science and cultural revolution *Sov. Stud.* **40** 245–65

[10] On the Soviet nuclear establishment, see ed Medvedev M 1992 *Legacy of Chernobyl* (New York: Norton)
Holloway D 1994 *Stalin and the Bomb; the Soviet Union and Atomic Energy, 1939–56* (New Haven, CT: Yale University Press)
R Norris, T Cochran and their colleagues at the NRDC and FAS produced important works on nuclear weapons, proliferation and other subjects over forty years that reveal the scale and ends of the enterprise, see ed Cochran T 1995 *Making the Russian Bomb: From Stalin to Yeltsin* (New York: Routledge)

[11] NTI Semipalatinsk Test Site *Nuclear Threat Initiative* https://nti.org/learn/facilities/732/

[12] Biriukov N 2010 Arkhipelag Osobogo Naznacheniia *Vozdushno-Kosmicheskaia Oborona* 26 October http://vko.ru/author/nikolay-biryukov
Khalturin V *et al* 2005 A review of nuclear testing by the Soviet Union at Novaya Zemlya, 1955–1990 *Sci. Glob. Secur.* **13** 1–42
Bukharin O (ed) 2001 *Russian Strategic Nuclear Forces* (Cambridge, MA: MIT Press)
Ispytaniia I 2000 *SSSR. Novozemelskii Poligon: Obespecheniye Obshchei i Radiatsionnoi Bezopasnosti Iadernykh Ispytanii* (Moscow: IzdAT) pp 75–8

[13] On various aspects of the Soviet space program, see ed McDougall W 1985 *…The Heavens and the Earth* (New York: Basic)
Siddiqi A 2003 *Sputnik and the Soviet Space Challege* (Gainsville, FL: University of Florida Press)
Andrews J 2009 *Red Cosmos: K E Tsiolkovskii, Grandfather of Soviet Rocketry* (College Station, TX: A&M University Press)
Bailes K 1985 Technology and legitimacy: Soviet aviation and Stalinism in the 1930s *Technol. Cult.* **17** 55–81
Jenks A 2012 *The Cosmonaut Who Couldn't Stop Smiling: The Life and Legend of Yuri Gagarin* (DeKalb, IL: Northern Illinois University Press)
Siddiqi A 2011 *The Red Rockets' Glare: Spaceflight and the Soviet Imagination* (Cambridge: Cambridge University Press)

[14] Kerber L 1996 *Stalin's Aviation Gulag: A Memoir of Andrei Tupolev and the Purge Era* ed Von Hardesty (Washington, DC: Smithsonian Institution Press)

[15] NTV 2018 *Putin Otmetit Den' Kosmonovtiki na Kosmodrome Vostochnyi* and n. a. Запуск исследовательской станции «Луна-Глоб» перенесен на 2020 год *Kosmicheskaia Lenta*, 25 January https://kosmolenta.com/index.php/1176–2018-01–25-luna-2025

[16] On the Soviet high energy physics program, especially that at the Budker Institute in Akademgorodok, see ed Josephson P 1997 *New Atlantis Revisited* (Princeton, NJ: Princeton University Press) ch 2

[17] On these cities and general themes of population migration control, see; Gentile M 2004 Former closed cities and urbanisation in the FSU: an exploration in Kazakhstan *Europe-Asia Stud.* **56** 263–78

Rowland R H 1996 Russia's secret cities *Post-Sov. Geogr. Econ.* **37** 426–62
See also; Siddiqi A 2012 ZATOs in view *Russian History Blog* 20 April
[18] Pospelov G L 1959 Bol'shaia nauka v sibiri *Sibirskie ogni* **12** 159
[19] There has been a precipitous decline in personnel through the year 2000 to around 35 000 persons, see Koptiug V (ed) 1997 *O Deiatel'nosti Sibirskogo Otdeleniia Rossiiskoi Akademii Nauk 1996 g. Nachalo 1997 g.* (Novosibirsk: SO RAN) p 13
[20] Gromyko F P 1949 *Energiia Rek—na Sluzhbu Kolkhozam* (Ivanovo: Ivanovskoe Oblastnoe Gosudarstvennoe Izdatel'stvo) p 3
See also ed Davydov M and Tsunts M 1955 *Rasskaz o Velikhikh Rekakh* (Moscow: Gosizdat Kul'turno-prosvetitel'noi Literatury)
Basov F G 1940 *Ispol'zovnie Vodnoi Energii Rek na Kolkhoznykh Gidrostantiiakh* (Voronezh: Voronezhskoe Oblastnoe Knigoizdatel'stvo)
[21] Vasilenko V A 2003 Ostrozhno, snova 'Povorot'! *Eko* 3–20
Zalygin S 1987 *Povorot* (Moscow: Mysl)
Micklin P 2011 The Siberian water transfer scheme *Engineering Earth: The Impacts of Megaengineering Projects* ed S D Brunn vol 3 (Dordrecht: Springer) pp 1515–30
Shishikin V Proekty Perebroski Stoka Severnykh Rek v Respubliki Srednei Azii http://www.cawater-info.net/review/pdf/shishikin.pdf
[22] On the social and cultural meaning of the Arctic for the USSR, see ed McCannon J 1998 *Red Arctic* (New York: Oxford)
On Big Science, engineering and the Far North, see ed Josephson P 2014 *The Conquest of the Russian Arctic* (Cambridge: Harvard University Press)
[23] Kola Science Center 2005 *Kola Science Centre* (Apatity: Kola Scientific Center) p 19
[24] Tropp E 2003 Petersburgskaia Nauka *Akademicheskaia Nauaka v Sankt-Peterbruge v XVIII–XX Vekakh: Istoricheskie Ocherki* ed Zh I Alferov (St Petersburg: Nauka) pp 550–51
Vodichev E G *et al* 2007 *Istoricheskii Ocherk. RAN. Sibirskoe Otdelenie* (Novosibirsk: Nauka) pp 323–6, 357
Josephson P and Dezhina I 1998 The slow pace of reform of science in Russia and Ukraine *Probl. Post-Communism* **45** 48–63
[25] Loren G and Dezhina I 2008 *Science in the New Russia* (Bloomington, IN: Indiana University Press) p 37
[26] Engvall J 2021 *Russia's Military R&D Infrastructure, FOI-R-5124-SE* (Stockholm: Swedish Defense Research Agency) p 4, 20, 25
[27] Luhn A 2013 Not just oil and oligarchs *Slate* 9 December http://slate.com/articles/technology/the_next_silicon_valley/2013/12/russia_s_innovation_city_skolkovo_plagued_-by_doubts_but_it_continues_to.html
[28] Genprokuratura R F 2015 Vkliuchila Fond Sorosa v Spisok Nezhelatel'nykh Organizatsii *Tass* 30 November http://tass.ru/politika/2483822
[29] Graham L 2013 *Lonely Ideas* (Cambridge: MIT Press)

IOP Publishing

Big Science in the 21st Century
Economic and societal impacts
Panagiotis Charitos, Theodore Arabatzis, Harry Cliff, Günther Dissertori, Juliette Forneris and Jason Li-Ying

Chapter 28

Big Astronomy: large telescopes and the dual narrative of impact

David Baneke

Astronomy has a long tradition of using ever-larger telescopes to explore the Universe. Today, large telescope projects define the structure of the discipline in terms of institutional organization and funding. But despite the long history of large telescope building, astronomy did not automatically adopt a 'Big Science' culture. In this essay, I argue that 'Big Science' ways of working were introduced into astronomy from the outside: via radio astronomy and space science. In recent decades, the different instrument cultures of these fields have gradually been integrated in to 'multi-messenger' and 'multi-instrument' astronomy.

Analyzing this history will help to understand a specific feature of Big Astronomy. Science and engineering are inseparable in big telescopes, but they are presented very differently to the outside world. Astronomy is often presented as the ultimate 'pure', curiosity-driven science, but its main instruments are also promoted (and funded) as technological development projects. These narratives are used side-by-side, depending on the context and audience. In this paper, I will analyze this 'dual narrative' of the impact of astronomy, and the implications for outreach, education, policy, and funding.

28.1 Introduction

The Very Large Telescope (VLT) in the Chilean Atacama desert has been observing the sky for over two decades. Several 'Extremely Large Telescopes' (ELTs) are under construction, and a $10 billion new space telescope was launched in December 2021. Clearly, astronomers have embraced bigness.

These instruments are used to investigate questions about the origins of the Universe and of the life in it: big scientific questions with strong popular appeal, but also quite abstract questions, with little direct application in daily life.

doi:10.1088/978-0-7503-3631-4ch28

Astronomy lost most of its practical applications with the improvement of clocks, compasses, and maps. Understanding the cosmos is often presented as the ultimate 'pure', curiosity-driven investigation. This endeavor has a strong cultural resonance, contributing in no small measure to the public appeal of astronomy.

Does this mean that the impact of astronomy is purely scientific and cultural? Are those large telescopes really funded only for scientific reasons, even in the current innovation-centered science policy environment? Are there no spin-offs? The scientific and cultural impact is definitely important, but looking at the curiosity-driven uses of telescopes reveals only half the story. Much of the funding for astronomy is not spent on using telescopes for astronomical research, but on building the telescopes and auxiliary instruments and developing the highly advanced technology they need[1]. That part of the discipline has a very concrete economic and technological impact. This has led to a 'dual narrative' about the impact of astronomy, emphasizing either pure science or technological innovation, depending on the context and audience.

Astronomy has a long tradition of pushing for ever-bigger telescopes to explore the Universe. Today, large instruments shape the structure of the discipline, and the future of astronomy is planned in terms of new telescopes of different kinds (figure 28.1). Astronomers did not automatically adopt a 'Big Science' culture, however. Until well into the twentieth century, even the largest telescopes were used mostly by individual researchers. In this paper, I argue that 'Big Science' ways of working were introduced

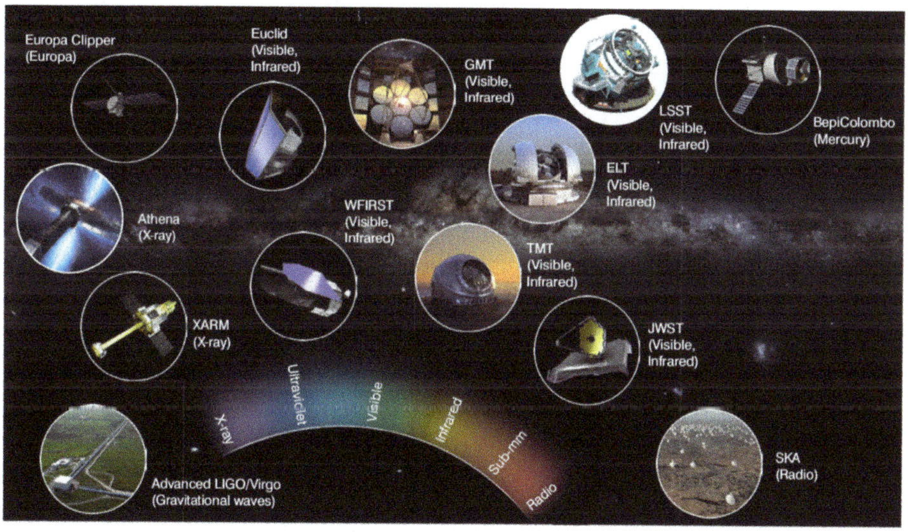

Figure 28.1. The future of astronomy is planned in terms of large instruments that can observe different parts of the electromagnetic spectrum: major instruments planned for the 2020s (International Astronomical Union 2020, p 20, reproduced with permission).

[1] In this chapter, I will mostly use the terms 'telescopes' and 'instruments' interchangeably; I will specify if I use 'instruments' in the more specific sense common in astronomy, for auxiliary instruments attached to telescope systems (for example cameras or spectrographs).

into astronomy from the outside: via radio astronomy and space science, both developed in Cold War contexts. It took time and effort to integrate the different instrument cultures into one discipline of 'multi-wavelength' and even 'multi-messenger' astronomy, since this always also involved 'multi-instrument' astronomy. In the first part of this essay, I will describe and analyze this process.

I will also describe how astronomical instruments gradually were institutionally separated from research. This is relevant for the second part of the essay, in which I analyze the 'dual narrative' of Big Astronomy's simultaneous pursuit of both cultural and technological impact.

I will start with a discussion of the importance of big telescopes for astronomy.

28.2 Why Big Astronomy?

Apart from missions within the Solar System, astronomers cannot investigate their research objects directly. All they can do is theorize, model, and, especially, observe. Astronomers have to work with whatever 'messengers' the Universe happens to send to their instruments. These can be cosmic rays, neutrinos, gravitational waves, but above all, light. For this simple reason, astronomers are always looking for ways to capture more light. That can be done by building bigger telescopes, by improving detectors to process the incoming light more efficiently, by improving data processing to filter out noise, and by extending observations to all parts of the electromagnetic spectrum. After the Second World War, new kinds of telescopes have expanded beyond visible light, by observing radio waves, infrared and ultraviolet light, x-rays, and gamma rays (Brandl *et al* 2010). This has radically changed our view of the Universe and the objects within it. The new telescopes revealed quasars, pulsars, interstellar molecules, and other things that had never been seen before, and in some cases not even been thought of before (Longair 2006, Dick 2013).

The idea that progress is strongly technology-driven plays a central role in astronomy (Smith 1997, Dick 2013, Harwit 2019). This started with the very first telescopes, which were such an improvement over the naked eye that Galileo discovered the moons of Jupiter, features on the surface of the Moon and the stellar nature of the Milky Way within months of first turning his instruments to the sky (Heilbron 2012). The argument that new instruments can 'uncover new phenomena not yet imagined' remains a strong and much-used argument for constructing ever-larger telescopes (Spitzer 1946, Smith 1989, 1997, cf. chapter 25 in this volume). The idea of 'Big Astronomy' is now widely embraced by astronomers, although the cost of some projects can still cause some to worry about consequences for funding of other projects (Smith 1989, McCray 2004, cf. chapter 24 in this volume).

Because of the importance of observation, access to large telescopes is one of the most important resources for any astronomer. Some large telescopes are open for all astronomers, who can apply for observing time. In other cases, some or all of the observing time is reserved for researchers from institutions that contributed funding or were otherwise involved in the development. These are highly coveted privileges (Smith 1997, DeVorkin 2000, McCray 2000, Baneke 2019, Williams and Mauduit 2010).

Large telescopes are not only important for researchers, but also for astronomical institutions. Institutions that develop (parts of) major instruments build up a lot of highly specialized expertise. To maintain that expertise—read 'staff', because expertise is embodied in people—after a project is finished, they need a new project that involves their specialty. In other words, they need a 'pipeline' of new instruments. Because each project needs a long time before it is approved and funded, astronomers, engineers, and managers are constantly thinking about future projects, in order to secure long-term institutional continuity (McCray 2004, Conway 2015, Baneke 2019). Having access to instruments also is an asset to attract new staff and students. All this explains why instruments are so central in the future planning of astronomy—also since the timescales of major instruments are much longer than the usual timescale of academic research grants.

This also means that new telescopes have an impact on astronomy long before 'first light', the start of scientific operations. Robert Smith and Patrick McCray have described beautifully how the run-up to the Hubble Space Telescope and the Gemini Telescopes have shaped telescope technology as well as the American astronomical community over several decades (Smith 1989, McCray 2004). Similarly, the Square Kilometer Array, a giant new radio telescope, to be built by a global consortium in South Africa and Australia, has already been influencing astronomy for several decades, even though construction has only just started (Baneke 2019).

28.3 Sources of instrument culture

The drive to build ever more powerful telescopes is a red thread throughout the history of modern astronomy, but that does not mean that every step was self-evident at the time, or that there was anything 'natural' about the evolution of large telescopes into 'Big Science'[2]. Every major new telescope was a risky project for its proponents and funders (Smith 1997, McCray 2004). In addition, major improvements involved new technologies, which often required new ways of working: how to do observations, how to process the data, and how to organize the operation of the telescopes. New kinds of instruments came with new 'instrument cultures'.

In this section, I will discuss three major sources of instrument cultures in modern astronomy: the optical tradition, radio astronomy, and space research. I will characterize how elements of all these instrument cultures were integrated into contemporary astronomy. I will also describe how the development of Big Astronomy led to an increasing separation between instruments and research, both in practice and institutionally.

28.3.1 Optical astronomy

Astronomical observatories were among the first institutions that were specifically created for scientific research. Ulugh Beg and Tycho Brahe's observatories had a significant staff and large instrument collections. Moreover, these observatories were

[2] Excellent overviews of the historiography of Big Science can be found in Capshew and Rader (1992) and Hallonsten (2016).

funded by monarchs. Team work, large instrumentation, and government funding are all commonly taken as characteristics of Big Science. But at the same time, these early modern observatories were personal enterprises that did not survive their founder. The funding by monarchs fits more into the pattern of early modern patronage than modern government funding. And while some instruments were large, they were operated by individuals. Collaboration mostly consisted in combining data by multiple observers (see chapter 22 in this volume).

This changed with the founding of national observatories, for example in Greenwich, Paris, and Washington. In the early nineteenth century, Pulkovo Observatory near St Petersburg became the model for a string of new observatories (Dick 1991). The very design of these observatories highlighted the importance of instruments in astronomy. They were literally constructed around the telescopes (especially meridian circles). But still, there was a clear division between designing, building, and maintaining instruments on the one hand, and using them on the other. Astronomers mostly purchased their instruments from specialized (commercial) instrument makers, who were not involved in using the instruments for observations (Smith 1997).

The telescopes were intended for long-term use, in a discipline that valued precision above everything else. So it was worth spending a lot of time on calibrating the instrument, and on getting to know all its remaining imperfections. 'Know your instrument' was a key part of every astronomer's work ethic (McCray 2004).

In the 1880s a large number of observatories agreed to jointly produce a giant photographic survey of the entire sky, the *Carte du Ciel* (Bigg 2000). This mammoth undertaking, which ran until the 1960s and was never completely finished, was certainly one of the most ambitious scientific projects of the era, even if it did not involve large instruments. It was triggered by the emergence of photography, a technology that fundamentally changed the instrument culture of astronomy (Lankford 1984). Instead of making measurements while looking through a telescope, astronomers increasingly took photos which they could investigate at leisure during the day. This made observations much more efficient. In addition, long exposure times brought much fainter objects within reach, and they provided a permanent record of the observations that could be revisited later. However, taking photos still required staying up in the telescope dome all night.

From the turn of the twentieth century, the largest telescopes were operated by privately funded observatories in the United States. The undisputed master of this 'large telescope era' was George Ellery Hale, who engineered the construction of the largest telescope in the world four times in a row, each time funded by a private philanthropist: at Yerkes Observatory near Chicago, Mt Wilson Observatory near Los Angeles (twice), and Mt Palomar in Southern California (Van Helden 1984, Wright 1994, Sandage 2004).

Hale did not build his new telescopes at existing observatories. Astronomical observations are not only limited by telescope quality and size, but also by clouds and atmospheric disturbances. Since astronomers could travel while big telescopes could not, it made sense to build telescopes in the places with the best 'seeing' conditions, which Hale found in California. This became standard practice in the twentieth century. In this way, telescopes got geographically separated from research institutions (although Hale made an effort to change a local college near Mt Wilson into a research university: this became Caltech).

Hale's telescopes feature in many publications about Big Science. Impressively large and expensive, they are regarded as direct predecessors of the large scientific instruments after the Second World War. But however big, the telescopes were used by individuals and operated by single institutions with powerful directors (Lankford 1997, cf. Sandage 2004). There was no team work, no government funding, and no 'big management'.

The era of large optical telescopes culminated in 1948 in the opening of Hale's masterpiece, the 200 inch (5 m) Mt Palomar telescope. But at the same time, this marked the start of stagnation. Palomar became the model for a string of 4 m sized telescopes that were built in the next few decades, but it was widely agreed that larger telescopes were impossible without significant technological innovations. The disappointing results of a Soviet 6 m telescope reinforced that idea (Keel 1992).

Apart from technological difficulties, cost was also an issue. No observatory could match the resources of the rich Californian institutes. In 1946 the International Astronomical Union (IAU) proposed to solve this by establishing international observatories (Blaauw 1994, Andersen *et al* 2019). In the next decades, international pooling of resources to develop large telescopes indeed became common practice, although without any central coordination by the IAU.

The Palomar telescope was only surpassed in size in the 1990s, although telescopes improved in the meantime because of more efficient detectors and active optics. In the decades after the Second World War, the main impulse in astronomical observations came from other directions, however: radio and space technology.

28.3.2 Radio astronomy

In the 1930s and 1940s radio and radar engineers noticed radio 'noise' coming out of the sky. This was a nuisance, but some of them realized that it could be scientifically interesting. After the end of the Second World War, groups of engineers and scientists started to investigate the radio noise systematically, using surplus wartime radar equipment as the starting point for their research (Agar 1998, Sullivan 2009, Munns 2012, Elbers 2017).

Radio receivers differed from existing telescopes. The most notable differences were the role of electronic technology and the format of the resulting data. Astronomers were used to look through telescopes themselves or to take photos, but radio observations required detectors which produced an electronic signal. It required a lot of work to translate this data into a visual image of the sky. Initially, it was also difficult to identify radio sources with known astronomical objects. This improved in the course of the 1950s, as more sources were identified. By that time, the receivers were called 'radio telescopes', indicating their gradual integration into astronomy (Sullivan 2009).

The emergence of radio astronomy is often described as the opening of a new 'window' on the Universe. Before, the Universe could only be observed in visible light, which revealed planets, stars, and nebulae. Radio waves revealed the existence of pulsars, quasars, black holes, and many other things that had never been observed before. Jan Oort, a pioneer of radio astronomy, commented in 1978: 'Before the advent of radio astronomy, astronomers had not realized at all how many things in the Universe are of an explosive nature'[3]. This explosive nature even included the Universe itself: the 'Big Bang' theory of the origins of the Universe was spectacularly confirmed by the discovery, with a radio telescope, of the cosmic background radiation in 1965. The Universe turned out to be a much more diverse place than anyone had thought.

Because of the differences, the instrument culture of early radio research also differed from traditional (optical) astronomy. Most notably, the boundary between science and technology was much less clear. While in optical astronomy, the instruments were built before they were used for observation, in radio astronomy the instrument kept being modified during observations. Engineers and scientists had to cooperate in every phase. Operating the telescopes, and even processing the data, required intimate knowledge of electronic technology. Radio astronomers were also among the first large-scale users of computers.

For all these reasons, it was not easy to integrate radio observations into astronomy. Another factor was the different background of the people involved: physicists and engineers had to learn the basics of astronomy, while astronomers had to familiarize themselves with electronic technology, and each had to appreciate the other's role in the process. This took time and effort (Agar 1998, Sullivan 2009, Munns 2012).

Radio astronomy was heavily based on military technology, especially radar, but the link was in the technology, not in the scientific research that was done with the instruments[4]. The 'dual use' nature of the technology was an asset: the expertise and experience with radio astronomy could easily be turned to military uses if needed, but until that time, nothing looks more peaceful than an astronomical observatory. Radio telescopes were also used as 'showcase' technologies, to display national technological prowess (Agar 1998).

Many radio telescopes were developed outside the institutional structure of astronomical observatories, and with direct government support. They were operated by newly created organizations, in many cases consortia of universities working together, for example the NRAO in the US and SRZM in the Netherlands (DeVorkin 2000, Baneke 2015). In Britain, the large Jodrell Bank radio dish was at the center of discussions about the relation between large instruments and academia (Agar 1998, ch 2).

From the 1950s, this model was also adopted for new large optical telescopes, which had become too expensive for most individual observatories. Several European countries joined the European Southern Observatory (ESO), which aimed to build a large telescope in the Southern hemisphere, while American universities collaborated to develop joint telescopes in both Arizona and Chile (Blaauw 1991, McCray 2004).

[3] Interview with W T Sullivan, 1978, NRAO.

[4] A high-profile exception was the use of the Jodrell Bank telescope to track Sputnik (Agar 1998).

The new organizations were explicitly modeled after the paradigmatic Cold War Big Science projects: high-energy physics. Brookhaven Laboratory was a model for the American National Optical and Radio Observatories, for example; some of the same people were involved in founding these institutions (DeVorkin 2000, McCray 2004). ESO was explicitly modeled after CERN, again with some of the same people involved. ESO even had an instrumentation department located at CERN (Blaauw 1991).

The emergence of 'facility' organizations to manage telescopes triggered discussions about access and division of labor. Should they have their own research staff, or should they only provide technical assistance? And who could use the joint instruments: only staff of the institutes or countries who contributed? Or could other researchers also apply? Would there be quota for specific groups? This were sensitive points, because they concerned the most important resource for astronomers: access to high-quality telescopes (DeVorkin 2000, McCray 2000).

By the 1960s, large radio and optical telescopes looked like 'Big Science' in the traditional, Cold War era sense. They were large and expensive instruments, funded by governments, operated by dedicated organizations with their own management structure, and used by many different researchers. Operating them required team work of scientists, engineers, and managers in all stages, from first proposal to data processing. Compared to high-energy physics, the teams were still small, however. In that sense, 'big instruments' did not automatically lead to a 'Big Science' instrument culture.

28.3.3 Space

Visible light and radio waves from outer space can be observed from the ground. Other wavelengths cannot: ultraviolet and infrared light, x-rays, and gamma rays are blocked by the Earth's atmosphere. Opening these 'windows' required going beyond the atmosphere, into space. It also required changing the ways astronomers worked (DeVorkin 2010). Even more than radio astronomy, the development of space research is intertwined with the politics of the Cold War.

From the early twentieth century, scientists used high-altitude balloons to investigate the upper atmosphere and to observe cosmic radiation. After 1945 they could also use 'sounding rockets', which left the atmosphere for a few minutes on a ballistic trajectory before falling back to Earth. The short time available, and the difficulty of pointing the instruments at a specific source, meant that they could not be used for most astronomical research, except solar physics and cosmic radiation (DeVorkin 1992).

Apart from the practical difficulties, the instrument culture of early space research also made it difficult for astronomers to participate. They were not used to the 'tinkering' style of the rocketeers, who built their own instruments. The fact that the instruments could be used for only a few minutes, after which they were destroyed on landing, was also very foreign to them (DeVorkin 1992, 2010).

The launch of Sputnik in October 1957, during the International Geophysical Year, changed the situation. Earth-orbiting satellites promised stable observing platforms outside the atmosphere. In the following years, space agencies initiated scientific programs that included astronomical projects. A major boost came in

1958, when the newly founded NASA offered to launch foreign scientific experiments. This was a diplomatic gesture, intended to re-gain the initiative after the blow to American prestige of having lost the first satellite launch to the Soviet Union (Krige *et al* 2013). The offer made it possible for scientists in many other countries to develop space instruments (Krige and Russo 2000, Baneke 2014).

Many governments were eager to support space initiatives, for scientific reasons but also to catch up with the technology and to support local high-tech industries. Developing astronomical instruments was an excellent way to do this. (Reinke 2007, Baneke 2014). Even if astronomical observations were not of any practical use, the technology was. Like radio astronomy, space technology is 'dual use', so any experience from developing scientific instruments could also be used for commercial and military applications. The 'arsenal of knowledge' argument worked in favor of 'pure science' (Krige 2010).

One would expect astronomers to jump at the opportunity, but many of them were initially cautious. Nancy Grace Roman, who was hired by NASA to develop its astronomy program, described the 'technical and social challenges' she encountered when approaching astronomers (Roman 2001, cf. Smith 1989, 2010). For one thing, satellite observations had to be done electronically, since the photographic technology that most astronomers were used to was not usable in space. Because the data had to be transmitted back to Earth, the amount of data was also very limited. This only changed later with advancing electronic and digital technology.

Astronomers were also not used to develop instruments to use them so briefly. The development of space instruments often took many years, while the actual missions were usually much shorter, especially in the beginning. According to Roman, this problem was solved in part by the influx of experimental physicists, who were more used to this way of working. Many pioneers of space research had a background in cosmic ray research (cf. Tucker and Giacconi 1985, DeVorkin 2019).

The short lifetime of the instruments, in combination with the difficulty of communicating with spacecraft in orbit, also meant that the observing programs of the instruments had to be planned before launch. In contrast to ground-based telescopes, each space instrument had a well-defined mission.

Because of their cost, complexity and political sensitivity, space projects could only be organized by government-supported space agencies. There were always many parties involved, including space agencies, scientific institutions, and industry. Each step, from design and construction to operation and data processing, required large, multidisciplinary teams.

If radio telescopes were big instruments but not necessarily Big Science, space telescopes were Big Science, but not big instruments. Satellites were small, and the astronomical instruments on board were even smaller. But precisely this small size made them even more complex. Each instrument had very tight limits regarding size, weight, energy use, temperature management, etc. Any design change immediately had repercussions for other components of the spacecraft. In addition, the quality requirements were extremely strict, because even the tiniest irregularity could endanger the whole mission, and fixing an instrument in space was impossible. All this required very strict management.

Roman commented that many astronomers were not used to the detailed management culture of space, and 'in fact tried to rebel against it [...] Part of the problem was that the astronomical community generally had no appreciation of the complexity of satellite projects'. For many space novices, the amount of paperwork was one of the biggest shocks (Roman 2001, p 507, Baneke 2014, cf. Johnson 2002).

A final complicating factor was the sensitive political context, which was new to most astronomers. The high cost of each project ensured political attention, and all technology was closely connected to military and intelligence uses. On the whole, this benefited astronomy though. Missiles were turned into rockets to provide access to space; Cold War diplomacy opened possibilities for scientists from smaller countries to develop space instruments; X-ray and gamma ray detectors developed to monitor nuclear testing proved to be useful for astronomical observations; and military infrared detectors boosted infrared astronomy.

Robert Smith has characterized space research as 'a gift of the Cold War' (Smith 2011). Groundbreaking observatories such as Uhuru (x-ray, launched 1970) and IRAS (infrared, launched 1983) really opened new windows on the Universe. They were followed by a string of ever more powerful observatories (Tucker and Tucker 1986, Bleeker *et al* 2001, Rieke 2006, Giacconi 2008).

Access to space has opened many new windows on the Universe. It has revolutionized our knowledge about the life cycle of stars, the origin of planets, the nature of the interstellar medium, and many other aspects of the Universe. But while astronomers could use space infrastructure, the way of working was shaped by space, not astronomy (DeVorkin 2010). This meant that they had to get used to Big Science in all its aspects, including 'Big Management', 'Big Money', and 'Big Politics'.

28.4 Merging instrument cultures

The opening of the electromagnetic spectrum created new subdisciplines like radio astronomy and high-energy astrophysics, each of which mostly used its own set of specialized instruments. In the last quarter of the twentieth century, astronomers increasingly started to combine different kinds of observations in their research: astronomy became 'multi-wavelength'. The bright supernova of 24 February 1987 in the Large Magellanic Cloud was a breakthrough. It was observed with many different kinds of telescopes, vastly increasing our understanding of stellar physics. Apart from light, several neutrinos from the supernova were also detected, the first observed neutrinos from any astronomical object other than the Sun. This was another addition to the astronomers' observational toolkit: not another wavelength but another kind of 'messenger' entirely (Jayawardhana 2014).

Interestingly, we can also see the multi-wavelength integration in the instrument culture. The differences in working with optical, radio, and space instruments diminished. Gradually, a new instrument culture of 'Big Astronomy' emerged. For one thing, all new telescopes were operated electronically, and electronic cameras and detectors became standard in all wavelength ranges. The era of chemical photography was over (McCray 2004, 2014). This meant that astronomers did their observations sitting behind a computer screen in a well-lit, well-heated control room, physically separated from the actual instrument, whether they were using optical, radio, or space instruments.

On the other hand, space observatories also adopted some elements of ground-based instrument culture. The International Ultraviolet Explorer (IEU), launched in 1978, was the first space observatory at which all astronomers could apply for observing time. Guest observers could do their observations together with the staff of the NASA and ESA ground stations. For all practical purposes, this was similar to ground-based telescopes. The IUE really popularized space astronomy: it was used by a significant share of the world's astronomers during its 18-year lifetime (Massey and Robins 1986, Roman 2001). This way of working was later also used at other space observatories.

From 1990, a series of large space observatories was launched to investigate all wavelength ranges, starting with the iconic Hubble Space Telescope, which had been in preparation for many decades (Smith 1989). This multi-billion-dollar instrument became a popular icon, in part because of the dramatic story of the human repair mission to fix its flawed primary mirror, and of course because of the spectacular images that the Space Telescope Science Institute produced based on its observations (Kessler 2012, Launius and DeVorkin 2014). It was followed by European, American, and Japanese space observatories in all wavelength ranges (Russia, and later China and India, focused more on Solar System missions).

The end of the twentieth century also saw a revival of ground-based optical telescopes. In the 1980s, a series of experimental telescopes tested new technologies that paved the way for a new generation of large telescopes[5]. The first to surpass the iconic Palomar telescope was the Keck telescope in 1993, with a 10 m segmented mirror. It was followed by a series of 8 to 10 m telescopes, including ESO's rather unimaginatively named Very Large Telescope (VLT), which consists of four 8 m telescopes (McCray 2004, Woltjer 2006, Madsen 2012, Leverington 2017).

The new telescopes all produced data in electronic (increasingly digital) format, increasingly following international standards (Giacconi 2008, McCray 2014). This made it easier to share data and link different archives. The resulting data archives could also be used as 'virtual observatories' for new research. An important side effect was that this improved research opportunities for people who did not have direct access to expensive facilities.

Another kind of virtual telescope emerged in radio astronomy: observations from different radio telescopes could be combined to reach the resolution of a telescope the size of the distance between them. Eventually this 'very long baseline interferometry' could span the entire Earth and even extend into space. This required communicating and processing extremely large amounts of data, making astronomers pioneers in 'Big Data' science.

Most telescopes have multiple cameras, spectrographs, and other auxiliary instruments attached to them[6]. By the end of the century, these had become as

[5] Most notably the American Multi-Mirror Telescope in Arizona, the European New Technology Telescope in Chile, and the Australian Advanced Technology Telescope in Australia.

[6] In fact, the term 'instrument' in astronomy usually refers to these auxiliary instruments, not whole telescopes.

complex and expensive as entire telescope systems of earlier decades. Some ground-based telescope organizations started to treat them as independent projects. The instruments would be proposed and developed by separate consortia of academic institutes and high-tech industry, sometimes with their own funding. This practice resembled the way space agencies such as NASA and ESA recruited instruments for their astronomical satellites. The required level of precision of some ground-based instruments also approached that of space technology. In fact, some ESO managers explicitly talked about importing the 'space culture' of management and quality control (Woltjer 2006, Giacconi 2008, Madsen 2012, cf. McCray 2004, p 233).

By the end of the century, the largest telescopes were all operated by specialized organizations, as research 'facilities' or 'infrastructure' (Cramer and Hallonsten 2020). Some organizations, such as ESO, operated multiple telescopes; others were *ad hoc* consortia for specific instruments, such as the Canada–France–Hawaii Telescope. Most telescopes involve international collaboration. The number of countries involved has grown and diversified beyond the traditional North American and Western European countries. Japan, Spain, and South Africa operate 10 m class telescopes; South Africa, China, India, and other countries also operate leading radio telescopes, while many more have joined international consortia.

Most of the major ground-based optical telescopes are now concentrated in a few places with the best 'seeing' conditions, most notably Hawaii, Northern Chile, and the Canary Islands. There, multiple telescopes share basic infrastructure such as roads, water, and energy supply, even while the individual telescopes can be operated by different organizations.

28.5 Megascience

The 1990s saw the start of a veritable 'telescope boom', both on the ground and in space. But of course, astronomers being who they are, they were already preparing the next generation of instruments. These would be even bigger, with even ground-based 'megascience' instruments costing in the order of a billion dollars (on megascience: Hoddeson *et al* 2008, Jacob and Hallonsten 2012).

The first of the new generation of instruments was ALMA in the Chilean Atacama desert, fully operational since 2013. It is an array of dishes to observe millimeter and submillimeter wavelengths, until then a relatively underexplored part of the electromagnetic spectrum. It combines characteristics of radio telescopes (multiple dishes linked by interferometry) and space science (some technology, and its operation in an environment that is inhospitable for both humans and instruments, in a desert at 5 km altitude). Its European component is managed by ESO, which until then only operated optical telescopes.

ALMA originated as a merger between planned American, European, and Japanese instruments, with other countries joining later (Vandenbout 2004). This is also characteristic of the latest generation of instruments: they are unique and global. The planned Square Kilometer Array, a radio telescope that will be divided over South Africa and Australia, will also be one of a kind (Ekers 2012, Garrett 2013). The James Webb Space Telescope, launched in December 2021 after many delays and cost overruns, will have unique capabilities. The telescope, which is regarded as the successor to the iconic Hubble Space Telescope, is mostly funded and developed by NASA, with several international partners[7].

The next generation of optical telescopes will have mirrors of 30–40 m. Three of these 'extremely large telescopes' are in preparation, although the timeline of two of them is unclear[8]. Several other large ground-based and space instruments are also in preparation, both general observatories and more specialized instruments such as the Vera C Rubin Observatory, which will survey the entire sky every few nights. It will produce enormous amounts of data, further pushing astronomy's role as a pathfinder in 'Big Data' processing.

In the meantime, other 'Big Science' instruments have become more relevant for astronomy as well. Neutrino detectors have increasingly been able to detect neutrinos from astronomical sources. And in 2015, the LIGO gravitational wave detector made the first confirmed observation of a gravitational wave. In 2017, the detection of another gravitational wave by LIGA and VIRGO could be identified with sources that could also be observed with other telescopes, marking a breakthrough in 'multi-messenger' astronomy. It has taken many decades to develop gravitational wave detectors; now they could be added to astronomy's toolkit (Collins 2017, Schilling 2017).

The instrument culture of neutrino and gravitational wave detectors resembles that of high-energy physics, with very large teams of scientists and engineers cooperating on very specific experiments. This is different from the dominant culture in astronomy, where even the largest instruments were still predominantly used by fairly small teams. In 2005, the average number of authors of astronomical papers was about 5 (Abt 2007). But also in astronomy, this was changing. A spectacular example was the creation of a picture of the immediate surrounding of a black hole with the Event Horizon Telescope in 2019. The EHT is not one instrument but a global consortium of many radio telescopes. Its publications had hundreds of authors (The EHT Collaboration *et al* 2019). 'Big Science' indeed, albeit not with a single big instrument.

28.6 The impact of astronomy: two narratives

The stream of new telescopes from the last three decades has vastly increased our understanding of the cosmos. Astronomers can look farther into the Universe than ever before, see more details, and investigate more different wavelengths. More giant instruments are in preparation.

But these new telescopes are being developed in a very different context than the ones from the Cold War era. The end of the Cold War not only changed geopolitics, it also changed science policy. Astronomy may have been one of the biggest

[7] I am looking forward to Robert Smith's forthcoming book about the James Webb Space Telescope.

[8] ESO's Extremely Large Telescope (ELT) is under construction in Chile; two US-led international consortia are working on the Giant Magellan Telescope (GMT) in Chile and the Thirty Meter Telescope (TMT) in Hawaii or possibly the Canary Islands.

beneficiaries of the Cold War, but the new policy context did not favor astronomical research. Research was supposed to stimulate economic growth via 'innovation' and connections with private enterprise (Elzinga 2012, Crease and Westfall 2016, Hallonsten 2016, see also chapters 23, 24, and 33 in this volume).

According to some scholars, 'Big Science' did not fit easily into the new policy context. Large instruments lost their exceptional status; they were increasingly subjected to 'normal' science policy criteria of relevance and accountability (Elzinga 2012, Baneke 2019, chapter 33 in this volume). The 'persistence of Big Science' after the Cold War has even been raised to the status of research problem (Jacob and Hallonsten 2012). The case for Big Astronomy would seem especially problematic. Solar physics has some practical value because of the potential impact of solar storms on electrical equipment, and there are high expectations of future Solar System mining, but the commercial value of research time on large telescopes is negligible, apparently leaving only 'pure science' arguments to legitimize their funding (Williams and Mauduit 2010).

How, then, to explain the telescope boom of the recent decades? To understand this, it is important to realize that the image of astronomy as a 'pure science' with few practical applications, is only one part of the story, albeit a very powerful part. In its Strategic Plan, the International Astronomical Union (IAU) published an image of the impact of astronomy in different fields (figure 28.2). Two of the main sectors relate to research, but the third addresses 'technology and skills'. These relate not to astronomical science, but to telescope technology and data processing, which

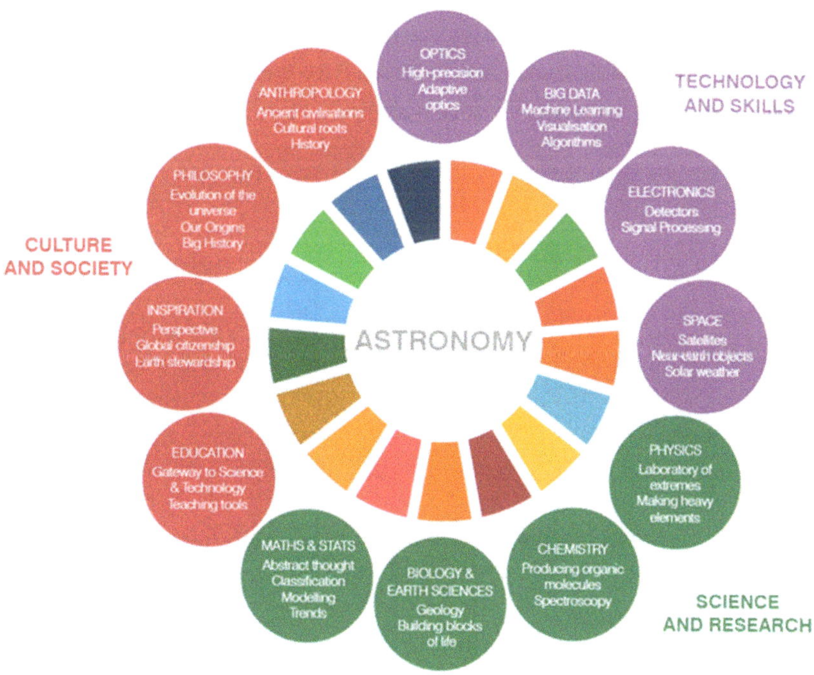

Figure 28.2. The 'wheel' image of the impact of astronomy (International Astronomical Union 2020, p 20, reproduced with permission).

have very clear connections to innovation and industry. So there are two complementary narratives about the impact of astronomy: that of pure, curiosity-driven science, and that of advanced applied technology. These two narratives reinforce each other in a unique way.

28.7 Research

The science and culture sections of the IAU's wheel image present the impact of astronomical research as a 'pure science'. It addresses big questions about the nature and origins of the Universe, life, and the fundamental laws of physics. The Universe provides extreme conditions that can never be replicated in any laboratory, revealing aspects of nature that could otherwise never be studied, which is of course interesting for scientists.

Outside academia, astronomy has a very strong cultural resonance, that sometimes borders on the religious ('inspiration' and 'philosophy' in the IAU's wheel). Astronomy is connected to fundamental ideas about humanity ('our origins' and 'cultural roots'), and realizing how small and fragile our Earth is in the vast cosmos ('perspective'). This has strong moral overtones, connecting to the unity of humankind ('global citizenship') and environmental issues ('Earth stewardship').

Astronomy is also connected to understanding the origins of the Universe, and increasingly the origins of life. This last topic has been boosted in recent decades by the study of exoplanets and the discovery of organic molecules in interstellar matter —and of course by the intense study of Mars (Lambright 2014). The cultural resonance of astronomy is also symbolized by iconic images such as the 'Pillars of Creation' and the 'Pale Blue Dot' (Kessler 2012). At the same time, despite the emphasis on 'origins', astronomy seems to have avoided major religious controversies in recent years, for example in comparison to evolutionary biology. The Vatican even operates a fully scientific observatory.

Because it has no obvious use, astronomical research has no ethical, political, commercial, environmental, or privacy-related downsides. While much astronomical technology has military connections, the link has always remained less visible than in the case of high-energy physics, for example. Together with its popular appeal and its universal image, this makes astronomy an appealing topic for international collaboration ('global citizenship' again). Even political opponents can agree on the value of astronomy, making it a great field for scientific diplomacy. The IAU is proud of the fact that North Korea and Syria joined recently (Andersen *et al* 2019). In addition, large telescopes are highly visible and prestigious. We should not underestimate the political importance of public appeal and the 'cool factor'.

All these effects make astronomy a perfect 'poster child' for science. Astronomy has a very strong popular presence in all kinds of popular science, educational activities, and public planetariums. New discoveries are frequently covered by mainstream news media. The International Year of Astronomy in 2008 was one of the largest science outreach events ever (Russo and Christensen 2010). Astronomers actively promote this image in their outreach, in particular emphasizing how astronomy can stimulate a more general interest in science. This is actually a way to connect to innovation, by arguing that attracting children to science and technology will increase the innovative potential of a community or nation ('gateway to science and technology' in the IAU wheel, cf. the Astronomy for Development initiative, (IAU Office of Astronomy for Development n.d.)).

28.8 Instruments

The third section of the IAU's wheel concerns 'technology and skills', referring to optics, electronics, space technology, and Big Data. These fields are crucial for the development of large telescopes, not least Big Data—modern telescopes produce unprecedented amounts of data, pushing astronomers to become pioneers in data processing on the largest scale, especially in filtering out noise. But these are also all fields of high economic and social value. Astronomers are quick to point out the links between astronomical and medical imaging technology, for example, or between radio astronomy and wifi (Downer *et al* 2019).

The message is that funding a telescope is not just an investment in astronomy, but also in technology and innovation. This of course provides an important legitimation for funding in an innovation-centered policy context. Since much of the funding for astronomy concerns instrument development rather than astronomical research, this is a major addition to the 'pure science' narrative. The technological development does not stop after the telescope starts scientific operations, because the auxiliary instruments can be renewed—most ground-based telescopes have multiple generations of cameras, spectrographs, and other instruments over their lifetime.

Telescope and instrument development and construction are well-defined projects with a clear timeline, matching the project-based structure of modern science funding. Major telescopes also involve major contracts for industry, so there are commercial interests involved as well—again, in contrast to 'pure' astronomical research.

Other than astronomical research, telescopes can be controversial. The Hubble and James Webb Space Telescopes have been the topic of fierce political debates, because of their sheer cost (Smith 1989 and forthcoming). There is also diplomatic competition for hosting large instruments, which are expected to add prestige and (economic and scientific) benefits to the host country.

Some of the controversies about the location of ground-based telescopes are more sensitive. Because they need remote locations, far from any (human) interference, telescopes are often built in protected natural areas. Some locations are also controversial for cultural and religious reasons. The most notable example is Mauna Kea in Hawaii, home to several of the world's largest telescopes, which is held sacred by the Hawaiian population (Swanner 2017).

28.9 The dual narrative of impact

The two complementary narratives of impact make building telescopes appealing projects from a political point of view. Just like the 'dual use' nature of astronomical technology was an advantage during the Cold War, the combination of fundamental scientific questions with innovative telescope technology is a benefit in current-day science policy.

Either narrative can be emphasized, depending on the audience: funding agencies, politicians, science diplomats, academic boards, or the broader public. Large telescopes can be legitimized as investments in fundamental science, but at the same time most of the funding will go to advanced technology. Or, conversely, telescopes can be 'sold' as stimulating innovation without them having to be directly useful as scientific instruments. This differs from the post-Cold War 'transformed Big Science' or 'megascience' that some authors have described, in which Big Science facilities are expected to provide more direct economic and practical benefits, including via use of the facilities by private parties (Hoddeson *et al* 2008, Hallonsten 2016). The dual narrative of astronomy provides an interesting case study for Hallonsten's analysis of the simultaneously grand and practical expectations connected to current-day Big Science (chapter 33 of this volume, cf Baneke 2019).

Another interesting example of the possibility provided by the dual narrative can be found in the controversies about telescopes on sacred sites. Leandra Swanner has described how astronomers invoke the universality and spiritual meaning of astronomy in their attempts to reach out to the opponents of the telescopes (Swanner 2017). In other words, they try to switch the debate from the instrument-focused narrative to the universal science one.

An excellent example of the combination of the two narratives can be found in South Africa, where the post-Apartheid regime invested strongly in astronomy. The choice for a 'pure science' was explicitly intended to underline that South Africa was not 'just' a developing nation that had to prioritize practical needs. But, at the same time, the decision to fund the construction of the 10 m South African Large Telescope (SALT) was also supported by the technological narrative and the benefits for local industry (Whitelock 2004). The IAU's Office of Astronomy for Development, based in Cape Town, explicitly uses the two narratives in a similar way to argue for investments in astronomy in all countries, regardless of their GDP (IAU Office of Astronomy for Development n.d.).

28.10 Conclusion

Astronomy is dependent on a constant supply of new large telescopes, both scientifically and institutionally. In the course of the twentieth century, astronomical instruments have become separated from research institutes. Hale famously moved telescopes away from universities, to places where the 'seeing' conditions were optimal. This has become common practice, so that most ground-based telescopes are now concentrated in a handful of locations on Earth. Space telescopes are even further away of course, but as I have shown, in practice, the difference in using ground-based and space instruments has all but disappeared. Their instrument cultures have merged into multi-messenger and multi-instrument astronomy. Telescopes have also become separated from research on the institutional level. We have seen how, since 1945, large telescopes have increasingly been built and operated by dedicated international organizations.

In the same period, another separation between instruments and research has emerged on the level of policy, funding, and legitimation. Much of the funding for astronomy is devoted to instrument development and construction, not use. This has given rise to a 'dual narrative' about the importance and impact of astronomy: a scientific and cultural narrative on the one hand, and a technological and economic one on the other hand. The two narratives do not contradict or obscure each other; they rather reinforce each other in a unique way. They provide flexibility in different policy contexts and for different audiences. Understanding the dynamics of the dual narrative of impact helps to understand the remarkable public, political, and financial support for Big Astronomy.

References

Abt H A 2007 The future of single-authored papers *Scientometrics* **73** 353–8

Agar J 1998 *Science and Spectacle: The Work of Jodrell Bank in Post-war British Culture* (Amsterdam: Harwood Academic)

Andersen J, Baneke D M and Madsen C 2019 *The International Astronomical Union: Uniting the Community for 100 Years* (Berlin: Springer Nature)

Baneke D 2014 Space for ambitions: the Dutch space program in changing European and transatlantic contexts *Minerva* **52** 119–40

Baneke D 2015 *De ontdekkers van de hemel: de Nederlandse sterrenkunde in de twintigste eeuw* (Amsterdam: Bert Bakker)

Baneke D 2019 Let's not talk about science—the normalization of big science and the moral economy of modern astronomy *Sci. Technol. Hum. Values* **45** 164–94

Bigg C 2000 Photography and the labour history of astronometry: the *Carte du Ciel The Role of Visual Representations in Astronomy: History and Research Practice* ed K Hentschel and A D Wittmann (Thun: Harri Deutsch) pp 90–106

Blaauw A 1991 *ESO's Early History: The European Southern Observatory from Concept to Reality* (Garching: ESO)

Blaauw A 1994 *History of the IAU: the Birth and First Half-Century of the International Astronomical Union* (Dordrecht: Springer)

Bleeker A M M, Geiss J and Huber M C E (ed) 2001 *The Century of Space Science* vol I (Dordrecht: Kluwer)

Brandl B, Stuik R and Katgert-Merkelijn J 2010 *400 Years of Astronomical Telescopes: A Review of History, Science and Technology* (Dordrecht: Springer)

Capshew J H and Rader K A 1992 Big Science: Price to present *Osiris* **7** 2–25

Collins H 2017 *Gravity's Kiss: The Detection of Gravitational Waves* (Cambridge, MA: MIT Press)

Conway E M 2015 *Exploration and Engineering: The Jet Propulsion Laboratory and the Quest for Mars* (Baltimore, MD: Johns Hopkins University Press)

Cramer K C and Hallonsten O (ed) 2020 *Big Science and Research Infrastructures in Europe* (Cheltenham: Edward Elgar)

Crease R P and Westfall C 2016 The new big science *Phys. Today* **69** 30

DeVorkin D H 1992 *Science with a Vengeance: How the Military Created the US Space Sciences After World War II* (New York: Springer)

DeVorkin D H 2000 Who speaks for astronomy? How astronomers responded to government funding after World War Two *Hist. Stud. Phys. Biol. Sci.* **31** 152–92

DeVorkin D H 2010 The space age and disciplinary change in astronomy *NASA's First 50 Years: Historical Perspectives* ed S J Dick (Washington, DC: NASA) pp 389–426
DeVorkin D H 2019 When to send your telescope aloft *J. Hist. Astron.* **50** 265–305
Dick S J (ed) 1991 National observatories issue *J. Hist. Astron.* 22
Dick S J 2013 *Discovery and Classification in Astronomy: Controversy and Consensus* (Cambridge: Cambridge University Press)
Downer B *et al* 2019 *From Medicine to Wi-Fi: Technical Applications of Astronomy to Society* (Paris: IAU) https://iau.org/static/archives/announcements/pdf/ann19022a.pdf
EHT n.d. *Event Horizon Telescope* https://eventhorizontelescope.org (Accessed: 21 April 2021)
Ekers R 2012 The history of the square kilometer array (SKA): born global *Proc. Sci. Resolving the Sky—Radio Interferometry: Past, Present and Future* 163 *(Manchester, 17–20 April)*
Elbers A 2017 *The Rise of Radio Astronomy in the Netherlands: The People and the Politics* (Basel: Springer)
Elzinga A 2012 Features of the current science policy regime: viewed in historical perspective *Sci. Public Policy* **39** 416–28
Garrett M A 2013 Radio astronomy transformed: aperture arrays—past, present and future *2013 Africon (Pointe aux Piments, Mauritius)* pp 1–5
Giacconi R 2008 *Secrets of the Hoary Deep: A Personal History of Modern Astronomy* (Baltimore, MD: Johns Hopkins University Press)
Hallonsten O 2016 *Big Science Transformed: Science, Politics and Organization in Europe and the United States* (New York: Palgrave Macmillan)
Harwit M 2019 *Cosmic Discovery: The Search, Scope, and Heritage of Astronomy. Re-Issue of 1981 Original* (Cambridge: Cambridge University Press)
Heilbron J 2012 *Galileo* (Oxford: Oxford University Press)
Hoddeson L, Kolb A W and Westfall C 2008 *Fermilab: Physics, the Frontier, and Megascience* (Chicago, IL: University of Chicago Press)
IAU Office of Astronomy for Development n.d. *OAD* http://astro4dev.org/ (Accessed: 21 April 2021)
International Astronomical Union 2020 Strategic Plan 2020–30 *IAU* https://iau.org/administration/about/strategic_plan/
Jacob M and Hallonsten O 2012 The persistence of big science and megascience in research and innovation policy *Sci. Public Policy* **39** 411–5
Jayawardhana R 2014 *The Neutrino Hunters: the Chase for the Ghost Particle and the Secrets of the Universe* (Richmond, VA: Oneworld)
Johnson S 2002 *The Secret of Apollo: Systems Management in the American and European Space Programs* (Baltimore, MD: Johns Hopkins University Press)
Keel W 1992 Galaxies through a red giant *Sky Telesc.* **83** 626
Kessler E A 2012 *Picturing the Cosmos: Hubble Space Telescope Images and the Astronomical Sublime* (Minneapolis, MN: University of Minnesota Press)
Krige J 2010 Building the arsenal of knowledge *Centaurus* **52** 280–96
Krige J and Russo A 2000 *A History of the European Space Agency 1958-87* (Noordwijk: ESA)
Krige J, Callahan A L and Maharaj A 2013 *NASA in the World: Fifty Years of International Collaboration in Space* (New York: Palgrave Macmillan)
Lambright W H 2014 *Why Mars: NASA and the Politics of Space Exploration* (Baltimore, MD: Johns Hopkins University Press)
Lankford J 1984 The impact of photography on astronomy *Astrophysics and Twentieth-Century Astronomy to 1950: Part A, The General History of Astronomy* ed O Gingerich vol 4 (Cambridge: Cambridge University Press) pp 16–39

Lankford J 1997 *American Astronomy: Community, Careers, and Power, 1859–1940* (Chicago, IL: University of Chicago Press)

Launius R D and DeVorkin D (ed) 2014 *Hubble's Legacy: Reflections by Those Who Dreamed It, Built and Observed the Universe With It* (Washington, DC: Smithsonian Institution Scholarly Press)

Leverington D 2017 *Observatories and Telescopes of Modern Times: Ground-based Optical and Radio Astronomy Facilities Since 1945* (Cambridge: Cambridge University Press)

Longair M 2006 *The Cosmic Century: A History of Astrophysics and Cosmology* (Cambridge: Cambridge University Press)

Madsen C 2012 *The Jewel on the Mountaintop: The European Southern Observatory through Fifty Years* (Garching: ESO)

Massey H and Robins M O 1986 *History of British Space Science* (Cambridge: Cambridge University Press)

Munns D P D 2012 *A Single Sky: How an International Community Forged the Science of Radio Astronomy* (Cambridge, MA: MIT Press)

McCray W P 2000 Large telescopes and the moral economy of recent astronomy *Soc. Stud. Sci.* **30** 685–711

McCray W P 2004 *Giant Telescopes: Astronomical Ambition and the Promise of Technology* (Cambridge, MA: Harvard University Press)

McCray W P 2014 How astronomers digitized the sky *Technol. Cult.* **55** 908–44

Reinke N 2007 *The History of German Space Policy: Ideas, Influences, and Interdependence 1923–2002* (Paris: Editions Beauchesne)

Rieke G H 2006 *The Last of the Great Observatories: Spitzer and the Era of Faster, Better, Cheaper at NASA* (Tucson, AZ: University of Arizona Press)

Roman N G 2001 Exploring the Universe: space-based astronomy and astrophysics *Exploring the Unknown—Selected Documents in the History of The US Civilian Space Program* ed J D Logsdon, J D Day and R D Launius vol 5 (Washington, DC: NASA) pp 501–43

Russo P and Christensen L L 2010 *International Year of Astronomy 2009: Final Report* (Paris: IAU)

Sandage A 2004 *The Mount Wilson Observatory: Breaking the Code of Cosmic Evolution* (Cambridge: Cambridge University Press)

Schilling G 2017 *Ripples in Spacetime: Einstein, Gravitational Waves, and the Future of Astronomy* (Cambridge, MA: Harvard University Press)

Smith R W 1989 *The Space Telescope: A Study of NASA, Science, Technology, and Politics* (Cambridge: Cambridge University Press)

Smith R W 1997 Engines of discovery: scientific instruments and the history of astronomy and planetary science in the United States in the twentieth century *J. Hist. Astron.* **28** 49–77

Smith R W 2010 Early history space astronomy: issues of patronage, management and control *400 Years of Astronomical Telescopes: A Review of History, Science and Technology* ed B Brandl, R Stuik and J Katgert-Merkelijn (Dordrecht: Springer) pp 447–59

Smith R W 2011 The making of space astronomy: a gift of the Cold War *Earth-Bound to Satellite: Telescopes, Skills and Networks* ed A D Morrison-Low, S Dupré, S Johnston and G Strano (Leiden: Brill) pp 235–49

Spitzer L Jr 1946 Astronomical advantages of an extra-terrestrial observatory *Exploring the Unknown—Selected Documents in the History of The US Civilian Space Program* ed J D Logsdon, J D Day and R D Launius vol 5 (Washington, DC: NASA) pp 546–52

Sullivan W T III 2009 *Cosmic Noise: A History of Early Radio Astronomy* (Cambridge: Cambridge University Press)

Swanner L 2017 Instruments of science or conquest? Neocolonialism and modern American astronomy *Hist. Stud. Nat. Sci.* **47** 293–319

The EHT Collaboration *et al* 2019 First M87 Event horizon telescope results. I. The shadow of the supermassive black hole *Astrophys. J. Lett.* 875 L1

Tucker W and Giacconi G 1985 *The X-ray Universe* (Cambridge, MA: Harvard University Press)

Tucker W and Tucker K 1986 *The Cosmic Inquirers: Modern Telescopes and Their Makers* (Cambridge, MA: Harvard University Press)

Vandenbout P A 2004 Origins of the Alma project in the scientific visions of the North American, European, and Japanese Astronomical Communities *The Dusty and Molecular Universe: A Prelude to Herschel and ALMA* ed A Wilson (Noordwijk: ESA) pp 23–6

Van Helden A 1984 Building large telescopes 1900–50 *Astrophysics and Twentieth-Century Astronomy to 1950: Part A, The General History of Astronomy* ed O Gingerich vol 4 (Cambridge: Cambridge University Press) pp 134–52

Whitelock P A 2004 Optical Astronomy in Post-Apartheid South Africa: 1994 to 2004 *Organizations and Strategies in Astronomy* ed A Heck vol 5 (Dordrecht: Kluwer) pp 39–60

Williams A and Mauduit J-C 2010 The access and return on investment dilemma in Big Science research infrastructures: a case study in astronomy *Big Science and Research Infrastructures in Europe* ed K C Cramer and O Hallonsten (Cheltenham: Edward Elgar) pp 198–216

Woltjer L 2006 *Europe's Quest for the Universe* (Les Ulis: EDP Sciences)

Wright H 1994 *Explorer of the Universe: A Biography of George Ellery Hale* (Woodbury, NY: American Institute of Physics)

IOP Publishing

Big Science in the 21st Century
Economic and societal impacts
Panagiotis Charitos, Theodore Arabatzis, Harry Cliff, Günther Dissertori, Juliette Forneris and Jason Li-Ying

Chapter 29

Big Meteorology and Big Climatology: history and prospects

James Fleming

29.1 Introduction

The term 'Big Science' typically refers to massive increases in the scale of basic research in the twentieth century and beyond. Historians have used the term to refer to large-scale, technologically intensive, and centrally managed research projects, supported by massive government funding and involving large numbers of technical and support staff. Conjuring up images of limited-access facilities, often surrounded by a chain-link fence, Big Science is described as being conducted by large teams of practitioners at national laboratories using one-of-a-kind machines, all of which require the massive funding that only governments or governmental consortia can provide. Examples of such efforts include the Manhattan Project, NASA's (or the Soviet) space program, and fundamental physical (or biological) investigations conducted at specially constructed facilities. Most scholars focus on the term as a twentieth century phenomenon, but the practice of Big Science is much, much older (Kinsella 2000, pp 79–80).

In 1987 the US Congressional Research Service published a 'World Inventory of 'Big Science' Research Instruments and Facilities' (US Congress, House of Representatives 1987). The term 'Big Science' was taken to mean what physicist and public policy scholar Harvey Brooks called 'a new scale of pure science carried on with complex and expensive equipment, and with a large supporting technological effort' (Brooks 1968). The world inventory focused on large-scale instruments and facilities constructed since World War II at a cost of $25 million or more (in 1984 dollars) for research in such fields as astronomy, aeronautical and space science, high-energy and nuclear physics, and materials science and engineering. Oceanography made the list largely due to a federal fleet of research vessels. Atmospheric science was mentioned only twice: the National Center for

doi:10.1088/978-0-7503-3631-4ch29

Atmospheric Research (NCAR) and its supercomputing facility. According to historians Bruce Hevly and Peter Galison, 'Big Science' remains an 'elusive category' with 'many faces'. Their edited volume (1992) includes a diversity of case studies about the Lawrence Radiation Laboratories, Los Alamos National Laboratory, CERN (Conseil Européen pour la Recherche Nucléaire), NASA's Hubble Space Telescope, DuPont's industrial research, and Silicon Valley (Galison and Hevly 1992).

This essay examines Big Meteorology and Big Climatology since antiquity. The atmosphere is big in scale, big in extent, big conceptually, and big in social import. It always has been. The goals of meteorology include explaining atmospheric dynamics and making precise and timely measurements and predictions of local, regional, and global phenomena at increasingly large spatial scales, from the surface to the top of the mountain, in clouds, in the free air, and in near and distant outer space, wherever human curiosity led and ingenuity allowed. Science, as a human activity, encompasses much more than basic research; it includes service. Weather services have always been desired, and have been sponsored by governments, including their military branches. Atmospheric science and services have expanded in scale from local observations to global coverage; in altitude from mountain top and tower observatories to radio balloons, aviation, satellites, and space probes; and conceptually from the whims of the gods and the deductions of natural philosophers to ideas about fluid dynamics, chaos theory, and climate change. The state and behavior of the atmosphere is linked to the economy, governance, military action, and, above all, the everyday life of everyone, everywhere, always. According to historians James Capshew and Karen Rader, 'Big is an adjective easily attached to modern science' (Capshew and Rader 1992), but ever since antiquity, it has been an adjective easily attached to meteorology and climatology. Meteorology has always been 'Big'—in scale, extent, concept, and import.

29.2 Antiquity

In the ancient world, in most cultures, the heavens were seen as the realm of the gods, and weather phenomena were believed to be manifestations of a divine will. Many religions invested natural forces and phenomena, such as Sun, storm, wind, and lightning, with divine characteristics. Traditional stories recount the war-like nature of the weather gods and the sacred power of storms. In many religions, lightning symbolizes spiritual enlightenment or perhaps divine judgment. The rainbow was often seen as the bow of god, a bridge between heaven and Earth. Many religious symbols derived from meteorological phenomena have expressed humanity's deep anxiety about the natural forces and sacred powers that sustain Earth, powers over which humans have had little or no control.

Ancient Greek philosophers, exemplified by the pre-Socratics, speculated on the primacy of substances and the nature of change. They found order in the cosmos, engaged in rational criticism and debate, and left the gods out of their naturalistic explanations. Still, a comprehensive theory of nature was lacking. According to Plato, Socrates once remarked, 'I believe that the Earth is very large and that we who dwell between the pillars of Hercules and the river Phasis live in a small part of it about the sea, like ants or frogs about a pond' (Plato 360 BCE, pp 109A–110A).

Aristotle, a student of Plato, dramatically expanded the purview of natural philosophy. His work, *Meteorologica*, is big, offering a systematic explanation of natural phenomena occurring in the sublunar realm, the realm of change. According to Aristotle, 'It is concerned with events that are natural, taking place in the region nearest to the motion of the stars, and all the affections we may call common to air and water' (Aristotle 350 BCE). *Meteorologica* presents explanations of such diverse phenomena as the Milky Way, comets, meteors, thunderbolts, whirlwinds, and earthquakes. Aristotle drew on his doctrines of natural place and natural motion, the four causes (formal, material, efficient, and final) in combination with the theory of the four basic elements (fire, air, water, earth), four primary qualities (hot, cold, moist, and dry), and two fundamental properties (opposition and affinity). Aristotle's goal was to provide a naturalistic explanation of aerial phenomena, the movement of the elements, their basic nature, and their mutual transformations. Aristotle drew on the work of Parmenides of Elea (c. 500 BCE) in describing the various climates of the known world. He believed that climate—from *klima*, meaning inclination—depended only on the height of the Sun above the horizon and the length of the longest day of the year. He also maintained that the meteorological conditions of various *klimata* (or climate regions) could change with time, since the interior parts of the Earth were changing cyclically and the sea and land were gradually changing places.

Medical climatology traces its roots to the treatises *On Airs, Waters, and Places* and *Epidemics* attributed to the legendary Greek physician, Hippocrates of Cos. These works emphasized the effects of climate and other geographical factors on human health. This tradition linked the physical situation of environments with the characters of their inhabitants. Overall, the relationship between health and lifestyle was under the direct influence, if not the control, of airs, waters, and places. In ancient Rome Claudius Galen elaborated the 'six necessary conserving causes' of health and disease that included air (that is, weather and climate), which influenced the balance of humors as well as character and mental development. Such ideas influenced countless generations of historians, ethnologists, and geographers.

29.3 The Renaissance

Renaissance meteorology remained close to its ancient roots, with Aristotelian and Hippocratic categories dominant (Heninger 1960). Serious researchers and the public at large were interested in questions of astronomical influences on weather and health. Many practitioners of astrology, searching for connections between the heavens and the Earth, also kept daily weather diaries. Almanacs, second in popularity only to the Bible, contained astrological tables and seasonal and yearly weather prognostications for better planting and harvests. The *Old Farmer's Almanack* (1792–present), is a descendant of this genre.

The use of meteorological metaphor was a powerful tool for Renaissance artists and writers. Leonardo da Vinci believed that the metamorphosis of water in the air produced perpetual and sometimes almost imperceptible shifts in atmospheric texture, color, and light. In the *Landscape of Santa Maria della Neve* (1473), one of the earliest landscapes in Western art, he suggests snowfall through his misty and sketchy lines. William Shakespeare invoked the folk tools of prognostication: 'Men judge by the complexion of the sky/The state and inclination of the day' (Shakespeare 1595, pp 194–95).

29.4 Bacon and the Scientific Revolution

The Scientific Revolution is traditionally seen as a period in which Aristotelian natural philosophy was decisively overthrown and replaced by the new 'mechanical philosophy' in the sixteenth and seventeenth centuries. In *The Great Instauration* (1620), Sir Francis Bacon announced his program to replace Aristotle's deductive natural philosophy with a new inductive observational and experimental program that involved 'collecting and perfecting' natural and experimental histories to ground philosophy and the sciences 'on the solid foundation of experience of every kind'. His wide-ranging catalog of particular histories included aerial and oceanic topics: lightning, wind, clouds, showers, snow, fog, floods, heat, drought, and the ebb and flow of the sea. The ultimate goal was to initiate a revolution in the sciences that would replace scholastic arguments based on the authority of a dominant natural philosopher such as Aristotle with a 'great instauration' dedicated to rapid progress in technology, and eventual control of nature (Bacon 1900).

In Bacon's utopian essay *New Atlantis* (1627), the scientists of Solomon's House practice both observation and manipulation of the weather: 'We have high towers … for the view of divers meteors—as winds, rain, snow, hail, and some of the fiery meteors also. And upon them in some places are dwellings of hermits, whom we visit sometimes and instruct what to observe … and engines for multiplying and enforcing of winds to set also on divers motions.' In great experimental spaces researchers imitate and demonstrate natural meteors such as snow, hail, rain, thunder, lightning, and 'some artificial rains of bodies and not of water'. Three so-called mystery men are in charge of expanding the repertoire of practices not yet brought into the arts, and three pioneers or miners try new experiments 'such as themselves think good'; that is, they manipulate nature without further review or oversight, a task requiring perfect virtue and vision in the experimentalist (Fleming 2010, p 50).

Bacon granted grudging respect to the ancients, denigrated the Middle Ages, and elevated modern accomplishments to equal or soon-to-be-greater status than those of antiquity. 'New discoveries', Bacon argued, 'must be sought from the light of nature, not fetched back out of the darkness of antiquity' (Bacon 1900). In his view the sciences were about to enter a period of great fertility because of his new method. Bacon's communitarian campaign was taken over by innumerable practitioners in the seventeenth century. His greatest legacy, without doubt, was institutional, in that his program was adopted by the Royal Society of London and by many other scientific societies. On the other hand, his greatest detractors criticize him harshly for his misogynist language involving masculine knowledge and power over a feminized 'Mother' nature (Fox-Keller 1985).

29.5 Meteorological instruments

In the seventeenth century, new instruments—the thermometer, barometer, hygrometer, and rain gauge—were developed for measuring, weighing, and describing the atmosphere (Middleton 1969). Air thermometers (or thermoscopes) are attributed to Galileo Galilei. The physician Santorio Santorre added a graduated numerical scale to the thermoscope for use in his clinical practice. In 1714, Gabriel Daniel Fahrenheit introduced a reliable mercury thermometer with a scale ranging in degrees from zero (ice, water, and salt) to 32 (ice and water) to 212 (boiling water) to 600 (boiling mercury). René de Reaumur and Anders Celsius also contributed standard thermometric scales of 80° and 100°. In 1848 William Thomson (later Lord Kelvin), proposed the absolute temperature scale that now bears his name. The multiplicity of scales and thermometric standards, however, meant that measurements taken with different instruments were difficult to compare.

Practical experience in hydrostatics led to the realization that water could not be raised more than 32 feet by a common suction pump. In 1642 Gasparo Berti conducted an experiment in which he inverted a long tube filled with water in a basin of water. The water column stood about 30 feet high with an empty space at the top of the closed tube. Two years later Evangelista Torricelli and Vincenzo Viviani conducted a similar experiment using a tube filled with mercury, a liquid almost 14 times denser than water. The column of mercury stood about 28 inches high above its reservoir, with an evacuated space at the top of the tube. Toriicelli and Viviani concluded that, contrary to Aristotle, they had created a vacuum by mechanical means and that the air had weight equal to the weight of the column of mercury. This was the first barometer. In 1648 Blaise Pascal and Florin Perier confirmed that both air pressure and the height of the barometric column decreased with altitude. In London in 1662, Robert Boyle developed the law that at constant temperature, the volume of a given quantity of air varies inversely as the pressure on it (Frisinger 1977).

Early hygrometers attempted to measure atmospheric humidity by weighing or stretching water-absorbing materials such as wool, hair, or catgut strings. In the closing decades of the eighteenth century, standard hygrometric scales were developed by Deluc, Lambert, and Saussure. In 1775 John Leslie used the principle of cooling caused by the evaporation of water to estimate the humidity of the air by measuring the temperature difference between thoroughly ventilated wet and dry thermometer bulbs. The instrument based on this principle is called the psychrometer.

Systematic records of rainfall exist from antiquity in India, Palestine, China, and Korea. In the 1690s Richard Townley weighed the rain he collected and published the results of 15 years of observations. Three decades later Samuel Horsley attached a 30 inch diameter funnel to the end of a graduated three-inch diameter glass cylinder, thereby improving the sensitivity of the gauge ten-fold and allowing direct measurements of the amount of rainfall as small as a hundredth of an inch. To address the problems caused by splashing, wind gusts, and local exposure, the American Francis Nipher devised an effective rain gauge shield in 1878.

Wind vanes are traditional throughout the ancient world and were installed on the steeples of Medieval churches. In about 1450 Alberti invented a pressure-plate anemometer to measure the force of the wind. Much later Robert Hooke designed a rotating windmill anemometer that recorded the wind speed on a strip of paper. Rotating cup and pressure-tube anemometers, especially as improved by William Dines in the 1890s, are still in use today (Middleton 1969).

29.6 Weather observations, maps, and charts

In the seventeenth century, the new scientific societies of Europe promoted the collection, compilation, dissemination, and discussion of meteorological observations by volunteers from remote locations and over widespread areas of the globe. The Accademia del Cimento in Florence led the way. Contemporaries Robert Hooke and Edeme Mariotte promoted similar projects in England and France. In 1686 Edmund Halley developed his theory of the trade winds by compiling observations in three oceans. He identified solar heating as the driving force of the trade winds, speculating that the air would rise in low latitudes and sink in high latitudes. Two years later, he published the first chart of the world's winds.

Results of eighteenth century projects to observe the weather over wide expanses are recorded in the scientific publications of Poland, England, Germany, France, Russia, and the United States (Fleming 1998). New instruments, practices, and perspectives meant that henceforth no atmospheric phenomenon, however seemingly insignificant, would be left unrecorded. As a result, a new culture of measurement emerged, linked to a new meteorological science of planetary proportions. In 1735 George Hadley explained tropical wind systems as the result of solar heating and the Earth's rotational motion. In his system, unequal heating from equator to pole produces a giant convection cell in each hemisphere, with heated air at the equator rising and flowing poleward before sinking and returning to the equator at lower levels.

Over the course of the nineteenth century, national weather services and international cooperation gradually replaced earlier volunteer systems and *ad hoc* organizations. The era of expanding meteorological horizons was an international phenomenon, and state meteorological observing systems in Europe, Russia, and North America followed roughly comparable paths of development. The lack of uniform standards and global and temporal coverage, however, remained a continuing challenge. Climatic surveys of various nations were widespread in the first half of the century, telegraphic experiments began in the 1850s, and storm warning services were established in the 1860s and 1870s. An international bulletin of observations began in 1875 (Fleming 1990).

Historian Cornelia Lüdecke has argued that the First International Polar Year (1882–83) was 'a big science experiment with small science equipment' (Lüdecke 2004). Each of the 12 participating nations was in charge of outfitting its own expedition as part of a big experiment to investigate the meteorology and magnetism of high latitudes.

29.7 Atmospheric dynamics

Beginning in the eighteenth century, a number of scientists promulgated the view that the atmosphere follows the laws of mechanics and thermodynamics, responding in a predictable fashion to external influences. In the 1750s experiments by Joseph Black on the freezing and melting of water and alcohol mixtures led to the concepts of heat capacity, specific heat, and latent heat. In 1851 William Thomson (later Lord Kelvin) provided the first expression of the second law of thermodynamics. Hermann von Helmholtz likened the behavior of the atmosphere to a number of flywheels, or rings of air, spinning around the Earth's axis at different latitudes. The energy for this circulation comes from seasonal and latitudinal differences in solar radiation that generate a slow meridional circulation. Surface friction and eddy viscosity serve to moderate the speed of the circulation. Von Helmholtz also defined sloping surfaces of discontinuity (later called fronts). Max Margules considered the atmosphere as both a mechanical and a thermodynamical system which generates the kinetic energy of storms (Kutzbach 1979).

William Ferrel employed his self-taught mathematical skills, physical intuition, and the latest observations compiled by others to explain, in 1858, wind flows on the rotating Earth. He used the equations of motion of Pierre-Simon de Laplace and the apparent 'force', identified by Gaspard-Gustave de Coriolis, that appears when a body moves in an accelerated system to explain the major wind regimes of the Earth. This was a big idea that went beyond Hadley's simple one-cell model of the atmosphere's circulation.

The 'disciplinary' period in the atmospheric sciences began rather late compared with parallel developments in other sciences. University and graduate education, well-defined career paths, and specialized societies and journals began in the second decade of the twentieth century (Fleming 1990). By 1920 Vilhelm Bjerknes' program to establish the theoretical basis of dynamic meteorology using the equations of hydrodynamics and thermodynamics had attracted considerable international attention and agreement. Accurate measurements of the atmosphere were woefully lacking, however, and the set of complex and non-linear differential equations he developed rendered prediction impossible. With his son Jacob and associates, notably Tor Bergeron, Bjerkenes developed graphical methods of analysis of weather maps that helped impose a semblance of order on the amorphous (and huge) collections of data (Fleming 2016).

29.8 Weather reports and forecasts

In 1856 Joseph Henry began to display weather messages received via telegraph on two large maps of the United States that hung in the Great Hall of the Smithsonian Institution and in the US Capitol. By 1870, the US Congress was supporting a federal telegraphic weather reporting service. Because it was well-funded (over \$400 000 in 1874), the US took the lead in international cooperation in meteorology by publishing the *Bulletin of International Simultaneous Observations*, which contained worldwide synoptic charts and summaries of observations recorded simultaneously at numerous locations around the world (Fleming 1990).

Radio broadcasts of weather conditions date to 1923 and televised weathercasts began in 1947. In 1982 the Weather Channel started round-the-clock cable operations. The earliest graphic displays, drawn by hand with chalk or grease pens, were gradually replaced by blue screen technology that allows the weathercaster to appear before a variety of images including radar, satellite, and analyzed surface observations.

Through the broadcast and print media, the products of the modern atmospheric sciences reach more people on a daily basis than those of any other science. Images of the weather generated by satellite photographs and radar networks are familiar to everyone. Moreover, recent social concerns such as air pollution, ozone depletion, climate change, and climate modification have placed the atmospheric sciences at the focus of national and international attention.

29.9 Atmospheric chemistry

The chemical identification of the constituents of air began in the eighteenth century, beginning with carbon dioxide (1756), nitrogen (1772), pure dephlogisticated air, later called oxygen (1774), and argon (1895) (Khrgian 1970). Ozone was identified as being naturally present in the Earth's atmosphere (1839) and reached its peak concentrations in the stratosphere (1917). By the 1970s stratospheric ozone production was found to be vulnerable to interference from chlorofluorocarbons and other anthropogenic chemicals. The discovery of an Antarctic ozone 'hole' in 1985 led to the enactment of the Montreal Protocol on Substances that Deplete the Ozone Layer (1987) (Brasseur 2020).

Air pollution has been a public issue for centuries. In 1306 King Edward I of England issued a proclamation banning the use of sea coal in London because of the smoke it caused. Over the next few centuries, additional efforts were made in Great Britain to reduce the amount of smoke in the air. The first attempt to control air pollution in the United States occurred during the industrial revolution. During the late 1940s, serious smog incidents in Los Angeles and Donora, PA, once again raised public awareness and concern about this issue and led to the air pollution control acts of 1955, 1963, 1970, and 1990. These 'Clean Air Acts' aimed to protect public health and welfare and set emissions standards (Fleming and Knorr 1998).

29.10 Technologies

The general-use technologies that were adopted by meteorologists in the twentieth century to transform their science include radio, aerospace, and nuclear (Fleming 2016). Radio's lineage involves wireless, continuous wave radio, radiosondes, radar, digital computers, and the telecommunications infrastructure of the space age. Aerospace—broadly interpreted to include the vertical dimension of the atmosphere—encompasses instrumented kites and balloons, commercial and military aviation, sounding rockets, and weather satellites. Aviation and radio-equipped sounding balloons allowed scientists, led by the theoretical work of Carl Gustaf Rossby, to investigate the dynamics of the middle atmosphere, above about five kilometers. The advent of electronic computers provided tools for scientists to experiment with mathematical models of the atmosphere and to compare their results with observations. As mentioned in the Congressional survey of Big Science, the US National Center for Atmospheric Research, its Cray supercomputer, and many new departments of atmospheric science date from the 1960s.

Nuclear issues date to the discoveries of 1895–96, followed by several decades of scientific interest in ionizing radiation in the atmosphere. After 1945 atmospheric detonations of nuclear weapons injected radioactive tracer particles that circled the planet in the wind regimes and descended to Earth as fallout, providing evidence for other studies of atmospheric chemistry and aerial pollutants.

The twentieth century transformation of transportation and communication changed our focus and our relationship to the atmosphere in fundamental ways. It changed how we travel, as innovations such as the airplane rendered new significance to visibility, winds, and weather far above the surface and far afield; and they changed how we send and receive information, via wireless, radio, television, and satellite surveillance—so much so that the twentieth century marked the dawn of a new electromagnetic social environment.

29.11 Climate and climatic change

Enlightenment thinkers, like their predecessors, believed that cultures were determined or at least strongly shaped by climate. They held that the climate of Europe had moderated since ancient times due to forest clearance and cultivation. The climate of the Americas, deemed to be harsher and more violent than in the Old World, would be improved by clearing of the forests, draining of the swamps, and cultivation of the land. Growing amounts of data, when analyzed, however, revealed no changes in the climate in recent times due to human influences (Fleming 1998). The discovery of the Great Ice Age in the 1830s initiated a new and robust discussion about its nature and possible causes. The two leading theories were periodic variations in the Earth's orbit and changes in the radiative properties of the atmosphere.

James Croll (1821–90) was the leading proponent of an astronomical theory of climate change. In 1864 he proposed that the 'true cosmical cause' of climate change 'must be sought for in relations of our Earth to the Sun', that 'geological and cosmical phenomena are physically related by a bond of causation', and that changes in the Earth's orbital elements, combined with physical feedbacks, were 'sufficiently great to account for every extreme of climatic change evidenced by geology' (Irons 1896, p 123). He called climate change, 'the most important problem in terrestrial physics… and the one which will ultimately prove the most far reaching in its consequences', asking, 'What are the physical causes which led to the Glacial Epoch and to all those great secular changes of climate which are known to have taken place during geological ages?' (Croll 1885, p 1). His insight that the orbital elements triggered feedbacks leading to complex changes—in seasonality, ocean currents, ice sheets, radiative forcing, plant and animal life, and climate in general—placed his theory at the nexus of celestial mechanics, terrestrial physics, geology, and evolution. Early in the twentieth century, the Serbian mathematician and geophysicist Milutin Milanković (1879–1958) set out to place the astronomical theory of climate on a firm mathematical basis for the inner planets of the Solar System and calculate the impact of Earth's secular orbital cycles on climate change. In 1920 he published, in French, his *Mathematical Theory of Thermal Phenomena Produced by Solar Radiation* and two decades later, in German, his *Canon of Insolation and the Ice-Age Problem* (1941) in which he calculated Earth's periodic orbital variations (eccentricity, obliquity, precession) along with their climatic influences. Today these are called Milanković cycles (Fleming 2021).

The theory of the Greenhouse Effect also has a long and distinguished intellectual ancestry that includes work done by Joseph Fourier, John Tyndall, and Svante Arrhenius, in the early, middle, and late decades of the nineteenth century (Fleming 1998). In 1856 Eunice Foote reported experiments on the heat of the Sun's rays as affected by air pressure, water vapor and carbonic acid gas (Ortiz and Jackson 2022).

Jean Baptiste Joseph Fourier (1768–1830) is remembered today as a mathematical physicist, but he also considered himself the Newton of heat: 'The principle of heat penetrates, like gravity, all objects and all of space, and it is subject to simple and constant laws' (Fleming 1998, p 153). For Fourier, solar heating and the temperature of outer space were the most significant factors controlling the Earth's heat budget. He did, however, write suggestive passages about what others later called the Greenhouse Effect, although he was not able to quantify it. In the middle decades of the nineteenth century John Tyndall (1820–93) experimented on the radiative properties of various gases and their ability to absorb and emit radiant heat. He established that the radiative properties of water vapor and carbon dioxide were of central importance in explaining meteorological phenomena such as the formation of dew, the energy of the solar spectrum, and, possibly, the variation of climates over geological time. His laboratory experiments with microphysical entities had informed a new view of the cosmos. Following Tyndall's suggestion, the Swedish scientist Svante Arrhenius (1859–1927) demonstrated that variations of atmospheric CO_2 concentration could have a very great effect on the overall heat budget and surface temperature of the planet and might trigger feedback phenomena that could account for glacial advances and retreats. Arrhenius's model relied heavily on the experimental and observational work of others, including Josef Stefan's new law that radiant emission was proportional to the fourth power of temperature, Samuel P Langley's measurements of the transmission of heat radiation through the atmosphere, Léon Teisserenc de Bort's estimates of cloudiness for different latitudes, Knut Ångström's early values for the bulk absorption coefficients of water vapor and carbon dioxide, Alexander Buchan's charts of mean monthly temperatures over the globe, and Arvid Högbom's carbon cycle, in which volcanic eruptions were the chief source of carbon dioxide in the Earth's atmosphere. The notion that humans can influence climate appeared in a new form in 1899 when Nils Ekholm, an associate of Arrhenius, pointed out that at present rates, the burning of pit coal could double the concentration of atmospheric CO_2 and could 'undoubtedly cause a very obvious rise of the mean temperature of the Earth' (Ekholm 1901, p 61). He further speculated about the possibility of deliberate climate modification to delay the onset of an ice age by burning coal exposed in shallow seams.

Guy Stewart Callendar (1898–1964) identified the links between the artificial production of carbon dioxide, enhanced sky radiation, and global warming. Today the theory of anthropogenic global warming (AGW) is also called the 'Callendar Effect'. He was one of Britain's leading steam and combustion engineers and a specialist in infrared physics who was keenly interested in weather and climate, taking measurements so accurate that they were used to correct the official temperature records of central England, and collecting a series of worldwide weather data that showed an unprecedented warming trend in the first four decades of the twentieth century. He formulated a coherent theory of infrared absorption and emission by trace gases, established the nineteenth century background concentration of carbon dioxide at 290 ppm, and argued that its atmospheric concentration was rising due to human activities, which was causing the climate to warm. He did this in 1938! (Fleming 2007).

During the International Geophysical Year of 1957–58, Charles Keeling began to analyze a series of accurate measurements of carbon dioxide taken at the summit of Mauna Loa volcano in Hawaii. Since then, this almost-continuous saw-toothed curve of rising carbon dioxide concentrations has become an environmental icon. Computer modeling of climate was beginning then, and although there was some discussion about increasing particulate pollution, concern soon focused on the rate and amount of warming that could result from a doubling of carbon dioxide in the atmosphere. In 1967 Syukuro Manabe and Richard Wetherald, scientists at the Geophysical Fluid Dynamics Laboratory in Princeton, published a paper presenting the results of a climate model that calculated the thermal equilibrium of the atmosphere with a given distribution of relative humidity. The authors used their radiative-convective model to compute the climate warming for a doubling of carbon dioxide, which had the effect of raising the temperature of the atmosphere by about 2.3 °C. This model constituted an important step toward the ultimate goal of designing three-dimensional climate models that account for additional factors such as oceans, ice sheets, and even weather patterns (Manabe 1997, Fleming 2008) (figure 29.1).

Beginning in the late 1970s, state funding for climate research grew rapidly. By the 1990s the US Global Change Research Program had an annual budget of over \$1.5 billion. By 2017 the US budget for climate change research was \$13.2 billion (figure 29.2). In this era, federal funding for all activities to understand and address

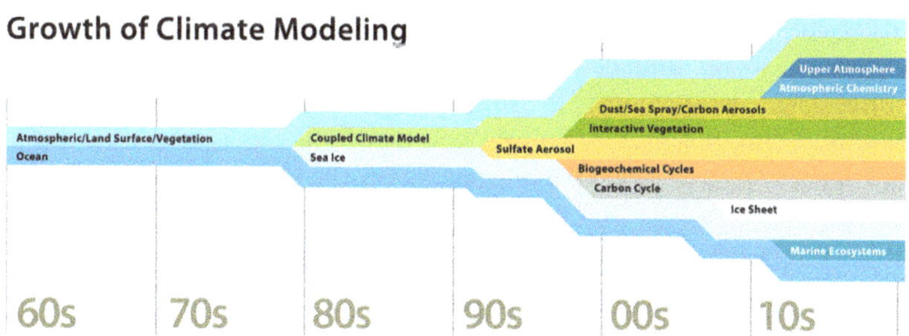

Figure 29.1. Growth of climate modeling—NCAR/UCAR, available at https://opensky.ucar.edu/islandora/object/imagegallery%3A2810. Copyright University Corporation for Atmospheric Research (UCAR)

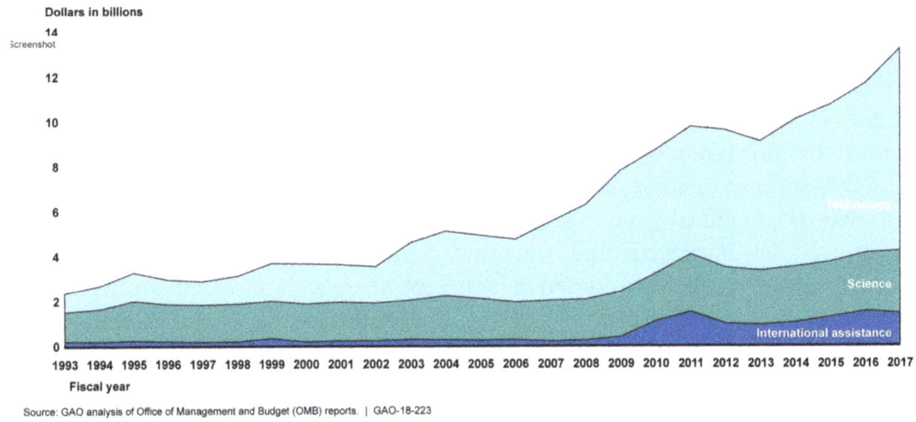

Figure 29.2. Government accountability office-reported federal climate change funding by category (in billions of nominal dollars), fiscal years 1993–2017 (GAO 2018).

climate change amounted to over $154 billion. Edward Lorenz, founder of chaos theory and one of the most influential figures in the modeling of dynamical systems, once observed, 'if by 1970 climate modeling [had] not become big science, it [had] certainly left the realm of little science' (Lorenz 1970, p 326).

In 1988 James Hansen, a climate modeler at the National Aeronautics and Space Administration, announced to the US Congress and the world that 'Global warming has begun'. He went on to report that, at least to his satisfaction, he had seen the 'signal' in the climate noise and we were in for a hell of a warming, perhaps in the form of a runaway greenhouse effect. Hansen later revised his remarks, but his statement remains the starting point of recent widespread concerns over global warming. That same year, the Intergovernmental Panel on Climate Change (IPCC) was formed as a joint program of the United Nations Environmental Programme, the World Meteorological Organization, and the International Congress of Scientific Unions. It has a mandate to prepare regular assessments of what is known and what should be done about anthropogenic climate change. The IPCC is widely considered to be the leading scientific authority on climate change. It represents the consensus of leading climate scientists, and its reports are published with the approval of 195 participating governments. The IPCC's original charge involved the science of climatic change, social and economic impacts, possible response strategies, and international legal instruments in support of the United Nations Framework Convention on Climate Change. Since 1988, the IPCC has issued six comprehensive assessment reports and several special reports on the state of knowledge on climate change.

The 1992 United Nations Conference on Environment and Development (the Earth Summit) in Rio de Janeiro produced the Framework Convention on Climate Change, which calls for a stabilization of atmospheric carbon dioxide concentrations at a level that would prevent dangerous human interventions in the global climate system. This process resulted in international climate change initiatives, including the 1997 Kyoto Protocol and the 2015 Paris Agreement (UNFCC 2021). These agreements represent geopolitical interventions in the climate system. Many more policy initiatives are under way. Economics has also begun to play a role, as taxes and incentives are being put into place to reduce unwanted emissions. Meanwhile, green social engineers are attempting to convince the general public to live sustainably, while 'geoengineers' are floating proposals for massive technical fixes for the climate system (Fleming 2010).

29.12 Interdisciplinary and international atmospheric science

Weather science and weather service established common ground in large cooperative projects such as the World Weather Watch (WWW) a project of the World Meteorological Organization, aimed at providing relevant and timely weather information to all of its member nations. It expanded and improved information sharing from land surface observational networks, civil aviation, merchant shipping, and special ocean weather vessels and ocean buoys. The US and Soviet weather satellites, TIROS-1 (1960) and Meteor-1 (1969), provided observational data from space, while helping to ease tensions between the world's two superpowers.

The Global Atmospheric Research Programme (GARP), coordinated by the World Meteorological Organization and the International Council of Scientific Unions from 1967 to 1982, was a massive international research program to measure the global circulation of the atmosphere, model its behavior, and improve predictions of its future state. It was big in scale, budget, and aspirations, arguing that scientists needed to view the atmosphere as a single, global system. An overarching goal of such activities is to monitor the world's weather through direct and remote sensing and feed the data instantaneously into computers running the most comprehensive atmospheric models in the quest for deeper understanding of atmospheric processes, and to issue the best possible short- and long-range forecasts and warnings.

New interdisciplinary problems, approaches, techniques, methods, and instruments now characterize the modern sub-disciplines of the atmospheric sciences. Cloud physics, atmospheric chemistry, climate dynamics, and radar and satellite meteorology have developed along with more traditional programs in weather analysis and forecasting. Weather and climate modelers use results from all of the sub-specialties, while those building new instruments, investigating particular phenomena, and administering observational networks provide theoreticians and modelers with observational data.

29.13 Conclusion

This essay aims to wrest the category of 'Big Science' away from its earlier focus on unique astronomical observatories and expensive, restricted-access nuclear physics experiments in order to examine 'Big Meteorology' and 'Big Climatology' from antiquity to today. While early conceptual framings, such as those of Aristotle, were big, involving worldwide changes in phenomena, more recent implications of weather and climate knowledge may be even larger, increasingly embedded as they are in societal affairs. 'Climate' has become a keyword in the vocabulary of culture and society. Its nature, history, and vicissitudes are key concepts organizing our ideas of the aerial environment and our relationship to it (Fleming 2015). Although immense effort has gone into mapping and modeling the climate, it cannot be directly visualized or forecast. The term 'atmosphere' as well, has changed in meaning since about 1900, expanding from largely surface weather phenomena to include mid-level tropospheric dynamics as measured by balloons and airplanes after the 1930s, to 'space weather' as measured by satellites and space probes after 1960. The relationships of climate to meteorology, meteorology to climate studies, and both weather and climate to society are dynamic and have expanded dramatically in meaning and import.

Progress in meteorology and weather forecasting over the past century has been threefold: technological innovation, observational data, and theoretical understanding. While progress in all three have been dramatic, all three need cultivation. An influential report noted in 1962 that 'The task of the atmospheric scientists is to make quantitative measurements of the properties that *describe* the atmosphere's successive states, to *understand* the physical processes by which the successive states are determined, to *predict* future states, and, if feasible, to *influence* future states in a beneficial manner' (Petterssen 1962). Atmospheric scientists have addressed the first two tasks and have generated robust progress; the final two tasks, (long-range) prediction and influence (or control) remain beyond their grasp. In the past century, the social aspects of meteorology have grown to include practical forecasting, severe weather warnings, and inter-governmental, diplomatic, and economic initiatives regarding climate's role in the health, prosperity, and future of the planet—that is, like the air itself, enveloping us all. Big Meteorology and Big Climatology indeed!

References

Aristotle 350 BCE Meteorology (transl. E W Webster) *The Internet Classics Archive* http://classics.mit.edu/Aristotle/meteorology.1.i.html

Bacon F 1900 *The Works of Francis Bacon* ed S James, L E Robert and D H Douglas (Boston, MA: Houghton Mifflin) http://onlinebooks.library.upenn.edu/webbin/metabook?id=worksfbacon

Brasseur G P 2020 *The Ozone Layer: From Discovery to Recovery* (Boston, MA: American Meteorological Society)

Brooks H 1968 *The Government of Science* (Cambridge, MA: MIT Press)

Capshew J H and Rader K A 1992 Big Science: Price to the present *Osiris* **7** 2–25

Croll J 1885 *Discussions on Climate and Cosmology* (Edinburgh: Adam and Charles Black)

Ekholm N 1901 On the variations of the climate of the geological and historical past and their cause *Quart. J. R. Meteorol. Soc* **27** 1–62

Fleming J R 1990 *Meteorology in America, 1800–1870* (Baltimore, MD: Johns Hopkins University Press)

Fleming J R 1998 *Historical Perspectives on Climate Change* (New York: Oxford University Press)

Fleming J R 2007 *The Callendar Effect* (Boston, MA: American Meteorological Society)

Fleming J R 2008 *Climate Change and Anthropogenic Greenhouse Warming: A Selection of Key Articles, 1824–1995, With Interpretive Essays* (Washington, DC: National Science Foundation) http://nsdl.library.cornell.edu/websites/wiki/index.php/PALE_ClassicArticles/GlobalWarming/Article10.html

Fleming J R 2010 *Fixing the Sky* (New York: Columbia University Press) p 50
Fleming J R 2015 Weather and climate as shape-shifting nouns: Gordian knots of understanding and prevision *Hist. Meteorol.* **7** 1–13
Fleming J R 2016 *Inventing Atmospheric Science* (Cambridge, MA: MIT Press)
Fleming J R 2021 Cosmic connections: James Croll's influence on his contemporaries and his successors *Earth Environ. Sci. Trans. Royal Soc. Edinburgh* **112** 239–44
Fleming J R and Knorr B 1998 History of the Clean Air Act: a guide to clean air legislation past and present *Interactive History of the Earth Sciences Project* (Boston, MA: American Meteorological Society)
Fox-Keller E 1985 Spirit and reason at the birth of modern science *Reflections on Gender and Science* (New Haven, CT: Yale University Press) pp 43–65
Frisinger H H 1977 *The History of Meteorology to 1800* (New York: Science History)
Galison P and Hevly B (ed) 1992 *Big Science: The Growth of Large-Scale Research* (Stanford, CA: Stanford University Press)
GAO 2018 *Climate Change: Analysis of Reported Federal Funding, GAO-18-223* (Washington, DC: GAO) p 33 https://www.gao.gov/assets/700/691572.pdf
Heninger S K Jr 1960 *A Handbook of Renaissance Meteorology* (Durham, NC: Duke University Press)
Irons J C 1896 *Autobiographical Sketch of James Croll with Memoir of His Life and Work* (London: Edward Stanford)
Khrgian A K 1970 *Meteorology: A Historical Survey* (Jerusalem: Israel Program for Scientific Translations)
Kinsella W J 2000 Big Science *Reader's Guide to the History of Science* ed A Hessenbruch (London: Fitzroy Dearborn) pp 79–80
Kutzbach G 1979 *The Thermal Theory of Cyclones: A History of Meteorological Thought in the 19th Century* (Boston, MA: American Meteorological Society)
Lorenz E 1970 Climatic change as a mathematical problem *J. Appl. Meteorol.* **9** 326
Lüdecke C 2004 The First International Polar Year (1882–83): a Big Science experiment with small science equipment *Proc. Int. Commission on History of Meteorology* **1** 55–64
Manabe S 1997 Early development in the study of greenhouse warming: the emergence of climate models *Ambio* **26** 47–51
Middleton W E K 1969 *Invention of the Meteorological Instruments* (Baltimore, MD: Johns Hopkins University Press)
Ortiz J D and Jackson R 2022 Understanding Eunice Foote's 1856 experiments: heat absorption by atmospheric gases *Notes Rec.* **76** 67–84
Petterssen S 1962 The atmospheric sciences, 1961–1971 *Weatherwise* **213** 185–7
Plato 360 BCE *Phaedo* 109A–110A http://classics.mit.edu/Plato/phaedo.html
Shakespeare W 1595 *Richard II (Folger Shakespeare Library)* https://shakespeare.folger.edu/shakespeares-works/richard-ii/
US Congress, House of Representatives 1987 Task Force on Science Policy, Committee on Science and Technology, Congressional Research Service Report *World Inventory of 'Big Science' Research Instruments and Facilities* 99th Congress, 2nd Session (Washington, DC: USGPO)
UNFCCC 2021 About the Secretariat *UNFCCC* https://unfccc.int/about-us/about-the-secretariat

IOP Publishing

Big Science in the 21st Century
Economic and societal impacts
Panagiotis Charitos, Theodore Arabatzis, Harry Cliff, Günther Dissertori, Juliette Forneris and Jason Li-Ying

Chapter 30

Big Biology

Niki Vermeulen, Rosalind Attenborough and Rodrigo Liscovsky

This essay will address the interplay between Big Science and the life sciences. Although the Human Genome Project is often seen as the start of Big Biology, (inter)national networks were key in much earlier forms of biology research, e.g. natural history efforts to explore, collect, and categorise life on earth. These two examples already indicate the diversity of scientific work contributing to the emergence of larger-scale organisational structures in the life sciences. Building on historian John Pickstone's conceptualisation of *Ways of Knowing* (2000), we will show how different forms of biology research have acquired different types of organisational structures, not only influenced by the type of work, but also crucially their social-cultural embedding. We will subsequently discuss 'natural history', 'analysis', 'experimentalism', and 'techno-science', arguing that these constitute different ways of collectively knowing and working in biology. Exemplary case studies can be found in ecology and biodiversity research, molecular biology and model organism research, integrative biology—such as synthesis research, systems, and synthetic biology—as well as biomedical research and space biology. We will show how these different forms of research into life combine different epistemic and societal agendas, including developments in science and innovation policy. Woven through these case studies are transformations in the funding of research, (re) constructions of disciplines, technology development including digitalisation, and shifting orientations towards application and public engagement. Together these examples of collective research arrangements in the life sciences will show a rich ecology of Big Biology and its development through time, while also providing food for thought for future forms of research and its politics.

30.1 Introduction

The idea of Big Biology is contested. Not only has the bioscience community debated the desirability of the growth of biology (Roberts 2001), also from an analytical point of view the question was raised: is biology big enough to count as a player in the Big Science league? (Knorr-Cetina 1999, Shrum *et al* 2007). When determining 'bigness' through the single dimension of instrument size, and comparing a particle accelerator to a genome sequencer, the answer might simply be no. However, such a simple comparison neglects the complexity of 'Big Science' as a modern concept, which captures transformation processes in the growth of science. Big Science has given rise to both quantitative and qualitative studies of increasing scale in science—quantitative studies mainly focus on an increase in numbers (of scientists, publications, co-authorships, funding, and economic impact), while historical and social studies attend to qualitative changes in the organisations of science. The latter take a more processual view to explore reasons and consequences of shifts towards larger-scale science (Vermeulen 2009, 2016). Centralisation around larger technologies is one of these processes which can also be observed in biology, but there is much more to say about the ways in which the organisation of research into life has changed to incorporate characteristics of Big Science (see also Kevles 1997). Most prominently, developments in biosciences culminating in the Human Genome Project showcase the role of information technologies in the growing global connectedness of research and its infrastructures (see also Glasner 1996, Garcia Sancho 2012, Hillgarner 2017). In addition, transformations in the socio-political environment have made biology a post-industrial form of Big Science in which issues such as the applications of research and ethical dimensions are more prominent than in modern forms of Big Science such as Big Physics. Although these societal transformations are enlightening for transformations in Big Physics as well (see Hallonston 2016, after Vermeulen 2009). This essay builds on more complex understandings of growth in science, attending to various forms of 'Big Biology'—from ecology to systems biology and the Human Brain Project—and how they relate to other types of large-scale research. Most importantly, the concept of 'Big Science' informs the connection between the content and form of research—the co-construction and mutual shaping of epistemic and organisational developments in science for which we develop the idea of 'collective ways of knowing'.

30.2 Collective ways of knowing

In his influential book *Ways of Knowing* (2000), historian John Pickstone has provided us with a long-durée analysis of the development of science, technology, and medicine in which he shows the entanglement of content and form of research over time. Using eggs as an example he lays out the different ways in which knowledge of an egg can be derived. One may describe and classify based on observations of the eggshell; analyse its reproductive function; look inside the shell; or intervene: what happens if you remove half of the egg? Do you get half an embryo or two? He uses this illustrative example to distinguish dominant phases in the history of science, from 'natural history' to 'analysis' and 'experimentalism'. Moreover, he connects these phases to specific forms of scientific work with corresponding social relations, nested in particular institutional and socio-political environments. As such, ways of knowing are connected to ways of working (see also

Agar 2012 who uses this perspective to write about the history of science in the twentieth century). From these ways of knowing and working we can easily derive the collective nature of these research endeavours—none of these phases involves just one person or one organisation; from the very start they involve networks of scholars and various institutions. As such we can also talk about 'collective ways of knowing' and by putting the spotlight on the collective nature of these research configurations, we relate analyses of ways of knowing to analyses of Big Science and its emergence.

A first insight that can be derived from this combination of perspectives is how natural history as a form of research is an early form of Big Science, perhaps even the most long-standing (Vermeulen 2009). Early collaborative networks in natural history formed to assemble the variety of life dispersed over the globe. Natural scientists were part of world explorations in the seventeenth century that drove cartography of the Earth and sky and widened knowledge of the various forms of life (Capshew and Karen 1992, Fernández-Armesto 2006). Natural historians joined expeditions to collect new species, giving rise to networks that exchanged objects and information. In this way of knowing the description and classification of objects or systems is key: 'From the seventeenth century, at least, natural history has been the study of "what we have"—in databanks, or public or private collections' (Pickstone 2000, p 11). This particular kind of scientific work has to be placed in the context of the Renaissance, trading nations, Empire building, cabinets of curiosity, and the establishment of scientific societies and national museums.

The beginning of the nineteenth century gave rise to a new way of knowing, the 'Age of Analysis': 'I include as analysis both the decomposition of "compounds" into their various elements, and the reduction of systems to the "flow" of single elements' (Pickstone 2000, p 85). Examples are analytical chemistry and thermodynamics which emerged together with a mechanical view, processes of rationalisation, the formation of disciplines and further development of research instruments such as the microscope. In biology this spawned a more reductionist view, including dissections which enabled research into tissues, organs, and cells as well as processes and mechanisms of life. In the late nineteenth century experimentalism and invention became more central as means to control nature and create novelty, as for instance exemplified by the shift from analytic to synthetic chemistry. In biology control of life became key, e.g. in the case of Pasteur experiments with bacteria led to the development of vaccines. Although this new experimental work in biology still encompassed collection, observation, and analysis, its proponents often portrayed themselves in opposition to older naturalists which they dismissed as 'stamp-collectors' (Maienschein 1993, Pickstone 2000). Thereby the laboratory became the dominant working place of research, with laboratory values subsequently influencing field biology practices through the creation of fieldlabs to carefully analyse and measure processes and relations of life (Kohler 2002).

It is in the context of analytical and experimental biology and the emergence of the laboratory that we can recognise the individual mode of working in biology as described by Knorr-Cetina (1999). She argues that unlike Big Physics, molecular biology is an 'individual, bodily lab-bench science' with scientists working on their own projects, and the laboratory represented by the individual laboratory leader. Even collaboration in biology takes place within this 'individual ontology', following a 'logic of exchange': 'Participants in the lab render services to each other in exchange for other services and co-authorship. Equally, the leader is often approached with requests for services by other labs, which he or she gladly renders, knowing that return favours will be forthcoming and might be needed' (Knorr-Cetina 1999, p 255). However, more contemporary studies of biology show how analytical and experimental ways of knowing in biology gradually became more integrated collaborations in interaction with the development of genetics, molecular biology, and genomics research (Parker *et al* 2010).

In addition, forms of collaboration between laboratories, government, and industry appeared in biology (Kwa 1987, Thackray 1998, De Chadarevian 2002, Strasser 2003). Connections were first built in the context of biomedical sciences before and during World War II. These developments comply with what Pickstone calls 'technoscience' or 'technological projects that are heavily dependent on science (or vice versa)':

> Techno-science, including techno-medicine, I take as first created at the end of the nineteenth century, when certain laboratory products (or processes) became commodities, so partially reconstructing the social relations of STM to include industrial research laboratories, linked to universities, to state laboratories and to other institutions, in ways which have since become more and more areas of STM. (Pickstone 1993, p 434)

This collaborative way of knowing is also described in theories of science and innovation, whereby connections are crucial for research and development, such as the 'Triple-Helix theory' which uses the structure of DNA to show the intertwining of academia, industry, and government (Leydesdorff and Etzkowitz 1998). And indeed, in analyses of large-scale biology projects such as the Human Genome Project to sequence human DNA, this intertwinement of public and private research is a main topic of discussion (Cook-Deegan 1994, Hillgartner 1995, Venter 2007).

So how should we think about contemporary forms of biology, such as genomics or systems and synthetic biology? What ways of knowing do they exemplify? Surely, they are all contemporary forms of techno-science, but at the same time they relate to earlier ways of knowing too—there is in fact a layering of ways of knowing in contemporary biology (Pickstone 2007). For instance, sequencing combines techno-science with analysis as it uses technology to analyse the human genome. It is also a form of natural history as it is about collecting and categorising human genome data. In turn, systems biology—which tries to understand life through the creation of digital models aiming to predict life—mainly combines techno-science with analysis. Synthetic biology more clearly aims to experiment with life, creating new or future organisms (Kastenhofer 2013, Schyfter and Calvert 2015). A similar argument about the layering of ways of knowing is made by Bruno Strasser (2019), showing how Big Data biology relates to natural history. Looking in reverse, we can see how contemporary natural history research has become techno-science too, e.g. marine biology becoming Big Science (Vermeulen 2013). As we outline next, the emergence of techno-science and the layering of ways of knowing in biology have also changed the organisation of research, resulting in more collective ways of working.

30.3 The growth of biology

If we take a closer look at the growth of biology, we can see how the globally dispersed nature of biological research material has been an important incentive for collaboration in the biosciences. In addition, technology development, research policy, and funding, and the shaping of biology *vis-à-vis* other scientific disciplines have contributed to increasing scales in the life sciences. Various research fields and communities have been actively arguing for and constructing Big Biology, most prominently the Human Genome Project and its international consortium, but they did this against the background of the development of model organism communities, and similar initiatives in proteomics, while giving rise to post-genomic endeavours such as systems and synthetic biology. At the same time, movements towards more integrated and larger-scale research can be observed in taxonomy and ecology. As such we see the development of an ecology of Big Biologies.

There is a clear connection between Big Physics and the emergence of Big Biology in the second half of the twentieth century. As investments in physics became controversial because of the devastating power of atomic bombs and subsequent debates on nuclear energy, the national laboratories in the United States started to incorporate biosciences in their strategy, looking into the biological effects of radiation on the one hand and environmental pollution on the other (Bocking 1997, Schloegel and Rader 2005, Creager and Santesmases 2006). In the context of their interest in the effects of radiation on the human genetic make-up, it was the Department of Energy which first envisioned large-scale research collaboration in molecular biology, using its experience with organising large-scale physics research in shaping and applying it to the Human Genome Project.

The Human Genome Project also built on increasing exchanges within the bioscience community itself. Similar large-scale aspirations were already present for some time in the community studying proteins, resulting in the 1965 Atlas of Protein Sequence and Structure (Strasser 2010). From the European side the creation of the European Molecular Biology Organization (EMBO) and the European Molecular Biology Laboratory (EMBL) as well as the EU framework programmes played a role in the increase of collaboration around technologies and model organisms, such as yeast (Garcia Sancho Sanchez *et al* 2021). Next to yeast, alliances began to form around the model organisms *Drosophila* and *Caenorhabditis elegans*, studying, respectively, the fly and the worm, involving exchange of research material and information and division of work across different laboratories (Kohler 1994, Gaudillière and Rheinberger 2004, Leonelli and Ankeny 2013). The type of materials these communities exchanged gradually transformed together with the development of information technologies, from the exchange of actual mutant organisms towards more abstract, and eventually digital, representations that also allowed for sharing (García-Sancho 2012, Kelty 2012, Attenborough 2020).

International coordination to sequence the human genome eventually resulted in a public consortium which consisted of 16 groups spread about 48 laboratories world-wide (Balmer 1996a, 1996b, Glasner 2002). With additional efforts in the private realm, the publication of the completed map of the human genome contained the names of some 520 scientists which constituted a major shift towards collective authorship in biology (Galison 2003). In opposition to more conventional Big Science arrangements, Peter Glasner describes sequencing efforts as 'conducted by mainly small groups of scientists contributing to a clearly defined long-term goal' (Glasner 1996, p 109). He used the term 'collaboratory' defined as 'technology, tools and infrastructure that allow scientists to work with remote facilities (co-laboratory) and each other (collaboratory)' (Glasner 1996, p 111, citing Lederberg and Uncapher 1989, p 3). Consequently, the Human Genome Project can be seen as an international infrastructure with different nodes—a networked form of Big Science.

In line with earlier reflections on the pros and cons of Big Science (Price 1963, Weinberg 1967) and Big Physics (see also Joseph Martin's contribution to this volume), Big Biology also gave rise to substantial debate. Proponents of Big Biology presented the Human Genome Project as the new and more effective way to perform research, while opponents claimed it undermined the very character of biological research, bureaucratising, and politicising it while diluting creativity (Davis 1990, Check and Castellani 2004). Genome sequencing was portrayed as 'massive, goal-driven and mind-numbingly dull' (Roberts 2001). Molecular biologist Sydney Brenner even joked that sequencing was so boring it should be done by prisoners: 'the more heinous the crime, the bigger the chromosome they would have to decipher' (Roberts 2001, p 1183). Unfortunately, the term Big Biology was mainly used as a rhetorical weapon, pictured as either good or bad, prematurely silencing the development of more nuanced understandings of the positive and negative sides of ongoing transformations in the biosciences.

In the slipstream of the Human Genome Project various collaborations have been set up. First of all, numerous sequencing projects targeted different species, from the microscopic to the gigantic (Núñez Casal 2019, Callaway 2021), producing enormous amounts of data, making storing and integrating data a key feature of modern biology (Groenewegen and Wouters 2004, Leonelli 2016, Wang *et al* 2020). In addition to genomes, other parts and processes of cells were charted within the contexts of so-called '-omics' research, including proteomics, metabolomics, and nutrigenomics (Lewis 2008, Penders 2010). Expertise came from different disciplines and sub-disciplines that were combined through the formation of multi-disciplinary collaborations, with the importance of technologies bringing the need for physicists, mathematicians, engineers, and computer scientists. In the case of creating computer models of living cells it was argued that the scale of these collaborations would need to surpass the size of the Human Genome Project as the task was more complex—involving the mapping of all the components of a cell and their interactions (Vermeulen *et al* 2012). However, the building of larger-scale projects in biology came with growing pains, as shifts in ways of working required shifts in research funding schemes which often proved difficult to achieve. Next to the increase in international networks and consortia—see also the analysis of the Human Brain Project as one of the more recent additions to Big Biology (Mahfoud 2021)—collaborations took the shape of research centres, as seen in systems biology and synthetic biology (Meyer and Molyneux-Hodgson 2016, Vermeulen 2018). Research funding to set up new centres working across departmental boundaries brought together researchers and instruments over longer periods of time, contributing to new ways of collaborative working and the integration of previously distinct disciplines and perspectives on life.

Such a combination of networks and centres can also be found in environmental biology. Early expeditions of natural historians gradually evolved into more coordinated multi-disciplinary research programmes, initially taking the form of thematic years or decades. Most notably the International Polar Year (IPY), which first took place in 1882–83 and was repeated in 1932–33, concentrated international research efforts investigating the North and South Poles. The success of the polar years led to the organisation of the International Geophysical Year (1957–58) and the International Biological Program (1968–74) advanced by the International Union of Biological Sciences (Pirie 1967, Dunbar 1971, Kwa 1987, NOAA 2007). Although originally genetics and human population studies were part of the core of the IBP, in light of emerging environmental concerns systems ecology, with its promise to control nature, became the central focus. As such the IBP can be seen as the first time in which ecology became Big Science, and since then it has also been analysed as such (Bocking 1997, Schloegel and Rader 2005, Parker 2010). The Long Term Ecological Research (LTER) network and National Ecological Observatory Network (NEON) are modern large-scale ecological research networks, while synthesis centres, such as the National Centre for Ecological Analysis and Synthesis (NCEAS), bring researchers together in one place to integrate work on similar topics and reach understandings of ecological systems on more global scales (Hackett *et al* 2008). At the same time taxonomists are actively trying to build a more coherent community to obtain funding for fundamental work in biodiversity on land and in the oceans (Vermeulen *et al* 2010, Wheeler *et al* 2012).

As a result, we see increasing scales of research reinforced by data infrastructures —in the form of (inter)national networks as well as centralisation of research in one place—in both environmental biology and in biosciences more orientated towards human health. On the one hand, these communities sometimes still operate at distance from each other: for instance, connections between systems ecology and systems biology communities are minimal and much less than expected given their joint emphasis on a systems approach to life. On the other hand, applications of systems and synthetic biology are often envisioned across biomedicine and environmental sciences, in line with the application of multi-species thinking in the development of the One Health approach (Schneider *et al* 2019). Moreover, different configurations of research mirror and develop in relation to one another, as in the case of lab and field studies, a boundary which is often crossed and transformed. This occurs not only with the development of lab-scapes (Kohler 2002), but also with the extension of the lab into society and the private sphere, e.g. through personalised medicine and its wearable devices (Nieroda *et al* 2015, Prainsack 2017). Moreover, there are relations between Big Biology and other members of the Big Science family. Not only has Big Physics integrated and inspired biology, they are currently merging in the advent of quantum biology (Marais *et al* 2018). Similarly, findings from deep-sea marine biology are relevant for life in the International Space Station, while research in the International Space Station (ISS) contributes to the understanding of experiments with life on Earth, e.g. the development of 3D bioprinting for human organs may be improved by an absence of gravity when printing living cells (Sims 2021).

30.4 The societal embedding of Big Biology

Big Science is also about politics and power and this is also the case with Big Biology. This becomes most clearly visible through competition between disciplines or countries—e.g. a recent race between the US and Russia to get the first bioprinter into space (Vaessen 2018). In contrast, international cooperation is often an objective, e.g. European integration was the main incentive for CERN and subsequent European framework programmes and infrastructure projects (Pestre 1992, Ulnicane 2020). As such, socio-political circumstances influence Big Science endeavours and this is relevant for Big Biology as the increase in scale took place in different times than the growth of Big Physics and space research. The emergence of Big Physics needs to be understood against the background of the modernisation of society—industrialisation and increasing need for energy; growth of the military-industrial complex; and the context of the Cold War in which big space research developed (Lenoir and Hays 2000, Creager and Santesmases 2006, Woods 2009, Agar 2012). However, the analysis of increasing scales of biology towards the end of the twentieth century needs to be situated in a different context—a knowledge society and economy in which science and technology replaced industrialisation as a foundational underpinning of society at large (Drucker 1969, De Wilde 2001). The network characteristics of Big Biology resemble the 'network society' as described by Manuel Castells (1996), with the development of information technologies and globalisation processes transforming the functioning of science as well as society. Moreover, the transition towards a knowledge economy shaped developments in biology research, as techno-science became a source of innovation and economic growth, thus knowledge about life became a commodity (Rose 2001, Sunder Rajan 2006, Waldby and Mitchell 2006).

The neo-liberal emphasis on innovation in science policy sets Big Biology apart from more classical forms of Big Science, because the requirement for applied research made the expansion of biology rather complicated. Being a source of medical, agricultural, and environmental innovation, life scientists are still always expected to know future (industrial) applications of their research, while observing that Big Physics and space research receive research funding without the societal usefulness of their projects being clear from the outset (Yeh 2019). As one of them commented: 'Even when these space shuttles explode, they still receive huge amounts of money'. The other way around, a physicist remarked that CERN could not have been built within contemporary science policy and funding regimes

(Vermeulen 2009). The solid organisational machinery of these classic forms of Big Science was built in times of the old social contract between science and society, with extensive lobbying experience acquired over time. These can be seen as reasons why Big Physics and space research managed to continue their funding. In contrast, life scientists have been trying to increase the size of their research within an environment that values application over more basic forms of science, while also having an organisational disadvantage vis-à-vis older forms of Big Science, as they do not have equivalent organisational and lobbying experience. Moreover, it has proven difficult to acquire long-term funding, as the end of the Census of Marine Life and the rise and fall of systems biology centres illustrates (Vermeulen 2013, 2018). In both cases, these research projects tried to survive in the form of (data) infrastructures, respectively, as the Ocean Biodiversity Information System (OBIS) or the Infrastructure for Systems Biology Europe (ISBE)[1].

While biology's potential for application can be used to legitimise large-scale biology research, this also draws it into important societal debates around health, food, and the definition of life. This means that biological research and its applications are often contested, as controversies about genetically modified food and stem-cell research have shown. Although Big Physics has given rise to some serious controversies too—for instance around atomic weapons and nuclear energy, giving an impetus to the philosophy of technology and social studies of technology—it is in the context of Big Biology that societal dimensions were first addressed in an integrated way. The Human Genome Project gave rise to an extensive programme on the ethical, legal, and social implications of genetic research, which set the stage for many more research programmes into the interaction between biology and society (Dreger 2000). An initial focus on downstream implications of biology research gradually shifted towards the integration of social sciences and humanities research into the start of scientific research, as is currently visible in synthetic biology centres funded by the BBSRC in the United Kingdom as well as the Human Brain Project (Balmer *et al* 2015, Aicardi *et al* 2018). As a result, Big Biology research has been an important factor in the broader movement towards the embedding of science in society or what now has been called Responsible Research and Innovation (Nowotny *et al* 2001, Owen *et al* 2021).

Perhaps the most prominent ways in which (big) bioscience and society are coming together is through citizen science and science activism. Natural history as a way of knowing has always relied on volunteers to collect specimens and consequently field biology has a long tradition of what now has been called 'citizen science' (Strasser *et al* 2019). Thereby, a variety of publics are able to contribute to scientific research, and digitalisation processes have played a role in the popularisation of these collection efforts, e.g. the Big Seaweed Search[2]. In addition, scientists seem increasingly active in their role as citizens, not only producing knowledge but also working on its societal impact. Some of this work might be driven through

[1] For more information see, respectively, the websites https://obis.org/about/ and https://cordis.europa.eu/project/id/312455.

[2] See https://www.bigseaweedsearch.org.

impact and valorisation agendas of funding and evaluation systems, but it is also a more bottom-up transformation with researchers feeling the need to act on their research findings (de Jong *et al* 2016). Not surprisingly, this is particularly visible within the environmental sciences, as issues of pollution and biodiversity loss are showing the urgency of such action and the need for collective action to have some impact. The combination of being both a scientist and an activist and the tension between these roles raises important questions regarding identity work and ways to successfully combine both activities[3]. This tension is also relevant in Big Science: do we need large-scale research to contribute to societal challenges and integrate societal actors?

30.5 Rethinking contemporary science through Big Biology

Since its conception, reflections on 'Big Science' have been evaluative (Vermeulen 2009). The growth of science was certainly not always perceived as progress and the characterisation of Big Science as a 'pathology' (Capshew and Karen 1992) was associated with Weinberg's critical outlook on Big Science as a phenomenon that covers the terminal diseases of mature science: 'journalitis, moneyitis and administratitis' (Weinberg 1961, p 162). This debate heated up again in the context of the Human Genome Project when some were against the making of Big Biology (Roberts 2001). Similar issues are currently debated in the context of biomedical research, e.g. the Wellcome Trust survey on research culture concluded that 'the system favours quantity over quality, and creativity is often stifled'[4]. Moreover, it states that while policy-makers often mention virtues of collaboration, the current research culture does not necessarily incentivise or facilitate collaborative behaviour, but rather (hyper) competition which can harm the quality of results. In an attempt to make sure their funding is used to reach its goal—the improvement of human health and wellbeing—the Wellcome Trust concluded that first and foremost the research system needs to be fixed. On the agenda are the responsible use of metrics through the implementation of the San Francisco Declaration on Research Assessment (DORA)[5] and a broader awareness that the science system is focussed too much on individual excellence and incentives, which harms research as a collective endeavour and makes young researchers increasingly vulnerable and precarious (Schönbauer 2019).

In line with current societal discussions on neo-liberal growth and its relation to pollution, biodiversity loss, climate crisis, and the current pandemic, these debates about research culture also hint at limits to growth. As such, reflections on Big Biology and the ways in which it has grown can provide input into contemporary discussions on (large-scale) research, and its opportunities and challenges. Here we will focus in particular on the ways in which Big Biology relates to current debates around decolonisation of science, global inequalities in research, and shifts towards

[3] See Sarah Schönbauer's project 'Choreographies of Environmental Engagement' funded by the Austrian Science Fund (FWF).

[4] See the Wellcome Trust website: https://wellcome.org/reports/what-researchers-think-about-research-culture.

[5] See the DORA website: https://sfdora.org/.

data sharing and open science. These three issues illustrate how the structuring of collaboration matters, because ways of knowing and working play a role in the creation of inclusive collectives and epistemic justice (see also Koch 2020, Valkenburg *et al* 2020).

First, Big Biology *avant la lettre* participated in empire building, through the development of global networks of exchange for biological materials as part of the natural history endeavour to collect and categorise life on Earth. The natural history museums established all over the world to house these collections often bear traces of the now controversial socio-political contexts in which the specimens were collected as well as presented. For example, the human skull collection of the Natural History Museum in Vienna has now moved behind the scenes and has become subject of extensive reflection on the history, ethics, and politics of European skull collections[6]. As such, these museums have the double challenge of preserving the huge quantity of specimens they inherited, while coming to terms with the value and meaning of this type of scientific heritage. This leads to instructive interventions and debates that can inform the decolonisation of science more broadly. For example, the Natural History Museum in Berlin has established a 'bureau for troubles' which successfully stimulates reflection on its own working practices and history including reflections on decolonisation[7]. However, epistemic injustices are not confined to past research practices, as issues of race and diversity have also come to the fore in relation to more contemporary (molecular) biology. In the context of the Human Genome Project and genomics notions of race and racism are questioned, e.g. how are categories of race constructed and reinforced through genomics (Rajagopalan and Fujimura 2012). More recently and as part of the Black Lives Matter movement, the synthetic biology community has started debates about its lack of diversity and the ways in which this influences the epistemic agenda of the field (Kuiken 2020). Decolonialisation, racism, and epistemic diversity are all themes that the scientific community at large needs to contend with too.

This relates very closely to global inequalities in research communities and the ways in which the current organisation of science concentrates around the Global North, as exemplified by model organism communities. With the advent of the Human Genome Project, granting bodies started recognising the value of model organism research by funding not only genome sequencing projects but also 'community resources' such as stock centres and cyberinfrastructures (Ankeny and Leonelli 2011, Leonelli and Ankeny 2012). Practices of collaboration and resource exchange (e.g. techniques, specimens, and data) common to all model organism communities have themselves become models for good behaviour in science (see Nelson 2013, Leonelli and Ankeny 2015). Although these studies are describing the transformative impact of internationalisation dynamics in the life sciences, they often decontextualise the very same international research

[6] This collection was part of the TRACES project and its exhibit 'Dead images: facing the history, ethics, and politics of European skull collections'. See http://www.traces.polimi.it/ and especially http://www.traces.polimi.it/download/3571/index.html.pdf.

[7] See https://www.museumfuernaturkunde.berlin/en/science/bureau-troubles.

communities they analyse and tend to ignore the power dynamics present between centre and periphery (Liscovsky 2021). By investigating the international dynamics of the zebrafish community from a Latin American perspective, Liscovsky shows how notions of asymmetry and dependency in practices of exchange, as well as structural configurations, determine the norms and expectations operating in these model organism communities. For instance, there is no zebrafish resource centre in Latin America, and there are barriers to the import of adult specimens and embryos. This means that scientists in Latin America must find ways to acquire and grow their own fish locally in order to sustain their contributions to the developing international body of knowledge on zebrafish. Liscovsky's work challenges the tendency to view model organism communities as uniform and harmonious international ecosystems governed by a strong collaborative ethos, opening the way for more critical reflection on model organism communities and the internationalisation of the life sciences.

Model organism communities and the Human Genome Project have also played an important role in the development of mechanisms to share data and consequently contributed to the construction of 'open science'. Twentieth century model organism communities typically operated within bounded social settings, depended on reciprocity and trust, which eased the sharing of ongoing work (Kelty 2012). Attenborough (2020) argues that the establishment of model organism communities are an important prehistory to the emergence of twenty-first century 'open science' on three grounds: they demonstrate that a cultural valuing of openness has been embedded in many biological research communities; model organism databases have been important sites for digital and online transformation towards public accessibility; and these communities trained many key figures in the public HGP effort, which established immediate, online data sharing as an internationally agreed norm for large-scale sequencing projects. As such, they were an important origin for more widespread open science movements in the 2000s, later framed as 'open data'. However, the sharing of data also lays bare inequalities in the current research system, such as the vulnerable position of early career scholars for whom data are their currency in a competitive system. Consequently, current proposals to promote open scientific cultures top-down by creating rewards for open practices (e.g. Wilsdon *et al* 2017, Ayris *et al* 2018) risk feeding into systematic societal inequalities, which likely make openness more accessible to scientists at intersections of power in gender, race, ethnicity, disability, sexual orientation, class, and so on (Attenborough 2020, p 321). Moreover, the sharing of unpublished ideas, or 'interpersonal openness'—'talking freely, sharing generously, with indirect reciprocal benefits; and a dependence on, or a creation of, trust' (Attenborough 2020, p 281)—is a crucial form of exchange for researchers which contributes to community building and flourishes in a collaborative and not in a competitive environment.

30.6 Conclusion

Following Alvin Weinberg's statement in *Science* 'Big science is here to stay' (1961, p 161), a *Nature* editorial stated: 'Like it or not, big biology is here to stay' (2001, p 545). As we have shown through the exploration of diverse collective ways of knowing in biology, its ways of working have indeed shifted towards larger-scale projects and infrastructures. This does not indicate a tendency towards a uniform size or configuration of 'Big Biology', but rather shows an ecosystem of various ways in which growth in biology has taken shape. The distinction between ways of working and ways of knowing attends to the interaction between the organisation of research and its epistemic focus, between form and content. In biology we can observe both the necessity to standardise and centralise in the laboratories of (molecular) biology orientated towards human health, and the introduction of this lab culture in field biology to acquire environmental knowledge, with both lab and fieldsites being increasingly connected through global research infrastructures. The types of knowledge generated in these networked research spaces aim to be more generalised and complete, e.g. integrating disparate theories, methods, and data across topics, specialties, and scales (see also Hackett *et al* 2016). As such, attention to collective ways of knowing shows how some types of knowledge can only be generated through collective work. Moreover, the ways in which collectives are structured influences the types of knowledge they produce, as shown by evaluations of Big Science in general and our own reflections on Big Biology, highlighting the importance of collectives which foster epistemic justice (see also Koch 2020).

The COVID-19 pandemic has emphasised the need to connect understandings of environmental and human health, which requires the development of organisational mechanisms that integrate different types of biological research. However, this type of synthetic research does not only require increasing scales, but an awareness of the pros and cons of different sizes of research—big, mezzo, and small—and how to find a balance between them (Vermeulen *et al* 2010). Synthesis centres have been able to find a solution in the integration of previously distinct but related research efforts by bringing researchers and their findings in conversation through a series of meetings over several years (Hampton and Parker 2011), and such smart configurations should be used and optimised to combine smaller and bigger forms of research and connect local efforts to make more global impacts. In line with our reflections on Big Biology, it also requires a changing relationship between centre and periphery to make the globalisation of knowledge production more equal, in contemporary work and in the rethinking and reworking of scientific heritage. In the face of increasing inequalities both within and outwith the research system, precarity should be replaced by care (Kerr and Garforth 2016), caring for a new generation of scientists and the creation of scientific cultures that foster community building and collective ways of working and knowing.

References

Agar J 2012 *Science in the Twentieth Century and Beyond* (Cambridge: Polity)

Aicardi C, Tyr Fothergill B, Rainey S, Stahl B and Harris E 2018 Accompanying technology development in the Human Brain Project: from foresight to ethics management *Futures* **102** 114–24

Ankeny R and Leonelli S 2011 What's so special about model organisms? *Stud. Hist. Philos. Sci. A* **42** 313–23

Attenborough R 2020 Finding virtue in open science? Biological scientists' constructions of openness in historical, advocacy and policy contexts *PhD Thesis* The University of Edinburgh http://dx.doi.org/10.7488/era/1150

Ayris P , López de San Román and Maes K 2018 Open Science and its role in universities: A roadmap for cultural change (The League of European Research Universities) Advice Paper no. 24

Balmer B 1996a Managing mapping in the human genome project *Soc. Stud. Sci.* **26** 531–74

Balmer B 1996b The political cartography of the human genome project *Persp. Sci.* **4** 249–82

Balmer A S, Calvert J, Marris C, Molyneux-Hodgson S, Frow E, Kearnes M, Bulpin K, Schyfter P, Mackenzie A and Martin P 2015 Taking roles in interdisciplinary collaborations: reflections on working in post-ELSI spaces in the UK synthetic biology community *Sci. Technol. Stud.* **28** 3–25

Bocking S 1997 *Ecologists and Environmental Politics: A History of Contemporary Ecology* (New Haven, CT: Yale University Press)

Callaway E 2021 Million-year-old mammoth genomes shatter record for oldest ancient DNA *Nature* **590** 537

Capshew J H and Karen A R 1992 Big Science: Price to present *Osiris* **7** 2–25

Castells M 1996 *The Rise of the Network Society* (Oxford: Blackwell)

Check E and Castellani F 2004 David versus Goliath *Nature* **432** 546–8

Cook-Deegan R M 1994 *The Gene Wars: Science, Politics, and the Human Genome* (New York: Norton)

Creager A N H and Santesmases M J 2006 Radiobiology in the atomic age: changing research practices and policies in comparative perspective *J. Hist. Biol.* **39** 637

Davis B 1990 The human genome and other initiatives *Science* **249** 342–3

de Chadarevian S 2002 *Designs for Life: Molecular Biology after World War II* (Cambridge: Cambridge University Press)

de Jong S, Smit J and van Drooge L 2016 Scientists' response to societal impact policies: a policy paradox *Sci. Public Policy* **43** 102–14

Dreger A 2000 Metaphors and morality in the human genome project *Controlling Our Destinies* ed P R Sloan (Notre Dame, IN: University of Notre Dame Press) pp 155–84

Drucker P F 1969 *The Age of Discontinuity: Guidelines to Our Changing Society* (London: Heinemann)

Dunbar M J 1971 Anatomy of an expedition *Geogr. Rev.* **61** 161–3

Fernández-Armesto F 2006 *Pathfinders: A Global History of Exploration* (Oxford: Oxford University Press)

Galison P 2003 The collective author *Scientific Authorship: Credit and Intellectual Property in Science* ed M Biagioli and P Galison (London: Routledge) pp 325–58

Garcia Sancho M 2012 *Biology, Computing, and the History of Molecular Sequencing: From Proteins to DNA, 1945–2000* (London: Palgrave Macmillan)

Garcia Sancho Sanchez M, Lowe J, Viry G, Leng R, Wong M and Vermeulen N 2021 Yeast sequencing: 'Network' genomics and institutional bridges *Hist. Stud. Nat. Sci.* **52** 361–400

Gaudillière J-P and Rheinberger H-J (ed) 2004 *From Molecular Genetics to Genomics: The Mapping Cultures of Twentieth Century Genetics* (London: Routledge)

Glasner P 1996 From community to 'collaboratory'? The human genome mapping project and the changing culture of science *Sci. Public Policy* **23** 109–16

Glasner P 2002 Beyond the genome: reconstituting the new genetics *New Genet. Soc.* **21** 267–77

Groenewegen P and Wouters P 2004 Genomics, ICT and the formation of R&D networks *New Genet. Soc.* **23** 167–85

Hackett E J, John N P, Conz D, Rhoten D and Parker A 2008 Ecology transformed: NCEAS and changing patterns of ecological research *Scientific Collaboration on the Internet* ed G M Olson, A Zimmerman and N Bos (Cambridge, MA: MIT Press) pp 277–96

Hallonston O 2016 *Big Science Transformed. Science, Politics and Organization in Europe and the United States* (London: Palgrave Macmillan)

Hampton S E and Parker J N 2011 Collaboration and productivity in scientific synthesis *BioScience* **61** 900–10

Hillgartner S 1995 The Human Genome Project *Handbook of Science and Technology Studies* ed S Jasanoff (Thousand Oaks, CA: Sage) pp 302–15

Hillgarner S 2017 *Reordering Life: Knowledge and Control in the Genomics Revolution* (Cambridge MA: MIT Press)

Kastenhofer K 2013 Synthetic biology as understanding, control, construction and creation? Techno-epistemic and socio-political implications of different stances in talking and doing technoscience *Futures* **48** 13–22

Kelty C M 2012 This is not an article: model organism newsletters and the question of 'open science' *BioSocieties* **7** 140–68

Kerr A and Garforth L 2016 Affective practices, care and bioscience: a study of two laboratories *Sociol. Rev.* **64** 3–20

Kevles D J 1997 Big science and big politics in the United States: reflections on the death of the SSC and the life of the Human Genome Project *Hist. Stud. Phys. Biol. Sci.* **27** 269–91

Koch S 2020 Responsible research, inequality in science and epistemic injustice: an attempt to open up thinking about inclusiveness in the context of RI/RRI *J. Respons. Innov* **7** 672–9

Kohler R E 1994 *Lords of the Fly: Drosophila Genetics and the Experimental Life* (Chicago, IL: The University of Chicago Press)

Kohler R E 2002 *Landscapes and Labscapes. Exploring the Lab-Field Border in Biology* (Chicago, IL: University of Chicago Press)

Knorr-Cetina K 1999 *Epistemic Cultures: How the Sciences Make Knowledge* (Cambridge, MA: Harvard University Press)

Kuiken T 2020 We must do better *Genetic Engineering and Society Center Blog* https://research.ncsu.edu/ges/2020/06/we-must-do-better/

Kwa C 1987 Representations of nature mediating between ecology and science policy: the case of the International Biological Programme *Soc. Stud. Sci.* **17** 413–42

Lederberg J and Uncapher K (ed) 1989 *Towards a national collaboratory. Unpublished report of a National Science Foundation invitational workshop* (New York: Rockefeller University)

Lenoir T and Hays M 2000 The Manhattan Project for biomedicine *Controlling Our Destinies* ed Philip R Sloan (Notre Dame, IN: University of Notre Dame Press) pp 19–46

Leonelli S 2016 *Data-centric Biology* (Chicago, IL: Chicago University Press)

Leonelli S and Ankeny R 2012 Re-thinking organisms: the impact of databases on model organism biology *Stud. Hist. Philos. Sci.* C **43** 29–36

Leonelli S and Ankeny R 2013 What makes a model organism? *Endeavour* **37** 209–12

Leonelli S and Ankeny R 2015 Repertoires: how to transform a project into a research community *BioScience* **65** 701–8

Lewis J T 2008 Computing genomic science: bioinformatics and standardisation in proteomics *PhD Thesis* Cardiff University

Leydesdorff L and Etzkowitz H 1998 Triple Helix of innovation: introduction *Sci. Public Policy* **25** 358–64

Liscovsky R 2021 Internationalisation dynamics in contemporary south American Life Sciences. The case of Zebrafish *PhD Thesis* The University of Edinburgh http://dx.doi.org/10.7488/era/2113

Mahfoud T 2021 Visions of unification and integration: building brains and communities in the European Human Brain Project *New Media Soc.* **23** 322–43

Maienschein J 1993 Why collaborate? *J. Hist. Biol.* **26** 167–83

Marais A *et al* 2018 The future of quantum biology *J. R. Soc. Interface* **15** 20180640

Meyer M and Molyneux-Hodgson S 2016 Placing a new science: exploring spatial and temporal configurations of synthetic biology *The Local Configuration of New Research Fields. Sociology of the Sciences Yearbook Book* ed M Merz and P Sormani vol 29 (Berlin: Springer) pp 61–77

Nature editorial 2001 Post-genomic cultures *Nature* **409** 545

Nelson N 2013 Modeling mouse, human, and discipline: epistemic scaffolds in animal behavior genetics *Soc. Stud. Sci.* **43** 3–29

Nieroda M, Keeling K and Keeling D 2015 Acceptance of mobile apps for health self-management: regulatory fit perspective *Proc. 1st Int. Workshop on Personalisation and Adaptation in Technology for Health, 23rd Conf. on User Modeling, Adaptation and Personalization* (Aachen: RWTH Aachen University)

NOAA 2007 Rockets, radar, and computers: the International Geophysical Year *NOAA* https://celebrating200years.noaa.gov/magazine/igy/welcome.html

Nowotny H, Scott P and Gibbons M 2001 *Re-thinking Science: Knowledge and the Public in an Age of Uncertainty* (Cambridge: Polity)

Núñez Casal A 2019 The microbiomisation of social categories of difference: an interdisciplinary critical science study of the human microbiome as the re-enactment of the immune self *PhD Thesis* Goldsmiths, University of London

Owen R, von Schomberg R and Macnaghten P 2021 An unfinished journey? Reflections on a decade of responsible research and innovation *J. Responsible Innov.* **8** 217–33

Parker J N 2010 Organising the field: collaboration in the history of ecology and environmental science *Collaboration in the New Life Sciences* ed J N Parker, N Vermeulen and B Penders (London: Routledge) pp 37–58

Parker J N, Vermeulen N and Penders B (ed) 2010 *Collaboration in the New Life Sciences* (London: Routledge)

Pender B 2010 *The Diversification of Health: Politics of Large-Scale Cooperation in Nutrition Science* (Bielefeld: Transcript)

Pestre D J K 1992 Some thoughts on the early history of CERN *Big Science; The Growth of Large-scale Research* ed P Galison and B Hevly (Stanford, CA: Stanford University Press) pp 78–99

Pickstone J V 1993 Ways of knowing: towards a historical sociology of science, technology and medicine *Br. J. Hist. Sci.* **26** 433–58

Pickstone J V 2000 *Ways of Knowing: A New History of Science, Technology and Medicine* (Manchester: Manchester University Press)

Pickstone J V 2007 Working knowledges before and after circa 1800. Practices and disciplines in the history of science, technology, and medicine *ISIS* **98** 489–516

Pirie N W 1967 The part the international biological programme will play in increasing world food supplies *Proc. Nutr. Soc.* **26** 125–8

Prainsack B 2017 *Personalized Medicine: Empowered Patients in the 21st Century* (New York: New York University Press)

Price D J de S 1963 *Little Science, Big Science* (New York: Columbia University Press)

Rajagopalan R and Fujimura J H 2012 Making history via DNA, making DNA from history: deconstructing the race-disease connection in admixture mapping *Genetics and the Unsettled Past: The Collision between DNA, Race, and History* ed K Wailoo, C Lee and A Nelson (New Brunswick, NJ: Rutgers University Press)

Roberts L 2001 Controversial from the start *Science* **291** 1182–8

Rose N 2001 The politics of life itself *Theory Cult. Soc.* **18** 1–30

Schloegel J J and Rader K A 2005 *Ecology, Environment, and 'Big Science': An Annotated Bibliography of Sources on Environmental Research at Argonne National Laboratory, 1955–1985* (Argonne, IL: Argonne National Laboratory)

Schneider M C, Munoz-Zanzi C, Min K-d and Aldighieri S 2019 'One health' from concept to application in the global world *Oxford Research Encyclopedia of Global Public Health* (Oxford: Oxford University Press)

Schönbauer S 2019 We are standing together in front. How scientists and research groups form identities in the life sciences *Sci. Technol. Stud.* **32** 60–80

Schyfter P and Calvert J 2015 Intentions, expectations and institutions: engineering the future of synthetic biology in the US and the UK *Sci. Culture* **24** 359–83

Shrum W, Genuth J and Chompalov I 2007 *Structures of Scientific Collaboration* (Cambridge, MA: MIT Press)

Sims J 2021 Why astronauts are printing organs in space *BBC Future; Health* https://bbc.com/future/article/20210601-how-transplant-organs-might-be-printed-in-outer-space

Strasser B J 2003 The transformation of biological sciences in post-war Europe. EMBO and the early days of European molecular biology research *EMBO Rep.* **4** 540–3

Strasser B J 2010 Collecting, comparing, and computing sequences: the making of Margaret O Dayhoff's Atlas of protein sequences and structure, 1954–1965 *J. Hist. Biol.* **43** 623–60

Strasser B J 2019 *Collecting Experiments: Making Big Data Biology* (Chicago, IL: University of Chicago Press)

Strasser B J, Baudry J, Mahr D, Sanchez G and Tancoigne E 2019 'Citizen science'? Rethinking science and public participation *Sci. Technol. Stud.* **32** 52–76

Sunder Rajan K 2006 *Biocapital: The Constitution of Post-Genomic Life* (Durham, NC: Duke University Press)

Thackray A 1998 *Private Science: Biotechnology and the Rise of the Molecular Sciences* (Philadelphia, PA: University of Pennsylvania Press)

Ulnicane I 2020 Ever-changing big science and research infrastructures: evolving European Union policy *Big Science and Research Infrastructures in Europe* ed K C Cramer and O Hallonsten (Northampton: Edward Elgar) pp 76–100

Vaessen S 2018 3D printer to be sent to space for human organ trials *Al Jazeera* https://aljazeera.com/videos/2018/12/2/3d-printer-to-be-sent-to-space-for-human-organ-trials

Valkenburg G, Mamidipudi A, Pandey P and Bijker W E 2020 Responsible innovation as empowering ways of knowing *J. Responsible Innov.* **7** 6–25

Venter J C 2007 *A Life Decoded. My Genome: My Life* (New York: Viking)

Vermeulen N 2009 *Supersizing Science; On Building Large-scale Research Projects in Biology* (Maastricht: Maastricht University Press)

Vermeulen N 2012 Growing a cell *in silico*; on how the creation of a bio-object transforms the organisation of science *Bio-objects: Life in the 21st Century* ed Niki Vermeulen, Sakari Tamminen and Andrew Webster (London: Routledge) pp 171–86

Vermeulen N 2013 The census of marine life; marine biology as big science *PLoS One* **8** e54284

Vermeulen N 2016 Big Biology; supersizing science during the emergence of the 21st century *NTM J. Hist. Sci., Technol. Med.* **24** 195–223

Vermeulen N 2018 The choreography of a new research field: aggregation, circulation and oscillation *Environ. Plan.* A **50** 1764–84

Vermeulen N, Parker J N and Penders B 2010 Big, small or mezzo? Lessons from Science Studies for the ongoing debate about 'Big' versus 'Little' science *EMBO Rep.* **11** 420–3

Waldby C and Mitchell R 2006 *Tissue Economies: Blood, Organs, and Cell Lines in Late Capitalism* (Durham, NC: Duke University Press)

Wang K, Shen X and Williams R 2020 Sequencing BGI: the evolution of expertise and research organisation in the world's leading gene sequencing facility *New Genet. Soc.* **40** 305–30

Weinberg A M 1961 Impact of large-scale science on the United States: big science is here to stay, but we have yet to make the hard financial and educational choices it imposes *Science* **134** 61–164

Weinberg A M 1967 *Reflections on Big Science* (Oxford: Pergamon)

Wheeler Q D *et al* 2012 Mapping the biosphere: exploring species to understand the origin, organization and sustainability of biodiversity *Syst. Biodivers.* **10** 1–20

Wilde R de 2001 *De Kenniscultus: Over Nieuwe Vormen van Vooruitgangsgeloof* (Maastricht: Maastricht University Press)

Wilsdon J, Bar-ilan J, Frodeman R, Lex E, Peters I and Wouters P F 2017 Next-generation metrics: responsible metrics and evaluation for open science. European Commission, Directorate-General for Research and Innovation

Woods B 2009 A political history of NASA's space shuttle: the development years, 1972–1982 *Sociol. Rev.* **57** 25–46

Yeh C-w 2019 Particle physics in public: legitimising curiosity-driven research on the Higgs boson and beyond *PhD Thesis* The University of Edinburgh

IOP Publishing

Big Science in the 21st Century
Economic and societal impacts
**Panagiotis Charitos, Theodore Arabatzis, Harry Cliff, Günther Dissertori,
Juliette Forneris and Jason Li-Ying**

Chapter 31

Turn and turn again: how Big Science both helped and hindered alternative energy in the 1970s

Thomas Turnbull and Cyrus C M Mody

By the 1970s the supposed benefits of big US federally-funded science had begun to invite criticism. In part, this was because federal mission-oriented scientific agencies had largely met their goals. This existential success created a degree of listlessness. Ironically, it was growing evidence of the impacts of science and technology on the environment that would renew justification for federal involvement in science and technology. At the center of these impacts was the energy system; increasingly unreliable, expensive, unproductive, land-hungry, and polluting, yet its growth seemed inexorable. This paper focuses on federally funded attempts to foster less impactful alternative energy technologies in response, specifically solar and wind. NASA, an agency in the midst of downsizing, and the National Science Foundation (NSF), traditionally a patron of small and 'basic' science, attempted to transform the work of marginal (even 'fringe') alternative energy scientists using the tools of prestigious Big Science. This injection of federal funding and rationales resulted in both successes and failures, created collisions between the prerogatives of small and big science, and gave impetus to Big Science's evolution toward a more collaborative and multi-stakeholder form in the 1980s. Moreover, this history offers notes of encouragement and caution with regard to contemporary state-led energy transitions.

31.1 Synecdoches of Big Science

Size is relative. In the annals of post-war US science, the term 'Big Science' has not always been clearly delineated. Ambitions and discoveries could be big and budgets small. One crude if obvious way of defining bigness, however, is to compare a research institution's annual budget with total government expenditure. By that

doi:10.1088/978-0-7503-3631-4ch31

metric, one of the biggest Big Science agencies of the post-war era was of course NASA. At its 1966 peak the agency received some 4.4% of the total federal budget (roughly 0.7% of US GDP). James Webb, NASA's chief administrator in the Apollo era (an oilman, both before and after his time with the agency), wrote that 'space age management' was the art of making 'constructive use of great aggregations of resources and power' [1]. However, if in early 1966 total employment in NASA programs stood at 420 000, by June 1970 the number fell more than 60% to 161 000 [2]. From 1966 to 1975, NASA's funding dropped to 1% of the federal budget, where it held steady until the late 1990s before falling once more. Today, NASA's budget hovers around 0.5% of the federal budget (or 0.1% of GDP) [3, p 149].

The 1969 Apollo 11 mission proved, in many senses, the organization's zenith. Despite this reversal in fortunes, parts of NASA still remained devoted to Big Science—just, fewer of them and they were less 'big' than in the agency's heyday. But other parts of NASA could no longer maintain a Big Science approach, and instead turned toward small projects, collaborating with new stakeholders and agencies. This collaborative model could be considered as the next phase in the evolution of classical Big Science[1]. As Catherine Westfall has argued, these kinds of collaborations involved 'interconnections of research enterprises of different types and scales' and 'formed a new kind of science', a 'mezzo' scale science that leveraged complex multi-organization collaborations as classical Big Science retreated [5].[2] While not wholly generalizable, NASA is representative of organizational changes across US Big Science institutions during the 1970s. We argue that one of the first areas where those organizational changes were attempted was in the field of alternative energy; and in doing so, we will show that NASA was somewhat representative of how Big Science agencies made this energy turn.

At NASA and elsewhere, the need for institutions to forge complex collaborations with other organizations opened the door for 'small science' institutions to contribute to the emerging mezzo approach in ways that departed markedly from the rationales of Big Science. In this new collaborative and parsimonious era, small science organizations were welcomed into large collaborations so long as they had resources to contribute; in return, they gained expertise and credibility, and most of all, utility. The prime example of such an organization is the National Science Foundation (NSF). In this chapter we look at how NASA and the NSF formed a variety of collaborations in this new mezzo mode in the 1970s. Those collaborations came to center on alternative energy; we therefore argue that the energy crises of the 1970s were central to the transition from 'big' to mezzo science. Looked at from the other direction, NASA and the NSF's involvement had a much more ambivalent impact on the nascent alternative energy scene—the mezzo mode both helped *and*

[1] In the 1990s, the International Space Station, for instance, was international for a reason: the scale of this project would be impossible without contributions of smaller pieces via complex transnational collaborations. For NASA's growing internationalization, especially after 1970, see [4].

[2] Elsewhere Lillian Hoddeson W and Adrienne Kolb dubbed the post-1970s evolution of Big Science 'megascience'—i.e. projects so costly that no one organization could support them, but rather they required large numbers of coordinated partners [6].

hindered efforts to stabilize alternative energy technologies and bring them to the market.

If NASA went from big to mezzo and NSF from small to mezzo during the 1970s, the Atomic Energy Commission's (AEC) National Laboratories underwent something like a mezzo-to-mezzo transition. The National Laboratories had always included a mix of small and big projects; as Peter Westwick argues, it was their decentralized yet coordinated *system* of institutions that allowed them to adapt to changing circumstances [7]. In that constellation, one of Big Science's first and most acute observers (who also coined the term 'Big Science') was Alvin Weinberg, head of the AEC's Oak Ridge National Laboratory (ORNL) from 1955 to 1973. Weinberg was ambivalent about Big Science's 'monumental enterprises', 'the huge rockets, the high-energy accelerators, the high flux research reactors' [8]. Indeed, he expressed serious misgivings about such things as early as 1961—Big Science was almost a slur from the start. He worried it substituted the spectacular in place of the useful, inhibited realistic appraisal, required outsized administrative support, and, at times, threatened to displace 'little science', i.e. small-scale specialized research, entirely [9].[3] By 1967 Weinberg's unease had become more insistent and more representative of his profession. By then, even Manhattan Project elders, such as Edward Teller and Hans Bethe, were suggesting that while Big Science had been successful, it was time to 'reaffirm the dignity of applied science' [10, p 156].

The dichotomy between big and small science did not precisely map on to the couplet basic/applied, but Weinberg and his peers articulated a growing call for a science that more directly improved society, rather than one that did so, or so it was claimed, only indirectly or symbolically; no one's life was saved and few were directly aided by Moon landings or particle accelerators. Yet as the audience for the *spectacle* of Big Science shrank, the *capacities* of its institutions remained. Organizations such as the AEC National Laboratories had simply become too big to fail. They were geographically big, a nation-spanning complex of laboratories and attendant technological apparatuses, built on the skeleton of New Deal and Second World War infrastructures. NASA's facilities also spread across the country, only one (!) of which (Cambridge, MA's Electronics Research Center) ever closed. Too many congresspeople had a stake in the employment, prestige, and pork-barrels Big Science provided to let it disappear entirely; but by the late 1960s, those same politicians could sense voters' desire for a science that more tangibly improved their lives.

31.2 Success at a cost

One way to understand Big Science's precarity is as a state of 'existential success'. By the 1970s, most of the US's major post-war scientific goals were considered to have been achieved: the space race was won, Communism largely contained, and some big unknowns in physics were now known [3, p 3–7, 11]. Science had proved an

[3] Weinberg was hardly alone. The sense of elegiac mourning for small science was common, even among some of those who had built the Big Science system in the first place, such as Donald Glaser, inventor of the bubble chambers widely used in high-energy physics. See [9], and also chapter 24 in this volume.

engine of post-war growth in population, affluence, and economic activity. Again, we rely on another crude but useful statistical proxy for that growth: per capita US energy demand increased by around 50%, from 184 million BTUs per person before the war in 1929 to 293 million in 1968 [12]. Population boomed from around 152 million in 1950 to 204 million by 1970 [13, p 3]. Consumer goods, from television sets to electric carving knives, were represented as spin-offs from federally-funded advances in science and technology [14]. Surveying this from his relatively privileged position, Weinberg wrote in 1967 that it appeared as if 'the historic central practical purpose of human life—economic survival' had been largely achieved in the US [10, p 37]. But at what cost? The impacts of this unprecedented post-war growth were readily apparent in terms of environmental pollution and destruction, and growing portents of material and energetic limits to growth, such as region-wide electrical grid blackouts and regular brownouts and rising electricity costs [15]. If scientists deserved credit for the nation's post-war success, then by the late 1960s the public also, logically, blamed them for the costs of this success. This disenchantment with science-driven development encouraged the idea that NASA and the like should now address more grounded problems.

As we shall see, partly due to a budget that sought to minimize inflation but perhaps also because of shifting sentiments, government and the scientific establishment responded by moving somewhat closer to the concerns of Big Science's critics [16] (see also chapter 24). By 1970, Weinberg, arguably the personification of establishment science, wrote of a 'yearning for science that is relevant' [17]. A group of arguably 'square' scientists and engineers of his ilk recognized the need for more obviously useful science, or just went with the flow as prerogatives changed, but did so within recognized career tracks [3]. Weinberg and many others would re-orient the research they carried out or administered, directing a greater proportion of their time and resources to 'relevant' or 'problem-oriented' topics such as pollution monitoring and control, disability and medical technologies, improvements to public housing and transportation, etc. Obviously, basic and Big Science are 'problem-oriented' in their way, but in the discourse of the day 'problem-oriented' meant research that aimed at solving defined challenges relevant to society. As we will show, the search for alternative sources of energy or means for reducing energy use was a prominent 'problem' for research to orient to from the off, but one that came to dominate after the 1973 OAPEC embargo.

31.3 Interdisciplinarity applied

By the late 1960s there were myriad reasons for federally funded research to move *away* from monumental projects as well as small-scale 'basic' research—and instead *toward* applied, problem-oriented research. Congress, which, as the NSF put it, provided the 'ambient legislative and financial environments' for science, was an important driver of this [18, p xi]. Skeptical of ever-growing federal science budgets, Democratic Congressperson from Connecticut, Emilio Daddario, as a member of the Committee on Science and Astronautics, had reviewed NSF funding allocation in 1966. Identifying considerable inefficiencies, two years later he proposed an amendment, Section 3.C, which would authorize the NSF to 'initiate and support scientific research, including applied research, at academic and other non-profit institutions'. With Congressional acceptance, historian of science Dian Belanger suggests, the NSF assumed a 'new character' [19, p 81]. An immediate outcome was the establishment of a new program to explore how the Foundation could support avowedly applied research, named Interdisciplinary Research Relevant to Problems of Society (IRRPOS)—'one of the least euphonious acronyms on the Washington science scene'—one journalist joked [20]. This awkward formulation was intended to side-step long-running debates over the distinction between applied and basic research, and to instead assess 'the contribution of fundamental scientific research in resolving major national problems' [20] (on the basic/applied debate see Kevles [21] and on the formation of this distinction, in administrative terms at least, see Godin [21]).

Why was 'research relevant to problems of society' framed as necessarily 'interdisciplinary'? The reason was twofold. First, the NSF considered the drive toward interdisciplinarity as partly driven by developments within the sciences. For one thing, according to preceding NSF Annual Reviews, advances in physics had provided a 'unified conceptual structure' within which other sciences had melded, creating fields such as bio-, astro-, and geophysics, which meant it was 'increasingly difficult to distinguish where the one discipline ends and the other begins' [22, p 39]. New computational techniques and technologies also encouraged interdisciplinarity, allowing the once 'esoteric' mathematics of partial differential equations to be used on real-world problems; a new age of 'applied mathematics' began in a range of fields, not least economics [23].

Interdisciplinarity was not only more common but it was also increasingly seen as *necessary*, particularly to Americans of an environmentalist orientation [24]. For instance, the NSF described how researchers in the life sciences had become aware that 'their findings become more meaningful when related to higher levels of biological organization'; researchers could now traverse from 'cracked' codes of DNA to entire ecosystems. Such studies were necessarily interdisciplinary, requiring biologists to work with meteorologists, geochemists, and physical chemists [22, p 66, 25]. In wider cultural terms, reductionist analysis was out, and holist, systemic understanding was very much in [26]. Such 'comprehensive ecosystem analysis' required the study of 'plants and animals, the structure and fluctuation of populations, analysis of production and diversity of organisms, study of the cycling of mineral nutrients, and analysis of the flow of energy and water through the system' [22, p 37]. At this scale, biology became 'problem-oriented' as the study of ecosystems bled into questions of food supply, disease control, and environmental quality [22, p 33]. Accordingly, early IRRPOS initiatives centered around biology [27, 33, p 238–9]. The same ideas were applied to energy, a field in which systems thinking provided a shared language for orientating diverse disciplines.

IRRPOS was headed by a 33-year-old solid-state physicist named Joel Snow [28]. In an interview he described the new organization as one addressed toward 'total systems', and one that accepted that society's problems may not have a solution so much as a 'partial or interim answer'. He hoped its discretionary budget of

$6 million dollars (for fiscal year 1970, set to rise to 13 million in 1971; i.e., from $42 to $86 million in today's dollars) would encourage existing institutions to 'modify their traditional departmental structures' to allow them to undertake the kind of 'interdisciplinary, problem-orientated research' of the kind that post-war 'think tanks' had done [20, 29]. His vision was that IRRPOS would engage in risky, innovative research, which more entrenched mission-oriented agencies, such as NASA, could not [30, p 176]. Yet the first recipient of IRRPOS's largesse was Weinberg's ORNL, which received nearly 1.5 million dollars (more than 10 million dollars today) to study 'The Environment and Technological Assessment', a project spanning environmental mutagenesis, human mutation, water hyperfiltration, regional modeling, and environmental information systems [31]. In total, 21 awards were made, including one to archetypal think tank the RAND Corporation's New York office, to investigate the city's solid waste flows [32, p 56].

IRRPOS was short-lived, but its successor, the Research Applied to National Needs (RANN) program, ensured that systemic, interdisciplinary approaches to problems of civil society would become steadily more attractive to the US's leading science policy makers. The transition from 'research relevant to problems of society' to 'research applied' was due largely to the NSF's new director William McElroy, who took over in 1970. A specialist in firefly bioluminescence, McElroy considered the formula 'research applied' a stroke of genius, because it once more side-stepped decades of disagreement over the supposed merits and demerits of scientific research designated either 'applied', directed toward specific ends, or 'basic', i.e., undirected, curiosity-driven, and free-roaming [19, p 96 ft 53]. Of course, this distinction was often more rhetorical than actual, a means for distinguishing a research initiative or a new idea from the old. But McElroy's demonstrative move toward applied research still raised hackles among more conventional members of the scientific establishment [33, p 239].

In part, this was because NSF had long presented itself as the patron of 'pure science'; this institutional mythos can and has been questioned, but it is true that until the late 1960s NSF administrators shied away from funding social or engineering sciences [34]. The NSF's grant review system ensured funds generally went to single investigators who aimed to solve generalizable topics of interest to their disciplines, rather than specific problems of interest to non-scientists, such as garbage collection or smog detection. By taking up such concrete problems, RANN represented a significant change of direction for the NSF. When it started up in March 1971, RANN initially took over IRRPOS' projects as well as those of NSF's 'Environmental Science' unit [32, p 57]. The roster of RANN projects quickly broadened; at its height, it absorbed around 10% of the NSF's total budget. Yet despite decades of historical interest in the boundary between pure and applied research, and in the NSF itself, RANN has largely gone undocumented by historians. Although mentions of RANN pepper histories of particular disciplines (such as ecology) during the 1970s, there has been no dedicated study of RANN's work, nor of its lasting influence [3, p 37, 19, 33, 35]. Meanwhile, historians of policy have cast it as a belated spurt of New Deal-style developmental science [36]. This is not quite right. RANN's trajectory shows how a program founded during the last gasps of the 'Big Science' era of the late 1960s gradually morphed into something stuck between the imperatives of state-funded Big Science and commercial industry. Here, RANN's engagement with NASA and the nascent alternative energy scene is used to exemplify this interstitial position.

31.4 Reach for the star

From the off, energy was a central RANN concern—although one inextricably linked to other topics. As the 1972 NSF annual report put it, RANN had three problem areas, of which:

> The first is energy, because of its importance to national productivity and economic growth, environmental concerns, and the decline in reserves of fossil fuels [37].[4]

As such, among RANN's first goals was to gain 'a complete picture of the present and prospective US energy problem'. The National Academy of Engineering sought to understand the energy system's contours and develop means for formulating national energy policies; while Weinberg's ORNL surveyed all ongoing US energy research at the time. The list was computerized, and around 4400 research projects were rendered interrogable. It was estimated that around a billion dollars was spent on energy related projects per year. California was the US hub of energy research with over 529 projects, with coal-rich Pennsylvania next with 446 projects [38].[5] Unsurprisingly research on fossil, chemical, and nuclear energy dominated [38, table 1].

Demonstrating the NSF's new openness to commercial research, a company called InterTechnology Corp, run by Hungarian-American solar advocate George Szego out of a converted Safeway store in Warrenton, VA, was tasked with producing an account of the scale of the US's energy problem. Its 15 000 page two-volume *The US Energy Problem* argued that the US would face an unmanageable growth in energy demand of 2.8% p.a. [39] Meanwhile the Truman-era non-profit Resources for the Future (RFF) worked with MIT to produce *Energy Research Needs*, a wide-ranging survey of possible areas of inquiry [37, p 56]. A second tranche of research supported specific developmental projects, such as Caltech's Environmental Quality Laboratory's work on pollution and power plant siting [40, p 62]. In line with RANN's origins, and as with the study of ecosystems, 'information from an exceedingly broad set of disciplines' was believed to be key to understanding the energy system [37, p 56]. As the RFF/MIT study attested, to address energy, the perspectives of science and engineering had to be informed by those of economics and policy. Let's not forget, all of this activity took place in the years *preceding* the conventional 'energy crisis'; as others have shown, the 1973 Middle Eastern oil embargo was more a symptom than a cause of the US's energy woes [41].

[4] The other two problem areas were 'regional environmental systems' and 'municipal systems and services' [37].
[5] Note that Pennsylvania's expenditure on energy research was greater at 63m USD versus California's 58m USD [38, pp xi–xviii].

Perhaps more than anything, RANN should be recognized as an important experiment in the history of energy R&D in the United States. Between 1971 and 1978 around 484 million dollars (somewhere around 2.5 billion in today's dollars) of federal funding was granted to applied research, with a focus on understanding the energy system and finding alternative energy sources that might prove less negatively impactful and, in the long run, less scarce [42]. Here, RANN's influence will be explored by addressing solar power, an expansive topic that covered many aspects of Earth's energy fluxes. In 1971, according to the ORNL's initial survey, there were just 72 solar projects of note in the whole country, with a cobbled-together budget of less than 3 million dollars ($20 million today) [38, table 1, p xiv]. RANN's 'Solar Energy Program' aimed to put solar R&D on a more substantial and coordinated footing. Led by SUNY Stonybrook physicist Richard Bleiden, the project convened an NSF/RANN-led workshop to assess the Sun's potential as a national energy resource. It involved 40 experts, including an economist and a sociologist [43], and considered solar in all forms, from building thermal regulation, to heat conversion, photovoltaics, and photosynthesis, to wind, and ocean currents—all were rightly considered energies of solar origin [44, p 69–70]. The final report, published in late 1972, informed Congress, the public, and industry of the potential of solar, but its initial request for three and a half billion dollars of investment over 15 years, later reduced to one billion, was not taken seriously [45, 64] (on the lack of impact see [67, pp 158–59]).

RANN's expansive approach toward solar was mirrored by other agencies. NASA, for instance, had projects involving wind energy, solar thermal collectors, solar ponds, a crewed solar power satellite (envisioned as beaming energy back to terrestrial cities), and so on. These projects were a product of both local and federal shifts. Locally, NASA facilities such as Cleveland, Ohio's Lewis Laboratory were spurred by the readily apparent impacts of nearby industry. The Cuyahoga River, east of the Lab, flowed into Lake Erie and was so heavily polluted it had caught on fire nine times in the 1960s [46]. In response to growing concerns, Lewis offered graduate courses on environmental topics and used its B-57 reconnaissance plane to monitor the local environment [47, p 205]. At the federal level, in 1973 Nixon's government reduced NASA's budget by 200 million dollars; Lewis faced cuts and lost over 800 personnel [48, p 204]. In pursuit of renewed relevance, not to mention funding streams—the annual NSF budget for solar research had quadrupled from $3 to $12 million in July 1973—Lewis and other facilities unsurprisingly pitched solar projects to the NSF, hoping their expertise might be profitably turned toward Earth [49, p 77].

In June that year, RANN was tasked with overseeing the coordination of federal solar research via a newly formed 'Interagency Panel for Terrestrial Applications of Solar Energy' [50]. The panel organized a major conference on terrestrial photovoltaics (PV) held in Cherry Hill, NJ in October 1973 [51, p 305].[6] In the meantime, the OAPEC embargo had accelerated concerns that solar might not just be

[6]The first specialist PV conference is remembered as that organized by the Institute for Defense Analyses (IDA) in Washington DC in 1961. It was distinctly space and defence oriented; see [51].

preferable to fossil fuels but vital if the oil supply remained vulnerable. And while the embargo heightened demand for solar, developments in solar technology and the emergence of a crop of solar entrepreneurs hinted at a sharp rise in the potential supply. For instance, the first commercial solar business, an Exxon subsidiary called Solar Power Corporation led by chemist Elliot Berman, had opened just a few years earlier, soon followed by Solarex, Day Star, and others [49, pp 58, 60, 52]. The time was ripe for RANN to catalyze the formation of a solar community and industry.

Co-organized by Caltech, the conference was intended to coordinate various scientists and industry figures working on the technology around a shared ten-year development plan. Bleiden later described how the 'aggressive targets' agreed at Cherry Hill fed directly into Nixon's proposed response to the embargo, the energy autarkic 'Project Independence' [53]. Researchers from Georgia Tech estimated if sunlight was used with a conversion efficiency of 10%, 1.5% of the nation's surface insolation could meet the nation's energy demand as of 1970 [54]. Central to the forecast plan was that government-backed R&D in PV could help this young industry to develop its product so as to provide power at 30 cents per watt by 1986, and this, it was hoped, would encourage an ambitious and commercially driven rollout of solar power technologies in the US and elsewhere up until the year 2000 [55].

Lewis researchers had been involved in PV research since 1961, for use on satellites, but because the space market was small and well-funded by the state there had been little incentive to lower costs. But at Cherry Hill researchers offered ideas for creating thinner and more efficient silicon arrays [56]. Hungarian émigré Joseph Lindmayer of the Solarex Corporation, who had previously worked for COMSAT, a Congress-founded satellite company, argued that cell cost was still 'controlled by the space market' and that further economies required the creation of thinner materials suitable for increasing the *area* over which insolation could be captured. He noted that this was the opposite of the notion of increasing device *density* that dominated the transistor industry [57].[7] Although a different technology, in a sense, the aim of Cherry Hill was to create a shared developmental target not unlike that of Moore's law but with thinner and more expansive substituted for the latter's target of denser and smaller [3, 58].

In 1979, with Department of Energy support, Lewis researchers developed a solar power system in a very sparsely populated area in Southern Arizona belonging to the Tohono O'odham nation [59].[8] There, in 1978, 96 residents formally dedicated 'the world's first solar powered village' in Schuchulik, thanks to the power granted by an array of 192 PV modules and a battery bank [47, p 207, 60].[9] As with many energy transitions, PV supplemented rather than replaced six on-site gas generators.

[7] On Lindmayer and COMSAT, see [71, pp 8–9]. Note that AMOCO were a shareholder in Solarex [71, p 111].

[8] This was not NASA-Lewis's only village solar project, the first was in Tangaye, Upper Volta (today, Burkina Faso) where in 1979 PV powered a water pump and grain mill [59].

[9] The Tohono O'odham (known at the time by the non-indigenous name 'Papago') were also participants and contributors in NASA's efforts in telemedicine [3, pp 173–4] The NASA/Lewis research was a collaboration with the DOE and the Indian Health Service.

By 1980, following a series of supply problems, anthropologists led by Arizona State's Donald Bahr noted how villagers referred to generator power as 's-gewk wepgi' or 'strong electricity' and PV as 'this thing on the other hand' or, implicitly, weak electricity [61]. A few years later, the village was connected to the grid, and the project was largely forgotten and left unmaintained [62]. Although not without problems, particularly reliability, the project was commendable in seeking to accommodate the interests of Indigenous people and it offers notes of both caution and encouragement that could help in better integrating and representing the interests of marginalized communities in present-day decarbonization efforts [63].

The relation between the 'thin' small-scale science of solar energy and the institutions of Big Science had clearer successes. One participant at Cherry Hill believed federal involvement had helped unify 'a diverse photovoltaic community' [64, p 785]; another remembered it as 'the most catalyzing event' in the technology's history [65, p 67]. Subsequently, the NSF supported research that allowed thin ribbons of silicon to be grown, markedly lowering costs and increasing efficiency [66, p 73]. But forecast cost reductions per unit of electricity were well off the mark, with 30 cents per watt only becoming feasible in recent years. The 1972 NASA/NSF prediction that in 2020 20% of US electricity would be solar-powered was way off; today the figure is only about 2.3% [67]. NSF's longest running solar researcher, Lloyd Herwig blamed this trajectory on the government having paid out less than the NSF requested in 1972 [64, p 789]. In fact, total federal outlay on PV would only exceed this amount some 26 years later [53]. Herwig believed Project Independence sidelined solar in favor of nuclear power, coal, and synthetic fuel. Indeed, the report's lead author, Dixie Lee Ray, had described solar as 'a flea on the back of an elephant' [64, p 799, 65, p 69]. Despite the involvement of mission-oriented agencies, and injections of federal funding, not to mention the outsized power the Sun offered, the inertia of the incumbent energy (and defense) industries subsumed solar energy's potential. To add further insult, the Project Independence 'blueprint' suggested solar only be used where it wouldn't compete with utility-supplied power, such as on 'ocean buoys' or at 'summer camps' [Kelly *et al* 1974, cited in 65, p 69].

NASA's involvement in terrestrial solar power would also prove fleeting. Despite initiatives such as the Low-Cost Silicon Solar Array (LSSA) taking off at JPL in the second half of the 1970s, it has been argued that NASA researchers developed no real allegiance to the technology. One observer dismissed their work as 'a conglomeration of minor activities', while Reagan's budget cuts ended any hope of more significant engagement [69, p 82]. In the 1980s, NASA turned back toward 'big' projects, notably the space shuttle and what would become the International Space Station. Yet in some sense these projects resembled the Schuchulik solar installation more than the Apollo Program. Not in terms of size, of course, but because of their comparable attempts at stakeholder representation and multilateral decision-making. Apollo-era NASA had simply done its own thing; but shuttle-era NASA had to listen to customers and foreign collaborators; to the Pentagon and other federal agencies; and congresspeople demanding greater diversity as well as free space rides for politicians [4, 70]. The Schuchulik solar village, alongside dozens of similar small-scale applied projects, arguably prepared NASA for the multipolar reality of the 1980s in ways that Apollo did not [3].[10]

Today, the solar industry's earliest innovators in the US can rightly look back with pride at having helped start an industry that has a global capacity of around 500 GW, equivalent to nameplate power of around 20 nuclear power plants [71].[11] But what happened to RANN-funded solar start-ups? Quite a few were purchased by larger conglomerates in the late 1970s, and specifically by oil firms. Solar Power Corp remained Exxon's, in 1983 Solarex was sold to the American Oil Company (AMOCO), a third start-up, Solar Technology International was sold to the Atlantic Richfield Company (ARCO) in 1978 [71, p 35, 72]. Beholden to the principles of financial capital, such start-ups were gradually broken up and sold as oil became cheaper [73]. Only in the second half of the 2000s, boosted by the German government's commitment to subsidized PV installations, and backed by the Chinese state's dramatic support for its domestic PV industry, was it possible for Liu Lifeng, a young electrical engineer from Tsinghua University, to set up a company, Beijing Sunlectric, that could produce power from the Sun at around the cost ($0.40 per watt) of which Cherry Hill participants had once dreamed [65, ch 7, p 155–7]. In a sense, it can be argued that state support, from the NSF's support of R&D to Germany's feed-in subsidy and China's state-led market creation, has bankrolled PV's shift from a rarified space technology to a common-place and increasingly affordable source of non-fossil energy.

31.5 Blew it

When it came to wind power, at the beginning of 1971 the ORNL identified just two extant projects, with a total budget of just over 100 thousand dollars [38, table 1, p xiv]. Of those projects, one was that of William Heronemus at the University of Massachusetts, Amherst, who was working on a 'National Network of Pollution-Free Energy Sources' which would allow the US energy system to gradually move from 'consumable fuel' to total reliance on the energy of the 'solar heat budget of Planet Earth' [74]. 'UMass Amherst' was a somewhat small-scale scientific institution; a former agricultural college with a modest student intake. At the time, the university was known for its chancellor, geologist Randolph Bromery, the second African-American to hold such a role [75]. Heronemus was another notable faculty member. A naval engineer who described having grown up during the 'embarrassing poverty of the Great Depression', he later served on the *USS Carolina* in WWII, and remained in the Navy for 17 years. Among other things, his warnings about the design of the nuclear-powered submarine, *USS Thresher*, had been disregarded. The

[10] Chapter 4 of [3] contains a number of other examples of NASA projects in the 1970s that were similarly multipolar but that were unrelated (or only partially related) to energy, in particular the Meal System for the Elderly (which combined NASA with several Texas government agencies and research institutions, Meals on Wheels, the Army, the Ford Foundation, and Martin Marietta) and the Modular Integrated Utility System (which combined NASA with ORNL and the Department of Housing and Urban Development) projects.

[11] For 2013, [71] estimates 100 GW p.a., whereas, as of 2018, global solar capacity stands at 512 GW (https://iea-pvps.org/publications).

boat failed to surface in 1963 and 129 crew died [76, pp 175–6, 79, p 48]. After a two-year spell as an engineer at the United Aircraft Corporation, he retired to teach ocean engineering at UMass Amherst.

By the late 1960s, 'Bill', as he was known, had begun warning of a coming energy crisis [77]. UMass Amherst researchers began studying how to increase the provision of electricity in the Connecticut River Valley. Bill became interested in working out how to meet increasing energy demand without causing ecological damage to this New England environment. He came to the conclusion that 'solar energy (including wind energy) was the only way out of the nation's energy (and pollution) dilemma' [78, p 196–203]. Fossil fuels were too 'valuable' to burn, given their potential petrochemical uses; moreover—just as Russian geochemist Mikhail Budyko had argued in the late 1960s—the risk was that their unchecked combustion might lead to a time 'when the temperature of the Earth is going to start increasing' [78, p 196–203]. While in the Navy Heronemus had received a grant to investigate the feasibility of placing nuclear reactors offshore, which had revealed the full cost of decommissioning nuclear plants [79, p 47]. That experience, and perhaps the *Thresher* before it, seem to have turned Heronemus against nuclear energy; but the idea of putting power plants offshore stuck with him. It became the kernel of a wide-ranging alternative energy proposal to the RANN Directorate that he submitted in 1971 [80]. His ideas were ambitious and imaginative, if not easily achievable. He proposed anchoring flotillas of wind ships out at sea, clusters of which could be used to produce liquid hydrogen fuel and even ammonia fertilizer offshore via electrolysis [77, p 336]. A year later he presented Congress with such 'gentle solutions' to the coming energy crisis, namely a network of off- and onshore alternative wind power systems, which in 1973, he argued to a Senate Subcommittee, could purportedly allow the country to be 'totally energized by solar energy and other renewable resources by the year 2000' [76, p 176].

Heronemus' grand claims attracted attention, and he received NSF support for a small-scale pilot project on campus, a 2.5 kW 'Wind Furnace' (WF-1) turbine that would generate heat for a family home. The project had not been allowed to generate electricity for the local utility company, Northeast Utilities, as they had claimed it could result in grid disruption [77, p 337, 79, p 52]. Whether true or not, such claims encouraged a suspicious mindset. Heronemus told a later interviewer that there was 'indeed a conspiracy. The utilities and energy companies are going to get us hooked on nuclear power, then they are going to drive the price up wherever it will go. And who are they? They are the same friendly souls who are selling us coal and petroleum products' [79, p 48]. It was not in the utilities' interest to provide low-cost solar (nor nuclear), as this might threaten the continued profitability of their staple fuels. Heronemus imagined bypassing the utilities entirely by instead converting the wind's kinetic energy into liquid hydrogen fuel, which could be piped around as needed.

Heronemus' students would form a company, US Windpower, to try and commercialize the WF-1 design. In many respects it was the prototype of the modern turbine, with three fiberglass blades, a pitch regulator, and computerized means of control [76, p 177, 82]. While his students sought to commercialize this design, Heronemus submitted further outsider proposals to the NSF, one for the beloved wind-driven DC generators that he imagined could be anchored off the coast of New England in strong winds. Sci-fi-like renderings of his plans were published in *National Geographic* and *Popular Science.* In a more grounded fashion, he also recommended replacing an onshore turbine at Grandpa's Knob, a hillock in Vermont, the largest of its kind at the time, that had been built in the 1940s, following a design by engineer Palmer Putnam. Towering above the landscape, the vast machine generated up to 1.5 MW in high winds [83, p 30]. But, ever idiosyncratic, Heronemus believed this monumental 185-foot diameter wind turbine should be replaced by a structural grid of 100 smaller turbines which, given their number, would be easier to maintain and less prone to intermittent breakdown [78, p 199]. His vision was that wind should be 'farmed' by such large arrays of smaller turbines, rather than one large showcase turbine [79, p 53]. That said, Heronemus still occasionally entertained more extravagant ideas, reimagining the windswept Great Plains as a vast forest of 70 story-high wind turbines for example [84].

Heronemus' ideas attracted Louis Divone, an MIT graduate working at the NSF and overseeing RANN's wind power work. By 1974, RANN was in its third year and had fully hit its stride; with the irruption of the OAPEC oil embargo in October of the previous year its focus was now clearly dominated by energy issues, alongside the related concerns of environmental protection and productivity. Divone convened a workshop, not unlike Cherry Hill, where Heronemus was joined by an earlier generation of wind power engineers, including Putnam, Marcellus Jacobs, and the German engineer Ulrich Hütter [76, p 116, 84]. It was Divone that suggested NASA's Lewis center participate. In 1970 Lewis engineers had been asked to build an experimental turbine on the Puerto Rican island of Culebra. Divone thought NASA's ambitions should be scaled up and that they should build a far larger 100 kW turbine at Plum Brook, a NASA-Lewis outpost that once housed a nuclear research reactor. By 1974, Lewis researchers were working on two-bladed turbines, with each turbine 62 feet in length, and 125 foot high turbines, second in size only to Putnam's creation—i.e., deviating very far from Heronemus' vision of large arrays of small, three-bladed turbines [48, p 64]. NASA's turbines were forecast to generate 180 000 kWh per year [47, p 206–7]. From 1974 onward, Divone's Federal Wind Energy Program continually pushed for the larger wind turbines. At one point he requested a ludicrous 400 foot long fiberglass rotor. Utilities were drawn to such technologies, and Westinghouse and GE, among others, invested in so-called 'behemoth' or megawatt-class turbines, with the idea being that the cost of wind power dropped rapidly with increasing size [79, p 54, 85, 87].

It seemed NASA's involvement had invoked the organization's old mantra of 'higher, faster, farther' but in a domain in which grass-roots engineering had suggested another approach [48, p xii, 86]. The Plum Brook turbine was launched in 1975 by NASA administrator James Fletcher, with the support of the newly formed Energy Research and Development Administration (ERDA), RANN's successor. However, it soon failed [83, p 159]. The blades oscillated severely as unexpected 'impulse' loads bore upon them. In the first eight months, the turbines generated power for just 57 h, as ERDA staff feared that otherwise their blades might succumb to metal fatigue

[87, p 971–2]. Not all money went in the direction of Big Wind, however; between 1973 and 1985, around 15% of the federal wind power budget was spent on smaller turbines, with Rocky Flats in Boulder, CO acting as a test-ground. Private industry also seemed to increasingly favor smaller turbines [76, p 178]. The windfarm idea took off at Crotched Mountain, New Hampshire in 1980 where 20 3kW turbines, descendant of Heronemus's WF-1, were installed by the UMass spin-off US Windpower. But wear and tear meant even these smaller systems began to lose blades. Just a year later the turbine test-site was bulldozed over [76, p 180, ft 17].

It appeared that both big and little wind faced operational problems at this time [87, p 973]. It was a symbolically inauspicious end to this period of federal support for alternative energy, but what looked like failure was arguably a learning process. Today's turbines combine Heronemus's three blades with the height of NASA's behemoths, and are often arranged in farms [76, p 154]. Thanks to federal involvement, the costs and risks involved in these early failures were largely absorbed by government, and on and offshore wind power is now widely seen as a viable source of power generation. Historian of wind power Brandon Owens has noted that the industry suffered from only receiving intermittent support from the US government. Once more, it was Reagan's administration that ended federal support, with his Office of Management and Budget questioning the need to subsidize such technologies if they were as commercially viable as their proponents suggested, and subsequently cutting funding for wind from a $60.5 million peak in 1980 to just $8.8 million by the decade's end [76, p 152]. Moreover, it was NASA's wind power research that informed private-sector efforts in the US and Europe, and once more definitively, China, where European (Vestas) turbines were first installed in Shandong Province in 1986. By the late 2000s, drawn to the vast scale of the Chinese market, legislation that required installed turbines to be built in China, and lower production costs, European and US wind companies formed joint ventures with the Chinese state-owned Big Wind enterprises Sinovel and Goldwind [76, ch 6]. Today's Big Science, as the recent Endless Frontier Act recognizes, is necessarily international [88].

31.6 Conclusion

As one account suggests, RANN, which had begun 'with aspiration', ended in 'dispersion' [89]. In 1977, under the directorship of Richard Atkinson, the NSF opted to 'kill' RANN and disperse its projects throughout the NSF. With ERDA's establishment in 1974, most of RANN's remaining energy projects were absorbed by this agency, and later became either private enterprises or administered by the DOE. On its face, then, RANN appeared to many observers to have been largely a failure.

Yet if we want to understand the evolution of Big Science into mezzo research, RANN looks more like a successful forerunner. RANN's alternative energy initiatives, only some of which are explored here, could be considered mezzo, both big and small. Big problems were framed at large scales but the point was 'to do more with less—less material, energy, and environmental impact' [66, p 9]. The involvement of Big Science institutions such as NASA, looking for their next paycheck, somehow both accelerated and hindered the development of these technologies. But it did so in a manner that was, in the case of solar, temporary. As soon as energy prices dipped, federal interest in solar waned, while the scientists involved lost interest in a technology that seemed further from feasibility than initially hoped. Federal investment in the development of solar expertise was subsequently lost to the market as funding disappeared under Reagan. For wind, NASA's involvement created a number of technological follies due to the inappropriate application of a Big Science approach. But similar technological flaws also plagued small-scale initiatives. If there is a lesson to draw from these case studies, it is that support for alternative energy must be sustained, and not dictated by fossil fuel prices.

It is clear that the energy crisis created a short-lived window of opportunity for research into alternative energy in the US, but one that swung shut as oil prices shrank back to pre-crisis levels under Reagan [90]. Yet if the intersection of alternative energy and the US R&D system in the 1970s had ambiguous results for energy technologies, its impact was considerably more certain and permanent for R&D. Between the late 1960s and the end of the Cold War, the hierarchical, spectacular, state-directed mode of Big Science largely collapsed, to be replaced by a public-private hybrid, multipolar, and applied mode—a transition captured nicely in Dan Kevles' contrasting of the fates of the Superconducting Supercollider and the Human Genome Project in the early 1990s [91]. At NASA, the Space Shuttle and International Space Station programs were necessarily multi-national and multi-organizational—built piece-by-piece from the contributions of each participating party. By contrast, rearguard attempts to preserve the hierarchical, lavishly funded ethos of Apollo days led—as Diane Vaughan has shown—to the disastrous *Challenger* launch decision [92].

At NASA the Big-to-mezzo transition is largely a story of decline. For the NSF, however, that transition led to the Foundation's growing influence over US R&D. The NSF came to Big Science late, as a result of projects transferred from military patronage, such as the Arecibo Observatory and the National Magnet Laboratory. Along with those transfers, NSF also received the Materials Research Laboratories (MRLs) network from ARPA in the early 1970s. As one of us (Mody, with Hyungsub Choi) has written elsewhere, the MRLs inspired the NSF to found other centers and networks, particularly associated with its growing Engineering Division (later Directorate) [93]. Those centers and networks largely support small science projects, but use a mix of active and passive coordination to ensure synergies that make the whole more than the sum of its parts—very much in the mezzo mode. Most of them also involve some level of public-private cooperation and some degree of public outreach—again, symptomatic of a multipolar way of organizing science. Most of them can also trace their origins to the onset of some new national 'crisis'—economic competitiveness, healthcare costs, pollution, etc, even if they also affirm allegiance to the primacy of basic research. The centers' approach is, thus, very much 'research applied' that follows the RANN template.

As we have tried to show, RANN helped the NSF learn how to operate in this problem-oriented manner. Although it lasted less than a decade, and many in the

NSF were happy to see it go, the fact is that RANN-style projects diffused widely across the Foundation during the 1970s. In a sense, the NSF became RANN, and so there was no longer a need for RANN within the NSF. Our argument has been that that organizational transformation took place because of 'existential success'. That is, organizations like NSF, NASA, and the AEC National Laboratories had achieved their mission of denying the Soviets an advantage on the literal and figurative battlefield of science and technology; thus, these organizations lost their once-lavish funding while simultaneously needing to find new missions that the public would support. That process occurred in two phases: an early, exploratory phase (corresponding to the IRRPOS era) of applying systems thinking to various progressive causes; and a second phase (corresponding to RANN) when the energy crises (and the associated economic recessions) catapulted to the top of the agenda. As the price of oil dropped in the 1980s, the connection between RANN-style mezzo science and alternative energy largely disappeared from view, but its legacy remains very much with us. If we wish to understand the unarticulated impulses guiding Federal R&D policy today, we should look to how R&D policy was deployed in the 1970s—when the environmental, economic, and energy crises of the day provoked quite explicit experimentation that echoes into the present.

References

[1] Webb J 1967 *Space Age Management: The Large-Scale Approach* (New York: McGraw-Hill) pp ix, 9
cited in; Capshew J H and Rader K A 1992 Big science: Price to the present *Osiris* **7** 25
Parker M 2009 Space age management *Manag. Organ. Hist.* **4** 317–32

[2] Low G M 1970 Statement before the Subcommittee on NASA Oversight, Committee on Science and Astronautics, US House of Representatives *Poindexter Reference Files* Box 6, Center Series, section 89 Johnson Space Center History Collection, UHCL Archives, Alfred R Neumann Library, University of Houston Clear Lake

[3] Mody C C M 2022 *The Squares: US Physical and Engineering Scientists in the Long 1970s* (Cambridge, MA: MIT Press)

[4] Krige J, Callahan A L and Maharaj A 2013 *NASA in the World: Fifty Years of International Collaboration in Space* (London: Palmgrave Macmillan)
Hersch M H 2013 Using the shuttle: operations on orbit *Space Shuttle Legacy: How We Did It and What We Learned* ed J Krige and J R Launius (Reston, VA: American Institute of Aeronautics and Astronautics) pp 191–211

[5] Westfall C 2003 Rethinking Big Science: modest, mezzo, grand science and the development of the Bevalac, 1941–1973 *ISIS* **94** 30–56

[6] Hoddeson L, Kolb A and Westfall C 2008 *Fermilab: Physics, the Frontier, and Megascience* (Chicago, IL: University of Chicago Press)

[7] Westwick P 2002 *The National Labs: Science in an American System, 1947–1974* (Cambridge, MA: Harvard University Press)

[8] Weinberg A 1961 Impact of large-scale science on the United States: big science is here to stay *Science* **134** 161–4

[9] Galison P 1997 *Image and Logic: A Material Culture of Microphysics* (Chicago, IL: University of Chicago Press) ch 5

[10] Weinberg A 1967 *Reflections on Big Science* (Cambridge, MA: MIT Press)
[11] Pickering A 1984 *Constructing Quarks: A Sociological History of Particle Physics* (Chicago, IL: University of Chicago Press)
[12] Nordhaus W 1973 Allocation of energy resources *Brookings Pap. Econ. Act.* **3** 540
[13] Bogue D 1985 *The Population of the United States* (New York: Free Press)
[14] Forman P 1987 Behind quantum electronics: national security as a basis for physical research in the United States, 1940–60 *Hist. Stud. Phys. Biol. Sci.* **18** 160–1
[15] Hirsch R 1989 *Technology and Transformation in the American Electric Utility Industry* (Cambridge: Cambridge University Press)
[16] Turnbull T 2017 From paradox to policy: the problem of energy resource conservation in Britain and America, 1865–1981 *PhD Dissertation* University of Oxford ch 7
[17] Weinberg A 1970 The axiology of science: the urgent question of scientific priorities has helped promote a growing concern with value in science *Am. Sci.* **58** 612–17
[18] NSF 1968 *Annual Report* National Science Foundation
[19] Belanger D 1998 *Enabling American Innovation: Engineering and the National Science Foundation* (West Lafayette, IN: Purdue University Press)
[20] Phelps J B 1970 IRRPOS looks for relevance to society's problems *Phys. Today* **23** 61–2
[21] Kevles D 1977 The National Science Foundation and the debate over postwar research policy, 1942–45 *ISIS* **68** 4–26
Godin B 2003 Measuring science: is there 'basic research' without statistics *Soc. Sci Inform.* **42** 57–90
[22] NSF 1967 *Annual Report* National Science Foundation
[23] On economics; Backhouse R and Cherrier B 2017 The age of the applied economist: the transformation of economics since the 1970s *Hist. Polit. Econ.* **49** 1–33
[24] Rome A 2003 Give Earth a chance *J. Am. Hist.* **90** 525–54
[25] Hammond A 1971 NSF prods scientists to coordinate bay research *Science* **172** 827–30
[26] Wood L S 2010 *A More Perfect Union: Holistic Worldviews and the Transformation of American Culture after WWII* (Oxford: Oxford University Press)
Belgrad D 2019 *The Culture of Feedback: Ecological Thinking in '70s America* (Chicago, IL: University of Chicago Press)
Heyck H 2015 *Age of System: Understanding the Development of Modern Social Science* (Boston, MA: Johns Hopkins)
Zangwill A 2021 *A Mind Over Matter: Philip Anderson and the Physics of the Very Many* (Oxford: Oxford University Press)
[27] On ecosystem analysis see ; Kwa C 1987 Representations of nature mediating between ecological and science policy: the case of the international biological program *Soc. Stud. Sci.* **17** 413–42
[28] Simon H A 1969 History of RANN's style, purpose, and criteria of managing research *Unpublished memorandum* Carnegie Mellon University, Pittsburgh, PA p 6 http://doi.library.cmu.edu/10.1184/pmc/simon/box00045/fld03519/bdl0001/doc0001
[29] On think tanks, see Medvetz T 2012 *Think Tanks in America* (Chicago, IL: University of Chicago Press)
[30] Squires A 1986 *The Tender Ship: Governmental Management of Technological Change* (Berlin: Springer)
[31] Greenberg D S 1970 Research priorities: new program at NSF reflects shift in values *Science* **170** 144–6

[32] NSF 1970 *Annual Report* National Science Foundation

[33] Appel T 2000 *Shaping Biology: The National Science Foundation and American Biological Research, 1945–75* (Baltimore, MD: Johns Hopkins University Press)

[34] Merton England J 1982 *A Patron for Pure Science* (Alexandria, VA: National Science Foundation)

[35] Egan M 2018 Survival science: crisis disciplines and the shock of the environment in the 1970s *Centaurus* **59** 26–39
Mata T and Scheiding T 2012 National Science Foundation patronage of social science, 1970s and 1980s: congressional scrutiny, advocacy network, and the prestige of economics *Minerva* **50** 423–49
Mark Solovey's work on federally-funded social science comes close, but understandably does not address the natural or engineering sciences, Solovey M 2020 *Social Science for What?* (Cambridge, MA: MIT Press)

[36] Green R J and Lepowski W 2006 A forgotten model for purposeful science *Issues Sci. Technol.* **22** 69–73

[37] NSF 1972 *Annual Report* National Science Foundation

[38] ORNL 1971 Inventory of Energy Research **vol 1** Congress, Committee on Science and Aeronautics, NSF

[39] Szego G C 1971 *The US Energy Problem* (Ontario: InterTechnology Corporation)

[40] NSF 1971 *Annual Report* National Science Foundation
Borelli P *et al* 1971 People, power and pollution: environmental and public interest aspects of electric power plant siting *Environmental Quality Laboratory* 1 September California Institute of Technology

[41] Lifset R D 2014 A new understanding of the American energy crisis of the 1970s *Hist. Soc. Res.* **39** 22–42

[42] Goodwin I 1988 Washington ins and outs: DOE's snow goes to Chicago, Marquet and Kerber leave DOD *Phys. Today* **41** 52

[43] Morse F NSF Presentation *National Science Foundation* https://core.ac.uk/download/pdf/42898361.pdf

[44] NSF 1973 *Annual Report* National Science Foundation

[45] NSF/RANN 1972 An assessment of solar energy as a national energy resource GI-32488, 73

[46] Stradling D and Stradling R 2008 Perceptions of the burning river: deindustrialisation and Cleveland's Cuyahoga River *Environ. Hist.* **13**

[47] Dawson V 1991 *Engines and Innovation: Lewis Laboratory and American Propulsion TechnologyNASA Technology Series* (Washington, DC: USGPO)

[48] Bowles M 2010 *The 'Apollo' of Aeronautics: NASA's Aircraft Energy Efficiency Program, 1973–1983* (Washington, DC: NASA)

[49] Perlin J 1999 *From Space to Earth: The Story of Solar Electricity* (Cambridge, MA: Harvard University Press)

[50] Bleiden R 1973 NSF/RANN plan for national photovoltaic conversion program *Workshop Proceedings: Photovoltaic Conversion of Solar Energy for Terrestrial Applications. Vol. 1. Working Group and Panel Reports* (Alexandria, VA: National Science Foundation) Research Applied to National Needs 84

[51] Forestieri A F 2011 The IEEE Photovoltaic Specialists Conference *Power for the World: The Emergence of Electricity from the Sun* ed W Palz (Stanford, CA: Pan Stanford)

[52] For a critical take on the NSF's supposed role in marginalizing small-scale solar in favor of mainstream corporate projects like the Solar Power Corporation, see Reece R 1979 *The Sun Betrayed: A Report on the Corporate Seizure of US Solar Energy Development* (Binghamton, NY: Maple-Vail)
[53] Bleiden R 1999 Cherry Hill revisited—a retrospective on the creation of a national plan for photovoltaic conversion of solar energy for terrestrial applications *AIP Conf. Proc.* **462** 796
[54] Williams J R and Clement J D 1973 Comparative evaluation of solar, fission, fusion, and fossil power systems *Part 1: Solar Energy Prepared by Georgia Tech. School of Mechanical and Nuclear Engineering on behalf of NASA* (Cleveland, OH: Lewis Research Center), p 1
[55] Ralph E L 1973 device development plan *Workshop Proceedings: Photovoltaic Conversion of Solar Energy for Terrestrial Applications. Vol. 2. Invited Papers* (Alexandria, VA: National Science Foundation) Research Applied to National Needs Fig. 1., 4
[56] Forestieri A F 1973 Photovoltaic terrestrial applications *Workshop Proceedings: Photovoltaic Conversion of Solar Energy for Terrestrial Applications. Vol. 2. Invited Papers* (Alexandria, VA: National Science Foundation) Research Applied to National Needs 150
[57] Bernatowicz D T 1973 The NASA-Lewis program on silicon solar cell and array technology *Workshop Proceedings: Photovoltaic Conversion of Solar Energy for Terrestrial Applications. Vol. 2. Invited Papers* (Alexandria, VA: National Science Foundation) Research Applied to National Needs 39–41
Lindmayer J Solarex Corporation, Silicon cells *Cherry Hill Conf. Proc.* vol 1 p 9
[58] On the imagined parallels between transistors and solar cells, see; Etkzkowitz H 1984 Solar versus nuclear energy: autonomous or dependent technology *Soc. Probl.* **31** 421–3
[59] National Research Council 1981 *Supplement to Energy for Rural Development* OST (Washington DC: Agency of International Development) 16
[60] D'Alessandro B 1979 Papago Indians light the way *Am. Indian J.* **5** 10–13
Greene J, Braitberg V and Maya Bernadett G 2020 Innovation on the reservation: information technology and health systems research among the Papago tribe of Arizona, 1965–80 *ISIS* **111** 443–70
[61] Bahr D, Garret B and Chrisman C 1978 Socioeconomic impact of photovoltaic power at Schuchulik Arizona *DOE/NASA* October, 115
[62] Tohono O'Odham Utility Authority 2019 *Solar Education and Case Study* April https://toua.net/wp-content/uploads/2019/11/TOUA-Solar-Education-and-Case-Study-04.2019.pdf
[63] On reliability problems, see; Stea D and Bugé C 1980 Cultural impact assessment: two cases in native American Communities *Environ. Impact Assess. Rev.* **1** 397–412
On the risk of marginalizing indigenous communities in contemporary energy transitions, see Zografos C and Robbins P 2020 Green sacrifice zones, or why a green new deal cannot ignore the cost shifts of just transitions *One Earth J.* **3** 543–6
[64] Herwig L O 1999 Cherry Hill revisited: background events and photovoltaic technology status *AIP Conf. Proc.* **462** 785–95
[65] Nemet G 2019 *How Solar Became Cheap* (London: Routledge)
[66] NSF 1974 (Alexandria, VA: National Science Foundation)
[67] Laird F N 2001 *Solar Energy, Technology Policy and Institutional Values* (Cambridge: Cambridge University Press) S157
For the 2.3 percent figure in 2020, see Electricity explained *EIA* https://eia.gov/energyexplained/electricity/electricity-in-the-us.php

[68] Bleiden R 1999 ‘Cherry Hill revisited—a retrospective on the creation of a national plan for photovoltaic conversion of solar energy for terrestrial applications’ *AIP Conf. Proc.* **462** 798
[69] Westwick P 2007 *Into the Black: JPL and the American Space Program, 1976–2004* (New Haven, CT: Yale University Press)
[70] Lewis C 2019 Arnaldo Tamayo Méndez and Guion Bluford: the last Cold War race battle *NASA in the Long Civil Rights Movement* ed B C Odom and S P Waring (Gainesville, FL: University Press of Florida) pp 145–66 , see also the other chapters in that volume
[71] Varadi P F 2014 *Sun above the Horizon: Meteoric Rise of the Solar Industry* (London: Routledge)
[72] Jones G and Bouamane L 2012 ‘Power from sunshine’: a business history of solar energy *Working paper* 12-105 Harvard Business School
[73] Jerneck M 2017 Financialization impedes climate change mitigation: evidence from the early American Solar industry *Sci. Adv.* **3** e1601861
[74] ORNL 1972 *An inventory of energy research: a report prepared for the National Science Foundation, 1971, Vol. 1* (Oak Ridge, Tennessee: Oak Ridge National Laboratory) 328. Project number 2175
[75] Cooperdock E *et al* 2020 Black Lives Matter: promoting diversity, equity, and inclusion in geochemistry *Elements* **16** 226–7
[76] Owens B 2019 *The Wind Power Story: A Century of Innovation that Reshaped the Global Energy Landscape* (Piscataway, NJ: Wiley/IEEE)
[77] Stoddard W 2002 The life and work of Bill Heronemus, wind engineering pioneer *Wind Eng.* **26** 335–41
[78] Heronemus W E 1973 A proposed national wind power R&D program *Wind Energy Conversion Systems* ed J Savino (Washington, DC: National Science Foundation) NSF/RA/W-73-006
[79] Asmus P 2001 *Reaping the Wind: How Mechanical Wizards, Visionaries, and Profiteers Helped Shape Our Energy Future* (Washington, DC: Island)
[80] Heronemus W E 1971 A National Network of Pollution-Free Energy Sources *Unpublished Proposal to the National Science Foundation* Research Applied to National Needs (RANN) Directorate
[81] Owens B 2019 *The Wind Power Story: A Century of Innovation that Reshaped the Global Energy Landscape* (Piscataway, NJ: Wiley/IEEE Press) p 176
[82] UMass WF-1, a retrospective *Wind Energy Center, University of Masschusetts Amherst* https://umass.edu/windenergy/about/history/windturbinewf1
[83] Righter R 2011 *Windfall: Wind Energy in America Today* (Norman, OK: University of Oklahoma Press)
[84] McCaull J 1973 Windmills *Environ., Sci. Policy Sustain. Develop.* **15** 6–17
[85] Heymann M 2001 Signs of hubris: the shaping of wind technology styles in Germany, Denmark, and the United States, 1940–90 *Technol. Cult.* **39** 654–5
[86] On the persistent folly of ‘aerospace-type thinking’ in wind power, see; Gandoin R 2019 A call to historicize wind and site studies *J. Energy Hist./Rev. d'Hist. de l'Énergie* **1** 8–9
[87] Metz W D Wind energy, large and small systems competing *Science* **197** 971–2
[88] Windham P, Hill C and Cheney D 2020 Improving the endless Frontier Act *Issues Sci. Technol.* https://issues.org/improving-the-endless-frontier-act/
[89] Sanderson J 1977 NSF kills RANN in aim to reunite agency *Chem. Eng. News* **55** 36

[90] Basosi D 2018 A small window: the opportunities for renewable energies from shock to counter shock *Counter Shock: The Oil Counter-Revolution of the 1980s* ed D Basosi, M Trentin and G Garavini (London: Bloomsbury)
[91] Kevles D 1997 Big Science and Big Politics in the United States: reflections on the death of the SSC and the life of the human genome project *Hist. Stud. Phys. Biol. Sci.* **27** 269–97
[92] Vaughan D 1996 *The Challenger Launch Decision* (Chicago, IL: University of Chicago Press)
[93] Mody C C M and Choi H 2013 From materials science to nanotechnology: institutions, communities, and disciplines at Cornell University, 1960–2000 *Hist. Stud. Nat. Sci.* **43** 121–61

IOP Publishing

Big Science in the 21st Century
Economic and societal impacts
Panagiotis Charitos, Theodore Arabatzis, Harry Cliff, Günther Dissertori, Juliette Forneris and Jason Li-Ying

Chapter 32

Reflections on the New Big Science: capabilities and challenges

Robert P Crease

The past few decades have seen a phase transition in large-scale science taking place not only throughout the US National Laboratory system but also at large scientific facilities elsewhere. This phase shift, which Catherine Westfall and I have dubbed the 'New Big Science', accompanies the replacement of high-energy physics accelerators by large-scale materials science facilities as the marquee projects at nationally supported laboratories. These facilities support a research ecology consisting mainly of relatively small instruments, experiments, and collaborations, and the research these support is growing not so much bigger as more complex. This phase shift has ushered in important developments in the character of research at these laboratories, along with a novel set of cultural, social, and economic impacts [1].

32.1 Big Science background

The term 'Big Science', promoted in the 1960s by Alvin Weinberg and Derek de Solla Price, named an undeniable, and what appeared to be irreversible, phenomenon: the rapidly increasing scale and expense of scientific projects—mainly physics—in the post-World War II era. The term designated both a condition and a dynamic.

32.1.1 Big Science: the condition

The condition was the seemingly inevitable reliance on large facilities, budgets, and large collaborations. Big Science was not a sudden development; nevertheless, the size of facilities, funding, and the scientific community made such a leap during World War II that it may not unjustifiably be used to date its beginning. Shortly after the War, the US, the national laboratory system was created to build and operate instruments—principally reactors and accelerators—that were of too large a

doi:10.1088/978-0-7503-3631-4ch32

scale for single institutions such as industries or universities to afford and operate. The first three national laboratories were the regionally distributed Argonne, Brookhaven, and Clinton (soon Oak Ridge) National Laboratories, but these were soon joined by other, still more widely distributed, national laboratories. New government agencies (the Atomic Energy Commission, or AEC) and congressional agencies (the Joint Committee on Atomic Energy, or JCAE), and several layers of expert panels, were created to support these efforts. In Europe, international collaborations sprang up to build laboratories and promote research; the founding document for CERN was signed by 12 member states in 1953, while the founding document for Euratom, the European Atomic Energy Community, was signed four years later. European space collaborations and the European Southern Observatory came into being in the 1960s, and at the end of the decade Germany and France collaborated to build the Institut Laue-Langevin.

Another aspect of the Big Science condition was Big Visibility. The size, cost, attention, and products of the now-government-supported facilities—not to mention the huge significance of such byproducts of the war's scientific efforts as the atomic bomb, radar, and the ability to produce penicillin on a large scale—inevitably made them conspicuously public places. Roger Pielke has pointed out that 'basic research' became a political symbol during the Cold War; so, one can add, did Big Scientific facilities [2].

32.1.2 Big Science: the dynamic

The dynamic involved certain ways of leading and managing large scientific facilities, and an important part of a manager's eye had to look upward. These facilities were not one-time commitments; their viability required continued increases in scale. To stay productive, a national laboratory had to be planning and building a new generation of instruments even as its researchers were working at existing ones. At Brookhaven, for instance, the Cosmotron accelerator (1952) was replaced by the Alternating Gradient Synchrotron (1961), and plans were to succeed it with ISABELLE (proposed in 1972); meanwhile, the laboratory's first reactor, the Brookhaven Graphite Research Reactor (1950) was replaced by the High Flux Beam Reactor (HFBR) in 1965. This dynamic had a wide range of consequences. One had to do with the source of funding, for laboratories were increasingly dependent on the government, with the concomitant dependence on its needs. Another consequence had to do with education and training. Larger collaborations meant that participants were generally restricted to performing more specialized roles. Graduate students often could not participate in an experiment all the way through. Another dynamic of Big Science relating to Big Visibility was the unavoidable need to develop community relations and political connections. Big Science facilities were equivocal symbols—of pride, political agendas, and sensationalistic media coverage—and could become targets of public activism about such things as pollution, cancer, meltdowns, and vulnerability. Big Visibility, therefore, came at a price. In the 1950s, during the Cold War, for instance, right-wing ABC News radio commentator Paul Harvey railed against Argonne National Laboratory for being easily infiltrated by enemy agents, at one point scaling its barbed wire fences to show how supposedly simple it was to penetrate a lab and potentially steal its atomic secrets. In the 1990s, the safely operated HFBR at Brookhaven was terminated following an antinuclear campaign by activists and catered to by politicians [3]. Oak Ridge National Laboratory, too, has experienced local suspicion and occasional demonstrations stemming from its role in the Manhattan Project and for its continued operation of a reactor.

32.2 Transition

The transition to New Big Science took place thanks to a surprisingly rapid overtaking of Big Science facilities by big materials science facilities as the most prominent projects at US national laboratories. A first factor in this development was the growing interest in materials science. The national laboratories, of course, had always pursued a certain amount of materials science, and some research had been carried out at 'first-generation' synchrotron light sources fashioned out of obsolete storage rings for high-energy physics accelerators. But the mission of the national laboratories was to build larger facilities than materials science required at the time. The initial second-generation synchrotron light source—built from the start for that purpose and with a new kind of lattice optimized for producing synchrotron light—was the National Synchrotron Light Source (NSLS), completed in 1983, at Brookhaven National Laboratory. The NSLS was built on a shoestring, with an order of magnitude less funding than ISABELLE, the lab's planned new state-of-the-art high-energy physics accelerator. Still, the NSLS quickly showed that light sources could serve a wide range of materials scientists, and demonstrated their value. The NSLS also brought in a new kind of culture, industrial participation, and diversity within collaborations to the labs, as well as new potential sources of funding. In 1984, a year after the NSLS opened, the growing interest in materials science facilities was ratified by two Department of Energy panels, the 'Planning Study for Advanced National Synchrotron Radiation Facilities' ('The Eisenberger-Knotek Report') and 'Major Facilities for Materials Research and Related Disciplines' ('The Seitz-Eastman Report') [4].

A second factor in the transition to the New Big Science was that, as Big Science facilities devoted to high-energy physics grew larger and more expensive—soaking up more and more funds for other projects—fewer were planned and constructed, they could be less geographically distributed, and they had to be more ambitious, thus taking more time to plan and build. Brookhaven's accelerator ISABELLE was a victim, as it was terminated in 1983 to free up resources for the much larger Superconducting Supercollider (SSC); the US high-energy physicists were in effect putting all their eggs in one basket. The SSC's huge expense then began to make materials scientists fear that it would divert support for their field and facilities, and for that reason some in the materials science community began to agitate against the SSC. One particularly vocal anti-SSC voice was materials scientist Philip Anderson, who took on high-energy physicists, including their prominent spokesman Steven Weinberg, to argue that fundamental knowledge was found throughout all scales of physics, not just high-energy physics, so that allowing the SSC to soak up physics funding would starve all but a small corner of the field [5]. The SSC, which was to be built at a new location in Texas, also made the directors of existing national laboratories elsewhere in the country concerned about the viability of their own facilities and labs.

In 1985 the DOE's director of the Office of Energy Research, Alvin Trivelpiece, faced that problem head-on. Trivelpiece forged a plan in which four national laboratories would each get a major new facility. Three of these projects focused on materials science: Argonne got the Advanced Photon Source (APS), the Lawrence Berkeley National Laboratory got the Advanced Light Source (ALS), Oak Ridge got a new reactor (the Advanced Neutron Source, or ANS, which was eventually replaced by the Spallation Neutron Source), while Brookhaven got a heavy-ion facility for nuclear physics, to be called the Relativistic Heavy Ion Collider, built in the tunnel and using other infrastructure that had been partially constructed for the terminated ISABELLE. To award specific facilities to specific laboratories required a significant deviation from the protocol. 'These senior management site decisions were made to maintain the technical viability among the Department of Energy laboratories' [6], as an internal memorandum put it. Plans for those four projects were then drawn up and got underway.

Then came a dramatic turn. In 1993, the US Congress terminated the SSC [7]. The decision left the four elements of the Trivelpiece plan intact as the key new big US projects. Each was poised to advance. The ALS was commissioned in 1993, the APS in 1996, in 2000 RHIC was commissioned, and that same year work on the Spallation Neutron Source (SNS) started. This gave a strong thrust to a new age of user facilities in the US. The national laboratories were no longer fenced-off facilities but academically public areas, with graduate and undergraduate students coming and going. Once upon a time, an outside user would physically bring an instrument to the lab along with a small group of collaborators, bolt it in place, take data, and then go back home. No more. Nobody would think of building another large facility without planning for a vast user community.

32.3 New Big Science

Thus around 2000, in the post-Cold War era and about half a century after the onset of Big Science, US national laboratories no longer followed the Big Science model marked by ever larger teams, facilities, and instruments. The New Big Science involved a new condition and new dynamic.

32.3.1 New Big Science: the condition

The new condition, again, was that large-scale materials science accelerators had replaced high-energy physics accelerators as the most prominent projects at most major US basic research laboratories. Their new generations of facilities were no longer gigantic leaps in scale or power over previous ones. Instead, the research that began to dominate at the national labs consisted of materials science facilities with smaller teams of researchers, more or less familiar instruments, and more or less stable funding. Instruments and collaborations did not get markedly bigger and bigger, but rather more diverse and complex, involving more disciplines and researchers with a broader range of affiliations working in teams that were more flexible and interactive. Some features of the New Big Science condition are:

- *Integrated industrial presence.* In Big Science, industrial users were essentially parasitic on the main basic research functions of the facilities. In the New Big Science, facilities and many beamlines and instruments support a spectrum of industrial or otherwise applied users who participate from the outset as intended users rather than as parasites.
- *Octopus-like networks.* A single beamline or instrument can serve as a hub whose research 'arms' reach out to different research projects; meanwhile, the arms of one research project can reach out to different beamlines.
- *Interdisciplinarity.* Some research projects may engage a spectrum of seemingly disparate disciplines in the sciences and beyond of a vastly larger scope than ever before; certain imaging projects, for instance, may include archaeologists or specialists in musical instruments.
- *Ecological complexity.* Research cultures at large-scale facilities now resemble ecosystems, comprised of complex and evolving interactions between individuals, institutions, and the overall research environment. The experiments are smaller, more numerous, and have more links with each another and to instruments, beamlines, industries, labs, funders, fields, and collaborators.
- *Intermural as well as intramural laboratory collaborations.* Instrument design and construction, or research projects requiring different tools such as x-ray, neutron, and other scattering methods, may depend on collaborations to a much larger extent than previously between different national laboratories.
- *Subfacilities.* Some facilities, such as synchrotron light sources, can support subfacilities that themselves support networks with the above properties, an example being x-ray lithography sources.
- *Integrative systems.* An integrative system is an organized and extended interdisciplinary network involving technologies planned and promoted for practical applications whose webs connect not just different scientific fields with each other but also with other webs involved with preparing and ushering technologies into the marketplace. An example of an integrative system is Biopolis in Singapore, which seeks to promote medical research and facilitate its clinical applications, including the construction of a legal and economic infrastructure to support these applications, or Fusionopolis, a research and development park to support engineering and materials science technologies. Though rare so far, these are also potential parts of the New Big Science.

32.3.2 New Big Science: the dynamic

The research dynamic at facilities with the above features is also new, more like the character and evolution of an ecology. Here are some features of the new dynamic:

- *Outreach.* One part of managing a New Big Science facility often actively seeking out and acquainting researchers in fields who may not be aware of the relevance and value of the research tools available at the facility for their own fields. Managers at Cornell's CHESS synchrotron light source, for instance, made botanists aware of the value of the kind of x-ray imaging obtainable at that facility for botanical studies.

- *Multistability*. Philosophers of science apply the term 'multistability' to the way artifacts and techniques developed for a single purpose can mutate to serve in other contexts for unforeseen purposes, the way that imaging tools have served in everything from engineering to biology and archaeology [8]. Mössbauer spectroscopy is one example of a technique that has morphed into several others.
- *Knowledge formation*. High-energy physicists generally use accelerators to add pieces to a single, coherent puzzle. In contrast, materials scientists often use large machines in concert with smaller, even table top, devices to piece together—with feedback loops—a mosaic of properties with an agenda that might change as the process proceeds and new properties of potentially useful materials emerge.
- *Krinsky effect*. This is named for the late Sam Krinsky, an NSLS accelerator physicist, who pointed out that x-ray users are more interested in keeping what they have than in envisioning—and waiting patiently for—new facilities, while the reverse is the case for high-energy physics users. At big high-energy physics machines, he noted, experimenters have a certain amount of patience when the machine breaks down or shuts for improvements, for they cannot move their detector, depend on collecting statistics, are rarely in danger of being scooped, and can use the time to upgrade. At synchrotron light sources, by contrast, the culture is different. Somebody *can* scoop you if the machine breaks down. Experiments are *not* limited by statistics. Detectors are rarely significantly upgraded. This cultural difference from Big Science facilities is a manifestation of a deeper phenomenon, the way a facility and community culture are bound up with each other.

The transition of Big Science to New Big Science in Europe did not exactly follow the US model; the European system was always more multinational with a greater variety of industrial national collaborations. Still, as in the US, European investment for light sources used primarily for materials sciences and biology has increased and is likely to continue [9].

32.4 Challenges

The new condition and dynamics of the New Big Science pose new challenges to science managers, educators, and scientists themselves.

- *Little visibility*. No more Big Visibility. Little visibility means that New Big Science facilities are less prone to acquire a symbolic valence, arousing fewer community concerns and less likely to become targets of environmental or anti-government activism. But little visibility, too, has a price, for New Big Science facilities are less politically exciting, meaning that the political case for them is harder to make. Congressmen like to build projects in their home districts, and see little compelling need to support those even in neighbouring districts. Oddly, New Big Science facilities have more users but are in danger of having less political support. That reality has prompted the Department of Energy to develop an interactive online map that, among other things, can reveal which users in a given congressional district have been to a particular DOE user facility elsewhere. What other solutions are possible?

- *Timely access.* The New Big Science has put stress on policies and management methods created during the era of Big Science. The traditional proposal approval process was designed for academic users and generally takes 9–12 months or so. But industrial users and small businesses generally need much quicker approval times, even a matter of weeks. Some industrial users, such as IBM or Exxon, may operate their own beamlines, and the approval process can be fast-tracked. But other industries or small businesses do not have that luxury, and they need quicker access the most. How can access to facilities be sped up in an equitable way?
- *Regulation.* When research is highly centralized, research regulation—of ethics and safety issues for instance—can be imposed by covering the biggest institutions. When the research is more spread out, as in the kind of networks typical of the New Big Science, regulations become more difficult to impose and enforce. This threatens to undermine the enterprise of such things as effective ethics and safety regulation. How can regulations be justly made and enforced given such research conditions?
- *Knowledge coordination.* How to coordinate relevant information from various kinds of facilities—neutron sources, synchrotron sources, and Cryo-EM in crystallography, for instance—when each of them provides different kinds of information with different uncertainties.
- *Management.* Effective management of New Big Science facilities requires being aware of rapidly changing developments and trends in research. How did the research patterns change from 5 and 10 years ago, and are they likely to continue? These patterns are difficult to spot; few people envisioned, for instance, the rise in biological research that has taken place at synchrotron light sources. The researchers at materials science facilities are diverse and transient, and therefore much harder to track; they include scientists from varied fields who receive money through universities, institutes, and industry, and medical doctors who obtain funding from NIH, hospitals, and pharmaceutical companies. How can one spot these patterns in a way to sustain and promote research? How can one extract, analyse, and display information for the benefit of managers, historians, and other researchers in academia and government when both methods and data are so dispersed? If the research climate is compared to an ecology, what are the keystone species? What elements (instruments, say) have to be protected?
- *Developing digital tools.* What digital tools can be developed for this task? [10] To develop it would require determining what key markers to use, and what the data set is one needs. One would have to treat each experiment station as

akin to a little stage: you'd want to input the actors (collaborators), their affiliations, publications, and awards—all the things that you find in the annual reports, the traditional or legacy database. But one would also want to list the props: the instruments, techniques, and audiences. One obstacle to developing such a database is the existence of proprietary research, and the EU's General Data Protection Regulation (GDPR). Such a database, however, would be valuable in the way it could detect patterns in the evolution of research—but valuable also because it could lead inquirers to refine their questions. For instance, such a database at a synchrotron light source might show that research at particular instruments shifted over time from general uses to a specific material, allowing the inquirers to approach the researchers and ask what developments led to that. Such a digital tool would be able to determine whether a case study was a good example of a trend in research or an anomaly.

32.5 Conclusion

The replacement of high-energy physics accelerators by large-scale materials science facilities at nationally supported laboratories has been accompanied by numerous shifts in the character of Big Science. High-energy and nuclear physics will not disappear, of course, at these labs. Still, the way the identity and mission of the national laboratories are characterized, and the character and culture of the research they support, is undergoing profound and long-term changes that affect their funding, visibility, leadership, administration, and responsibilities to their local communities. Historians and other scholars of the New Big Science have to respond to these changes by crafting new tools to extract, analyse, and display information about research at these labs, and that information will be essential to maintaining the robustness of these labs.

References

[1] Crease R P and Westfall C 2016 The new big science *Phys. Today* **69** 30

[2] Pielke R Jr 2012 'Basic research' as a political symbol *Minerva* **50** 339–61

[3] Crease R P with Bond P N 2022 *The Leak: Politics, Activism, and Loss of Trust at Brookhaven National Laboratory* (Cambridge, MA: MIT Press)

[4] Eisenberger P and Knotek M L 1984 *Planning Study for Advanced National Synchrotron Radiation Facilities* Printed March 14 Department of Energy, Office of Basic Energy Sciences (the 'Eisenberger-Knotek Report') https://aps.anl.gov/files/APS-Uploads/APS-Science/1984/eisknot.pdf

Eastman D and Seitz F 1984 Major facilities for materials research and related disciplines *Major Materials Facilities Committee Commission on Physical Sciences, Mathematics, and Resources, National Research Council* (Washington, DC: National Academy Press) https://aps.anl.gov/files/APS-Uploads/APS-Science/1984/rf002.pdf

[5] Martin J D 2018 *Solid State Insurrection: How the Science of Substance Made American Physics Matter* (Pittsburgh, PA: University of Pittsburgh Press)

[6] Salgado J F 1986 The record *United States Government Memorandum*

[7] Riordan M, Hoddeson L and Kolb A W 2015 *Tunnel Visions: The Rise and Fall of the Superconducting Super Collider* (Chicago, IL: University of Chicago Press)
[8] On multistability see, in particular: Ihde D 1990 *Technology and the Lifeworld: From Garden to Earth* (Bloomington, IN: Indiana University Press)
Also: Rosenberger R 2022 Technological multistability and the trouble with the things themselves *The Oxford Handbook of Philosophy of Technology* ed S Vallor (Oxford: Oxford University Press)
[9] For an example of recent Big Science in Europe, see: Kaiserfeld T and O'Dell T 2013 *Legitimizing ESS: Big Science as a Collaboration Across Boundaries* (Falun: Nordic Academic Press)
[10] For an example of an attempt to develop such a digital tool, see: Crease R P, Elyse G and Folson J 2019 Database thinking and deep description: designing a digital archive of the National Synchrotron Light Source *Digit. Scholarsh. Humanit.* **34** i46–57

IOP Publishing

Big Science in the 21st Century
Economic and societal impacts
Panagiotis Charitos, Theodore Arabatzis, Harry Cliff, Günther Dissertori, Juliette Forneris and Jason Li-Ying

Chapter 33

Selling and over-selling Big Science in the knowledge society

Olof Hallonsten

Big Science originated during World War II and was allowed to expand during the Cold War era much because of the cultural and political prestige enjoyed by nuclear energy. During the last decades of the twentieth century, however, research policy in the United States and Western Europe shifted towards more explicit demands of (economic) relevance and 'innovation policy', and other forms of Big Science also grew to prominence, of multidisciplinary character and more oriented to practical applications. In this chapter, public marketing campaigns of contemporary Big Science are analysed and put into relevant historical and theoretical context. With the help of an institutional perspective and a case study, the European Spallation Source (ESS) in Lund, Sweden, the analysis connects Big Science marketing efforts to current trends in society of prioritizing the spectacular but superficial over the truly or deeply useful or meaningful. The analysis brings to life the concept of 'ideographs', relates the development of Big Science into its current state to wider shifts in research policy, in particular competition for resources and attention, and conducts a theoretical discussion on the sociology of expectations and promises in policymaking in general, and in advertisement campaigns of Big Science in particular.

33.1 Introduction

The use of very large infrastructure for scientific research—'Big Science'—is a twentieth and twenty-first century phenomenon, although historians and sociologists have also used the concept for some research-related endeavors in previous times. In the literature, the military-lead research programs during World War II are usually identified as the origins of Big Science as we know it, and in the aftermath of the war Big Science became an inalienable part of the public science systems (PSSs) that were built up partly on basis of existing academic institutions and partly on a new policy of generous public funding as well as new organizations and missions to fund, steer, and carry out research in the interest of society (e.g. Galison and Hevly 1992, Westwick 2003, Krige 2006, Hiltzik 2015). A lot of the rationale for this ambitious buildup could be found in the strained military/geopolitical situation of the Cold War, and consequently, research areas that had proven their worth for weapons development and other national security applications during World War II gained a prominent place in the postwar regime. Most notably, nuclear physics and its 1950s offspring particle physics (or high energy physics) took a crown position not only in the United States but also in Western Europe, and to some extent in the Soviet Union (Kevles 1995/1977, Krige 2006, Martin 2015).

Today, particle physics has lost most of its previous glory and the use of very large infrastructure for scientific research has spread to many disciplines, including structural biology and materials science. In this 'transformed Big Science' (Hallonsten 2016, Cramer *et al* 2020), nuclear reactors and particle accelerators are used for purposes for which they were originally not invented, namely the production of x-rays and neutron beams for the multidisciplinary study of various materials, including biomaterials. These facilities attract several tens of thousands of users in the world annually, and have become indispensable experimental resources for many areas of the natural sciences. They are significantly more oriented to applications and to scientific work to meet the perceived challenges or stated needs of contemporary society, such as innovation for sustainability and economic growth and national or regional competitiveness in a globalized world, than were the Cold War era counterparts (Hallonsten 2016, Cramer 2020, Cramer and Hallonsten 2020).

This chapter analyses public marketing efforts of contemporary Big Science, by placing them in proper historical context and within a theoretical framework of institutionalist analysis of research policy and public marketing campaigns. It thus discusses the meaning of promises and expectations in political campaigning, and how this relates to the public image of science and technological development in current society. Empirically, the analysis focuses on the European Spallation Source (ESS), a state-of-the-art neutron scattering facility built in Lund in Southern Sweden by a consortium of 13 European countries, and its marketing campaign that was at its height in the years 2008–09 (Hallonsten 2020). Theoretically, the analysis makes use of an institutional view of research policy and insights from the work of Harro van Lente (1993, 2000) on the role of promises and expectations in the promotion of scientific and technological projects; the various conceptualizations of a doctrinal shift in research policy in the United States and Western Europe from the late 1970s onwards (Irvine and Martin 1984, Smith 1990, Stokes 1997, Krige 2006, Berman 2012, 2014, Hallonsten 2021); the early work of Niklas Luhmann (1968) on trust and expectations as a mechanism for the reduction of (social) complexity; as well as the concept of 'ideographs' in rhetoric studies, launched by McGee (1980) and revived by van Lente (2000).

The ambitions of the chapter are mainly theoretical and critical. An important rationale for social science is its critical capacity and its ability to uncover features of contemporary social life that are unnoticed or obscured by mainstream discourse, and specifically engaging with topics where powerful interests reign and are allowed to set and control the agenda, to the detriment of society in a broader or deeper sense (Berger 1963, Habermas 1971/2002). Contemporary Western society is characterized by an obsession with innovation, grandiose promises and expectations, and short-term rewards (Lasch 1979/2018, Roberts 2014, Vinsel and Russell 2020), which makes organizations and institutions engage in excess 'window dressing' and the staging of 'pseudo-events', often at the expense of real and broad quality and meaning (Boorstin 1961/2012, Alvesson 2013). A key task of social science is therefore, arguably, to critically examine such delusory and adverse features of organizations and society, and expose their incoherencies and deficiencies, with the long-term aim of contributing to better policy and priorities, more efficient and responsible resource utilization, and more harmonic relationships between society's various institutions and between the public and society's various elites. While such approaches are easy to dismiss as quibbling, fault-finding, or even sabotage of the prospects for reaching political goals, it has been proven many times that critical thinking and impartial analysis of current and historical events as well as of the public communication of powerful individuals and organizations in society are crucially important for the well-being of democracy in the long run.

The organizations that propose, develop, and take on stewardship for contemporary Big Science projects consistently engage in selling and over-selling these projects, with the help of grandiose and superficial promises of spectacular scientific breakthroughs and vast socio-economic benefits. This is allowed to continue even though the professionals at the helm of these organizations—and, conceivably, also some of the communication officials in charge of crafting and communicating the public messages—quite obviously must be aware that Big Science, like any science, is inherently uncertain and its outcomes are impossible to predict. Therefore, both silly promises of 'smoother cream cheese' and more serious (and thus perilous) assurances that Big Science will bring solutions to the climate crisis (Hallonsten 2020, p 249) are blatantly unscientific promises that give unrealistic and even counter-productive impressions of the capabilities of contemporary Big Science, among politicians, industrialists, and the general public. This rather obvious incoherence in the public message of the marketing campaigns, and the broad gap between advertising and reality, has not really been critically examined by social scientists, who otherwise are rather well-equipped to analyse its sociological basis. The chapter makes a contribution towards remedying this, by connecting the study of contemporary Big Science with historically oriented science policy studies and sociological theory on the role of expectations and promises in society, in the hope that more voices of reason enter the discussion in the future.

33.2 Transformed Big Science and the doctrinal shift of research policy

The World War II experience of the mobilization of technology and scientific talent for mission-oriented work with military purposes not only made Big Science a natural part of the postwar research systems of several countries, it also provided a firm ground for the formation of a research policy doctrine in the Western world where fundamental research was designated the true origin of societal and economic development as well as long-term military superiority, through the 'linear model of innovation'. Governments began investing heavily in both academic and mission-oriented R&D efforts and the contours of a 'social contract for science' also began to emerge. Generous public funding would secure active and healthy research programs but the governance of these would be left to expert rule, i.e. the scientists themselves, and thus would useful outcomes also be secured (Guston 2000, pp 37–45, Greenberg 1999/1967, pp 107, 112–14).

The economic boom of the postwar era enabled a continuous growth of spending on R&D in most Western countries, and physicists became used to a privileged situation of having seemingly unlimited funds available (Greenberg 1999/1967). With the 1953 launch of the 'Atoms for Peace' policies by the Eisenhower administration, Big Science gained a natural role to play also outside the military realm, through particle physics, a fundamental and entirely peaceful branch of Big Science (Hewlett and Holl 1989, pp 257). Soon, particle physics also became another area of superpower competition, as the larger and larger accelerator complexes and also the continuous discovery of new subatomic particles and forces had an easily graspable logic and an element of grandiosity, proving technological edge on behalf of nations just like, e.g., space programs (Greenberg 1999/1967, pp 218–19). The typical cost of cutting-edge particle accelerators grew from hundreds of thousands of dollars in the 1940s, to millions in the 1950s, hundreds of millions in the 1960s and 70s, and eventually billions in the 1980s (Hoddeson and Kolb 2000, p 308). Most of these big machines were built and operated within the frameworks of national laboratories in various countries, which meant that Big Science was also institutionalized in specific organizational fields and provided with a solid organizational basis where it could eventually renew itself and expand beyond the initial purposes and missions, when contextual factors were altered and priorities in research policy shifted (Westfall 2003, 2010, Hallonsten and Heinze 2012, 2016, Hallonsten 2016).

The 'postwar consensus' of US and Western European research policy (Smith 1990, pp 36–7) came to an end in the 1960s, when claims were heard that the linear model of technological innovation was not foolproof and that basic science was not transformed into technological innovations and economic and social development at the rate expected (Guston 2000, p 78, Westwick 2003, p 296). As the postwar economic boom can to a halt in the early- to mid-1970s and austerity took hold on several economies, not only did the growth of R&D investments in the United States and Western Europe slow down (and in some countries even turn into decline) (Herman 1986, pp 18–19, Smith 1990), new ideals also began to spread that preached strategic priority, accountability, and demands of applicability of R&D efforts (Irvine and Martin 1984, Stokes 1997, p 58ff, Guston 2000). The public sector reforms of the 1970s and 1980s have often been identified with 'neoliberal' ideology (Rose and Miller 1992, Wacquant 2012) but in the case of research policy reforms in the same period, state-interventionist agendas were also behind the 'economization' of research policy (Berman 2012, 2014, Hallonsten 2021). Governments began to view the main purpose of their policymaking as seeking to positively affect the larger economy, and research policy was consequently reformed to make PSSs most of all deliver innovations for the economy to grow. Put differently, the organizations and institutions that produce and disseminate scientific knowledge became viewed primarily as elements of an 'innovation system' (Lundvall 1992). As a consequence, a greater number of actors and actor groups in society started to impose their particular interests and ideas about the role of knowledge in society, and how knowledge production should be managed, on the organizations and institutions of PSSs. In practice, therefore, although the 'knowledge society' is usually meant to denote a society permeated by knowledge in all its parts, it can also be conceptually understood as a society where a multitude of interests demand, and are allowed, a say in defining what knowledge is, or ought to be, how it shall be produced and disseminated, and by whom (Sörlin and Vessuri 2007, pp 1–2, Hallonsten 2021, pp 15–16). In this context, 'economization' also means that the market economy takes precedence over other interests so that science undergoes 'commodification', which means that economic criteria are allowed to dominate the evaluation of quality, relevance, and success (Kleinman and Vallas 2001, Radder 2010), and accountability becomes a watchword, not meaning responsibility for conduct but rather capability to be counted, i.e. evaluated and ranked with the help of metrics (Dahler-Larsen 2012, Hallonsten 2021). Strategic priority becomes an inalienable part of this new research policy regime where science is no longer viewed primarily as a general public good but instead as a financial good and/or a deliverer of specific solutions to society's current problems (Pestre 2005, Elzinga 2012).

But for Big Science, the shift was slightly different. Due to its symbolic connection to the arms race and the superpower competition, particle physics retained a privileged position throughout the whole Cold War period, with new accelerator complexes built and planned in Europe and the United States according to the same logic as before, although their sizes grew to a level that demanded the concentration of resources at a limited number of locations (Hoddeson *et al* 2008, p 281). Outside of the particle physics hotspots in Europe and the United States, other labs on both of these continents experienced a 'squeeze' already in the 1970s when their plans for new Big Science facilities were turned down as part of the concentration of particle physics funding to a few prioritized labs (Westfall 2008, Hallonsten and Heinze 2012, p 454). Most of them managed to redirect priorities and reinvent themselves so that in the mid-1990s new large infrastructures of the 'transformed Big Science' (Hallonsten 2016) were in operation at several former nuclear and particle physics labs—foremost synchrotron radiation and neutron scattering facilities for the multidisciplinary study of materials (Hallonsten and Heinze 2015).

Thus while particle physics continued to grow in terms of investment and operating costs of accelerator complexes at least until the early 1990s, and hence appeared untouchable by the cycles of the economies of the United States and Western Europe and by 'economization' and 'commodification' of research policy, other uses of similar large infrastructures by a broader community of natural scientists grew in its shadows. Materials science became an area of priority of many national research policy systems in the 1970s and on (Bensaude 2001, Mody 2011,

Gribbe and Hallonsten 2017), and the life sciences experienced a similar growth in importance somewhat later (Kevles 1997, Kastor 2010). Both areas benefited tremendously from the progress in synchrotron radiation and neutron scattering that produced the 'transformed Big Science', and both also contributed greatly to a rebranding of large-scale research facilities that made them resources for R&D work with greater societal relevance and industrial potential, such as the molecular and atomic study of advanced materials and the mapping of biomolecules with relevance for drug development and other life science research (Doing 2015, Hallonsten and Heinze 2015, Hallonsten 2016). When in 1993 the United States congress canceled the Superconducting Super Collider, the largest scientific infrastructure project ever launched and the most recent in the succession of US particle physics machines, it put a halt to a previously uninterrupted progress and created (perhaps unknowingly) a symbolic event that marked the end of the Cold War era Big Science (Riordan *et al* 2015). Several factors contributed to the demise of the Super Collider, not least dramatic cost overruns (taking the estimated total construction costs beyond the $10 billion mark), but the end of the Cold War and thus the end of an obvious motive for spending enormous amounts of money on fundamental physics, should not be underestimated as a cause for its canceling (Hoddeson and Kolb 2000, p 309, Riordan *et al* 2015, p 77). As discussed above, a research policy system of post-Cold War character had started to develop long before the end of the Cold War in 1989–91, and the emergence and growth of the 'transformed Big Science' cohered with this development.

'Economization', 'commodification', strategic priority, and the increased demands for relevance and accountability in PSSs can all be interpreted as functions of an increased public and/or political *impatience* with the institution of science and a demand for directness and swiftness in its contributions to economic and social development. Although the private sector is also part of the 'innovation system' and would in principle have to take some of the blame for a perceived inefficiency in the turnover from investments in R&D, the focus for research policy reform was mainly on PSSs, since these are within the boundaries of the jurisdiction of the state—'being a public institution, it is fair game for public intervention' (Widmalm 2012, p 40). However, since some organizations, actors, and institutions within PSSs are generally less susceptible to reform, intervention will not be even across the systems. In many countries, universities are not only old and strongly reliant on conservative and inert institutional practices, but also have some of their fundamental organizational principles protected by constitutional law, such as academic freedom, which comes in many flavors but which also places practical limitations on organizational reform. It is therefore only reasonable to expect that economization, commodification, and the implementation of managerial reforms that seek to emphasize relevance, accountability, and fast turnover of results becomes stronger in organizations with less institutional precedent or with a hybrid status or role in the innovation system, such as governmental institutes or national laboratories. If such organizations also live off the opportunity to make big discontinuous investments in physical infrastructures with a naturally high visibility, that can play important roles as symbols of determination or strategic priority and decision making, or simply political 'pork', they will quite logically be exposed to policies built on expectations and demands of what they can or should deliver, and organizational/administrative schemes that are designed to follow up the delivery on expectations, for example through continuous (quantitative) performance assessment and management. This means, among other things, that while all sciences—i.e. also 'small science'—have been the targets of reforms and expectations of the type summarized above with the help of the concepts 'economization' and 'commodification' and growing impatience regarding its payoffs, Big Science seems to be a particularly easy target given its visibility and potential political contentiousness.

33.3 Selling and over-selling Big Science: a case in point

One of the current and most prestigious publicly funded Big Science facilities in Europe is the European Spallation Source (ESS), a state-of-the-art accelerator-based neutron scattering laboratory where intense bursts of neutrons will be used to penetrate materials to reveal important information about their properties and qualities, much like x-rays but with slightly different special characteristics that allow other types of studies. The project was first proposed by the European neutron-user community in the early 1990s, as a long-term replacement of the very successful reactor-based neutron source Institute Laue Langevin (ILL) in Grenoble which started operation in 1972 (Kaiserfeld 2013, p 29ff, Rush 2015, p 140). After nearly a decade of scientific planning and technical design work, in 2002 the ESS project entered the political level as an assumed intergovernmental collaboration (although outside the EU framework), but the three obvious candidates for hosting it and taking the lead initiative for its finalization—France, Germany, and the UK—all backed out and left the future of the ESS in uncertainty. Meanwhile, the governments of Japan and the United States had pressed ahead with decisions and could start operation of their own spallation sources, very similar to the planned ESS, in 2007 and 2009, respectively (Hallonsten 2015, p 416; Rush 2015, p 149). Discussions continued in Europe, on lower levels of research policy, and in the years 2007–09 the campaigns on behalf of Hungary, Spain, and Sweden to become the future host of the ESS intensified. In May 2009 a meeting between representatives of interested countries decided to propose Lund in Sweden as a site, and five years later the Swedish government presented a solution for the funding of the project realized in Sweden, with contributions from 13 countries, who could become future members and shareholders in the facility, and construction could start (Hallonsten 2020, p 180ff).

Parallel with the government's lobbying campaign on the European political level, a notable marketing effort was undertaken on the local and national level to mobilize public support for the Swedish ESS bid and the ESS in general. As Agrell (2012) concludes in his analysis of this campaign, the promotion of *expectations* was a key part. What stood out in the advertisement material were the *beliefs* and *anticipations* of what the future ESS would deliver, not only in terms of scientific breakthroughs that would lead to technological innovations, but also in a broader and more comprehensive sense—the future ESS facility was presented as a motor of economic growth and regional competitiveness in the globalized knowledge economy. The advertisement material produced by the local ESS organization, with particular intensity in 2008 and 2009, presents the future ESS as a producer of 'important new knowledge with practical implications in fields like energy, climate, environment, chemical products for everyday life, materials and health' (Agrell 2012, p 433), and as the perfect match for an already dynamic and growing knowledge-intensive cross-border region with several universities and innovative companies (Hallonsten 2020, p 249ff). This material claims that locating the ESS in Lund will mean a boost to existing capabilities in R&D and it will add new ones, extending far beyond basic research and to the forming of hi-tech spinoff companies and scientific clusters. Importantly, the rhetoric around the facility and the experimental techniques it will provide to scientific use is one where curiosity, uncertainty, and the mystery of unexplored worlds take precedence (Agrell 2012, p 434). The material, hence, shows a remarkable imbalance between uncertainty as something genuinely and inherently positive in the realm of the future scientific breakthroughs at the ESS, and a striking *certainty* when it comes to the broad socio-economic benefit of the facility for the region, the Swedish nation, and the world (Hallonsten 2020, p 248). Although nobody knows what will come out of the ESS once it is built and put into operation, the advertisement materials are clear that 'there are only winners' (Agrell 2012, p 434). The science of the ESS is, in other words, described as 'incomprehensible and yet predictable' which is an image very much like the linear model of technological innovation (see above)—the content and nature of future discoveries are unknown, but there is certainty that they will lead to sellable products and other improvements of society (Agrell 2012, pp 434–7).

But advertising was not limited to colorful brochures and (over-)selling slogans. It also featured spectacular promises by official representatives of national and regional governments, of an eight- to nine-fold return on investment for the region hosting the ESS, in the shape of dynamic socio-economic effects, and that the gross regional product of the Southern Swedish province Skåne would increase by SEK 214 billion (approximately 21 billion euros) because of the ESS, in its first two decades of scientific operation (Hallonsten 2020, pp 258, 261). Moreover, in 2010 the local ESS organization released a 10 min video (on a freely distributed DVD, and on Youtube) with the actor Patrick Stewart, best known for his role as Captain Jean-Luc Picard in *Star Trek*. Although in 2010 Stewart was Chancellor of the University of Huddersfield in the United Kingdom, and rightly identified thusly in the video, it was clear that the video was produced to exploit his iconic look and acting skills to invoke a science fiction aura or framing to the whole ESS endeavor. Stewart's narration contained a 'tirade of exaggerations' and was combined with 'imaginative imagery and visual effects' to present the ESS as 'a unique and absolutely vital tool for furthering humanity's understanding of our world' (Hallonsten 2020, pp 251–2).

The exaggerations that this advertisement campaign produced can be explained by several historical circumstances. They may very well have been the result of an unnaturally large underdog mentality on behalf of the campaign to locate the ESS in Lund, caused both by the unlikeliness that Sweden would win the competition for a major collaborative European Big Science facility, and the fact that Sweden can muster only a very small community of neutron users which means that the arguments for Sweden to invest in the ESS must be other than scientific (Hallonsten 2020, pp 278–9). It should also be noted that the campaign took off for real during the heyday of theories of regional innovation systems, clusters, learning regions, and the like (e.g. Castells 1996, Maskell *et al* 1998, Florida 2002), and it is clear that the campaign officials were strongly inspired by these (Hallonsten 2020, p 292). Given all of this, it should probably be acknowledged that both the people in charge of campaigns such as this one (politicians, administrators, directors), and their communication staff, face tremendous challenges in the shifting public sphere. The sometimes ostensibly unscientific or even dishonest campaign promises that they made when (over-)selling the ESS can perhaps be explained, at least in part, by the *Zeitgeist*, unrealistic expectations, and a lack of proper preparation.

But while the example of the ESS is spectacular, it is not unique. A minimum of soul-searching on behalf of directors and other employees of other similar organizations will show quite clearly that the promotion efforts for most Big Science projects invoke promises and expectations of tangible and measurable impacts both in the scientific realm and, by extension, in terms of innovation, economic growth, and global competitiveness, that are exaggerated to the limit of what can be scientifically justified (and sometimes far beyond that limit). In order to 'cut to the chase' and avoid boring the reader with an abundance of grandiose and colorful yet tediously standardized and predictable examples of slogans and promises from other similar organizations (these can easily be found online, at laboratory websites), this chapter will now proceed to the theoretical discussion that is its key purpose.

33.4 The sociology of expectations

Harro van Lente (1993, pp 10–11) has argued that the doctrinal shift in research policy in the last decades of the twentieth century, briefly outlined in the previous section, brought an increase in the importance of *expectations* and *promises*, and an increase in their use in the promotion of certain projects. The reason for this is quite logical: when the need for strategic priority grows, projects will have to be promoted in order to receive priority over other projects, and hence the marketing effort also grows in importance. Moreover, it is only possible to consider a specific project or investment 'strategic' if there is some kind of expectation that it will have a benefit or utility. Thus also if the utility lies several years ahead and is shrouded with uncertainty, it must be part of the advertisement campaign, otherwise the project will not be viewed as strategically important and thus not worth prioritizing. This goes to the core of what analysts of strategic research and strategic research policy have concluded about its nature: It is the expectation(s) of future benefit(s) that characterizes 'strategic' research and singles it out from the purely basic and applied (Irvine and Martin 1984, pp 3–5, Stokes 1997).

It is also reasonable to suggest that both economization and commodification further bolsters the importance of expectations and promises by promoting strong beliefs in over-simplified measures of productivity, excellence, and relevance. When there is (perceived) harsh prioritization between areas of policy and public expenditure, and within the area of research policy, between projects, it would also strengthen the role of promises and expectations of (preferably measurable) results and benefits. This means, among other things, that expectations and promises of future results and benefits of specific projects need to be used as assets in political campaigning for investments. In addition to 'economization', which means that investments in science are made with the ultimate end of stimulating economic growth, market logic has taken over current research policy also by imposing its marketing procedures on the game of prioritization. Projects are 'advertised' and 'sold' on a market of public and political trust and support that seems to be oversaturated with projects but undersaturated with funding. In line with how marketing typically works, the promises and expectations of scientific and technological breakthroughs that projects will deliver therefore need to be made real and very tangible, regardless of the fact that uncertainty is essential to these projects and the policy area—the ultimate motivation for investment in scientific research can be none other than the (well-grounded) hopes that it will produce things unexpected or even inconceivable.

Interestingly, by the use of expectations and promises, this inherent uncertainty of projects is turned into a direct advantage. Promises have built-in imperatives that can give just the right momentum for a political campaign, because once a promise is presented and expectations of its fulfillment are spread, the politically responsible course of action is to act to (try to) fulfill the promises and (try to) meet the expectations (van Lente 2000)—a kind of political 'self-fulfilling prophecy' (Merton 1948). Promises and expectations are thus no longer prospects to be fulfilled or not, but are transformed into assets in the mobilization of action and support; when they are adopted by the public or a significantly influential interest group they swiftly lead to demands of action to reduce the uncertainty they carry, by their fulfillment (van Lente 1993, 2000, Brown and Michael 2003, Borup *et al* 2006).

In his early work on trust as a mechanism to reduce social complexity, Luhmann (1968) puts forward a similar argument: expectations are varieties of trust, because the inherent uncertainties of expectations necessitate some trust that they will be fulfilled. As a consequence, argues Luhmann (1968, p 79), expectations become stronger political assets the more grandiose they are, because trust is not bred on basis of what is justified by observation or reason, but instead on basis of uncertainty. Whereas trust in the certain 'collapses at the first disappointment', trust in the uncertain is continuously reinforced by the threats that the alternative to the expected, or hoped-for, poses. In the view of the public, therefore, uncertainty necessitates (political) action to fulfill the expectation (Luhmann 1968, p 79).

But expectations can grow in importance not only by being grandiose; they may also become *institutionalized* by their reproduction and continuous use as political assets and thus produce a lock-in. This lock-in not only keeps the proponents of the specific project tied to the promises they once made and the expectations they bred, but also pressures the proponents of other similar projects to take part in the same game, not unlike *isomorphism* within social or organizational fields. Such isomorphism can take coercive, normative, and mimetic forms, but overall it amounts to a proliferation of ideas and practices among organizations in the same field, that become articulated or subliminal points of reference for all organizations and actors that seek (social) legitimacy (DiMaggio and Powell 1983). However, there is also a dire risk that such points of reference and models for proper conduct become 'rationality myths' that displace substantively rational procedures and means of learning and improvement (Meyer and Rowan 1977). Instead of making use of the knowledge and competence residing in an organization to work purposefully towards goal fulfillment with reasonable expectations and mechanisms for learning and re-evaluation of priorities, executive branches and their communication offices push advertisement campaigns towards messages and images that embody promises and expectations with little or no anchoring in reality. The sense of urgency in grand narratives conveyed by politicians, including 'progress' but recently also 'globalization' and the advent of the 'knowledge society', supports the uncritical embrace of promises and expectations for spectacular future benefits of investing in Big Science, because it allows the marketing efforts of these projects to tap into society's collective understanding about the most favorable course of action given the challenges and opportunities that lie ahead. McGee (1980, p 15) named this rhetorical and social devices 'ideographs', defined as 'high order abstraction, representing collective commitment to a particular but equivocal and ill-defined normative goal'. Ideographs are colloquial terms or concepts that are imported from ordinary language into political discourse and thus are laden with meaning that goes way beyond plain semantics. They can be used as a political asset in the promotion of a specific cause, because they warrant the use of power and guide 'behavior and belief into channels easily recognized by a community as acceptable and laudable' (McGee 1980, p 15).

Some ideographs carry a highly ambiguous meaning and can therefore be used for vastly different purposes in different contexts; the most extreme examples of this interpretive flexibility are the uses of the words 'liberty' and 'equality' on the different sides of the Iron Curtain during the Cold War. Although these words evidently had very different practical meanings in the East and the West, they (and similar expressions) could be forcefully used to motivate political (and military) action in both types of societies (McGee 1980, p 6). Throughout the past few hundred years, as van Lente (2000, p 44) points out, classic watchwords of modernity such as 'freedom', 'equality', and 'brotherhood' have been complemented by 'rationality', 'progress', and 'technology'—all of them constituting ideographs that surely unite people around a common normative goal more or less impossible for anyone to renounce but with highly 'equivocal and ill-defined' practical meanings (McGee 1980, p 15). 'Technological progress', writes van Lente (2000, pp 46–9), is exactly the type of 'abstraction that expresses a social commitment' that makes it qualify as an ideograph, and recent history is also filled with examples of scientific and technological projects that have been promoted in part by a direct or indirect branding of its opponents as being against progress or 'irrational' or similar—the advocates for a specific scientific or technological project are fast in connecting their campaign to the ideograph of 'technological progress' or 'progress' and defining all opponents as followers of a condemnable counter-ideograph. There are certainly some examples in history of when 'dangerous technology' or similar has been used as a successful counter-ideograph to 'technological progress'; let's for the sake of simplicity mention only three very obvious ones: Frankenstein's Monster, Rachel Carson's book *Silent Spring*, and the Chernobyl disaster. Although these have risen to a level of public awareness enough to mobilize successful protest movements or critical debates over technology, in greater perspective they have only accomplished minor perturbations on the surface, and 'progress' and 'rationality' retain their overall superiority. A lot of the explanation to this lies in the nature of the word 'progress' and what it means. Among ideographs, 'progress' has the rare quality of signaling inevitability in a very long historical perspective—past achievements, current states, and future prospects are woven together in a sequence that becomes an unmistakable part of our society and its capability to produce wealth, fight disease, and make cultural advances.

Thus can, for example, the inventions of the wheel, the compass, the steam engine, the penicillin, nuclear energy, space exploration, and so on, be invoked as past advances on which our society and civilization are built and to which a scientific or technological project currently in need of support is added as the natural next step. The chain should not be broken, argue proponents of the project (directly or indirectly); this is the 'next generation' and it is only logical to make it happen (van Lente 2000, pp 46–7). Importantly, events of the past can be chosen rather freely in order to obscure or ignore failures, harmful side effects, and false starts. Other historical events or achievements of non-technological nature, such as social movements, political action, or the long-term shaping of a cultural identity among nations or across regions can also be added to the mix so that tradition is invented to contextualize the contemporary and identify it as the natural consequence of a (successful) historical development, although its place in this sequence or its relation to the past events invoked is not entirely historically accurate—(part of) tradition is thereby *invented* (Hobsbawm and Ranger 1983, Hanel and Hård 2015).

It is quite evident from the history of Big Science in the Cold War era, and the accelerators for particle physics that succeeded each other in the 1960s, 1970s, and 1980s, that the ideograph 'technological progress' had a role to play in the promotion of new projects, albeit perhaps not as articulated. The major Fermilab machine outside of Chicago, which started operation in 1974 as the flagship particle physics lab of the United States, was very much framed in these terms. Its founding director, Robert Wilson, likened the role of the lab in pushing the frontier of knowledge with the westward expansion of the United States in the nineteenth century and also worked purposefully to make the lab's architectural style reflect his vision of frontier exploration (Hoddeson *et al* 2008, p 95ff). In the 1980s, the proponents of the Superconducting Super Collider used similar metaphors of a 'frontier outpost', comparing the accelerator complex and its mission to explore the inner secrets of matter and the origins of the Universe with the explorers' missions that discovered the New World in the fifteenth and sixteenth centuries. The machine itself was pictured as a vessel for exploration, in other words a successor or direct descendant of the ships that brought Columbus and Vespucci to the Americas (Riordan *et al* 2015, p 79).

Less grandiloquence is used in today's campaign rhetoric for new Big Science facilities, but the message is similar. Although more oriented to practical needs and not so much appealing to cultural self-images, investments in large-scale research infrastructures are still, at least in public advertisement material, largely motivated by their capabilities to contribute to (technological) progress and thus to creating a better world.

It is worth returning to the historical context and the difference between old or Cold War era Big Science and the transformed Big Science of today, as distinguished by Hallonsten (2016), not least because it is quite evident from historical studies that the spirit of the times of the early Cold War era was very important for the creation of the 'postwar consensus' (Smith 1990, pp 36–7) around the social contract and the linear model of technological innovation as the cornerstones of research policy in the United States and Western Europe. Science, and especially Big Science, had a demonstrated capability to contribute to technological progress and to translate into the military might that was necessary in the bipolar geopolitical world order. Nuclear physics was a mere apex of a pyramid but scientists in this and adjacent fields had an apparent *carte blanche* from their governments until at least the beginning of the 1960s. As Del Sesto (1986) has shown, the promises of nuclear energy were indeed great enough to build a powerful vision of the future, in which nearly all of mankind's problems would be solved by the careful and purposeful exploration and exploitation of the inner forces of the atom. The combination of this optimism and the horrendous prospects of 'mutually assured destruction (MAD)' by nuclear war between the two superpowers produced a mighty cultural and political *raison d'être* for huge investments in science and technology.

The transformed Big Science is sustained by expectations and promises that similarly combine the abstract and the tangible. It should be noted, however, that the most striking difference lies not in the subject of the expectations (nuclear energy was in fact expected to contribute to the development of new materials, also in the area of the life sciences), but rather in the way in which results are supposed to be transferred to areas of application. The 'knowledge society', here taken to mean both a society permeated by knowledge in all its parts, and a society whose organizations and institutions devoted to the production and dissemination of knowledge are increasingly penetrated by all kinds of actors and interests, has evolved into a concept very much reminiscent of an ideograph. As Välimaa and Hoffman (2008, p 266) have pointed out, the concept carries a wide interpretive flexibility and has become 'an imaginary space' that can include almost anything related to the concept of knowledge and how it is produced, spread, and used. The connection between the 'knowledge society' as a rhetorical concept and images of enlightenment was, furthermore, established already in its first documented use in the mid-1960s, where the vision of a 'knowledgeable society' was launched as a redemption of a society plagued by corruption and moving in the wrong direction (Lane 1966). 'Knowledge' is very much like 'liberty' and 'equality' (cf. McGee 1980, p 6), as it can hardly or never at all be interpreted negatively *per se* but evidently be given many different practical meanings, especially when turned into policy. The ubiquitous presence of the concept 'knowledge society' in current research policy rhetoric and in wider society is part of the foundation for the advertisement campaigns for new Big Science projects from which these draw some of their strength.

As noted in connection with the short case study on the ESS above, the 'innovation paradigm' is very present in current research policy and was especially dominant a decade ago. This sentiment turns all R&D—regardless of its organizational or institutional belonging—into means for achieving innovation, which in turn supposedly drives economic development and societal progress. The impatience with the speed and scope of practical turnover of R&D results that produced 'economization', 'commodification', strategic priority, and the increased demands for relevance and accountability in PSSs (see above) perhaps had as an ultimate consequence the emergence of 'innovation' as a higher or highest cause. Only by declaring 'innovation' the purpose of all R&D, and devising policy reforms and management schemes that make all R&D oriented to 'innovation', will the impatience be calmed. Obviously, there was no such master plan to calm impatience and it is thus very difficult to point at specific actors behind the shift, but with an institutional view and in retrospect, the historical development doubtlessly makes more sense when viewed in light of such a logic.

33.5 Conclusions

The doctrinal shift in research policy in the last decades of the twentieth century has had a broad set of consequences for the institution of science that have been commendably analysed by several authors (see e.g. Guston 2000, Pestre 2005, Radder 2010, Berman 2012, Elzinga 2012, Hallonsten 2021), but among the most interesting consequences for the public image of science and technological development in current society are van Lente's (1993) claim that the importance of expectations and promises, and their use in the promotion of certain projects, have increased. In this article, sociological theory has been used to analyze and discuss the reasons for such a development, and in wider perspective the meaning of promises and expectations in political campaigning for Big Science projects.

Logically, a research policy of strategic priorities enforces a market-like model on projects and their proponents. Promotion campaigns are necessary to get ahead, and campaigns need to focus on promises and expectations of future results and impacts to brand their projects 'strategic'. Political action is necessary once promises are made and expectations take root, because inaction in the light of these promises and expectations will be viewed as irresponsible (van Lente 2000, p 59). Organizations within the same field imitate each other, either on purpose or in more subtle ways, by the proliferation of models for proper conduct, and marketing efforts begin to live lives of their own, as rationality myths that give social legitimacy to the organizations (Meyer and Rowan 1977, DiMaggio and Powell 1983); critical voices that spot the poor grounding of slogans and promises in reality are easily quieted by the lure of the messages and the sense of urgent need to get ahead in competition.

The result of the doctrinal shift of research policy, to simplify somewhat, is therefore not a drift away from basic science and towards applied science, but instead a reinforced status for basic science as long as the performing agents of PSSs agree to be the subjects and figureheads of society's expectations. Strictly speaking, there is nothing in the above to suggest that basic science needs to fulfill expectations and promises, only that its potential to do so is highlighted enough in advance of decisions to grant a certain project priority and make an investment. Both the Cold War era Big Science and the transformed Big Science are also products of their times, and represent the *Zeitgeist* by reflecting the society's expectations. While 'technological progress' seems to be intact as overarching ideograph between these two eras, it is quite clear that both 'innovation' and the 'knowledge society' are the most important contemporary watchwords. Both of them are ideographs that join 'technological progress' in a mix whose materialization in policymaking is a market or quasi-market of expectations and promises. The consequences of this turn in science policy, and especially the handling of investments in Big Science, should be analysed with attention to the evident drift of motives away from national security concerns to regional and national competitiveness in a globalized economy, the mobilization of promises and expectations in the marketing efforts around Big Science, and the risks of goal displacement that this brings.

The perspective in this chapter has been entirely institutional, and the empirical material used has of course been limited to one case (although, as noted, it is easy for the reader to find other very similar cases on their own). Further studies should adopt an agency perspective, and attempt to identify both the role of different actors in the (over-)selling of Big Science, and their relative weight, and also study how the audiences of the advertisement campaigns react, and to what extent they are actively mobilizing and perpetuating the expectations to which the campaigns respond. Such studies, preferably also looking at other cases, would yield important insights and advance the state of knowledge on this important topic far beyond the very preliminary and conceptually oriented analysis accomplished in this chapter.

References

Agrell W 2012 Framing prospects and risk in the public promotion of ESS-Scandinavia *Sci. Public Policy* **39** 429–38

Alvesson M 2013 *The Triumph of Emptiness. Consumption, Higher Education, and Work Organization* (Oxford: Oxford University Press)

Bensaude V B 2001 The construction of a discipline: materials science in the United States *Hist. Stud. Phys. Sci.* **31** 223–48

Berger P L 1963 *Invitation to Sociology. A Humanistic Perspective* (New York: Anchor Books)

Berman E P 2012 *Creating the Market University. How Academic Science Became an Economic Engine* (Princeton, NJ: Princeton University Press)

Berman E P 2014 Not just neoliberalism: economization in US science and technology policy science *Technol. Human Values* **39** 397–431

Boorstin D J 1961/2012 *The Image. A Guide to Pseudo-Events in America* (New York: Vintage)

Borup M, Brown N, Konrad K and van Lente H 2006 The sociology of expectations in science and technology *Technol. Anal. Strateg. Manag.* **18** 285–98

Brown N and Michael M 2003 A sociology of expectations: retrospecting prospects and prospecting retrospects *Technol. Anal. Strateg. Manag.* **15** 3–18

Castells M 1996 *The Rise of the Network Society. The Information Age: Economy, Society and Culture* vol 1 (Oxford: Blackwell)

Cramer K 2020 *A Political History of Big Science: The Other Europe* (London: Palgrave Macmillan)

Cramer K, Hallonsten O, Bolliger I and Griffiths A 2020 Big Science and research infrastructures in Europe: history and current trends *Big Science and Research Infrastructures in Europe* ed K Cramer and O Hallonsten (Cheltenham: Edward Elgar)

Cramer K and Hallonsten O (ed) 2020 *Big Science and Research Infrastructures in Europe* (Cheltenham: Edward Elgar)

Dahler-Larsen P 2012 *The Evaluation Society* (Redwood City, CA: Stanford University Press)

Del Sesto S L 1986 Wasn't the future of nuclear energy wonderful? *Imagining Tomorrow: History, Technology, and the American Future* ed J J Corn (Cambridge, MA: MIT Press)

DiMaggio P and Powell W W 1983 The iron cage revisited: institutional isomorphism and collective rationality in organizational fields *Am. Sociol. Rev.* **48** 147–60

Doing P 2015 Epistemic politics at work: national policy, an upstate New York synchrotron, and the rise of protein crystallography *The Local Configuration of New Research Fields On Regional and National Diversity* ed M Merz and P Sormani (Berlin: Springer)

Elzinga A 2012 Features of the current science policy regime: viewed in historical perspective *Sci. Public Policy* **39** 416–28

Florida R 2002 *The Rise of the Creative Class* (New York: Basic)

Galison P and Hevly B (ed) 1992 *Big Science—The Growth of Large-Scale Research* (Redwood City, CA: Stanford University Press)

Greenberg D S 1999/1967 *The Politics of Pure Science* 2nd edn (Chicago, IL: University of Chicago Press)

Gribbe J and Hallonsten O 2017 The emergence and growth of materials science in Swedish universities *Hist. Stud. Nat. Sci.* **47** 459–93

Guston D H 2000 *Between Politics and Science: Assuring the Integrity and Productivity of Research* (Cambridge: Cambridge University Press)

Habermas J 1971/2002 *Knowledge and Human Interests* (Cambridge: Polity)

Hallonsten O 2015 Unpreparedness and risk in Big Science policy: Sweden and the European Spallation Source *Sci. Public Policy* **42** 415–26

Hallonsten O 2016 *Big Science Transformed. Science, Politics and Organization in Europe and the United States* (London: Palgrave Macmillan)

Hallonsten O 2020 *The Campaign. How a European Big Science Facility Ended up on the Peripheral Farmlands of Southern Sweden* (Lund: Arkiv Academic)

Hallonsten O 2021 Stop evaluating science: a historical-sociological argument *Soc. Sci. Inform.* **60** 7–26

Hallonsten O and Heinze T 2012 Institutional persistence through gradual adaptation: analysis of national laboratories in the USA and Germany *Sci. Public Policy* **39** 450–63

Hallonsten O and Heinze T 2015 Formation and expansion of a new organizational field in experimental science *Sci. Public Policy* **42** 841–54

Hallonsten O and Heinze T 2016 'Preservation of the laboratory is not a mission'. Gradual organizational renewal in national laboratories in Germany and the United States *Innovation in Science and Organizational Renewal. Historical and Sociological Perspectives* ed T Heinze and R Münch (London: Palgrave Macmillan)

Hanel T and Hård M 2015 Inventing traditions: interests, parables and nostalgia in the history of nuclear energy *Hist. Technol.* **31** 84–107
Herman R 1986 *The European Scientific Community* (London: Longman)
Hewlett R G and Holl J M 1989 *Atoms for Peace and War, 1953–1961* (Berkeley, CA: University of California Press)
Hiltzik M 2015 *Big Science. Ernest Lawrence and the Invention that Launched Military-Industrial Complex* (New York: Simon and Schuster)
Hobsbawm E and Ranger T (ed) 1983 *The Invention of Tradition* (Cambridge: Cambridge University Press)
Hoddeson L and Kolb A 2000 The superconducting super collider's frontier outpost, 1983–1988 *Minerva* **38** 271–310
Hoddeson L, Kolb A W and Westfall C 2008 *Fermilab: Physics, the Frontier and Megascience* (Chicago, IL: University of Chicago Press)
Irvine J and Martin B R 1984 *Foresight in Science. Picking the Winners* (London: Pinter)
Kaiserfeld T 2013 ESS from neutron gap to global strategy: plans for an international research facility after the cold war *Legitimizing ESS: Big Science as a Collaboration Across Boundaries* ed T Kaiserfeld and T O'Dell (Falun: Nordic Academic)
Kastor J 2010 *The National Institutes of Health 1991–2008* (Oxford: Oxford University Press)
Kevles D J 1995/1977 *The Physicists: The History of a Scientific Community in Modern America* (Cambridge, MA: Harvard University Press)
Kevles D J 1997 Big Science and big politics in the United States: reflections on the death of the SSC and the life of the Human Genome Project *Hist. Stud. Phys. Sci.* **27** 269–97
Kleinman D L and Vallas S P 2001 Science, capitalism, and the rise of the 'Knowledge Worker': the changing structure of knowledge production in the United States *Theory Soc.* **30** 451–92
Krige J 2006 *American Hegemony and the Postwar Reconstruction of Science in Europe* (Cambridge, MA: MIT Press)
Lane R E 1966 The decline of politics and ideology in a knowledgeable society *Am. Soc. Rev.* **31** 649–62
Lasch C 1979/2018 *The Culture of Narcissism. American Life in an Age of Diminishing Expectations* (New York: Norton)
Luhmann N 1968 *Vertrauen* (Stuttgart: Lucius and Lucius)
Lundvall B Å (ed) 1992 *National Systems of Innovation* (London: Anthem)
Martin J D 2015 Fundamental disputations: the philosophical debates that Governed American Physics, 1939–1993 *Hist. Stud. Nat. Sci.* **45** 703–57
Maskell P, Eskelinen H, Hannibalsson I, Malmberg A and Vatne E 1998 *Competitiveness, Localised Learning and Regional Development: Specialisation and Prosperity in Small Open Economies* (Abingdon: Routledge)
McGee M C 1980 The 'Ideograph': a link between rhetoric and ideology *Quart. J. Speech* **66** 1–16
Merton R K 1948 The self-fulfilling prophecy *Antioch Rev.* **8** 193–210
Meyer J W and Rowan B 1977 Institutionalized organizations: formal structure as myth and ceremony *Am. J. Sociol.* **83** 340–63
Mody C 2011 *Instrumental Community. Probe Microscopy and the Path to Nanotechnology* (Cambridge, MA: MIT Press)
Pestre D 2005 The technosciences between markets, social worries and the political: how to imagine a better future? *The Public Nature of Science under Assault: Politics, Markets,*

Science and the Law ed H Nowotny, D Pestre, E Schmidt-Aßmann, E Schultze-Fielitz and H H Trute (Berlin: Springer)

Radder H (ed) 2010 *The Commodification of Academic Research. Science and the Modern Universities* (Cambridge, MA: Harvard University Press)

Riordan M, Hoddeson L and Kolb A 2015 *Tunnel Visions. The Rise and Fall of the Superconducting Super Collider* (Chicago, IL: University of Chicago Press)

Roberts P 2014 *The Impulse Society. What's Wrong with Getting What We Want?* (London: Bloomsbury)

Rose N and Miller P 1992 Political power beyond the state: problematics of government *Br. J. Sociol.* **431** 173–205

Rush J J 2015 US neutron facility development in the last half-century: a cautionary tale *Phys. Perspective* **17** 135–55

Smith B 1990 *American Science Policy since World War II* (Washington, DC: Brookings)

Sörlin S and Vessuri H 2007 *Knowledge Society vs Knowledge Economy: Knowledge, Power, and Politics* (London: Palgrave Macmillan)

Stokes D 1997 *Pasteur's Quadrant: Basic Science and Technological Innovation* (Washington, DC: Brookings)

Välimaa J and Hoffman D 2008 Knowledge society discourse and higher education *High. Educ.* **56** 265–85

van Lente H 1993 Promising technology. The dynamics of expectations in technological developments *Dissertation* University of Twente, Enschede

van Lente H 2000 Forceful futures: from promise to requirement *Contested Futures* ed N Brown, B Rappert and A Webster (Farnham: Ashgate)

Vinsel L and Russell A L 2020 *The Innovation Delusion. How Our Obsession with the New Gas Disrupted the Work that Matters Most* (Redfern: Currency)

Wacquant L 2012 Three steps to a historical anthropology of actually existing neoliberalism *Soc. Anthropol.* **20** 66–79

Westfall C 2003 Rethinking Big Science: modest, mezzo, grand science and the development of the Bevalac, 1971–1993 *ISIS* **94** 30–56

Westfall C 2008 Surviving the squeeze: National laboratories in the 1970s and 1980s *Hist. Stud. Nat. Sci.* **38** 475–8

Westfall C 2010 Surviving to tell the tale: Argonne's intense pulsed neutron source from an ecosystem perspective *Hist. Stud. Nat. Sci.* **40** 350–98

Westwick P J 2003 *The National Labs: Science in an American System 1947–74* (Cambridge, MA: Harvard University Press)

Widmalm S 2012 Innovation and control: performative research policy in Sweden *Transformations in Research, Higher Education and the Academic Market: The Breakdown of Scientific Thought* ed S Rider, Y Hasselberg and A Waluszewski (Berlin: Springer)

IOP Publishing

Big Science in the 21st Century
Economic and societal impacts
Panagiotis Charitos, Theodore Arabatzis, Harry Cliff, Günther Dissertori, Juliette Forneris and Jason Li-Ying

Chapter 34

Big Science, scientific cosmopolitanism, and the duty of justice

Michela Massimi

For very long time, since the time of Roosevelt, science policy has been based on what is often referred to as a 'linear model', whereby the assumption was that national governments could invest into science and innovation, and the scientific outcomes would eventually cascade down linearly onto society and foster societal progress. Clearly, the pitfalls of such irenic picture became evident during WWII and the Cold War years.

What is needed is a way of bridging the gap between scientific and technological innovations on the one hand, and societal progress on the other hand. This often means rethinking how large-scale scientific investments may offer ways of redistributing scientific knowledge and its benefits, well beyond the traditional cascade-down model of science policy, or some kind of principle of charity or philanthropy. In this chapter I sketch a version of what I shall refer to as 'scientific cosmopolitanism', which I see as promising in delivering on this challenge.

34.1 The problem with Big Science

On 14 April 2021 a letter to the American President Joe Biden signed by 100 Nobel Laureates and 75 Heads of State was published by the People's Vaccine Alliance. The letter raised concerns about the 'very slow progress in scaling up global COVID-19 vaccine access and inoculation in low- and middle-income countries'. It called for a temporary waiver of the World Trade Organization (WTO) intellectual property rules as a 'vital and necessary step to bringing an end to this pandemic. It must be combined with ensuring vaccine know-how and technology is shared openly' against narrow nationalism that risks leaving nine out of ten people

doi:10.1088/978-0-7503-3631-4ch34

in the poorest countries 'without a vaccine this year'[1]. Hence the plea for an emergency waiver of COVID-19 patent rules that will 'give people around the globe a chance to wake up to a world free from the virus. We need a people's vaccine'.

Discussions about Big Science have often focussed on large financial-economic investments and capacity building in some sectors. They have often highlighted disparities in resource allocation and research income across different scientific fields, much to the disgruntlement of those who work in self-avowedly Small Science. But if the COVID-19 pandemic has taught us something at all, it is that in addition to all those disparities and inequities, more profound issues of justice run deep in Big Science. For it is not just how resources, capital investments, and capacity building are allocated in some areas that make some scientific fields 'Big', it is also the associated oligopoly in know-how, scientific practices, and experimental tools necessary to deliver on the scientific outcomes. The debate on the waiver for COVID-19 vaccine is a case in point—made even more glaring by the awareness that peddling vaccine nationalism can only be self-harming and self-defeating in a global pandemic.

Much as a global TRIPS[2] waiver is an important first step in this debate, I think it would nonetheless be a mistake to assume that—in and of itself—it could function as a panacea for all the current problems. A much bigger challenge still lies ahead—namely, to create the necessary *local* infrastructures to manufacture and freely distribute the vaccine in *all* countries; and to meet demands and accelerate the successful vaccination rollout world-wide. A people's vaccine requires a people's science. And how to think about a 'people's Big Science' is my topic here. Or more provocatively, *who* is Big Science for?

The question easily invites the answer: 'Big Science is for everyone'. But such an easy answer would fly in the face of a glaring asymmetry between *who* overwhelmingly benefits from scientific advancement and innovations (in the Northern Hemisphere) and *who* should also be benefitting from them rather than being left to the mercy of philanthropic efforts on behalf of high-income countries. That scientific advancements and innovations (more broadly speaking) have historically been a commodity for mostly Global Northern consumption has been well-documented by sociologists of science, historians of science (e.g. Cooper 2007), STS scholars, and environmental philosophers (e.g. TallBear 2014, Wylie 2020, among others).

The problem with Big Science runs even deeper than this. Historians (Galison and Hevly 1992) have documented how Big Science was born in North America in the aftermath of WWII and brought to new high during the Cold War years (from 1939 to 1960) where collaborations among physicists and engineers—from the Manhattan Project to the national space programme—were often instrumental to military and

[1] https://peoplesvaccinealliance.medium.com/open-letter-former-heads-of-state-and-nobel-laureates-call-on-president-biden-to-waive-e0589edd5704

[2] TRIPS stands for 'Trade Related Aspects of Intellectual Property Rights' and it is a key legal agreement that regulates the relationship between intellectual properties and trade. For more info see: https://www.wto.org/english/tratop_e/trips_e/trips_e.htm (Accessed: 27 August 2021).

industrial interests. The biomedical and pharmaceutical sciences quickly followed suit with the International Biological Program (1964–74). As Aranova *et al* (2010, p 183) have documented:

> This new, potentially complementary model of Big Science encompassed the field practices of ecologists and suggested a model for the natural historical sciences to achieve the stature and reach of the experimental physical sciences. …By 1974, when the program ended, many participants viewed it as a failure. However, this failed program transformed into the Long-Term Ecological Research program. Historical analysis suggests that many of the original incentives of the program (the emphasis on Big Data and the implementation of the organisational structure of Big Science in biological projects) were in fact realized by the program's visionaries and its immediate investigators.

Issues about fairness and equity in the use of Big Data in biomedical and health sciences remain a much-debated topic in contemporary history and philosophy of science (see Leonelli *et al* 2021 and Krige and Leonelli 2021). And increasing legal work—such as the UN Convention on Biological Diversity and the Nagoya Protocol—have highlighted the need to protect the resources, intellectual property rights and access to benefits for local communities, especially in the Global South, who are often at the receiving end of unfair practices of bioprospecting in the use of the local flora, fauna, and, more in general, genetic resources on behalf of Big Pharma (see Kolopenuk 2020, Mgbeoji 2006, Gerber *et al* 2021).

In my work on perspectival realism (Massimi 2022a) I have endeavoured to contribute to this debate by going right to the heart of it: namely, by investigating how scientific knowledge is usually portrayed in philosophical narratives and scientific canons as well. Discussions about fairness and equity in Big Science, and science more broadly, require a good look at how scientific knowledge is in fact historically produced and the epistemic injustices lurking behind particular narratives and images of it.

The epistemology of science I have advocated treats all scientific knowledge as *perspectival knowledge* and shows how scientific knowledge production is always the outcome of *intersecting and interlacing scientific perspectives*, each of which is historically and culturally situated. Perspectival realism is realism that places centre-stage epistemic communities, whose ontology of phenomena and natural kinds—qua open-ended groupings of historically identified phenomena—reflects the multicultural nature of scientific knowledge (see Massimi 2022a, ch 8 and ch 11, and also Massimi 2022b).

In this chapter I briefly sketch an important consequence of this epistemological view that bears on the topic of Big Science: namely, how to think about scientific progress (more on this topic in Massimi 2022c and 2022d, on which I draw in what follows). I argue that a philosophical account of scientific progress ought to be able to answer the question '*Who* is Big Science (and science more broadly) for?' beyond the dichotomy of scientific nationalism and philanthropic solidarity.

34.2 Scientific progress, societal progress, and science policy

In the seminal 1945 report *Science, the Endless Frontier* the then Director of the Office of Scientific Research and Development for the American President Roosevelt, Vannevar Bush, laid out an influential vision for science policy. Hailing scientific progress as 'one essential key to our security as a nation, to our better health, to more jobs, to a higher standard of living and to our cultural progress' (Bush 1945/1960, Letter of Transmittal, p 2), in the summary of the Report Bush went on to outline how what he called 'basic scientific research' was pivotal to the prosperity of the nation:

> Progress in the war against disease depends upon a flow of new scientific knowledge. New products, new industries, and more jobs require continuous additions to knowledge of the laws of nature, and the application of that knowledge to practical purposes. Similarly, our defence against aggression demands new knowledge so we can develop new and improved weapons. This essential, new knowledge can be obtained only through basic scientific research.
>
> Science can be effective in the national welfare only as a member of a team, whether the conditions be peace or war. But without scientific progress no amount of achievement in other directions can insure our health, prosperity, and security as a nation in the modern world (Bush 1945/1960, p 5).

The document left little doubt about the importance of scientific research. However, it left much more unclear how to understand the notion of scientific progress: or better, how the seamless flow of new knowledge was meant to translate into better public health and nation-wide prosperity. As the philosopher of science Philip Kitcher (2001, p 141) remarks,

> As a consequence of these omissions, the institutional framework the report constructed is one in which scientists are left free to pursue their own curiosity but simultaneously saddled with the task of advertising their research as potentially satisfying the untutored preferences of the public. It is easy to see how the stage was set for later struggles (only partly acknowledged) between scientific efforts to manipulate public funding and suspicious outsiders.

We live in a society that is obsessed with metrics about progress in science. Usually, progress is measured in terms of milestones towards a target: how many vaccinations per day; how much CO_2 emissions have to be cut to reach intended targets; how many phases (and related risk assessment) in the delivery of a large-scale infrastructure. But what is scientific progress? What makes science a progressive kind of enquiry, beyond targets, milestones, and metrics?

Philosophers of science have engaged with this question for long time and offered a number of different answers to the question. A traditional view has linked scientific progress with getting closer to the truth (Niiniluoto 1987, 2014). For example, one can say that a scientific theory like Copernican astronomy made progress compared to ancient Greek Ptolemaic astronomy by being closer to the truth in its analysis of planetary motion. We now know that planetary motion does not require crystalline spheres and does not happen in epicycles and deferents with the Earth at the centre. This is an example of scientific progress understood as getting closer to the truth about the nature of planetary motion.

Other accounts, for example Thomas Kuhn's (1962), have linked scientific progress with puzzle-solving. In Kuhn's view, Copernican astronomy was progressive not because it was more likely to be true than Ptolemaic astronomy but because it had a higher puzzle-solving power than the latter. For example, Copernican astronomy could explain phenomena such as the phases of the planet Venus or the changing dates of the Moon calendar associated with the date of Easter, which were puzzling within Ptolemaic astronomy.

More recently, the philosophical debate has continued with Alexander Bird (2007) championing his epistemic account of scientific progress, whereby progress is identified not with the accumulation of truths but of knowledge over time. And Finnur Dellsén (2016) has in turn proposed to link scientific progress with the notion of understanding rather than knowledge. In brief, the debate has revolved around the question as to whether scientific progress should be measured in terms of how close (or far) we are from the truth; or whether or not we have accumulated over time a sufficiently bigger stock of scientific knowledge claims; or whether without necessarily appealing to either truth or knowledge (qua justified true belief) a better way of thinking about progress is in terms of gaining more understanding, i.e. better grasping explanatory and predictive aspects of theories. The debate is familiar among philosophers of science and I shall not rehearse it here[3].

Here I want to take my cue from an observation made by Heather Douglas (2014) in a paper where she discusses Kuhn's view of progress and his emphasis on pure science and neglect of applied science. In her words: 'If WWI created doubts about the equivalence of scientific and societal progress, WWII and the atomic bomb demolished it, science and the powers it unleashes could indeed make society better, or it could end civilization completely' (Douglas 2014, p 63). The Cold War and Vietnamese War made glaringly evident the limits of what Vannevar Bush still in 1945 was hailing as the linear model of scientific progress. It is simply not enough to invest in science and expect a linear cascade-down return in increased well-being and wealth in society, given the often dangerous industrial and military uses of scientific and technological advancements. Scientific progress—in and of itself—does not

[3] Philosophers of science make a distinction between scientific knowledge and scientific understanding. Knowledge is defined classically as beliefs that are true and justified. Understanding remains more elusive to capture but it departs from knowledge in not appealing to beliefs or truth but rather to the ability of scientists to better grasp how to correctly explain 'or reliably predict more aspects of the world than they did before' (Dellsén 2016, p 7).

automatically translate into societal progress. Rightly in my view, Douglas stresses therefore the need for a 'more socially and ethically mediated conception of progress, one that takes into account all of science, both pure and applied' (Douglas 2014).

Philip Kitcher (forthcoming) has spearheaded such 'more socially and ethically mediated conception' of scientific progress by taking inspiration from the American Pragmatist tradition. What he calls non-teleological pragmatic progress draws attention to particular types of inquiry that would be able to identify problems and allows to recognise situations as 'problematic'.

Douglas' and Kitcher's remarks are my starting point for asking whether there is any hope to restore a link between scientific progress and societal progress—or, perhaps more pressingly, whether such link ought to be expected at all (needless to say, found). Can we navigate the complexity of this landscape where questions of justice arise at every corner while looking at science at large—pure and applied—progress in scientific theories no more than progress in large-scale scientific investments and innovations? Is not there a risk of harking back to a view of scientific progress that with Roosevelt's advisor, Vannevar Bush, was designed to offer a linear cascade-down model of investments in basic scientific research with the expectation that societal progress would *naturally* result from it?

I believe that a link between scientific progress and societal progress can be found. I also believe that such a link should be restored. But it would require a deep rethinking of a host of issues about scientific research, scientific advancements, and science policy models that can hopefully avert the worry of harking back to a governance model that ultimately proved defective.

34.3 Scientific advancements as transnational 'public goods'?

One might think at this point that a more promising way of establishing a connection between scientific progress and societal progress is not in terms of truth, knowledge, or understanding, but in terms of the Kuhnian original idea of problem-solving. Science often does progress by solving problems and some kinds of problems are not purely theoretical but have far-reaching societal implications. How to go about identifying the problem in the first instance? And how to rethink of scientific progress along these lines?

Consider the following example again inspired by the COVID pandemic. In August 2020 the Kayapo Mekragnotire people blocked highways in the Amazon forest, near the city of Novo Progresso, to draw attention to how roadworks were spreading COVID-19 virus in their population causing the death of elders and with not enough masks, personal hygiene supplies, or health care workers available in the community[4]. Recognising when a situation is problematic (and what the problem is) becomes often a highly contextual matter. For the same airborne mechanism of COVID-19 virus transmission leads to dramatically different public health scenarios

[4] See the Washington Post article: https://www.washingtonpost.com/world/indigenous-protesters-in-brazil-demand-covid-19-protection/2020/08/17/280e3e32-e0a0-11ea-82d8-5e55d47e90ca_story.html.

depending on whether the population in question has (or does not have) access to basic supplies and health care.

There is more. Recognising when a situation is 'problematic' (and what the problem is) is also a highly contextual matter in *who gets mostly affected* by it. Consider as a foil the virus Ebola, which is highly contagious and lethal, with a case fatality rate ranging from 50% to as high as 90% in the Republic of Congo outbreaks of 2003 and 100% in Uganda in 2011[5]. Case fatality rate measures the proportion of people who died of the disease among those who contracted the virus in a certain period. The Ebola case fatality rate is much worse that the COVID-19 case fatality rate registered across countries world-wide at the time of writing this chapter (in 2021). Even for some of the worst affected countries the COVID-19 case fatality rate peaks at 4.9% max for Peru, among the countries most affected to date[6].

Yet the global response to COVID-19 in terms of fast-tracking vaccine research has been of an order of scale, magnitude, and timely response that is unprecedented and has no equal in the fight against Ebola. Why is that? Well, despite being a much more lethal virus than COVID-19 with a much higher case fatality ratio, Ebola outbreaks have so far occurred mostly in equatorial Africa and with only very minor incidences outside of it. Had Ebola spread as a virus at a global scale as COVID-19 has done, the international response in terms of efforts in vaccine research would have been entirely different from what it has been.

This does not mean that Ebola has disappeared as a virus. Or that it is not a problem any more and that there have not been any more outbreaks: there have been, even recently in 2018–2020 in the Democratic Republic of Congo with an estimated case fatality rate of 66%. And yet no similar international efforts in the treatment and eradication of Ebola have remotely paralleled what we witnessed with COVID-19. Although Ebola was discovered in 1976 and caused severe outbreaks in West Africa in 2014–16, it was only in 2019–2020 that vaccines for it were eventually approved.

To sum up, despite the *prima facie* very similar problems, scientific progress in fighting Ebola has been nowhere near progress in fighting COVID-19. And it is hard to resist the conclusion that the reason for this asymmetry is that COVID-19 is regarded as *everyone's problem* whereas Ebola seems to have been perceived by the international community as *somebody else's* problem because of its geographically confined spread to date. Yet, a problem is still a problem no matter whether it is *somebody else's* problem rather than *everyone's* problem (and no matter either whether *somebody's* problem is ever likely to become *my own* problem).

It is in response to situations such as this one that I think scientific progress—broadly understood along Kuhn/Kitcher lines in terms of problem-solving—can be fruitfully expanded in a certain direction, a direction which is congenial to the kind of perspectival realism I have articulated (Massimi 2022a). Here I can only very briefly sketch a possible way of expanding on this particular take on scientific progress so as to better articulate its link with societal progress and science policy.

[5] See data from https://www.who.int/news-room/fact-sheets/detail/ebola-virus-disease.

[6] See data from Johns Hopkins Coronavirus Resource Centre: https://coronavirus.jhu.edu/data/mortality.

The underlying issue concerns distributive justice about scientific advancements and how to think of it beyond philanthropy delegated to individual nation states. Are scientific advancements such as a vaccine a 'public good' once in the hands of nation-state governments, despite being patented, produced, and commercialised by private companies? And how should governments treat such 'public goods'? Are they essential or discretionary public goods? Most importantly, how to ensure scientific advancements qua public goods are distributed fairly?

In political philosophy 'public goods' are typically defined as being non-rival (one person's benefitting from it should not decrease the value of the good to others) and non-excludable (they must be provided without excluding anyone in a given population—see Miller and Taylor 2018). Most of the discussion in economics or political philosophy has concentrated on issues such as cost-benefit analysis, or how the ensuing costs of providing such public goods have to be allocated via taxation systems.

But my focus here is on a different yet no less relevant aspect of this debate, an aspect that is upstream to these other kinds of concerns and regards more directly the link between scientific progress and societal progress. Namely, can one treat scientific advancements as 'public goods' in the first instance? Can one accordingly treat scientific progress not just as something internal to science—i.e. how to go from one scientific theory to a better scientific theory, from Copernican astronomy to Ptolemaic astronomy—but as something that speaks to science's ability to produce knowledge that serves the needs of society (in whatever form that might take) and ultimately improves society?

To be perfectly clear, the question is not an invitation to reflect on the (often dreaded) 'impact' of scientific research for society—a narrative that has historically benefited more applied areas than blue-sky types of research. I am taking here for granted that such divisions are unhelpful and often artificial ones: there are plenty of examples of blue-sky research that have turned out to be highly applicable (including, more recent striking examples such as research on mRNA that—once considered blue-sky —has in fact proved pivotal for the new generation of COVID-19 vaccines).

There are two parts to my original question. First, in what sense (if any) is scientific knowledge itself a 'public good'? Second, how can scientific advancements be counted as 'public goods'—especially those that often require the intervention of private companies (as in the example of a vaccine) and are administered by individual nation states through systems of taxation of their own citizens?

The main barrier in answering these two sub-questions and especially the second one, as I see it, is *not only* the divide between public and private in the production of scientific knowledge, the use of patents, and TRIPS, but also the fact that the epistemology of science is not very well equipped to think about the *transnational value* of scientific knowledge and its advancements as 'public goods'. For they are not like national parks, local art galleries, or national defence systems—they are not commodities that act as public goods for specific communities with a clear national identity and well-defined national governments. They are *transnational public goods*, exportable and moveable across countries although they are mostly administered by national and international bodies rather than transnational ones.

This has profound consequences. For even if often there are international organisations whose remit is to cater to specific scientific targets/goals for a consortium of nations (think of Fraunhofer-Gesellschaft in Germany, or ESA or CERN in Europe, or similar institutions elsewhere), catering to specific targets/scientific goals that go beyond any well-defined national or international boundaries (defined either by national or international law)—and often involving even areas beyond national jurisdictions (think of the use of marine genetic resources in areas beyond national jurisdictions)—require transnational institutions. This task is often left to the United Nations—think of the aforementioned Convention on Biological Diversity, the Nagoya Protocol, and the March 2023 Treaty for Biodiversity Beyond National Jurisdiction (BBNJ)—whenever there are aspects of scientific knowledge and its advancements that impinge upon broader issues concerning human rights. But we lack transnational institutions that can regulate and govern in a systematic way the use, distribution, and consumption of scientific advancements as transnational public goods.

This is, I think, the origin of some of the difficulties in thinking about '*who* Big Science *is* really *for*', *who ought to be benefitting* from scientific knowledge and its advancements, and associated problems about distributive justice. That scientific progress is linked—and ought to be linked—to societal progress is neither wishful thinking nor weary 'research impact' agenda. It is a way of realigning epistemological narratives about science with the social responsibility of scientists. But it is also first and foremost a way of revising science policy models that—despite their best intentions—did not necessarily deliver on this important link; and a way of reflecting on institutional infrastructures and legal frameworks that often are not designed to cater for scientific advancements as transnational public goods.

34.4 Scientific cosmopolitanism and the duty of justice

Where to go from here? I can only sketch the general lines of an argument here with more work to be done in this direction[7]. Often philosophers of science have been thinking about science in terms of goals or aims. What is the aim of science? Is it to save the phenomena? Or is it to tell a true story about nature? Is it to accumulate knowledge or to deliver better understanding? Or does it have to do with increasing puzzle-solving? I would like to realign some of these narratives in a slightly different direction, where instead of asking about the aims and goals of science, we consider instead the duties and obligations that come with scientific knowledge. When thinking about duties and obligations in science, familiar ethical questions come to mind: integrity in the production and replication of data; accuracy in measurements; non-biased algorithms; and countless other examples.

But here I am not referring to these daily concerns of bioethics committees and ethics of AI groups. Instead, I'd like to draw attention to what I shall call *cosmopolitan obligations* in science, which—in legal lingo—are obligations pertaining to 'world citizenship'. The term 'cosmopolitan' means 'citizen of the world' in a

[7] The material that follows is reproduced from Massimi (2022a, ch 11), to which I refer the interested reader for more details. I expand on these points in Massimi (2022c).

long-standing tradition in social, political, and ethical philosophy. The general philosophical idea is that 'all human beings, regardless of their political affiliation, are (or can and should be) citizens in a single community' (Kleingeld and Brown 2019). One of the questions that this thousand-year old tradition has however left glaringly untouched concerns how to best understand the role of science. *What (if any) are the cosmopolitan obligations associated with scientific knowledge and its advancements?*

Traditionally, there have been various strands of cosmopolitanism. For my purpose here, the one that matters most is the tradition that begins with the Kantian idea of a *Weltbürgerrecht* (cosmopolitan right)[8] and continue with philosophers like Karl Jaspers, Hannah Arendt, and more recently Seyla Benhabib, who has drawn attention to the UN Declaration of Human Rights[9] in 1948 as a watershed in the transition from international to cosmopolitan norms of justice. Indeed the UN Declaration of Human Rights (UNDHR), in its Article 27 (1), says something important about science and technology:

> (1) Everyone has the right freely to participate in the cultural life of the community, to enjoy the arts and to *share in scientific advancement and its benefits* (emphasis mine).

This human rights approach to scientific advancements signals an important shift: *from* scientific knowledge as an individualistic achievement whose economic benefits can be commodified *to* scientific knowledge as a human activity without which, in the words of the Mexican drafting members of the UNDHR, 'no social progress would be possible' (UN 1950, ch 4, p 71). Treating 'sharing in scientific advancement and its benefits' as a human right implies democratising scientific knowledge. It makes it possible to think of science as carrying cosmopolitan obligations. For where there are rights, there are also duties to respect those rights.

An overarching right 'to share in scientific advancement and its benefits' ramifies into a number of others, e.g. the rights to education, the right to live in a sustainable environment, and so on. The emphasis in the UN Declaration is on '*Everyone* has the right *freely*…' as if there were indeed a genuine, cosmopolitan world citizenship to which everyone partakes and in virtue of which everyone enjoys these rights.

But *who* is 'everyone'? The vague formulation of Article 27 (1) leaves unspecified what the terms 'the cultural life of the community' or 'scientific advancement and its benefits' refer to. The risk of assuming some dominant form of 'community' qua repository of cultural life and scientific advancement remains (see Donders 2008, p 3). Unsurprisingly, scholars in critical theory, post-colonial studies, sociology, and anthropology point out that 'everyone' is often a disguised proxy for the Western-centric 'we'.

[8] Kant introduced the notion of a 'cosmopolitan right' (*Weltbürgerrecht*) in Kant (1795/2006). Following Kant, Hannah Arendt argued for example that 'crimes against humanity' are not just violations of moral norms, but violations of cosmopolitan rights that everyone enjoys. See on this point also Waldron (2006) and Kleingeld (2011, 2016). For an analysis of the normative foundations of Kant's notion of cosmopolitan right and its relevance to discussions about the right to science, see Massimi (n.d.).

[9] See https://www.un.org/en/about-us/universal-declaration-of-human-rights.

They have called for varieties of cosmopolitanism that do not advance universalistic globalising Western-centric claims disguised under cosmopolitan lingo. I want to echo some of these scholars (from Homi Bhabha to Michael Onyebuchi Eze among others) in making a plea for a non-classist, non-elitist kind of scientific cosmopolitanism as a way of starting a conversation about *who* scientific knowledge is really *for*, *who produces* scientific knowledge over time, and *who* ought to be benefitting from it.

One of the main consequences of this shift of narrative about scientific progress as linked to a duty of justice concerns science policy and what governance models this kind of scientific cosmopolitanism might deliver. In particular what kind of transnational institutions and tools can be created and called to deliver on this model—recall, after all, that it was a shift of legal focus to 'cosmopolitan obligations' that historically prompted the creation of institutions such the International Court of Justice of Le Hague for crimes against humanity.

I think that to overcome scientific nationalism and self-serving nation-state interests, to genuinely deliver on the promise of distributive justice in science, a similar sort of shift is needed, going beyond national and international institutions. What is needed is the jurisdicial space where transnational institutions can be created to protect scientific knowledge as a 'transnational public good' so that cosmopolitan obligations in distributing scientific advancements can be met.

Acknowledgements

I am very grateful to Theodore Arabatzis and Panagiotis Charitos for inviting me to contribute to this volume. The material presented here draws on presentations I gave at Philip Kitcher's Descartes Lectures (Tilburg 2020) and at a Royal Society of Edinburgh workshop on 'Perspectives on Big Science and the Question of Justice' (April 2021). I am very grateful to the audiences for their questions and comments. Section 34.4 is reproduced in abridged form from Massimi (2022a) with permission from Oxford University Press. Section 34.2 is an expanded version of material presented in Massimi (2022d), first published in *Nature Physics* **18** 604–6. Reproduced with permission from Springer Nature. This article is part of a project that has received funding from the European Research Council (ERC) under the European Union's Horizon 2020 research and innovation programme (grant agreement European Consolidator Grant H2020-ERC-2014-CoG 647272 *Perspectival Realism. Science, Knowledge, and Truth from a Human Vantage Point*).

References

Aranova E, Baker K S and Oreskes N 2010 Big Science and Big Data in biology: from the International Geophysical Year through the International Biological Program to the Long Term Ecological Research (LTER) Network, 1957–Present *Hist. Stud. Nat. Sci.* **40** 183–224

Bird A 2007 What is scientific progress? *Noûs* **41** 64–89

Bush V 1945/1960 *Science, the Endless Frontier* (Washington, DC: National Science Foundation) https://hdl.handle.net/2027/heb.01131

Cooper A 2007 *Inventing the Indigenous. Local Knowledge and Natural History in Early Modern Europe* (Cambridge: Cambridge University Press)

Dellsén F 2016 Scientific progress: knowledge versus understanding *Stud. Hist. Phil. Sci.* **56** 72–83
Donders Y 2008 Cultural life in the context of human rights *UN Committee on Economic, Social and Cultural Rights* E/C.12/40/13 9 May
Douglas H 2014 Pure science and the problem of progress *Stud. Hist. Phil. Sci.* **46** 55–63
Galison P and Hevly B (ed) 1992 *Big Science: The Growth of Large-Scale Research* (Redwood City, CA: Stanford University Press)
Gerber K *et al* 2021 Ethical concerns about Psilocybin intellectual property *ACS Pharmacol. Transl. Sci.* **4** 573–7
Kant I 1795/2006 *Toward Perpetual Peace and Other Writings in Politics, Peace and History* ed P Kleingeld (New Haven, CT: Yale University Press)
Kitcher P 2001 *Science, Truth, and Democracy* (Oxford: Oxford University Press)
Kitcher P 2021 *Moral Progress.* (Oxford: Oxford University Press)
Kitcher P forthcoming *Descartes Lectures*
Kleingeld P 2011 *Kant and Cosmopolitanism* (Cambridge: Cambridge University Press)
Kleingeld P 2016 Kant's moral and political cosmopolitanism *Philos. Compass* **11** 14–23
Kleingeld P and Brown E 2019 Cosmopolitanism *Stanford Encyclopaedia of Philosophy* ed E N Zalta, Winter 2019 edn https://plato.stanford.edu/archives/win2019/entries/cosmopolitanism/
Kolopenuk J 2020 Provoking bad biocitizenship *Hastings Center Rep.* **50** S23–9
Kuhn T S 1962 *The Structure of Scientific Revolutions* (Chicago, IL: University of Chicago Press)
Krige J and Leonelli S 2021 Mobilizing the transnational history of knowledge flows. COVID-19 and the politics of research at the borders *Hist. Technol.* **37** 125–46
Leonelli S, Lovell R, Wheeler B, Fleming L and Williams H 2021 From FAIR data to fair data use: methodological data fairness in health-related social media research *Big Data Soc.* **8**
Massimi M 2022a *Perspectival Realism* (New York: Oxford University Press)
Massimi M 2022b Perspectival ontology: between situated knowledge and multiculturalism *The Monist* **105** 214–28
Massimi M 2022c A human rights approach to scientific progress. The deontic framework. In *New Philosophical Perspectives on Scientific Progress* ed Y Shan (New York: Routledge)
Massimi M 2022d Perspectives on scientific progress *Nat. Phys.* **18** 604–6
Massimi M n.d. Normative foundations of Kant's cosmopolitan right: the overlooked legacy of Kant's metaphysics of nature *Kantian Rev.* (in press)
Mgbeoji I 2006 *Global Biopiracy: Patents, Plants and Indigenous Knowledge* (Vancouver: UBC Press)
Miller D and Taylor I 2018 Public goods *The Oxford Handbook of Distributive Justice* ed S Olsaretti (Oxford: Oxford University Press) pp 556–74
Niiniluoto I 1987 *Truthlikeness* (Dordrecht: Reidel)
Niiniluoto I 2014 Scientific progress as increasing verisimilitude *Stud. Hist. Phil. Sci.* **46** 73–7
TallBear K 2014 Standing with and speaking as faith: a feminist-indigenous approach to inquiry *J. Res. Pract.* **10** N17 http://jrp.icaap.org/index.php/jrp/article/view/405/371
UN 1950 *These Rights and Freedoms* (Lake Success, NY: United Nations, Department of Public Information)
Waldron J 2006 Cosmopolitan norms. In *Another Cosmopolitanism. The Berkeley Tanner Lectures* ed S Benhabib and R Post (Oxford: Oxford University Press) pp 83–98
Wylie A 2020 The Indigenous/Science project: collaborative practice as witnessing *George Sarton Memorial Lecture* https://hssonline.org/about/honors/george-sarton-memorial-lecture/

Part IV

Big Science in culture and education
[Editor: H Cliff]

Panagiotis Charitos, Theodore Arabatzis, Harry Cliff, Günther Dissertori, Juliette Forneris and Jason Li-Ying

Chapter 35

Introduction to part IV—Big Science in culture and education

Harry Cliff

Science is commonly thought of as a discipline directed towards the acquisition of knowledge that leads to tangible, technological benefits, many of which are discussed in the other chapters of this book. However, science also has tremendous power to inspire, lift the spirits, and change how we look at the world or imagine our place within it. This is particularly true of fundamental physics, astronomy, and space exploration, which grapple with profound questions about the cosmos, from its origins and history to the physical laws that govern reality at the deepest level.

Increasingly, answers to such fundamental questions are sought at large international facilities such as the LHC, LIGO, and SKA. Such Big Science projects are themselves a source of inspiration—their sheer scale and ambition can produce a sense of awe, as can the apparently noble spectacle of hundreds or thousands of scientists, engineers, and technicians from different nations working in concert to understand the world around us.

It is therefore unsurprising that Big Science has long been fertile ground for filmmakers, writers, artists, and curators. The first half of this chapter features essays from academics and cultural practitioners who have either explored the influence and depiction of Big Science in wider culture or who have themselves directly engaged with Big Science in their own creative work. What becomes apparent from their collective accounts is that the relationship between Big Science and wider culture is a two-way exchange. The case of the Hollywood blockbuster *Interstellar* presented by Oliver James is a prime example of how close engagement between scientists and artists can bring tangible benefits to both. For the filmmakers, working with a leading theoretical physicist resulted one of the first, and certainly the most widely seen, scientifically accurate depictions of a black hole on the silver screen, while engagement with the film-making process led to a deeper understanding of black holes for the consulting scientist.

A more direct and obvious benefit to Big Science accrues from documentary film-making, which for decades has been one of the primary ways in which such endeavours are shared with the public. Documentaries play an important role in making the case for Big Science, while also providing ample fodder for documentarians looking for scientific topics that they know will grab the audience's interest, as Jean Baptiste Gouyon describes in his essay on BBC science programming in the post-war period.

Furthermore, as the artist Ariane Koek reflects, Big Science institutions increasingly recognise the importance of engagement with wider culture, as exemplified by the long-running Arts at CERN programme, through which artists are able to spend extended periods based at the laboratory, inspiring new creative directions in their own work, while also giving scientists the opportunity to engage with artists and see their scientific practice in a new light.

Another growing trend identified by Matias del Campo and Sandra Manninger is the use of technology developed for Big Science directly in the production of artistic works. Machine learning algorithms developed for pattern recognition, background rejection, and data analysis in Big Science projects have been used to generate new forms of computer rendered artworks, blurring the lines between artist, artwork, and technology.

But there are also difficulties associated with engagement between Big Science and culture. As curator Alison Boyle reflects, Big Science poses particular challenges for museums: the artefacts in question and often far too large to exhibit, and even when displayed are unfathomable to most of the general public. Worse still, the scientific questions being explored by Big Science projects are often extremely challenging to convey in a meaningful way. Overcoming these hurdles, in the particular case of the *Collider* exhibition at the Science Museum in London, led to the development of innovative new interpretive techniques, which both helped meaningfully convey the science and engineering on display to visitors, while advancing curatorial practice down new and productive avenues.

Big Science also presents particular opportunities and challenges in more formal educational settings, both within the classroom and without. Sibel Erduran and Wonyong Park note the opportunities presented by engaging school students with Big Science, which can be an opportunity to learn both scientific principles as well as the nature of scientific practice itself and in particular the ways in which Big Science projects function. However, they also acknowledge the intellectual hurdles presented to many students by the advanced scientific topics under study by such projects, requiring considered syllabus design, teaching methods, and importantly, evaluation of their efficacy in the classroom.

Meanwhile, Steven Goldfarb describes the programmes created by Big Science organisations to engage the public and educate students both inside and outside the classroom, through the production of classroom materials and activities, as well as visits to the experimental sites and direct engagement with working scientists. Such programmes have great potential to engender a broad interest in science, as well as to develop critical thinking skills. However, as Stephen Pompea and Pedro Russo argue, there is much that Big Science institutions can do to improve their current activities, which often fall short of what might be achieved, particularly in the educational sector. They encourage such organisations to avoid conflating education with communication, public engagement, and outreach, and to take a less self-centred approach to their activities in order to better serve the educational section and their local communities. Their essay amounts to a thought-provoking and often challenging manifesto for how best Big Science organisations can deliver the greatest benefits to their local communities, rather than taking short-termist and self-serving approaches to engaging with the public.

Together these essays paint a vivid picture of the many and varied interactions between Big Science and wider society through culture and education, as well as pointing to new and productive directions in which these mutually beneficial interactions may develop in future.

IOP Publishing

Big Science in the 21st Century
Economic and societal impacts
Panagiotis Charitos, Theodore Arabatzis, Harry Cliff, Günther Dissertori, Juliette Forneris and Jason Li-Ying

Chapter 36

Picturing Big Science on television in the post-war period. The case of BBC science programmes

Jean-Baptiste Gouyon

36.1 Introduction

'The Knowledge Explosion' (21 September 1964), the sixth episode in the BBC science series *Horizon*, became famous for the sequence it contains of science fiction author Arthur C Clark depicting the future as a technocratic dream come true. Driven by technoscientific innovation it would be 'fantastic'. The same programme also contains an interview that has not achieved the same degree of fame, with then historian of Science Derek de Solla Price. There, Price reprises the ideas contained in his book *Little Science, Big Science*, published the year before, and the notion that the exponential growth of science presented new challenges to policymakers and that scientists entering politics would be a positive development [1, pp 114–5]. As Price concluded in his book:

> The ranks of senior scientists and key administrators of science have now swelled to the point where I think it will not be long before some of the good ones begin to enter politics rather more forcibly. We need such men, on the national scene and on the international scene.

The interview with Price probably is one of the rare occasions when the phrase 'Big Science', which Alvin Weinberg coined in 1961 and Price popularised in his 1963 volume, was used to discuss the concept in a science television broadcast [2]. But even without using the phrase, many programmes over the years have dealt with the large-scale, state-sponsored, resource intensive scientific endeavours grouped under the umbrella term 'Big Science', which have flourished in the West in the second half of the twentieth century. As shall appear, these programmes can be

doi:10.1088/978-0-7503-3631-4ch36

interpreted as, each in its own way, making the case for technocracy, government by experts [3].

This essay alights on a few of these programmes, broadcast between 1959 and 2016, to sketch an overview of the coverage of Big Science on the BBC. Many branches of what has come to be named Big Science, such as space exploration, high energy physics, nuclear research, military research, and so forth, have provided material for numerous science television programmes. Science broadcasters are always on the lookout for compelling narratives. These come in the guise of breakthroughs, ambitious or uncommon ventures, and very esoteric research programmes—all things Big Science is replete with. At this stage, however, a caveat is necessary. As media scholar Roger Silverstone remarks, each programme is the outcome of a unique process of construction which is at the same time contingent and creative [4, p 162]. Any attempt at constructing an overall narrative would therefore be doomed to be contradicted by individual examples [5]. However, despite this variety, a common thread seems to run through all the programmes discussed below; a concern for the questions that Big Science research projects raise about the relationship between science and politics.

If one defines politics as a mode of collective action to maintain or challenge existing power relations between social groups, it appears that each of the programmes discussed in this chapter can be interpreted as an attempt, on the producer's part, to come to terms with social change and the role of science in relation to it. What varies as we move closer to the twenty-first century is the position ascribed to science in the process. Beginning in the 1960s with a depiction of science and scientists as necessary actors in the elaboration of a common future increasingly seen as dominated by science and technology innovation, it ends up in the early 2000s with a portrayal of Big Science as a kind of counterfactual 'pure' science [6, p 725], supposedly freed from external influence, as a microcosm whose functioning in terms of cooperation and collaboration could serve as a model to organise and run society.

Representations of science on television relate to the public culture of science in two ways. They help to shape it, through the production and diffusion of representations of science and of cultural meanings about it [7]. At the same time, television programmes are an index of the public culture of science. Embedded in specific social contexts and relations, broadcasters produce representations for mass consumption that must relate to their intended audiences. Broadcasters are more likely to conform to dominant ideologies than to contest them [8, 9]. I approach the television programmes discussed below from this double perspective. I look at them as propositions to their audiences about how to make sense of Big Science. I also interpret them as indices of the meanings already attached to large technoscientific projects in the cultural context where TV programmes originate, and how these meanings change. But for a start, it is necessary to say a word about the beliefs broadcasters associate with the production of science television programmes.

36.2 History of science TV

Although scientific research featured on television as early as 1936, the presentation of science on TV is for the most part a phenomenon which developed in the second half of the twentieth century [5]. Producers of science television programmes in the post-war period did not conceive of science broadcasting as a means of further education. What mattered to them instead was to try and develop a televisual discourse about science. This understanding of science television broadcasting is the product of a two-decade evolution. In the 20 years that followed World War Two, British television broadcasters developed a medium-specific professional culture, a syncretism of radio broadcasting, filmmaking, and documentary production [10]. In 1966 Aubrey Singer, who was then the Head of Science and Features at the BBC, spelled out the key tenet of this new professional culture, in a famous BBC lunchtime lecture:

> The televising of science is a *process of television*, subject to principles of programme structure, and the demands of dramatic form. Therefore, in taking programme decisions, priority must be given to the medium rather than scientific pedantry [11].

Singer's pronouncement, as much a *vademecum* to upcoming television producers as a signal to 1960s British scientists lobbying to control the BBC's science output, also invites us to recognise that television stands outside the cultural space of professional science. As such it puts television in the position to 'play the stranger' in relation to science and approach it as 'a questionable topic of investigation, not an instrument for disentangling problematic situations but a problematic situation itself and one hard to master' [12]. Standing outside the cultural space of professional science enables television to question features of science that are invisible to scientists and to propose cultural meanings for science that professional scientists might not recognise as adequate, but which nonetheless will make sense to viewers because they relate to wider social debates about the fate of our societies and the role science might play there.

Scientists on the BBC's Science Consultative Group challenged Aubrey Singer over his claim to television's autonomy in relation to science. This advisory panel had been created by the BBC in 1966 as a concession to scientists who were incessantly lobbying the broadcasting institution, and following the publication of the Pilkington Report on the future of broadcasting [5]. In a meeting of the Consultative Group that followed Singer's lecture, scientists took issue with his contention, expressed in his lunchtime talk, that fundamentally it was unimportant whether BBC TV producers in charge of science programmes had been scientifically trained. Scientists on the board thought this was essential if the programmes were to convey accurate renditions of scientific knowledge. But Singer considered it only 'a bonus'. What mattered most of all was 'a genuine interest in the broadcasting of science' [13]. To him television producers were not meant to 'convey other people's images' but to speak their own mind, 'utter [their] own statements' [14, p 305].

Instead of looking at science as a body of knowledge, a collection of facts, broadcasters approached it as an activity taking place in society, something which scientists do as participants in the social body. As Robert Reid, then the Head of Science and Features at the BBC, noted during a later exchange with the BBC Science Consultative Group, 'It would be wrong to broadcast a discussion on genetic engineering, for example, without exploring the social consequences' [15]. More than in an epistemic account of science, they were interested in providing viewers with a social one [16]. Through examining science's impact on society and vice-versa, 1960s British science broadcasters were set on providing viewers with keys to understand science's relationship with 'the shifting face of society and the fading importance of environment' [14, p 305].

The main outcome of Singer's vision for science TV broadcasting was the series *Horizon*, first broadcast in 1964. No television series has been more influential in post-war Britain in its continuous contribution to placing science in British culture [16]. The way in which television producers have engaged with science, through *Horizon* episodes and other science programmes should be seen as an index of theirs and television audiences' relationship with science. As such, we need to pay attention to what is a 'substantial body of work, and to understand its content and form' [16, p 87]. Since 1964 approaches to presenting science in this series have evolved from celebratory—science as the path to progress—in the 1960s, to a more critical outlook from the 1970s onwards. In the 1990s a narrative approach—a use of science to tell stories—developed [5]. In the first decades of the twenty-first century science television took a cinematic turn, whereby visual aspects of science television programmes became central to programmes' contents [17]. Despite these variations, a constant remains: the constraints of television take precedence. The necessity to attract and hold audiences' attention and the obligation to tell a story by visual means determine television programmes' form and content as much if not more than their subject matter.

36.3 Between Manhattan and Apollo. Making the case for depoliticised Big Science

Although the phenomenon of Big Science, as money-hungry scientific research on a big scale centred around large pieces of equipment, predates the Second World War, the Manhattan Project is considered to have set the tone for Big Science in the second half of the twentieth century [18]. In 1964 Robert Reid, a science television producer at the BBC, began work to produce a filmed documentary to tell the story of the Manhattan Project. Looking at this programme provides a good entry point in an examination of the coverage of Big Science on British television.

Broadcast on 2 March 1965 on BBC1, *The Building of the Bomb* traces the history of the Manhattan Project back to 1930s Germany, and the research that physicists conducted there at the time into the atom as a source of energy. It follows these scientists, many of whom identified as Jews, as they escaped Germany in the wake of the Nazis' rise to power. The film recalls how they then successfully lobbied British and US governments to take interest in nuclear research, which eventually led to the start and completion of the Manhattan Project. The film concludes on a discussion of the ethics of dropping the bombs on Hiroshima and Nagazaki on the 8 and 9 August 1945, which caused the death of 200 000 people, and the impact this had had on science as it is conceived of by both scientists and non-scientists. *The Building of the Bomb* is a work of public history, presenting evidence from the past, to retrace past events and human actions, in a public setting—a television programme—outside of academia [19]. It is also intended as a means of prompting reflection on the relationship between science and politics in (what was then) the present (1965).

With *The Building of the Bomb* Reid does not merely tell the story of the Manhattan Project, but uses it as a vehicle to explore moral and ethical dilemmas scientists face when 'theoretical scientific research may be used to produce new weapons' [20]. Ultimately, the film interrogates the consequences for science of getting ensconced in geopolitics. Adopting scientists' perspective, the programme introduces 'political and military background only when this impinges on the decisions and the work of the scientist' [21]. Providing this context is delegated in the film to contemporary newsreel, reconstitutions, and archival footage, introduced and commented upon by a narrator, Rene Cutforth, who delivers a script authored by Antony Jay. Interspersed with this material are interviews with nuclear physicists, some of whom worked for the Manhattan Project whilst others were part of the German nuclear research effort during the Second World War.

The Building of the Bomb characterises the Manhattan Project as evidence of the potentially devastating consequences of merging scientific interests with industrial, political, and military ones, and then letting politicians take the final decision. The overall narrative for the documentary was intended to be balanced and 'objective' when it comes to the punctual decision of dropping the bomb in August 1945. As Aubrey Singer noted, 'considerable care was taken to balance the statements by the protagonists' [22]. But from a larger perspective, that of the on-going issue of the association of science with political, military, and industrial interests, the film can be interpreted as a warning of the sinister consequences that might come from political decisions driving developments in science.

During the research phase of production, Robert Reid met with nuclear physicist Edward Teller, on 29 October 1964. Teller, who at the time had quit his academic positions and had gone fully political, actively lobbying against a test-ban treaty, famously maintained that 'the balance of terror is a terror of terror and that his "baby", the H-Bomb, provides this balance at the present time' [23]. Reid's account of his meeting gives a flavour of how worrisome he perceived the association between politics and science when scientists have all but left the room:

> I've just returned from dinner with Edward Teller. Never have I been so terrified of the views of a scientist, politician, political scientist, scientific politician or any other conceivable combination of these sinful professions. On my walk back here I made a note of the nearest fall-out shelter in each street in case Teller's Washington equivalent was at that moment testing the button. For a time I suspected that he was Peter Sellers imitating Dr Strangelove [24].

Robert Reid's early treatment for *The Building of the Bomb* reveals the foundational principle of the film to be the notion that politics was the driving force of the developments in science which resulted in the use of an atom bomb. The political context in 1930s Germany is presented as the origin point of the atom bomb:

> The story will open as being that of a group of men all working on aspects of nuclear physics, either as visitors, or as permanent residents in the country which was the scientific centre of the world: Germany: Germany—the centre of racial prejudice, the centre of hate, and the centre of fear. All of these men have a common characteristic: they are all Jews. ...
>
> It was fear of Germany that drove the Jewish intellectuals, one by one, from German Universities. It was fear of Germany that sent them to America, to Britain and to France: there to work with scientists with non-Aryan names, such as Compton, Rutherford and [J]oliot-Curie [25, p 1].

But eventually, 'Japan—the nation against whom the emigres never intended it should be used—by a fantastic stroke of irony, was to bear the brunt of what had been a terrible fear of Germany' [25, p 5]. To Reid, the atom bomb was fear materialised, the perverted outcome of a pathological form of knowledge production, the result of negative external political interferences with the development of science.

An important point of debate in 1945, after the bomb had been produced, was the decision to use it. As one of the interviewees in the film put it, Fritz Houtermans (1903 66), who was a nuclear physicist in Germany during the War, the fabrication, and the decision to drop the bomb 'prostitute[d] science' [26]. The decision was taken by politicians which the film implies were ignorant of the true nature of the device. As Reid noted in his early treatment:

> Truman, having apparently no knowledge whatever of the meaning of an atomic bomb became President. The committee called upon to advise the new President found itself asked not whether the bomb should be dropped, but how it should be dropped [25, p 4, 27].

Starting with Leo Szilard and Albert Einstein's warning note to Roosevelt, which was put on his desk the day he died, the film shows physicists' failed efforts to convince politicians not to drop the bomb over civilian targets. To Edward Teller, politicians' decision not to follow scientists' advice disrupted the relationship between nuclear physics and society, leading to what he saw as misconceptions of the former by the latter:

> the dropping of the bomb has distorted our view, has changed our whole outlook, we are not looking on the accomplishment of atomic explosives

> as progress which can and should be used in the right way. We've started at that time to look at it as something horrible, something that should not be continued. This general theme took its very beginning, its very start from the mistaken decision to drop the bomb [26].

Teller, in other words, thought that an outcome of the Manhattan Project had been to sully the public image of nuclear physics, and the very idea of scientific progress. This could have been avoided if control of the Bomb had remained with scientists. The film can be interpreted not as a criticism of Big Science, but as a cautionary tale of the risks involved with taking the control of science innovation out of the hands of scientists.

The Building of the Bomb, produced as the Cold War was well underway, when nuclear arsenals assume a prominent role on the geopolitical stage, alights on the history of Big Science at the moment when, following the discovery of nuclear fission in 1938, nuclear physics becomes one of the driving forces in 'the amalgamation of military, industrial and scientific interests' [28, p 231]. The film interrogates the consequences of this amalgamation. It concludes that the intrusion of politics in fundamental research to innovate in weaponry production places scientists in a position which should not be theirs. 'We were all aware of the fact that in one way or another, we were intervening, explicitly and heavy-handedly in the course of human history. That is not for a physicist a natural professional activity' [26]. Robert Oppenheimer's dramatic closing statement says as much. The film postulates an earlier state of purity for science, when science and scientists were 'innocent', as Fritz Houtermans interviewed for the film put it, which ended with the Manhattan Project. In other words, the film frames the building of the bomb as a demonstration that science is best left to scientists and that they should not associate too closely with politics.

In the late 1960s, a *Horizon* episode on space exploration developed a similar argument, making the case that scientists-led innovation might offer better prospects to humanity than one led by politicians and the military [29]. 'After Apollo' (22 May 1969), produced by Peter Goodchild, was broadcast as the Apollo 10 mission was orbiting the Moon, 'at almost exactly the time when the lunar module was making its descent to the surface' [30], in what was a dress rehearsal to the Apollo 11 Moon landing. The programme takes stock of the achievements of the US space programme thus far and considers its future. In particular it examines its possible links with the arms race:

> The American space programme is being pursued with the enthusiasm of a war; … Yet the real objectives remain obscure. Is it purely adventure and excitement … or is it the quest for scientific knowledge, or are there military reasons? These are the questions this programme tries to answer and in doing so looks at the future possibilities of Man in Space [31].

But this gesturing towards potential negative military applications of Big Science is mostly rhetorical. In fact, space research is presented in the programme positively, as a valuable derivative from lethal military endeavours. The conquest of new worlds, starting with the Moon, is discussed as able to potentially satisfy the appetite of aspiring empire building nations, and thereby avoid the devastation caused by wars. As *The Times* television critic put it, reviewing the programme, 'If man is by nature a *conquistador* he can do less harm, and may possibly pay his fare if he turns his attention to the skies' [32]. In this case, Big Science, left in the hands of scientists, would bring peace rather than destruction.

The two programmes discussed so far can be interpreted as attempts on the part of their respective producer to live up to Aubrey Singer's ambition that science television should be a vanguard participant in social and cultural debates. Singer, then the head of Science and Features at the BBC, was adamant, as mentioned earlier, that television should aim to 'create out of its own potentialities its own genuine statements' and thereby should 'claim equal responsibility with those who create the values of society. With architects, authors, scientists, designers, film-makers, with all those who create and communicate original work' [14, p 305]. The producers of *The Building of the Bomb* (1965) and 'After Apollo' (1969) both engage with Big Science as the embodiment of the military–scientific–industrial complex in relation to larger contemporaneous social debates. And both argue in favour of a counterfactual pure state of science which could be restored, or preserved, if science kept away from the military.

Robert Reid produced *The Building of the Bomb*, shortly after the Cuba missile crisis, and as the world was preparing to mark the twentieth anniversary of the bombing of Hiroshima and Nagasaki. At the height of the Cold War, the fear of a nuclear Armageddon was ever present. In May 1965, shortly after the broadcast, he noted in the rough outline of a programme on the H-Bomb:

> The lesson of Hiroshima and Nagasaki has been that no atomic weapon has been used in anger in 20 years. The hydrogen-bomb has never once been used and scientists who work on its production now stand in the same position as they did before Hiroshima, as political crises alternately tighten and remove the finger from the trigger. Would they, the midwives of the weapon, in spite of the legacy of Hiroshima, agree to any circumstances in which the H-bomb might be used? [33, p 2]

Producing a programme discussing the 'political responsibilities of scientists before, during and after the 1939 to 1945 War' [34] and the ethical and moral dilemmas they faced at a time when scientists again faced the same interrogations was a means for television to weigh in the debate. Stephen Black, the interviewer for *The Building of the Bomb*, asked several of his interviewees, all involved in the building of the bomb, what their war time activities led them to think about 'the role of the scientist in politics of the future' [35]. The last question Black asked Professor Joseph Rotblat is evidence of Reid's belief in the relevance of his film to reflect on the relationship between science and politics in the mid-1960s geopolitical context: 'What political–scientific lessons did you learn as a result of your war-time work on the Bomb which you consider are important to apply to today's situation?' [36]. Several scientific institutions in the US, among which the Atomic Energy Commission, requested viewing copies of the film [37], suggesting that it reached at least a part of its intended audience.

Besides the fear of a nuclear apocalypse, the second half of the 1960s was also marked by the escalation of US military operations in Vietnam, which is the backdrop of the production of the *Horizon* episode 'After Apollo' (1969). The war in Vietnam provided further illustration of the amalgamation of scientific, military and industrial interests. From the use of napalm to that of agent orange, several of the weapons the US army deployed in Vietnam had been developed by chemists working in military–academic labs in US universities. The anti-war movement questioned once again the value of the military–scientific–industrial complex [38]. A sequence of footage from the war in Vietnam, showing bombs dropped from B52s and people burned on the ground by napalm, is explicit in contrasting space exploration as a science-led project for the peaceful conquest of new territories, to chemical warfare as a case of politicians (ab)using science to commit atrocities. Central to both *The Building of the Bomb* and 'After Apollo' is the question of the balance of power between science and politics in relation to science and technology innovation, and the notion that science having the upper hand on politics is the most desirable situation for the future of humanity.

36.4 CERN: Big Science for peace

Programmes about CERN, in particular, and high energy physics in general, contrast with the picture painted so far as they offer a view of Big Science detached from politics. However, despite their seemingly apolitical outlook, these programmes, by far the most numerous on Big Science, still participate in an attempt at inserting a televisual discourse on Big Science in broader debates about the fate of our societies. Yet, instead of depicting politics as a negative influence on science, they tend to picture science as a potentially positive influence on politics.

The European Centre for Nuclear Research (CERN) has become the post-war epitome of western Big Science, in its pacific version. As historian of science Jürgen Renn notes: 'CERN demonstrates the possibility of large-scale international cooperation on knowledge production under very special boundary conditions, that is, in the absence of immediate political, military, or economic implications' [28, p 252]. CERN was established in the early 1950s as an international institution dedicated to promoting scientific collaboration in nuclear physics and thereby promote world peace [39]. (On the public presentation of nuclear research see [29].) A British scientist explained as much to Raymond Baxter, the series host, in the 1959 episode of the series *Eye on Research*, 'The Particle Hunters' (on the BBC series *Eye on Research*, see [10]):

> The work we conduct at CERN is in no way connected to atomic armament, or even with atomic power. It is entirely directed to peaceful uses and any result we find is immediately made public and can be used by anybody who cares to make it [40].

This CERN version of Big Science stands in contrast with depictions of large, multinational science projects emanating from the military–scientific–industrial complex. CERN, a result of multi-states cooperation in fundamental physics research is shown to be putting science to work for diplomatic purposes and international peace, rather than producing destructive weapons. In terms of public relations for science, the establishment of CERN in the early 1950s and the visible publicization of the research taking place there were part of scientists' efforts to compensate for the negative consequences of the Manhattan Project on public attitudes to research in physics.

'The Particle Hunters', a live outside broadcast from Geneva, was the first British TV programme about CERN. This strand of broadcasts introducing viewers to particle physics and the international research facility located in Switzerland have become the quintessential 'hard science' programmes. They are intended to attract a specific audience of 'hard-core' science viewers, and their esoteric subject matter serves to establish the identity of the broadcaster as a trustworthy and authoritative source of knowledge [5, 17]. In the following years, BBC cameras visited CERN for several *Horizon* episodes, approximately once every decade. In 1969 an episode soberly titled 'CERN' (27 October 1969) examined 'the significance and need for the big machine at CERN' [41] at a time when a new accelerator was being built there. 'The Hunting of the Quark' (6 May 1974), whose soundtrack contains large extracts from The Pink Floyd's album *Meddle*, places the emphasis on the international aspect of particle physics, showing how researchers at CERN collaborate with physicists at Stanford in the search for sub-atomic particles, the quarks. 'The Geneva Event' (24 January 1983) presents the work of Italian Nobel Prize laureate Carlo Rubbia that led to the findings of the W and Z bosons. Filming began in 1978 as work was underway to construct the Super Proton–Antiproton Synchrotron and follows Rubbia and his team until the announcement by CERN on 21 January 1983, of the discovery of a new sub-atomic particle, the W boson. The final sequence, showing Rubbia's announcement of his team's finding in a large amphitheatre in CERN, had been shot four days before the programme transmission, allowing viewers to witness for themselves, almost as it was happening, a major announcement in the small world of particle physics. In 2007 'The Six Billion Dollar experiment' (1 May 2007), presented the Large Hadron Collider, nicknamed 'the big bang machine' after the experiment it had been built for: recreating the conditions of the birth of the Universe. Five years later 'The Hunt for the Higgs' (9 January 2012) told the story of the discovery of the Higgs boson, and like the 'Geneva Event' concludes with the public announcement of the discovery. Finally, in 2016 'Inside CERN' (10 August 2016) again introduced viewers to the build-up of the discovery of a new particle, again highlighting the international networks of particle physics through interviewing of researchers from different countries. This episode is slightly different from earlier ones in that it ends on an anti-climax. What was thought to be the signature of a new particle was in the end recognised as an artefact of the measurement system.

These programmes all about pacific Big Science, follow more or less the same pattern and suggest that, left to its own devices, science is harmless. The abstruseness and relative invisibility of the subject matter means that rather little in terms of epistemic content is featured. Particles and what physicists do with them is mostly evoked through animations, visual signatures of collisions, or visual effects. Most of the programmes' content concentrate on the context of knowledge production, bringing forward the cost, the gigantism of the hardware, the exceptionalism of CERN as a research facility. Interviewed scientists explain in plain terms what they do and how their equipment works. By contrast with interviews from the 1960s, they are not asked about their views on science's relationship with politics and society at large, evoking instead their hopes, their fears, all in relation to their research. Coming out of these programmes is a vision of science as radically isolated from politics and the rest of society. When the LHC was switched on in 2008, headlines in the popular press evoked the possibility that this could create a black hole that would destroy the Earth [42]. Those rumours had been dismissed with a smile in the 2007 *Horizon* episode 'The Six Billion Dollar experiment', as irrational fears.

The programmes listed above, set at CERN, paint a picture of Big Science as an autonomous pursuit taking place in a location slightly separate from the mundane world. David Sington, a science television producer of several documentaries depicting international scientific collaboration projects is explicit about the narrative purpose of sequences showing how scientists co-exist together in such large research environments as CERN:

> I think that you know, scientific cooperation, internationally, is kind of a model for how we ought to run society. The most successful international collaborations that we have are the international scientific collaborations. So I think that in all sorts of ways, science is a model for human society, and so I want to sort of show that and make people understand. (Interview with David Sington, 7 March 2017. Quoted in [9, p 453].)

Television pictures research workers at CERN free from most everyday considerations and only driven by the quest of truth. Sequences showing them congregating in conference halls or conversing in open plan offices, or during parties or dinners, contribute to constructing on screen the image of a utopia, dedicated to knowledge production and whose functioning could stand as a model of how to run society at large. Historian of science Jeff Hughes noted, 'the History of Big Science is in important ways the story of relationship between Science and the State' (2002, p 152). As part of the history of our societies' relationship with Big Science, television programmes about it are expressions of a scientistic belief in science as a rational model for organising society.

36.5 Conclusion

From the Manhattan Project to the latest experiment in high energy physics conducted at CERN, Big Science in the post-war period has provided science television producers with countless stories. The scale of projects, their potential consequences for humanity, and their (often) esoteric nature make them prime candidates for narratives likely to attract what broadcasters might refer to as 'hard-core science' audiences. The programmes reviewed in this chapter approach Big Science from different perspectives, but they all point in the same direction. Whether it is *The Building of the Bomb* (1965) which can be seen as a cautionary tale about the sinister consequences of ignorant policymakers overruling scientific experts on important decisions, or *Horizon* episodes about CERN depicting international scientific research as a model society run on the basis of scientific expertise, these programmes can be interpreted as advocating a technocratic view of society. In this, they may reflect their producers' views, but more broadly they might also be expressions of enduring normative—and comforting—beliefs in public culture about science as the ultimate source of morality, knowledge, and certainty.

References

[1] de Solla Price D J 1963 *Little Science, Big Science* (New York: Columbia University Press)

[2] Weinberg A M 1961 Impact of large-scale science on the United States *Science* **134** 161–4

[3] For a critical examination of technocracy from the perspective of the social study of science see for instance ed Bucchi M 2009 *Beyond Technocracy: Science, Politics and Citizens* (Berlin: Springer)

[4] Silverstone R 1985 *Framing Science: The Making of a BBC Documentary* (London: BFI)

[5] Boon T and Gouyon J B 2014 The origins and practice of science on British television *The Routledge Companion to British Media History* (Abingdon: Routledge) pp 488–501

[6] Rappert B, Balmer B and Stone J 2008 Science, technology, and the military: priorities, preoccupations, and possibilities *The Handbook of Science and Technology Studies* ed E J Hackett, O Amsterdamska, M Lynch and J Wajcman (Cambridge, MA: MIT Press), pp 719–39

[7] Kirby D A 2017 The changing popular images of science *The Oxford Handbook of the Science of Science Communication* K H Jamieson, D M Kahan and D A Scheufele (Oxford: Oxford University Press)

[8] Curtis R 1994 Narrative form and normative force: Baconian story-telling in popular science *Soc. Stud. Sci.* **24** 419–61

[9] Gouyon J B 2018 From engaged citizen to lone hero: Nobel Prize laureates on British television, 1962–2004 *Public Understand. Sci.* **27** 446–57

[10] Boon T 2014 Formal conventions in British science television, 1955–1965 *Actes d'Hist. Cièn. Tècn.* **7** 51–69

[11] Singer A 1966 Science broadcasting *BBC Lunch-Time Lectures: Series 4* (London: British Broadcasting Corporation) p 13

[12] Schutz A 1964 Collected papers, vol II *Studies in Social Theory* A Brodersen (Nijhoff: The Hague) p 104
cited in Shapin S and Schaffr S 2011 *Leviathan anthe Air-Pump* (Princeton, NJ: Princeton University Press) p 6

[13] BBC 1966 Note of a meeting held on Thursday, 2 June 1966 at Broadcasting House, W.1 *Minutes* WAC R78/76/1 Science Consultative Group p 5

[14] Singer A 1966 Television: window on culture or reflection in the glass? *Am. Scholar* **35** 303–9
[15] BBC 1969 Note of a meeting held on Thursday, 17 November 1969 at Broadcasting House *Minutes* WAC R78/76/1 Science Consultative Group p 4
[16] Boon T 2015 'The televising of science is a process of television': establishing Horizon, 1962–1967 *Br. J. Hist. Sci.* **48** 87–121
[17] Kantor B J 2021 Performing 'the authoritative account': how the BBC's Horizon produces epistemic authority *Soc. Stud. Sci.* **51** 463–83
[18] Morus I 2017 Experimental cultures *The Oxford Illustrated History of Science* I R Morus (Oxford: Oxford University Press) pp 213–45
For a discussion of the Manhattan Project in relation to the history of the concept of Big Science see, for instance, Hughes J 2002 *The Manhattan Project: Big Science and the Atom Bomb* (New York: Columbia University Press)
Thorpe C 2017 The political economy of the Manhattan Project *The Routledge Handbook of the Political Economy of Science* (Abingdon: Routledge) pp 43–56
[19] Kelley R 1978 Public history: its origins, nature, and prospects *Public Hist.* **1** 16–28
[20] BBC Anonymous *Radio Times Billing* WAC T14/1,915/1
[21] Reid R 1965 *The Incomplete History of the Hydrogen Bomb* BBC WAC T14/1,915/1
[22] BBC 1966 *The Building of the Bomb*—use in Japan *Memo by Aubrey Singer to Controller of Programmes Television* WAC T14/1,915/1
[23] On Teller lobbying against a test-ban treaty see ed Herken G 2002 Brotherhood of the bomb: the tangled lives and loyalties of Robert Oppenheimer *Ernest Lawrence, and Edward Teller* (New York: Henry Holt)
The quote is from; BBC 1964 The Manhattan Project *Anonymous* WAC T14/1,915/1, 10 April
[24] BBC 1964 *Letter from Robert Reid to Aubrey Singer* WAC T14/1,915/1, 29 September
[25] Reid R 1964 *The Manhattan Project* BBC WAC T14/1,915/1
[26] BBC 1965 *The Building of the Bomb* (documentary programme)
[27] For historical context see ed Agar J 2012 *Science in the 20th Century and Beyond* (Cambridge: Polity)
[28] Renn J 2020 *The Evolution of Knowledge: Rethinking Science for the Anthropocene* (Princeton, NJ: Princeton University Press)
[29] For a discussion of other BBC science programmes about space exploration, see ; Gouyon J B 2014 Making science at home: visual displays of space science and nuclear physics at the Science Museum and on television in postwar Britain *Hist. Technol.* **30** 37–60
[30] Goodchild P 1969 *Letter to Dr A Krafft Ehricke* BBC WAC T14/2,311/1, 2 June
[31] Reid R n.d. Horizon: after Apollo.... *Transmission promotion* BBC WAC T14/2,311/1 Editor *Horizon*
[32] Critchley J 1969 Watch this space *The Times* 23 May 23, p 9 *The Times Digital Archive* https://callisto.ggsrv.com/imgsrv/FastFetch/UBER2/0FFO-1969-MAY23-009?legacy=no&crop=643+2649+580+1737&format=jpeg (Accessed: 14 September 2021)
[33] BBC The incomplete history of the hydrogen bomb *Anonymous [Robert Reid]* WAC T14/1,915/1, May
[34] Reid R 1964 *Letter to Professor Klaus Fuchs* BBC WAC T14/1,915/2, 14 August
[35] See for instance; Anonymous 1964 *Questions for Professor Rotblat* BBC WAC T14/1,915/1, 14 August

Anonymous n.d. *Professor Dr Heisenberg* BBC WAC T14/1,915/1 p 2
Anonynous n.d. *Professor C F von Weizsäcker* BBC WAC T14/1,915/2 p 2
[36] Anonymous 1964 *Questions for Professor Rotblat* BBC WAC T14/1,915/1, 14 August
[37] e.g.; Mendelsohn R 1965 *Atom bomb prints, memo to Robert Reid* BBC WAC T14/1,915/1, 17 May
[38] Bridger S 2015 *Scientists at War* (Cambridge, MA: Harvard University Press)
Zierler D 2011 *The Invention of Ecocide* (Athens, GA: University of Georgia Press)
Moore K 1999 Political protest and institutional change: the anti-Vietnam war movement and American science *How Social Movements Matter*vol 10ed M Giugni, D McAdam and C Tilly (Minneapolis, MN: University of Minnesota Press) pp 97–116
[39] For more on the Atoms for Peace movement in the 1950s, see; Krige J 2006 Atoms for peace, scientific internationalism, and scientific intelligence *Osiris* **21** 161–81
Hamblin J D 2006 Exorcising ghosts in the age of automation: United Nations experts and atoms for peace *Technol. Cult.* **47** 734–56
[40] BBC 1959 The particle hunters *Eye on Research* (documentary programme)
[41] BBC n.d. Transmission promotion: Horizon 'CERN' *Memo from Editor, Horizon* BBC WAC T14/2,319/1, Week 44, p 2
[42] For example, see ; Daily Mail Reporter 2008 Scientists receive death threats over Big Bang experiment from critics who fear end of the world *Daily Mail* 8 September https://dailymail.co.uk/sciencetech/article-1053162/Scientists-receive-death-threats-Big-Bang-experiment-critics-fear-end-world.html (Accessed: 24 September 2021)

Chapter 37

Bringing Big Science to the Big Screen: lessons from the film industry

Oliver James

Science and science fiction in film have been intertwined ever since Georges Melies' *Le Voyage dans la Lune* helped popularise the idea of sending man to the Moon. Early films were, of course, inspired by the science fiction novels of the time, but cinema magnified the audience these ideas reached, and now that film has become the dominant art form of our century, it is no surprise that it has a pervasive impact on the perception of science and technology.

Science provides a rich pool of ideas for fiction writers to draw inspiration from. As fundamental science has progressed, every breakthrough discovery seems to create as many new questions as it answers, and opens the doors to even more curious possibilities. As Philip Pullman said, 'It's a gift to a novelist to be told that here is an intriguing scientific fact, and no one knows what the truth is …' [1], and he uses the mystery of dark matter and the many-worlds interpretation of quantum mechanics as springboards to launch into the fantasy worlds of *His Dark Materials*.

Science fiction can also influence science. Fantastical ideas presented in film can seem so real, they can prompt the question 'why can't this be real?'. Miniature Communicators in *Star Trek*, and Dick Tracey's watch were fantastical devices when they first appeared, but they inspired Martin Cooper and his team at Motorola to create a real version with the first handheld cellular phone [2].

Science in film can increase public engagement and interest with the subject. Interestingly, scientific accuracy does not always seem to be a factor here: Roland Emmerich's *2012*, where mutating neutrinos boil up Earth's core, was voted 'the most scientifically flawed and absurd science fiction film ever made' by NASA experts [3], but that didn't stop worldwide Google searches for 'neutrinos' spiking in November 2009, coinciding with the release of the film. This ties in with a shift in science communication from public *understanding* to public *engagement* [4]. Public engagement may sound like a vague term, but it matters and can have far-reaching

doi:10.1088/978-0-7503-3631-4ch37

consequences. Much as we might hope that scientific funding is determined solely on scientific merit, the reality is that funding a large-scale physics project becomes a political decision and therefore influenced by public perceptions. The US Superconducting Super Collider failed to gain enough support from Congress to maintain its increasing costs, whereas after broadly similar cost overruns, plans to cancel the James Webb Space Telescope were reversed. Could the difficulty in communicating the abstract nature of particle physics, versus the immediate beauty of space telescope images we've come to expect from Hubble, have played a factor in these decisions?

In 2013 Alfonso Cuaron put the topic of space debris centre stage in the blockbuster *Gravity* which went on to dominate the subsequent Academy Awards, winning seven of its ten nominations. It would be difficult to prove the movie influenced the decision to fund the European Space Agency's ClearSpace-1 proposal in 2019, but the movie is referenced in the team's PR video for the 2017 Conference on Space Debris [5]. It's hard to imagine the opposite conclusion, that such a graphic and high-profile depiction of the danger space debris made it *more* difficult to communicate the importance of the project to the various funding committees.

I imagine there are few people working in the physical sciences today who wouldn't include science fiction movies as one of the sparks that lit a fire of curiosity in them, leading them down their career path. That was certainly the case with me, although my love of physics was also paired with a love of photography, and this eventually led me to a career in visual effects (VFX), where we create virtual photography using physics. I'm Chief Scientist at DNEG visual effects, so when Christopher Nolan needed someone to create images of black holes and wormholes for his movie *Interstellar*, I was fortunate enough to be in the right position at the right time to lead the techincal challenge.

Interstellar was described by Richard Roeper in the *Chicago Sun-Times* as 'a sci-fi adventure of the greatest magnitude' and 'one of the most beautiful films I have ever seen'.

So what did *Interstellar* do differently than other science fiction movies?

It has many aspects in common with any successful movie: a director with a strong vision, a compelling story brought to life by engaging actors, and backed by a tremendous musical score. But this movie had another aspect which made it resonate strongly in the scientific community, which was the involvement of the movie's scientific advisor, Kip Thorne. Kip Thorne is something of a legend in the world of physics. He is an authority on Einstein's general theory of relativity and has a reputation for probing fundamental questions in physics that lie close to the fringe of science fiction, including the possibility of wormholes and their potential use for time-travel. Many movies have a scientific advisor, but their role is typically focussed on narrow aspects of the production, for example advising on key plot points, or providing visual reference for phenomena. Their involvement is rarely woven through the fabric of the movie as thoroughly as Kip Thorne's was; from the initial concept for the movie, an eight-page treatment he co-authored with Lynda Obst; coaching the actors on the finer points of quantum mechanics; through to post-production. I'm going to focus on the latter part, in which I believe *Interstellar* was unique: the collaboration of the science advisor with the visual effects team to create images of a black hole and a wormhole.

This collaboration could easily have been limited to Kip giving the visual effects artists reference images of these phenomena. These would have been approximated, re-imagined, and beautified before appearing on screen, and I doubt they would have had lasting impact. However there were some pivotal moments during post-production that led to the images being created directly from Einstein's equations, and this process seemed to lead to these images creating a story of their own, and having an impact well beyond the scope of the movie.

I'm going to describe what those pivotal moments were, how they came about, and the effect they had. This also gives a concrete example of how art and science can complement each other—science being used to answer creative questions, and ultimately, art influencing science.

The first pivotal moment was Christopher Nolan directing Kip towards the visual effects supervisor, Paul Franklin, to explain what a black hole might look like. Note that he didn't send Paul a picture he'd pre-selected; he wanted a direct interaction between Kip and the visual effects team. Paul loves science, and has an encyclopedic knowledge of science fiction, but his training is in fine art, so he invited me and a few of the more scientifically minded members of DNEG's R&D team to the discussion, 'in case things get technical'.

Kip gave us an overview of black hole and wormhole science that frankly had our heads spinning and made the task suddenly seem even more daunting. Kip described how the intense gravity of a black hole bends light, much like light is bent by a lens. In fact, it's called gravitational lensing. The light isn't interacting with an optical medium though—it is the warping of space and time itself that bends the light. He described having some colleagues in mind who might be able to help us, which gave me the impression he wasn't too confident about us being able to make much progress with the physics. However, I remember asking one question that seemed to change the tone of the conversation: could he provide us with the equations which described the path of a ray of light from a distant star, around a black hole and into the eye of an observer?

The simplicity and contained form of this question must have struck the right note as twenty-four hours later we received a personalised, typeset paper with exactly these equations.

Kip's wisdom was to simplify the physics to match the level of my naive question. I later learnt that the motion of the observer and the rotation of the black hole both have a big effect on the path, and the colour of light, but since I hadn't mentioned those, Kip was free to choose values which made the equations relatively straight-forward. I was then able to create a crude rendering from those equations within a day. Had he hit us with the full formulation that day, there's a good chance we would have been overwhelmed and not been able to make those vital first steps.

It took a further nine months to progress from these rough images to code capable of producing the renders seen in the movie. It's probably extremely difficult for anyone outside the VFX industry to understand how it could possibly take so long just to polish up these pictures. Outlining some of the steps involved might help in explaining how committed we all were in getting these images to look right, and how deeply embedded Kip became in our development team. It might also give some insights into the VFX process: it's never a linear journey, there are always detours and dead-ends we go down, and it's often these detours that are the most rewarding part of the job. Developing software for visual effects is also a race against time: we will always have a deadline by which we have to deliver final images for the movie. Our producers will have worked back from that and created multiple internal deadlines by which rough animation, look-development, lighting, final-animation, and compositing will have been completed. It's also a race against 'Plan B': if production loses confidence that the code will create images that are going to fit in the movie, both technically and creatively, then digital artists will be put to the task, using every tool at their disposal to make it happen. This adds a salesmanship dimension to the process, where we need to build confidence in the evolution of our 'product'. I've found a good way of doing this is to show results early and often and to get people engaged in these results; explain the defects in them; create a set of milestones that will eliminate the shortcomings and get you to the final goal; then start to hit some of these milestones.

The first step, getting people engaged and excited by the images, was very easy. Even though we were creating crude renders, they were intriguing and full of complex details, unlike anything we'd made before. This built support internally for the work, but immediately diverted our effort as we needed to make our prototype available for artists to explore the world around a black hole. The artists' renders were soon in front of Christopher Nolan, who liked what he saw. This certainly solidified the support, but also committed us to delivering a polished product.

One of the interesting detours we took was in imaging stars, and this illustrates some of the lengths we went to in the project. Our initial images were of distant stars seen through the gravitational lens of a black hole. There's a fractal quality to these images: the closer light rays get to the black hole, the more they are bent, which results in more and more complex patterns the closer you get to the black hole's shadow. Each star will appear multiple times in these images, and when we animate the viewpoint changing, to simulate an observer slowly orbiting around the black hole, the stars appear to move around in marvellous swirls (figure 37.1).

We created these images from the equations Kip had given us describing the path of light from a star to an observer's eye. In place of an eye, we constructed a virtual camera and we ran the equations in reverse, from the camera backward around the black hole and out into distant space. Each pixel in the film-plane of the camera represents a direction to start tracing backwards. Sufficiently far from the black hole, light travels in a straight line and we could map this new direction to a point on a chart of the night sky. If that part of the chart contains a star, then the pixel becomes bright; if there's no star, then the pixel remains dark. Every pixel in the final image represents a slightly different direction to trace back from, and as we sample all these directions, we slowly build up the full picture. The chart we initially used was simply a high resolution image of the constellations with hundreds of thousands of bright points, each one representing a different star. The effect of the gravitational lensing was to stretch and squeeze this chart as it was imaged by our virtual camera. In some

Figure 37.1. Starfield image distorted by a black hole. (Image courtesy of DNEG © 2015 DNEG.)

areas the chart was stretched massively in one direction, so that a single star might be smeared across a large part of the image (figure 37.2). This stretching wasn't what we'd expect to see with a real camera imaging real stars; it was an artefact caused by representing the stars as dots on a limited-resolution canvas, so something we wanted to correct. Our solution was to replace the limited-resolution star chart with a database of stars. We downloaded the publicly available Tycho2 catalogue of two-and-a-half million stars from the European Space Agency. Now, when we traced the ray of light back through space, we just needed to see if the ray intersected one of these points in our database. Alert readers will notice there's a problem here: we're trying to intersect a line with a point which is a fruitless task—there's no chance of them intersecting. Instead we construct a little cone of light directions and search for stars within this cone. At the camera, the cone represents a bundle of light rays incident on a small circular patch on the camera's focal plane with a radius of one pixel. Kip's equations told us how to trace back the central axis of the cone, but how would the boundary of the cone evolve as we trace it back through curved spacetime? Again, I had a simple, self-contained question for the professor, so emailed it off and waited. A few days later, a reply arrived:

'The equations that you need are actually rather complicated.'

When the professor who literally wrote the textbook on gravitation describes the problem as *complicated*, I should have been very concerned, but I wasn't. Kip uses very precise language and there are plenty of (maybe unprintable) adjectives he could have used instead. I figured *complicated* was exactly what he excelled at. Days turned to a week and a week into a month, and then the answer finally arrived: five pages of equations, again beautifully typeset and with references. It took time to implement these equations in our code, but when eventually this was done, our stars were no longer smeared. Several person-months had been devoted to solving this

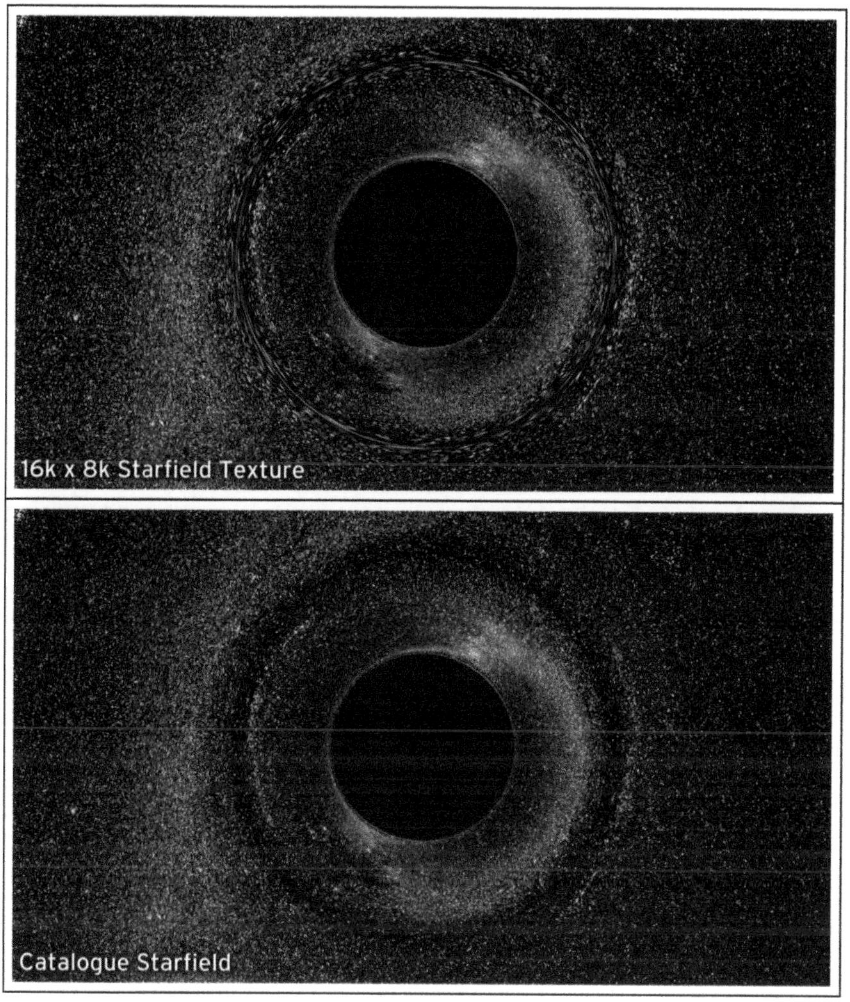

Figure 37.2. Images of starfields distorted by a black hole. In the upper figure, the starfield is stored as a 16 384 × 8192 pixel image. In areas of high distortion the images of stars become smeared because of this limited resolution. In the lower image, we store the stars as points in a database and compute the intersection of light cones with these points, leading to clearer images. (Image courtesy of DNEG © 2015 DNEG.)

relatively minor problem, and this might seem like overkill, but this obsessive attention to detail is characteristic of high-end visual effects.

This investigation into how the shape of a cone of light was affected by gravitational lensing had an interesting side effect. It turns out that the gravitational lensing shears, stretches, and focuses the cross section of our little cone of light as we trace its history back in time. Occasionally, the lensing causes the bundle of rays within the cone to cross over each other, forming an imaginary crease in space. Adjacent creases in neighbouring cones join together to form a two-dimensional ribbon in three dimensional space. This ribbon is called a caustic surface and as a star passes through it, two images of the star come together and massively brighten, or, out of nothing, a very bright image appears and splits into two. These caustics are of interest to astrophysicists because the particular way they magnify distant stars is characteristic of the geometry of the black hole. They could infer properties of a black hole by studying the patterns of starlight affected by these imaginary surfaces in space. The methods we came up with to reduce the smearing of pixels in our images were easily re-purposed to analyse these patterns.

This interaction was representative of many of the technical discussions I had with Kip. We would typically spend a few days exchanging emails exploring a problem and clarifying it to the point where Kip could focus on it in isolation for a week, then summarise the physics in a short paper, or updates to previous papers. We would frequently be working on several problems at once, each one in its own email thread and ending in another paper.

We were based in different time zones (UK and Pacific USA), and this separation actually helped with communication: it forced me to ask very precise questions as answers would only be received the next morning, and responses could be carefully thought through over the whole day for the other to receive their morning. Over the first seven months of the project, we exchanged around 250 emails—an average of almost two a day—and accumulated 14 papers (figure 37.3). I mention these numbers because they point to an unprecedented involvement with DNEG's software team.

These interactions illustrate how we solved highly technical problems by working very closely with the scientific advisor. We were also able to use science to solve creative problems: the black hole images we were creating were fascinating, but lacked any sense of depth. The hole itself appeared as a flat disc with no hint of its

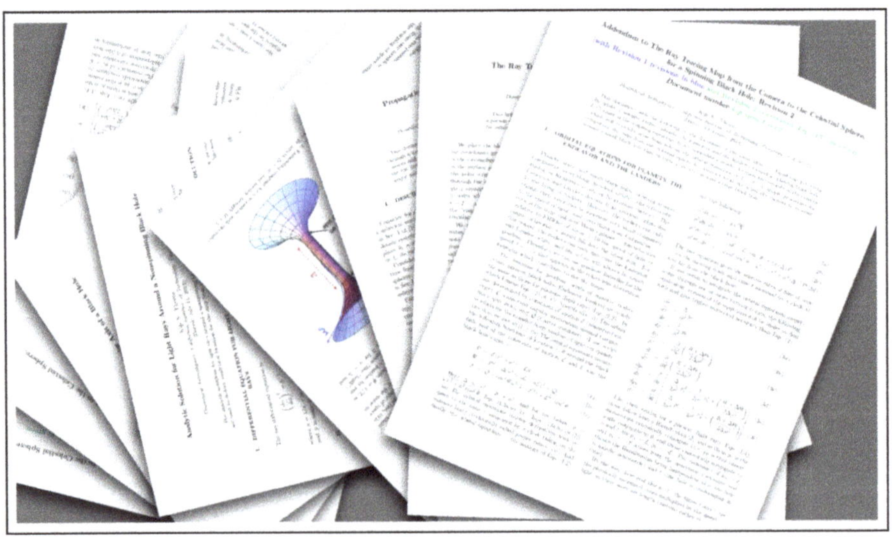

Figure 37.3. Some of the papers written by Kip Thorne for the visual effects team during the post-production of Interstellar. (Reproduced with kind permission from Kip Thorne © 2015 K S Thorne.)

spherical shape. To help carry the story, the black hole needed to have a more substantial and menacing presence.

'Can Kip think of any phenomenon/effect that would help us to establish the spherical nature of the hole?' This was the slightly frantic message sent to me and my colleague Eugenie von Tunzelmann, just before we were due to have lunch with Kip. We were in Los Angeles, attending the Siggraph conference and Kip had graciously juggled his diary to meet us a few hours before our flight back to London.

Over lunch, Kip described how a star that passed too close to a black hole could get ripped apart by tidal forces: the side of the star nearest to the black hole would experience a greater gravitational pull than the far side, and the effect of this difference would be to stretch out the star, further amplifying the difference until the star is eventually stretched apart. Fragments of the star will spiral into the black hole, but other remnants can end up orbiting the black hole much like the rings of dust around Saturn, only much hotter and circling much faster. The orbiting matter is called an accretion disc, and this fiery disc could frame and emphasise the spherical nature of the black hole. It could give us a sense of scale and danger, solving our creative problem. But how would this disc appear under gravitational lensing?

I don't think I slept on the flight home, trying to imagine the answer to this question. Modelling the curved passage of light through a flaming gas cloud was going to take time, but returning to our mantra of showing rough results quickly, I was able to insert a flat, textured plane, painted to look like a disk, into our computer model within a couple of days to get an indication of the effect (figure 37.4).

If the earlier images were intriguing, maybe mysterious, these new ones added a bold, graphic element which went on to define the look of our black hole. With the camera positioned slightly above the plane of the disc, the area of the disc in front of the black hole was relatively undistorted. However the part located behind the hole appeared twice, folded up above the sphere and down below it. This representation of a black hole evolved into the iconic images used in the movie (figure 37.5) and has replaced the idea of a whirlpool in space (figure 37.7). To me, it encapsulates how science, technology and art combined creatively on this project: Science introduced what could be possible and how it might behave; technology made it possible to turn these ideas into images and create tools artists could then use to explore the strange

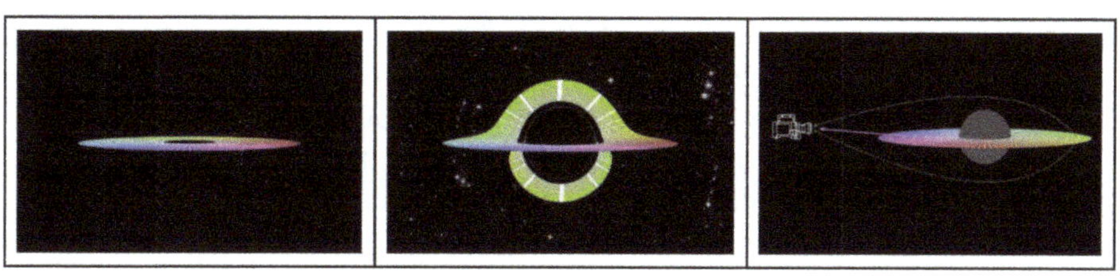

Figure 37.4. (a) Coloured disc in space without gravitational lensing. Panel (b) shows the effect of gravitational lensing. Panel (c) illustrates the approximate paths of light from different parts of the disc to the observer. (Images Courtesy of DNEG © 2015 DNEG.)

Figure 37.5. A black hole image similar to the one used in the movie. Reproduced from [8]. (© 2015 IOP Publishing Ltd under a Creative Commons Attribution 3.0 Unported (CC BY 3.0) license.)

warped space around a black hole, creating environments within which the story in the film could play out.

In addition to the black hole, we were also developing a model for a wormhole. For a long time it was thought that black holes were little more than a mathematical glitch in Einstein's equations[1], but over time, work by Chandrasekhar, Oppenheimer, Snyder, Penrose, Hawking, and others [6] built up the theoretical framework showing how a black hole can form as a star depletes its source of energy, and now it is thought that space is littered with stellar-mass black holes and that supermassive black holes with the mass of millions or billions of suns lurk at the centre of every galaxy. The idea of wormholes arise from another oddity in Einstein's equations, where somehow the topology of the Universe is split into two halves; one half outside a spherical mouth of the wormhole and the other half inside. If someone was to venture into a wormhole, like an insect swimming into a droplet of water, they would see something like an air bubble in the centre, and if they swam into that bubble, instead of finding themselves in a small enclosed space, they would find themselves emerging from the droplet into another Universe. Looking back at the surface they had just emerged from, instead of seeing the concave interior of the air bubble, they would see the convex surface of the interior of the droplet, appearing just like the sphere they originally entered. They would have discovered a whole new Universe located in a bubble at the centre of a drop of water. Some theorists imagine the two halves of the Universe, connected through the wormhole, may also be connected by a gentler curving of space, over a much greater distance with the wormhole creating a shortcut through space. If a driver on a motorway wants to get to the opposite carriageway, they usually need to take a circuitous route via slip-roads and roundabouts to eventually get to the other side; by analogy, the wormhole is the wilder route, smashing straight across the central reservation.

[1] Freeman Dyson describes Einstein as 'not merely skeptical, he was actively hostile to the idea of black holes'.

As with black holes, the wormhole is allowed by Einstein's equations, but unlike black holes, nobody has worked out how one might come into being, given our current knowledge of the rest of physics [7]. Despite this pessimistic outlook for wormholes, we can still model what they might look like. Many of the problems we solved for the black hole can be applied to the wormhole and the software we developed creates images of either phenomena. In the early stages of the project I thought the wormhole would create more interest than the black hole, it is certainly a more unusual concept and the images we could produce were equally, if not more mesmerising (see figure 37.6). However we hit a creative problem: in the movie, as the hero travels through the wormhole, he is supposed to be making an epic voyage, but when we visualised it, the transition from one Universe to another was rather underwhelming. We spent several weeks trying to work out ways to make it more dramatic within the constraints of our mathematical model, but with little success.

Despite all the mind-bending science in *Interstellar*, *science* is not the story being told. At its heart, it is a story about a father and daughter and separation. Science is the backdrop, and it provides the motivations and mechanics to separate father and daughter—the warping of time near a black hole means the father and daughter age at different rates. The closer he gets to the black hole, the further he is being separated from his daughter in *time*. Hundreds of years could pass for her, while he only ages a few months. The wormhole separates them in *distance*, and if the images derived from the equations can't convey that with sufficient emotion, then we have to use a different strategy, unconstrained by physics. This is what our artists thrive at, and they did a fantastic job in splicing together a sedate, but physically correct, arrival at, and exit of the wormhole with a much more dramatic middle section which was created using more traditional CGI methods. I believe this deviation from science is a key reason that compared with the black hole, there was little discussion about the wormhole in the media storm that followed the release of the film. Overall,

Figure 37.6. View into a wormhole. (Image courtesy of DNEG © 2015 DNEG.0

the wormhole had limited impact outside the scope of the movie. In contrast, the work we did on the black hole caused more interest.

The description of the methods we used to create our black hole images formed a paper we published in the journal *Classical and Quantum Gravity* [8]. The paper has now been downloaded over 330 000 times and cited by 142 other articles (this contrasts with a second paper we produced describing the methods we used to visualise the wormhole [9], which has only been downloaded 46 000 times and cited 66 times).

I prefer to think of the impact in cultural terms, although it is impossible to isolate our work on the black hole from its connection with a high-profile movie. If you searched online for images of a black hole before the release of *Interstellar*, you would most likely find an image like figure 37.7: a vague whirlpool in space. After the release of the movie you would have only found images like figure 37.5. These sorts of images existed before the publication of [10], but the combination of high quality, dramatic images, and their role in a Christopher Nolan movie seemed to change the public perception of black holes overnight.

On its own, this apparent change in public perception might be of little significance—why should people care whether black holes look like spheres surrounded by flaming discs, or whether they look like whirlpools in space? A good reason surfaced in the months following the movie's release. By fortuitous timing, the movie coincided with an era of rapid development in black hole science, and this connection brought some benefits. The developments were from the Laser Interferometer Gravitational-Wave Observatory (LIGO) and the Event Horizon Telescope (EHT).

Figure 37.7. This image represents typical pre-Interstellar depictions of a black hole. It combines ideas of whirlpools in space and embedding diagrams, but probably hinders the viewer's understanding more than it helps. (Image courtesy of DNEG © 2020 DNEG.)

LIGO is the project Kip co-founded in 1984 to detect gravitational waves, the ripples in spacetime created by the collision of black holes. Shortly after the release of *Interstellar*, this 30-year project went live, detected the collision of two black holes and thereby produced the most direct evidence of the existence of black holes. By popularising a more realistic idea of black holes, the movie made the concept that they might collide and generate some sort of shock wave a little easier to digest. We made this connection between our work and LIGO more explicit in a concert at the Starmus Festival. This concert, *The Warped Side of the Universe*, celebrated the discovery of gravitational waves, with Hans Zimmer's orchestra performing pieces from *Interstellar* to a video backdrop produced by DNEG, and Kip introducing the concert with a short lecture on black hole science.

In April 2017 EHT captured the first image of a black hole. A common theme in the press at the time was the comparison between that image and the image in *Interstellar*. I felt that by referencing a familiar picture, they could focus the public's attention on the amazing science and engineering behind EHT, without having to explain what the image represented.

What started as a short question eight years ago: 'how could we best depict a black hole?' just kept growing. Having unprecedented access to a world authority on the subject allowed us to push the boundaries in visual effects by simulating general relativity in our visualisations. A visionary film director gave us time and freedom to experiment. By following where science led us, we were rewarded with images of striking complexity that we had not anticipated, and discovered unexpected solutions to creative problems. The technical demands of creating images for film led to a new approach of tracing cones through warped spacetime and subsequent contributions to scientific journals. Art and Science truly enriched each other in this project.

The work contributed to *Interstellar* winning the Academy Award and BAFTA for Best Special Visual Effects; it has helped connect students with applications of STEM subjects they hadn't previously considered; but I believe the most important consequence was bringing black holes back into the public eye with a more realistic depiction, increasing engagement with the real black hole science of LIGO and EHT.

References

[1] Q&A 2002 with Philip Pullman *The Guardian (Online)* 18 February https://www.theguardian.com/books/2002/feb/18/fiction.philippullman

[2] Marty Cooper 2015 interview *Scene World Magazine* 12 February https://youtu.be/B6OKTJMavtw

[3] Edwards L 2011 NASA's Silly Sci-Fi Film List—2012 the Most Flawed *Phys.org* 7 January https://phys.org/news/2011-01-nasa-silly-sci-fi-flawed.html

[1] Schäfer M S 2008 From public understanding to public engagement *Sci. Commun.* **30** 475–505

[5] Dealing with Space Debris 2017 *ESA* 11 April https://www.esa.int/ESA_Multimedia/Videos/2017/04/Dealing_with_space_debris

[6] Penrose R 1996 Chandrasekhar, black holes, and singularities *J. Astrophys. Astron.* **17** 213–31
[7] See ; Everett A and Roman T 2013 *Time Travel and Warp Drive: A Scientific Guide to Shortcuts through Time and Space* (Chicago : University of Chicago Press) for a discussion of this topic
[8] James O *et al* 2015 Gravitational lensing by spinning black holes in astrophysics, and in the movie *Interstellar Class. Quantum Grav.* **32** 065001
[9] James O *et al* 2015 Visualizing interstellar's wormhole *Am. J. Phys.* **83** 486–99
[10] Luminet J P 1979 Image of a spherical black hole with thin accretion disk *Astron. Astrophys.* **75** 228–35

IOP Publishing

Big Science in the 21st Century
Economic and societal impacts
Panagiotis Charitos, Theodore Arabatzis, Harry Cliff, Günther Dissertori, Juliette Forneris and Jason Li-Ying

Chapter 38

Distracted enchantment and public attitudes to space exploration

Simon Stewart

Space exploration has the characteristics of Big Science when it involves large-scale, long-term, and costly research that employs a range of scientists, technicians, engineers, and administrators, and is coordinated across a range of institutions including government, industry, and the university sector (Lambright 2013). As Parker and Bell (2009) argue in their introduction to a special issue on space travel and culture, space exploration is an under-researched area in the social sciences. This is in great part due to the scepticism among academics about the vast resources deployed and the wider power games associated with the 'space race'. However, space exploration continues to have a hold over the imagination of the wider public. With reference to public attitude studies conducted in Europe and the United States of America, this essay formulates sociological explanations for the broad though unfocused interest and support among the populace for space programmes. In doing so, I put forward the notion of *distracted enchantment*. Space exploration can be understood as an example of what the classical social theorist Georg Simmel (2011[1907]) terms objective culture in that it opens up a body of knowledge so vast and expansive that it is deemed largely incomprehensible and inaccessible by most people. This notwithstanding, space programmes appeal to subjective culture in that they open up vistas of enchantment, connect with wider stories that form part of the extensive repertoire of popular culture, and, perhaps most significantly, provide a sense of meaning and future-oriented purpose even if the precise nature of the purpose is elusive. The Big Science of space exploration can thus be understood as a collective endeavour that people are able to feel connected to, albeit in a relatively distracted and unfocused way.

This essay commences with the findings of two recent studies on public attitudes to space exploration/activities conducted in 2018. Both studies are based on representative samples of the adult population aged 18 or over. The Pew

doi:10.1088/978-0-7503-3631-4ch38

Research Center (PRC) report (2018) draws on the results of the American Trends Panel which is 'a nationally representative panel of randomly selected US adults living in households recruited from landline and cellphone random-digit-dial (RDD) surveys' (Pew Research Center 2018). The survey that informs this report included 2541 respondents. The Harris Interactive study, conducted for the European Space Agency, was based on a representative sample of 5227 Europeans including representative samples from Germany (1051), France (1059), UK (1068), Italy (1037), and Spain (1012) (Lévy *et al* 2019). Both studies present quantitative data and so it is not possible to gain access to deeper reflections on public attitudes to space exploration. Furthermore, many of the questions are designed to elicit responses associated with the purposes of space exploration. The questions do not offer much scope for critical responses. Regrettably, there is a lack of qualitative sociological research on public attitudes to space exploration, although Dickens and Ormrod (2007a) have been able to reflect on answers to questions included in the 2005 Mass Observation Project (MO). They found, for example, that the vast majority of MO respondents had no burning desire to travel into space and those who did tended to be middle-aged men in professional occupations in possession of high volumes of the various forms of capital (Bourdieu 2020). In contrast, those with fewer resources at their disposal found the increasingly abstract nature of scientific knowledge of the Universe to be alienating (Dickens and Ormrod 2007b, p 620):

> Well what is a Universe and what is space? I may read more about it, but my understanding of it doesn't increase—it's all just too immense. When I look at the Milky Way and understand what it means and then relate it to infinity, my mind just shrivels and my brain gets tangled.

The data discussed below has limits inasmuch as respondents were asked to select from a pre-given set of potential survey responses. Nevertheless, the findings provide a useful basis on which to develop the notion of distracted enchantment as means of understanding the positive and yet unfocused interest that runs through public attitudes to space exploration.

38.1 Public attitudes to space exploration: survey findings

In the PRC study there is overwhelming support among Americans for the central role of NASA (Pew Research Center 2018). This is despite the claims made in recent years by pro-space libertarian groups that NASA is an outmoded, excessively bureaucratic organization that stifles innovation (Parker 2009). Libertarian groups argue that the way forward for space exploration is by means of private sector innovation. In contrast to this perspective, the PRC report demonstrates high levels of support for NASA and for the continuing involvement of the United States government in space exploration (Pew Research Center 2018). 65% of respondents state that it is essential that NASA continues to be involved, while 35% say that the private sector can ensure that progress is made even without NASA's involvement. Of those who declare a keen interest in space matters, a slightly higher proportion are in favour of the private sector, with 55% advocating NASA's essential role and 45% believing in private sector-led progress without NASA. Meanwhile, 72% state that it is essential for the USA to be a leader in space exploration. This widespread support from the American public is 'widely shared across gender, educational and political groups', and finds similar levels across the generations (71% among 'Baby Boomers' and 70% among 'Millennials'). At the same time, 80% of respondents view the International Space Station as a good investment. The only slight discrepancy in the views is that Republican supporters are more divided on the issue of whether space exploration should be led by the public or private sector, with 53% maintaining that NASA should continue to play a leading role. Meanwhile, many respondents are confident in the private sector's ability to make profit from space exploration but are sceptical about the extent to which private companies would minimise human-made space debris. Interestingly for the purpose of this essay, only 7% of respondents state that they are attentive to space news. Meanwhile, those who are more attentive tend to be those who are more supportive of the US leading space exploration (88%, as compared with 60% who proclaim to have heard nothing about NASA and/or space exploration) and consider that the International Space Station has been a good investment (94% as compared to 67% who do not follow space news) (Pew Research Center 2018).

In the PRC report it is notable that respondents identify as a top priority for NASA the monitoring of key parts of the Earth's climate system (63%). In addition to this, monitoring asteroids/objects that could hit the Earth (62%), scientific research (47%), and the development of technologies that can be adapted for other uses (41%) are also considered to be significant. In contrast, plans to send astronauts to Mars (18%) or to the Moon (13%) are not considered to be top priorities. Interestingly, the survey respondents were not asked about whether military and defence-related innovations should be prioritised by NASA, and so it is not clear how important these would be for those surveyed. It is worth noting that there are some modest demographic differences in the responses: for example, Republican voters are less likely than Democrats to prioritise environmental concerns; and those with a postgraduate qualification and those who pay more attention to space news are more likely to prioritise scientific research than those with high school or less education and those who do not follow space news (Pew Research Center 2018).

The public attitude to space exploration in relation to environmental concerns can be traced back to the impact of the Apollo programme. Henry and Taylor (2009) draw attention to the significance of Apollo 8's *Earthrise* and Apollo 17's *Whole Earth* images in raising environmental consciousness both about space exploration but also, perhaps more significantly, about the Earth. The photographs, taken by crew members, convey a strong sense of Earth's majesty and yet also its fragility. They became hugely popular because they 'sparked a radical rethinking of our dependence on, and responsibility for, sustaining Earth's ecosystems and the diversity of life on our fragile biosphere' (Henry and Taylor 2009, p 190). For example, *Whole Earth*, according to the former US Vice President Al Gore, became the most published photo in history. Moreover, Gore argues that the image played a key role in the development of a number of progressive environmental policies in the

US, including the Clean Water Act and Clean Air Act, as well as the Natural Environmental Policy Act (Henry and Taylor 2009). Apollo 8 astronaut Bill Anders remembers the genesis of the *Earthrise* image as follows:

> When I looked up and saw the Earth coming up on this very stark, beat-up lunar horizon, an Earth that was the only color we could see, a very fragile looking Earth, a very delicate Earth, I was immediately almost overcome with the thought that here we came all this way to the Moon and yet the most significant thing we are seeing is our home planet, the Earth (Henry and Taylor 2009, pp 192–3).

Whilst space exploration enables humans to envisage new frontiers to explore and conquer, the images of Earth captured by Apollo were taken up as powerful symbols for environmental concerns, highlighting our interconnectedness and collective responsibility to protect the planet. However, in contrast to the reflective mode of global consciousness enabled by *Earthrise* and *Whole Earth*, the recent 'private turn' in space exploration, which has seen billionaires at the forefront of the newly developing space tourism market, reinvokes the Western imperialist imagination. At present, these activities are the preserve of the super-rich. Nevertheless, 50% of the respondents in the PRC survey envisage that space tourism will soon be a routine activity within the next 50 years and 32% think that colonies on other planets on which humans can live for long periods of time will be developed before 2068. Meanwhile, 42% would like to orbit the Earth in a spacecraft (as opposed to 58% who would not like to do so). Eagerness to travel in space differs by age category and, perhaps predictably, there is a higher proportion of Millennials (63%) than Baby Boomers or older generations (27%) interested in space tourism. Those most interested in space news are more likely to be keen on space travel than those who do not follow such news (74%, as compared to 30%). Those not interested in space tourism cite prohibitive factors such as the high cost (28%) and/or the scariness of the venture (28%), and health or age reasons (28%). Those keen on the possibility of space tourism say that they would be keen to experience something unique (45%), see Earth from space (29%), and learn more about the world (20%) (Pew Research Center 2018).

The Harris Interactive Study for the European Space Agency (ESA) report, in common with the PRC report, finds that public attitudes to space activities are favourable. For example, the ESA report highlights that 90% of respondents have a positive view of space activities and of these, 33% have a very positive view (Lévy *et al* 2019). This view is shared across the five countries surveyed albeit with minor variations: in Italy, the positive view is held by 95% of respondents as compared to 86% in the UK. In terms of the indirect benefits of space activities (for example), 91% refer to the benefits of space activities for research and scientific purposes; 85% say that these activities give people a reason to dream; 84% state they inspire young people; and 83% think that they enable corporation between countries. Less prominent but significant nevertheless are the material or everyday concerns of creating economic wealth (64%) and making everyday life easier (53%). The enabling aspect of space activities is also emphasised by the respondents with reference to (for example) furthering our knowledge of the Universe (93%; 54% 'high priority'); exploring the Solar System (92%; 50%); facilitating transport, e.g. through satellite navigation (89%; 40%); gaining a better understanding what is happening on Earth (90%; 45%); developing the means of communication (85%; 37%); and identifying the effects of climate change (85%; 36%). Looking to future research, identifying the effects of climate change (91%; 53% 'high priority'); and furthering our knowledge of the Universe (87%; 44%) are among the priorities (Lévy *et al* 2019).

While exploring the effects of climate change is considered to be a significant priority, the main threats identified are space debris, asteroids, solar flares and, above all, data hacking. The ESA report did not ask about the public/private distinction in leading space exploration. However, it did ask respondents about the relative significance of various actors in space activities. Despite the development of space programmes in India, Japan, and China, the respondents unanimously highlight the US and Russia as the major superpowers in space, with the US identified as important by 93% ('very important' by 56%) and Russia at 90% ('very important' by 46%). Other powers mentioned include Europe and China. The majority of respondents from each country within Europe believe that the various countries should pool their resources, though the level of support for this is slightly lower in the UK (perhaps explainable due to the post-Brexit political climate). At the same time, and perhaps related to this, respondents overstate the estimated expenditure on space activities as part of their countries' public spending. In the five European countries surveyed, the costs of space travel through public expenditure is around 10 euros per person each year. In contrast to this, the respondents estimate that the cost per person is 245 euros per year, more than 20 times the actual amount (Lévy *et al* 2019).

38.2 Political and economic considerations

Discussion of the role of the military–industrial–space complex is conspicuous by its absence in both surveys. Dickens and Ormrod (2007a) and Dickens (2009) provide key insights into the political and economic dimensions of space exploration. The foundation of NASA, over 60 years ago, 'required something of an unholy alliance between military hawks, big research institutions, "defence" contractors and politicians' (Dickens 2009, p 85). It was enabled in great part by the state funnelling money through lucrative contracts to a range of organizations including Boeing, North American Rockwell, and McDonnell Douglas. In the 1960s expenditure on NASA reached its peak with an allocation of 5.3% of the total Federal budget (Dickens 2009). Project Apollo cost around $25.4 billion, which is approximately $156 billion in today's money (Congress House Committee on Science and Astronautics 1973). Since Project Apollo, advocates of space programmes have had to make persuasive arguments to the president and Congress to gain funding, in particular where there has been less political will for such projects. For example, President Nixon (in 1976) argued that 'space expenditures must take their place within a rigorous system of national priorities' (cited in Lambright 2013, p 132). Nevertheless, numerous Big Science space exploration projects have been funded since the Apollo programme. These have included, for example, the Viking project (1975–83), which sent a robotic probe to Mars in order to advance knowledge of the planet and ascertain whether life exists there. This was a huge project involving 70 principal scientists, including government and academic researchers, and '[a]t its height, Viking employed over ten thousand people in a government, industry and university team' (Lambright 2013, p 133). Other similar projects have included the Hubble Telescope (and later the James Webb Space Telescope) which commenced in 1990, and the Cassini project, which was enabled in great part by NASA's collaboration with the European Space Agency. The Cassini project sought to orbit Saturn and land on one of its moons, Titan. It ran from 1997 to 2017 and the mission 'involved over 250 scientists, 5000 engineers and other professionals, and 19 nations' (Lambright 2013, pp 143–4).

In 2019 the global turnover of the space industry was estimated to be around $423.8 billion (as compared to $216.6 billion in 2009). Government expenditure worldwide on space programmes is estimated to have been $82.5 billion in 2020 as compared to $42.4 in 2014 (Statista 2020). The US leads the way with expenditure of $48 billion dollars in 2020. In the same year, the nearest competitors are China ($9 billion), France, Russia, and Japan ($3–4 billion). The US Government spent $10.8 billion on space exploration in 2017 with the figure projected to rise to $12.8 billion by 2027. In 2021 France (1.065 billion euros) and Germany (968.6 million euros) made the most significant contributions to the European Space Agency's budget. In 2021 US has the most significant presence in space, with 1897 satellites in orbit. The nearest competitors are China (412) and Russia (176). NASA is the leading space agency, with a government budget of $19.5 billion. The nearest rivals are CNSA (China) with a budget of $11 billion, ESA (Europe) with $6.3 billion and Roskosmos (Russia) with $3.3 billion (Statista 2020). NASA's leading contractors include the California Institute of Technology, Boeing Company, Lockheed Martin Corporation, Northrop Grumman Corporation, and Jacobs Engineering Group Inc. In 2021 the global leading aerospace and defence manufacturers (in terms of revenue) include Lockheed Martin ($66.01 billion), Airbus ($60.5 billion), Raytheon Technologies ($60.48 billion), Boeing ($56.97 billion), Northrop Grumman ($37.34 billion), and BAE Systems ($26.94 billion) (Statista 2020).

Dickens argues that the considerable expenditure that space exploration involves is legitimised by the state and by key organizations such as NASA in terms of 'general, universal values rather than the competing values of factional and warring interests' (Dickens 2009, p 78). This means that space exploration is justified with recourse to the pursuit of scientific knowledge and the search to explain the fundamental mysteries of the Universe. Also emphasised are the benefits to humankind such as the ability to monitor the Earth and as a means of identifying and thus resolving environmental crises. These benefits are referenced in the PRC and ESA surveys. A further benefit emphasised is that of defence where space exploration provides means of protecting the citizens of a nation. A final universalising justification put forward is that space exploration is an expression of an inherently human curiosity to explore new horizons. In fact, the potential consequences of exploring the cosmos, envisioned as capitalism's 'outside' are often detrimental. They include, for example, the erosion of the 'outside' in pursuit of new resources, and new forms of galactic imperialism stimulating conflict between various nation states and private corporations for control over new territories. For example, Dickens (2009) draws attention to the intensifying competition between the military–industrial–space complexes in the USA, Russia, Europe, and China. At the same time, he argues that investment that could be better used to counteract social inequalities and environmental degradation is funnelled towards space exploration as one of a series of temporary 'fixes' designed to further prolong the life of capitalism. Drawing on the work of David Harvey's notion of 'spatial fixes', Dickens (2009, p 67) argues that '[c]apitalism now has the cosmos in its sights, an outside which can be privately or publicly owned, made into a commodity, an entity for which nations and private companies can compete'. In this sense, the exploration of outer space is another example of capitalism's tendency to construct outer transformations: new, fresh zones that have up until now not been invested in or fully exploited. Dickens (2009) argues that space is the new frontier for imperialistic conquest and the pursuit of raw materials. Meanwhile, the deployment of space technologies also enables extensive surveillance on Earth for military purposes and to better coordinate warfare on the planet.

Outer space is also a site of a new form of conspicuous consumption which Dickens and Ormrod term cosmic narcissism (Dickens and Ormrod 2007b). They argue that the contemporary figure of the 'have it all' narcissistic individual has its roots in the notion of the 'universal man' that came to prominence during the Renaissance. In the current era, billionaire space tourists are cosmic narcissists for whom 'the Universe is very much experienced as an object; something to be conquered, controlled and consumed as a reflection of the powers of self' (Dickens and Ormrod 2007b, p 617). Billionaire space tourism creates strong waves of publicity and public awareness. However, the extent to which it helps individuals to further their understanding of space exploration is debatable.

38.3 Distracted enchantment

In common with the American respondents in the PRC survey, the majority of the European respondents in the ESA survey state that they are not particularly well-informed about space activities (Lévy *et al* 2019). Only 8% of respondents consider themselves to be very well informed, with the majority not well informed (and 13% of these 'not at all well informed'). Interestingly, when it comes to specific space exploration-related activities, the respondents are—broadly speaking—aware of them, but are not so well informed on the specifics of the programmes. For example, whilst 83% have heard of the European Space Agency, only 37% know exactly what it is. This extends out to knowledge of European astronaut missions to the International Space Station, the Galileo navigation satellite system, the Ariane rockets, Copernicus—the Earth observation programme, and the Rosetta Mission. Even if respondents are aware of these activities only a minority know precisely what they are about (e.g. 15%, in the case of the Copernicus programme and 17% in the case of the Rossetta Mission) (Lévy *et al* 2019). A recent IPSOS study of attitudes to space exploration has similar findings in relation to levels of engagement: whilst 21% of respondents are very interested in space exploration, 54% are moderately interested (IPSOS 2019).

In considering the results of the PRC and ESA surveys, I put forward the notion of *distracted enchantment* to help explain the positive attitude towards space exploration despite limited knowledge and the distracted manner in which information about space activities is consumed. I argue that even without the detail and focus that comes with a more sustained level of engagement, space exploration stimulates a sense of enchantment through collective endeavour and shared structures of meaning. In other work, with reference to the temporal dimension of aesthetic appraisal, I have distinguished between *sustained* and *distracted aesthetic engagement* (Stewart 2015). Sustained aesthetic engagement refers to the ability to become wholly absorbed in an engagement with an object so that we lose track of time. It involves, for example, becoming totally immersed in an artwork, film, novel, or piece of music. The ability to find the time to do this regularly depends on the extent to which individuals are able to possess time. Stella Chatzitheochari and Sara Arber distinguish between pure free time and contaminated free time to account for this distinction, with highly paid men the most likely to be able to generate unconstrained free time even if they lead busy lives (Chatzitheochari and Arber 2012). In contrast, those experiencing time poverty, with domestic and childcare duties impinging on free time, will find it particularly difficult to free up time for sustained aesthetic engagement. The more common form of aesthetic engagement, possible for almost everyone, is distracted aesthetic engagement. This is where individuals engage with cultural objects in a distracted manner, whether they are carelessly flipping through the pages of a magazine or moving from one song to another on a playlist. In this mode, individuals are not fully absorbed and their minds might be on other things. Along these lines, I argue that public attitudes to space exploration are, by and large, characterised by an orientation of distracted enchantment.

In order to understand *distracted* enchantment, we can learn much from Walter Benjamin's notion of exhibition value, which, he argues, is replacing cult value in how objects or images are consumed. Only a small number of people can simultaneously view a unique work of art such as the *Mona Lisa*, hence it acquires cult value. This lends the original art work an aura of uniqueness and when people view it, there is a sense of reverential awe. In contrast, any number of people can simultaneously watch a film shown in cinemas because there is a plurality of copies. Moreover, film assembles multiple fragments of reality 'assembled under a new law' that can be consumed in a distracted manner (Benjamin 1968, p 234). Images of space exploration that are consumed by the public are typically broadcast to the nation and/or are available retrospectively on sites such as YouTube. They possess exhibition value because they are (at least in principal) available for anyone to see. The value of space exploration thus has democratic potential inasmuch as footage can be viewed simultaneously, or different times, by any number of people. Even when viewed in a distracted manner, contemplation of the properties of space and time and the potential existence of life on other planets has the potential to provide collective structures of meaning through which people can make sense of life

(Carroll 2008). In contrast, the experience of space tourism, which is only available to the super-rich, retains cult value because of the aura associated with a unique experience that is only available to a select few. The cult value of space tourism is enhanced by the fact that it is a brief and intense experience, e.g. of a few minutes of weightlessness in space. The vast majority of people are excluded from having a stake in these activities that flagrantly display the principle of conspicuous waste (Veblen 1994[1899]).

Another aspect of distracted enchantment is that it is characterised by individuals' lack of detailed knowledge about space exploration. To help understand why positive attitudes persist despite this lack of knowledge, I refer to Max Weber's (1946[1918]) argument about rationalization and disenchantment and Georg Simmel's (2011 [1907]) observations about the limits of subjective culture (the development and enrichment of an individual's mind) in relation to the exponential growth of objective culture (the accumulation of mental labour across generations as expressed in technology, literature and so on). Broadly speaking, Weber argued that rationalization is a process that has taken hold in the modern West over the past few centuries that has involved the increasing mastery of reality through technical procedure, planning, and calculation. Rationalization, understood as increasing mastery over empirical reality, extends to all societal fields, from the field of law to the field of sport, and of course, to knowledge about outer space. However, as Weber was aware, the impact of rationalization processes on individuals does not necessarily lead to an increase in knowledge of reality. Weber was acutely aware that 'increasing intellectualization and rationalization do *not* ... indicate an increased and general knowledge of the conditions under which one lives' (Weber 1946, p 139). He goes on to discuss that while the knowledge is available to understand how a streetcar works, most of us who ride on one will have no idea how the car achieves motion. Those riding the streetcar are happy to 'count' on the expectation that the streetcar will take them where they want to go.

Disenchantment, the decline of a magical orientation to reality that comes about because 'one can, in principle, master all things by calculation' (Weber 1946, p 139) has been accompanied by a parallel process of (re)enchantment where individuals seek meaning in (for example) forms of spirituality, neo-paganism, religious fundamentalism, psychoanalysis, consumerism, celebrity culture, tourism, and science fiction. These interests conjure up and are 'rooted in, understandings and experiences of the world in which there is more to life than the material, the visible or explainable' (Jenkins 2000, p 29). Richard Jenkins (2000) argues that (re)enchantment is as integral to modernity as is disenchantment. It is my contention that an interest in space exploration offers, for many people, a form of distracted *enchantment*. They do not necessarily become immersed in details of space-related activities, nor are they preoccupied by them. However, they are mildly interested in them and as such, the activities imbue their lives with a sense of meaning. This means that whilst science can be seen as 'modern nihilism's most authoritative and dispiriting guise' inasmuch as it informs the view that human life has no meaning beyond its occurrence within a history of chance evolutionary mutations (Carroll 2008, p 129), it can also serves a means of enchantment, such as when physicists find something akin to spiritual awe in their contemplation of the Universe or when members of the public are captivated by popular science that introduces them to cutting edge debates in fundamental physics and cosmology.

Simmel's (2011) discussion of the tension between objective and subjective culture also enables insight into an interest in space exploration as a form of distracted enchantment. According to Simmel (2011, p 488), objective culture is 'the concentrated mental labour of a cultural community' which is accumulated over countless generations. It is 'is embedded in language and custom, political constitutions and religious doctrines, literature and technology as objectified spirit from which everyone can take as much of it as they wish to or are able to, but no single individual is able to exhaust it all' (Simmel 2011, p 449). As part of the modern division of labour with its associated specialisms, everyone involved contributes—to some extent—to this accrual of knowledge. However, Simmel argues that the tragedy of culture is that there is a lag in the development of subjective culture. The wealth of objective culture grows apace, spurred on by the increasing sophistication of the modern division of labour. At the same time, the individual can only absorb a limited amount of this objective culture. The gap between the two widens every day, such is the relentless pace of growth that characterises objective culture. The individual is likely to feel a sense of inadequacy when confronting this totality, and 'the individual mind can enrich the forms and contents of its own development only by distancing itself still further from that culture and developing its own at a much slower pace' (Simmel 2011, p 449). There is a seemingly endless supply of information and reading material. It is, in principal, possible to read any of the books or articles available. However, there is only so much that an individual will have the time or inclination or ability to absorb. The same logic applies when considering technological, linguistic, scientific, or militaristic development. For example, Simmel refers to the example of the soldier (subjective culture) as compared to modern methods of warfare (objective culture):

> The work of the individual soldier has essentially remained the same for a long time, and in some respects has even been reduced through modern methods of warfare. In contrast, not only the material instruments but, above all, the completely impersonal organization of the army have become extremely sophisticated and a real triumph of objective culture. (Simmel 2011, p 449)

The lack of specific knowledge about space exploration revealed in the PRC and ESA surveys highlights that individuals do not have the time, resources, or inclination to understand it in great detail. In Simmel's terms, the totality of knowledge deriving from space exploration is a form of objective culture from which individuals can only take so much. Therefore, they might engage with space news when something such as a much-publicised mission strikes a note with the public and creates a sense of collective effervescence or a sense of heightened consciousness of the fragility of the planet. However, as way of coming to terms with the limitations of subjective culture and knowledge about such matters, individuals adopt a mode of distracted enchantment. It is *distracted* because of the time, resource, and knowledge constraints that individuals face. It is a form of *enchantment* because it offers a sense of compensation for the lack of knowledge when confronting objective culture. Space exploration offers the contemplation of fundamental questions relating to the origins of the Universe. As a form of public science, it has the appeal of a collective endeavour, even if it participation occurs in distracted mode. It has an integrative function that extends beyond the nationalistic and imperialistic ambitions of the day. The notion of distracted enchantment might potentially be applied to public attitudes to other Big Science projects such as the James Webb Telescope and the Large Hadron Collider. A glance at the survey results of a public attitudes in relation to fundamental particle physics research at CERN suggests that there is broad support for this publicly funded science even if detailed knowledge of CERN's activities is limited (Florio 2018). The researchers surveyed a sample of 1005 French citizens and find that whilst public awareness of CERN's activities is lower than awareness of other organisations (such as UNESCO, NASA, WHO, and IMF), the general perception of CERN, among those who are aware of its activities, is positive, with 84% agreeing that 'scientific research carried out by CERN is important for everybody in the world, confirming the fact that knowledge advances in physics … can be considered as a "public good"' (Florio 2018, p 11).

38.4 Conclusion

In this chapter, I have argued that the Big Science of space exploration can be perceived as a collective endeavour that people are able to feel connected to, albeit in a relatively distracted and unfocused way. Space exploration involves a paradox: on the one hand it is a means to the opening up of new market segments and frontiers for imperialistic conquest. On the other, it has something about it that is akin to radical politics: 'the ends of space exploration rest on a certain orientation to the future, a possibility that what comes next might be substantially different than what happens at the present time' (Parker 2009, p 83). For example, the Apollo 11 mission, created out of American capitalism, promised 'something else … Perhaps even something sublime … The idea that our world might be different, both larger and smaller than we normally imagine, and that human beings can do extraordinary things' (Parker 2009, p 96). In this context, space exploration is appealing precisely because its purpose and outcomes are largely beyond comprehension, and so it is quite reasonable to nod approvingly, expressing a modicum of interest, before returning to the tasks at hand. However, distracted enchantment also relies on the sense that space exploration is something that in an intangible way, everyone owns. Strangely, despite its embeddedness in the military–industrial–space complex, space exploration as public science offers a sense of collective enchantment, especially during the moments when a specific space programme captures the public imagination. At other times, distracted enchantment ensures its popularity and the sense that as an intangible aspect of objective culture it somehow belongs to everyone. However, it remains to be seen whether the recent activities of billionaire space tourists such as Richard Branson and Jeff Bezos, with their ultra-conspicuous consumption, will change the public mood in relation to space exploration by projecting their narcissistic fantasies and making all too apparent the wealth and status inequalities that characterise social life in the 21st century.

References

Benjamin W 1968 *Illuminations* (New York: Schocken)

Bourdieu P 2020 The forms of capital *The Sociology of Economic Life* (London: Routledge)

Carroll J 2008 *Ego and Soul: The Modern West in Search of Meaning* (Melbourne: Scribe)

Chatzitheochari S and Arber S 2012 Class, gender and time poverty: a time-use analysis of British workers' free time resources *Br. J. Sociol.* **63** 451–71

Congress House Committee on Science and Astronautics 1973/1974 *NASA Authorization Hearings, Ninety-Third Congress, First Session* H.R. 4567 (Superseded by H.R. 7528) Washington, DC http://hdl.handle.net/2027/mdp.39015084762718

Dickens P 2009 The cosmos as capitalism's outside *Sociol. Rev.* **57** 66–82

Dickens P and Ormrod J 2007a *Cosmic Society: Towards a Sociology of the Universe* (London: Routledge)

Dickens P and Ormrod J S 2007b Outer space and internal nature: towards a sociology of the Universe *Sociology* **41** 609–26

Florio M 2018 Scientific research at CERN as a public good: a survey to French citizens *Future Circular Collider Publication* CERN-ACC-2018-0024 CERN http://cds.cern.ch/record/2635861

Henry H and Taylor A 2009 Re-thinking Apollo: envisioning environmentalism in space *Sociol. Rev.* **57** 190–203

IPSOS 2019 Attitudes toward space exploration *New C-SPAN/Ipsos Poll* https://www.ipsos.com/en-us/news-polls/cspan-space-exploration-2019

Jenkins R 2000 Disenchantment, enchantment and re-enchantment: Max Weber at the millennium *Max Weber Stud.* **1** 11–32 https://www.jstor.org/stable/24579711

Lambright W H 2013 Big Science in space: Viking, Cassini, and the Hubble Space telescope *Exploring the Solar System* ed R D Launius (London: Palgrave Macmillan) pp 129–48

Lévy J-D, Lancrey-Javal G and Gautier A 2019 Europeans and space activities: how do Europeans perceive issues related to space? https://harris-interactive.fr/wp-content/uploads/sites/6/2019/11/Harris-Results-T1-Europeans-and-space-activities-ESA.pdf

Parker M 2009 Capitalists in space *Sociol. Rev.* **57** 83–97

Parker M and Bell D 2009 Introduction: making space *Sociol. Rev.* **57** 1–5

Pew Research Center 2018 Majority of Americans believe it is essential that the US remain a global leader in space *Report* 6 June https://www.pewresearch.org/science/2018/06/06/majority-of-americans-believe-it-is-essential-that-the-u-s-remain-a-global-leader-in-space/

Simmel 2011 *The Philosophy of Money* (London: Routledge)

Statista 2020 Space industry worldwide https://www.statista.com/study/59030/space-exploration/

Stewart S 2015 Aesthetics and time: sustained and distracted modes of engagement *Cult. Sociol.* **9** 147–61

Veblen T 1994 *The Theory of the Leisure Class* (New York: Dover)

Weber M 1946 *From Max Weber: Essays in Sociology* ed C W Mills and H Gerth (New York: Oxford University Press)

IOP Publishing

Big Science in the 21st Century
Economic and societal impacts
Panagiotis Charitos, Theodore Arabatzis, Harry Cliff, Günther Dissertori, Juliette Forneris and Jason Li-Ying

Chapter 39

Preserving and displaying Big Science

Alison Boyle

Alvin Weinberg famously predicted that 'when history looks at the 20th century … she will find in the monuments of Big Science—the huge rockets, the high-energy accelerators, the high-flux research reactors—symbols of our time just as surely as she finds in Notre Dame a symbol of the Middle Ages' (Weinberg 1961).

For museums of science and technology concerned with the preservation and popularisation of the recent past, Big Science offers great opportunities for engaging visitors. Subjects such as space exploration, particle physics, and astronomy already carry a high public profile. The political and economic significance of major projects and specialised facilities is often a source of national and local pride.

But Big Science also poses curatorial challenges. Practically, its monuments are far too large to preserve and display in their entirety, but it is the intellectual challenges that are particularly pressing. In museums that have traditionally tended towards memorialising individuals and moments of 'discovery', what sorts of artefacts can capture the range and reality of today's large-scale scientific enterprises—hundreds or even thousands of personnel with specialised subdivisions of labour, projects which can span generations, and the increasingly physically distributed nature of work?

This essay will explore some examples of how public and private institutional collections have been tackling the preservation and interpretation of Big Science, with particular reference to CERN's collecting practices and the London Science Museum's 2013–17 touring exhibition on the Large Hadron Collider.

39.1 Introduction

At the Musée des Confluences in Lyon, France, the *Societies: human theatre* gallery explores how knowledge is organised and exchanged by societies via a selection of objects drawn from different world cultures. Amongst these an object of Big Science stands out amongst the displays—a Cockcroft–Walton generator made by Phillips in

1958 (Derolez 2020). Cockcroft–Walton generators are frequently seen in museum galleries. Prominently displayed examples can be found in Edinburgh as a highlight of the National Museum of Scotland's *Grand Gallery* (refurbished in 2011), in London in the Science Museum's flagship *Making the Modern World* gallery (2000), and in Milan as the centrepiece of the Museo Nazionale Scienza e Tecnologia Leonardo da Vinci's *Extreme* particle physics gallery (2016). Their ubiquity is not surprising: they are common in physics practice, of an age when items discarded by practitioners are available for collections, visually imposing with a large ladder-like structure, but of a size that can still be fit into a large museum gallery.

In many cases, the details of manufacture and operation of the particular generators on display take second place in the interpretation to what they are used to symbolise—the original Cockcroft–Walton 'atom splitting' experiment (despite that being a different experimental set-up), and the Manhattan Project development of the atomic bomb. The challenge for curators is to square grand narratives of Big Science with what its artefacts can tell us.

39.2 Big Science: advantages and challenges for museum storytelling

As Ludmilla Jordanova has noted, science loves its heroes, and even the most sceptical curators understand the power of name recognition and compelling stories in attracting visitors to museums (Jordanova 2015). For large-scale enterprises such as space exploration or particle physics, this heroism often transcends the many individuals involved—the monumental technology itself is a draw (Jordanova 2015).

Of the science and technology museums which consistently top the world's most-visited attractions, space exploration is a prominent part of the exhibits (AECOM 2021)[1]. Narratives of progress and national exceptionalism have tended to dominate, from the 'right stuff' and 'red stuff' of the US–Soviet space race, to China's twenty-first century missions (Ellis 2020, Lewis 2005, Schäfer and Song 2017). However in recent years many museums have been adopting a more internationalist approach and seeking to explore a wider range of roles from the thousands of people involved in space science and technology (Small 2017, Science Museum Group 2019). The challenge for science museums, particularly those that retain a collection-focused rather than science centre approach, is how to incorporate the artefacts of Big Science into the story.

One key obstacle is of course bigness—spacecraft, rocket first stages, and particle detectors simply dwarf most museum buildings, and it can be difficult for curators to find a component that can meaningfully speak for the whole. On its own, a single photomultiplier tube from Japan's Super-K neutrino detector lacks the eloquence of Andreas Gursky's famous *Kamiokande* photograph of the half-filled underground tank with two small human figures in dinghies against a wall of thousands of tubes (Gursky 2007). Toyko's Miraikan museum attempts to capture some of this sense of scale with a tube displayed in front of a recreation of a segment of the detector

[1] China Science and Technology Museum and Shanghai Science and Technology Museum have dominated the AECOM lists in recent years, while the US National Air and Space Museum, usually a perennial favourite, has been partially closed for refurbishment.

(Miraikan n.d.). Many twentieth and twenty-first century artefacts also tend to suffer from being 'black boxes', and while arguably these are not necessarily more confusing for the average museum visitor than the more visually attractive scientific instruments of earlier times (Boyle 2020a), space in museum galleries and stores is limited and so curators must make strong arguments for preserving items in the long term.

39.3 Preserving Big Science: why?

When considering the preservation of the material culture of Big Science, the first question might be: why bother? If there is one thing the great machines of Big Science generate it's paperwork—every aspect from initial scoping to construction of the large-scale scientific and technological projects of the twentieth and twenty-first century tends to be exhaustively documented, and during operational life there are thousands of published academic works. The situation has been amusingly summed up by curator and historian of computing Doron Swade as: 'one of the best definitions I have heard of a "high tech" object is one for which the documentation occupies more space than the object' (Swade 1991). So, do the objects of Big Science need to be preserved for the historical record?

Curators, by nature, prefer objects to documents, and Swade has previously called for 'high tech' objects to be preserved for 'indefinite interrogation in the light of unforeseen circumstance', wary of consigning objects to the scrap heap when future study methods might elicit information that the documentation cannot provide (Swade 1991, p 22). However, it is simply impossible to preserve everything—large or small—so museums and other repositories must make pragmatic choices as to which objects should survive for the future. David DeVorkin, longstanding curator at the US National Air and Space Museum, is also no stranger to wrestling with the practicalities of preserving large volumes of material, often very large things. Considering both the intellectual and symbolic value of museum objects, he proposed five functions an object can perform. They are validation: material proof that something happened; celebration: providing a sense of transcendence from physically encountering an object that made history, and lending weight to memorial events; inspiration: providing evidence of challenges overcome, and potentially useful for illustrating principles of science and technology; illumination: preserving something about a historical event that would not be available otherwise, and exploration of the reasons behind an object's survival acting as a prompt to historical research; and stimulation: the very act of preserving an object prompts efforts to learn about it, and therefore to preserve relevant documentation (DeVorkin 2006).

In terms of validation and illumination, it is rare for an object of Big Science (of any size) to provide crucial historical evidence that is unavailable elsewhere (Agar 2013). Historians of material culture have long recognised that objects with such primacy in historical research tend to be those from much earlier periods where no documentation survives and the objects are our only sources (Corn 1996). But paying close attention to objects of recent make can allow us to pay what DeVorkin terms ‘attention to the fine structure of nuts-and-bolts history’, encouraging historians to interrogate documents even more carefully and acting as a stimulus to research and writing that enhances scholarship on broader themes (DeVorkin 2006, Boyle 2020b).

And for many institutions—particularly museums—historical research is just one facet to an object’s storytelling potential. Celebration and inspiration are important functions, both in institutions of Big Science and in museums, particularly those with a pedagogical function. Institutional and even disciplinary identities are often built on the folk memories of what went before, even if these may differ from the perspective of historians (Pestre 1999). And, as Roland Wittje argues, professional archivists and curators often follow behind the scientists, secretaries or technicians who make the first decisions about what is kept—their practices and motivations will often influence the stories that historians and curators can find about objects in the future (Wittje 2013). Once equipment is no longer useful for research or teaching it may be kept for a variety of reasons: plans for repurposing, emotional attachment, a sense that it may be historically important, or simply being forgotten and not taking up needed storage space (Wittje 2013). The example of CERN is a case in point.

39.4 Creating collections: the case of CERN

At CERN a collection which was initially amassed for the sorts of reasons Wittje identifies has begun to become more formalised. In recent decades collecting has been moving from what Susan Pearce terms ‘souvenir’ mode—nostalgic recollections of past experience of individuals and groups—to a more ‘systematic’ mode with a role for wider public display (Pearce 1992). This is no accident—CERN’s public engagement team consciously used the start-up of the Large Hadron Collider as an opportunity to reboot public engagement with particle physics, and has historically had close links with the exhibition team (Gillies 2006, Dorey 2016). Interest from museums in covering the LHC, including the London Science Museum’s 2013 international touring exhibition covered later in this essay, helped to reinforce the value of the collection as part of CERN’s public output.

Objects can be found in buildings all over the site; they include those made and owned by CERN, and those which officially belong to the experiments but are stored at CERN. Ad hoc survival due to an individual’s nostalgia (or an object having been simply overlooked) is attested to by the presence on CERN’s database of several ‘mystery objects’ whose function has been forgotten (CERN 2017)[2]. Other objects have survived due to practicalities—during a visit to CERN in 2012 I saw the DELPHI detector from the Large Electron Positron Collider (LEP) accompanied by a display panel. This is a rather curious display, as it can only ever be seen by relatively few members of the public (such as VIP guests and CERN Open Day tour groups) during operational shutdowns when it is possible to access the underground shaft where it sits. The display is a result of the object being too difficult and

[2] The list was initially assembled in 2017; as of October 2021 some objects had been identified but others remained ‘unknown’.

expensive to extract to ground level; instead it was moved off to the side of the beam line (the space is now occupied by the Large Hadron Collider's LHCb detector). It is in what Lourenço and Gessner term the 'limbo' stage of scientific instrument life, lingering half-forgotten (Lourenço and Gessner 2014). It has at least survived; another LEP detector, L3, was not spared, with some parts being re-used for the LHC's ALICE detector and most scrapped.

Some larger 3D objects which have survived have become semi-public display items, moved to concrete stands in an outdoor area visible from CERN's administration building. Including the Gargamelle bubble chamber, a hero of CERN mythology for its role in the 1973 discovery of neutral currents, these objects are displayed as 'fossils ... of venerable dinosaurs' (CERN 2014).

At the time of writing this chapter, the public audience at CERN could access the collection in two permanent display areas, Microcosm near the visitor reception and Universe of Particles in the Globe of Science and Innovation, a striking wooden dome repurposed from the Expo 2002 site at Neuchatel to provide a centrepiece for CERN's public activities. Microcosm closed in late 2022 and a new flagship Science Gateway education and outreach facility was due to open in 2023. CERN's small professional exhibition and outreach team has oversight of these areas. In recent years less formal exhibition display spaces, mostly created by physicists and engineers, also sprang up across the CERN site, for example at the entrances to the four main experiments and in the magnet testing hall. While the level of interpretation and design polish varies between display sites (although part of the role of CERN's Heritage Committee is to create more internal coherence), the role of all is largely to host education groups and political stakeholders to create enthusiasm and support for CERN's work.

Increasingly, CERN's policies of collecting objects are becoming more formalised. The year 2017 saw the establishment of the Heritage Committee and the advertising of a Heritage Collection for use in displays at sites around CERN and for loan to museums and science centres (Warakaulle 2017, Raynova 2017). Notable objects include the NeXT computer used by Tim Berners-Lee as the first World Wide Web server, on loan to the London Science Museum's *Information Age* gallery since 2014. CERN's Heritage Action Plan principles of openness and collaboration, disseminating its research to a wider audience, and 'maximising the positive impact for CERN' (Sanders 2016). Thus, in the nascent collection plan we can see the transition from ad hoc collecting motivated by nostalgia and pragmatism to a 'corporate museum' type of collection—an artefactual organisational memory that is actively deployed in the development of image and identity, for the organisation's own staff and for a wider public audience (Nissley and Casey 2002). A search of the database indicates that the collection mainly seems to consist of hardware and a few models, without the more everyday or ephemeral material that a social history approach to collecting might capture, although this may change as the heritage policy develops. A blend of hardware and ephemera was used by the Science Museum in its touring exhibition *Collider: step inside the world's greatest experiment* (2013–17), which influenced subsequent displays at CERN's *Microcosm*.

39.5 Elements of Big Science: exploring the Large Hadron Collider at the Science Museum

Museums, which have traditionally tended towards memorialising individuals and 'discovery' objects, might seem particularly ill-suited to Big Science. Capshew and Rader remark that the prevalent late twentieth century view of Big Science 'entails a drama of scale that juxtaposes the huge machines, large organisations, and massive expenditures found in some contemporary research projects with the stereotyped lone investigator of the past, working with the proverbial sealing wax and string in a private laboratory' (Capshew and Rader 1992). But the string and sealing wax model was no more an accurate reflection of the realities of laboratories in J J Thomson's day than it is now (Hughes 1998). Avoiding lone genius narratives is all very well, but decades of museum evaluation show that if there's one thing museum visitors want it's stories about people, so the challenge of covering Big Science is to bring out individuals amongst the large teams (including those with managerial and administrative skills) and multi-authored papers of today's large-scale science projects. Centring on a major facility such as CERN provides an opportunity to consider different communities who coalesce around the lab—onsite and around the world—as 'a sort of collaborating organism' (Knorr Cetina 1992) and to examine some of the different subcultures of physics—engineers, experimenters, computer programmers, theorists (Galison 1997).

Collider: step inside the world's greatest experiment ran at London's Science Museum from November 2013 and was followed by a world tour. Led by the author and one of the editors of this volume (Harry Cliff), the exhibition team was tasked with improving the Museum's coverage of fundamental science topics, creating an offer that would appeal to independent adults, and launching a new programming strand of high-profile ticketed special exhibitions (Science Museum 2012). The global brand recognition of CERN and a climate of public interest in the Large Hadron Collider made this a subject likely to appeal to visitors around the world, but the usual selling point of museum blockbusters—being able to show real objects—was not so pronounced this time: with the LHC operational during 2012–13 only a handful of prototypes or spare components was available for loan from CERN. While some of these were visually attractive in their own right, others looked more like humdrum pipes or pieces of electronics; it would also be difficult to explain their operation and how they fit into the much larger LHC and detector infrastructure. Meanwhile, the object most closely associated with the LHC in the public mind, the elusive Higgs boson, was intangible. Overall, this was an exhibition featuring technology much too large to fit in a museum, seeking particles much too small to be seen in a museum.

The lack of available objects, and a bank of audience research indicating that previous exhibitions on the subject had failed to capture the overall ambition and scale of the LHC (Hobson 2012), made it clear from the outset that a traditional museological approach of showcased objects with panels and labels would not be successful. The Science Museum team therefore sought a different approach, with input from professionals in other creative sectors. Thus, in addition to the usual specialisms such as content, design, and construction that contribute to a large exhibition project, *Collider* included a creative team who had experience of theatre rather than museums. Research visits to CERN interviewing dozens of personnel and taking in a range of sites and functions (including detector caverns, building services management, and the canteen) convinced all professional specialists that the experience of visiting the large-scale CERN site was of more interest than considering the physics phenomena in isolation. From this author's point of view it would allow the exhibition to be informed by work in the history and sociology of particle physics, while for Cliff it allowed a science communication based take on his present-day experiences of working at CERN (where, in his specialised role as an experimental physicist working on one of the detectors, he was almost as unfamiliar with the accelerator engineering as the average museum visitor would be). The creative team was enthused by the possibilities of capturing aspects of CERN's distinctive architecture and design.

Although this was not an artefact-led exhibition in the traditional sense, ultimately the visitor experience was strongly shaped by considering the LHC as a whole artefact. On one of the team's research visits Control Room shift leader Rossano Giacchino talked about the cycle of a typical eight-hour shift with beams in the collider (going through a cycle of injection, acceleration, collision, and dumping) and describing how some of the more experienced shift leaders tend to rely on intuition as well as data from their control decks—'feel the beam'. This became the basis of the exhibition storyline, with visitors following the LHC beam around different parts of CERN. In addition to providing a way for the available objects to play strong roles in the narrative rather than merely illustrating concepts, the approach gave visitors a logical sequence to follow and the ability to build on knowledge as they went around the exhibition.

The exhibition creative team was led by Nissen Richards Studio (NRS), a design and architectural firm with expertise both in museum and theatre scenography; a particularly strong element of NRS's bid was the inclusion of several theatre professionals including playwright Michael Wynne and video designer Finn Ross, both Olivier Award winners. The exhibition opened with a dramatised film scripted by Wynne, starting in CERN's Lecture Theatre on the day the Higgs discovery was announced (this having conveniently occurred just in time to be incorporated into the exhibition planning) and moving onto the LHC Control Room where the stages of the beam were introduced. The characters in this film were composites of people at CERN that the team had interviewed, and played by actors. After viewing the film, visitors could walk around recreated parts of the LHC tunnels, with objects from the LHC accelerator chain embedded in the displays. These included a hydrogen bottle canister used for injecting protons into the LHC, segments of the LHC's dipole and quadruople magnets, and also—most familiar to visitors—a bicycle from CERN's pool, used to get around the LHC ring during construction.

The creative team recreated these spaces using large wall panels printed with detailed photographs. In an attempt to create an authentic 'CERN visit' feel, the graphic designers Northover & Brown deliberately avoided traditional museum signage and labels: scientific concepts were delivered via diagrams on whiteboards—

ubiquitous items at CERN—while object labels were in the style of equipment data sheets (for more on the recreation of spaces, see Boyle and Cliff 2014).

The dramatic high point of the visitor journey was provided by Ross, with a large-scale 'Collision' animation recreating the aspects of the LHC too large or small to display—visitors could zoom down through the CMS detector (based on technical drawings provided by the CMS team) and into the invisible particle world; the animation then showed how data from CERN are distributed around the world for processing. Here the visuals were accurate as far as could be described by known physics, but Ross had plenty of artistic licence; the team ultimately agreed that particles could be yellow, in the absence of any available evidence to the contrary. After the collision, another object display area showed detector components—here the scenery was a less literal recreation than the tunnel spaces, but was heavily informed by CERN's equipment testing labs, with the object showcases set into stylised workbenches. Visitors then moved onto a depiction of the work of data analysis, with a faithful recreation of a corridor in CERN's Building 2 (home to Tim Berners-Lee and Robert Cailliau during the development of the first web interface) and an office space with another film scripted by Wynne featuring a PhD student who had done crucial software analysis for the Higgs discovery. The 'CERN visit' ended in a more abstract-style display area discussing future directions for particle physics after the Higgs discovery and the next generation of collider to follow the LHC. Throughout the exhibition, in addition to the composite characters in Wynne's scripts, visitors encountered a range of real CERN personnel, including magnet engineer Marta Bajko and CMS experiment leader Jim Virdee, via film and audio. These snippets, based on longer interviews, covered topics such as developing instruments from the prototype stage, and different working styles at CERN.

Post-opening visitor evaluation showed that the exhibition met the Museum's aims of illustrating the large-scale and international nature of Big Science. The interpretation was particularly successful in communicating the engineering challenges and the global scale of the LHC project, and the human side of CERN—the corridor and recreated office were particularly popular, although some visitors were disappointed to realise that CERN is quite shabby (Boyle and Cliff 2014). This seems to have been a result of the overall exhibition environment, rather than being reactions to any particular objects in the space, not surprising with a 'keypiece' approach where objects are part of a suite of interpretation (Maas 2013).

In terms of long-term preservation, the Science Museum did not manage to retain as many objects from CERN for its permanent collection as initially hoped, since the objects were now required for CERN's corporate collection. However the exhibition has had a longer-term legacy as many aspects of its interpretation style were adopted for CERN's own displays in the *Microcosm* display area. This 'people and process' approach, contrasting with the more phenomena-led approach in CERN's *Universe of Particles* displays, shows that it is possible to cover some of the nature of Big Science in museums and displays, although it's best not to rely solely on technical objects.

39.6 Curating Big Science: lessons

An ethnographic approach to collecting, capturing experiences and ephemera as well as technological artefacts, could be deployed across many Big Science projects. In astrophysics, the global effort of the Event Horizon Telescope group to capture the first direct image of a black hole has a great story arc from data capture by global network of telescopes, to transmission of data—including physical shipping of hard drives—to data processing, to media splash (Boyle 2020a). The Square Kilometre Array, which will also involve sites around the world and significant data requirements, has particularly interesting political and social issues around the project, including debates over appropriate land use in the post-colonial (and in South Africa post-apartheid) landscapes of the telescope sites, where any local activities causing radio interference would be restricted. While the ethnographic approach is well suited to Big Science, it is also applicable to smaller-scale enterprises that still involve multidisciplinary groups concentrating around equipment (Westfall 2003). Such an approach, detailed but also highly personal, could perhaps address the challenges noted by Derolez in preserving and presenting Big Science as heritage—to capture the singularity and complexity of each individual artefact (with its own particular scientific, technical, and social status) with what it symbolises about politics, or the aesthetic response it elicits (Derolez 2020).

Ultimately, Big Science has many important ingredients of a compelling museum visitor experience: large visually imposing objects, compelling story arcs of vastly ambitious projects with the challenges and successes these bring, and global stories. But, as noted earlier, the single most important ingredient for engaging museum visitors is stories about other people, so we must balance the Big with the human scale. For the *Collider* exhibition, the everyday human interactions in CERN's offices and corridors were as important as the LHC's giant superconducting magnets. It is a balance well illustrated by *Cosmonauts: Birth of the Space Age* at the Science Museum in 2015. Considered the Museum's most successful temporary exhibition of recent years due to a combination of critical acclaim and commercial success, the exhibition framed the Soviet and Russian exploration of space with a strong focus on the stories of people on the ground. Amongst striking technical artefacts such as Valentina Tereshkova's Vostok 6 capsule or the LK-3 lunar lander were personal items. One of the most affecting objects in the whole exhibition was Chief Designer Sergei Korolev's aluminium drinking mug from the Siberian Gulag. In the end, Big Science is one of the symbols of our time, and it's all about us.

Acknowledgements

Thanks to the Collider project team and colleagues at the Science Museum, CERN and the Department of Science and Technology Studies, University College London.

References

AECOM 2021 AECOM theme index *AECOM* https://aecom.com/theme-index/

Agar J 2013 What's your CHOICE? *UCL STS Observatory* https://blogs.ucl.ac.uk/sts-observatory/2013/11/26/whats-your-choice/ (Accessed: 1 July 2018)

Boyle A 2020a *Festschrift:* of mice and myths: challenges and opportunities of collecting contemporary science in museums *Sci. Mus. Group J.* **13**

Boyle A 2020b Stories and silences in modern physics collections: an object biography approach *Doctoral Thesis* University College London https://discovery.ucl.ac.uk/id/eprint/10113201/

Boyle A and Cliff H 2014 Curating the collider: using place to engage museum visitors with particle physics *Sci. Mus. Group J.* **2**

Capshew J H and Rader K A 1992 Big Science: price to the present *Osiris* **7** 2–25

CERN 2014 CERN Courier Archive 1971 *CERN Document Server* https://cds.cern.ch/record/2064550 (Accessed: 8 June 2018)

CERN 2017 DE mystery objects *CERN Document Server* https://cds.cern.ch/yourbaskets/display_public?bskid=23340&ln=en (Accessed: 8 June 2018)

Corn J J 1996 Object lessons/object myths? What historians of technology learn from things *Learning from Things: Method and Theory of Material Culture Studies* ed W D Kingery (Washington, DC: Smithsonian Institution Press) pp 35–44

Derolez S 2020 Instrument de la 'big science' et biographie culturelle *Cult. Mus.* **36** 191–6

DeVorkin D H 2006 Space artifacts: are they historical evidence? *Critical Issues in the History of Spaceflight* ed J D Steven and D L Roger (Washington, DC: National Aeronautics and Space Administration, Office of External Relations) pp 573–600

Dorey J 2016 When public relations and particle physics collide: an ethnographically informed account of life in the CERN Communications Group *PhD Thesis* The Open University http://oro.open.ac.uk/47760/

Ellis T 2020 Curating the space race, celebrating cooperation: exhibiting space technology during 1970s détente *Eur. J. Am. Cult.* **39** 275–95

Galison P 1997 *Image and Logic: A Material Culture of Microphysics* (Chicago, IL: University of Chicago Press)

Gillies J 2006 Letter from the Guest Editor: LHC *Symmetry Magazine* 8 January https://www.symmetrymagazine.org/article/august-2006/letter-from-the-guest-editor

Gursky A 2007 2007 Selected works—Kamiokande *Andreas Gursky* https://www.andreasgursky.com/en/works/2007/kamiokande

Hobson M 2012 LHC and CERN: a summary of the public's knowledge, understanding and interest *Internal report* Science Museum

Hughes J 1998 Plasticine and valves: industry, instrumentation and the emergence of nuclear physics *The Invisible Industrialist: Manufactures and the Production of Scientific Knowledge* (Basingstoke: Macmillan) pp 58–101

Jordanova P L 2015 On heroism *Sci. Mus. Group J.* **1**

Knorr Cetina K 1992 The couch, the cathedral, and the laboratory: on the relations between experiment and laboratory in science *Science as Practice and Culture* ed A Pickering (Chicago, IL: University of Chicago Press) pp 113–38

Lewis C S 2005 The birth of the Soviet Space Museums: creating the earthbound experience of space flight during the golden years of the Soviet Space Programme, 1957–68 *Showcasing SpaceArtefacts: Studies in the History of Science and Technology* ed M Collins and D Millard (London: Science Museum) pp 142–58

Lourenço M C and Gessner S 2014 Documenting collections: cornerstones for more history of science in museums *Sci. Educ.* **23** 727–45

Maas A 2013 How to put a black box in a showcase: history of science museums and recent heritage *Stud. Hist. Philos. Sci. Part* A **44** 660–8

Miraikan n.d. Understanding the universe through neutrinos *Miraikan—The National Museum of Emerging Science and Innovation* https://www.miraikan.jst.go.jp/en/exhibitions/world/neutrino/ (Accessed: 31 October 2021)

Nissley N and Casey A 2002 The politics of the exhibition: viewing corporate museums through the paradigmatic lens of organizational memory *Br. J. Manag.* **13**

Pearce S M 1992 *Museums, Objects, and Collections: A Cultural Study* (Leicester: Leicester University Press)

Pestre D 1999 Commemorative practices at CERN: between physicists' memories and historians' narratives *Osiris* **14** 203–16

Raynova I 2017 Enriching CERN's heritage object collection *CERN Document Server*

Sanders E 2016 CERN heritage—an action plan *Internal document* CERN

Schäfer D and Song J-O 2017 Interpreting the collection and display of contemporary science in Chinese science museums as a reflection of science and society *Challenging Collections: Approaches to the Heritage of Recent Science and Technology Artefacts: Studies in the History of Science and Technology* ed A Boyle and J-G Hagmann vol 11 (Washington, DC: Smithsonian Institution Scholarly Press), pp 88–103

Science Museum 2012 Strategic Ambitions 2012–22 https://www.sciencemuseumgroup.org.uk/wp-content/uploads/2017/06/Science-Museum-Strategic-Ambitions-2012-22.pdf

Science Museum Group 2019 Culture-space: towards a better understanding of space exploration rationale *Science Museum Group* https://www.sciencemuseumgroup.org.uk/project/culture-space-towards-a-better-understanding-of-space-exploration-rationale/

Small L M 2017 Space museums: technical and cultural considerations *Thesis* University of Toronto https://tspace.library.utoronto.ca/handle/1807/76674

Swade D 1991 Napoleon's Waistcoat button: modern artefacts and museum culture *Museum Collecting Policies in Modern Science and Technology: Proc. of a Seminar Held at the Science Museum, London, 3 November 1988* (London: Science Museum) pp 19–26

Warakaulle C 2017 Mandate and terms of reference of the CERN Heritage committee *Internal document* CERN

Weinberg A M 1961 Impact of large-scale science on the United States *Science* **134** 161–4

Westfall C 2003 Rethinking big science: modest, mezzo, grand science and the development of the Bevalac, 1971–1993 *ISIS* **94** 30–56

Wittje R 2013 The Garching nuclear egg: teaching contemporary history beyond the linguistic turn *Stud. Hist. Philos. Sci.* A **44** 683–9

IOP Publishing

Big Science in the 21st Century
Economic and societal impacts
Panagiotis Charitos, Theodore Arabatzis, Harry Cliff, Günther Dissertori, Juliette Forneris and Jason Li-Ying

Chapter 40

Sublime giant machines—artistic reflections between Big Data and a new Grand Tour

Matias del Campo and Sandra Manninger

40.1 A *ménage à trois*—science, data, and the sublime

The Grand Tour was considered the ultimate rite of passage for the cultured gentleman or lady of the eighteenth century. This journey could last up to several years and, most importantly, it included roaming the Italian peninsula to visit, admire, and absorb the remnants of classical antiquity. One significant aspect of these travels comprised scaling the Alps on the passage to Italy (figure 40.1). This experience was sublime and terrifying at the same time—the giant dimensions of nature dwarfing the traveler's own existence [1]. The idea of the sublime became a mainstay in the discussion of the relationship between humans and nature [2]. Today, the sublime has been extended by the presence of Big Science, Big Data, and the colossal machines that interrogate nature at the most intimate levels. It is almost ironic that CERN, the home of the largest machines in service of science, is deep within the mass of the Alps, as if to remind the visitor to these gargantuan machines about the history of the intellectual relationship between the efficacy of science and the overwhelming scale of nature [3]. This new condition is highly inspiring as an educational setting, provoking questions about the position of nature, science, and the role of the arts in this *ménage à trois*. In effect, it constitutes a new Grand Tour, a pilgrimage of sorts, to the Alpine region in order to engage with the dualism between scientific insight, able to shed light on the origin of the Universe, and fundamental artistic questions that explore the nature of human agency, creativity, and authorship. In this frame of thinking, the existence of Big Science—and thus Big Data—opens new opportunities for the arts to reflect on the role of humans in an ecology where agency is shared with additional agents that emerge through the intelligent processing of massive amounts of data.

doi:10.1088/978-0-7503-3631-4ch40

Figure 40.1. *The Grand Tour. St Gotthard Pass, The Devils Bridge*, J M W Turner 1803. This [Joseph Mallord William Turner 028.jpg] has been obtained by the authors from the Wikimedia website, where it is stated to have been released into the public domain. It is included within this article on that basis.

40.2 Scaling the Alps and the quest for data

The Grand Tour was not just the beginning of tourism in the eighteenth and nineteenth century, it also presented an educational opportunity. It represented a method of data gathering by experience. We are talking about a time of data scarcity. Science had to rely on methods of data compression in order to be able to think through a problem, and applications were limited to the ones that could be rationalized for fast performance with a small set of data [4, p 19]. Think, for example, about the vast number of mathematical tables collected in books (figure 40.2). These solutions of trigonometric functions and tables of logarithms aided, for example, in nautical navigation and the calculation of artillery's ballistic trajectory [5], until the middle of the twentieth century—their most significant detriment: calculation errors (*errare humanum est*!). These faulty mathematical tables led, in the early nineteenth century, directly to the development of the difference engine, most notably by Charles Babbage [6]. Astonishingly, around the same time German philosophers[1] and artists [7] were scrutinizing the idea of the Sublime; developing their own intellectual frameworks for the observation and understanding of nature. While Babbage used brass and steel, Schopenhauer used words to ask the same questions. The difference engine, able to interpolate or tabulate functions by applying a limited set of polynomial coefficients, represents a first step towards the ability to operate with large datasets—the emergence of Big Data. First, however, it is necessary to obtain the data in order to be able to process it.

[1] In particular Arthur Schopenhauer in [3].

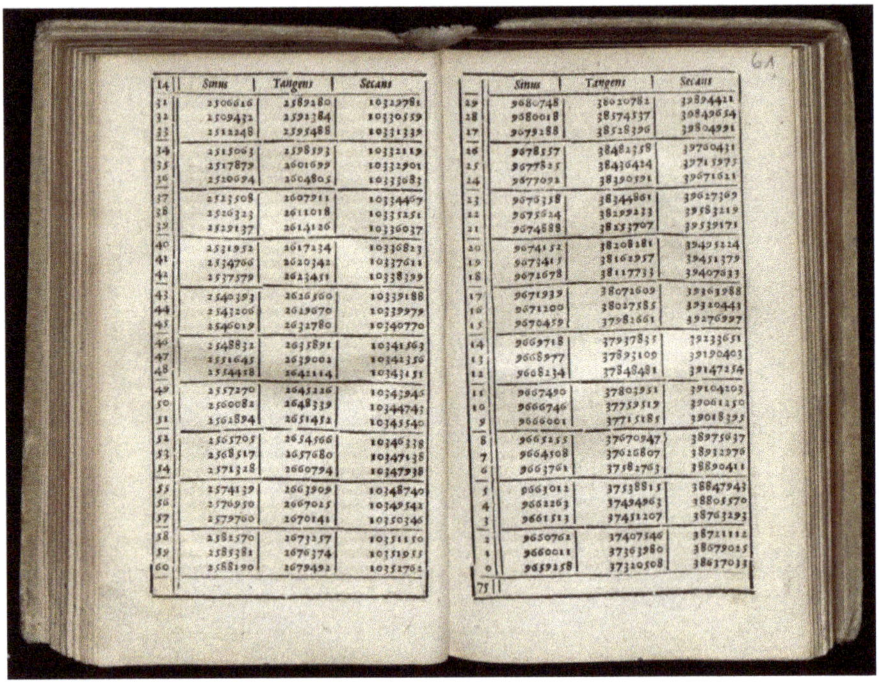

Figure 40.2. An example of a mathematical table collection from the seventeenth century. Matthias Bernegger 1619 *Manuale Mathematicum* (Strassburg: Anton Bertram). This [Bernegger Manuale 136.jpg] has been obtained by the authors from the Wikimedia website, where it is stated to have been released into the public domain. It is included within this article on that basis.

As Mario Carpo, architecture historian at the Bartlett UCL, put it: 'Writing, print, and other media have made information more easily available.... Yet until a few years ago, the culture and economics of data were strangled by a permanent, timeless, and apparently inevitable scarcity of supply: we always needed more data than we had' [4, p 10].

At this point, a brief account of the meaning of Big Data. Big Data is a relatively novel field of inquiry that interrogates ways to systematically analyse, organize, and mine information from large-scale datasets that are too big to be explored with conventional data processing methods, such as balancing and shuffling [8]. Volume, variety, and velocity can be considered vital concepts that can be applied to Big Data. These aspects allow harnessing the potential of Big Data to perform, for example, user behavior analytics, predictive analytics, and other analytic methodologies that extract value from Big Data. In this role, the algorithms applied to search for correlations in the data can be used to 'spot business trends, prevent diseases, combat crime and so on' [9]. There are multiple examples as of how Big Data can contribute to deeper knowledge, resulting in paradigmatic shifts in areas such as diagnostics [10], meteorology [11], genomics [12], connectonomics [13], and numerous other areas.

Fortunately for all these areas, the number of available datasets has grown at an almost exponential rate in recent years, collected by mobile devices, software logs, cameras, microphones, radio-frequency-identification (RFID) readers, WIFI networks, and many other data harvesting methods. Using the ever-so-popular Internet scraping methods, it has become easy to harvest data. By the year 2025, the International Data Group (IDC), estimates that the global data volume will reach 163 zettabytes(!) [14]. However, the wild harvesting of Big Data also contains a tale of caution, as there are profound ethical implications involved in data harvesting, particularly in data labeling. As Kate Crawford points out in her book *Atlas of AI* [15] there are several pitfalls. One example is the way that images are labeled in ImageNet [16], the world's largest labeled image dataset and the bedrock for multiple machine learning (ML) tasks in operation today. The artist Trevor Paglan demonstrated these underlying problems with his project ImageNet Roulette [17] (2019) that dared visitors to upload an image of themselves to an app that would show how they would be classified based on ImageNet's person categories. This project exposed the underlying inclusion of racist and sexist terms in the labeled images. Consequently, ImageNet removed 1593 of the 2832 categories describing a person's traits in an attempt to reduce the inherent bias in their labeling algorithm [18].

Apart from the opportunity to use Big Data as a canvas for critical reflection, artists have a broad experience with the processing of data and methods to extract information from large amounts of data—even if it is done unconsciously. Let me illustrate this notion with an example. Traditional forms of art such as oil painting can teach us something about Big Data. Oil paint consists of granular solids called pigments, suspended in a binder, commonly linseed oil. The pigments themselves can be described as a form of data, which is laid onto the canvas in an informed way by the artist. The result can be the realistic representation of a smiling person, such as in the *Mona Lisa* or the tangled hypnotic jungle of whirls and splashes present in a Jackson Pollock painting. This is what we as humans extract as information from a canvas. It is surprising that despite the fact that artistic expression can vary so much, there is only a limited number of ways we can put oil paint on a canvas. If we expand this thought, the information created by artists in the form of artwork can itself serve again as data. In doing so, contemporary artists can mine the vast storage vaults of art history in order to supply data to neural networks able to recognize edges, corners, and pixel blobs. The laying of pigments on a canvas transforms into the way pixels are organized by neural networks, which learn the salient features from a set of images. Neural networks such as generative adversarial networks (GANs) [19] are then able to generate artworks—sometimes strange, sometimes alien, almost always provoking (figure 40.3).

40.3 The allure of latent spaces and dark caverns

Like the example of the oil paint, the vaults of the arts—the collection of all known artworks—are vast, yet finite. These data are not endless, but the number is high enough that machines can be trained to produce images. The question is how to use this ability, which can be translated to any other art form, be it music, poetry, or

Figure 40.3. 'Strange Family Members' A set of Neural Portraits based on the portrait collection of the UMMA, SPAN (Matias del Campo, Sandra Manninger) 2021. Copyright SPAN - Matias del Campo & Sandra Manninger 2020.

architecture. We can demonstrate a possible response to this question with the example of the project *Latent Carceri.* Giovanni Batista Piranesi was an artist, architect, and one of the first examples of what we understand today as a classical archeologist. Born in 1720 in Venice, he emerged as one of the leading figures in the arts of the Baroque era. He became primarily famous for his etchings of Rome and the atmospheric renderings of fictitious brooding cavernous 'prisons', i.e. *Le Carceri d'invenzione* (1749 and 1761). Befitting to the central motif of this essay, he became known outside Italy for providing travelers of the Grand Tour with souvenirs in the form of portfolios with his illustrations and etchings. The authors of this essay created a dataset of 1427 samples (admittedly a relatively small dataset) of Piranesi's *Carceri* and ran it through a StyleGAN2 [20] algorithm that reveals the hidden connections between the salient datapoints of these images (figures 40.4 and 40.5).

Figure 40.4. Giovanni Battista Piranesi, The Smoking Fire, 1761, etching and engraving printed in black ink on laid paper. Detroit Institute of Arts, Founders Society Purchase, Hal H. Smith Fund and Elizabeth P. Kirby Fund, 42.66. Copyright DIA, the Detroit Institute of the Arts.

The neutral network creates variations of the central motifs of the *Carceri*, such as stairs, bridges, piers, wells, and chains. None of these are as accurate as Piranesi's originals—*a priori*, not the point—but they contain the inherent atmospheric quality of the original etchings. The resulting images are strange but familiar enough to be recognized as the distant, AI-mutant siblings of the etchings by Piranesi. This ability for *defamiliarization* [21] and *estrangement* [22] is what makes these results so provocative as a basis for debate on art today. Resulting questions about the agency, authorship, and aesthetics emerging from Big Data are part of a hot debate about contemporary art and culture at large [23]. Applying these novel methods in architecture design has the potential to unleash potentialities dormant in the enormous amount of data gathered by the architecture discipline over several millennia of production. The sheer scale of this architectural repository is currently invisible to the profession, as we lack the means and organization to uncover these

Figure 40.5. Latent Carceri, SPAN (Matias del Campo, Sandra Manninger) 2021. Copyright SPAN - Matias del Campo & Sandra Manninger 2021.

potentials. It is not done just by collecting data; it has to be correctly labeled and processed for ML applications. In a proof-of-concept project, the authors designed, in collaboration with the robotics department of the University of Michigan, the Robot Garden, a test facility for robots with legs (figure 40.6). This is the first architecture project that used neural networks throughout the entire design process. Using satellite images, the design used methods such as Neural Styletransfer [24] and Deepdreaming [25] to populate the site with various geological terrains. These novel design methods mark a paradigm shift, representing the first genuinely twenty-first century methods for architecture design.

This essay started with the Grand Tour as a metaphor for the relationship between humanity and the sublime scale of nature. Today, humanity faces another sublime entity: Big Data. The vast data collections gathered by Big Science can serve as a source of inspiration for the arts. At the same time, the ML methods developed to sift through these giant repositories of potential knowledge provide the arts with novel tools to interrogate the human condition, the state of the world, or questions of agency in the age of (almost) intelligent machines. The critical interrogation of these emerging effects, the ontology as well as the epistemological considerations arising from this field, has barely started. It will provide a stage for further discussions in the foreseeable future.

Artists engaging with Big Data today are climbing many Alps.

Figure 40.6. The Robot Garden, SPAN (Matias del Campo, Sandra Manninger) and Alexandra Carlson 2019. Copyright Matias del Campo 2021.

References

[1] Burke E 1757 *A Philosophical Enquiry into the Origin of Our Ideas of the Sublime and Beautiful* (London: R and J Dodsley)

[2] Kant I 1764 *Observations on the Feeling of the Beautiful and the Sublime* ed Frierson P and Guyer P (Cambridge: Cambridge University Press)

[3] Schopenhauer A Fullest feeling of sublime—immensity of universe's extent or duration. (Pleasure from knowledge of observer's nothingness and oneness with Nature) *The World as Will and Representation* (Cambridge: Cambridge University Press) § 39

[4] Carpo M 2017 *The Second Digital Turn* (Cambridge, MA: MIT Press)

[5] Bernegger M 1619 *Manuale Mathematicum, darinn begriffen Die Tabulæ Sinuum, Tangentiu [m], Secantium,* (Strassburg: Anton Bertram)

[6] Dasgupta S 2014 *It Began with Babbage: The Genesis of Computer Science* (New York: Oxford University Press)

[7] Pinto P 2021 The experience of the sublime in the work of Caspar David Friedrich *medium.com* https://medium.com/@pspinto/the-experience-of-the-sublime-in-the-work-of-caspar-david-friedrich-d9f06ffad358 (Accessed: 21 September 2021)

[8] Leevy J L *et al* 2018 A survey on addressing high-class imbalance in big data *J. Big Data* **5** 42

[9] Kukier K 2010 Data, data everywhere *The Economist* 25 February 2010 (Accessed: 15 September 2021)
[10] Dash S *et al* 2019 Big data in healthcare: management, analysis and future prospects *J. Big Data* **6** 54
[11] Guo X 2016 Application of meteorological big data *2016 16th Int. Symp. on Communications and Information Technologies (ISCIT)* pp 273–9
[12] Tadist K *et al* 2019 Feature selection methods and genomic big data: a systematic review *J. Big Data* **6** 79
[13] Lichtman J W, Pfister H and Shavit N 2014 The big data challenges of connectomics *Nat. Neurosci.* **17** 1448–54
[14] Shirer M and Rydning J 2020 IDC's global datasphere forecast shows continued steady growth in the creation and consumption of data *IDC* https://www.businesswire.com/news/home/20200508005025/en/IDCs-Global-DataSphere-Forecast-Shows-Continued-Steady-Growth-in-the-Creation-and-Consumption-of-Data (Accessed: 21 September 2021)
[15] Crawford K 2021 *Atlas of AI* (New Haven, CT: Yale University Press)
[16] Deng J, Dong W, Socher R, Li L, Li K and Fei-Fei L 2009 ImageNet: a large-scale hierarchical image database *2009 IEEE Conf. on Computer Vision and Pattern Recognition* pp 248–55
[17] Crawford K and Paglen T 2021 Excavating AI, The Politics of Images in Machine Learning Training Sets https://excavating.ai/ (Accessed: 25 September 2021)
[18] Yang K, Qinami K, Fei-Fei L, Deng J and Russakovsky O 2020 Towards fairer datasets: filtering and balancing the distribution of the people subtree in the ImageNet hierarchy *Proc. of the 2020 Conf. on Fairness, Accountability, and Transparency* (New York: Association for Computing Machinery), pp 547–58
[19] Goodfellow I J, Pouget-Abadie J, Mirza M, Xu B, Warde-Farley D, Ozair S, Courville A C and Bengio Y 2014 Generative adversarial networks, arXiv:1406.2661
[20] Karras T, Laine S, Aittala M, Hellsten J, Lehtinen J and Aila T 2020 Analyzing and improving the image quality of Stylegan *2020 IEEE/CVF Conf. on Computer Vision and Pattern Recognition (CVPR)* pp 8107–16
[21] Shklovsky V 1991 *Art as Device: Theory of Prose* (Funks Grove, IL: Dalkey Archive)
[22] Young M 2016 The affects of realism—or the estrangement of the background *Architect. Design* **86** 58–65
[23] Dahlstedt P 2019 Big data and creativity *Eur. Rev.* **27** 411–39
[24] Gatys L A, Ecker A S and Bethge M 2015 A neural algorithm of artistic style, arXiv:1508.06576
[25] Mordvintsev A, Olah C and Tyka M 2015 Inceptionism: going deeper into neural networks https://ai.googleblog.com/2015/06/inceptionism-going-deeper-into-neural.html (accessed 20 February 2023)

Chapter 41

The art of Big Science

Ariane Koek

In the twenty-first century, Big Science laboratories are being transformed beyond being simply spaces for scientific exploration. They are becoming places which generate sources of creativity and ways of knowing other than science and technology—namely the arts. Big Science laboratories are becoming what I call 'cultural generators'—dynamic places which demonstrate how science and technology don't exist in a vacuum, but together with the arts, form the richness of human existence, experience, and ingenuity known as culture. The seeds of this change can be traced back to the birth of Big Science in the previous century, and in particular to the founding in 1967 of Fermi National Accelerator Laboratory, the US Department of Energy's particle accelerator laboratory for particle physics in Batavia, IL, known as Fermilab. Under its visionary founding director, Robert R Wilson, the arts—unusually—were a core founding principle in the creation of the laboratory. He imagined an ideal laboratory which would be architecturally impressive and artistically inspiring, contributing to the well-being of the scientists and all those employed there. This was reflected in every aspect. The architecture of the site included the iconic and dizzying twin-towered Wilson Hall, which was inspired by French gothic architecture; the colours of the buildings, design of tables and chairs, and distinct Fermilab logo (replete with symbols, bosons and bisons) were created by the laboratory's first artist in residence Angela Lahs Gonsalez, who for 30 years ruled the aesthetics of the laboratory including the colour of the buildings (Biron 2017); the grounds of the laboratory housed large-scale public sculptures, including *Alle Funi* (1978) and *Tractricious* (1988), designed by the founding director himself (Fermilab n.d.).

Thus the arts were a cornerstone of the existence of Fermilab since it began. Even the auditorium, essential for sharing new experimental and theoretical information to the international scientific community, was designed to also offer cultural events for lab employees and users who could experience them with members of

doi:10.1088/978-0-7503-3631-4ch41

surrounding communities. The first public performance in the auditorium was by the Fine Arts Quartet in 1972.

This weaving of the arts into this major science laboratory right from its beginning, although unique, reflected a cultural shift in the US during the late 1960s towards the creation of arts and science programmes and initiatives. This includes programmes such as Experiments in Art and Technology, commonly referred to as EAT, at Bell Telephone Laboratories created by the engineer Billy Kluger and the artist Robert Rauchensburg; the setting up of The Exploratorium in San Francisco as the museum of the science and art of human perception by the particle physicist Frank Oppenheimer, the brother of the so-called father of the atomic bomb; and the founding of the Center for Advanced Studies at the Massachusetts Institute of Technology (MIT) by Hungarian artist Gyorky Kepes. These initiatives at the time were then perceived as being cutting-edge, radical, and avant-garde.

Today in the twenty-first century, another arts/science explosion is happening, except it is now a worldwide cultural phenomenon and mainstream, rather than avant-garde. New institutions, festivals and programmes are being setting up globally, from Nigeria and Senegal to across Europe, Asia, and Australasia. Today there over 660 worldwide arts/science initiatives alone, including the international Science Gallery International network—the global network for universities dedicated to the collision of science and art (Karalambeva n.d., Gorman 2020). These initiatives all explore in very different ways the interface and interactions between art and science as two 'ways of knowing', as the title of the Exploratorium landmark conference in 2011 posited (Exploratorium 2011).

So it is into this twenty-first century cultural framework and significant societal shift that Big Science laboratories are now creating their own arts programmes in ever growing numbers and becoming cultural generators. This development is partly driven by Big Science organisations wishing to let the public know what they are doing, and they see art as a way of reaching the general public by touching the public's hearts and not just their minds. It is partly due to global challenges, such as the climate emergency, ageing populations, and refugee crises, which need creative thinking to solve these issues. Science is seen as a solution, but not the entire solution. Cross-disciplinary work in arts, science, and technology is seen as key in trying to solve these crises. In addition, the growing interaction of arts and science is also partly driven by economics—money is leaving the humanities, including at the university and cultural level, and being concentrated more in science and technology, so the arts and the humanities are forging new alliances within the sciences in order to survive. Equally there is an increasing need by Big Science to engage with the public to demonstrate what it is doing, particularly in the case of publicly funded institutions, which need to show governments and tax payers how money is being spent. Art is increasingly being used as a public communications and relations tool for doing this effectively. Whatever the reasons, what is emerging is a fundamental understanding of the importance of curiosity and critical enquiry, as well as new ways of looking at complex issues in order to accelerate innovation. After all, exploration and discovery are the *modus operandi* of both science and art—they just express them in different ways.

This essay will look at this increasing movement of Big Science creating its own art engagement by focusing on three disciplines: physics, biology, and space/aeronautics. The laboratories discussed are all at different points on their journey of introducing art into their institutions. It should be emphasised that these are examples, and they are by no means exhaustive or representative of the entire scope of this ever-expanding field. By looking at the work of key laboratories—what they are doing, examining their different motivations, challenges, and results of these endeavours—what emerges are some interesting new cultural developments. These have exciting implications for the future of arts/science interactions, the future of Big Science laboratories, as well as science and art as disciplines in their own terms.

41.1 Physics

To this day, Fermilab is one of the only laboratories to have engaged with the arts since its foundation and to weave it into the very fabric of life at the laboratory, including the surrounding communities. It is the USA's premier particle physics laboratory on a 6800 acre site, with 1750 employees from around the world, collaborating with more than 50 countries on experiments based in the USA and elsewhere. It has always worked with art in manifold ways, reflecting the complexity of what is often referred to as the art/science movement or sci-art, but which in reality is not one movement but an ever-expanding field with many different tributaries, interests, and approaches (Ball *et al* 2020, p 19). Fermilab engages with art as illustrator and explainer of the science, art as public engagement of science, art as collaborator with the scientist, and also contemporary art. Since 1974 Fermilab has held an extensive programme of cultural talks and performances open to the public as well the scientific community. The programme—which now includes its own official artist in residence—ranges from magicians, dancers, and singers, and has included Suzanne Vega, Stephen Hawking, and Stephen Jay Gould, all of whom appeared in the science auditorium which holds 800 people.

As Kurt Riesselmann, Creative Services Manager says 'cultural activity builds connection with the community. It is another way to connect with neighbours, people of all ages, and bring new people onto our site who are obviously coming for the art itself and for our artists. Art is a good conversation starter. Rather than going straight into the science of quarks and leptons, the public get interested in the art, and then want to know about the science which lies behind it' (Riesselmann 2022).

Fermilab also has always had its own art gallery located in the main buildings of the laboratory with a dedicated curator, Georgia Schwender, who is herself a photographic artist. Each year the art gallery hosts approximately four exhibitions, typically showcasing the work of between one and four artists. The laboratory's arts programme also extends to the bi-annual Employee Art Show which hosts as many as 50 people. In 2021 the gallery launched a global project for an online virtual exhibition called pARTicles, billing it as a Fermilab Virtual Family Open House for anyone and everyone to submit art connecting with Fermilab scientists, whether an artist or not.

Schwender also runs the Fermilab artist in residency (AIR) programme which she founded in 2014, partly inspired by the programme at Canada's national laboratory for physics, TRIUMF, held with Emily Carr University, Vancouver, and partly by Arts at CERN. The Fermilab AIR, like all residencies, gives artists access to people and spaces which are not open to the general public on the site, and gives them the opportunity to create a signature body of work which is displayed in the gallery or performed in the auditorium. The programme invites both visual artists and guest composers, the latter working with Janet Mackay-Galbraith, who manages Fermilab's performing arts programme. For example, in 2021 the composer in residence was the American musician David Biedenbender, who experimented with different sounds and instruments to musically interpret the science. His aim was very close to art as the illustrator and explainer of the science, which is the traditional help-mate role that art has played in the history of science, and which like any collaboration can produce exciting and novel results. At the beginning of the residency in 2021 he made his motivation clear:

> My hope is to use music to help those who may not know much about the subatomic world learn about and reflect on some of the amazing phenomena, questions and experiments that happen at Fermilab (Magnuson 2021a).

California-based visual artist Mark Hirsch is Fermilab's 2021–22 artist-in-residence. He uses computer models and coding for his art, and will draw on his data visualisation background to make Fermilab science more accessible and intriguing to the public. His aims are similar—art as communicator of the science:

> 'The work I do can play a role in communication', Hirsch said. 'And one of the goals of that communication through data visualisation is to hopefully broaden the conversation around some of those scientific concepts and bring in some folks who maybe would otherwise not have the experience to see what scientists are working on' (Magnuson 2021b).

Due to the COVID pandemic, both artists had to complete their residencies online rather than in person. But Georgia looks on the positive side, pointing out that with public arts discussions moving online, they are now attracting a global and more diverse audience, with attendance up 90%, again proving the power of art to attract new audiences to Fermilab science.

The resources which Georgia relies on to run each year-long residency programme are very limited. The programme does not cover housing or travel costs, which are paid for out of a small monetary award given by the laboratory. The access and support from Fermilab staff and scientists for a year both remotely and in person are seen by all the artists to be invaluable as shown by the growing number of international artists applying for the open call, for a residency which can only offer minimal financial support. In addition, the artists can separately fundraise for additional support during the residency, and also raise funds for production and exhibition following their time at Fermilab, using the name of this world-famous laboratory as leverage.

Despite the budgetary constraints, Georgia has ambitions for the AIR programme, and she cites the positive effects of the AIR programme in creating ambassadors for Fermilab science who reach audiences outside the traditional science communications field. For example, the first artist in residence, Lindsay Olson, is still an ambassador for the programme seven years later and continues to do outreach work mentioning Fermilab science and scientists. Indeed, the benefits flow in both directions. Being artist in residence at a well-known science laboratory opens doors in wider culture too—Lindsay was invited to give talks at institutions she wouldn't normally have access to due to the profile of the laboratory, including to the Field Museum's conference on Women in Science. The ripple effects of the small but successful AIR programme continues to this day, combined with the highly established public engagement programme which works directly with the arts. All are driven by the skill, dedication, and commitment of the AIR part-time curator, other colleagues, and the scientists who participate and enjoy the benefits of learning to talk about their passion and interests to a committed audience who will share their work and inspiration.

41.2 The creative collider—Arts at CERN, Geneva, Switzerland

In 2011 CERN formally announced its first open call for its own artist in residency programme, Collide, at the Ars Electronica Festival in Linz, Austria. The Collide residency open call was for two editions—Geneva and International—and for artists to win a fully funded residency including travel subsistence and accommodation as well as prize money of 10 000 euros, for up to a three-month residency at the laboratory. As programme curator, I was responsible initiating the programme in 2009 and for raising the necessary funding, and went on to design and direct the programme for its first five years. By the time I left CERN in January 2015 I had raised 4.1 m CHF both internally and externally, in cash and in-kind support to develop the programme and secure its future. This is an example of how an outsider with cultural skills and know-how can develop an arts programme at a Big Science laboratory. In this task I received crucial support from CERN's newly appointed Director General, Rolf Dieter-Heuer, who was committed to showing the effects of science in society and breaking down the walls between the laboratory and the outside world.

From the outset, the Arts at CERN programme made it clear that this Big Science laboratory was situating its programme directly in the contemporary art and new media field, rather than being driven by communications and illustration of science. It placed fundamental research for artists on the same level as that for scientists in a public laboratory dedicated to fundamental research. This was indicated by the curator's manifesto for the programme being published in *The Art Newspaper*, the world's leading art journal (Koek 2017b). This, together with my creation of CERN's first cultural policy in 2010—*Great Art for Great Science*—made it explicit that the programme would select artists for the different strands of Arts at CERN, with artists and scientists working at the same level of excellence, achievement, and potential (Koek 2017b). This also signalled that the programme was at its heart a cultural change programme, introducing ideas of equal excellence of artists to the scientists, which was enforced by a selection process which included leading curators as well as scientists in the open calls for both the Collide residencies, and later the Accelerate programme. Significantly too, the artists were entirely fundraised for and paid for their work—thus putting them economically on a less unequal footing.

What also was unusual was that from the first it was made clear that the science of CERN was to be the springboard for the artists' imaginations, and that it was not the purpose of the artists to explain, communicate, illustrate, or even demonstrate they fully understood the science. The selection of the artists for the three different strands of Arts at CERN in its first five years—Collide, Accelerate, and Guest Artists—reflected this. In addition, in the first five years artists who had never had a residency in a science laboratory were selected for the Collide residencies in order to bring something fresh and new to arts/science interactions and into culture.

What was equally unusual was that during its first five years the programme concentrated solely on the process of research by the artist in the laboratory, offering artists a fundamental research space (including the physicist's classical office with desk, table and blackboard on the main Meyrin site) with no expected outcome. This was a deliberate provocation to draw attention to the importance fundamental research has as much for scientists as the artists—with no expected outcomes and deadlines. However, it was also a tongue in cheek statement, because it was known that by careful curation by the programme's curator, every artist who came in contact with the scientists and science of CERN would create a piece without fail, which was exactly what happened (Koek 2017a). The first Swiss artist, Gilles Jobin, created the dance piece *Quantum* two years after he won the residency, which was first premiered at CERN, and then won the New Settings award of the Foundation d'Entreprise Hermès enabling it to open in New York and Paris the following year. Even artists who just came for one day visits to CERN on the Guest Artist's programme, in which the curator selected artists such as, for example, Goshka Macuga and Iris van Herpen, created pieces that were showcased at the leading contemporary art festival, Documenta 13 in Kassel, Germany, and even putting CERN science on the catwalks of Paris in 2014 with Iris van Herpen's Magnetic Motions collection.

Very quickly Arts at CERN attracted the attention of the European Commission. I was invited to give a keynote presentation about creative collisions, innovation, and the Arts at CERN programme at the European Parliament in Brussels in April 2012, and thus Arts at CERN became involved in the very early days of the European Commission's Technology Directorate, DG-CNect, workshopping and developing their own arts programme. This became the now celebrated Science Technology and the Arts (STARTs) programme, which began in 2016, with a mission to support collaborations between artists, scientists, engineers, and researchers in order to develop more creative and inclusive technologies.

What had been spotted was the potential of art and science interactions as providing a new benefit to the European sustainable research and innovation agenda, including the power of art as a complementary innovation practice extending scientific and technological knowledge into the societal, cultural, and subjective realm. STARTS is now a highly influential landmark European Commission project, and was initiated and is still run by Austrian former physicist turned European Commission bureaucrat, Ralph Dum as part of the DG-CNect directorate, beginning within the framework of Horizon 2020. Out of the early brainstorming for its formation, with which Arts at CERN was involved, came the Future and Emerging Technologies (FEAT) programme in 2015 (also funded under Horizon 2020), which included research and artists' residencies exploring the implications of new technologies. The SciArt programme, initiated and run by the Joint Research Centre—the European Commission's Environmental Science Research and Policy Institute, which has seven science laboratories across Europe and 1500 staff—was also directly inspired by the success of the Arts at CERN programme and that of the STARTs initiative. The JRC SciArt programme was launched in 2016, and includes the Resonances art science summer camp which matches artists with environmental scientists and policy makers at the laboratory, exhibiting the results of this matching at leading cultural centres in Europe. The programme is headed by European Commission member of staff Adrian Eeckels.

Today Arts at CERN is recognised as one of the pre-eminent arts/science programmes in the world, with its Collide residency programme internationally recognised (Muñoz-Alonso and Rea 2017). Over 178 artists have passed through its doors since the official programme began in 2010. Since March 2015, under the direction and curatorship of Mónica Bello, the original three strands of the programme have developed even more significantly. For example, the programme has moved from its focus from being process orientated to becoming an extensive formalised international production, commissioning, exhibition, and touring programme. There is a new programmatic framework, with residencies which still include the original Collide, Accelerate, and Guest Artists strands, alongside the new Connect programme with Pro Helvetia. Since 2018 the programme also includes art commissions, exhibitions, and events which are held with other institutions and public programmes. Through consortium building with leading international contemporary new media and arts organisations such as FACT Liverpool, Le Lieu Unique Nantes, Ical Brussels, and CCCB Barcelona, Arts at CERN has built an international touring and production art science network of its own, with co-production funds from the consortium for new commissions. The touring exhibition *Quantum/Broken Symmetries* are part of a European network established by Bello in collaboration with leading cultural partners in Europe (ScANNER, Science and Art—Network for New Exhibitions and Research). These exhibitions have toured to Liverpool, Barcelona, Tallinn, Taichung, and Brussels, attracting well over 100 000 visitors, showing the significant public relations value of touring and the attraction of the work.

Scientists at CERN who have been involved in the Arts and CERN programme, either as inspiration partners or by interacting more informally with the visiting resident artists, speak eloquently about the benefits of artists engaging with Big Science. CERN theorist, James Wells, talked to the German radio station Deutsche Welle about how engaging with artists reminds a scientist of their humanity and how the work they do exists in the social framework of a world outside that of Big Science, which can become all absorbing (Mitzman 2012). Others talk about how it gives them new ways of looking at their science, and new skills in explaining and describing it.

For the artists, the benefits come through new and profound sources of inspiration, including new technology and cutting-edge science which a Big Science laboratory such as CERN affords, together with its scale of personnel and resources, which are so much more than a smaller laboratory. This means artists have more opportunities to engage and have more sources of inspiration, including in the production and application of their ideas through having direct access to a large pool of different disciplines which go into realising science as a process: theorists, experimentalists, engineers, technicians, and fabricators all feed into work.

In 2018 there was clear confirmation of Arts at CERN as a contemporary art force. The installation at Art Basel of HALO by the British artistic duo Semiconductor, who had won the Collide Residency in 2015, was a significant moment in the programme's development. Audemars Piguet invited Mónica Bello to commission and curate a piece to be shown at the prestigious contemporary art fair. HALO, which was inspired by Semiconductor's residency at CERN in 2015–16, signalled Arts at CERN's increasing influence and importance in the contemporary art field as well as the shift towards new media, digital art, and art which engages with science being finally recognised as being a key part of contemporary art (Dafoe 2018).

The effects and the impact of the CERN residency programme are clear to see in the case of the artists who come through the programme, for example Semiconductor. Even though they are highly experienced in working in science laboratories, the Collide programme gave them the opportunity to extend their practice, and the Collide residency 'definitely attracted the Audemars Piguet commission and enabled us to make our artwork HALO which has raised our profile significantly as artists, in certain fields of art and among the broader general public where our practice wasn't as well known. It grabbed the attention of the international and national press in the UK which will have reached audiences our work hasn't previously reached' (Semiconductor 2022).

The Arts at CERN programme continues to develop and grow to this day, both in scope and influence. There is a new strategy to connect with other arts and science programmes around the world, as shown by the new Connect programme with the Swiss Arts Council, Pro Helvetia, announced in 2021 (Swiss Arts Council n.d.a). The Connect programme includes new annual residency formats in collaboration with prestigious scientific and cultural organisations in Chile, South Africa, Brazil, China, and India. Pro Helvetia have been inspired by Arts at CERN and are now focussed on art and science as their lead policy for the next 4 years 2021–24 (Swiss Arts Council n.d.b).

Externally, the Arts at CERN programme is judged as a success, thriving well beyond its conception in 2009. This is thanks in part to the three-month feasibility study I carried out in 2009, which gave it clear positioning, branding, and communications, allowing it to establish itself on the international scene in its first five years. Twelve years on, it has developed extensively, with CERN recognising the success of the programme as signalled by the creation of a formal staff position for the curator Mónica Bello. However, challenges remain:

> The interest and the demand is huge, so responding to the expectation and numerous number of proposals is a big challenge. Sustaining the

> programme financially is an important challenge too… Working with scientists (and artists) requires a continued and sustained dialogue and interaction, which is exciting and challenging for the insufficient time—due to staff, resources (Bello 2021).

With such a wide-ranging programme, with multiple international partners, which now includes commissioning, production, and exhibiting, Arts at CERN still has only one full-time curator and director—Bello—one coordinator/administrator, and more recently a communications and content producer for the website and social media, with one or two interns a year. With such a small team, commanding an extensive international programme with many partners and deliverables, it is very hard to know how the team manages its work across research residencies, commissioning, production, and touring. Nevertheless, the programme continues to evolve, expand, and grow, and Charlotte Lindberg Warakaulle, Director for International Relations at CERN, says that the programme will continue to be open to change and evolve according to the times because 'Arts and science are both dynamic areas of constant change, which is what makes them so exciting; Arts at CERN needs to react to, and take on board, that change' (Warakaulle 2021).

One way in which it will certainly evolve is with the development of the new Science Gateway at CERN—the flagship project for science outreach and education—designed by the internationally renowned Italian architect, Renzo Piano, and due to open in 2023 (CERN n.d.). Charlotte Lindberg Warakaulle says 'Arts will be integral to the CERN Science Gateway. The gallery devoted to the purely creative end of the science spectrum—theory—will be combined with the arts. It is a way of bringing scientific and artistic creativity together in a tangible way' (Warakaulle 2021). Collide artists in residence Julius von Bismarck (2012) and Ryoji Ikeda (2014/5) have been commissioned to make new art works for the Gateway. The new public centre will also provide opportunities to showcase some of the results of the Arts at CERN programmes on site, at last bringing the art home to the place that inspired it. Thus, it will complete the circle of creativity, inspiration and production, and will crucially give CERN scientists the opportunity to see the fruits of the programme at its point of inception.

41.3 Biology

41.3.1 The community builder—Wellcome Sanger Institute, Cambridge, UK

The world-famous Wellcome Sanger Institute in Cambridge, UK was set up to undertake the most ambitious project in biology to date—the sequencing of the human genome. It is based on the Wellcome Genome Campus in Cambridgeshire, employing 1200 people from over 70 different countries. The campus is also home to some of the world's foremost institutes and organisations in genomics and computational biology, including EMBL-EBI, Wellcome Connecting Science, and the BioData Innovation Centre. There are public art pieces on the campus, such as *Inaugural Window* by Kathy Shaw Urlich, which was commissioned for the opening of the Sulston Building in 1997, and *Fingerprint* 2011 by the poet Fiona Sampson, in which some of her published poems about genomics were installed on the walls within one of Sanger's buildings. Despite the ongoing interest in public art from the beginning, there is no formal public art strategy in place.

However, there is a history of arts engagement on the campus, with the arts channelled through the public engagement programme. This is one of the traditional ways in which science laboratories engage with the arts, using the arts as a public engagement and science communication tool, rather than as a force and discipline in its own right.

Two recent projects at the Wellcome Sanger Institute show a different, blended direction, combining socially engaged artistic practice with public engagement and science communication. Socially engaged practice describes art that is collaborative, often participatory, and involves people as the medium or material of the work, with a mission to change or effect the perceptions of the people in some way. There has been a steady rise in this socially engaged artistic practice, accelerated by the COVID pandemic, with artists seeking to make change happen by working directly with communities and people. The lines between public engagement with science, which often uses art to engage the public and change their perceptions of science, and contemporary art socially engaged practice, are becoming increasingly blurred, even fused.

#flowcellular was devised with the Sanger Institute's Cancer, Ageing and Somatic Mutations team (CASM). The team came with the germ of an idea themselves to the public engagement team of wanting to creatively explore the delicate subject of cancer. As Becky Gilmore, the Public Programmes Manager of Wellcome Connecting Science at the Wellcome Genome Campus says, the 'holy grail' of arts/science public engagement is when the scientists themselves approach you with an idea' (Gilmore 2021).

The Saturday Museum is a collaborative art project run by artists Lucy Steggals and George Moustakis. Due to the lockdowns, #flowcellular was run on Zoom, and the artists invited everyone to meet them in their kitchens with the rationale that that whether you were a scientist, artist, or member of the general public, the kitchen is the common space every person has. The artists used food to explore themes around cancer and ageing, using culinary metaphors for scientific concepts, and setting up experiments that turned kitchens into homemade laboratories. For example, how cancer develops from a cell was explored through the making of bread. These simple domestic projects in a home setting opened up neutral spaces for communicating about cancer and ageing, enabling levelling conversations between diverse groups of people each with different experiences of cancer, whether lived or witnessed second hand, with some going through cancer treatments themselves. *#flowcellular* began in February 2020, and ran for a year on during the global pandemic. The project also had the additional benefit of giving the laboratory scientists direct contact and interaction in a safe neutral space with people that have lived experience of cancer or even as a carer, enabling them to better understand the people they are trying to find treatments for, which Gilmore described as very revealing. For her, this had all the trademarks of a successful project—having both internal and external value—and at the end of the year the group also published a cookbook.

Another much larger project with socially engaged artists is *One Cell at A Time*, which is a Sanger Institute led public engagement project. Under the artistic direction and project management of art curator Suzy O' Hara in 2020 and 2021, *One Cell at a Time* brought together arts and communities, patients, and researchers to explore the biological and cellular make-up of the human body. The project took place in four cities—Cambridge, Oxford, London, and Newcastle Upon Tyne—with performances also screened online. Artists from different arts modalities, from digital to dance, were commissioned to make pieces in the cities with Human Cell Atlas scientists and the communities in the chosen cities. For example, dance/moving artist Anna Macdonald worked for six months with people who are experts in giving and receiving; a group of transplant recipients and a group of young care-givers. What emerged from these conversations were layers of images and movement that sit playfully together, gently provoking thought about issues of consent, donation, and what might be saved and lost within processes of mapping. During the performance, which is experienced by one viewer at a time, people are invited to make patterns, matches, and shapes, in response to the performer's hands, and donate these gestures to the work. Each of these gestures is added, with the viewer's consent, to a growing map of other people's gestures, leaving an online legacy of different ways of doing things.

Both projects show how the Sanger Institute is approaching art/science and the way in which they would like to continue—creating projects which have meaning and relevance for the participants from the local communities, interest groups and community of scientists and blending socially engaged artistic practice with public engagement. As Becky Gilmore eloquently says:

> For our scientists, it finds new ways of thinking and talking about their science, and challenging some of them that art is just not an accompanying way of illustrating and explaining the science. Working with an artist brings their own practice into the mix and opens up ways of thinking and feeling, taking some scientists out of their comfort zones and gives them a way of bringing new meaning for themselves back into the laboratory. It also brings meaning for the artist, for us, and the public—making us thinking how we can all be brought into the creative development as well. (Gilmore 2021).

Funding these projects is not an issue for the laboratory. Both *One Cell At a Time* and #flowcellular are funded by Wellcome which provides funding for public engagement to Wellcome Connecting Science and the Sanger Institute on a five year cycle. Wellcome, which is the fourth largest endowment foundation in the world, spent £7 billion on funding medical research in 2020, including a small proportion on public engagement, and has now changed its approach to the arts. Wellcome has moved away from its pioneering decades of funding and commissioning of cutting-edge arts/science initiatives which began in the 1990s with the progressive SciArt fund, and continued until the early twenty-first century. In the twentieth century, Wellcome made one of the biggest and most significant contributions to setting the standards and inspiring the arts/science field in Europe, but the trust has now moved towards funding almost exclusively much more traditional public engagement of science projects which focus on its mission to solve the urgent global health challenges. It has also disbanded its Culture and Society department.

However, the increased influence in contemporary art by socially engaged artistic practice, as evidenced by both the nominees and the winner of the UK's prestigious Turner Prize in 2021 (Array Collective, the Belfast-based collective initiate collaborative actions and street interventions on social justice issues and Northern Irish politics) has meant that art and public engagement initiated by Big Science which works with communities and people are now increasingly aligned in a meaningful and impactful way. The work is motivated not only by communication but a recognition of the essential role that well-being and social value play in our culture, even more so in a society which is experiencing COVID, together with the climate emergency and inequality.

This alliance between the arts and science for social purpose also enhances the scientific work of laboratories by giving scientists valuable access to real people's experiences, feelings, and the implications of their research. The challenge of getting artistic practice into a scientific laboratory with a very specific funding focus, such as at Wellcome, can be skilfully and meaningfully met as the Sanger Institute shows, for the good of all both internally and externally. The programme also shows the way for awareness of social value and wellness now becoming a core value of work in culture and society which will be an increasing priority over the coming years (Future of Arts & Culture 2022).

41.4 Space

41.4.1 NASA—the space race

When the National Aeronautics and Space Administration (NASA) rang up the performance artist Laurie Anderson in 2002 offering her a residency, she thought initially it was a prank. She said, sure, what do I do? And the response she got from NASA was anything she liked because they had never done anything like this before. They were offering to make her NASA's first artist in residence, with a 20 000 US dollars commission with complete creative freedom to make and perform a piece about NASA.

She was not the obvious choice to work for a government agency. Her work has always been fiercely critical of the status quo, the country's wars, and its repression on the domestic front. The agency was taking a risk and opening itself potentially up to criticism.

In the end she created a piece called *End of the Moon*, a music-theatre journey that examines, among many other compelling themes, twenty-first century perceptions of beauty and time, and the stories we exchange to help us along the way. Part travelogue, part personal philosophy, part daydream, using the mediums of music, film, song, and storytelling for which she is well-known, *The End of the Moon* looks at the relationships between war, aesthetics, spirituality, and consumerism.

The residency was not accepted as a positive by everyone. After her residency, Indiana Republican Chris Chocola successfully amended the Science, State, Justice and Commerce annual appropriations bill ‘to prohibit federal funds from being used to employ an “artist in residence” at NASA’ (Cowing 2005). The discussion in Congress includes many of the arguments for and against art in science laboratories—including that such programmes were a waste of money that could have been given to science (even though the amount is tiny in comparison to a science project budget) versus the need to educate and inspire the public by means beyond simply science.

Laurie Anderson’s residency was a development and continuation of the well-established NASA arts programme which had been initiated by the NASA Administrator, James Webb, in 1962, just four years after agency has been founded. The intention of the programme was to document different aspects of the space agency because ‘an artistic record of this nation’s programme of space exploration will have great value for future generations and may make a significant contribution to the history of American art’ (Ulrich n.d.). To date well over 350 artists were commissioned, for as little as 800 US dollars and up to 2500 US dollars, and the 3000 plus piece collection including sketches, photographs, compositions, films, and finished works by the artists Robert Rauchensberg and Andy Warhol, the photographer Annie Liebovitz, and even a composition by the musicians the Kronos Quartet. All were attracted by the unprecedented access and the opportunity to be part of history being made. The NASA Art collection is one of the most unique art collections in the world, charting the space race and parts of the collection are exhibited at the Kennedy Space Centre. The programme now is minimal and has all but stopped.

In addition, NASA’s first officially run homegrown artist in residency is no longer run due to the congressman’s amendment. But NASA still has informal involvement in artists in residence, which are often generated by the artists themselves or often in collaboration with other cultural institutions, with private artists finding their own grants to fund their activities in different parts of NASA. For example, in 2016 the artist and photojournalist Justin Guarglia embedded himself working with geoscientists on the NASA’s Oceans Melting Greenland Mission led by the agency’s Jet Propulsion Laboratory, California Institute of Technology. However, he funded the work himself. The results were showcased at a the Norton Museum of Art in West Palm Beach, FL as part of a major exhibition on the effects of the Anthroprocene on climate.

41.4.2 ESA and ESTEC, Noordwijk, Netherlands

Today the entire space and aeronautics industry is the largest component of the Big Science world, accounting for a reputed 40.5% of investments in Big Science projects (BBC Research 2021). In 2020 NASA employed over 17 000 people and had an annual budget of over 22.6 billion US dollars. It is the biggest of the Big Science space programmes. In comparison, the European Space Agency (ESA), which is funded by 22 member states and is the European gateway to space, has 2200 staff and a budget (for 2021) of 6.49 billion euros. The largest of the ESA establishments is the European Space Research and Technology Centre (ESTEC) in the Netherlands. ESTEC is the technical heart of ESA, the main technology development and test centre, as well as being home to most of the scientists responsible for guiding ESA's space science missions.

Founded in 1975, ESA has a history of working with artists, including musicians, writers, and film makers, on an ad hoc basis, like many science laboratories, including CERN, before it created its own official programme. Despite the interest clearly demonstrated by artists there has never been a coherent, ongoing programme at agency level to foster interaction between the arts and the sciences nor consistent logging of the number or names of artists passing through. Artists have typically found their way into ESA through personal contacts, often via scientists they knew, and their interactions were on an informal, ad hoc, unstructured, and unselected basis. Artists visited the laboratories at the invitation of individual scientists they knew, while other colleagues across the agency would support access and information exchange on a best effort basis.

Even so, despite these informal channels, ESA has seen direct benefits from artists engaging with its science. Consider the following two recent examples. In 2014 the exploits of ESA's Rosetta spacecraft as it encountered comet 67P Churyumov-Gerasimenko and deployed a lander on the comet's surface captured the attention of visual artist Ekaterina Smirnova and inspired her to create a series of works exploring the comet through watercolour paintings, sculpture, music, and augmented reality. Ekaterina exhibited her work at the 50th ESLAB Symposium 'From Giotto to Rosetta' in Leiden, the Netherlands, beginning a dialogue with many of the participants about the relationship between science and art. Ekaterina continues this dialogue with scientists by participating in scientific conferences as well as exhibiting her Rosetta-inspired work in galleries across the globe.

Artist Lisa Pettibone, who in 2018 won an Arts Council of England funded artistic residency at the Mullard Space Science Laboratory (MSSL), has been following the progress of the ESA Euclid Mission, an initiative to explore the dark Universe. Lisa visited ESTEC several times to talk to the scientists and the wider community about what she was doing. The conversations with scientists at ESA, driven by MSSL and the enthusiastic reception of her work have culminated in a collaborative artwork called *Fingertip Galaxy*, made with over 250 scientists, that will be placed on the Euclid spacecraft before it finally launches in 2023.

In 2016 ESA made its first significant step towards creating a structured programme by beginning a collaboration with Ars Electronica—the internationally renowned institute founded in 1979, based in Linz, Austria, and active in the field of new media art. This was the first time ESA had participated as an official partner in a structured programme, although as early as 2013 Karen O'Flaherty, then working as a science communicator at ESA, had contacted Arts at CERN in order to learn how within just two years of being publicly launched, the programme was already attracting international attention and acclaim. She had been inspired by the Collide residencies of Arts at CERN programme as well as a smaller one at University College Dublin (UCD), called *Tunneling Art and Physics.* This latter programme was introducing artists into the physics department at UCD as an experiment and thanks to its success has led to the establishment of a residency programme firmly embedded across the University.

One of the reasons Karen was given for the successful implementation of the Arts at CERN programme was the unique three-month in-depth feasibility study which had looked at the people, place, and culture of CERN, as well as looking at art/science programmes across the world in depth. Normally this would have been a significant investment, but in this case was sponsored by the Clore Fellowship, underwritten by the UK government's Department of Culture, Media and Sport (Koek 2017).

ESA's collaboration with Ars Electronica in 2016 was a way of learning and experiencing what a designed residency would involve without having to find the investment needed for a feasibility study and with the additional benefit of exposure to experienced people. The collaboration was the final partnership of Ars Electronica's arts residencies programme with major science laboratories in their European Digital Art and Science Network, whose format was itself based on the Collide programme created by Arts at CERN which I had invited Ars Electronica in 2011 to join as partners for the first 3 years. In total, 139 artists responded to the ESA-Ars Electronica open call, called art&science@esa, and the winning artist was chosen by a jury comprising members of ESA, Are Electronica, and the European Digital Art and Science Network.

The winner was the internationally recognised multidisciplinary artist, Aoife van Linden Tol, whose practice spans photography, sculpture, performance, film, drawing, and installation and who since 2000 had been working with explosives to explore her interests in nature, cosmology, chemistry, and physics. She spent two months at ESTEC and one month at Ars Electronica Futurelab, following exactly the Collide residency model. The result was *Star Storm*, a site specific explosive performance inspired by the processes of stars and premiered at the Ars Electronica Festival 2017 which that year had a record-breaking 100 076 visitors. *Star Storm* emerged from Aoife's research into ESA's space science missions, in particular focusing on the composition, life cycle, magnetic behaviours, and light production within stars, including the Sun. Aoife described this residency as 'life-changing' and she has become a passionate advocate for dialogue between scientists and artists in various fora. She is now the vice-chair of the Committee for the Cultural Utilisation of Space (ITACCUS), which seeks to promote and facilitate the innovative utilisation of space (data, systems, applications) by organisations in the cultural sector internationally, covering all areas of the arts and humanities, including the fine arts, entertainment, popular culture, and tourism.

The success of this art&science@esa residency in terms of public awareness raising, public audience, and an outstanding art piece, along with the accumulating evidence of past successful, albeit informal and sporadic, collaborations with artists, created the momentum to build on these great experiences and to explore the possibility of ESA having its own, sustainable, open programme. Karen and ESA CHEOPS mission scientist and astrophysicist, Kate Isaak, prepared a proposal for a pilot programme to explore different models of ArtScience collaborations and to demonstrate the feasibility of establishing an ArtScience programme. The proposal built on the knowledge accumulated by Karen, Kate, and colleagues in the Science Directorate and across the Agency, and included costings, structure, mission, vision, and values, along with a plan for how it could be carried out over a three-year period in the first instance. The proposed programme has three different models of ArtScience collaboration: one is a residency at ESA-ESTEC, another a joint residency with an industrial partner, and finally special commissions for a mission or an event which would demonstrate another role of art, that of commemorating the extraordinary.

The proposal was approved in 2019, championed by ESA's Senior Advisor for Science and Robotic Exploration, Mark McCaughrean, a long-time advocate of cultural engagement, and also the co-founder of Space Rocks which celebrates space exploration and the art, music, and culture it inspires through public events and more. However, due to the COVID-19 pandemic these plans have been put on hold for the foreseeable future. As a result, in order not to lose momentum, in September 2021 Karen and Kate launched an open call, called Mission: Art-Space Exchange, for a modified version of the programme in collaboration with Science Gallery Rotterdam. It is a two-month research residency at ESTEC and is described as a pilot project for an ArtScience programme in the space science domain at ESA. To account for restrictions on international travel arising from the COVID-19 pandemic, the residency this time was only open to artists resident in the Netherlands, although future editions of the residency will be open to international participation.

The winner was Dutch visual artist and filmmaker Anna Hoetjes whose artistic practice focusses on seeing and being seen in a technologically mediated world. She has made it clear that in her residency she will focus on spectroscopy—the study of the interaction of matter with light across the electromagnetic spectrum. Spectroscopy is a vital component of space exploration, and in addition she will also have an opportunity to explore its use in medical research at the Erasmus MC in Rotterdam which is the home of Science Gallery Rotterdam. On winning, she stated 'My aim is to re-imagine the history of space science as one that is more inclusive and diverse. At ESTEC, I hope to get a deep understanding of the hardware and software that are used in space spectroscopy today, and discuss with scientists how techniques can be applied to develop an immersive optical installation' (European Space Agency 2021).

Karen O'Flaherty is convinced about the positive aspects of artist residencies and what they offer. Artists and scientists are driven by curiosity about a topic but come to it from different perspectives. Exploring these perspectives can bring new insights and this is especially the case when there is the time and space both to dive deep into a subject and to stand back and reflect on it. On-site residencies foster fruitful collaborations and 'serendipitous meetings on site maximise the opportunities for interaction with scientists and transform the often stereotyped view of an "artist" to a recognition of a fellow professional with whom one can have fruitful discussions' (O'Flaherty 2022). Equally, when scientists have so many demands on their time, adding an extra activity by asking them to engage with an ArtScience collaboration when it is not highly regarded by some in management can make it difficult, so the on-site presence of artists over lunches and other common activities can also facilitate engagement and make it much easier to happen.

The astrophysicist and project scientist at ESA Kate Isaak is very clear about the benefits of art and science colliding:

> In some ways it is positive interference if you want to think about it in terms of physics terms—the two perspectives looking at a common problem and coming up with an interpretation that is greater than the individual contributions. It's more than PR. It's communications and engagement for sure but it's also two way because both sides gain from the experience. I wouldn't see it as contemporary art or as communication. I would see it as a hybrid. (Isaak 2022)

For both, the marks of success for the new programme will be providing a fruitful and inspiring residency for an artist, inspiring and motivating scientists in their work with new ways of looking at the world, and ultimately the programme becoming an integral part of the ESA activities. Crucially, they are being advised and supported in the development of their programme at every step of the way by curator Dr Claudia Schnugg, advocate of art/science collaborations, the author of *Creating ArtScience Collaborations—Bringing Value to Organisations*, and former Head of the Art Electronica Residency Network.

Karen sees that one of the ways to make the case for this even stronger would be to have longitudinal studies of every artist who comes through ESA, tracking the lectures, exhibitions, and activities they do subsequently, and how the artists as well as the scientists feel years down the line after the encounters.

What is clear is the enthusiasm and dedication of both Karen and Kate working with Mark to make the full pilot programme happen and launch it. As is often the case with major projects in Big Science laboratories, it takes years of persuasion and dedication to achieve lift off for engaging with the arts. Once this is done, the benefits soon become apparent, as the ESA experience shows.

41.5 Conclusion

In the twenty-first century, Big Science is engaging with the arts like never before. This development is an important part of the mainstreaming of arts and science today, and the Big Science laboratories which have been examined in this essay demonstrate a multiplicity of approaches rather than a homogeneous one. From Arts at CERN, which from the first worked with contemporary art and new media, placing the art of contemporary art on equal footing as an equally valid way of knowing as science; the Sanger Institute merging socially engaged artistic practice with public engagement; through to NASA with its arts programme driven by documentation of the space race, and ESA's new residency programme and its hybrid arts programme to come; and Fermilab's multidimensional arts programme which covers everything from communications, arts development for the staff, to contemporary art done even by the founding director himself.

Whereas in the twentieth century, the impetus for science engaging with arts mostly came from outside the laboratories, today the impetus is increasingly coming from within the institutions themselves. This is a massive cultural shift, and Big Science laboratories are becoming cultural generators—generating different responses and contributions to culture other than just science, technology, and engineering. By including the arts, Big Science is creating significant new spaces for creativity, discovery, and knowledge, demonstrating how all forms of knowledge create and are a fundamental part of human culture.

This engagement by Big Science with art also enhances the understanding and interest in science and has a vital communication role. As experimental physicist Michael Doser, who was a founding member of the CERN Cultural Programme for Arts at CERN, says:

> As much as we might wish that all audiences are as fascinated by our science as we are (and this is the rationale for outreach), that is not necessarily the case. Artists (either directly, through their work, or indirectly, through the possibility of making the dialogue with them public) reach other audiences, and some of those come to see the beauty of science and discovery through the artists eyes. This naturally enhances the visibility and reputation of the labs, as does being seen to be open to dialogue with non-experts in the field of science. Ideally it also encourages a more open and broadly curious atmosphere among the younger scientists. Particularly Big Science laboratories, with their international and wide mix of cultures and backgrounds, can thus help foster an open, curious approach to alternative, but equally rigorous, approaches to thinking about the world. (Doser 2022).

All the Big Science arts programmes—except for Arts at CERN—have been initiated by insiders who are already working at the laboratories. However, today all the curators and initiators of the arts programmes in the Big Science laboratories all share a common concern—how to ensure their programme's long-term continuation once started and established. This anxiety is not just driven by funding needs in many of their cases but is significantly driven also by the regular changes of directors at the top of the laboratories, who have different strategies and politics which may in turn affect the arts programmes. Engagement with the arts in Big Science laboratories relies ultimately on the political savvy and adroitness of the people inside the institutions who are supporting and administering the arts programmes.

One of the ways to make the case for continuing programmes may be, as Karen O' Flaherty identified, the creation of long-term tracking of artists well after they have left the laboratory—tracking their careers, their work, the impact their connection with the laboratory, and the audiences they reach when exhibiting and talking about the work they have done in Big Science. Scientific organisations understand data and numbers—it is an easy tool for evaluation—rather than qualitative data which is seen as subjective and open to interpretation. But this involves yet more investment to enable it to happen.

In the meantime, the creation of arts programmes by Big Science laboratories is an important trend which has cultural implications for the future of science laboratories and art. This trend is the emergence of internationally renowned scientists who are leaders in their field, dedicating themselves not only to their science and leading their scientific teams, but also simultaneously investing their time and energy into creating and directing art programmes, curatorial teams, and artists in their own laboratories. This can be seen in the examples of the Horizon prize-winning physicist Dr Suchitra Sebastian who also leads the Cavendish Arts Science Programme at the University of Cambridge which she founded in 2016, and the new residency programme initiated in 2020 by leading epigeneticist Professor Maria-Elena Torres-Padilla who is Director of the Institute of Epigenetics and Stem Cells (IES) at the Helmholtz Centre in Munich.

What has made this possible is the blurring of the boundaries between what used to be science communications and what I have called creative science communications (ScienceComm 2014). This form of science communications is dedicated to creating new entry points for new audiences through enlisting the senses—sight, touch, hearing, smell, and taste—as well as the imagination, rather than just using accurate scientific description which appeals only to the scientifically literate and the already interested. Thus, classic science communications are increasingly complemented in science laboratories by creative science communications. These are more dynamic and fluid, and touch the heart, body, and soul of the audience, rather than their cerebral interests. They also includes arts programmes as a form of creative communications and public engagement.

This is also part of another interesting shift, as shown by the Sanger Institute, towards engaging different communities where the laboratory is located in ways which blend traditional public engagement and science communications with what is known as artistic social practice. This reflects a trend within the arts themselves, which has increased during the current COVID pandemic, for the arts to have socially applied relevance, working directly with communities to create new understanding, expressions, change, and connections.

Finally, this emerging trend in the twenty-first century for Big Science laboratories themselves to create their own arts programmes shows a return to the historical roots of the laboratory as a place for radical and daring experimentation. One of the earliest associations of the laboratory was to the medieval practice of alchemy—an ancient branch of natural philosophy and a forerunner to chemistry that was historically practiced in China, India, the Muslim world, and Europe (Schmidgen 2014). It was a place for experimentation of the visible and the invisible phenomena in our lives, with the possibility to create the unknown whilst searching to find the philosopher's stone which could turn base metal to gold. The introduction and development of scientific protocols and methodology which the Enlightenment in Europe ushered in, were formalised in the laboratory which subsequently became the designated hygienic space for objective experimentation, measurement, and observation under controlled conditions, from which the practice of alchemy was excluded and banished.

Today, with Big Science laboratories engaging with the arts they become once again spaces that welcomes disruptive and alien discoveries and events, which defy objective understanding, yet expand our knowledge of our world by working across disciplines rather than by being locked solely within the post Enlightenment methodologies and objectivity of science.

This shift towards Big Science laboratories increasingly welcoming the arts by increasingly initiating their own programmes may also be due to another important cultural change—the growing acknowledgement and acceptance the role of creativity and the imagination has in the sciences themselves. Previously, these were barely talked about, let alone acknowledged, because creativity and imagination were seen as subjective, not objective, and therefore not scientific. This was despite Albert Einstein and Richard Feynman famously turning to artistic practice, most particularly music as well as their dream worlds, for developing theories and ideas for furthering their scientific enquiry and knowledge. However, recently there have been increasing investigations by scientists themselves into the role of creativity and imagination in science. Take for example the Institute of the Imagination's four-day seminar in Cambridge in 2016. The seminar, which featured some of the top physicists of our age including Sir Martin Rees and Professor Melissa Franklin, concluded that imagination was key and critical to science (Robison 2016). The report began with the following quote: 'If you can't imagine something marvellous, you are not going to find it. The barrier to discovering what can be done is actually imagination.' (Haldane 2016).

With this growing open acceptance and acknowledgement of the imagination in science, combined with the obsession with interdisciplinarity and the wish to accelerate new ways of thinking in order to find new ways of solving the United Nations Global Challenges such as the climate emergency and increasing urbanisation, the age of science laboratories instigating and supporting arts initiatives is truly in motion. What is needed now, for their sustainability, are the continued expansion of modes of funding which are no longer siloed, in order to ensure that these projects continue, develop, and thrive. The former Director General of CERN, Professor Rolf Dieter-Heuer, called for this in 2011 with the launch of the Arts at CERN programme at Ars Electronica.

Certainly the European Commission's Horizon 2020 programme and other funding programmes, such as Creative Europe, go some way to begin changing this, and the European Commission has continued building on funding interdisciplinary activity. What can result is an enriched culture, new ways of seeing the world, unexpected results, and potentially the acceptance of artistic research as being as valid a form of discovery, knowledge-making, and policy making as that of scientific research itself.

We are truly on the threshold of a new era for art and science interactions. The European Commission is now discussing creating the European Arts Science Institute, which would work across its different directorates. The Big Science laboratories are significant drivers of this change, with their unique highly multicultural workforces and distributed networks, and they are influencing this development. So much so, that Big Science laboratories are increasingly becoming cultural generators which produce new ways of knowing not just in science and technology, but also in contemporary arts and the world at large.

Ariane Koek, December 2021.

References

Ball P, Koek A, Parkinson G, Rovelli C and Triscott N 2020 *Entangle: Physics and the Artistic Imagination* (Berlin: Hatje Cantz)

BCC Research 2021 What is big science, and why are so many companies investing in it? *Market Research* https://blog.marketresearch.com/author/bcc-research

Bello M 2021 personal communication

Biron L 2017 Fermilab's 11th employee *Symmetry Magazine* https://symmetrymagazine.org/article/fermilabs-11th-employee

CERN n.d. CERN Science Gateway http://sciencegateway.cern/project

Cowing K 2005 NASA's first and last artist in residence? *NASA Watch* http://nasawatch.com/archives/2005/06/nasas-first-and-last-artist-in-residence.html

Dafoe T 2018 How artist duo semiconductor turned subatomic data into a 'chiming time sculpture' for Art Basel *Artnet* https://news.artnet.com/art-world/semiconductor-audeMars-piguet-1300586

Doser M 2022 personal communication

European Space Agency 2021 Artist wins residency at ESA's technical heart *ESA* https://esa.int/About_Us/ESTEC/Artist_wins_residency_at_ESA_s_technical_heart

Exploratorium 2011 Art as a Way of Knowing https://exploratorium.edu/knowing/index.html

Fermilab n.d. Fermilab architecture and sculpture *Fermilab National Accelerator Laboratory* https://news.fnal.gov/wp-content/uploads/architecture.pdf

Future of Arts & Culture 2022 Drivers and impacts for the next decade *Future of Arts & Culture* https://futureofartsandculture.org/

Gilmore B 2021 personal communication

Gorman M 2020 *Idea Colliders: The Future of Science Museums* (Cambridge, MA: MIT Press)

Haldane D G 2016 Six O'Clock News *Radio Broadcast* BBC Radio 4, London

Isaak K 2022 personal communication

Karalambeva N n.d. Art and science platforms in the beginning of XXI century: European Digital Art and Science Network *Doctoral Dissertation (Unpublished)* (Sofia: National Academy of Art)

Koek A 2017a In/visible: the inside story of the making of Arts at CERN *Interdiscip. Sci. Rev.* **42** 345–58

Koek A 2017b Arts and science are similar in that they are expressions of what it is to be human in this world *Medium* https://luclalande.medium.com/arts-and-science-are-similar-in-that-they-are-expressions-of-what-it-is-to-be-human-in-this-world-b5624a2ffe2a (originally published in *The Art Newspaper*)

Magnuson M 2021a Classical composer David Biedenbender named Fermilab's 2021 guest composer *Fermilab* https://news.fnal.gov/2021/06/classical-composer-david-biedenbender-named-fermilabs-2021-guest-composer/

Magnuson M 2021b Visual artist Mark Hirsch Named 2021 Fermilab Artist-in-residence *Fermilab* https://news.fnal.gov/2021/06/visual-artist-mark-hirsch-named-2021-fermilab-artist-in-residence/

Mitzman D 2012 At CERN, art and science collide *Deutsche Welle* https://dw.com/en/at-cern-art-and-science-collide/a-15838136

Muñoz-Alonso L and Rea N 2017 What is the best artist residency in Europe? What about the weirdest? We asked the experts *Artnet* https://news.artnet.com/art-world/best-artist-residencies-europe-1060478

O'Flaherty K 2022 personal communication
Riesselmann K 2022 personal communication
Robison R 2016 Physics imagination retreat: exploring the role of imagination in the work of scientists *Imagination Institute* https://cpb-us-w2.wpmucdn.com/web.sas.upenn.edu/dist/8/471/files/2018/06/Physics_Imagination_Retreat_Report-wl9qnd.pdf
Schmidgen H 2014 Laboratory *Encyclopedia of the History of Science* Carnegie Mellon University https://ethos.lps.library.cmu.edu/article/id/450/
ScienceComm 2014 *ScienceComm'14 (Sursee/Beromünster, 18–19 September 2014)* https://sciencecomm.ch/images/SC_ScienceComm14_Programm_Abstracts_f_EV3.pdf
Semiconductor 2022 Personal communication
Swiss Arts Council n.d.a Connect: at the intersection of art and science *Prohelvetia* https://prohelvetia.ch/en/press-release/connect-at-the-intersection-of-art-and-science/
Swiss Arts Council n.d.b Art, Science and Technology—Overview *Prohelvetia* https://prohelvetia.ch/en/dossier/art-science-and-technology/
Ulrich B n.d. NASA and the arts *NASA* https://nasa.gov/50th/50th_magazine/arts.html
Warakaulle C L 2021 personal communication

IOP Publishing

Big Science in the 21st Century
Economic and societal impacts
Panagiotis Charitos, Theodore Arabatzis, Harry Cliff, Günther Dissertori, Juliette Forneris and Jason Li-Ying

Chapter 42

A systemic approach to the integration of Big Science in science education

Sibel Erduran and Wonyong Park

Secondary students' engagement in science lessons has been a concern to educators for many decades. Various strategies have been proposed to motivate students in science lessons and ultimately to ensure that the science professions are equipped with graduates with rigorous scientific skills. One way of enthusing students in science careers is through their exposure to Big Science where students get a first-hand understanding of how scientists actually do science through transnational large-scale projects. While this proposal has a common-sense appeal, from an educational point of view, it raises significant questions. First, are secondary students equipped with sufficient cognitive skills to understand frontier science, including its conceptual and methodological dimensions? Second, what exactly are students supposed to learn through exposure to Big Science? What are the potential learning objectives that can be set to clarify how Big Science can be integrated into secondary science? Third, how exactly can engagement in Big Science be structured so as to ensure that students are supported in their learning? In other words, what are the educational approaches that can facilitate students' appropriation of how science works? In this chapter, we address such questions by highlighting the various dimensions of Big Science including its cognitive, epistemic, social, and institutional components. We take a broad and systemic perspective in order to ensure that a comprehensive framework can be conceptualised not only for theoretical justification but also for practical utility in educational applications. We illustrate LIGO as an example for how educators can disentangle the various learning goals related to Big Science for inclusion in science education. For example, we illustrate how the organisational aspects of Big Science can be modelled to students in order to highlight the importance of peer review in the validation of knowledge claims. Implications for future research and development are discussed.

doi:10.1088/978-0-7503-3631-4ch42

42.1 Introduction

Big Science is characterised by large and complex instrumentation, and organisational arrangements, supported by funding from government or other international agencies, where teams of scientists and technicians are involved in the accumulation of knowledge (Dennis 2017). Physics has been closely associated with Big Science since the early twentieth century mainly through its interactions with the military, although practices of Big Science later developed in other disciplines as in the example of the Human Genome Project (Vermeulen 2016). Given the dramatic shift in the practice of physics since the advent of Big Science, there seems to be an urgent call to incorporate the recent changes into the secondary curriculum so that students can develop an authentic image of physics. According to Hallonsten (2016), Big Science consists of three major dimensions: big machines, big organisations, and big politics. First, it requires big machines such as reactors, accelerators, and huge telescopes. Moreover, the discovery of gravitational waves by the LIGO–Virgo Collaboration and the first visual imaging of a black hole with the Event Horizon Telescope were made possible with contributions from various observatories scattered around the globe. Second, Big Science requires a large group of people and long experiment runs, often taking thousands of individuals and decades of time to achieve its aim. Knowledge of Big Science is collectively generated by thousands of scientists, engineers, technicians, and administrators across the globe, many of whom have not met or even do not know each other well. This is in sharp contrast with scientific research conducted within small groups within institutional and geographical boundaries. Third, Big Science influences and is influenced by politics. The term 'Big Science' has been often used strategically to win national R&D budgets for scientific research for military applications and economic growth (Westfall 2003).

The changing focus in scientific practice from 'little' to 'Big' science had a number of consequences for science education. The secondary physics curriculum is often criticised for focusing on 'old' physics knowledge, most of which was discovered up to the early twentieth century, with few exceptions such as quantum mechanics and relativity theory. This is partly due to the fact that physics has enjoyed the status of mature science (Kuhn 1962) for much longer than, for instance, chemistry, biology, and geology have, which made the physics curriculum already too crowded to accommodate the inclusion of recent achievements in the field. However, education based on old physics practices can communicate a misleading image of physics to learners in some respects. One notable example is that to a large extent, science textbooks and popular science still perpetuate the 'lone geniuses' image of science (Charney 2003, Hadley 2004). In Newton's time, most natural philosophers wrote treatises as a single author to disseminate their discoveries. In contrast, a recent scientific paper reporting the measurement of the mass of the Higgs boson included 5154 authors from institutions in dozens of countries (Aad *et al* 2015). Science textbooks tell students how individual scientists achieved their scientific discoveries, but they speak little about the role of collaboration, funding and organisation that have become essential elements of scientific practice in today's science. They are told that science should be open to all, but a significant portion of Big Science, particularly in the context of the military and space research, is carried out in secrecy. These examples show that the traditional 'textbook' image of science fails to explain the practice of science.

With such a mismatch between professional science and school science in mind, the aim of this chapter is to explore some implications of Big Science for science education at the secondary school level. We begin by considering emerging themes in science education research that support the introduction of Big Science in the curriculum. After situating Big Science in the current landscape of science education research, we propose the 'family resemblance approach' (FRA) as a systematic model to frame various aspects of Big Science for educational purposes. FRA is a framework that characterises the nature of science (NOS) as a cognitive, epistemic, social, and institutional system (Erduran and Dagher 2014, Irzik and Nola 2014), and it has recently garnered interest in the science education research community (Akgun and Kaya 2020, Cheung 2020, Couso and Simmaro 2020, Mohan and Kelly 2020, Peters-Burton *et al* 2023, Petersen *et al* 2020). Based on FRA, we illustrate how institutional examples such as LIGO may offer some guidelines for how educators can disentangle the various learning goals of secondary science education.

42.2 Why Big Science should be addressed in secondary science education

It is worth considering some rationales for introducing Big Science into secondary science. If Big Science has little presence in schools today, why would one argue for introducing it into the secondary science curriculum? The question of what to teach in schools can be approached in several ways. One may consider that a curriculum should mirror the academic discipline it stemmed from (Bruner 1960, Peters and Hirst 1970). In this view, teaching Big Science can be justified on the grounds that it constitutes an important part of contemporary physics. Alternatively, we can think of a curriculum as a means to equip pupils with life skills and competencies (OECD 2005), according to which learning Big Science should contribute to pupils' abilities to solve problems and carry out tasks successfully. Furthermore, it can be argued that students' exposure to Big Science may improve their motivation although evidence on the impact of authentic science experiences on students' interests and attitudes is limited. For example, Hellgren and Lindberg (2017) reported that students' motivation remained unchanged following joint projects with scientists. However, the incorporation of Big Science in lessons would constitute a different context that may result in more positive outcomes, an empirical question that needs further investigation. In addition to these considerations, in this chapter we have a particular interest in the role of the curriculum in promoting democratic citizenship (Dewey 1916, Erduran and Kaya 2016). In other words, science education should aim for cultivating the 'willingness to be responsible for the state and to engage at all levels in the decisions that chart its course' (Engle and Ochoa 1988, p 18). In this view, while school science shares some goals with physics education at the university and postgraduate levels, it needs to accommodate broader educational goals besides preparing future physicists. Accordingly, a curriculum need not be a microcosm of academic disciplines in every aspect; rather, curriculum makers often engage in selecting and prioritising among different aspects of the discipline to best serve the goal of democratic citizenship education. Here, let us consider several trends in science education research that are coherent with the introduction of Big Science into the curriculum.

First, Big Science has an important overlap with teaching the nature of science (NOS) and scientific practices. The US *Next Generation Science Standards* (NGSS Lead States 2013) highlights that pupils should not only acquire scientific knowledge but also 'engaging in the practices of enquiry and the discourses by which such ideas are developed and refined' (NGSS Lead States 2013, appendix, p 48) and 'understanding the enterprise of science as a whole—the wondering, investigating, questioning, data collecting, and analysing' (NGSS Lead States 2013, appendix, p 96). There is a broad consensus among science educators that the process in which scientific knowledge is generated, in addition to the end product, should be taught to students (Hodson 2014, Erduran 2015). In addition, historical, philosophical and sociological research on the practice of Big Science carried out in the past century offers fruitful insights into how science as a collective endeavour operates (e.g. Galison and Helvy 1992, Collins 2004, 2017, Karaca 2017). Scientists now know that gravitational waves and the Higgs boson exist, but it is also known that these facts were not discovered by a single group consisting of a few scientists. Big Science raises interesting questions about science and its practice. Such questions are essential to understanding the role of science in society. For example, how are the research topics decided in Big Science? How do different types of expertise of individual scientists contribute to collaborative enquiry? How does science interact with technological advances? These questions are of central importance if we take the aim of school science to be cultivating responsible citizens.

Second, Big Science resonates with the increasing emphasis on socioscientific issues (SSI) as an aim of science education over the past two decades. The idea of SSI is that traditional science teaching focused on decontextualised content knowledge fails to cultivate active participants in democratic decision-making (Zeidler *et al* 2019). Recent educational policy documents have highlighted cultivating pupils' abilities to discuss social and ethical issues by analysing, evaluating, and synthesising various scientific evidence (Zeidler 2014). Several key properties of Big Science make it an inherently socioscientific topic. For example, the Manhattan Project and other weapon science projects contributed significantly to the advancement of knowledge but also raise questions about the societal impact and ethics of science, along with the political and financial contexts of scientific research. Similarly, space exploration represents humankind's curiosity about nature, but it has levelled considerable criticisms in terms of its utility and cost, which reached a high point with the 1986 *Challenger* disaster (Krauss 2011). These science-related issues in society call for the public's attention to the role, function, and impact of science in broader society, in particular from the view of citizenship education.

Third, scientists, educators, and science communicators have also been active in teaching Big Science in various informal, out-of-school settings. Pupils and adults engage in physics learning across their lives through exposure to visits to museums and science centres, and TV, films, books, magazines, and various social media platforms (Falk and Dierking 2012). Although Big Science has yet to establish a strong presence in formal education (see Begalli and Bilow 2019, Park *et al* 2019 for exceptions), Big Science organisations provide online and in-person experiences from which learners can develop an understanding of how physics is practised today. They publish summaries of research for the general public in plain language, provide instructional resources for teachers and pupils, and offer guided tours and internship opportunities (see, e.g., LIGO's public resources at https://www.ligo.org/public.php). Nowadays, it is not unusual to find films and documentaries about space exploration and black holes made with consultations with renowned physicists (e.g. *Interstellar*, produced in collaboration with 2017 Nobel laureate Kip Throne) that can be readily watched from home. The public's as well as students' interest in contemporary physics heightened by these recent efforts provides an ideal environment for addressing Big Science at the school level.

42.3 Integrating Big Science in school science

If the case for the inclusion of Big Science in secondary education receives endorsement by the relevant stakeholders including curriculum designers and teacher educators, then how can Big Science be framed for it to be effectively embedded in secondary science education? How can the science curriculum be structured to be inclusive of Big Science content and what models of science can accommodate for such inclusion? There is a long tradition in science education of framing science from the perspective of scientific knowledge, or what Richard Duschl calls the 'final form science' (Duschl 1990) where students are taught the outcomes of scientific knowledge at the expense of the processes and broader contexts of the scientific endeavour. In more recent years, more emphasis has been placed on the inquiry processes (Hodson 2014) and the social context (Zeidler 2014) of science. However, the reconsideration of the science curriculum from an emphasis on knowledge, processes, and context to more nuanced aspects of science such as the aims and values, the social ethos, the community practices as well as the financial and political dimensions requires a coherent and systemic framework to ensure that students are introduced to a well-balanced and justified account of how science works.

A recent framework within science education research is the so-called 'family resemblance approach' (FRA) to characterising the nature of science (NOS). Numerous studies have now incorporated FRA as a theoretical as well as an empirical framework in science education research (e.g. BouJaoude *et al* 2017, Cheung and Erduran 2022, Erduran and Kaya 2018, Erduran *et al* 2019, 2020, McDonald 2017). The relevance of FRA for science education (Erduran and Kaya 2019) as well as other school subjects (Puttick and Cullinane 2021) have been demonstrated, and practical learning resources for secondary education have been generated (Erduran *et al* 2020). FRA considers the nature of science as a cognitive, epistemic, social, and institutional system, and as such it provides a holistic account of the various dimensions that characterise science. In Erduran and Dagher's (2014) characterisation of FRA, within the cognitive–epistemic aspects there are four sub-categories as follows: aims and values, scientific practices, methods and methodological rules, and scientific knowledge. The social–institutional aspects consist of seven sub-categories as follows: professional activities, social values, scientific ethos, social certification and dissemination, social organisations and interactions, political power structures, and financial systems. The descriptions of the FRA categories are provided in table 42.1. The table also provides examples from gravitational-wave physics to illustrate content that would be relevant and appropriate for the secondary science curriculum. The case of the first-ever detection of gravitational waves in 2016 is used as a Big Science example and framed from the FRA approach to indicate potential approaches to teaching and learning through some reflective prompts that can be used to discuss various aspects of Big Science with secondary students.

One factor that makes the teaching of Big Science in secondary schools promising is that it is not necessary to teach the complicated mathematics and physics to pupils. As highlighted by Park *et al* (2019) with the examples of general relativity and gravitational-wave physics, there are many aspects of Big Science that do not require advanced mathematical knowledge on the topic. The ideas in table 42.1 can be introduced and discussed without a mathematical understanding of general relativity, which would be implausible to acquire for secondary school pupils. In the classroom, the teacher and students can discuss whether scientists can believe their observations, and how the LIGO scientists established the accuracy of their findings by comparing the results from two identical interferometers. The cutting-edge apparatuses and technologies used in the interferometers and the role of computer simulations can be highlighted as key features of Big Science. The aim is not to remember the technical details but to grasp how these diverse, interrelated features collectively comprise the practice of Big Science. In the case of the examples provided here, students would be introduced to some of the epistemic dimensions of science including accuracy of data and observations. Moreover, the activities of LIGO provide a fruitful source for teaching how Big Science operates in the social, ethical, political, and financial contexts. The story of the detection tells us much about the dissemination of scientific findings, the social functions of Big Science, as well as how it is influenced by organisational structures, funding, and politics.

There are several potential benefits of using FRA as a framework for Big Science. By foregrounding the epistemic features of Big Science and how they relate to the social, political, and cultural contexts, this approach can offer a holistic understanding of contemporary physics that exceeds a mathematical and technical understanding. Understanding the content and the context of physics should be pursued simultaneously in developing pupils' physics learning (Driver *et al* 1997). For example, the teacher can teach about not only how the interferometer works scientifically but also how and when it was developed, how much it costed to build the LIGO interferometers, which naturally leads to the financial and institutional aspects of gravitational-wave physics. Such coverage ranging from the epistemic to the institutional aspects of science is likely to engage a broader and more diverse group of students. In other words, for example, some students who may not necessarily be motivated by the epistemic aspects of science may potentially be

Table 42.1. Definitions of FRA (adapted from Erduran and Dagher 2014) and application of FRA in the context of the detection of gravitational waves (adapted from Park *et al* 2019, pp 1074–5).

FRA category	Definition	Examples from gravitational-wave physics	Sample reflective prompts
Aims and values	• Science is guided by a set of values that are often implicit, and they may include accuracy, objectivity, consistency, scepticism, rationality, simplicity, empirical adequacy, and commitment to logic.	• To increase the accuracy of the detection and isolate real signals from noises, two identical interferometers were built separately in Livingstone, Louisiana, and Richland, Washington. • When the detection of signal GW150914 was first shared by a LIGO scientist, many LVC scientists were initially sceptical about it, suspecting that it was a blind injection, software bug, etc, and later tried to eliminate such possibilities (Collins 2017, p 253).	• Discuss the diverse techniques LIGO scientists use to reduce errors.
Methods	• Scientists use a variety of observational, investigative, and analytical methods to generate reliable evidence and construct theories, laws, and models in a given science discipline, which are guided by particular methodological rules.	• Different detecting technologies are used for different frequency ranges of gravitational wave. • Early scientists used a 'bar detector', while later projects used a 'laser interferometer detector', each based on different theoretical principles.	• Investigate how the observational methods for gravitational waves differ from the ones for electromagnetic waves.
Scientific practices	• Science has cognitive, epistemic, and discursive practices. Scientific practices such as observation, classification, and experimentation utilise a variety of methods to gather observational, historical, or experimental data. Cognitive practices, such as explaining, modelling, and	• The production, processing, and visualisation of observation data were performed by computer, and LIGO scientists had to be alert to malicious hacking (Collins 2017, p 109).	• Write about whether LIGO scientists' work is an example of observation or experimentation. Why do you think so?

(*Continued*)

Table 42.1. (*Continued*)

FRA category	Definition	Examples from gravitational-wave physics	Sample reflective prompts
	predicting, are closely linked to discursive practices involving argumentation and reasoning.		
Scientific knowledge	• Theories, laws, and models (TLM) are interrelated products of the scientific enterprise that generate and/or validate scientific knowledge and provide logical and consistent explanations to develop scientific understanding.	• The detection of a gravitational wave, in addition to confirming Einstein's theory, provided detailed knowledge about the collision of two black holes, e.g. what their respective masses are and how far we are from where it happened.	• Write a short essay that describes how our knowledge about black holes has evolved from the eighteenth century to now.
Social certification and dissemination	• Scientists present their work at conferences and write manuscripts for peer-reviewed journals. Scientists' work is reviewed and critically evaluated by their peers. Peer review ensures social quality control aids in the validation of new scientific knowledge by the broader scientific community.	• The paper was submitted to the journal *Physical Review Letters* and was reviewed by anonymous referees. Most referee comments were positive, but some changes had to be made in order for it to be certified by the international physics community (Collins 2017, p 220).	• Think about why scientific papers are subject to the peer-review process and why it is often done anonymously.
Scientific ethos	• Scientists are expected to follow a set of norms both within their own work and during their interactions with colleagues and scientists from other institutions. These norms may include organised scepticism,	• When the signal GW150914 was first reported, LIGO imposed an embargo about the detection, following its 'procedure for making a first discovery' and asked all members to maintain secrecy. This was to prevent	• Assume you are a LIGO scientist and think about whether to immediately disclose the discovery or keep it secret until it is

	universalism, communalism and disinterestedness, freedom and openness, intellectual honesty, respect for research subjects, and respect for the environment.	unnecessary rumours and misunderstandings.	sure enough, and defend your position.
Social values	• The scientific enterprise embodies various social values including social utility, respecting the environment, freedom, honesty, addressing human needs, and equality of intellectual authority.	• The technical challenges for setting up the LIGO interferometer eventually led to many advances in technology, providing benefits to the society (Gravitational Wave International Committee (GWIC) 2010).	• Compare the way gravitational-wave physics and cancer research respectively contribute to better lives for people?
Professional activities	• Scientists engage in a number of professional activities including conference attendance and presentation, writing manuscripts for peer-reviewed journals, reviewing papers, developing grant proposals, and securing funding.	• Key contributors in the detection of gravitational waves such as Kip Thorne, Rainer Weiss, and Barry Barish gave lectures and television interviews to the public, and some of them wrote books about the discovery.	• Search for scientific papers about a recent historical discovery (e.g. the detection of Higgs boson) and compare the title and abstract to the gravitational-wave detection case.
Social organisations and interactions	• Science is socially organised in various institutions including universities and research centres. The nature of social interactions among members of a research team working on different projects is governed by an organisational hierarchy	• Projects for gravitational-wave detection have become a 'Big Science', done in intimate collaboration among hundreds of universities and research groups worldwide and by other detectors in Italy and Japan.	• Compare how Newton communicated and interacted with other scientists and how gravitational-wave scientists did so.

(*Continued*)

Table 42.1. (*Continued*)

FRA category	Definition	Examples from gravitational-wave physics	Sample reflective prompts
Financial systems	• Economic factors may influence how science is done. Scientists often receive funding for their research from the state- and national-level governing bodies. Funding is regulated through institutions such as universities and research centres.	• As physicists' interests in gravitational waves increased in mid-1980s, LIGO received a large amount of money for its construction. However, the astronomy community was against the funding on the grounds that LIGO was less proven and thus a risky project (Collins 1999, p 170).	• Think about why astronomers opposed the construction of LIGO. Are physicists and astronomers in a cooperative or competitive relationship?
Political power structures	• The scientific enterprise operates within a political environment that imposes its own values and interests. The outcomes of science are not always beneficial for individuals, groups, communities, or cultures.	• When LIGO was striving for funds from Congress, Joseph Weber wrote a series of letters to politicians stating that LIGO was a total waste of money in an attempt to interrupt the interferometer project (Collins 2004, pp 360–1).	• Search and discuss the ways political power has an impact on scientific research in science today.

drawn in by the social and contextual aspects. In short, FRA allows for *teaching about* Big Science in addition to *teaching* the Big Science. In practice, which Big Science topic to teach, what aspects of the selected topic to highlight and in what sequence should be carefully decided by curriculum makers and teachers, in consideration of the education system, the developmental stage of pupils, and the teacher's instructional goals (Erduran and Dagher 2014, Kampourakis 2016).

42.4 Conclusions and discussion

We have presented a case for the inclusion of Big Science in secondary science education, in particular by drawing on the FRA framework (Erduran and Dagher 2014) to illustrate how the various aspects of Big Science, ranging from its aims and values to the institutional aspects can be framed for the science curriculum and science learning. Some concerns can be raised by educators about the cognitive demands that might be placed on school-age children in understanding Big Science issues as well as the range of aspects of science that are inherent in the FRA framework. While we recognise such potential concerns, it should be noted that the identification of age-specific aspects of content to teach is a general problem of education and it is a problem of empirical testing through classroom-based research. There is already a vast amount of research in science education that illustrates that some of the concepts that are taken for granted as part of conventional science curricula (e.g. energy, atom) can pose significant learning challenges for students (Neidorf *et al* 2020). In other words, it is plausible to argue that many difficult concepts are already part of secondary science education and thus the inclusion of Big Science ideas is no different. To the contrary, situating the coverage of scientific knowledge in Big Science may potentially help students make sense of these difficult concepts when they are articulated in a coherent manner in relevant and authentic science contexts. In the past decade, empirical research conducted by an international group of physics educators has shown that some modern physics concepts underpinning Big Science can be understood by students as young as 10 years old when taught appropriately and can intrigue young students (Pitts *et al* 2014). Furthermore, it is anticipated that broadening the characterisation of Big Science from a narrow focus on scientific knowledge to other aspects including the social context may potentially raise students' motivation and interest in science because it would humanise science from students' point of view (Matthews 2014).

Future research and development efforts could focus on designing age-appropriate teaching and learning interventions that would transform proposals such as our example of gravitational-wave frames from FRA into classroom practice. Classroom-based research can be carried out to test the effectiveness of introducing Big Science ideas to students' understanding of nature of science. The incorporation of Big Science ideas in secondary education will have significant challenges for secondary science teachers who may not have had much exposure to Big Science themselves. There is considerable research on science teachers' learning and professional development that point to the challenges to educational reform (e.g. Minstrell and VanZee 2000, Supovitz and Turner 2000). Teacher education initiatives have highlighted factors that impact teacher change which is often a long-term process (e.g. Fullan 2001). Early approaches to teacher learning which had little sustained impact were underpinned by beliefs that teacher learning is a linear process, and that educational change is a 'natural consequence of receiving well-written and comprehensive instructional materials' (Hoban 2002, p 13). A more complex view of professional development of science teachers is required, incorporating professional learning in a supportive and collaborative community, where sustained change can occur over a long period of time by teachers developing a sense of ownership of Big Science ideas themselves. Hence, teachers' engagement with Big Science will require more than their being provided with teaching and learning resources. It will require an evidence-based approach to teachers' learning. Comprehensive and inclusive accounts such as the FRA can frame the content for educating teachers about Big Science.

References

Aad G *et al* 2015 Combined measurement of the Higgs boson mass in pp collisions at and 8 TeV with the ATLAS and CMS experiments *Phys. Rev. Lett.* **114** 191803

Akbayrak M and Kaya E 2020 Fifth-grade students' understanding of social-institutional aspects of science *Int. J. Sci. Educ.* **42** 1834–61

Akgun S and Kaya E 2020 How do university students perceive the nature of science? *Sci. Educ.* **29** 299–330

Begalli M and Bilow U 2019 Sharing LHC research and discovery with high school students *Upgrading Physics Education to Meet the Needs of Society* ed M Pietrocola (Cham: Springer) pp 60–84

BouJaoude S, Dagher Z R and Refai S 2017 The portrayal of nature of science in Lebanese 9th grade science textbooks *Representations of Nature of Science in School Science Textbooks: A Global Perspective* ed C V McDonald and F Abd-El-Khalick (New York: Routledge) pp 79–97

Bruner J S 1960 *The Process of Education* (New York: Vintage)

Charney D 2003 Lone geniuses in popular science: the devaluation of scientific consensus *Writ. Commun.* **20** 215–41

Cheung K K 2020 Exploring the inclusion of nature of science in biology curriculum and high-stakes assessments in Hong Kong *Sci. Educ.* **2** 491–512

Cheung K K C and Erduran S 2022 A systematic review of research on family resemblance approach to nature of science in science education *Sci. Educ.* https://doi.org/10.1007/s11191-022-00379-3

Collins H M 1999 Tantalus and the aliens: Publications, audiences and the search for gravitational waves *Soc. Stud. Sci.* **29** 163–97

Collins H 2004 *Gravity's Shadow: The Search for Gravitational Waves* (Chicago, IL: University of Chicago Press)

Collins H 2017 *Gravity's Kiss: The Detection of Gravitational Waves* (Cambridge, MA: MIT Press)

Couso D and Simmaro C 2020 STEM education through the epistemological lens: Unveiling the challenge of STEM transdisciplinarity *Handbook of Research in STEM Education* ed C C Johnson, M J Mohr-Schroeder and T J Moore (Abingdon: Routledge) pp 17–28

Dennis M A 2017 Big science *Encyclopaedia Britannica* https://britannica.com/science/Big-Science-science

Dewey J 1916 *Democracy and Education* (New York: Macmillan)

Driver R, Leach J, Millar R and Scott P 1997 A study of progression in learning about 'the nature of science': issues of conceptualisation and methodology *Int. J. Sci. Educ.* **19** 147–66

Duschl R 1990 *Restructuring Science Education: The Importance of Theories and their Development* (New York: Teachers College Press)

Engle S H and Ochoa A S 1988 *Education for Democratic Citizenship: Decision Making in the Social Studies* (New York: Teachers College Press)

Erduran S 2015 Introduction to the focus on … scientific practices *Sci. Educ.* **99** 1023–5

Erduran S and Dagher Z R 2014 *Reconceptualizing the Nature of Science for Science Education* (Dordrecht: Springer)

Erduran S, Dagher Z and McDonald C 2019 Contributions of the family resemblance approach to nature of science in science education: a review of emergent research and development *Sci. Educ.* **28** 311–28

Erduran S and Kaya E 2016 Scientific argumentation and deliberative democracy: an incompatible mix in school science? *Theory Pract.* **55** 302–10

Erduran S and Kaya E 2018 Drawing nature of science in pre-service science teacher education: epistemic insight through visual representations *Res. Sci. Educ.* **48** 1133–49

Erduran S and Kaya E 2019 *Transforming Teacher Education through the Epistemic Core of Chemistry: Empirical Evidence and Practical Strategies* (Dordrecht: Springer)

Erduran S, Kaya E, Cullinane A, Imren O and Kaya S 2020 Designing practical learning resources and teacher education strategies on nature of science *The Nature of Science in Science Education: Rationales and Strategies* ed W McComas (Dordrecht: Springer) pp 377–97

Falk J H and Dierking L D 2012 Lifelong science learning for adults: the role of free-choice experiences *Second International Handbook of Science Education* ed B J Fraser, K G Tobin and C J McRobbie (Dordrecht: Springer) pp 1063–80

Fullan M 2001 *The New Meaning of Educational Change* 3rd edn (New York: Teachers College Press)

Galison P and Helvy B (ed) 1992 *Big Science: The Growth of Large-Scale Research* (Stanford, CA: Stanford University Press)

Gravitational Wave International Committee (GWIC) 2010 The gravitational wave international committee roadmap: The future of gravitational wave astronomy https://gwic.ligo.org/roadmap/Roadmap_110726_WEB.pdf

Hadley T 2004 Lone geniuses in popular science: the devaluation of scientific consensus *Tech. Commun.* **51** 328–9

Hallonsten O 2016 *Big Science Transformed: Science, Politics and Organisation in Europe and the United States* (London: Palgrave Macmillan)

Hellgren J M and Lindberg S 2017 Motivating students with authentic science experiences: changes in motivation for school science *Res. Sci. Techn. Educ.* **35** 409–26

Hoban G F 2002 *Teacher Learning for Educational Change: A Systems Thinking Approach* (Buckingham: Open University Press)

Hodson D 2014 Nature of science in the science curriculum: origin, development, implications and shifting emphases *International Handbook of Research in History, Philosophy and Science Teaching* ed M R Matthews (Berlin: Springer) pp 911–70

Irzik G and Nola R 2014 New directions for nature of science research *International Handbook of Research in History, Philosophy and Science Teaching* ed M R Matthews (Dordrecht: Springer) pp 999–1021

Kampourakis K 2016 The 'general aspects' conceptualization as a pragmatic and effective means to introducing students to nature of science *J. Res. Sci. Teach.* **53** 667–82

Karaca K 2017 A case study in experimental exploration: exploratory data selection at the Large Hadron Collider *Synthese* **194** 333–54

Krauss L 2011 The space shuttle programme has been a multi-billion-dollar failure *Guardian* 21 July https://theguardian.com/science/2011/jul/21/space-shuttle-programme

Kuhn T S 1962 *The Structure of Scientific Revolutions* (Chicago, IL: University of Chicago Press)

Matthews M R 2014 *Science Teaching: The Contribution of History and Philosophy of Science* 2nd edn (London: Routledge)

McDonald C V 2017 Exploring representations of nature of science in Australian junior secondary school science textbooks: a case study of genetics *Representations of Nature of Science in School Science Textbooks: A Global Perspective* ed C V McDonald and F Abd-El-Khalick (London: Routledge) pp 98–117

Minstrell J and Van Zee E (ed) 2000 *Teaching in the Inquiry-based Science Classroom* (Washington, DC: American Association for the Advancement of Science)

Mohan A and Kelly G J 2020 Nature of science and nature of scientists *Sci. Educ.* **29** 1097–16

Neidorf T, Arora A, Erberber E, Tsokodayi Y and Mai T 2020 Review of research into misconceptions and misunderstandings in physics and mathematics *Student Misconceptions and Errors in Physics and Mathematics* (IEA Research for Education vol 9) (Berlin: Springer)

NGSS Lead States 2013 *Next Generation Science Standards: For States, by States* (Washington, DC: National Academies Press)

OECD 2005 *The Definition and Selection of Key Competencies: Executive Summary* (Paris: OECD)

Park W, Yang S and Song J 2019 When modern physics meets nature of science: the representation of nature of science in general relativity in new Korean physics textbooks *Sci. Educ.* **28** 1055–83

Peters R S and Hirst P H 1970 *The Logic of Education* (London: Routledge and Kegan Paul)

Peters-Burton E, Dagher Z and Erduran S 2023 Student, teacher, and scientist views of the scientific enterprise: A comparative epistemic network analysis using two nature of science frameworks *Int. J. Sci. Math. Educ* **21** 347–75

Petersen I, Herzog S, Bath C and FleiBner A 2020 Contextualisation of factual knowledge in genetics: a pre- and post-survey of undergraduates' understanding of the nature of science *Int. J. Environ. Sci. Educ.* **16** e22115

Pitts M, Venville G, Blair D and Zadnik M 2014 An exploratory study to investigate the impact of an enrichment program on aspects of Einsteinian physics on Year 6 students *Res. Sci. Edu* **44** 363–88

Puttick S and Cullinane A 2021 Towards the nature of geography for geography education: an exploratory account, learning from work on the nature of science *J. Geogr. Higher Educ.* **49** 343–59

Supovitz J A and Turner H M 2000 The effects of professional development on science teaching practices and classroom culture *J. Res. Sci. Teach.* **37** 963–80

Vermeulen N 2016 Big biology: Supersizing science during the emergence of the 21st century *NTM Zeitschrift für Geschichte der Wissenschaften Technik und Medizin* **24** 195–223

Westfall C 2003 Rethinking big science: modest, mezzo, grand science and the development of the Bevalac, 1971–1993 *ISIS* **94** 30–56

Zeidler D L 2014 Socioscientific issues as a curriculum emphasis: theory, research and practice *Handbook of Research on Science Education* ed N G Lederman and S K Abell vol 2 (New York: Routledge) pp 697–726

Zeidler D L, Herman B C and Sadler T D 2019 New directions in socioscientific issues research *Discip. Interdscip. Sci. Educ. Res.* **1** 11

IOP Publishing

Big Science in the 21st Century
Economic and societal impacts
Panagiotis Charitos, Theodore Arabatzis, Harry Cliff, Günther Dissertori, Juliette Forneris and Jason Li-Ying

Chapter 43

Education and public engagement: why the science is Big for everyone

Steven Goldfarb

43.1 Introduction

Human beings, like other species, have certain basic needs tied to our survival. We eat, we make shelter, and seek mates, and we try to understand the world around us. These needs are integral to our existence and, as such, bring us joy when attained.

For this reason it should not have been surprising in July 2012 that over one billion people[1]—14% of the human inhabitants on our blue planet (perhaps a few dogs and cats, as well)—watched video of a scientific seminar at CERN [1] reporting the findings of the search for a subatomic particle called the Higgs boson (figure 43.1). One of the talks featured well over 100 slides, the other was carefully crafted in Comic Sans font, and both featured formulae, selection criteria, histograms, and statistical analyses well beyond the typical mathematical grasp of the audience. Yet, the deep and profound importance of the human endeavour they summarised grabbed the world's attention.

The search, which had lasted 48 years from the first proposal of the particle's existence, was an important advancement in human understanding of our Universe. Discovery of the particle meant that we now had a grasp on the mechanism that gives mass to all other elementary particles. It also provided scientists with an important new tool to search for explanations to the many puzzles that remain. Humanity had improved its understanding of the world and it brought us all joy. Well, at least a billion of us.

[1] Based on estimates from journalists of approximately 5000 TV shows on 1000 stations that rebroadcast the webcast in part or in whole, debriefing report of CERN-IT, August 2012.

doi:10.1088/978-0-7503-3631-4ch43

Figure 43.1. ATLAS Spokesperson, Fabiola Gianotti, reports on the Higgs boson search during an LHC seminar held at CERN on 4 July 2012. Copyright CERN.

43.2 Big questions

Bringing this story and others like it to classrooms, lecture halls, and festivals gives enormous pleasure to those of us involved in science education and public engagement. It is an easy sell and a story with many important lessons: the scientific process, international collaboration, technological advancement, and human perseverance, to name a few. Moreover, it is fun and interesting. The science is Big because it affects pretty much every particle in the entire Universe and, well, the Universe is about as Big as Big gets.

But, that kind of big is difficult for humans—even students—to comprehend. It comes more as a feeling than a realisation. And students want more than that. They want concrete answers to concrete questions. For example, 'Why do we need to build such large devices to measure the smallest things in the Universe?' 'What else are we looking for?' 'What will this do to improve our lives?'

Such questions give us an opportunity to introduce basic concepts to the students (or any audience) without necessarily going into the details of the theory or underlying mathematics. For example, we can describe the equivalence of energy and mass, the transformation of energetic low mass particles to those of higher mass, the transformation of high mass particles to those of lower mass and energy, the need for large accelerators to produce the more massive particles, the need for large detectors to measure the trajectories of high-energy charged particles in electromagnetic fields, etc.

Simply explaining the construction and operation of accelerators and detectors is rich with physics concepts that are well within the grasp of nearly any audience (figure 43.2).

Figure 43.2. Physicist Howard Gordon describing a particle physics detector during a classroom visit. (Image courtesy of Brookhaven National Laboratory.)

Only the level of detail changes, based on its scientific background and the amount of time available. Yet, the physics concepts are only a taste of the lessons to be shared and discussed.

Once the basic concepts of experimental particle physics are understood, it becomes clear that further exploration at the high energy frontier requires the construction of very large and complex accelerators and detectors. The successful planning, construction, and operation of these devices requires human expertise and financial resources that are well out of reach of any one single institution or country. Hence the need for international collaboration.

43.3 Big collaborations

The Large Hadron Collider (LHC) [2] at CERN represents the current high energy frontier for experimental particle physics. The 27 km circular machine accelerates protons in opposite directions at energies up to 7 TeV, allowing the beams to pass through each other, and thus interact, at four different locations (figure 43.3). Large, complex, multi-layered detectors surround these interaction points, capturing the output from the collisions, which occur around every 25 ns.

Each of the very large and complex detectors has been planned, built, and is currently being operated by international teams that include thousands of physicists, engineers, technicians, and support staff. The very fact that these teams exist and are working together can be important news for various audiences. In many cases, scientists from countries often seen in strong disagreement—even war—on the political scene, are working together closely to try to answer some of the most difficult and fundamental questions.

Figure 43.3. Underground installation of the Large Hadron Collider at CERN. Copyright CERN.

Similarly large international teams have been assembled for the gravitational wave experiments of LIGO (1200 scientists from over 100 institutions in 18 countries) [3] and VIRGO (650 members, 119 institutions, 14 countries) [4] experiments exploring gravitational waves. The size, scope, and complexity of these experiments effectively require that the efforts be global in scale.

The case of CERN is somewhat unique. The laboratory was founded in the decade following World War II, in part to bring together scientists from across Europe, with the goal of performing scientific research for peace. That motivation was made explicit in its original convention [5]. Yet, this fact is not known by many of the classrooms being introduced to the science of the laboratory. It can also be news to policy makers who control the funding of the efforts, yet who did not realise the global impact of the effort beyond the scientific goals.

For this reason, scientists who are active in science education and public engagement often dedicate significant time in classrooms, public events, and congressional offices to describe the importance of international collaboration to society. As an example, the members of these collaborations, around one third of whom are post-graduate students, gain invaluable experience working in an international environment. They learn to thrive in that environment and they also learn the value that environment brings to the science.

Scientists with different national, cultural, social, and linguistic backgrounds necessarily think differently, even when it comes to science. They attack problems from different points of view and often come up with solutions that would not necessarily have occurred to their colleagues. This not only creates a work environment that is more rich and inventive, but it also leads to more robust analyses and results.

These lessons, embraced by the collaboration participants, are also made evident when the scientists visit classrooms or host visits at the laboratory. Even if the host scientists are of the same nationality as the students, they will inevitably discuss their international adventures in the field, the teams they work with, conferences they attend, schools where they have studied. Students are typically impressed and fascinated, but also receive a positive impression of international collaboration, and view it as an attractive aspect of the field.

43.4 Big impact

There is no doubt that improving our understanding of the Universe has a vital impact on humanity. It is a fundamental component of our nature. Exploring the world around us and sharing our findings with each other provides future generations with the tools they need to survive. We have been practicing science since prehistoric times and it is impossible to imagine us living without it. There will forever be new challenges facing the survival of our species and we will forever be finding solutions to meet these challenges. Well, at least we hope so.

The major findings of the LHC are plentiful. Beyond making a giant leap in our understanding of the phenomenon of mass, we have also discovered a variety of new composite states of particles, put limits on the proposed existence of others, enhanced the precision with which we understand current models in the field, and eliminated others. In short, we have already travelled many rivers, found El Dorado in some, and mapped the details of others. And the physics coming from the experiments is still only in its early stages.

Given that, however, it can be difficult to predict the benefits to society such discoveries and measurements might bring. In many cases, the gains will only become evident well into the future, in the form of advanced technologies or applications. In other cases, we might never be able to clearly trace the path from scientific advancement to application. Imagine Thomson or Maxwell developing an electric car or Born and Einstein coming up with a smart phone. There are just too many steps of development between the initial idea and the application.

Yet, there are quite a few important applications that have come directly from the research of Big Science experiments, such as those of the LHC. Work at the host laboratory, CERN, has led to such applications as the PET scan, MRI, touchscreen, mouse, solar panel, hadron therapy, and, yes, the World-Wide Web. There have been significant advancements in vacuum technology and superconductivity driven by the construction of the accelerator, as well as important breakthroughs in micro-electronics, high-speed computing, and networking pushed by the demands of the experiments.

These advancements were not the stated goals of the experiments and, in many cases, they might have eventually come out of other research. However, it is important to note that they are necessary products of large international scientific collaborations pushing the limits of technology in order to solve some of the most complex problems of humankind. This is another important message being delivered to audiences, from classrooms to government offices.

43.5 Big engagement

In the list of important applications and developments produced directly as a result of the pursuit of Big Science, one should not forget the messenger. A number of significant advancements in science communication, education and public engagement have been achieved in recent decades. While some are designed to satisfy an attentive and curious public, others have charted new paths to spark new interest in science and to encourage a new generation of potential researchers to pursue their dreams.

Large scientific laboratories, such as CERN, DESY [6], GSI [7], KEK [8], and FNAL [9], have run successful communication and education programmes for many years now. With significant sources of public funding dedicated to their science programmes, each has a responsibility to inform the public on plans, progress, and findings. Each of these labs supports communication programmes that include the maintenance of news-oriented websites and social media, as well as press offices that author releases for the media, and coordinate communication with contributing institutes and national offices.

Many of the larger experiments, such as those of the LHC, which comprise thousands of collaboration members from around the globe, also run their own communication programmes. These are typically modest in size and resources, ranging from one or two volunteer scientists, to teams that include dedicated writers, media creators, and web developers. The experiment programmes rely on support from the host laboratories for infrastructure and common content, but choose to produce their own written summaries, images, and video in support of the experiment's research programme.

For the LHC experiments, a periodic outreach and communication meeting was set up well before the original accelerator start up, to coordinate their activities, and to seek common support from CERN. These meetings of the LHC Outreach Group (LOG) continue to this day, allowing the experiments the freedom to run their own programmes, while sharing common material and best practices.

Beyond communication, the laboratories, experiments, and participating institutes also host and contribute significantly to educational programmes. Busloads of students, typically from secondary schools visit the large laboratories, every day. Perhaps the most interesting part of the tours, at least to the students, is the fact that they are typically led by active scientists.

More intensive opportunities are offered to undergraduate students in physics, engineering, and computing, seeking experience in research. The CERN Summer Student programme [10] hosts around 300 students each year, providing an opportunity for the participants to embed in international teams, and to contribute to leading-edge research in the experiments and laboratories on campus (figure 43.4). Semester programmes, supported by national institutes and organisations such as the US State Department [11], provide longer-term in-depth activities, and are able to attract students from under-represented communities to embed in research teams at CERN.

In addition to the student programmes, CERN hosts high school teachers several times during the year. Over the summer, there is an international programme involving the participation of teachers from around the world; throughout the year,

Figure 43.4. CERN Summer Student Programme 2018. Copyright CERN.

there are national programmes, hosted by scientists in their native language. Participating teachers discover material and activities developed by CERN, and have an opportunity to share ideas and best practices with their colleagues, often leading to lasting friendships and connecting classrooms around the world.

While the facilities at CERN draw attention, arguably the most important aspect of the educational programmes are the interactions between the participants and the researchers. Students and teachers get a chance to ask questions pertaining to the science, experimental activities, educational possibilities, and even the personal lives of the researchers, learning the human aspects of science. The researchers learn how to clarify and express their views, improving their own communication skills and broadening their own understanding of their scientific contributions.

Tours of the laboratories also include members of the public, many of whom travel long distances to view the accelerator complex and experiments; members of the media, to get first-hand information, interview footage, and b-roll for their shows; policy makers, curious to witness the effectiveness of their decisions to fund research and a few selfies for their social media; and even an occasional VIP, including heads of state, musicians, artists, and influencers. All leave with deep impressions of the value of fundamental research and a better understanding of humanity's quest to understand our Universe. Well, OK, some just leave with more followers on TikTok or Instagram. Regardless, the results are positive for everyone. However, for those six or seven billion who have not yet made the trip to visit in person (and now for times of pandemic) the experiments and laboratories sought means to provide remote views of the apparati and activities, and to offer engagement with the scientists involved in the research.

In the years leading up to the LHC start up, much of outreach activities offered unidirectional communication: books, brochures, images, and video on tape or disk. For major events or on special occasions, video conference or webcast links could be set up, and live conversations could be held between remote audiences and the scientists or educators.

As an example, for the start up of the LHC, an elaborate webcasting system was set up, allowing for presentations to be made from the CERN accelerator and experiment control rooms. Questions prepared ahead of time could be answered, and there was a good opportunity to show off the equipment and the people behind the success of the project. Such events, however, were costly, labour intensive, and thus rare.

Social media offered a possibility for further engagement with remote audiences. Questions and comments could be posed online and, when the authors were attentive and had the time and inclination, responded to. Yet, such interactions were rare and clear communication was not always the result. (Author's note: In the many years that have followed, it appears that the problem of poor communication through social media could be with us for a very long time.)

Yet, the development of tools combining social media and video conferencing technology gave the scientists and educators an affordable method to reach remote audiences. The concept of Virtual Visits was first championed by the ATLAS experiment [12], where the hardware left over from the LHC start up was combined with video conferencing equipment to create a platform that was accessible for most remote participants. Given only a personal computer, microphone, speaker, camera, and screen, classrooms around the world were able to click on a link in their web browser and connect with scientists at the experiment.

Public websites, social media, and word of mouth were used to quickly spread the availability of Virtual Visits from ATLAS and, soon after, from major experiments around the world. During a Virtual Visit, a classroom might connect directly to scientists standing in front of their experiment, control room, or other research facility. They are given brief introductions and short tours, then spend most of the time (if they are good hosts) answering the questions of the teachers and students.

Now, with the proliferation of social video sites and the ubiquitous nature of Zoom™ around the world, such opportunities are capable of reaching classrooms anywhere within reach of the Internet. The LHC experiments make these connections all year long, now. ATLAS hosted more than 150 visits this past year alone, reaching thousands of viewers, as hosts could take the visitors underground to see the detector during its upgrade and, despite the pandemic, students were able to connect from the classroom or home.

But visits are not the only way audiences around the world are being introduced to the fundamental research of particle physics. Scientists involved in the research often speak and give demonstrations of their work in classrooms, public exhibitions, science festivals, and art and music festivals, around the world. There are a tremendous number of projects under development and in use, often in conjunction or coordination with education departments, universities, consortia, foundations, or funding agencies hoping to improve the science, technology, engineering, and mathematics (STEM) literacy of their national population or constituents.

But, how to organise and coordinate all these efforts? Without some kind of structure there can be inefficiency in the effort, including redundancy, missing or inaccurate material, poor methodology, and repeated mistakes. Fortunately, this was taken into consideration by the CERN Directorate in the early preparations for the LHC, under the guidance of CERN Director General Chris Llewellyn Smith. He proposed the creation of a group of scientists, educators, and communicators who would oversee the development, maintenance, and coordination of outreach material, and the sharing of best practices across the member states.

43.6 Big reach

43.6.1 EPOG, EPPOG, and IPPOG

The European Particle Physics Outreach Group (EPOG) was formed in 1997 under the joint auspices of the European Committee for Future Accelerators (ECFA) [13] and the High Energy Particle Physics Board of the European Physical Society (EPS-HEPP) [14]. The primary tasks of EPOG (later rebranded as EPPOG) included: providing a forum for the exchange of information on outreach activities, pooling resources, identifying media contacts, and identifying other possible collaborating partners. The original group comprised one delegate from each CERN member state, one additional CERN and DESY member, and a Chair and a Deputy Chair appointed by ECFA and EPS-HEPP.

The group agreed to meet twice a year, to exchange ideas and best practices in particle physics education and outreach, to define common activities, and to develop and share material supporting their activities. Following agreement by the collaboration board and subsequent approval by ECFA in 2010, EPPOG evolved to IPPOG, the International Particle Physics Outreach Group [15]. The first new member of IPPOG was the USA in 2012, quickly followed by Israel, Ireland, Slovenia, Australia, and South Africa.

In 2016 the IPPOG adopted a Memorandum of Understanding (MoU) [16] defining it as a scientific collaboration, with an organisational structure, terms of membership, and operation. The MoU specifies agreements with member countries, collaborations, and international laboratories. Countries participate in the IPPOG

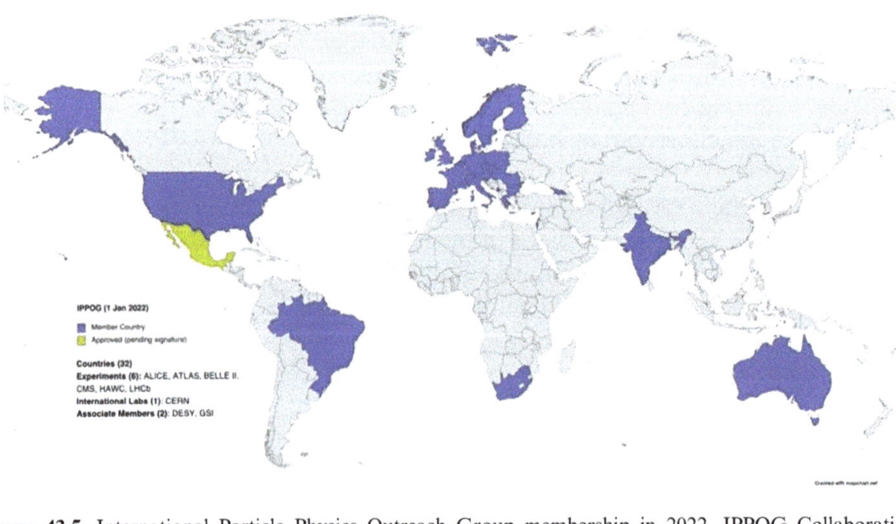

Figure 43.5. International Particle Physics Outreach Group membership in 2022. IPPOG Collaboration. copyright CERN.

programmes, appoint a representative, and provide monetary contributions for the core infrastructure.

International collaborations agree to recognise and support outreach efforts, provide access to data, tools and expertise for IPPOG activities, and appoint a representative. CERN is currently the only member international laboratory, providing masterclass coordination, a monetary contribution, and access to administrative, legal and financial services. National laboratories and other scientific bodies can join IPPOG as associate members, allowing for participation in the Collaboration Board, but without a vote. There are currently 39 IPPOG members: 32 countries, six international collaborations, one international laboratory (CERN), and two associate members (DESY and GSI) (figure 43.5).

43.6.2 Particle physics masterclasses

The best known of the IPPOG educational programmes are its Particle Physics Masterclasses (figure 43.6) [17]. These were first developed in the UK in 1997 by Ken Long of Imperial College and Roger Barlow of the University of Manchester. A Masterclass differs from a typical class, as it is taught by an expert in the field, who is not necessarily a professional educator. Particle physics masterclasses are run by scientists active in current research to classes of secondary students, outside of their usual lessons. The students get a chance to work with current tools on real data from active experiments, and have a chance to engage in conversation with experts in the field. They learn what it is like to be a particle physicist for a day, but the influence is immediate and long-lasting.

Figure 43.6. Particle Physics Masterclass with students in Colombo, Sri Lanka. IPPOG Collaboration. copyright CERN.

The idea of organising a world-wide programme of particle physics masterclasses was put forward by Michael Kobel in October 2003. The programme would be hosted at universities and research centres around Europe, inviting local classrooms to participate. The morning featured talks and demonstrations about the LEP [18] experiments and local activities at the host institute. The afternoon agenda included a masterclass on measuring Z branching ratios at LEP, as well as other hands-on exercises, including cosmic-ray measurements. The day concluded with a video conference joining classrooms who had taken part in the programme, so the students and mentors could share and discuss their results.

The proposal was enthusiastically embraced by EPOG members and by March 2005, particle physics masterclass learning packages had been developed and translated into 16 different languages. These were used in conjunction with the event visualisation tools provided by the participating experiments. CERN organised video conferencing connections for the schools and helped to operate and moderate the end-of-day wrap up sessions. These sessions instilled students with an appreciation for international collaboration and a stronger understanding of the scientific process, and how it is used in modern research.

Since then, International Masterclasses (IMCs) has become a flagship programme of IPPOG, reaching students in all populated continents around the globe. In 2019 IMC included the participation of over 14 000 students at 220 institutions in 60 countries. The content has evolved and grown in physics scope as well, featuring authentic data from the four major LHC experiments and a neutrino experiment in FNAL. New Masterclasses in 2020 included analysis of data from the Belle II experiment at KEK [19], particle therapy research at the GSI/FAIR collaboration [20], and a direct dark matter search based on the DarkSide Experiment at LNGS [21]. The 2020 IMCs were unfortunately cut short due to the global COVID-19 pandemic, but new lessons, including the Big Analysis of Muons in CMS [22] (BAMC), allowed some of those courses to continue remotely, with students participating from home.

In 2021 student participation rebounded to nearly half the 2019 peak, despite the fact that most were held remotely due to restrictions posed by the pandemic. It is hopeful that adjustments made to reach out to students connecting remotely from home or distant classrooms will increase the size and breadth of the programme in the years to come. In particular this adaptation allows us to reach locations far from scientists and institutions involved in current research. It also allows for participation by individual students whose classrooms might not normally subscribe. An overview of the IMC programme [23] has been written up by coordinators Uta Bilow (Technical University of Dresden) and Ken Cecire (University of Notre Dame and QuarkNet[2]) on behalf of IPPOG.

43.6.3 The global cosmic rays portal

Much of our early understanding of high-energy particle physics came from the study of cosmic rays [25]. The subsequent discoveries of massive elementary particles

[2] QuarkNet is an NSF-funded partnership between FNAL and the University of Notre Dame supporting particle physics education and outreach [24].

outside of the lighter, stable collection in the existing models motivated physicists to develop accelerators and detectors, eventually leading to the large-scale programmes that define the field today.

Fortunately, construction and operation of the instruments needed to detect and make simple measurements of cosmic rays are inexpensive, making them excellent tools for the classroom. Students learn basic particle physics concepts and methods, while setting up experiments, operating hardware, taking data, and performing analysis. As a result, cosmic-ray experiments, employing a variety of detector types, have been installed in classrooms around the world. Many of these have been connected together into networks, allowing students to share data and produce scientific results in a manner similar to our much larger-scale international collaborations.

To better exploit the potential of cosmic-ray experiments for particle physics outreach, IPPOG has developed a common umbrella for these globally separate programmes [26]. Current global activities include International Cosmic Day (ICD) [27], organised by DESY, and International Muon Week [28], organised by QuarkNet. Both events are promoted and supported by IPPOG and its members.

43.6.4 Public events and cultural festivals

With limited, but sustainable funding from its members, IPPOG is now able to support a broader reach of activities outside of the classroom, as well. This includes public engagement at events ranging from science, art, and music festivals (figure 43.7) to film screenings, open days, pubs, cafes, libraries, and more. Example venues featuring IPPOG activities in recent years include the 2021 FICA Science Festival in Portugal, 50 Years of IN2P3 in France, a portable exhibition

Figure 43.7. Festival participant enjoying virtual reality views of the LHC experiments. Copyright RATIO, Bulgaria.

called Urknall Unterwegs in Germany, the WOMAD and Blue Dot Music Festivals in the UK, the Ostrava Festival in the Czech Republic, the Pohoda Festival in Slovakia, RATIO in Bulgaria, MoogFest in the USA, and the Montreux Jazz Festival in Switzerland. Again, these are only a few of the ongoing activities supported by IPPOG, its members, and their institutes around the globe.

The originality and quality of the talks, exhibitions, and demonstrations prepared by the researchers are impressive, and these novel ideas are often shared during IPPOG meetings. Dedicated IPPOG working groups, such as 'Exhibitions and Public Events', 'Outreach of Applications for Society' and 'Explaining Particle Physics Hot Topics to a Lay Audience' provide forums in for discussion, the development of material and sharing of best practices.

43.6.5 International competitions

In addition to providing support for local events, projects, and exhibitions, IPPOG has hosted international competitions with the goal of reaching younger and more diverse audiences. Particles4U [29] was a competition held in 2018 that challenged groups of students or classrooms to create their own physics-themed educational material to be shared with their fellow students. We received a total of 77 entries from 15 countries spanning the globe: Austria, Brazil, Bulgaria, Colombia, France, Greece, Iran, Italy, Japan, Philippines, Poland, Slovakia, South Africa, Spain, and the United States. Two winners were selected from two different age groups, and their classrooms were awarded educational material and visits from local researchers.

Girls Do Physics [30] was a competition held between 11 February 2019 (United Nations Day for Women and Girls in Science) and 8 March 2019 (International Day of Women) challenging girls to share their physics projects and activities on social media. Participants joined from classrooms around the world, with winning efforts coming from Iran, Bulgaria, Romania, and Spain. This effort helped IPPOG to raise awareness within participating countries and to encourage girls around the globe to consider future careers in science.

43.7 Big partners

Large international scientific collaborations dedicated to education and outreach, such as IPPOG, are not common. One recent addition in the world of Big Science is the International Gravitational Waves Outreach Group (IGrav) [31], founded by members of the LIGO [3] and VIRGO collaborations. This group formed not long after the remarkable discovery of gravitational waves and has been active ever since. There is some overlap between IPPOG and IGrav in both content and membership, making for a natural synergy in our activities.

In addition, there are a number of national and international collaborations, consortia, societies, and networks that are investing important effort and resources in that direction. A number of them, including the Astro-Particle Physics European Consortium (APPEC) [32], the Nuclear Physics European Collaboration Committee (NuPECC) [33], the American Physical Society (APS) [34], the European Physical

Society (EPS) [35], and the European Particle Physics Communication Network (EPPCN) [36] partner with IPPOG, sharing attendance in meetings and contributing to common projects. Another recent partner is the International Association of Physics Students (IAPS) [37], an association of physics students and student societies from around the globe.

43.8 Countering big lies

Science education also has an important role to play in protecting us from misinformation. The invention of the World-Wide Web at CERN by Tim Berners-Lee [38] has been widely celebrated across the planet as an important advancement in communication, tying together people in remote lands and allowing for the instant propagation of information far and wide. The impact of this communication revolution has been vital to education and science, and is frequently cited as a key example of the direct positive impact of fundamental research on society.

Yet, while the web was invented as a means for scientists to propagate and share scientific findings around the globe, it has rapidly become a very decisive and dangerous means for others to push lies and misinformation to vast audiences for less than noble reasons. Whether it be for fame, fortune, or political gain, some of the world's most visible leaders have abused this powerful means of communication for personal gain, and the effect has been devastating.

Lies, misinformation campaigns, and conspiracy theories propagate quickly on social media and, unfortunately, *argumentum ad authoritatum* (argument by authority) is still confused by much of the public as legitimate justification. This is not surprising. Not so long ago, when the words of an authority were published, they had usually passed through a series of editors and peer-review, with the goal of attaining accuracy. Social media and unscrupulous media outlets have done away with much of that, with their primary goal being profit and fame. This is achieved by taking advantage of a public that appears more and more thirsty for outrage.

Many of the solutions being sought to address these issues are aimed at curbing the onslaught of misinformation through legislation and social media regulation. Many focus on the algorithms that drive viewership, through likes, shares, and other activity, hoping a tweak here and there could improve things. But, economists learned some time ago that supply-side efforts are not the most efficient. For every bad player whose account is blocked, several new ones appear. Where there is demand, there will be supply.

This is where science education and public engagement can make a difference. Our work helps to reduce the demand for misinformation by teaching the methods used to find real information. Although the topics we bring to the classroom or the public arena are focused on particle physics and our efforts to understand the basic building blocks of the Universe, the methods we present are common to all science and, in fact, many of our life decisions. We observe, we measure, we pose questions, we remeasure, we analyse, and we transparently report our findings.

Arguably the most important lesson we bring to their desktop is that of the inherent messiness of our work. Unlike rolling a ball down a ramp, what the

Figure 43.8. Particle Physics Masterclass students differentiating signal from background. ATLAS Collaboration @ CERN.

students measure does not always fit pre-supposed hypotheses. Sometimes two oppositely charged particles seen by a detector will have come from the decay of another particle. Sometimes they won't have been produced in this way, making them what we term 'background'. All a student can do is calculate the mass of the particle they might have come from and look for a peak (figure 43.8). There is signal, there is background, and there is uncertainty. And these values do not come from the supposition of an authority figure. They are measured.

43.9 Big future

The recent publication of the European Strategy for Particle Physics Update (ESPPU) [39] includes big plans for science research in Europe and around the world. It followed several years and many contributions from experts in all areas of our field to plot a sustainable course for its future. Its implementation will have a major long-term impact on the continent and our planet, as it not only presents priorities for infrastructure, but also a timeline and overall strategy for improving our understanding of the Universe.

As an example, one of the priorities put forth in the ESPPU would be an important expansion of our high energy frontier, including construction of a Future Circular Collider (FCC) [40] that could take us to proton collision energies as high as 100 TeV. Just as the recent launch of the James Webb Space Telescope [41] promises to greatly extend our understanding of the distant expanses of our Universe, the FCC promises to dramatically improve our knowledge of its most fundamental internal structure.

As these opportunities unfold, tens of thousands of students will seek advanced degrees in physics, engineering, computing, and other related topics. They will also learn how to thrive and compete in an international setting, creating lasting ties with their colleagues around the world and becoming drivers for world peace. They will make discoveries, build devices, and develop concepts that will shape our society in the decades to come. In short, they will continue the essential, genetic drive of our species to explore the world and to pass on their findings to future generations.

If we want to ensure the success of these generations, we need to be active now, providing them with the tools they need to reach their goals in a healthy and positive manner. The impact of our efforts will not be limited to those who plan to enter the field, but also apply to those who choose a different path. It is essential that these new generations grow up with the ability to differentiate fact from fiction, signal from background, so they all become better citizens and better people. After all, these students will choose and become the leaders who decide the future of science and of our planet. We are counting on them and they are counting on us.

References

[1] European Organization for Nuclear Research (CERN) http://home.cern

[2] Large Hadron Collider (LHC) https://home.cern/science/accelerators/large-hadron-collider

[3] LIGO Scientific Collaboration https://ligo.org

[4] The VIRGO Collaboration https://www.virgo-gw.eu

[5] The history of CERN *CERN* https://home.cern/about/who-we-are/our-history

[6] Deutsches Elektronen-Synchrotron (DESY) https://desy.de

[7] Helmholtzzentrum für Schwerionenforschung (GSI) https://gsi.de

[8] High Energy Accelerator Research Organization (KEK) https://www.kek.jp/en/

[9] Fermi National Accelerator Laboratory (FNAL) https://fnal.gov

[10] Summer Student Programme *CERN* https://home.cern/summer-student-programme

[11] Goldfarb S and Schwarz T 2021 University of Michigan CERN Semester Research Program *Proc. Sci.* **390** 943

[12] ATLAS Experiment *CERN* http://atlas.cern

[13] European Committee for Future Accelerators (ECFA) *CERN* https://ecfa.web.cern.ch

[14] European Physical Society—High Energy Particle Physics Board (EPS-HEPP) https://eps-hepp.web.cern.ch

[15] International Particle Physics Outreach Group (IPPOG) https://ippog.org

[16] Beck H P, Bardeen M and Bruant Gulejova B 2016 Memorandum of understanding establishing the International Particle Physics Outreach Group (IPPOG) collaboration *Memorandum of Understanding* IPPOG-MOU-2021-001 CERN https://cds.cern.ch/record/2789139

[17] Particle Physics Masterclasses *IPPOG* https://physicsmasterclasses.org

[18] Large Electron–Positron Collider *CERN* https://home.cern/science/accelerators/large-electron-positron-collider

[19] Belle II Experiment https://belle2.org

[20] GSI/FAIR International Accelerator Facility https://gsi.de/en/researchaccelerators/fair.htm

[21] Direct dark matter search based on the DarkSide experiment *Darkside Masterclass* https://sites.google.com/unisa.it/ds-20k-masterclass/masterclass-authors-and-past-events

[22] CMS Experiment *CERN* http://cms.cern

[23] Bilow U and Cecire K 2020 Current status of international particle physics masterclasses *Proc. Science* **ICHEP2020** 937

[24] QuarkNet https://quarknet.org
[25] Pacini D 1912 La radiazione penetrante alla superficie ed in seno alle acque *Nuovo Cimento* **3** 93–100
[26] Global Cosmic Rays Portal *IPPOG* https://ippog.org/global-cosmic-rays-portal
[27] International Cosmic Day *DESY* https://icd.desy.de
[28] International Muon Week *QuarkNet* https://quarknet.org/content/international-muon-week
[29] Particles4U Competition *IPPOG* https://ippog-static.web.cern.ch/ippog-static/particles4u_en.html
[30] Girls, Do Physics Competition *IPPOG* https://ippogorg.wixsite.com/girlsdophysics
[31] International Gravitational Waves Outreach Group (IGrav) https://www.igrav.org/
[32] Astroparticle Physics European Consortium https://www.appec.org/
[33] Nuclear Physics European Collaboration Committee http://nupecc.org
[34] American Physical Society https://aps.org
[35] European Physical Society https://www.eps.org/
[36] European Particle Physics Communication Network (EPPCN) *CERN* https://espace.cern.ch/EPPCN-site/_layouts/15/start.aspx
[37] The International Association of Physics Students https://www.iaps.info
[38] Berners-Lee T 1990 Information Management: A Proposal *CERN* May https://cds.cern.ch/record/369245/files/dd-89-001.pdf
[39] European Particle Physics Strategy Update 2018–2020 *CERN* https://europeanstrategyupdate.web.cern.ch
[40] The Future Circular Collider (FCC) *CERN* https://home.cern/science/accelerators/future-circular-collider
[41] The James Web Space Telescope *NASA* https://jwst.nasa.gov

Chapter 44

Big Science and science education: steps towards an authentic partnership

Stephen M Pompea and Pedro Russo

Big Science projects and facilities can move towards a less self-centred frame of reference as they strive to better identify and serve educational audiences. By doing this, their science education efforts will be more productive in general, and their service to local schools will be more effective. By developing an enlarged awareness of local educational needs, they will become better stewards and partners in their roles in the science education system. They will also become more valued and trustworthy neighbours to their local and cultural communities. We propose a practical way for large science organisations to organise their budgets and their allocation of staff time to greatly increase the effectiveness of their organisation in its contribution to local science education.

44.1 Introduction

Big Science transforms our understanding of the natural world. It brings new data, knowledge, and wonder not only to scientists but to the broader science-interested public. It mobilises entire sectors of society intellectually as well as commercially as it develops new projects and the new facilities that support them. The scientific results are conspicuous and consistent successes: Earthscope, the National Ignition Facility, the Large Hadron Collider, and the Human Genome Project have all transformed their branches of science. The largest projects involve tens of thousands of scientists from over a 100 countries. These projects, and the tools they have created for international collaboration, have transformed modern science by rapidly making enormous amounts of data available to large groups of scientists worldwide. The era of Big Science has transformed our relationship to science data. The title of one data-based project provides a very apt description of the era of Big Science: 'From Petabytes to Megafolks in Milliseconds' (National Science Foundation 2022).

doi:10.1088/978-0-7503-3631-4ch44

These Big Science initiatives are very familiar to the authors; we have communications and educational leadership roles in astronomical research institutes that are conducting large, ambitious projects such as the Event Horizon Telescope, the Dark Energy Survey, and the Legacy Survey of Space and Time. We have also been involved with the teams that are creating the next generation of highly capable and extremely large ground-based telescopes. These include optical-infrared telescopes with apertures of over 1500 square metres and radio telescope arrays whose collecting area is over a million square metres. We have also led projects in a different type of Big Science: enormous worldwide science education ventures involving over 100 countries. These projects include the International Year of Astronomy 2009, the IAU100 projects, and many other ambitious national and international science education initiatives (Pompea *et al* 2010, Russo and Christensen 2010, van Dishoeck and Elmegreen 2018).

This article describes how Big Science projects and facilities can enlarge their outlook on science education and move towards a less self-centred frame of reference. By doing this their science education efforts will be more focused and productive in serving local schools. They will also become better stewards and partners to the science education system, as well as more trustworthy neighbours to their local and cultural communities. We propose a way for large science organisations to organise their budgets and staff time to greatly increase their effectiveness in supporting local science education.

The first step to an enlarged perspective on what organisations can accomplish in science education is to explore the general area of science communication, community engagement, outreach, and science education. These sectors are often unified in research organisations as a functional department with a name that is a combination of these words. We believe that this combination or blurring of duties is inadvisable, as it muddies the distinctions between the goals and results of the different efforts, often leading to a neglect of science education, and especially of local science education efforts. We also suggest that science organisations can benefit from adopting the strategy of making several specific commitments of money and staff time to educational efforts. Finally, we advise that Big Science organisations and their scientists begin to view science education efforts from a deeper ethical and cultural stewardship perspective, rather than from a more limited 'business' model that tends to emphasise a quick return on educational efforts and investments.

We use the term 'Big Science' to mean the lead and subsidiary science organisations and institutions that create and implement these projects. This includes the employees of these organisations: the scientists, engineers, project managers, and the leadership of these organisations. It also includes the funders and the supporting companies that make products used in doing science. In short, we are addressing everyone who works for these organisations that have an orientation towards enabling large science projects and their associated science advancements. This includes the individual scientists who work independently or at small organisations as well as university scientists who are part of a larger project. It also includes the support staff such as those who work in accounting, purchasing, safety, and human resources at these national labs, private labs, universities, and non-profits.

44.2 Communications: a success story for Big Science

Big Science projects have advanced and transformed the field of science communication. They do a terrific job of communicating the excitement of the scientific process. They are highly successful at publicising new results to science stakeholders and to the public that play an important role in the support of their private or government funding. Science communicators are often among the most important first hires for a Big Science project under development. Science communication experts are essential players in helping outline and communicate a project's scientific objectives and in elucidating the exciting science questions that will be addressed. Science communication specialists are also needed for their professional skills during the lengthy and often convoluted odyssey of aligning stakeholders, participating institutions, foundations, and government agencies during the quest for funding and support.

The science of science communication has mature proven best practices which are essential elements in the creation of Big Science projects, organisations, and facilities (Christensen 2007, National Academy of Sciences 2017, 2018, Jamieson *et al* 2017). Science communication professionals are highly creative in teaming with scientists to create exciting and successful wide-reaching projects designed to interest children and adults in science. For example, a worldwide project was hosted in many countries to allow the public to name a unique exoplanet and its host star. This IAU100 project was very successful in stimulating interest in exoplanets, with 113 national campaigns that involved nearly 780 000 people (Mamajek *et al* 2020).

44.3 Science education: a story with diffident results

If the science research and science communication efforts of Big Sciences are outstanding successes, their education efforts often fall short. There are many reasons for this. The initial educational efforts often have a significant activation energy and may start slowly or not at all. The international nature of these large projects can lead to a committee mentality where no one part of the project or organisation takes responsibility for the science education program. There is often a question of what could and should be done in education, with perhaps no one person on the initial team with solid answers or experience in this area. With distributed projects, there may be no mandate or marching orders towards doing science education, since the funding source and project scope are still under discussion. While science communication efforts are perceived as an essential need in a project's early phases, science education efforts are perceived to lack a compelling rationale and may be viewed as a bridge too far or a bridge to be crossed later.

The remote and rural nature of some of the large science facilities can lead to a sense of geographic disembodiment and a lack of identification with the local community, the region, or even the country in which the facility is placed. User facilities have an additional identity problem. They have many people coming and going and their operations managers may feel that they have enough difficulty in keeping the facility running continuously and in responding to the needs of the diverse users who cycle through the facility or its data. Thus, day-to-day operational concerns overwhelm a longer-term perspective.

The newest research facilities produce enormous amounts of data for a worldwide community of scientists who may never visit the physical location or understand the local situation. Related to this, in data intensive projects there often is a diminished relationship to the source of the data. The actual place that takes the data might be viewed more as a set of sensors or as a 'peripheral' connected to the computers and its pipelines. There may be little concern for place or a sense of place as long as the data keeps flowing. Finally, with tight budgets, facilities are trying their best to focus on their core mission: doing science and providing science data. They work hard to eliminate mission creep, and to look for opportunities to divest themselves of any stranded tasks or requirements that no one seems to miss. They also try to minimise or ignore any new demands that don't arrive with additional money.

All of these factors influence the ability of an organisation to maintain an educational mission. If educational efforts are not well-defined and built in at the beginning of a project, they tend not to be treated seriously and they may be forsaken later.

These observations apply to the many players in the Big Science structure and not just to the national, federal, or state research labs, user facilities, and university science departments and institutes. Other participants are the industrial contractors that make the steel and concrete, aerospace companies, engineering and architectural firms, and the support companies in information and data services. Many of these players have adopted a 'not my job' approach to helping the science education ecosystem, even as they have solid communication, community engagement, and occasionally outreach efforts. The distinctions between these areas are important and will be discussed in detail later.

44.4 Why the education mission matters

The lack of a solid, well-defined, and measurable educational mission in many Big Science organisations has some serious long-term consequences for society and for the science enterprise. Even if the organisation is doing a great job of generating press releases, adding social media followers, making inspiring movies, enjoying the publicity from popular articles, and creating well-written science and funding justifications, they may not be perceived as a good neighbour or as a core member of their local community. Does this actually matter?

We believe that, in the longer term, it does matter, as there are very direct feedback loops that affect the organisation. It is often a surprise to science organisations when their trustworthiness and legitimacy to operate, expand, or to procreate is questioned by geographic or cultural communities nearby. This has happened in many places. In Hawai'i the building of a billion-dollar telescope on Maunakea (the TMT) has been challenged, delayed, and possibly stopped by the efforts of Hawai'ians with strong concerns about colonial attitudes by scientists and how the culture of science has allowed scientific advances to be made all too often at the expense of marginalised people (Kahanamoku *et al* 2020, Prescod-Weinstein *et al* 2020).

Partially in response to this outcry against the TMT telescope, the US astronomy community's once-each-decade report that recommends what facilities and missions should proceed has called for the astronomy community to pivot towards 'community science' (National Academies 2021). Community science encourages a larger involvement of the research community with the many other local and cultural communities participating in and affected by science projects. It also calls for greater inclusion and equity in science endeavours, in alignment with other calls for greater social justice and equity in science (O'Donnell and Scott 2022).

The science community is increasingly asking questions about which communities it belongs to and if it is a good neighbour to them. This is forcing science organisations to examine whether they have empathy for a community's problems and knowledge of its history and concerns, or if interest is only ramped up when the science institution needs something from them or encounters resistance to its plans.

44.5 Communications, community engagement, outreach, and education: differences that makes a difference

These are actually four quite different areas, though they are often thought of as quite similar. At many research centres they are also often combined into one functional department with one set of resources to handle all of these areas. This unification is done with the notion that the four areas are closely related, can use the same people, and require the same skill sets. These assumptions are mostly incorrect; this can be best seen by an examination of each field and its audiences.

Communications is essential to research organisations and consists of the creation of press releases, graphics, and social media posts, and many other creative strategies to reach different audiences. Their target audience can be scientists, funding agencies, the larger organisation in which most research centres are embedded, sister institutes, the local community, the science-interested public, the government, or the public at large, to name a few.

Community engagement often means encouraging a broader engagement with the science community. This can be done through expanded communications efforts including workshops, meetings, panel discussions, town halls, and strategy sessions, thus aligning scientific and non-scientific communities to support new initiatives. The beauty of the term 'community engagement' is that it shifts the focus from the organisation and its output (e.g. the number of press releases) to engagement and dialogue with external communities, though with a goal in mind. All community engagement efforts must first decide which community is being targeted. Then these efforts must be sensitive to the needs of that community and to the engagement strategies that may be effective. Community engagement is usually about choosing, tailoring, and understanding the medium, message, reception, and response, and the encouragement of listening and dialogue, often to further a communication goal.

Community engagement for education is a sound and genuinely good goal but is not often done. When was the last time your favourite research organisation met with the local school district superintendent, head of local school science programs, or the indigenous leader of a nearby tribal Nation or with the First Nations community to discuss education? We strongly encourage sustained and productive interactions with the local, regional, or national cultural and educational communities to gain an understanding of their needs. In practice this is not usually done on a regular and long-term basis. Thus, community engagement efforts at research organisations have largely focused either on their own science community or on local and regional community leaders who can influence general public opinion or policy.

Outreach or educational outreach is often used synonymously with education when describing the efforts of research organisations. However, it is quite different as it usually refers to handling requests from outside the organisation for advice, guidance on where to look for materials, requests for tours, or requests for scientists to visit or speak. It also includes working with those outside the organisation who need guidance for their science-interested children and talking with those who want to contribute their own idea on how the Universe (or their private Universe) works.

Outreach covers education-related actions that require the research centre to advance outside of its walls, either physically or virtually (e.g. by video conferencing). Outreach is often justified by both its educational and public relations benefits. It is viewed as a form of social responsibility, and as such it is often viewed as an affable but not essential action. If people from the outside want to speak to someone on the inside, the public outreach department is the place to refer them to.

The priority of outreach in an organisation varies widely, from viewing it as a welcoming and much needed interface with the public to a barrier that shields and insulates the scientists from the public. A figurative example of this latter case is this: the research institute on the hill occasionally opens its formidable door and gates (and lowers its drawbridge over its moat!) and allows some lucky few from below to ascend and enter, to see the wonders that have been created. Or the institute sends its trained emissaries beyond its walls into the village below, usually bearing gifts of posters, PowerPoint slides for the Rotary Club, or props that can be used in school talks. Thankfully, most organisations realise the value of outreach and extend their efforts beyond this minimal level with great passion and enthusiasm, even if they are limited by resources

However, many organisations still conduct educational outreach with paternalistic and neo-colonial components that reflect their sense of being above or better than the community. The research institute is not in a dialogue with its local community about what the community or culture wants or needs but has decided for them on what they should want or deserve. In this case, the institute is mainly serving its own needs for connection and control at the expense of helping the community develop.

In this, there are parallels to 'parachute science', where researchers from areas with more resources go to a less developed area to gather data or to use the local resources for their research. The researchers then publish the results with little or no consideration or involvement of the local population, local experts, local culture, or development of local capacity and infrastructure (DeVos 2020). The danger is that educational outreach activities may function in this same way, dropping briefly into educational settings with little awareness of educational or cultural needs, and with no sense of how to help build capacity (Heenatigala 2022).

These outreach activities are often initiated by demands that are presented or funnelled to the outreach department by schools, clubs, or the organisation's own employees. Examples of outreach activities include a request to provide mentors or judges for a student science fair, a request for someone to participate in a career day at a school, a request for a class visit to the research facility, or an inquiry for someone to answer questions from a student or teacher. There can also be requests to provide a speaker to a group, to give a demonstration or a talk, to make a presentation to the city government, or to provide a science activity at a camp or special event. Requests are often local but can be from anywhere in the world.

These requests are taken seriously by the outreach workers, especially if they were forwarded from upper management. Responding to them takes significant time that cannot be used for other broader or more productive initiatives. Unfortunately, outreach is not usually viewed as an activity requiring special skills; the perception is that it needs only minimal knowledge coupled with abundant patience and civility. It might also be viewed as a tertiary mission of the communications department. Unfortunately, the efforts are usually viewed from the perspective of management and non-outreach professionals as 'anything we do in this area will be fine and should be appreciated by those we do it for'. Thus, outreach is highly reactive and relies heavily on triage. It is perceived as a favour that should be appreciated and that does not need to be evaluated or examined for its opportunity costs.

Education and outreach are quite different. It is tempting, and not too inaccurate to say that quality education efforts are usually everything that outreach is not. Responding to educational outreach requests is like answering your door when your next-door neighbour needs a spare rubbish bag. You should do it to show that you heard the knock and have a willingness to help. In contrast, working in education is the equivalent to mobilising the entire block to discuss, plan, find resources, and do the things that may keep your entire neighbourhood from being bulldozed by a developer. Education relies heavily on evaluation along the way as to how you are doing, and at the end as to what level you succeeded. Education is goal-oriented (usually SMART goal-oriented) and evaluation is done against the goals. In education, the time-on-task, that is, the level of effort directly applied to something is a good indicator, as efforts are usually designed to be more prolonged and in-depth in order to be effective. Thus, educational efforts to explore a topic deeply are complementary to communications efforts designed to excite interest in a large number of people. Education programs are always designed with a sense of the opportunity costs–what cannot be done if the resources are used in particular way.

Education requires a sophisticated understanding of research and practice, and how best to interact with the complexities of the educational system (Pompea and Russo 2020, 2021a, 2021b). Education efforts that are ad hoc or poorly planned are often unproductive since project success relies on the successful completion of the many parts of each project. At its core, deciding what to do in an educational context and what can be done best must rely on science education professionals with the general knowledge of how children learn and who have experience with learners of different ages. These practitioners must also have detailed knowledge of the science concepts and their accompanying misconceptions and of the pedagogical knowledge of how best a concept can be approached and explored. Thus, a strong knowledge and experience base in science and in science education are essential. As one expert in both fields described science education: 'It is not rocket science, it is harder' (Education World 2021, Pompea and Russo 2021a).

44.6 Considerations in science education

We are mainly discussing 'formal' education, the education that takes place in classrooms with teachers and students. Informal, or free-choice and out of school education is also important but has been addressed in detail elsewhere (Pompea and Russo 2020). Educational efforts are always intentional; they are designed to address specific problems for well-defined educational audiences. This is necessary as schools have requirements and standards of which subjects and topics can be taught at each grade level. It makes little sense to encourage a teacher to teach a topic that shouldn't even be addressed in their class.

Education efforts are largely 'operational' rather than 'research and development' (R&D). By operational we mean working with schools and teachers to help them in their teaching that happens every day. R&D efforts are designed to develop or to try out a new approach and are conducted by university researchers or by expert practitioners. These R&D efforts are often funded by competitive government or foundation grants, in much the same way as science R&D gets its resources. We won't be discussing the many interesting new approaches being researched here. Instead, we focus on some basic foundational issues that form the basis of how Big Science organisations can be working on operational projects with local schools. Here are some basic considerations:

44.6.1 Short-term efforts are not needed

The activation energy for collaboration is significant so it makes sense to partner for longer-term efforts that are more important (and more challenging) to the community. One-off activities create visibility and public relations stories, but do not satisfy the deeper educational needs of the community. Giving real value back to the community by conducting quality education projects requires long-term commitment and support. If only public relations is desired, choose easy and short projects since these project can appear successful in a press release.

44.6.2 Teachers need guides on the side

The maxim 'Be a guide on the side, not a sage on the stage' is applicable. Teachers need human guidance and support as much as they need teaching equipment for science. Many teachers would like to have a science resource person they can ask question to about a science topic that they don't know very well. These trust-sensitive relationships take time to develop.

44.6.3 Teachers need and want additional professional development

Most elementary teachers get very little science training so this makes it very difficult for them to teach it effectively. Programs to give them this professional development and support afterwards are invaluable. However it takes experience to create a good program. The good news is that the best practices for professional development in science are understood and well described in books like Loucks-Horsley *et al* (2009).

44.6.4 Polite acceptance doesn't mean success

When evaluating an educational activity provided by others, many teachers will politely accept and perhaps even praise what is of only incremental value to them. This is partly because of the culture of education with its collegiality and civility, and partly because teachers are used to having their professional skills undervalued, questioned, and attacked. Because of this, further clarity is needed on whether an 'outreach' effort is really of educational value to their classes or is just useful to the research organisation.

44.6.5 Without local partnerships, little can be done

The close cooperation and collaboration of local stakeholders is essential to get anything done. Teachers need the permission and support of their department heads, principals, and superintendents to participate in many professional development programs. Without that support, the time for the teachers to participate may not be made available and the resources to implement additional training may be lacking, all but guaranteeing the failure of the effort. Teachers, like many other professionals, do not like programs imposed on them and especially resent programs that are poorly planned or that do not meet their needs. Detailed discussions and long-term relationships play an important role in creating understandings and in fulfilling expectations.

44.6.6 Where did the project come *from*?

Ask honestly whether the program originated from the community and its needs or was pitched by the science organisation to the community. There is a stark difference between finding a venue for a program that you want to do (e.g. a science lecture series by scientists) and finding out whether your needy non-scientific community really wants and needs that particular program. Projects that focus on, and mesh with the school curriculum are often the most valuable to teachers.

44.6.7 Educational audiences are not all equal

There is an essential distinction to make about choosing your educational audience. Educational programs from research institutes often go the schools where their better paid employees send their children. This is often in quite a different area than the facility. For example, the children of employees might be in schools in a suburban area nearby or in a distant borough with already excellent schools. Or perhaps the employees send their children to an already excellent private school. Conversely, if the organisation operates a research station in a rural area or on indigenous land, it is unlikely that employees send their children to the schools near the research station. When choosing an audience, ask 'Where will the most value be created?' Then, aim the program at that audience. Helping the wrong school is gilding the lily. There is no need for excess, when many are lacking the most basic.

44.6.8 Some educational projects are hard

An educational program, if well designed, will have the most value by going to a place where it is needed desperately. However, going to such a place is not easy. For example, many city-centre schools and schools in core neighbourhoods have started to resemble prisons in the United States, and access by outside visitors is restricted. These programs may be more frustrating and take more effort. Some programs are designed to pluck the easy to reach, low-hanging fruit. But don't confuse that kind of success with that from programs that are aimed at larger and harder to reach goals, and which may have great rewards if they succeed.

44.6.9 Projects rely on co-creation to meet participant needs

Developing authentic partnerships for education is much harder than creating your own outreach events or educational programs independently of your audience or users. For example, a research institute's program that creates teacher training workshops without direct involvement of teachers and the school district is sending this message:

> We think there is a need for what we want to do. We have decided by ourselves to proceed to do this. We have designed something by ourselves, since this was easier. We think highly of our own approach and choices, and we hope you like it, too. Who wants to come and get this training? When you finish, we hope it was helpful, but we would rather not know too much. We will do it again sometime, as we feel that is was useful to you. Thank you for telling us that it was useful. (See section 44.6.4 for the translation of polite statements by teachers.)

44.6.10 Elementary schools are the most needy

Teaching at the elementary school level is the most challenging type of teaching (Pompea and Russo 2021a) and incorporating science into these classrooms has always been particularly difficult. Spend time in these classrooms if you are sceptical. Elementary teachers often have large classes and many different kinds of learners who are at many different levels. The teacher is responsible for many areas of instruction and is under extreme pressure in many cases. Encouraging robust science education programs in these classrooms is critically important (National Academies of Sciences 2022).

44.7 The situation of science education today

Research labs often interact most strongly with upper-middle class suburban and private schools and draw their conclusions on the educational system from those experiences. The needs of those schools and their students may not be typical. Here are some common concerns from more typical public schools worldwide:

- Public schools have inadequate technology, a lack of books, and poor or no science lab equipment.
- There is a shortage of trained teachers in the science disciplines at all grade levels. Many science teachers are teaching outside their field of expertise.
- There is a strong disconnection between science classrooms and how science is done, with many science teachers having little practical experience in doing science (Pompea and Russo 2021a).
- Elementary and primary grade teachers have little training in science content or in science teaching pedagogy and may not teach very much about science.
- Teaching has not attained the full status of a profession or guild (Tobias and Baffert 2010). Disrespect for teachers and their lack of autonomy seems to be getting worse each year.

In the United States there are additional problems. School segregation and unequal school financing is still ubiquitous. The emergence of charter schools and the expansion of school voucher systems have undermined a once strong public education system. Government leaders such as former Secretary of Education Betsy DeVos and her predecessors have also contributed to the undermining of public schools by promoting the contrived virtues of private schools over public schools (Ravitch 2014, 2016, 2017). In the United States, the number of teaching certificates awarded for secondary school teaching in science, technology, engineering, and math is dropping precipitously (Feder 2022).

Another large problem in the US, but that also occurs worldwide, is the increased attack on science and science education, often by the same groups attacking the value and autonomy of public education. The 'Scopes Monkey Trial' of 1925 convicted a science teacher of violating a Tennessee law prohibiting the teaching of evolution. Although this law was taken down in recent years, there is very little teaching of biological evolution in many states. Organisations such as the US National Academy of Sciences have worked hard to counteract such teaching prohibitions (National Academy of Sciences 1998) but many issues remain.

Laws are now being passed to hinder teaching about the ancient Earth, the Big Bang, and climate science. The argument is that these are inappropriate 'controversial' topics and should not be taught in schools. There are numerous proposals to require schools to teach 'creation science' or 'intelligent design' in science classrooms. Most US textbook publishers have significantly altered school science textbooks to make them more acceptable to anti-science textbook committees. This situation will not solve itself.

The current attacks on science and the science of global climate change have tremendous repercussions for our planet. The next generation of children must have the opportunity to develop an understanding and appreciation of how science works and the value of scientific data and evidence in order to address this important issue.

44.8 The business model perspective

A research institute can help itself short-term (i.e. achieve a good return on investment) by vigorously communicating its successes: its science discoveries, service to its users, and the relevance of its research to society (Entradas *et al* 2020). Similarly, from a short-term business perspective, research institutes can be effective in generating positive community responses from short-term, ad hoc, self-serving, or unwanted educational efforts. Is this the right way to evaluate a research institute's efforts? That value can be obtained by these educational efforts perhaps illustrates that the community is grateful for the smallest efforts made, especially when none have been made before.

However, the gains obtained through self-serving approaches rarely help develop the deeper community trust or critical mass of supporters in Big Science projects (and their sponsoring institutions) that is essential today for many ambitious Big Science projects. The process of building an iterative and equitable relationship with a community is much more difficult (Kahanamoku *et al* 2020).

44.9 Adding an ethical perspective

We feel that it is time for research institutes to move beyond basing their educational actions on the direct return for the organisation and its stakeholders (Entradas *et al* 2020). This type of justification can be convincing, but it is also rather narrow. The case for promoting diversity in organisations has often been justified in this same narrow way, namely that diversity improves the productivity of the organisation. While diversity does improve the productivity, adding ethical considerations can strengthen the argument for encouraging a diverse work force (Haacker *et al* 2022).

Ethical considerations are a key part of many decisions and adding it to strategic planning for education broadens the perspective. Ethical considerations allow us to examine the moral questions and dilemmas associated with the choices of action or non-action that can be made by research organisations as they begin to notice the local education arena and its needs. Hard questions need to be addressed:

> Do individuals and their organisations have an ethical obligation to address and repair the problems of the world, or at least the ones closely related to their profession, culture of science, and support for science?
>
> If the local science education situation is seriously unwell, is there a duty to help or rescue it, even if the rescue effort is hard?

Can science organisations behave like the ploughman in Auden's 1940 poem 'Musée des Beaux Arts', who chose not to notice the fall of Icarus? These questions frame the ethical considerations that need to be considered in addition to considering the direct and short-term value that comes back to the organisation from its education efforts.

44.10 Educational stewardship: preserving the educational landscape

A second perspective that can be added to the ethical perspective is that of 'stewardship' of science education and of the public school system. A culture of stewardship and partnership can frame how research organisations support and improve the landscape of science education.

Just as environmental stewardship is a key element in the maintenance of fragile places, the adoption by research institutes of an educational stewardship approach will help them work with others to maintain the public schools and to reinforce their important role in our society. A stewardship approach also can help support the culture of scientific research and discovery so essential to the future of research. With some reflection, we should realise that there are strong long-term connections between science teaching and science research, that both are vulnerable, and that by working together they can both flourish.

The protection and caring for land is one of the original meanings of stewardship. This caring perspective includes all elements of the present land, and the cleaning up and repairing of damage from earlier times. In true stewardship, this responsibility exists even if the present stewards didn't cause the damage or realise its extent. Stewardship implies a responsibility to preserve for future generations and includes making decisions that avoid short or long-term harm. The research literature on land-use planning and stewardship is instructive in its descriptions of how land can be protected and nurtured in the face of diverse threats (Wallace *et al* 2005).

For educational stewardship, the same general principles of land stewardship apply. Educational stewardship can preserve the landscape of the public education system and its main inhabitants: the teacher, students, and administrators. It involves maintaining and improving the parts of the educational landscape that have not been taken care of previously. This culture of stewardship is not about replacing public education with private versions. As in land stewardship, stewardship demands a commitment to preserving what is there, even if has been degraded under the guise of educational reform or improvement.

Research labs and large user facilities are often based in rural areas or on indigenous land. This is particularly true for astronomical observatories such as the Square Kilometer Array in South Africa and Australia and for high energy physics facilities, such as the Pierre Auger Observatory in Argentina. Rural and indigenous communities often struggle with poverty, youth unemployment, low teacher retention, poor Internet service, and overwhelmed health facilities. The Covid-19 pandemic has accentuated many of these problems even further. In particular it has highlighted the importance of computers and the internet in providing equitable educational access.

At first glance, research institutes seem ill-prepared to become stewards of either the public school system or the science education ecosystem since they have few education professionals on their staff. However, 'Big Science' projects and the facilities that support them are actually quite rich in resources that are valuable for educational stewardship and even for the broader stewardship of our planet. In particular, the interest and time of their employees is a valuable resource.

Science organisations have a large number of highly educated scientists, engineers, and facility users as well as a diverse and often international work force. They also have a strong supporting infrastructure of management, facilities, purchasing, and accounting that can support organisational missions. Finally, their core values are highly congruent with the core values in education. These attributes can be used to great advantage in education and community engagement efforts. The longevity of research facilities also can contribute to developing strong community roots and allegiances, which are essential to stewardship.

44.11 What research and development organisations can do: a 3 + 5 proposal

Our starting (and ending) point is that research institute resources should be used in the schools and other situations where the need of learners are the greatest. In our experience this means the elementary or primary public schools, i.e. the schools that serve students who are 11 years old or younger. Educationally vulnerable students are often found in the postal code areas that are less affluent. Students at private schools or a public school situated in an upper-middle class neighbourhood are usually much better served. Public schools that serve inner city, rural, and indigenous residents usually have the most needs and a clear case for equity and fairness can be made. Where there is a dire situation for the students, that is where the help is needed, and also where the help can do the most good, in the long-run.

We propose a 3 + 5 program to describe a moderate level of educational support that can be productive. If research organisations and the companies that provide support to them could commit to 3% of their budget to education and 5% of any employee's time to work on the organisation's educational programs, this would be a step in the right direction. This is the essence of the program: a minimum of 3% of the budget and 5% of the time for some employees.

In this plan for action, 3% of the regular or core budget would be solely for education staff and the resources they directly need. Not included in this 3% would be money, people, or time for organising, creating, and supporting booths, meetings, and conferences; town halls; press releases; social media production; graphics production; web site improvements; working with film crews; arranging interviews with media; rotary club talks; or for answering communications inquiries. It would be used to hire or support education professionals who can organise and direct the broader range of truly educational activities for the staff who are interested. The money can also be used to provide the resources needed for these efforts. Grant money on top of the 3% can be used to organise special projects or educational experiments.

The second number of 5% represents the amount of regular time allocated for staff who are motivated to participate in these organised and productive educational activities. These are the activities that were created in a dialogue with the local stakeholders. If even 5%–10% of the organisation's employees participate in this program, 5% of their time is a significant effort that can supplement the efforts of the educational specialist. An employee who has 5% of their time available for education work means that they can spend one full day a month preparing and actually working with schools or teachers. This represents a great improvement over having one or a few education professionals in a large organisation trying to work with schools or students.

The staff volunteers can start that day by not coming to work and spending time in a school, learning their needs. Note that this 5% of time is not allocated for working at their own children's school (unless it is a highly needy one), or in some out of school activity that they feel strongly about. We don't discourage these volunteer activities, but this time is meant to be in direct service of their parent organisation in a well-planned program. This time is for employees to participate in some solid activities or projects organised and supervised by their own education professional(s). This might be a tutoring program, or it might be to support a professional education and training program for teachers, for example.

Paid employee time is a powerful resource and represents a deep organisational commitment that can augment the expertise of the organisation's education professionals. Allocating and freeing the time for employees who wish to participate to do educational community service is critical, and represents a serious commitment. Encouraging employees to volunteer time in the evenings and on weekends for educational causes is not ineffective, but it gives the message that these efforts are not a priority.

If these 3 + 5 programs are implemented, they need to run long-term, and not be subject to budget cuts or attempts to subvert them through merging of budgets and other common ways that shift operational priorities.

44.12 Moving from local support to national support

The allocation of resources to contribute locally to schools and the local educational ecosystem is a most important part of the picture. Many organisations have created successful local educational partnerships and have then gone beyond these local contributions in order to contribute on a larger scale. The national and international character of Big Science projects often pushes a project to contribute beyond the local level as the organisation evolve educationally and partner more widely (Pompea *et al* 2015).

There are many examples from our own experiences of how research organisations may contribute to larger regional, national, or international efforts to promote science education. We will use a few from our field of physics and astronomy that may serve as models for other fields of science. We especially encourage the development of programs that encourage science interest and develop science skills in younger and pre-secondary school children.

For example, *Universe Awareness* (UNAWE) is an international program for elementary school students (Miley *et al* 2019) that trains teachers and develops inexpensive innovative materials for classrooms. The *Hands-On Optics* program was a United States program funded by the National Science Foundation for 11–13 year old school students. It developed innovative curricular materials and kits and trained museum educators and teachers across the country (Pompea *et al* 2005).

The *Galileoscope* program (Pompea *et al* 2010) was an international program to develop and teach with small, high-quality telescope kits. It was also used in Arizona as a state-wide program for 10 year old students who built their own telescopes and used them at star parties (Pompea *et al* 2013, 2015). Some projects, such as the *Colors of Nature* project, focus specifically on new approaches that encourage girls in science activities (Tzou *et al* 2014, Conner *et al* 2017).

Other innovative projects with global reach that were developed at research organisations include the International Year of Astronomy 2009 (Russo *et al* 2009) and the International Year of Light 2015 (Dudley and González 2015, Curticapean *et al* 2016). Additionally, there are many educational initiatives such as the IAU100 and Pale Blue Dot led by professional societies such as the International Astronomical Union (van Dishoeck and Elmegreen 2018, van Dishoeck 2019). A number of highly successful programs for older students have emphasised research using observations and data from research telescopes on the ground and in space (Buxner 2010, Doran *et al* 2012, Fitzgerald *et al* 2014, Rebull *et al* 2018a, 2018b). Many other exemplary programs are described in Pompea and Russo (2020). These programs have been highly effective either to encourage science interest or to suggest new approaches that can be incorporated into current education programs.

44.13 Conclusion and summary

We have described the resources Big Science has to offer to its local community and to the larger science education ecosystem. These resources are valuable and need to be allocated and committed in an intentional way for effective results. The efforts need to be given to those who can benefit the most from them, namely public elementary school students and their teachers. Educational projects will only succeed if they are based on the best practices elucidated by science education research. A key best practice is to form strong, authentic partnerships and to co-create programs with educational stakeholders and organisations. Stronger educational engagement with the local, regional, national, and international education community is not only possible, but highly desirable from an ethical and stewardship perspective.

We've suggested a sensible allocation of money and time to implement strong local educational partnerships in our 3 + 5 approach. For some research organisation this appears to be a large stretch to implement such a program. These organisations feel vulnerable working in this new direction and worry about failing. This fear should not be an impediment to getting started. Perhaps they should take heart from the words of a doctor who worked under very difficult circumstances. Paraphrasing the Talmud, he said: 'The day is short. The task is difficult. It is not possible to complete it. But we are forbidden not to try' (White 2013).

We encourage Big Science organisations to try, to commit themselves to the difficult task of supporting science education. Although there are easier ways for an organisation to look good to their community in the short-term, the long-term benefits are powerful and important to the science organisations, to their people, and to the new communities they now will serve.

Acknowledgements

The work of SMP is supported by NSF's NOIRLab, which is managed by the Association of Universities for Research in Astronomy (AURA) under a cooperative agreement with the National Science Foundation.

The opinions expressed by the authors are theirs alone and not that of NSF's NOIRLab, AURA, Inc., the National Science Foundation, the University of Arizona, or Leiden University.

References

Buxner S R 2010 *Exploring the Impact of Science Research Experiences for Teachers: Stories of Growth and Identity* (Tucson, AZ: The University of Arizona)

Christensen L L 2007 *The Hands-On Guide for Science Communicators: A Step-by-Step Approach to Public Outreach* (Berlin: Springer)

Conner L D C, Tzou C, Tsurusaki B K, Guthrie M, Pompea S and Teal-Sullivan P 2017 Designing STEAM for broad participation in science *Creat. Educ.* **8** 2222

Curticapean D, Vauderwange O, Wozniak P and Mandal A 2016 The International Year of Light 2015 and its impact on educational activities *Proc. SPIE* **9946** 994608

deVos A 2020 The problem of 'Colonial Science' *Sci. Am.* 1 July https://scientificamerican.com/article/the-problem-of-colonial-science/

Doran R, Melchior A L, Boudier T, Ferlet R, Almeida M L, Barbosa D, Gomez E, Pennypacker C, Roche P and Roberts S 2012 Astrophysics datamining in the classroom: exploring real data with new software tools and robotic telescopes arXiv:1202.2764

Dudley J and González J R 2015 Shining light on sustainability: a legacy for the international year of light *Opt. Photon.* **10** 1

Education World 2021 Former astronaut George 'Pinky' Nelson champions science literacy https://www.educationworld.com/a_issues/chat/chat048.shtml

Entradas M *et al* 2020 Public communication by research institutes compared across countries and sciences: building capacity for engagement or competing for visibility? *PLoS One* **15** e0235191

Feder T 2022 The US is in dire need of STEM teachers *Phys. Today* **75** 25

Fitzgerald M T, Hollow R, Rebull L M, Danaia L and McKinnon D H 2014 A review of high school level astronomy student research projects over the last two decades *Publ. Astron. Soc. Aust.* **31** E037

Haacker R, Burt M and Vara M 2022 Moving beyond the business case for diversity *Eos* **103** 9 February

Heenatigala T 2022 Decolonizing astronomy *EDU INAF* https://edu.inaf.it/in-english/Universe-world/thilina-heenatigala-decolonising-astronomy/

Jamieson K H, Kahan D and Scheufele D A (ed) 2017 *The Oxford Handbook of the Science of Science Communication* (Oxford: Oxford University Press)

Kahanamoku S, Alegado R A, Kagawa-Viviani A, Kamelamela K L, Kamai B, Walkowicz L M, Prescod-Weinstein C, Reyes M A D L and Neilson H 2020 A native Hawaiian-led summary of the current impact of constructing the thirty meter telescope on Maunakea arXiv:2001.00970

Loucks-Horsley S, Stiles K E, Mundry S, Love N and Hewson P W 2009 *Designing Professional Development for Teachers of Science and Mathematics* (Thousand Oaks, CA: Corwin)

Mamajek E, Meloy Elmegreen D, Lecavelier des Etangs A, Lindberg Christensen L, Monfardini Penteado E, Yamaoka H, Williams G and Anglada-Escudé G 2020 Public naming of exoplanets and their stars: implementation and outcomes of the IAU100 Name-ExoWorlds Global Project *Commun. Astron. Public J.* **28** 22

Miley G, Ödman C and Russo P 2019 The inspiring universe *Seeking Understanding* (Leiden: Brill) pp 119–35

National Academy of Sciences 1998 *Teaching About Evolution and the Nature of Science* (Washington, DC: National Academies Press) http://nap.edu/catalog/5787.html

National Academy of Sciences 2017 *Communicating Science Effectively: A Research Agenda (2017)* (Washington, DC: National Academies Press)

National Academy of Sciences 2018 *The Science of Science Communication III: Inspiring Novel Collaborations and Building Capacity: Proceedings of a Colloquium* (Washington, DC: National Academies Press)

National Academies of Sciences, Engineering, and Medicine 2021 *Pathways to Discovery in Astronomy and Astrophysics for the 2020s* (Washington, DC: National Academies Press)

National Academies of Sciences, Engineering, and Medicine 2022 *Science and Engineering in Preschool Through Elementary Grades: The Brilliance of Children and the Strengths of Educators* (Washington, DC: The National Academies Press)

National Science Foundation 2022 BIGDATA: F: DKM: Collaborative research: making Big Data active: from petabytes to megafolks in milliseconds *NSF Award* 1447720 Michael Carey P.I., University of California Irvine https://nsf.gov/awardsearch/showAward?AWD_ID=1447720

O'Donnell C and Scott K A 2022 Culturally responsive astronomy education: using a critical lens to promote equity and social justice https://caodonnell.github.io/files/culturally_responsive_astronomy_preprint.pdf

Pompea S M *et al* 2010 International Year of Astronomy 2009 Cornerstone Projects: what's available for you *ASP Conf. Series* **431** 86

Pompea S M, Fine L W, Sparks R T, Walker C E, Dugan C L and Dokter E F 2013 The Galileoscope project: community-based technology education in Arizona *Education and Training in Optics and Photonics* (Washington, DC: Optical Society of America) p EWC3

Pompea S M, Johnson A, Arthurs E and Walker C E 2005 Hands-on optics: an educational initiative for exploring light and color in after-school programs, museums, and hands-on science centers Proc. Ninth Int. Topical Meeting on Education and Training in Optics and Photonics (Marseille)

Pompea S M, Walker C and Sparks R 2015 The evolution of optics education at the US National Optical Astronomy Observatory *Proc. 12th Int. Conf. on Education and Training in Optics and Photonics* vol 9289M F P C Martins Costa and M Zghal (Washington, DC: Optica), ETuC1

Pompea S M and Russo P 2020 Astronomers engaging with the education ecosystem: a best-evidence synthesis *Annu. Rev. Astron. Astrophys.* **58** 313–61

Pompea S M and Russo P 2021a Improving science education: it's not rocket science—it's harder! *Phys. Today* **74** 26

Pompea S M and Russo P 2021b Ten simple rules for scientists getting started in science education *PLoS Comput. Biol.* **17** e1009556

Prescod-Weinstein C, Walkowicz L M, Tuttle S, Nord B and Neilson H R 2020 Reframing astronomical research through an anticolonial lens—for TMT and beyond arXiv:2001.00674

Ravitch D 2014 *Reign of Error; The Hoax of the Privatization Movement and the Danger to America's Public Schools* (New York: Vintage)
See also About Diane *Diane Ravitch* https://dianeravitch.com/

Ravitch D 2016 *The Death and Life of the Great American School System: How Testing and Choice Are Undermining Education* (New York: Basic)

Ravitch D 2017 Don't like Betsy DeVos? Blame the Democrats *New Republic* pp 109–18

Rebull L M, Fitzgerald M, Roberts T, French D A, Laurence W, Gorjian V and Squires G K 2018a The NASA/IPAC teacher archive research program (NITARP) arXiv:1804.08743

Rebull L M, French D A, Laurence W, Roberts T, Fitzgerald M T, Gorjian V and Squires G K 2018b Major outcomes of an authentic astronomy research experience professional development program: an analysis of 8 years of data from a teacher research program *Phys. Rev. Phys. Educ. Res.* **14** 020102

Russo P, Cesarsky C and Christensen L L 2009 SpS2-The International Year of Astronomy 2009 *Proc. Int. Astron. Union* **5** 559–609

Russo P and Christensen L L (ed) 2010 *Final Report* (Paris: International Astronomical Union) http://astronomy2009.org/resources/documents/detail/iya2009_final_report/

Tobias S and Baffert A 2010 *Science Teaching as a Profession: Why It Isn't, How It Could Be* (Arlington, VA: National Science Teachers Association Press)

Tzou C, Conner L, Pompea S and Guthrie M 2014 *Colors of Nature: Connecting Science and Arts Education to Promote STEM-Related Identity Work in Middle School Girls* (Boulder, CO: International Society of the Learning Sciences)

van Dishoeck E F and Elmegreen D M 2018 The IAU strategic plan for 2020–2030: OAO *Proc. Int. Astron. Union* **14** 546–8

van Dishoeck E F 2019 The IAU and education: introduction *Proc. Int. Astron. Union* **15** 2–7

Wallace G N, Barborak J R and MacFarland C 2005 Land use planning and regulation in and around protected areas: a study of legal frameworks, best practices and capacity building needs in Mexico and Central America *COPA* http://copa.acguanacaste.ac.cr:8080/handle/11606/153

White W L 2013 A life in addiction psychiatry: an interview with Dr Herb Kleber https://archive.org/details/2013-dr.-herb-kleber/page/n3/mode/2up
Also; Seelye K Q 2018 Herbert D Kleber, pioneer in addiction treatment, dies at 84 *New York Times* 17 October https://www.nytimes.com/2018/10/15/obituaries/herbert-d-kleber-dead.html

IOP Publishing

Big Science in the 21st Century
Economic and societal impacts
Panagiotis Charitos, Theodore Arabatzis, Harry Cliff, Günther Dissertori, Juliette Forneris and Jason Li-Ying

Chapter 45

From Big Science to open science

Kamran Naim, Jelena Brankovic, Anne-Gentil Beccot, Sunje Dallmeier-Tiessen, Micha Moskovic, Pamfilos Fokianos, Alexander Kohls, Antonios Papadopoulos and Milosz Artur Zielinski

CERN has been known for its large-scale projects, experiments, and collaborations to study the foundations of the Universe. The laboratory has been a driver for innovations to conduct Big Science. Since the foundation of CERN, open science has been part of the CERN research practices and its culture. CERN continues to push the boundaries of open science on a global scale. This chapter provides an overview of the major pillars of CERN's open science activities and achievements.

The community at CERN has been disrupting traditional publishing practices with its preprint culture and continues to drive innovation in scholarly communication. The community-wide solutions for open access—from SCOAP3 to repository technologies—have been serving as role models or solutions for other communities' challenges. Similarly, CERN has also been taking a lead on open data and delivered large-scale solutions for big open data releases on the web. Lessons learned for other Big Science disciplines are discussed, including one of the oldest open science pillars, which is increasing in importance, open software. In this chapter the history and links to the community challenges and solutions are explained. Although less well-known, since 2008 CERN has been a driving force in open hardware initiatives with the result of standardized licensing and repositories that have become indispensable to this movement. This chapter sheds more light on the developments and specific needs of this open science pillar. Finally, the chapter concludes by summarizing the lessons learned from CERN's large-scale open science activities for other stakeholders and communities.

45.1 Introduction

On 19 June 2020, following almost two years of deliberations, the highest governing authority of the European Organization for Nuclear Research (CERN), the CERN

Council, announced its updated strategy to guide the future of particle physics in Europe within the global landscape. This document, the European Strategy for Particle Physics (ESPP), represents the most important strategic document for CERN, proposing a vision for the future of the field in both the near- and long-term, with an ambition to maintain Europe's leading role in advancing particle physics and its associated technologies. While setting the future research agenda for CERN, the recommendations of the ESPP further highlighted not just the scientific impact of particle physics, but also its contributions towards technological advancement, human capital development, and broad societal impacts.

Significantly the practice of open science was given strong endorsement by the CERN Council. According to the ESPP,

> 'European science policy is quickly moving towards Open Science, which promotes and accelerates the sharing of scientific knowledge with the community at large. Particle physics has been a pioneer in several aspects of Open Science. The particle physics community should work with the relevant authorities to help shape the emerging consensus on Open Science to be adopted for publicly funded research, and should then implement a policy of Open Science for the field.' (European Strategy Group 2020)

Although this statement provided for the formalization of open science through policy efforts, the values underlying the practice of open science have been embedded in CERN since it was first founded.

Established in 1954 with the objective to 'increase international scientific collaboration' by twelve European nations to rebuild their cooperation in the spirit of 'science for peace' following the devastation of World War II, CERN has made significant contributions to our understanding of the world and the Universe. The mission of the organization is to provide a unique range of particle accelerator facilities that enable research at the forefront of human knowledge, perform world-class research in fundamental physics, and unite people from all over the world to push the frontiers of science and technology, for the benefit of all. Supported through a global partnership of 23 member states who jointly finance and contribute towards the governance of the organization, CERN is home to the world's largest scientific instrument, the Large Hadron Collider (LHC), and hosts over 13 000 scientists and engineers from over 70 countries. The LHC is the epitome of Big Science: a particle accelerator of 27 km circumference buried 100 meters underground, protons accelerated to almost the speed of light, beams colliding billions of times per second observed by four detectors each the size of a cathedral, hundreds of petabyte of data recorded, and collaborations of several thousands scientist analysing them.

The core scientific goals of CERN's research program are to attain a fundamental understanding of the laws of physics, to explain the origin of mass, and to understand the dark matter in the Universe. Insights into any of these key areas of exploration would dramatically change our view of the world. The pursuit of this scientific agenda is undertaken by experimental collaborations of thousands of experimental particle physicists who, using machines like the LHC, study collisions between highly accelerated particles to get insights into the constituents of matter. They are supported in their work by theoretical particle physicists, who formulate hypotheses and theories to accommodate and predict experimental findings. As stated in CERN's mission, the scientific goals of the organization are intended 'for the benefit of all', implying that broader social benefit is pursued in tandem with its research agenda. The practice of open science is increasingly recognized as one of the key mechanisms on delivering this broader social benefit, and was even enshrined in CERN's founding convention back in 1953, which states that '... the results of its experimental and theoretical work shall be published or otherwise made generally available' (CERN 1953).

Open science at CERN is not pursued, however, as a matter of obligation. It rather reflects the social imperative of Big Science, representing the collective moral and financial responsibility, not only to member states but to the global scientific community, to advance the boundaries of human knowledge in a manner that delivers optimal scientific and social outcomes. This recognition of the importance of open science is gaining in momentum, particularly among political entities, funding bodies and research agencies, seeking to optimize their investments in accelerating scientific progress. Deriving benefits from the practice of open science requires implementing technologies, practices and policies to change the way in which we conduct and perceive research. The challenges posed therefore are not merely technological but sociological, and affect all aspects of the research enterprise: from how we conceptualize research, to how we collect and store data, preserve analysis elements to ensure reproducibility, share and communicate findings, conduct peer review, and assess the results of projects and grants.

The community at CERN has helped to pioneer many elements of what is described today as open science: we helped build the open Internet, ushered in the preprint culture, pioneered initiatives to allow discipline-wide open access to scientific publications, and facilitated the openness of experimental data. Yet, the complexity of the research undertaken at CERN makes the comprehensive pursuit of open science an ongoing and ever-evolving effort. The thousands-strong research community at CERN is dynamic, with scientists from across the world engaged in multiple experimental collaborations that operate largely independently of each other and have different tools and practices. The high-energy physics research conducted at CERN is one of the most data-intensive branches of science: particle collisions at CERN produce about one petabyte of collision data per second (Gaillard 2017). The complexity, scale and value of these data presents unique and unprecedented challenges for the research community at CERN. The responsible stewardship, analysis and preservation of these data requires a supporting organizational research culture, but also a range of tools and services to optimize how data are used, verified, and built upon to advance knowledge.

This chapter describes the ecosystem of initiatives, projects, and technologies that have been developed at CERN to maximize the impact of our 'Big Science' research agenda, through building an open science infrastructure that is effective, collaborative, and responsive to the needs of the scientific community. We aim to demonstrate that despite the complexity of the research undertaken at CERN, open science can be advanced through concerted efforts and, as such, the CERN example could serve as an inspiration for the global scientific community.

Open science is defined by the United Nations Educational, Scientific and Cultural Organization as

> an inclusive construct that combines various movements and practices aiming to make multilingual scientific knowledge openly available, accessible and reusable for everyone, to increase scientific collaborations and sharing of information for the benefits of science and society, and to open the processes of scientific knowledge creation, evaluation and communication to societal actors beyond the traditional scientific community. (UNESCO 2021).

For the purpose of this article we focus on open access to research publications, open data, open software, and open hardware, which are all key pillars of open science. It is important to note, however that other relevant open science movements and concepts such as open tools, open research assessment, open peer review, and citizen science also contribute to the progress and transformation of modern scholarly communication and how we preserve, share, and publish research results. In particular for Big Science and its diverse and large-scale research outputs it is essential to scale and link such initiatives early on.

45.2 Open access at CERN

The best-known of the movements within the umbrella of open science is undoubtedly open access (OA), the movement that aims to grant anyone, worldwide with free and immediate access to the results of scientific research. Although OA was formalized as a global movement with the Budapest Open Access Initiative in 2002[1], and the Bethesda Statement on Open Access Publishing[2], and Berlin Declaration on Open Access to Knowledge in the Sciences and Humanities[3] in 2003, the research community at CERN had been practicing a form of OA through its culture of sharing of preprints for over four decades. Physicists actually mass-mailed manuscripts intended for publication in hundreds of copies for maximum circulation to colleagues around the world for comment prior to publication. Librarians at CERN then compiled a catalog of these documents which became a backbone of scholarly communication in the discipline. The practice of sharing preprints was significantly and radically reformed with the advent of the Internet and the establishment of the first research repository, arXiv at Los Alamos Laboratory in 1991. The wider distribution of knowledge was achieved following the invention of the World Wide Web in 1989 by Tim Berners Lee while working at CERN, and its underlying source code subsequently being placed into the public domain by CERN in 1993. Thanks to the speed with which they made results available, repositories re-enforced the role of

[1] https://www.budapestopenaccessinitiative.org/read/
[2] http://nrs.harvard.edu/urn-3:HUL.InstRepos:4725199
[3] https://openaccess.mpg.de/Berlin-Declaration

preprints—what is now termed 'green OA'—as the lifeblood of HEP scientific information exchange.

Having laid the groundwork for the digital revolution in scientific communication, CERN began to play a central role in emphasizing the importance of science in the information society. In particular, the organization recognized that the cost of traditional scientific journals were becoming a great burden for library budgets at CERN and other scientific institutions (the 'serials-crisis'), and began to develop some ideas on how to shape an open future for scholarly communication in the discipline of high-energy physics and beyond.

45.2.1 Achieving discipline-wide open access

Although the discipline of HEP had already pioneered green OA through the self-archiving of over 90% of HEP preprint articles in the arXiv repository, there remained a robust consensus of the value and contribution provided by the high-quality journals in the discipline, for the quality control delivered through their processes of peer review and editing, and for their role as vehicles for the career development of individuals and for assessing the outputs of research groups (Schimmer 2013). As such, the scientific community at CERN recognized the need for action to convert the final published versions of the approximately 7000 articles published each year in the discipline to open, i.e. achieving so-called 'gold OA' for HEP. With the support of the Director-General at the time, Dr Robert Aymar, and international leading actors from the particle physics community, a working party was formed in 2005, and a report entitled 'Towards open access publishing in high energy physics' (Bianco *et al* 2007) was published by the group in 2007.

In this report, the working party proposed the launch of an initiative with two goals: first, to provide open and unrestricted access to all HEP literature in its final, peer reviewed form; and second, to contain the overall cost of journal publishing while ensuring sustainability. In the spirit of 'Big Science' and echoing the LHC experiments' models of international collaboration, the proposal of the working party was to form the Sponsoring Consortium for Open Access Publishing in Particle Physics (SCOAP^3), a coordinated collective action of funding agencies (including governments) and laboratories involved in the field, alongside leading national libraries and library consortia, that would use its joint leverage to achieve its specified goals (figure 45.1).

The funding concept of SCOAP^3 is based on the understanding described in the seminal white paper of the Max Planck Digital Library, 'Disrupting the subscription journals' business model for the necessary large-scale transformation to open access' (Schimmer *et al* 2015), that there are sufficient financial resources in the subscription system to sustainably fund open access, and to deliver significant savings to libraries. Facilitating this process requires that journal subscription packages including the leading journals in the discipline be replaced by subscriptions to individual journals (a practice known as 'unbundling'), and that subscription income for these leading titles instead be replaced by a single revenue stream from SCOAP^3 that would underwrite the open access publication of the global HEP literature. This represents

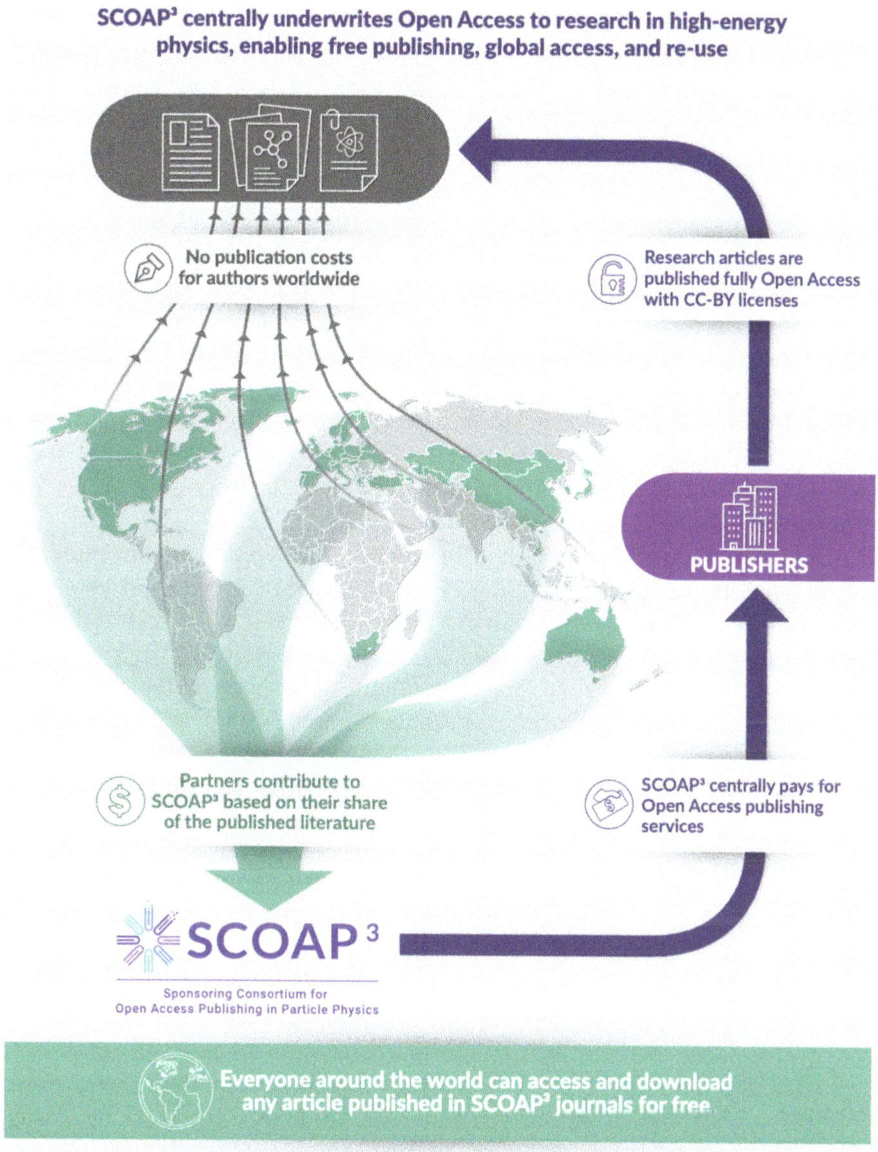

Figure 45.1. The SCOAP^3 model. This image is licensed by the SCOAP3 Consortium under a Creative Commons Attribution 4.0 International License.

a global redirection of funding by thousands of subscribing entities around the world paying for access to read content in HEP, to instead reinvest into a single fund that would arrange for the payment of open access publishing services to be delivered by the leading outlets in the discipline. By centrally covering the costs involved in providing open access, SCOAP^3 would effectively remove subscription fees for individual journals and any expenses scientists might normally incur to publish their articles openly (i.e. author processing charges (or APCs)). This way, authors from anywhere in the world would be able to publish their work in participating journals without any financial burden, with licensing rights arranged so that they would retain copyright of their work. To coordinate this activity, a steering committee was established and tasked with developing an innovative tender process, while simultaneously working to build a coalition of stakeholders across research agencies, funders, and libraries.

Following an extensive process of consultation across global stakeholders, SCOAP^3 launched an open tender in June 2012, inviting bids from commercial, society (non-profit), subscription as well as born-OA journals. Bids from publishers were evaluated on the basis of delivering the 'best value for money', determined by three factors: price per article; quality (determined by impact metrics as well as licensing and format criteria); and an agreement to reduce subscription prices, according to the volume of articles that would be published OA. Through this mechanism, contracts would be awarded that would establish links between quality and price as well as the total volume of articles published and price; features that did not previously exist in the subscription model.

The initiative started operation in 2014 entering into contracts with publishers of leading journals in the discipline, funded and supported by representatives in each of several dozen countries who each worked at the national level to secure the funding of their contribution to SCOAP^3. Initially predicated on being supported by the redirection of subscription reductions issued by publishers to their customers in SCOAP^3 partner countries for the amounts commensurate to the price of the journals in each subscription package into the central SCOAP^3 fund, the funding contribution from each partner is determined relative to their total share of publishing in the HEP journal literature, termed their 'fair-share'. Partners also contribute an additional 10% to underwrite publications from scientists from countries that cannot be reasonably expected to contribute towards the consortium.

SCOAP^3's governance is modeled on large research collaborations, similar to those of the large CERN experiments. SCOAP^3 partners enter into collaboration agreements with CERN through memorandums of understanding, and their participation in the initiative grants them a seat on the governing council (GC) of SCOAP^3, the collective governance entity of the partnership that steers and votes on major strategic matters related to the consortium. The GC also appoints committees that oversee and support the regular operations of the initiative, and appoints working groups of members from across the consortium to work on new strategic initiatives. As the host organization, CERN provides for the operation of the consortium, ensures the faithful execution of the contracts, and provides in-kind support for the financial, legal, and technical aspects of the operation of the initiative.

SCOAP^3 is currently in its third phase of operation (2020–24) and is the world's largest open access initiative, with the collaboration consisting of over 3000 libraries, national funding agencies, and research institutions from 45 countries and three intergovernmental organizations. Since its launch in 2014, SCOAP^3 has made almost 50 000 scientific articles published by authors from over 120 countries freely accessible to everyone. Through the partnership between libraries and the world's leading outlets in high-energy physics, today, nearly 90% of scientific articles in this field are available to everyone without restriction and authors from anywhere in the world can publish their papers without any financial barriers.

The program delivers significant benefits across a diverse range of stakeholders. First, researchers across the world are able to publish in SCOAP^3 journals without any cost, delivering an equitable approach to open access that does not disadvantage authors from economically disadvantaged communities as other predominantly author-fee based approaches to open access (such as the payment of APCs, etc) do. This approach further delivers dramatic benefits for institutions, who save on subscription spending as well as administrative expenses (on article processing workflows, etc) that are typically required for arranging open access. Publishers also benefit from the SCOAP^3 structure, which arranges for stable financing from a single contractual partner for an entire discipline's publishing output. The introduction of a 2.5 million euros growth account in phase 3 of the initiative further financially incentivizes publishers whose journals are increasing in popularity and total article output. Finally, readers of HEP literature worldwide benefit from free and unrestricted access to the literature in the discipline, previously only accessible to institutions with the financial means to pay for subscriptions. The SCOAP3 model and its key benefits are summarized in figure 45.1.

Having successfully converted the majority of the journal literature in the discipline to open, in 2019, SCOAP^3 decided to expand the scope of the initiative to include academic books in high-energy physics and related disciplines (such as accelerator physics and instrumentation). Working in partnership with leading publishers of books in the discipline, the SCOAP^3 for Books initiative was formally launched in 2021, and will convert over 100 important texts in the discipline to full OA, enabling unrestricted and immediate access to important texts in HEP for students, faculty, and researchers around the world. The books will all be made available via the OAPEN library[4], which will support the dissemination of the books to a wide range of organizations such as academic libraries, library intermediaries, research institutions, and discovery services (including Google Scholar), to ensure optimal global visibility for these open access books.

45.2.2 CERN open access policy and its impact

To formalize and further support open access to the research output from its experimental collaborations, CERN issued its first formal open access policy in October 2014, requiring OA publishing of all original high-energy physics results, whether experimental or theoretical. In 2017 the policy was expanded further to all publications in related fields such as instrumentation, scientific computing, and accelerator physics (CERN Scientific Information Policy Board 2017). Strengthened by historical community practices of research sharing, the CERN open access policy enjoys high levels of compliance, and the overwhelming majority of CERN research is now published OA. In 2020 CERN authors published a total of 982 scientific articles, and 85% of scholarly articles from CERN were published immediately OA,

[4] oapen.org

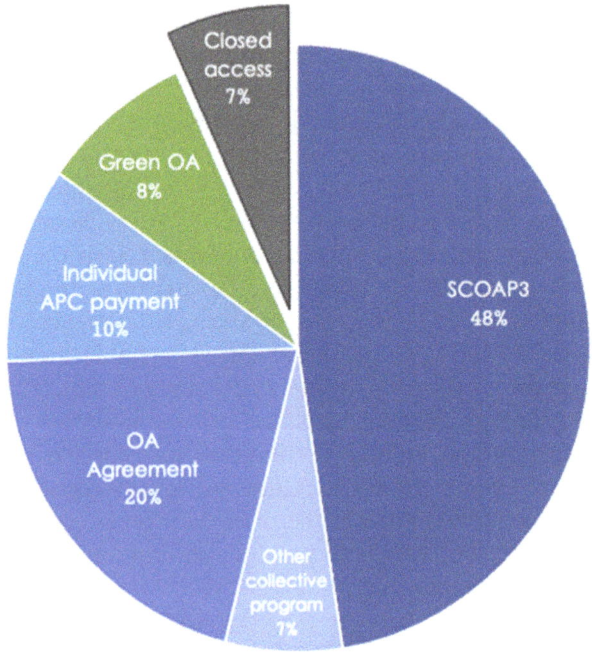

Figure 45.2. Open access status of CERN publications (2020).

with an additional 8% of the articles deposited as a preprint in arXiv. A total of 93% of CERN research articles during 2020 were freely accessible and reusable by researchers worldwide (CERN Scientific Information Service 2020).

As demonstrated in figure 45.2, the majority of CERN's research output is published OA through its participation in the SCOAP^3 project. However, SCOAP^3 only supports high-energy physics articles. The research at CERN is significantly more diverse. Therefore, additional measures are needed towards the goal to publish all CERN articles OA. The next largest contributor to CERN's share of OA articles is the so-called transformative OA agreements the CERN Scientific Information Service is systematically entering into with major publishers. These are primarily 'read and publish' agreements, which seek to shift the contracted payment from libraries to publishers away from subscription-based reading and towards OA publishing. Through these arrangements the CERN affiliation of corresponding authors is automatically detected by the publishers, and articles submitted to journals covered by the agreements are automatically published OA. As a consequence of the diversity of CERN research, a number of scientific articles are also published outside of SCOAP^3 and the transformative agreements. This includes publishing in 'other collective programs', such as SciPost, a non-profit publishing platform that is supported by consortial contributions without the reliance on author fees. To facilitate the policy of achieving full OA to the organization's research output, CERN continues to pay individual article processing charges on an article-by-article basis for research published in outlets beyond the mentioned agreements. To facilitate the administration of this wide variety of OA approaches, CERN is a pilot partner of the OA Switchboard initiative[5] which aims to simplify the exchange of metadata to facilitate OA funding requests at the article level.

45.2.3 Open access services at CERN

CERN maintains a range of services for the more 'traditional' outputs, such as papers, proceedings, multimedia, presentations, etc. INSPIRE is the core HEP information system operated and curated by a consortium consisting of a number of high-energy physics labs worldwide, including the German Electron Synchotron (DESY), the Institute of High Energy Physics (IHEP) of the Chinese Academy of Sciences, the French National Institute of Nuclear and Particle Physics (IN2P3), the Stanford Linear Accelerator (SLAC) and the Fermi National Accelerator Laboratory (Fermilab) in the United States, and CERN. The service focuses on aggregating scholarly works from the global HEP community from various sources ranging from publishers to arXiv. To support the fair assessment of research and researchers, INSPIRE also computes transparent citation metrics for scholarly articles including citations from and to preprints. To further foster this important cause, the INSPIRE collaboration is in the process of signing the Declaration on Research Assessment (DORA)[6].

The CERN Document Server (CDS)[7] is CERN's institutional repository which gives access to CERN works (preprints, articles, etc), as well as administrative and multimedia content. Through CDS, CERN also publishes various CERN reports, most notably the CERN Yellow Report series[8], which was started in 1955. It provided a medium for communicating the work done at CERN where publication in a journal was not appropriate. The Yellow Report series includes the proceedings of schools and workshops having a large impact on the future of CERN, as well as reports on detectors and technical papers from individual CERN departments.

45.2.4 Open access at scale—lessons learned

Open access for CERN publications has become a reality and the vast majority of CERN's scientific output today is published immediately OA. This success is of course to a large extent based on SCOAP^3 covering the majority of HEP papers worldwide and was supported by the wide range of additional commercial OA arrangements as well as the availability of dedicated OA services. However, based on anecdotal evidence, talking to hundreds of researchers at CERN, a key success factor was the chosen approach. Even though it might not seem so from the outside, CERN publishing more than 90% of its articles OA is the result of a long-lasting step-by-step approach that started more than 15 years ago. Even at an organization such as CERN, which was founded based on the principle of openness, researchers did not perceive publishing OA as a key priority. As such, CERN was not different

[5] https://www.oaswitchboard.org/
[6] https://sfdora.org/
[7] https://cds.cern.ch/
[8] https://cds.cern.ch/collection/CERN%20Yellow%20Reports

to other institutions. Several studies, for instance Tenopir *et al* (2016) have shown that the ability to publish OA is regularly ranked at the end of factors authors take into consideration for journal decisions. While this has meanwhile improved (Rowley *et al* 2020), it is still a reality that many researchers do not see publishing OA as a priority. CERN's approach has been to gradually introduce OA measures aiming for a pragmatic balance between what is achievable at a given point in time and being progressive and innovative. Each step was accompanied by a central OA fund as well as tools and processes that help researchers in complying with OA requirements. Therefore, it has always been reasonably easy for researchers at CERN to comply with an increasingly strict OA policy. On this aspect, SCOAP^3 can be considered as an ideal solution from a researcher's perspective. Sometimes advertised as the 'invisible infrastructure', it has helped to establish OA as a norm in high-energy physics because researchers could continue to publish in popular established journals, where their articles were published OA automatically.

Soon to enter into the ninth year of continuous operation, SCOAP^3 has served as a powerful demonstration of the sustainability and effectiveness of global collective action in delivering sustainable open access and the importance of making an OA transition easy for authors. The initiative has been cited as the source of inspiration for other global initiatives in open access, including the OA2020 initiative of the Max Planck Digital Library, Plan S, and the European Commission. While inspirational for the global OA community, in the increased development of collective approaches towards OA that do not rely on author fees (such as the Subscribe to Open model, the Open Anthro project by Libraria, and the Community Action Publishing Model by PLOS), SCOAP^3 has yet to be replicated at a global scale in another discipline. A possible reason for is that a such global collective efforts have large coordination costs, i.e. CERN's role as a large and well-connected central organization to act as coordinator, financial guarantor, and host organization was critical to the success of SCOAP^3. Nevertheless, some smaller initiatives have succeeded indeed in replicating the model. An example is the transition of several Norwegian Social Science and Humanities journals in 2017 (Open Access Norway 2019) based on a central consortium.

45.3 Open data at CERN

The four particle detectors of CERN's LHC are home to the experiments ALICE, ATLAS, CMS, and LHCB, operated by international collaborations of scientists and engineers with up to several thousand members each. During the so-called 'run-periods' of the LHC, i.e. active periods of the collider's operation, these experiments generate tens of petabytes of data, even after applying highly de-selective triggers, likely making the LHC the world's most prolific source of experimental data. It is estimated that as the machine is progressively updated over time, that annual data volumes will increase to result in total data sample volumes of between 10 and 100 exabytes by the end of active operation of the LHC (currently planned for around 2035–40) (Shiers *et al* 2016). Members of the LHC experiments perform their research using these data, studying the fundamental particles and forces of nature, and testing various theories and models that have been proposed to explain their behavior. The reuse of this data within the collaborations is fundamental to the discipline, including the ability to reproduce past analyses. CERN experimental data are unique and the result of vast long-term human and financial investment by the international community, and form an important part of the scientific legacy of the LHC.

CERN advances the principle that openness of data and the adoption of FAIR standards in making data available will, in the long term, optimize investments and allow the maximum realization of their scientific potential. As such, it represents a realization of the collective moral and fiduciary responsibility to member states and the global scientific enterprise more broadly. Opening experimental data from CERN presents unique opportunities but also unprecedented challenges, due to the complexity, value, and scale of these data. The data are accompanied by software, supporting data and metadata, reference data etc that need to be taken into consideration when opening up the research results to the public. This is in parts unique to the size of the experiments and the scale of data taking, but is in principle a challenge that is transferable to other disciplines too. Hence, making data available responsibly—at different levels of abstraction and at different points in time—ensures that it can be appropriately used, verified and built upon to advance knowledge. Preserving associated software and analyses supports reproducibility and can underpin further research.

In conclusion, this means that the data generated by the CERN experiments and that gets published can take many forms. Starting from either raw experimental or simulated data through to reconstructed data and the datasets of higher abstraction generated by analysis workflows, and finally all the way to data represented in scientific publications. Each of these layers has the potential to afford different opportunities for long-term reuse and poses different challenges for openness and preservation.

45.3.1 Open data policies

To ensure that LHC data can be fully exploited for their scientific value by researchers and communities beyond the CERN collaborations, the four LHC experiments adopted policies for preservation of significant subsets of the data and their release to other scientists, as well as the general public, following appropriate embargo periods. To align these efforts and empower the experiments by adopting a consistent and progressive approach towards the openness and preservation of experimental data to maximize its overall value to the discipline and society more broadly, in December 2020, CERN announced a new progressive framework for the openness and preservation of experimental data from the LHC Experiments (CERN 2020).

The CERN open data policy for the LHC experiments describes how the LHC experiments believe in making data publicly available—at different levels of complexity—and where practical to enable their reuse by a wide community and a diverse set of use-cases including for: collaboration members long after the data are taken, experimental and theoretical HEP scientists who were not members of the collaboration, professional science, educational and outreach initiatives, and citizen scientists in the general public. The LHC experiments will provide open access to their data after a suitable embargo period, allowing experiment collaborators to fully exploit their scientific potential.

The data generated by the CERN experiments take many forms. Starting from either raw experimental or simulated data through to reconstructed data and the datasets of higher abstraction generated by analysis workflows, and finally all the way to data represented in scientific publications. Each of these layers has the potential to afford different opportunities for long-term reuse and poses different challenges for preservation.

The Data Preservation in High Energy Physics (DPHEP) study group identified four general levels of complexity of data in HEP which serve varying audiences, and imply a diversity of openness solutions and practices, as described in figure 45.3. This categorization has been facilitating a more strategic discussion of the different levels of data in the community and could be considered an enabler for the open data developments in particle physics.

The first level of data, i.e. level 1, comprises data that are directly related to publications which provide documentation for the published result. This includes additional documentation to put results in context and understand the analyses procedures, or additional numerical data which did not or could not appear in the publications (e.g. cross sections as a function of multiple variables, data behind figures, etc). In compliance with the CERN open access policy, all scientific publications from the LHC experiments are published OA, and so are available to the public. To allow for the maximum scientific value of their publications, the LHC experiments make public additional information and data with the publications. For example, this is done to allow a re-interpretation of search analyses in new BSM models, to allow combinations between analyses, or to allow new theoretical

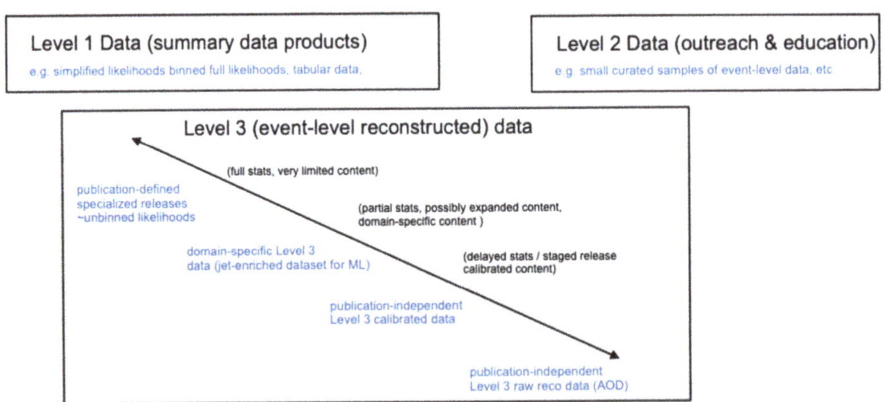

Figure 45.3. Towards a common policy. A spectrum between unbinned likelihoods and open data (Heinrich 2020).

models to be used for spectroscopy results. CERN collaborates with the Durham High-Energy Physics Database (HEPData) for storage of these data.

In recognition of the vital role of outreach and education, CERN actively participates in and encourages associated activities and allows selected data to be made available for them. This category of data, termed level 2, includes simplified, portable and self-contained formats for analysis in outreach and education/training exercises. For these purposes, dedicated subsets of data are used, selected and formatted to provide rich samples to maximize their educational impact, and to facilitate the easy use of the data. Typical use-cases for this data range from analysis programs and event displays for high-school and university education, to releasing selected subsets of simulated data events for data science related contests, etc. Often lightweight computational environments to allow for the easy exploration of these data are also provided alongside the data through the CERN open data portal.

The most transformative policy measure of the CERN open data policy relates to the release of so-called level 3 scientific data, the type required to make scientific studies, collected by the LHC experiments. This level 3 data consists of reconstructed data and simulations, the type used as input to most physics analyses, and will be released together with the associated software, analysis workflows and documentation required to access the data, understand them, reproduce published analyses, and perform new analyses not requiring re-reconstruction of the data or new simulations.

Under this policy, the LHC experiments have committed to releasing the fully calibrated reconstructed data with the level of detail used for most physics results released by the collaborations. In exercise of the 'as open as possible, as closed as necessary' principle for openness of scientific data, level 3 data will be released following an appropriate latency period to allow thorough understanding of the data, the reconstruction and calibrations, as well as to allow time for the scientific exploitation of the data by the respective experimental collaboration. Up to 50% of the data will be released five years after collection (although some experiments may exceed this release volume and on an accelerated timeframe), and the aim is for full datasets to be publicly available ten years after their collection, or by the close of the experiment concerned.

The foreseen use-cases of level 3 include professional physics data analysis, analysis by interested members of the public, as well as technical and algorithmic developments for example improving data reconstruction, analysis techniques or data management techniques. In order to facilitate the latter, the experiments may release dedicated smaller sets of data at different levels of detail, for example to allow development of reconstruction algorithms. The release of level 3 data is intended to contribute to scientific research in particle physics, as well as research in the field of scientific computing, for example to improve reconstruction or analysis methods based on machine learning techniques, an approach that requires rich datasets for training and validation. As such, and to support the maximum scientific benefit from these data, the level 3 policy requires that data be accompanied by a concurrent release of associated analysis artifacts, appropriate simulated data samples, and metadata and software to allow reusability and reproducibility (through the REANA) and further high-quality analysis of the data. The dataset size is expected to be of the order of a few petabytes of data per year (at least for the data from LHC run-2 and run-3), and as such should be a manageable data volume for analysis. The release of level 3 data in this manner is intended to allow high-quality analysis by diverse groups: non-CERN scientists, scientists in other fields, educational and outreach initiatives, and the general public.

Level 3 scientific datasets will be released through the CERN open data portal under the emerging standard Creative Commons CC0 waiver. Data will be available using FAIR standards, a set of data guidelines that ensure the data are findable, accessible, interoperable, and reusable. This will include the application of persistent identifiers to allow discovery and citation.

The final level of data identified by DPHEP and described in the CERN open data policy is the level 4 raw data. Data at this level is not considered practical to be made available in a meaningful way outside the LHC experimental collaborations. This is due to the complexity of the data, metadata, and software, the required knowledge of the detector itself and the methods of reconstruction, the extensive computing resources necessary, and the access issues for the enormous volume of data stored in archival media. For these reasons, general direct access to the raw data is not even available to individuals within the collaboration, and so instead the production of reconstructed data (i.e. level 3 data) is performed centrally.

45.3.2 Open data services at CERN

The practice of open science for Big Science, particularly in a field with the complexity (and associated cost) of high-energy physics, demands more than simply providing OA to research data. Instead, it needs to concern itself with providing users of data with the information they need to repeat or verify an analysis over given datasets, or to produce results that can be reused by others for comparison, confirmation, or simply for deeper understanding and scientific inspiration. This requires runnable examples of how the research was performed, accompanied by the software, documentation, runnable scripts, notebooks, workflows, and computational environments to enable reuse and reproducibility (Dallmeier-Tiessen and Šimko 2019).

With the practice of releasing data following embargo periods, this means that much of the associated contextual information required to understand, repeat, or reuse research data is typically also developed prior to the release of the data. Thus, the process of preparing for openness of scientific data needs to start at the very inception of an analysis itself. Associated analysis artifacts need to be captured throughout the regular iterative, closed research cycle, but preserved in such a manner that they complement open data releases to ensure their optimal use, reuse, and reproducibility. Rather than thinking about open data from its inception, it is more useful to consider data to be FAIR, i.e. findable, accessible, interoperable, and reusable (Wilkinson *et al* 2016), throughout the scientific process, from being initially closed to being made open later in meaningful ways that allow for the full scientific value to be extracted from the data by those outside of the experimental collaborations. This suggests the adoption of a set of scientific practices and technological solutions that can be readily integrated into existing research processes and procedures, making them a seamless part of a researcher's daily work.

The CERN Analysis Preservation and Reuse framework (Chen *et al* 2019) consists of a set of services and tools developed together with the LHC collaborations, that enable physicists to preserve, document, organize, and share the datasets, code, and tools used during analyses. The core services that constitute the CERN Analysis Preservation and Reuse Framework include: the CERN Analysis Preservation (CAP) service, the CERN open data portal (COD), and the reproducible research data analysis platform, REANA. Each of these services, built on free and open source software, are built on flexible metadata structures that allow researchers to define everything from experimental configurations to data samples, from analysis code to software libraries and computational environments used to analyse the data, accompanied by documentation and links to presentations and publications. For all these services, capturing and preserving data provenance is a key design feature. Research provenance facilitates reproducibility and data sharing as it provides a formal model for describing published results. The result is a standardized, yet flexible, way to describe and document an analysis for the purposes of discoverability, reusability, and reproducibility.

CAP is a digital repository designed to aggregate high-level physics information about individual physics analyses, in order to facilitate preservation, discoverability, and reproducibility. With the use of flexible JavaScript Open Notation (JSON)-based metadata schemas, the service allows for the definition of components of analyses ranging from experimental configurations, data samples, analysis code, presentations, and publications, supporting their identification and preservation. The CAP service is interconnected with collaboration databases and workflows in ways to allow seamless integration of analysis preservation activities into the researchers' workflows. It features both a 'push' protocol that enables researchers to deposit material via a user interface or automated command-line client; as well as the ability to 'pull' information from existing databases within the information infrastructure of the LHC collaborations. For software and intermediate datasets, it can also completely ingest the material referenced by the researcher.

The COD was launched in 2014, to allow CERN experiments to share their data —currently over two petabytes of HEP experimental collision and simulated datasets, software, configuration files, virtual machines, etc—with a double focus: for the scientific community, including researchers outside the CERN experimental teams as well as citizen scientists, and for the purposes of training and education through specially curated resources (CERN 2014). The COD service provides access to open experimental collision and simulated datasets, example software, virtual machines with the suitable computational environments, together with associated usage documentation that were released to the public by the HEP experiments. It is implemented as a standalone data repository on top of the open source Invenio digital repository framework developed at CERN, which powers numerous CERN open science tools and initiatives. Data are published on the COD platform following embargo periods, first specified by data policies of each individual LHC collaboration, and since December 2020 specified by the CERN open data policy for

LHC experiments. The service captures extensive metadata to describe the complex materials on the portal to ensure preservation and discoverability, and includes usage instructions, related software and other supplementary materials along with the release of data to enable reuse. All data on the COD platform are shared under a Creative Commons CC0 public domain dedication; data and software are assigned unique DOI identifiers to make them citable in scientific articles; and software is released under open source licenses.

The final core service of the CERN Analysis Preservation and Reuse Framework is the reusable analysis service, REANA[9]. REANA allows researchers to instantiate preserved research data analyses on remote compute clouds, as well as run to 'live' analyses that are still ongoing and have not yet been published or preserved. Using containerized information about the input datasets, the computational environment, the software framework, the analysis code, and the computational workflow steps to run the analysis, REANA permits researchers to submit parameterized computational workflows to run on remote compute clouds. As such, REANA provides a systematic mechanism for the research outputs generated at the LHC to be made reproducible. Linked with CERN Analysis Preservation and CERN open data and their preserved content, REANA enables the community to practice open and reusable research at scale.

45.3.3 Open data at scale—lessons learned

The CERN open data portal was released in 2014 amid a discussion as to whether particle physics data, due to its large volume and complexity, would find any use outside of the LHC collaborations. Some skeptical voices questioned the feasibility of any open data sharing in the community, despite the long tradition of HEPdata as a resource of the community. With several data releases over the past years by LHC experiments, the feasibility has clearly been shown. Furthermore, exploitation of the released open content has been demonstrated both on the educational side and for research purposes. Some examples follow. A team of researchers, students, and summer students reproduced parts of published results from the CMS experiment using only the information that was released openly on the CERN open data portal. The developed code produced plots comparable to parts of the official CMS Higgs-to-four-lepton analysis results (Jomhari *et al* 2017). In 2017 Thaler and colleagues (Tripathee *et al* 2017) confirmed their jet substructure model predictions using the open data from the Compact Muon Solenoid (CMS) experiment that were released on the portal in 2014, demonstrating that research conducted outside of the CERN collaborations could indeed benefit from such open data releases.

The examples underline that practicing open science does have an impact on the community and its research practices. Transforming research culture in Big Science over time, however, does demand sustainable services that connect to the individual researchers, teams, and groups. For Big Science in particular this includes scalable and evolving services that can be trusted over the course of a researcher's career.

[9] reana.io

Service must continuously be enhanced and supported to maintain the trust that is being put into them—by researchers and the public.

Furthermore, it deems necessary to create visibility and benefits for the effort put into capturing research assets beyond an article, i.e. data, code, metadata, etc. Competition in Big Science is fierce and provision of high-quality research outputs (e.g. data, code) needs to be rewarded. Up to now (2022) practicing open science in Big Science relies in part on a few champions of the community who are driving the way forward. For a larger cultural shift, administrative and funding frameworks need to include open science practices in their work programs too.

Within the CERN Analysis Preservation and Reuse framework a range of defined use-cases from closed to open can now be addressed. With this foundation, a continued monitoring and evaluation of the community landscape, practices, and demands can happen to further develop and sustain the services. In particular in Big Science with enormous complexities and scalability challenges, open science needs to meet certain technological maturity standards to connect to community tools, services, as well as the spirit of the research projects.

Taking a step away from the high-energy physics community, CERN has worked to leverage its expertise to the benefit of the global research community. Zenodo, a data repository that is also based on the Invenio Digital Library software framework, is a free data repository service for everyone to preserve their research assets in. Zenodo is hosted by CERN and commissioned by the European Commission (EC) through the OpenAIRE project to support the open data and open access movements in Europe. The OpenAIRE project, in the vanguard of the open access and open data movements in Europe was commissioned by the EC to support their nascent open data policy by providing a catch-all repository for European Commission (EC) funded research. CERN, an OpenAIRE partner and pioneer in open source, open access, and open data, provided this capability and Zenodo was launched in May 2013. Zenodo quickly became a successful product worldwide and got additional project funding from the Alfred P Sloan Foundation and Arcadia Fund. Zenodo accommodates all kinds of research artifacts that can be grouped into community collections. Zenodo played a key role making software a first class citizen in scholarly publications. Thanks to its integration with GitHub (GitHub 2016), which allows automatic archival of software releases, Zenodo hosts 80% of the world's software DOIs (Fenner *et al* 2018). Zenodo provides leading features such as versioning (to support reproducibility) and metrics as usage statistics (following Make Data Count and COUNTER guidelines).

Researchers in many disciplines and countries are still, unfortunately, lacking critical infrastructure when it comes to storage and computing as well having a safe place to store their research assets for the long term. Having services such as Zenodo that could cover such requirements, while still keeping the barrier to share research at a comfortable level. One important example arose very recently—as the service is easy to use, Zenodo played an active role during the recent scientific efforts to fight COVID-19, by creating a central open access sharing space for related research artifacts between disciplines (Ioannidis 2020).

45.4 Open software

The oldest of the movements to make science and technology more open, the 'open source' software movement was the first to articulate the values of 'open' science, and inspired subsequent movements such as open access publishing. While based on the premise that recipients of technologies should have access to their building blocks and source code in order to study it, modify it and redistribute it to others, many consider the open source movement a natural extension of academic freedom and knowledge sharing for the digital age.

The roots of open software date to the free software movement of the 1980s which rose in response to restrictions on proprietary software and in response to increased monopolization over intellectual property rights by corporations. The fundamental value is facilitating open collaboration, enabling peers to freely, openly, and collectively build and develop software. A second value is to enable attribution, openly providing recognition and credit for the individuals or institutions who make valuable contributions to the development of software. A third value pertains to the so-called four freedoms that are granted to users, namely to use software for any purpose, to study the source code and understand it, to share and redistribute it, and to contribute towards its improvement and share these improvements with the wider community. As such a virtuous circle is fostered, in which software is continuously improved upon and shared towards the common good, while minimizing corporate control of software.

Perhaps the most significant and impactful example of CERN's embrace of the open source philosophy was the development and ultimate release into the public domain of the software underlying the World Wide Web. This software was initially created to serve the need of CERN's scientific collaborations to share information between the experiments at CERN as the host laboratory and scientists in universities around the world. CERN and sister institutes instantiated web servers, and put information online, including at SLAC in California whose preprint server SPIRES (predecessor of INSPIRE) became the first web server in the USA. Having experienced first-hand its potential to connect physicists around the globe, in 1993 CERN released the web software into the public domain so that developers could collaborate and improve on it. The following year, CERN released the next web-server version under an open source license with the explicit goal of preventing private companies from turning it into proprietary software. Within a few years the concept spread from physics institutes to all of science, and on to society, bringing businesses online and transforming life as we know it, making the WWW the most impactful example of the societal impact of open source software, but a key exemplar of CERN's embrace of open science philosophy.

Ever since releasing the WWW software under an open source model in 1994 (Smith and Flückiger 2020), CERN has continuously been a pioneer in this field, and makes use of free and open source software (FOSS) where possible. One of the most prominent examples—that is also being used to run CERN open Data and Zenodo—is Invenio, a free and openly-developed integrated digital repository system that was born at CERN around 20 years ago. The software, built on the

Python-flask framework, covers all aspects of digital repository management. It complies with the next generation repository (NGR) standard and supports multiple bibliographic standards for metadata. Its flexibility and performance make it a comprehensive solution for the management of document and data repositories of small to very large size. Invenio is a framework and toolbox for building and managing a digital repository server. Together with 18 partner institutions, CERN is currently building a turn-key open source research data management platform called InvenioRDM, whose main purpose is to enable reproducibility and reuse of research artifacts (Nielsen 2019).

There are many other prominent examples at CERN, e.g. Indico, EOS, Geant4 and Root that have been transforming the way research is being organized, presented, stored, simulated, or analyzed in the community. All of them have for example been used to prepare or accompany open data releases on the CERN open data portal. This underlines that it requires a suite of excellent services to push the open science boundary in one Big Science discipline.

45.5 Open hardware

Initiatives at CERN to enable researchers to collaborate and use, study, share and co-design hardware components through open practices similar to software date back to 2008, when CERN scientists identified the need for a technology to synchronize the timing systems of LHC particle accelerators with sub-nanosecond accuracy and picoseconds precision (Wilson Center 2021). CERN engaged with private companies and organizations to develop the open source hardware, software, and peripheral technologies. This engagement was not based out of a profit motive, but a mutual interest in the knowledge that would ensue from the open exchange of this information. The result of the work was a shared mechanism for 'distributing clocks' ensuring that the timing systems of networks of particle accelerators shared nanosecond scale accuracy, playfully named 'White Rabbit' after the time-obsessed character from Lewis Carroll's *Alice in Wonderland*. White Rabbit was first used in 2012, and since has demonstrated diverse application beyond particle physics labs and in industry more widely. Subsequently, the Institute of Electrical and Electronics Engineers (IEEE) incorporated the White Rabbit into Precision Time Protocol (PTP) extension into the industry standard, facilitating global adoption by industry and other partners (Lapka 2016).

The successful example of White Rabbit spawned an initiative for open hardware, formalized at CERN in 2013 to support hardware engineers working for particle physics labs to collaborate on hardware designs, consequently distributing the load of developing research infrastructures collaboratively with high levels of quality control and assurance. This initiative is supported by a set of licenses for open hardware (the CERN Open Hardware license)[10], which provides a sound legal foundation for designers to study, modify, and republish designs and associate hardware, and the CERN Open Hardware Repository, which is intended as a place

[10] https://ohwr.org/project/cernohl/wikis/Documents/CERN-OHL-version-2

on the web for electronics designers at experimental physics facilities to collaborate on open hardware designs. This repository further facilitates community norms through a dedicated manifesto stating how collaborations should take place, and aims to stimulate collaboration by organizing meetings and workshops. An example of one meeting was the Gathering for Open Science Hardware, which served as a forum to address the needs of the global open science hardware community through convening meetings, publications, and other activities.

Open hardware has been proven, thus far, to be flexible and adaptable enough to cover a wide range of use-cases: from educational and hobbyist activities to infrastructure solutions operating in research labs around the globe. A more recent example is the development of a washable 3D printed mask with replaceable filters, developed for the CERN Fight COVID-19 initiative and released via the CERN Open Hardware Repository[11] in June 2020. This mask, designed by a team of researchers and engineers at CERN, is designed to offer improved face contact over standard tissue masks, offering wearers a higher level of protection from airborne pathogens if the correct filter is applied. The washable mask is made available as an open source solution, allowing anyone around the world to 3D print it using standard 3D printing filaments (Jakobsson 2020).

45.6 International engagement

As examples such as the multidisciplinary open science platform Zenodo demonstrate, CERN's engagement in open science—consistent with the tenet to pursue our mission for 'the benefit of all'—is intended to benefit and accelerate scientific progress beyond disciplinary boundaries. CERN actively engages in international programs to promote and advance elements of open science, including many EC projects, Horizon 2020, and Horizon Europe Projects, particularly focusing on science, research, and e-infrastructures at the European level. Some recent examples include OpenAIRE, EOSC-hub[12], THOR[13], FREYA[14], and ESCAPE[15]. This engagement has allowed CERN and the particle physics community to play an active role in shaping the global discourse in open science and support international open science policy developments. This international engagement is also considered an opportunity to learn from the community of multidisciplinary external partners and practitioners operating in this innovative and progressive domain.

CERN has been at the forefront of supporting and adopting open science related guidelines, most recently the UNESCO Recommendation on Open Science[16] (adopted by the UNESCO General Assembly in November 2021), as well as the FORCE11 Joint Declaration of Data Citation Principles (Martone 2014), the FORCE11 Software Citation Principles (Smith *et al* 2016), the FAIR Data Principles

[11] https://ohwr.org/project/3dmask/wikis/home
[12] https://www.eosc-hub.eu/
[13] https://cordis.europa.eu/project/id/654039
[14] https://www.project-freya.eu/
[15] https://projectescape.eu/
[16] https://en.unesco.org/science-sustainable-future/open-science/recommendation

(Wilkinson *et al* 2016), the Plan S Principles (European Science Foundation 2020), the Code of Practice for Research Data Usage Metrics (COUNTER 2020) (in Zenodo), and the recommendations of the Declaration on Research Assessment (San Francisco Declaration on Research Assessment 2012) (in INSPIRE).

CERN also engages actively with partners in the domain of open science in multidisciplinary projects of global scientific importance. Examples include the Asclepias project (CERN 2018) and the Biodiversity Literature Repository (BLR)[17], a joint project of Plazi, Pensoft, and Zenodo that opens data from biodiversity literature. Information such as taxonomic treatments, images, and the associated metadata is being extracted from closed access publications, and made publicly available. The data are then immediately distributed to global biodiversity platforms, such as the Global Biodiversity Information Facility, which makes original taxonomic information accessible that would otherwise be hidden behind paywalls (Pensoft 2019).

45.7 CERN Open Science Policy

On October 1st, 2022, CERN approved its inaugural Open Science Policy (Open Science Strategy Working Group 2022) that aims to make all CERN research fully accessible, inclusive, democratic and transparent, for both other researchers and wider society. The policy development process was entrusted to a multi-stakeholder group consisting of representatives across CERN departments and the experiments. Drawing on existing initiatives, the working group designed comprehensive guidelines for the CERN community on sharing its research within a new framework for open science. Published alongside the policy document is a dedicated website explaining all the open science initiatives at CERN.

The new policy encompasses the existing policies for open access, open data and open source software and hardware, which make all research papers, experimental data and research software and hardware publicly available. It also brings together other existing elements of open science–research integrity, open infrastructure and research assessment, which make research reliable and reproducible–and training, outreach and citizen science, which aim to educate and create dialogue with the next generation of researchers and the public.

CERN's new Open Science Policy heralds a new era in knowledge sharing. The OSWG will continue to assess the practice of open science at CERN, and will revisit the policy frequently to ensure it remains at the forefront of the dynamic open science landscape. Alongside this, a new open science report will be published every two years, to demonstrate CERN's continued commitment to the initiative. Furthermore, as directed by the European Strategy for Particle Physics, SIS will initiate engagement with other particle physics labs across Europe (and beyond) to advance the adoption of open science policies and practices.

[17] https://zenodo.org/communities/biosyslit

45.8 Conclusion

Given CERN's extensive engagement in the domain of open science, we hope to extend our experiences, knowledge, and tools to further accelerate the global adoption of open science policies and practices, and to support scientific communities beyond the discipline of high-energy physics to embrace this vision.

The COVID-19 global health pandemic has brought into sharp focus both the inadequacies of traditional science, but also the embrace of open science practices as the most effective mechanism for addressing global crises. CERN with its open science vision and program does serve a truly global community that is active in Big Science. The lessons learnt from delivering the programs and services at scale hopefully help other disciplines or communities to do the same: be ambitious to support a community's steady transition into a more open environment. With the continued advancement in research and the technological changes, the transition perhaps never stops, but will always demand a close link between research practitioners and the supporting services.

That is why, it is our firm belief that we are at an inflection point, where open science should go beyond an aspiration, to become the default standard for global research. This will not only equip the scientific community to better respond to current and future challenges, but will further accelerate the advancement of the research enterprise itself, making it more efficient, transparent, responsive, and dynamic.

The COVID-19 pandemic has brought open science to the forefront of global science policy. At CERN we have mobilized our resources and expertise to aid the global fight against COVID-19. As this fight continues, it is incumbent upon the scientific community and the administrative and funding frameworks around it to recognize the potential for open science, not only in addressing global crises such as climate change, but in building a future scientific ecosystem that is optimized to achieve scientific progress for the benefit of all.

References

Bianco S *et al* 2007 Towards Open Access publishing in high energy physics *Report of the SCOAP3 Working Party* CERN-OPEN-2007-015 CERN https://cds.cern.ch/record/1034368/files/open-2007-015.pdf

CERN 1953 *CERN Convention for the Establishment of a European Organization for Nuclear Research* https://council.web.cern.ch/en/content/convention-establishment-european-organization-nuclear-research

CERN 2014 CERN makes public first data of LHC experiments *CERN* https://home.cern/news/press-release/cern/cern-makes-public-first-data-lhc-experiments

CERN 2018 Asclepias Broker: scholarly link broker for the Asclepias Project https://asclepias-broker.readthedocs.io/en/latest/

CERN 2020 CERN announces new open data policy in support of open science *CERN* https://home.cern/news/press-release/knowledge-sharing/cern-announces-new-open-data-policy-support-open-science

CERN Open Science Strategy Working Group 2022 Open Science Policy CERN-OPEN-2022-013 https://cds.cern.ch/record/2835057/files/CERN-OPEN-2022-013.pdf

CERN Scientific Information Policy Board 2017 Open access policy for CERN publications *Report* CERN-OPEN-2021-009 http://cds.cern.ch/record/1955574

CERN Scientific Information Service 2020 CERN Scientific Information Service: Activity Report 2019 *Report*

Chen X *et al* 2019 Open is not enough *Nat. Phys.* **15** 113–9

Dallmeier-Tiessen S and Šimko T 2019 Open science: a vision for collaborative, reproducible and reusable research *CERN Cour.* **59** 25–6

European Strategy Group 2020 *2020 Update of the European Strategy for Particle Physics* https://home.cern/sites/home.web.cern.ch/files/2020-06/2020%20Update%20European%20Strategy.pdf

Fenner M *et al* 2018 DOI registrations for software *DataCite Blog*

Gaillard M 2017 CERN Data Center passes the 200-petabyte milestone *CERN* https://cds.cern.ch/record/2276551?ln=en

GitHub 2016 Referencing and citing content *GitHub Docs* https://guides.github.com/activities/citable-code/

Heinrich L 2020 ATLAS open data remarks *Presentation to the CERN Open Data Working Group*

Ioannidis A 2020 Fighting the coronavirus disease (COVID-19) *Zenodo Blog* https://blog.zenodo.org/2020/04/02/fighting-the-coronavirus-disease-covid-19/

Jakobsson E 2020 CERN develops washable 3D-printed face mask *CERN* https://kt.cern/article/cern-develops-washable-3d-printed-face-mask

Jomhari N Z, Geiser A and Bin Anuar A A 2017 Higgs-to-four-lepton analysis example using 2011–2012 data *CERN Open Data Portal*

Lapka M 2016 White Rabbit, a CERN born open-source technology sets new global standard empowering world innovators *CERN* https://kt.cern/article/white-rabbit-cern-born-open-source-technology-sets-new-global-standard-empowering-world

Martone M (ed) 2014 Data Citation Synthesis Group: Joint Declaration of Data Citation Principles (San Diego CA: FORCE11) https://doi.org/10.25490/a97f-egyk

Nielsen L H 2019 InvenioRDM: a turn-key open source research data management platform *Invenio Blog* https://inveniosoftware.org/blog/2019-04-29-rdm/

Open Access Norway 2019 Norwegian open journals in the social sciences and humanities: a consortium for Norwegian SSH OA journals https://openaccess.no/english/humsam/

Pensoft 2019 Data mining applied to scholarly publications to finally reveal Earth's biodiversity *Pensoft blog* https://blog.pensoft.net/2019/10/21/data-mining-applied-to-scholarly-publications-to-finally-reveal-earths-biodiversity/

Rowley J, Sbaffi L, Sugden M and Gilbert A 2020 Factors influencing researchers' journal selection decisions *J. Inf. Sci.* **48** 321–35

Schimmer R 2013 A road long travelled: is SCOAP 3 now arriving? *Insights* **26** 135–40

Schimmer R, Geschuhn K K and Vogler A 2015 Disrupting the subscription journals' business model for the necessary large-scale transformation to open access http://hdl.handle.net/11858/00-001M-0000-0026-C274-7

Shiers J, Berghaus F O, Cancio Melia G, Blomer J, Dallmeier-Tiessen S, Ganis G and Simko T 2016 CERN services for long term data preservation *Internal Note* No. CERN-IT-Note-2016-004 https://cds.cern.ch/record/2195937?ln=en

Smith A M, Katz D S and Niemeyer K E 2016 FORCE11 Software Citation Working Group. Software Citation Principles *PeerJ Comput. Sci.* **2** e86

Smith T and Flückiger F 2020 Licensing the Web *CERN* https://home.cern/science/computing/birth-web/licensing-web

Tenopir C, Dalton E, Fish A, Christian L, Jones M and Smith M 2016 What motivates authors of scholarly articles? The importance of journal attributes and potential audience on publication choice *Publications* **4** 22

Tripathee A, Xue W, Larkoski A, Marzani S and Thaler J 2017 Jet substructure studies with CMS open data *Phys. Rev.* D **96** 074003

UNESCO 2021 UNESCO recommendation on open science *Programme and Meeting Document* SC-PCB-SPP/2021/OS/UROS UNESCO https://unesdoc.unesco.org/ark:/48223/pf0000379949.locale=en

Wilkinson M D *et al* 2016 The FAIR guiding principles for scientific data management and stewardship *Sci. Data* **3** 1–9

Wilson Center 2021 Standing on the shoulders of giants: CERN https://wilsoncenter.org/open-sourced-science/stories/standing-on-the-shoulders-of-giants-cern/

Part V

Mapping a global outlook on the role of Big Science organizations [Editors. G Dissertori and P Charitos]

IOP Publishing

Big Science in the 21st Century
Economic and societal impacts
Panagiotis Charitos, Theodore Arabatzis, Harry Cliff, Günther Dissertori, Juliette Forneris and Jason Li-Ying

Chapter 46

Introduction to part V: Mapping a global outlook on the role of Big Science organizations

G Dissertori and P Charitos

In the fifth part of this book we aim to present lessons from the development of Big Science facilities around the globe. As international collaboration in science continues to grow at a remarkable rate, we felt that this volume would profit from a collection of contributions discussing the socio-economic impact of big scientific projects beyond Europe and reflecting on the challenges of building synergies across borders. Big Science projects generate common interests between otherwise unrelated parties, including academia and industry, and motivate the development of rich networks for knowledge transfer.

The spread of scientific workers and research and development activity around the world, outside the North American and Western European poles, has facilitated the increase in international collaborations. The growing number of science and engineering PhDs in developing countries, some of whom are international students and post-docs returning to their country of origin, has expanded the supply of potential collaborators and fostered the investment in large-scale research infrastructures to tackle some of the most fundamental questions in modern physics. Coupled with a rising trend of government and industry R&D spending in developing countries, a certain number of grant policies by the European Union and other funding agencies favor international cooperation. At the same time, the lower cost of travel and communication as well as the rise of teleworking favor collaborating with persons across different geographic locales.

Despite tensions emerging around the world, collaborative research in the framework of Big Science projects has remained resilient. The essays in this volume remind us that the joint efforts for the development of Big Science facilities are not only important for accelerating scientific advances, but for strengthening ties between nations. At a time that concerns are growing that opportunities for large-scale, international projects could be curtailed, it is important to look at the benefits

doi:10.1088/978-0-7503-3631-4ch46

that investments in Big Science projects bring in different geographical areas, allowing the establishment of new networks and empowering the young scientific workforce in these areas.

The first essay in this section, chapter 47, by Fahmida N Chowdhury, William Sims Bainbridge, and Frederick Kronz, reflects on the critical role that the National Science Foundation (NSF) has played in establishing US leadership in science and engineering, creating innovations that drive the nation's economy and educating the next generation of scientists and engineers. The authors discuss the NSF's approach for catalyzing interest and directing investment in fundamental research, as the basis for discovery, invention, and innovation. NSF maintains a broad portfolio of investments that also require collaborations with industry, private foundations, other agencies, science academies, and universities. Over the last few decades the NSF has developed a comprehensive, convergent approach and has modified and adapted its Merit Review Criteria for submitted grant proposals to increasingly emphasize societal impacts. Three examples of NSF's Big Science investments are presented, and their societal impacts are explored, along with discussions of NSF's various mechanisms for advancing the prosperity and welfare of society.

Chapter 48, by Hiromi M Yokoyama, Yuki Akimoto, and Yuko Ikkatai from the University of Tokyo, focuses on Japan, another key player in Big Science over the past decades. The authors present some of the major Big Science facilities including Japan's proposed or planned infrastructures in high-energy physics and neutrino physics. Moreover, the decision-making system and its different pillars are introduced, helping readers to understand the types of criteria and the power play in the decision process for Big Science projects. The essay captures the transition from medium-scale to large-scale research projects such as Hyper-Kamiokande and the proposed ILC and identifies the challenges for Japan's particle physics society; bringing in useful lessons from the science and technology theory.

In chapter 49, Chen Guang, from the Chinese Academy of Sciences, offers an account of China's investment in large scientific facilities and their role, beyond scientific progress, in providing better driving roles for economic and social development. The paper focuses on the FAST radio telescope and uses it as a case study to demonstrate the role of large scientific facilities in the region's economy and societal progress.

Beyond Japan and China, chapter 50 by E Tsesmelis and Leong-Chuan Kwek offers an overview of how Big Science and the scientific community have been developing in Southeast Asia in the recent past and an outlook for science in the region for the twenty-first century. An active and dynamic scientific research landscape has been realized in academia, government, and industry, which can lead to exciting opportunities for the future. After a brief introduction to the historical development of science in southeast Asia, prospects for growth in the area of Big Science are depicted.

Developments in one field of Big Science drive collaborations in other fields, such as India's participation in CERN activities in the field of experimental high-energy and particle physics. In chapter 51 D K Srivastava and V S Ramamurthy from the National Institute of Advanced Studies discuss how the participation in the construction of CERN's Large Hadron Collider opened new opportunities for academia, industry, as well as for young Indian researchers who benefited from the stimulating international environment offered by this project. Moreover, India's participation in CERN has also become a model for several other international collaborations in which India is taking part, such as the LIGO-India Project, planned to build and commission a Laser Interferometer Gravitational-wave Observatory (LIGO) detector in India. In chapter 52 Sukanta Bose, Project Coordinator of LIGO-India, discusses the benefits stemming from India's investment in this project and the broader cultural and societal changes it can bring.

Moving from Asia, we also try to highlight the impact of Big Science efforts in the African content where important steps have been made in the past years. The first essay, in chapter 53, by Dr Daniel Adams (Department of Science and Innovation, Ministry of Research South Africa) and Dr Adrian Tiplady (South African Radio Astronomy Observatory (SARAO)) describes a strategy of investment in Big Science infrastructure over the last decade to expand and exploit pools of excellence and areas of particular strategic advantage. Their chapter looks at the motivation for South Africa's policy decisions, as well as demonstrable examples of the value derived from some of these investments. This includes investments in the Southern African Large Telescope (optical astronomy), the MeerKAT radio telescope and the Square Kilometer Array (radio astronomy), as well as supporting access to research infrastructures such as the LHC at CERN, the Joint Institute for Nuclear Research (JINR) in Dubna, and the European Synchrotron Facility (ESRF) in Grenoble. Chapter 54 by Simon Connell, Katharina C Cramer, Edward Mitchell, Sekazi K Mtingwa, and Prosper Ngabonziza focuses on the prospective merits of an AdLS in Africa; an ambitious international flagship project that could create a high-impact multi-disciplinary and trans-cultural hub of science and technology of world impact. It would represent a major step of science diplomacy towards the Pan-African vision and play an important role in stopping the African scientific diaspora, and perhaps even reversing it as the diaspora returns.

Finally, in chapter 55, we turn our attention to Latin America, and specifically the efforts to contribute to the LHC experiments with the support of the EU-funded HELEN and EPLANET programs that facilitated the scientific interchange between Europe and Latin America and the establishment of a Latin-American community working at the forefront of a Big Science project. Despite myriads of problems in many countries, pockets of excellence continue thriving in South American science. Recent academic literature captures the full extent of different efforts as well as the historical perspectives. Clearly one of the open challenges remains the design of regional science budgets that can address the vast economic differences and social inequalities; recognizing that public investments in Big Science projects also have a redistribution character and a justice perspective.

The complexity of this landscape and the interconnectivity of modern Big Science is reflected in the essay by Dr Mark Robinson (SOAS, London) in chapter 56. Exasperatingly the world is beset by an inability of countries to cooperate via international institutions to address these policy problems that span borders. This international relations gridlock encompasses both a deadlock or dysfunctionality in existing organizations and the inability of countries to come to new agreements as issues arise. Big Science communities have been remarkably effective in evolving complex processes and mechanisms to enable international collaborations. Major powers, who are fierce rivals in other domains, form lasting alliances. Three of the most relevant case examples are the Conseil Européen pour la Recherche Nucléaire (CERN) community, the International Thermonuclear Experimental Reactor (ITER) nuclear fusion project community, and the International Space Station (ISS) community. Extensive field work in each of the three case study communities reveals winning combinations of pathways for Big Science collaborations.

Finally, the essay by Simone Turchetti (University of Manchester) in chapter 57 offers a framing analysis of the term Big Science, charting its transition in meaning from a notion advertising the national laboratory in the 1960s to one now by-and-large promoting international collaborations, and, contingently, publicizing science diplomacy. The essay reminds us that Big Science collaborations not only present a promising model for large-scale scientific research, but also a global governance exemplar instigated through science diplomacy.

IOP Publishing

Big Science in the 21st Century
Economic and societal impacts
Panagiotis Charitos, Theodore Arabatzis, Harry Cliff, Günther Dissertori, Juliette Forneris and Jason Li-Ying

Chapter 47

A convergent approach for achieving societal impacts

Fahmida N Chowdhury, William Sims Bainbridge and Frederick Kronz

This essay describes how the consideration of broader, societal impacts informs and shapes the scientific research funding processes at the United States National Science Foundation (NSF), which supports research and education in all fields of science and engineering. Over the last few decades NSF has developed a comprehensive, convergent approach and has modified and adapted its Merit Review Criteria for submitted grant proposals to increasingly emphasize societal impacts. In this essay, three examples of NSF's Big Science investments are presented and their societal impacts are explored, along with discussions of NSF's various mechanisms for advancing the prosperity and welfare of society.

47.1 Introduction

The term 'Big Science' indicates scientific/technical enterprise undertaken by large teams of researchers supported by large-scale facilities and instrumentation, funded by governments or international organizations (American Academy of Arts and Sciences 2021). It is generally recognized that before the Second World War most scientific research was accomplished by individual scholars, perhaps with graduate students or assistants, working on problems/issues of their own choosing (curiosity driven), supported by universities, professional societies, private benefactors, or philanthropies. During and immediately after the war, the nature of scientific research establishment underwent a significant change: governments and militaries became the biggest sponsors and patrons of science, and many successful scientists started playing the role of research managers and fundraisers. Big Science was born, although it did not have that name yet. The first use of the term Big Science can be traced back to an article written by Alvin Weinberg (Weinberg 1961). Weinberg, who was the Director of Oak Ridge National Laboratory from 1955 to 1973, expressed a level of nostalgia for the previous

doi:10.1088/978-0-7503-3631-4ch47

style of 'little science' as compared to the (then) newly emergent Big Science, and raised a few crucial, fundamental questions. He wrote:

> The emergence of Big Science and its tools as a supreme outward expression of our culture's aspirations has created many difficult problems, both philosophic and practical. Some of the problems concern science itself, some the relation between science and our society. I shall address myself to three specific questions, all of which arise from the growth of Big Science: first, is Big Science ruining science?; second, is Big Science ruining us financially?; and third, should we divert a larger part of our effort toward scientific issues which bear more directly on human well-being than do such Big Science spectaculars as manned space travel and high-energy physics? These questions are so broad, and so difficult, that I cannot do more than raise them here.

Sixty years later, these questions have not yet been fully answered; we may note, however, that the third issue Weinberg raised, regarding diverting a larger part of our scientific efforts toward human well-being, has resonated well with the US National Science Foundation (NSF), and in this paper we explore that specific issue in some detail and depth.

47.2 About the National Science Foundation (NSF)

Established in 1950 by the US Congress, NSF is an independent federal agency with a budget of $8.5 billion (in the fiscal year 2021) and poised to receive a substantial increase in 2022. NSF funds research and education in all fields of science and engineering and strives to promote the progress of science and to advance national health, prosperity, and welfare; quoting from its website: 'we support basic research and people to create knowledge that transforms the future'. Thus, supporting science that creates or drives societal benefits is woven into the very fabric of NSF. Table 47.1 provides some numbers from the latest available 'NSF Fact Sheets'[1] and 'NSF by the numbers'[2] at the time of writing.

Table 47.1. Basic NSF information.

94%	The percentage of NSF's budget for research, education, and related activities
307 000	The total number of people supported by NSF in 2020
43 000	The number of proposals submitted and evaluated in 2020
12 000	The number of awards NSF-funded in 2020
2000	The number of institutions funded by NSF
$200M	NSF annual funding for small businesses to move discoveries to the marketplace
$1.4B	NSF annual spending on STEM education and workforce development
248	The number of Nobel Prize winners who received NSF funding

[1] https://nsf.gov/about/congress/factsheets.jsp

[2] https://nsf.gov/news/factsheets/Factsheet_By%20the%20Numbers_05_21_V02.pdf

Proposals submitted to NSF are typically reviewed by at least three (usually more) independent external reviewers, using specific (publicly available) 'Merit Review Criteria'. The results of these merit reviews are used as the basis of funding decisions. Over the last few decades, NSF has modified and adapted its Merit Review Criteria to increasingly emphasize societal impacts. In 1997 the National Science Board (NSB) used recommendations from the Committee on Equal Opportunities in Science and Engineering (CEOSE) to simplify the merit review criteria for proposals from four to two—intellectual merit and broader impacts (NSB/MR-97-05, NSF News Release 97-028). While intellectual merit generally refers to the scientific (or engineering, as the case may be) quality of the proposed project, the term broader impacts refers to social aspects with the following five general subcategories:

1. Integrating research and education.
2. Broadening the participation of under-represented groups.
3. Enhancing the infrastructure for research and/or education.
4. Disseminating project results broadly to enhance understanding of science and technology.
5. Describing potential benefits to society at large.

In 2011 the National Science Board issued a report in which one of the three merit review principles was stated as follows:

> NSF projects, in the aggregate, should contribute more broadly to achieving societal goals. These 'Broader Impacts' may be accomplished through the research itself, through activities that are directly related to specific research projects, or through activities that are directly supported by, but are complementary to, the project.

In this essay we are focusing on Big Science, but we must note that evaluating potential societal impacts is a fundamental part of NSF's award making process. All projects funded by NSF must be highly rated on intellectual merit as well as broader impacts on science, society, or both.

47.2.1 Societal impacts via NSF's broader impacts criterion

It is clear from the characterization of broader impacts that the term applies to both impacts on the scientific community as well as societal impacts (Roberts 2009). The guide lists five sub-questions in characterizing the meaning of the broader impacts question: What are the broader impacts of the proposed activity? Four of these sub-questions focus on specific activities and one is open ended pointing only to 'benefits to society', or societal impacts. The four specific concerns are pedagogy, broadening participation, enhancing research and education infrastructure, and dissemination to promote scientific understanding. The first (pedagogy) and the third (infrastructure) are primarily oriented towards benefits to the scientific community, with possible benefits to society, whereas the second (broadening participation) and the fourth (promoting scientific understanding) are more oriented towards societal benefits with possible benefits for the scientific community.

Benefits to the scientific community are predominant in the pedagogy question. Proposals that address it typically plan or describe impacts on teaching, training, and learning; corresponding activities mainly include curriculum development and research assistantships for undergraduates, graduates, and post-doctoral fellows. The infrastructure enhancement question is similar; proposals that address it often plan or describe enhancements to facilities or instrumentation, or the development of research networks or partnerships with industry. In both cases there may be broader societal impacts, such as mentoring high school students or partnering with industry to develop applications or extending networks internationally. In contrast, broadening participation is mainly a societal impact. Enhancing participation in science, technology, engineering, and mathematics (STEM) by members of under-represented groups is based on gender, ethnicity, disability, or geographic location; these social constructs are pertinent to social justice concerns. Of course, the recruitment and retention of members of these groups serve to enhance the scientific community, and may also serve to enhance STEM disciplinary or interdisciplinary innovations (Intemann 2005). Finally, proposals that plan or describe broad dissemination of results to promote scientific and technological understanding are mainly addressing societal concerns. They characterize activities that provide opportunities for mutual learning between scientists and members of the public. Such activities include public engagement at high schools, museums, science fairs, or science cafes; they also include engagement with policymakers, which could lead to the integration of research results into policy and practice at the federal, state, or local levels.

Further elaboration and clarification of the broader impacts criterion are provided for applicants in NSF's *Proposal and Award Policies and Procedures Guide*, NSF 20-1. In section II.C.2.d.i, the guide indicates that the project description must contain a separate section in the narrative on 'broader impacts'. That section then provides the following list of examples of what it refers to as 'societally relevant outcomes' and explicitly indicates that the list is not intended to be comprehensive or prescriptive:

- Full participation of women, persons with disabilities, and under-represented minorities in STEM.
- Improved STEM education and educator development at any level.
- Increased public scientific literacy and public engagement with science and technology.
- Improved well-being of individuals in society.
- Development of a diverse, globally competitive STEM workforce.
- Increased partnerships between academia, industry, and others.
- Improved national security.
- Increased economic competitiveness of the US.
- Use of science and technology to inform public policy.
- Enhanced infrastructure for research and education.

Similar instructions may be found in previous versions of the guide for the last 20 years, first occurring in NSF 00-2. The list of examples has evolved somewhat over time. The current list, which is indicated above, is an amalgamation of goals indicated in prior NSF guides and the term 'broader impacts' is used to refer to impacts on the scientific community as well as societal impacts, and each can be distinguished as being primarily one or the other in its main thrust. No single project, even a large-scale engineering research center, could be expected to engage with each of these impacts; the extensive range of broader impact activities is clearly meant to be addressed in aggregate.

The reviewers and panelists who evaluate submitted proposals are given clear instructions on the merit review criteria (intellectual merit and broader impacts) and their definitions. In the 'Instructions for Proposal Review' page, reviewers are given this message regarding the broader impacts criterion[3]:

> What are the broader impacts of the proposed activity? Potential considerations: How well does the activity advance discovery and understanding while promoting teaching, training, and learning? How well does the proposed activity broaden the participation of underrepresented groups (e.g. gender, ethnicity, disability, geographic, etc.)? To what extent will it enhance the infrastructure for research and education, such as facilities, instrumentation, networks, and partnerships? Will the results be disseminated broadly to enhance scientific and technological understanding? What may be the benefits of the proposed activity to society?

Reviewers are expected to separately address the two criteria in their written review report. Across NSF, many if not most programs provide review templates as well as templates for writing summaries of panel discussions in which the intellectual merit and broader impacts criteria must be addressed specifically. The panel summaries must be read and approved by NSF Program Directors before the review panel can be concluded. After the funding decisions are made, written reviews and panel summaries are released to the proposal submitters for the sake of process transparency. Thus, the consideration of both scientific merit and broader societal impacts are incorporated into NSF's routine operations.

In addition to embracing societal impacts as a core review criterion for submitted proposals, NSF has invested in a center on the topic. Societal impacts of the scientific enterprise is the theme of a recently established NSF-initiated center, the Center for Advancing Research Impact in Society (ARIS), which was founded in 2018 with the support of a $5.2 million grant from NSF's Office of Integrative Activities; each of NSF's eight research supporting directorates contributed funding for the grant[4]. ARIS is based at the University of Missouri, and its primary objects are to work with scientists and engagement practitioners to build capacity, advance scholarship, grow partnerships, and provide resources to help them engage with and demonstrate the impact of research in their communities and society. It provides workshops and consultation on broader impacts related topics for a variety of

[3] www.fastlane.nsf.gov/a5/A5Crit_Conflict_Conf.htm
[4] www.nsf.gov/awardsearch/showAward?AWD_ID=1810732

audiences, and it offers a range options from one-hour to multi-day sessions. It also offers a program to enhance institutional capacity for supporting broader impacts activities with the aim to establish an office or center at the institution for enhancing those activities. It has also developed a Broader Impacts Wizard, a web-based tool that facilitates the development of a broader impacts plan that will satisfy NSF's broader impacts requirements for submitted proposals.

47.2.2 Societal impacts via research collaboration and industrial engagement

In addition to achieving societal impacts by implementing the broader impacts merit review criterion to all proposals, NSF has established mechanisms for supporting research collaborations between scientists and institutions. Such collaborative efforts are supported in multiple ways. In principle, all NSF proposals can include collaborative research, and these collaborations can be with academic institutions, research laboratories, non-profit organizations, museums, or any other type of NSF-eligible entity. Some NSF programs are specifically designed for supporting collaborative work, for example, the Research Coordination Networks (RCN) program, which supports the development of new collaborative networks. Investigators constituting a network jointly plan to create new and innovative directions in research, training, or education[5]. RCN grants support communication and coordination efforts of investigators to engage in such activities across disciplinary, organizational, geographic, and international boundaries. There is another NSF program which specifically mandates international collaboration: Accelerating Research through International Network-to-Network Collaborations (AccelNet)[6]. This program, administered by the Office of International Science and Engineering (OISE), seeks to accelerate the process of scientific discovery and prepare the next generation of US researchers for multiteam international collaborations. In addition to funding programs, NSF supports and facilitates international collaboration via agreements with scientific research funding agencies in other countries. NSF currently has memoranda of understanding with counterpart funding agencies of the European Union and several countries including the United Kingdom, Israel, the Czech Republic[7]. The purpose of each memorandum is to facilitate bilateral international collaborative research.

NSF launched the innovative Small Business Innovation Research (SBIR) program in 1977, designed by Roland Tibbetts, that was soon adopted by other US government agencies and became one of the components of NSF Engineering directorate (founded in 1981). Tibbetts conceptualized SBIR primarily as a way through which academics could create start-up companies, usually commercializing hardware ideas developed through prior research, in three phases: (i) a small NSF grant to get started, (ii) a second somewhat larger grant to develop really promising prototypes, and (iii) completing development with funding entirely from investors.

---

[5] beta.nsf.gov/funding/opportunities/research-coordination-networks

[6] www.nsf.gov/publications/pub_summ.jsp?WT.z_pims_id=505584&ods_key=nsf21511

[7] www.nsf.gov/od/oise/IntlCollaborations/index.jsp

The multi-agency SBIR program reports it 'has resulted in 70 000 issued patents, close to 700 public companies, and approximately $41 billion in venture capital investments'[8]. The continuing NSF component has recently operated within the Division of Industrial Innovation and Partnerships that

> supports programs to accelerate NSF-funded and other federally-funded fundamental research into market opportunities, and fosters public–private partnerships to advance technological innovation. IIP invests in high-tech small businesses and collaborations between academia and industry to transform discoveries into innovative commercial technologies with societal benefits.

The 2020 budget for the IIP division was $291 460 000, of which SBIR invested $196 040 000[9].

NSF's Innovation Corps (I-Corps) program promotes collaborations between academia and industry to transform discoveries into innovative commercial technologies with societal benefits, and it supports experiential education to help researchers gain valuable insight into entrepreneurship, into starting a business, or into industry requirements and challenges[10]. I-Corps supports teams, hubs, and the national innovation network. The teams have three primary members: a technical lead, an entrepreneurial lead, and an I-Corps mentor. A hub is a regional network of universities that helps researchers learn how to translate fundamental research to the marketplace. The national innovation network is designed to be a tightly connected, highly functional organization of I-Corps™ Hubs, Nodes and Sites spanning the nation. Other federal agencies such as the National Institutes of Health (NIH) and National Aeronautics and Space Administration (NASA) have their own versions of I-Corps programs.

Another set of industrial engagement is led by NSF's Directorate for Computer and Information Science and Engineering (CISE). Since its emergence in the late 1980s, CISE has maintained close connections to industry (Freeman *et al* 2019). Recently, its Division of Information and Intelligent Systems has emphasized Big Science investment in artificial intelligence:

> Led by NSF, and in partnership with the US Department of Agriculture National Institute of Food and Agriculture, US Department of Homeland Security, Google, Amazon, Intel and Accenture, the National AI Research Institutes will act as connections in a broader nationwide network to pursue transformational advances in a range of economic sectors, and science and engineering fields—from food system security to next-generation edge networks[11].

---

[8] www.sbir.gov/birth-and-history-of-the-sbir-program

[9] www.nsf.gov/eng/iip/about.jsp

[10] www.nsf.gov/news/special_reports/i-corps/index.jsp

[11] www.nsf.gov/news/news_summ.jsp?cntn_id=303176

In 2020–21 $360 000 000 was invested in 18 institutes, each of which is an extensive partnership, linking in total nearly 400 organizations across 40 of the 50 US states, including additional companies such as Dell, Facebook, IBM, Microsoft, Netflix, and Qualcomm. At the same time, the Computer and Network System division has been closely connected in several ways to the network communication industry, for example:

> The Platforms for Advanced Wireless Research (PAWR) program is an NSF-led public–private partnership to advance the development of next-generation wireless technologies and services. In 2016, NSF convened a consortium of over 30 leading companies and industry associations in the wireless sector, to include networking vendors, device manufacturers, and wireless carriers. Beginning in 2017, NSF is investing $50 million over 7 years, together with an additional $50 million in cash and in-kind contributions from the industry consortium, to design, develop, deploy, and operate four wireless research platforms. (NSF 2018, p 20)

Overall, collaboration and partnerships play a large role at NSF, and their societal benefits are explicitly pointed out on NSF's Partnerships webpage:

> Partnerships can accelerate discovery by expanding the kinds of questions that can be addressed; enabling access to expertise, infrastructure, or sites; and building broader communities of researchers. This, in turn, can accelerate translation of research results to products and services, and enhance preparation of the future workforce to benefit society and grow the American economy.[12]

47.2.3 Evolution of Big Science and convergent approach at NSF

Over the several last decades, the definition of Big Science has evolved significantly, and now includes large but decentralized projects sometimes at least partially supported by the private sector: one prominent example is the Human Genome Project, although many other, more recent, examples can be cited. NSF has funded a large number of projects that would fit into various descriptions of Big Science, expanding beyond the initial physics/astronomy based efforts into other fields such as engineering, education, and human resources, and the social sciences; these projects may showcase the societal connectivity more directly. It is also noteworthy that through the integration and collaboration of various disciplinary parts of NSF, a convergent approach has taken shape. Convergence is defined as a means of solving complex problems focusing on societal needs. It entails integrating knowledge, methods, and expertise from different disciplines and forming novel

[12] www.nsf.gov/about/partners/

frameworks to catalyze scientific discovery and innovation[13]. Convergence research has two primary characteristics, the first of which incorporates societal needs:

- *Research driven by a specific and compelling problem.* Convergence research is generally inspired by the need to address a specific challenge or opportunity, whether it arises from deep scientific questions or pressing societal needs.
- *Deep integration across disciplines.* As experts from different disciplines pursue common research challenges, their knowledge, theories, methods, data, research communities, and languages become increasingly intermingled or integrated. New frameworks, paradigms or even disciplines can form sustained interactions across multiple communities.

We note that the convergent approach applies not only to scientific research, but also to designing, implementing, and evaluating the broader societal impacts of the scientific enterprise. To illustrate that point, before considering three Big Science initiatives in some depth, we present brief sketches of three others—these map the general structure of NSF involving convergence across all fields of science and engineering. NSF often invests significantly in a unified project designed to explore a well-defined scientific question, such as IceCube in Antarctica that observes cosmic neutrinos, but other, no less important, efforts focus on expanding the scope of discovery and invention, through large sets of much smaller investments. Dating from 1978, the Established Program to Stimulate Competitive Research (EPSCoR) partners with a wide range of NSF programs to jointly support projects in geographic areas of the United States 'that had historically received lesser amounts of federal R&D funding, and had demonstrated a commitment to developing their research bases and improving the quality of S&E research conducted at their universities and colleges'[14]. Based on NSF's success, these other agencies also developed extensive EPSCoR programs: Department of Energy, National Aeronautics and Space Administration, National Institutes of Health, US Department of Agriculture, and Department of Defense. Table 47.2 shows the distribution of recent EPSCoR awards across the main units of NSF, called directorates, stopping with 2015 because later grants may not yet have received all their funding. These data are available for anyone to analyse—they were downloaded as a set of spreadsheets from the public database of abstracts of NSF awards[15].

Over the decades, NSF has experimented with Big Science composed of many small projects working in parallel. In 2008, on the basis of experience developing earlier transformative research initiatives, NSF invented Early-Concept Grants for Exploratory Research (EAGER), small investments regular programs could make in research 'considered especially "high risk-high payoff" in the sense that it involves radically different approaches, applies new expertise, or engages novel disciplinary or interdisciplinary perspectives'[16]. The two other columns of the table focus on just one year, 2020, and NSF's response to the COVID-19 pandemic: 'RAPID is a type

[13] www.nsf.gov/od/oia/convergence/index.jsp
[14] www.nsf.gov/od/oia/programs/epscor/nsf_oiia_epscor_epscor_eicc.jsp
[15] www.nsf.gov/awardsearch/advancedSearch.jsp
[16] www.nsf.gov/about/transformative_research/submit.jsp

Table 47.2. Examples of small grants that collectively achieve innovative Big Science.

NSF Directorate	EPSCoR 2001–15	EAGER 2008–15	CARES RAPIDS 2020	Regular RAPIDS 2020
Biological Sciences	559	400	116	52
Computer and Information Science and Engineering	392	1204	127	32
Education and human resources	309	151	0	58
Engineering	815	849	146	74
Geosciences	570	399	0	73
Mathematical and Physical Sciences	780	152	38	26
Office of the Director	236	13	1	12
Social, Behavioral, and Economic Sciences	224	98	22	189
Total	3885	3266	450	516
Total funding	$1 661 338 509	$537 366 069	$73 707 143	$71 294 510

of proposal used when there is a severe urgency with regard to availability of, or access to, data, facilities or specialized equipment, including quick-response research on natural or anthropogenic disasters and similar unanticipated events'[17]. According to data from the public abstract system, 450 grants were funded through an emergency government response, The Coronavirus Aid, Relief, and Economic Security Act, also known as the CARES Act, and 410 of them were RAPIDs. But NSF also funded about an equal number of pandemic-related grants using its regular funding, here tallied along with a few RAPIDs responding to other sudden situations of scientific significance. Already, these brief grants are nearing their conclusions, reporting findings related to the pandemic itself that may help us deal with future virus emergencies, but also seeking to understand indirect effects and the best ways to reduce many kinds of harm from this societal disruption.

47.3 Big Science investments at NSF

NSF supports research funding for 'little science' (individual investigators working on problems of their own choice, typically with graduate students supported by the grant), 'Big Science' (large-scale instrumentation, facilities, multi-disciplinary science driven by big ideas and involving large teams), and everything in between, in all fields of science and engineering (excluding medical/clinical), including education and the social sciences. NSF has made many Big Science investments over the last

[17] www.nsf.gov/pubs/policydocs/pappg20_1/pappg_2.jsp#IIE1

half-century, including several large, major research facilities, for example, the world famous Laser Interferometer Gravitational-Wave Observatory (LIGO), the National Ecological Observatory Network (NEON), and National Center for Atmospheric Research (NCAR). From this rich repertoire, we have chosen to focus on three specific examples to illustrate cross-domain collaboration within NSF and show how such collaboration strengthens the societal impacts of each program.

1. The Digital Library Initiative
2. The National Nanotechnology Initiative
3. Three large-scale social science surveys

47.3.1 Digital libraries

The Digital Library Initiative (DLI) that NSF launched officially in 1993 is a good example of how the social organization and effects of Big Science can vary from one field of research to another, because it extended across Internet rather than being housed in a single observatory (Lesk 1997, 2012). The foundation of the DLI was six grants that totaled about $27 million, which would be about $48 million in today's money, but there were also smaller DLI investments made in parallel, such as the grant from the Sociology program, Prototype Internet Services for the General Social Survey[18], that still today provide students and researchers free access including online analysis tools for the survey and for the American National Election Studies, two of the most significant NSF social science datasets[19]. The Defense Advanced Research Projects Agency and The National Aeronautics and Space Administration were partners with NSF from the beginning of the DLI, and the six main initial awards covered a very wide range of sciences and engineering applications, to be developed at these leading educational institutions: University of Michigan for a testbed for textual, video, still image, and datasets, from both primary and secondary information suppliers, in the Earth and space sciences[20]; Carnegie-Mellon University to develop intelligent, automatic mechanisms to populate a digital video library based on knowledge-based search and retrieval[21]; University of Illinois for fundamental research on the scalable organization of large digital collections, using engineering journals and magazines as the empirical challenge[22]; University of California, Santa Barbara, for design, development, and evaluation of a distributed digital library of spatially indexed information that includes collections of maps and digital images[23]; University of California, Berkeley, for a prototype digital library focused on environmental information, with a concentration on the utilization, evaluation, and preparation of environmental impact reports, statements, and related documents[24]; Stanford University to develop

[18] www.nsf.gov/awardsearch/showAward?AWD_ID=9422785
[19] sda.berkeley.edu/archive.htm
[20] www.nsf.gov/awardsearch/showAward?AWD_ID=9411287
[21] www.nsf.gov/awardsearch/showAward?AWD_ID=9411299
[22] www.nsf.gov/awardsearch/showAward?AWD_ID=9411318
[23] www.nsf.gov/awardsearch/showAward?AWD_ID=9411330
[24] www.nsf.gov/awardsearch/showAward?AWD_ID=9411334

the enabling technologies for a single, integrated, and 'universal' library, proving uniform access to the large number of emerging networked information sources and collections; and one result was the Google search engine (Page *et al* 1999)[25].

NSF program director who managed these grants, Stephen Griffin (1998), reported that the initiative: 'was the result of a community-based process which began in the late 1980s with informal discussions between researchers and agency program managers. These discussions progressed to planning workshops designed to develop research values and agendas'. Thus, it was a collective effort across many organizations, rather than the inspiration of one individual leader or even a small committee. An NSF historical webpage describes what followed: 'In 1999, the $55 million Phase 2 (DLI-2) included 36 projects supported by NSF, DARPA, the National Library of Medicine, the Library of Congress, NASA, and the National Endowment for the Humanities, with participation from the National Archives and the Smithsonian Institution, to extend and develop innovative digital library technologies and applications'[26].

Participation by the National Endowment for the Humanities was especially significant, because the Internet allowed computer and information science to open new possibilities for the arts and humanities. In the second phase of DLI, Gregory Crane, who had already founded the Perseus Digital Library of ancient Greek and Roman literature, received a substantial grant titled A Digital Library for the Humanities, 'both to improve the ways that humanists can perform their intellectual work and to design materials that are more accessible to the vastly expanded audience already reached by the World Wide Web'[27]. Another large grant focused on recordings to create A National Gallery of the Spoken Word, 'a significant, fully searchable, online database of spoken word collections that span the twentieth century. A collaborative project among the humanities, engineering, education and library science, this gallery will provide the first large-scale repository of its kind'[28]. Another major DLI-2 grant explored innovative online alternatives to traditional academic publication, Re-inventing Scholarly Information Dissemination and Use: 'to develop tools and technologies that support highly improved models of information dissemination and access. A goal is to facilitate moving from the current centralized, discrete publishing model, to a distributed, continuous, self-publishing model, while at the same time preserving and enhancing the best aspects of the current model'[29].

Near the end of the formal initiative, the DLI emphasized international connections, for example in a three-year International Symposium on Music Information Retrieval[30]. In 1999 a few grants were made in collaboration with the United Kingdom, such as Integrating and Navigating Eprint Archives through

[25] www.nsf.gov/awardsearch/showAward?AWD_ID=9411306
[26] www.nsf.gov/news/news_summ.jsp?cntn_id=103048
[27] www.nsf.gov/awardsearch/showAward?AWD_ID=9817484
[28] www.nsf.gov/awardsearch/showAward?AWD_ID=9817485
[29] www.nsf.gov/awardsearch/showAward?AWD_ID=9817353
[30] www.nsf.gov/awardsearch/showAward?AWD_ID=9905842

Citation-Linking, and Emulation Options for Digital Preservation: Technology Emulation as a Method for Long-Term Access and Preservation of Digital Resources[31]. From 1994 to 2009, a total of 158 research awards received funding from the Digital Library Initiative, for a sum just over $100 000 000, although the formal competition was completed before the new century, and research of this type quickly found support across NSF. The Directorate for Education and Human Resources funded its own efforts, beginning in 1995 with A Digital Library for Middle Schools: Supporting Authentic Science Investigations[32].

47.3.2 From nanotechnology to large-scale convergence

The National Nanotechnology Initiative (NNI) is one of the largest investments in research and development, with great implications for human welfare, yet concerns many of the smallest physical structures in nature and technology (Roco 2011). A January 1999 workshop report observed that the potential 'for nanotechnology to transform so many aspects of human existence is almost without precedent', capable of shaping the future, based on recent experience from the past:

> In the last few years, applying fundamental discoveries related to nanotechnologies has already developed multibillion-dollar businesses. These latter include giant magnetoresistance (for hard disks), nanolayers (for data storage and the photographic industry), nanoparticles (for drugs in the pharmaceutical field and colorants in printing), confinement effects (for optoelectronic devices and lasers), nanostructured materials (for nanocomposites and nanophase metals), and chemical and biological detection (for national security and the food industry). Fundamentally novel phenomena and processes have led to new, high-value-added technologies. Investment in enabling basic research and infrastructure for nanoscience and engineering promises extraordinarily high economic and societal returns; it is due primarily to this fact that the need to establish a nationally coordinated nanotechnology initiative is so compelling. (Roco *et al* 2000, p xv)

Thus, from the very beginning, the benefits for the general public were paramount in justifying this huge effort. But the possible harms were also in mind, and the proper balance of innovation and caution was the focus of a second conference in September 2000 that produced another book-length report, Societal Implications of Nanoscience and Nanotechnology, with 64 academic and government contributors (Roco and Bainbridge 2001). There was no naive assumption that ethics and values were simple concepts, but rather awareness that the results of both scientific research and public discussion would raise new questions, as well as achieving solid

[31] www.nsf.gov/awardsearch/showAward?AWD_ID=9907892; www.nsf.gov/awardsearch/showAward?AWD_ID=9905935

[32] www.nsf.gov/awardsearch/showAward?AWD_ID=9554205

discoveries (Roco 2003). Therefore, a second conference was held in 2003, resulting in a pair of book-length reports that extended and updated the findings of the first (Roco and Bainbridge 2006a, 2006b). In both cases, immediate publications were shared electronically, and later physical printings preserved them. As illustrated from its 2004 text, the annual NSF grant competition, Nanoscale Science and Engineering, stressed: 'Innovations in science and technology both require societal support and influence social structures and processes, sometimes in unexpected ways. Examining the ethical and other social implications of these societal interactions is necessary, in order to understand their scope and influence and to anticipate and respond effectively to them'[33].

In 2004–05 NSF-initiated several important programs to engage with the societal and ethical aspects of nanotechnology (Roco 2011). It established two Centers for Nanotechnology in Society (CNS), one at Arizona State University (ASU) and the other at the University of California at Santa Barbara (UCSB). On the education front, it established the National Centers for Learning and Teaching in Nanoscale Science and Engineering, which focuses on high school and undergraduate nanotechnology educations; it also established the Nanoscale Informal Science Education Network, which focuses on nanotechnology science education in informal settings, such as museums or science cafes. In pursuing socially responsible research with regards to environmental, health, and safety (EHS) concerns, NNI's Nanoscale Science, Engineering, and Technology Subcommittee of the National Science and Technology Council formed the Nanotechnology EHS working group, which coordinated efforts across multiple federal agencies including NSF.

The establishment of two centers for nanotechnology in society, ASU CNS and UCSB CNS, is a very important development for NSF's Directorate for the Social, Behavioral and Economic Sciences (SBE). Each is a large-scale center that conducts SBE research on ethical and societal aspects of nanotechnology research and development; they were made possible by funding provided by multiple directorates across NSF. That contrasts with the large-scale surveys, which are discussed in section 47.3.3; the surveys constitute large-scale infrastructure for SBE research, and they are funded solely by the SBE directorate.

ASU CNS was funded for ten years by NSF with an initial award for five years[34], and a renewal award for five more years[35]. The center evolved over the ten-year period; a snapshot will suffice. In 2013 the key elements of the center consisted of four Real Time Technology Assessments, two Thematic Research Clusters, which create linkages across the assessments, one group dedicated to nanotechnology education at all levels and another group dedicated to public engagement. The Research and Innovation Systems Assessment uses bibliometric and patent analyses to understand the evolving dynamics of the nanotechnology science and engineering (NSE) enterprise; the research group has constructed a large-scale set of global databases of nanotechnology research publication and patent records. The Public

[33] www.nsf.gov/pubs/2004/nsf04043/nsf04043.htm
[34] www.nsf.gov/awardsearch/showAward?AWD_ID=0531194
[35] www.nsf.gov/awardsearch/showAward?AWD_ID=0937591

Opinion and Values Assessment uses surveys and quasi-experimental media studies to understand changing public and scientists' perspectives on developments in nanotechnology. The Anticipation and Deliberation Assessment uses scenario development and other techniques to foster deliberation on plausible NSE applications, thereby operationalizing the Centers predominant unifying theoretical framework, anticipatory governance. The Equity, Equality and Responsibility Cluster studies the ways in which NSE contributes to increasing or decreasing social and economic inequalities in different national contexts, and it explores ways to ensure that NSE can contribute to equity, equality, and responsibility as public values. Finally, the Urban Design, Materials, and the Built Environment Cluster investigates the nano-enabled city of the future, addressing the links among NSE, the built environment, social structures, and sustainability. The sketch above shows that ASU CNS is pertinent to a broad range of societally relevant outcomes (such as those listed in section 47.2.1 that were abstracted from NSF 20-1).

The UCSB CNS was also funded for ten years by NSF, initially for five years[36], and renewed for five more years[37]. In 2013 the key elements of the center consisted of three Interdisciplinary Research Groups, six Cross Group projects, and education and public engagement components. The Nanotechnology in Historical Context Group focuses on historical and contextual aspects of nanoscale science and engineering using both qualitative and quantitative methods, including archival research and ethnography. The Nanotechnology and Globalization Group strives to develop a better understanding of the importance of both state policies and international collaboration in fostering research, development, and commercialization of nanotechnology through a comparative study of the US, China, Japan, India, Korea, and selected Latin American countries. The Risk Perception and Nanotechnology Group engages in research to generate new knowledge about the perceived risks and benefits of nanotechnology among multiple stakeholders in the nanotechnology enterprise; it develops and documents methods for public engagement in the US context and contributes to work within the Center to disseminate the knowledge gained to an array of critical stakeholders groups. Cross Group work is on a variety of topics including solar nanotechnology, spatial and global-value-chain analyses of nanotechnology in California, and ethnographic studies of nanoscience and nano-toxicology labs; ten senior participants are directly involved in six Cross Group projects. The sketch above shows that UCSB CNS addresses a broad range of societally relevant outcomes.

Initial discussion of societal implications tended to focus on the harm that nanoparticles might do to the natural environment, or to biological systems like the human body, if they were accidentally released, but there turned out to be another issue that caused considerable confusion (McCarthy and Kelty 2010). Much of the early public discussion of nanotechnology was based on unscientific notions of what it might accomplish, thus adding impossibilities to the risks people were worried

[36] www.nsf.gov/awardsearch/showAward?AWD_ID=0531184
[37] www.nsf.gov/awardsearch/showAward?AWD_ID=0938099

about (Smalley 2001, Bainbridge 2007). Subsequently another more reasonable issue arose, how to assess and improve the inclusion of women and other under-represented groups in nano-related professions (Villanueva-Felez *et al* 2015). Throughout the history of the NNI, a significant challenge was how to accomplish coherent science and engineering when it is being done by large numbers of individuals at multiple institutions. One solution was establishment of communication hubs and networks, such as the 2015 announcement that NSF would 'provide a total of $81 million over five years to support 16 sites and a coordinating office as part of a new National Nanotechnology Coordinated Infrastructure'[38].

The National Nanotechnology Coordinated Infrastructure (NNCI) has four main sites that do work researching, training, and communicating some of the social and ethical issues raised by and inherent to Nanoscale Science and Engineering (NSE): North Carolina State University, Arizona State University, University of Texas at Austin, and Northwestern University. These sites constitute a support system for the rest of the NNCI sites that want to work societal and ethical implications of NSE programs into their sites. Their collective goal is to integrate research on societal, ethical, and environmental concerns with nanotechnology research and development, and ensuring that advances in nanotechnology bring about improvements in quality of life for all Americans.

A decade into extensive convergence developments that began with ethical aspects of the NNI and expanded to include not only all the physical sciences and fields of engineering, but also the social sciences, a major series of conferences combined retrospectives with more advanced perspectives on the future (Roco *et al* 2013). Working groups deeply explored five broad areas in which convergence could have major implications: (i) Human health and physical potential, (ii) human cognition and communication and the emergence of the cognitive society, (iii) societal collective outcomes including manufacturing, (iv) people and physical infrastructure, and (v) convergence of knowledge and technology for a sustainable society.

47.3.3 Three large-scale social science surveys

The American National Election Studies is often cited at the first 'Big Science' project for the social sciences at NSF, although it actually began before NSF was founded, and shifted from support by the Carnegie and Rockefeller foundations to NSF beginning in 1970 (Larsen 1992, pp 105–11, Solovey 2020, pp 191–6). A progress report on its first fifty years stated:

> Although there is an unbroken series of national election studies covering all thirteen presidential elections, and 10 midterm elections, since 1948, a crucial turning point came in 1977 when NSF established NES as the first *national social science resource*, a model NSF later used with the General

[38] www.nsf.gov/news/news_summ.jsp?cntn_id=136211

Social Survey and the Panel Study of Income Dynamics. (Sapiro and Bartels 1998, p 64)

The survey's website describes its mission:

Why does America vote as it does on Election Day? The mission of the American National Election Studies (ANES) is to inform explanations of election outcomes by providing data that support rich hypothesis testing, maximize methodological excellence, measure many variables, and promote comparisons across people, contexts, and time. The ANES serves this mission by providing researchers with a view of the political world through the eyes of ordinary citizens[39].

Thus it is Big Science not merely in terms of the size of the investment, or its multi-decade longevity, but in that it seeks to understand the huge American voter population, that well exceeded 150 million in 2020. A recent study based on analysis of 1163 articles published in 11 journals central to the field concluded: 'the ANES is a critical investment for the scientific community and a main driver of political behavior research' (Robison *et al* 2018, p 1).

From its website, the Panel Study of Income Dynamics (PSID) links to an overview article that describes it as 'the world's longest-running household panel survey' that achieves 'breakthroughs in understanding the connections between economic status, health and well-being across generations and over the life course' (McGonagle *et al* 2012, p 268). A panel study recruits respondents for a continuing research project that will repeatedly interview them or administer questionnaires to them, tracking changes over time. The ANES sometimes did a panel study, most often interviewing the same people before and after a presidential election, but given the many years of the PSID, methodological concerns were raised about whether the sample of respondents remained representative of the general population, or attrition had introduced biases. However, careful statistical analysis and proper recruitment of new respondents has preserved the validity of the PSID over the decades (Becketti *et al* 1988, Fitzgerald *et al* 1998). It is important to note that it studies families, rather than just individuals as most surveys do, although collecting data by interviewing the 'head of household' or 'reference person', and can add children as respondents when they reach adult independence. It began in 1966 as a study of poor families conducted by the Census Bureau, then added a sample of the general population in 1968. When NSF began supporting it in 1979, it was called The Panel Study of Family Income Dynamics, but the word 'family' was dropped for its fourth grant in 1984, when the simple PSID name became familiar[40].

The prominence of the word 'income' in the PSID name connects directly to the main NSF program through which funding has been provided, the Economics

[39] electionstudies.org/about-us/

[40] www.nsf.gov/awardsearch/showAward?AWD_ID=7818535, www.nsf.gov/awardsearch/showAward?AWD_ID=8311866

program. Among the core research questions have been the stability versus volatility of family income, such as what fraction of the population is trapped in poverty rather than being poor only temporarily, and the wider economic question of social class mobility across generations (Moffitt and Zhang 2018, Smeeding 2018). Yet over the years other programs have also contributed, indicating the convergent value of the PSID for social science, including Sociology, Political Science, Geography, and Spatial Sciences, and an interdisciplinary program, Methodology, Measurement, and Statistics. For example, sociologists have been especially interested in PSID data for studies on the changing structure or composition of families, racial and ethnic inequality, and even the effect on families of the characteristics of their neighborhood (Pfeffer *et al* 2020). Additional funding from the National Institute of Child Health and Human Development supported supplementary data collection on child development, permitting analysis of issues such as how the long-term impact of mass media, such as extensive TV watching or computer use, varied by gender and ethnicity (Hofferth 2010).

The General Social Survey (GSS) began with NSF funding in 1971, rather than having initially been supported by private foundations or other government agencies as was the case for ANES and PSID. It was proposed by James A Davis, who had just become head of the National Opinion Research Center at the University of Chicago, and in 1984 it established an alliance called the International Social Survey Program (ISSP) with colleagues in Great Britain, West Germany, and Australia (Alwin 2017). The GSS consists of core items that are included each year the questionnaire is administered, and *topic modules* that appear just one year or occasionally, to focus on particular themes and to test significant hypotheses. The partners in the International Social Survey Program often develop a topic module they all will include, even as some of their core questions are also identical. For example, the 1992 ISSP module focused on 'beliefs and perceptions regarding social inequality in general and income differences in particular', while the 1994 module focused on 'attitudes toward environmentalism and on women, work and the family'[41]. Currently, 42 countries participate in the ISSP[42], and its data have been used worldwide to teach social science from a global perspective (Lauer and Yolandis 2004). In addition, many social scientists use sets of questions from the GSS in their own questionnaires, both because these items have been verified and in order to compare distinctive groups with the general population (Bainbridge 2002).

47.4 Concluding remarks

The examples of NSF's Big Science investments that we provided in this paper show the convergent approach in action; although the long-term societal impacts of these investments have not been formally assessed, there is plenty of evidence of the numerous ways these projects have touched the lives of people. Just noting that Google was the result of one of the first six projects supported by the Digital Library

[41] www.nsf.gov/awardsearch/showAward?AWD_ID=9122133, www.nsf.gov/awardsearch/showAward?AWD_ID=9318653

[42] issp.org/members/member-states/

Initiative is sufficient in terms of its revolutionary global societal impact. Many authors have written about the various aspects and impacts of the DLI; we cite one example which explicitly considered long-term societal impacts. Karen Calhoun (2014) explored the social roles of digital libraries and proposed a conceptual framework by which these roles could be analysed. Calhoun proposes that the digital library initiative has impacted the following ten inter-related societal roles that build on or reinforce one another: broad access to content, public infrastructure, free flow of ideas, individual empowerment, informed citizenry, formal education, scholarly collaboration and partnerships, progress of knowledge, economic benefits, preservation, and curation. Among other things, she pointed out the potential role of open-access digital libraries a mechanism to bridge the digital divide.

A recent retrospective for the 50th anniversary of the General Social Survey (GSS) reports that it

> has prospectively recorded the current characteristics, backgrounds, behaviors, and attitudes of representative cross sections of American adults covering more than two generations and more than a century of birth cohorts. As a foundational resource for contemporary social science, the data that it produces and disseminates enable social scientists to develop broad and deep understandings into the changing fabric of US society, and aid legions of instructors and students in teaching and learning. (Marsden *et al* 2020)

The GSS contributed greatly to development of social indicators that policymakers could use to assess the changing social and economic situation, and it measured the changing nature of American culture and the personal opinions of the general population.

All seven directorates of NSF participated actively in creating and expanding the National Nanotechnology Initiative (NNI), as did other parts of the US federal government. For example, in 2013 the US invested over $1.55 billion in the NNI across 11 agencies, of which four invested over $100 million each: National Science Foundation, Department of Health and Human Services, Department of Energy, and Department of Defense[43]. From practically the very beginning, the fact that many separate sciences and engineering professions were involved encouraged creative thinking about how they could best cooperate. This immediately launched a series of large conferences about what came to be called NBIC Convergence, combining nanoscience with biology, information science, and cognitive science, and especially the technologies based on them (Roco and Bainbridge 2003, Bainbridge and Roco 2016a). The first extensive report identified five main areas of societal benefits: (i) expanding human cognition and communication, (ii) human health and physical abilities, (iii) enhancing group and societal outcomes, (iv) national security, and (v) unifying science and education.

[43] 'The National Nanotechnology Initiative, Supplement to the President's 2015 Budget,' p 8, obamawhitehouse.archives.gov/sites/default/files/microsites/ostp/NNI_FY15_Final.pdf

Organized differently, with a large audience for public presentations by a few deep thinkers, the second NBIC conference illustrated convergence at the level of individual scholars who presented challenging ideas that linked well together. For example, George Khushf (2004) explained why many of the great ethical questions about new technologies cannot be answered without close collaboration between specialists in the given field and philosophers or others with deep understanding of the human issues, because both modern technology and society are complex systems interacting with each other. Sonia Miller (2004) outlined how the legal system would need to adapt to the rapidly changing system that includes intellectual property rights, social impacts, and the ethical balance between rights and duties. James Canton (2004, p 186) observed: 'The emergence of NBIC convergence will challenge us in new ways to balance risk and return, threat and opportunity, and social responsibility'. The third and fourth major reports also were collections of contributions from individuals and small teams, often focused on how convergence might transform a particular field of endeavor. For example, a team from IBM led by Jim Spohrer *et al* (2006) explained how the emphasis of information technology corporations may shift from selling hardware to providing information services. Philosophers Anders Sandberg and Nick Bostrom (2006) from Oxford University considered the range of options for human cognitive enhancement. A recent overview suggested, 'Convergence of knowledge and technology for the benefit of society (CKTS) is the core opportunity for progress in the twenty-first century' (Bainbridge and Roco 2016b). It can be safely stated that analysing and quantifying the societal impacts of science and technology is a complex task; moreover, it is difficult, if not impossible, to separate the impact of 'Big Science' from the overall impact of science and science funding by government agencies and other organizations. Despite that complexity, we can point out one significant societal impact of NSF's funding programs: the development of future scientific workforce; quoting from 'NSF by the Numbers'[44]:

- Annually, NSF funding directly impacts 307 000 researchers, technical professionals, post-doctoral students, graduate students, undergraduates, K-12 teachers, and students.
- Since 1952 NSF has supported more than 61 000 students through its flagship Graduate Research Fellowship Program.
- In the fiscal year 2020, approximately 42 000 graduate students received funding through NSF programs as part of research and education grants.

NSF continues to support basic research by individuals and teams, both small and large, in every scientific and engineering discipline including social, behavioral and economic sciences. In addition, NSF leads, supports and facilitates large-scale endeavors in rapidly emerging fields such as quantum methods, artificial intelligence, advanced manufacturing, and biotechnology, sometimes in cooperation and collaboration with the private sector. NSF welcomes partnerships with private industry,

[44] nsf.gov/news/factsheets/Factsheet_By%20the%20Numbers_05_21_V02.pdf

foundations, nonprofits, other federal agencies, state and local governments, and international organizations. Consideration of broader societal impacts remains one of NSF's core values and permeates all its activities. For example, the opening paragraph on the NSF fact sheet entitled 'NSF Partnership Opportunities'[45] states: 'Partnering can accelerate scientific discovery as well as the translation of research into products and services that benefit society'.

To complete the story, we draw attention to the fact that in addition to requiring that all funded projects contribute to societal benefits, NSF funds social science research that may carry out in-depth exploration of various aspects of the topic itself, mainly, but not exclusively, through its program on Science and Technology Studies (STS). This program supports research that uses historical, philosophical, and social scientific methods to investigate the intellectual, material, and social facets of science and technology. The program encompasses a broad spectrum of topics including interdisciplinary studies of ethics, equity, governance, and policy issues that are closely related to STEM disciplines. The STS program can be considered as an added dimension in NSF's convergent approach to not only achieving but also exploring societal impacts of science.

Many questions regarding societal impacts of technological advances in general, and of Big Science in particular remain open: in-depth exploration may reveal not only great societal benefits but also unintended consequences. The scientific approach demands that we remain open and unbiased as we delve into the topic. Social responsibility requires that we remain cognizant of the continuous need to find better ways to mitigate the potentially negative impacts of science even as we acknowledge and celebrate its tremendous benefits.

47.5 Disclaimer

This article was produced while the authors were employed by the National Science Foundation. Any opinions, findings, and conclusions or recommendations expressed in this material are those of the authors and do not necessarily reflect the views of the National Science Foundation.

References

Alwin D F 2017 Dedication: James A Davis, Master of social surveys *Sociol. Methodol.* **47** ix–xv

American Academy of Arts and Sciences 2021 Bold ambition: international large-scale science *Report* American Academy of Arts and Sciences, Cambridge, MA

Bainbridge W S 2002 *The Endtime Family* (Albany, NY: State University of New York Press)

Bainbridge W S 2007 *Nanoconvergence* (Upper Saddle River, NJ: Prentice-Hall)

Bainbridge W S and Roco M C (ed) 2016a *Handbook of Science and Technology Convergence* (Dordrecht: Springer)

Bainbridge W S and Roco M C 2016b Science and technology convergence: with emphasis for nanotechnology-inspired convergence *J. Nanopart. Res.* **18** 211

[45] https://nsf.gov/news/factsheets/Factsheet_Partnerships_v8_P.pdf

Becketti S, Gould W, Lillard L and Welch F 1988 The panel study of income dynamics after fourteen years: an evaluation *J. Labor Econ.* **6** 472–92

Calhoun K 2014 *Exploring Digital Libraries: Foundations, Practice, Prospects* (New York: Neal-Schuman)

Canton J 2004 Designing the future: NBIC technologies and human performance enhancement *The Coevolution of Human Potential and Converging Technologies* ed M C Roco and C D Montemagno (New York: New York Academy of Sciences) pp 186–98

Fitzgerald J, Gottschalk P and Moffitt R 1998 An analysis of sample attrition in panel data: the Michigan panel study of income dynamics *J. Human Resour.* **33** 251–99

Freeman P A, Adrion W R and Aspray W 2019 *Computing and the National Science Foundation, 1950–2016* (New York: ACM)

Griffin S M 1998 NSF/DARPA/NASA digital libraries initiative; a program manager's perspective *D-Lib Magazine* July/August www.dlib.org/dlib/july98/07griffin.html

Hofferth S 2010 Home media and children's achievement and behavior *Child Develop.* **81** 1598–619

Intemann K 2005 Why diversity matters: understanding and applying the diversity component of the National Science Foundation's broader impacts criterion *Technol. Soc.* **27** 249–66

Khushf G 2004 Systems theory and the ethics of human enhancement: a framework for NBIC convergence *Ann NY Acad. Sci.* **1013** 124–49

Larsen O N 1992 *Milestones and Millstones: Social Science at the National Science Foundation, 1945–1991* (New Brunswick, NJ: Transaction)

Lauer S R and Yodanis C L 2004 The International Social Science Program (ISSP): a tool for teaching with an international perspective *Teaching Sociology* **32** 304–13

Lesk M 1997 *Practical Digital Libraries* (San Francisco, CA: Morgan Kaufmann)

Lesk M 2012 The digital library initiative *Leadership in Science and Technology* ed W S Bainbridge (Thousand Oaks, CA: Sage) pp 703–11

Marsden P V, Smith T W and Hout M 2020 Tracking US social change over a half-century: the General Social Survey at fifty *Annu. Rev. Sociol* **26** 1–25

McCarthy E and Kelty C 2010 Responsibility and nanotechnology *Soc. Stud. Sci.* **40** 405–32

McGonagle K A, Schoeni R F, Sastry N and Freedman V A 2012 The panel study of income dynamics: overview, recent innovations, and potential for life course research *Longitud. Life Course Stud.* **3** 268–84

Miller S E 2004 How the legal system should change as a result of converging technologies *Ann NY Acad Sci.* **1013** 178–85

Moffitt R A and Zhang S 2018 Income volatility and the PSID: past research and new results *NBER Working Paper Series* Working paper 24390 www.nber.org/papers/w24390

National Science Foundation 2018 Building the future: investing in discovery and innovation *NSF Strategic Plan for Fiscal Years (FY) 2018–2022* National Science Foundation, Alexandria, VA

Page L, Brin S, Motwani R and Winograd T 1999 The PageRank citation ranking: bringing order to the web *Technical Report* 1999-66 Stanford Digital Libraries http://ilpubs.stanford.edu:8090/422/

Pfeffer F T, Fomby P and Insolera N 2020 The longitudinal revolution: sociological research at the 50-year milestone of the Panel Study of Income Dynamics *Annu. Rev. Sociol* **46** 83–108

Roberts M 2009 Realizing societal benefit from academic research: analysis of the national science foundation's broader impacts criterion *Soc. Epistemol.* **23** 199–219

Robison J, Randy T S, Druckman J N, Jackman S, Katz J N and Vavreck L 2018 An audit of political behavior research *SAGE Open* **8** 1–14

Roco M C 2003 Broader societal issues of nanotechnology *J. Nanopart. Res.* **5** 181–9

Roco M C 2011 The long view of nanotechnology development: the national nanotechnology initiative at 10 years *J. Nanopart. Res.* **13** 427–45

Roco M C R, Williams S and Alivisatos P (ed) 2000 *Nanotechnology Research Directions* (Dordrecht: Kluwer)

Roco M C and Bainbridge W S (ed) 2001 *Societal Implications of Nanoscience and Nanotechnology* (Dordrecht: Kluwer)

Roco M C and Bainbridge W S (ed) 2003 *Converging Technologies for Improving Human Performance* (Dordrecht: Kluwer)

Roco M C and Bainbridge W S (ed) 2006a *Nanotechnology: Societal Implications - Individual Perspectives* (Berlin: Springer)

Roco M C and Bainbridge W S (ed) 2006b *Nanotechnology: Societal Implications: Maximizing Benefit for Humanity* (Berlin: Springer)

Roco M C, Bainbridge W S, Tonn B and Whitesides G (ed) 2013 *Convergence of Knowledge, Technology and Society* (Dordrecht: Springer)

Sandberg A and Bostrom N 2006 Converging cognitive enhancements *Ann. NY Acad. Sci.* **1093** 201–27

Sapiro V and Bartels L M 1998 The American National Election Studies: a progress report *PS: Polit. Sci. Polit.* **31** 64–8

Smalley R E 2001 Nanotechnology, education, and the fear of nanobots *Societal Implications of Nanoscience and Nanotechnology* ed M C Roco and W S Bainbridge (Dordrecht: Kluwer) pp 145–46

Smeeding T M 2018 The PSID in research and policy *Ann. Am. Acad. Pol. Soc. Sci.* **680** 29–47

Solovey M 2020 *Social Science for What? Battles over Public Funding for the 'Other Sciences' at the National Science Foundation* (Cambridge, MA: MIT Press)

Spohrer J, McDavid D, Maglio P P and Cortada J W 2006 NBIC convergence and technology-business coevolution: towards a services science to increase productive capacity *Managing Nano-Bio-Info-Cogno Innovations: Converging Technologies in Society* ed W S Bainbridge and M C Roco (Berlin: Springer) pp 227–53

Villanueva-Felez A, Woolley R and Cañibano C 2015 Nanotechnology researchers' collaboration relationships: a gender analysis of access to scientific information *Soc. Stud. Sci.* **45** 100–29

Weinberg A 1961 Impact of large-scale science on the United States *Science* **134** 161–4

IOP Publishing

Big Science in the 21st Century
Economic and societal impacts
Panagiotis Charitos, Theodore Arabatzis, Harry Cliff, Günther Dissertori, Juliette Forneris and Jason Li-Ying

Chapter 48

Big Science and Society in Japan: Kamiokande Series, policy systems and ILC

Hiromi M Yokoyama, Yuki Akimoto and Yuko Ikkatai

The success of Japan's particle experiments has symbolized the growth in neutrino facilities and B-factories, two of which have medium-sized cores. However, in recent years, Japan has tried medium-scale to full-scale Big Science, such as Hyper-Kamiokande and the International Linear Collider (ILC). The following are discussed in this chapter: the success of the Kamiokande series regarding local relations; the Master Plan and Road Map, which have been central to Big Science in Japan over the past decade, as well as its problems; activities undertaken to attract ILC and public awareness; and other topics of science and society.

48.1 Chain of success: Kamiokande series

48.1.1 Japan's strategic medium-sized experiment

The Kamiokande series (i.e., Kamiokande, Super-Kamiokande, and Hyper-Kamiokande), which comprises devices used in neutrino research, began with a proposal by Masatoshi Koshiba of the University of Tokyo (2002 Nobel Prize in Physics and distinguished university professor, 1926–2020). The Kamiokande, which was installed in a mine in the town of Kamioka, Hida-shi, Gifu Prefecture, Japan, was initially proposed as an experimental facility for proton decay. However, Koshiba's team succeeded in reducing detector noise in early 1987. Immediately after that, they detected neutrinos from the supernova explosion SN1987A in a large magellanic cloud on February 23. Yoji Totsuka (1942–2008, distinguished university professor and director of the Kamioka Observatory at the University of Tokyo; 2003–2006 director of the KEK) asked the researchers in Kamioka to send the tapes by courier to the Hongo Campus in Tokyo, where Koshiba's laboratory analyzed it

doi:10.1088/978-0-7503-3631-4ch48

and submitted a paper based on the findings. As a result, Koshiba won the Nobel Prize in Physics in 2002 for his achievements in pioneering neutrino astronomy.

From the 1960s to the 1970s, large accelerators were built in Japan (e.g., the proton synchrotron). However, at that time, the most advanced pioneers in high-energy physics were in the United States and Europe. Koshiba's laboratory aimed to be a leader in high-energy physics research both by participating in large-scale experiments overseas and while doing middle-sized research in Japan. The Kamiokande series and the B factory series (Belle and Belle II experiments) also began as medium-scale experiments.

In 1996, a larger Super-Kamiokande started observation. At a conference held in Takayama in 1998, Takaaki Kajita (distinguished university professor, The University of Tokyo, and Director of the Institute for Cosmic Ray Research, University of Tokyo (ICRR)) reported the oscillation of atmospheric neutrinos, which led to his receiving the 2015 Nobel Prize in Physics. Totsuka was expected to receive the award with Kajita, but unfortunately, he passed away in 2007.

Koshiba, Totsuka, Kajita, and other leaders in neutrino physics won the Breakthrough Award in 2016. They are esteemed by the Japanese people. In addition, the respect and acceptance of leaders in Japanese society was a major factor in the continuation of Kamiokande as a series. After receiving the Nobel Prize, Koshiba established the Heisei Basic Science Foundation and aimed to popularize basic science by giving lectures to high school students.

48.1.2 Huge accident in the Super-Kamiokande

In 2001, when Totsuka was the director of the Kamioka Observatory, a huge tragedy struck. There was a roaring sound for about five seconds in the Super-Kamiokande. When the tank was opened, about half of the 6000 photosensors, which are called photomultiplier tubes (PMTs), were shattered. Developed for neutrino detection by Hamamatsu Photonics, Super-Kamiokande PMTs are about 50 cm in size. When the accident occurred, the researchers were in the process of storing water in the tank. Before the accident, the researchers had been engaged in replacing ~100 PMTs, and they were in the process of filling water to restart observations.

A photo depicting the shattered PMTs was immediately released. This photo was posted on the Yomiuri Shimbun website on the night of the accident. The cause of the accident was not immediately known. However, because of the prompt disclosure of information immediately after the accident, social reactions were surprisingly good.

Two days later, Totsuka sent a powerful message to his collaborators: 'We will rebuild the detector. There is no question.' Totsuka stated that it was his role to rebuild and bring Japan's neutrino physics back to an internationally competitive level. In a brief email message, he not only stated his determination to rebuild, but also highlighted two strategies. He planned to deploy the remaining PMTs at half density and restart the K2K experiment which detect artificial neutrinos with Super-Kamiokande. He also planned to remake the PMTs in time for the opening of the

Japan Hadron Facility Kamioka (JFK) experiment (currently the T2K experiment). Koichiroh Nishikawa (professor, KEK; spokesperson, K2K experiment), who was the representative of K2K, sent an email in full support of Totsuka and the Super-Kamiokande Group. Subsequently, the cause of the accident was investigated.

After several months of research, it was speculated that one of the two PMTs attached to the bottom of the tank exploded during the PMT replacement efforts because it could not withstand the water pressure. The explosion was thought to have resulted from a crack in the PMT during PMT exchange work. PMTs can usually withstand water pressure, but PMT implosion can occur due to cracks. It was thought that after water flowed into the vacuum part of the 20-inch tube, a shock wave was generated, and the adjacent PMTs were destroyed in sequential order.

Totsuka released the information quickly and maintained the trust of Japanese society. He had always interacted with the media. In addition, during the summer of the following year, university students from all over the country gathered to help rebuild the Super-Kamiokande. Criticism arises after major accidents, but in this case, the media supported these efforts. As planned, the reconstruction of the Super-Kamiokande was completed in one year and immediately after that, Koshiba won the Nobel Prize. Everyone involved was doubly excited.

48.1.3 Overcoming pollution: Super-Kamiokande with local communities

The Kamiokande series has continued to be regarded positively by the local community. The Kamioka Mine was opened in the 1700s. In 1874, the Mitsui Group assumed the successful operation of the mine. However, cadmium, which was contained in the zinc ore, flowed into the river, and caused large-scale pollution. The ailment caused by this exposure was named *itai-itai* disease because its symptoms are bone fractures and pain, even at the slightest touch. This pollution, which lasted from the 1910s to the 1970s, is remembered as the most damaging in Japan during this period.

The Kamiokande series employs a huge tank of purified water. Clean water is supplied by water-rich mines. The series name 'Kamiokande' was based on the town name Kamioka, and Kamiokande series uses clean, pure water. It was a source of great joy and pride for the locals that Kamioka town is known with an image of pure water when they had suffered the effects of polluted water. Thus, good relationships have long existed between neutrino researchers and local residents.

Researchers at Super-Kamiokande began experimenting with gadolinium in 2020 to capture low-energy supernova neutrino events. Ongoing polite communication is maintained between the researchers and the local community. KamiokaLab, situated in Hida City, serves as a popular tourist attraction where visitors can casually explore an exhibit that replicates the interior of Super-Kamiokande. Moreover, a donation system called "Furusato Nozei" has been established to support KamiokaLab and young researchers, with the participation of local governments. These initiatives receive strong backing from the local governments and communities, as the physicists' humble endeavors foster trust.

48.1.4 The Hyper-Kamiokande budget and U Tokyo bonds

The success of Super-Kamiokande and the Nobel Prize were well known in Japan when the construction of Hyper-Kamiokande began in 2020. One problem involved the increased costs of constructing this device, the effective mass which was eight times that of Super-Kamiokande. The construction costs of Super-Kamiokande amounted to 10 billion yen, and Hyper-Kamiokande was expected to cost around 64.9 billion yen. The budget for Big Science, including astronomy and nuclear fusion, which is managed by the Ministry of Education, Culture, Sports, Science, and Technology of Japan (MEXT), is about 35 billion yen annually, which includes a supplementary budget. Thus, the construction of Hyper-Kamiokande could not be guaranteed.

However, the University of Tokyo, home of the Kamiokande series, devised a bold solution. In 2018, in the face of increased international competition, the University of Tokyo announced that it would issue a 20 billion yen university bond. Some of these monies would be used for facilities such as Hyper-Kamiokande. The University of Tokyo's bond was well received and secured a budget that could be used at the university's discretion. This was made possible because the University of Tokyo was one of the first groups to be recognized under the designated country system, which was established in 2017, to ensure a considerable amount of freedom in terms of management. Meeting the budget requirements for constructing Hyper-Kamiokande was a major step forward for the university system.

However, it has also been pointed out that in the future, Big Science may also require the investment of universities. However, the number of universities that can raise part of their own budget will be very limited. In Japan, new big science that can be used for commercial purposes has begun to change significantly from the time when the government paid the full amount. For example, Nanoterasu, a synchrotron radiation facility under construction on the premises of Tohoku University, which is scheduled to start operation in 2024, has requested investment from the private sector from the beginning, and the government, the private sector, and the university are cooperating in construction. IBM partnered with the University of Tokyo in 2019 to pursue multiple quantum computer projects. Funding for large-scale basic science will become an even more difficult problem in the future, but new solutions must be considered.

Hyper-Kamiokande has the highest rating on the roadmap described below. The fact that the best Big Science projects can no longer be covered by national funds, even with international cooperation, poses a major issue for Japan's future large-scale science research.

48.2 Policy system for Big Science in Japan (2010–2021)

48.2.1 The history of the Master Plan and the Road Map

Since 2010, Big Science projects in Japan have been funded in two stages. Every three years, the Science Council of Japan (SCJ) creates a Master Plan that lists 200 projects. Only a dozen projects are placed on a priority list. Next, based on the

Master Plan, the MEXT Examination Committee announces the top 20 projects. This list, which is prepared by the MEXT Committee, is called the Road Map. In 2022, the SCJ decided they would not update the Master Plan with the same style because the three-year term was too short and there are several political issues. In this chapter, we describe Japan's Big Science policy system from 2010 to 2021 and review relevant research in science and technology studies (STS).

The Master Plan and Road Map system for Big Science in Japan was initiated in 2010. In 2009, the Democratic Party of Japan began a budget review activity called business sorting. At that time, Big Science, including supercomputers and Super-Kamiokande, was also targeted. In particular, a supercomputer (the K computer, operated from 2012–2019) was judged to be temporarily frozen. Over 100 post-doctoral researchers were assigned to this project, and there was a chance they could be suddenly fired. This posed a major issue for talent development. There was a strong backlash from scientists, and a press conference was organized by Nobel laureates.

This event had a significant impact on scientists in Japan. In Big Sciences other than space science, there was no protocol for officially deciding on future projects. Therefore, scientists began to list future Big Science projects according to their scholarship value. They examined several reports: the European Strategy Forum on Research Infrastructures 2006 [1]; UK Research Councils 2008 [2]; Department of Energy 2007 [3]; and National Science Foundation 2008 [4]. At that time, the SCJ took the initiative regarding Big Science. Some distinguished Japanese scientists who were leaders in Big Science served on committees and increased communication between the stakeholders. However, a few years later, it became difficult to bridge the two because the membership had changed, and a joint meeting between the SCJ and MEXT was set up in the middle of the discussions. This brought the SCJ to the attention of the community. Scientists in not only physics and astronomy, but also the biological sciences and humanities, which had not previously been considered big sciences, submitted proposals to the SCJ. This was a good opportunity for everyone to consider future plans for the field and to reduce the sense of unfairness felt by nontraditional Big Science scientists. However, there was a big gap in the degree of completion of the proposed projects between fields that were used to large-scale science and those that were not.

One reason that the MEXT did not conduct independent evaluations is that it was difficult to collect and organize bottom-up proposals from a wide range of academic fields. In the SCJ, three fields—Part 1 (Humanities and Social Sciences), Part 2 (Biomedical Sciences), and Part 3 (Science and Engineering). The SCJ can also appeal to other ministries and agencies, as well as the MEXT, for funds based on the budget in the Master Plan. Thus, the Master Plan has a strong effect on academia. Dozens of projects are selected as priorities in the Master Plan, and they are posted on the Road Map committee. At this time, the MEXT launched a budget for a large-scale frontier promotion project. The committee that deliberates on this is called the Working Group for Large-scale Academic Projects, which is under the Research Environment Infrastructure Committee. In addition, under this committee is the Research Infrastructures (RI) Evaluation Subcommittee (University Joint Use

Evaluation subcommittee), which operates large-scale facilities and promotes university research.

Thirty Meter Telescope (TMT) was the first to be funded by the SCJ and the MEXT. A Japanese literature project was also funded at an early stage. The project aimed to develop a system for automatically reading Japanese books up to the 19th century, searching for pictures, and creating a database. The adoption of the first project in the humanities and social sciences symbolized the success of this evaluation system.

48.2.2 Evaluation of item priority awareness

In creating Master Plan 2010 [5], the SCJ outlined the following: 1) definitions; 2) budgets; 3) scientific goals; 4) international standards and international cooperation; 5) researcher–community consensus; 6) core organizations; 7) the evaluation index (i.e., the shared-use system); and 8) the validity and transparency of the plan. In 2014 [6] and 2017 [7], the examination criteria included the following: 1) academic value; 2) consensus of the scientific community (including the existence of duplication with other proposals); 3) implementation of the plan, validity of the plan, and fulfillment of the shared-use system; 4) social value (e.g., laypersons' understanding, intellectual value, economic and industrial value, etc.); and 5) appropriateness as a Big Science research plan.

The MEXT Committee prioritizes the Master Plan prepared by the SCJ and develops it as a Road Map. Based on the SCJ evaluation, three basic evaluation axes were fixed in the 2010 Road Map: 1) scientist's agreement, 2) sharing responsibilities, and 3) joint use [8]. Before the revision of the index by SCJ, MEXT Committee added four more axes: 4) feasibility, 5) urgency (6) strategy, and 7) social support. The MEXT Committee used the same index for the Master Plan in 2014 [9] and 2017 [10]. The seven items were treated in the same way. No different weighting calculations were performed for the items. In this way, the SCJ and the MEXT Committee, while independent of each other, decided on evaluation indicators while referring to each other.

Is each item similarly prioritized by the evaluating scientists and the general public? In recent years, responsible research innovation (RRI) has been recommended in European science and technology policy, and the public is encouraged to be involved in upstream policymaking. Each time MEXT creates a roadmap, it uses a public comment system to solicit public input on major science decisions. It is good to ask society widely, but it doesn't attract much social attention. If there is no difference in the evaluation of each item between scientists and citizens, judgments made by scientists may be evaluated as reflecting the intentions of citizens to some extent. Dr Yokoyama presented the results at the MEXT meeting on November 11, 2014, which are summarized as follows[11].

The respondents to the survey included both Big Science scientists and the general public. Between March 19, 2013 and March 29, 2013, 274 Big Science scientists responded to the survey. The breakdown of researchers who qualified as valid respondents is as follows. 92 (40.2%) in space science, 87 (38.0%) in physics and

Table 48.1. Responses to the question, 'In what order do you think the seven evaluation items should be emphasized to select and promote better research?'. Professor Kazuya Nakayachi at Doshisha University, Associate Professor Mia Takeda and Professor Hiromi Yokoyama conducted the survey.

	Priority order			
	Big Science researchers (N = 235)		Citizens (N = 700)	
	Average rank	Rank	Average rank	Rank
(1) Scientist's agreement	2.64	1	3.60	1
(2) Sharing responsibilities	3.95	4	4.10	5
(3) Joint use	4.58	6	4.34	6
(4) Feasibility	3.33	2	3.67	2
(5) Urgency	4.27	5	3.80	3.5
(6) Strategic	3.75	3	3.80	3.5
(7) Social support	5.47	7	4.70	7

engineering, 31 (13.5%) in material and analytical science, 12 (5.2%) in life science, 3 (1.3%) in informatics, 2 (0.9%) in energy, environment, and earth sciences, 12 (5.2%), 3 (1.3%) in informatics, 2 (0.9%) in energy, environment, and earth sciences, and 2 (0.9%) in others. The survey was distributed to the respondents from the general public on February 16, 2013 by Macromill, Inc. The survey was conducted between February 16 and 18, 2013. The respondents were adults aged 20 to 60 years old, and the ratio of males to females and the age structure were set for consistency with the Japanese population.

Both groups of respondents answered the same question: 'In what order do you think the seven evaluation items should be emphasized to select and promote better research?' In both cases, scientists' agreement (1) and feasibility (4) were ranked first and second, respectively. However, society support (7) was ranked the lowest (table 48.1).

Of the seven items, the scientists and the citizens ranked the same top three items and the lowest items, which was a high level of congruence, suggesting that the general public's perspective is similar to that of scientists.

48.2.3 Master Plan difficulties

In 2010, 43 projects were listed by the SCJ in the Master Plan, and the MEXT committee decided that they were all suitable for inclusion in the Roadmap. A small revision was made a year later, and the list was updated every three years. However, in the 2014 Master Plan, the number of projects was increased dramatically to 207, of which 27 were the SCJ priority list. In 2017, 179 cases were posted in the Master Plan, 28 of which were selected as priority plans.

The SCJ emphasizes the equality of fields, and the lists have become enormous. However, the physics fields, which were originally high quality projects, have not always been included in the SCJ priority list. For example, in 2017, the High-Luminosity Large Hadron Collider (HL-LHC) was not included in the priority list.

Therefore, the MEXT committee reselected the examination targets based on the list in the Master Plan rather than in the priority list. Even though HL-LHC was not listed as a priority list in the SCJ Master Plan, it was ranked as the top project on the MEXT Roadmap. However, this suggests that the assumption of the equality of scientific fields within the SCJ makes it difficult to cover projects that are rated highly in academia.

In addition, the MEXT committee's decision to support the frontier budget for the period of 10 years had an enormous impact on the research community. The MEXT committee thought that the 10-year deadline would allow projects to be replaced and new projects to be introduced, but this was not easy because many large projects evolved with the continued operation of the facility.

In the 2020 Roadmap, the MEXT committee added 'academic significance' to the assessment. The decision by the MEXT committee to judge independently without relying on the SCJ's evaluation caused anxiety regarding the future compliance of the Master Plan and the Road Map. Finally, in February 2022, the SCJ declared that they will change the strategy of the Master Plan. In the future, Japan's approach to promoting Big Science policy system will undergo further changes.

48.2.4 Budget community 'hype' in Japan

In science and technology studies (STS), the separation of science and non-science has been debated [12–14]. The era in which science and non-science could be separated according to Popper's disprovability has passed. Recently, the traditional idea that science and non-science can be separated has been questioned, and many have put forward the idea that how people talk about science is important [15].

Moreover, it has been pointed out that the academic 'journal community' plays an important role in determining whether a science can be included in a field or not [16]. However, before journals review, a 'budget community' decides which projects should be funded. Thus, the journal community is powerful in the process of research [17]. Because the Master Plan and Roadmap committees have authority in deciding Big Science project priorities, they can be played as the 'budget community' of Big Science.

The European Strategy Forum on Research Infrastructure (ESFRI) roadmap committee is also a budget community. Because Big Science budgets are huge, they attract attention. Such attention generally raises expectations that lead to investment. However, the larger the investment, the more complex and time-consuming the process. And, the gap between expectations and disappointments has a greater impact. This type of 'hype' is common in the budget community, similar to the criticism regarding the development of the ESFRI roadmap [18]. The Master Plan and Roadmap committees can also elicit hype that involves both researchers and policy systems. Long review periods involving the entire community are overly focused and time-consuming, which burdens researchers. The Japanese big science frontier budget is limited to 35 billion yen in these days. In the existing project, the budget is tight, and the government cannot afford to add multiple new projects. Nevertheless, the number of proposals is enormous compared to the number that is accepted, which invokes hype. Projects listed on the roadmap have also been followed up in the MEXT. Ready projects start with other budgets. At a minimum, the projects listed on the Roadmap are guaranteed to be academically recognized by the SCJ and the MEXT committee.

Scientists in the community believe that Master Plans and Roadmaps will catalyze the development of research. However, in the current economic situation, the budget scale will not increase, and some projects will not be budgeted, even after repeated reviews. For these reasons, it is better to reduce the numbers of projects which are listed in the SCJ Master Plans and Roadmaps. Instead, Japanese scientists could spend time discussing high-quality projects in harmony with the MEXT Committee. Japan's Master Plan and Roadmap system is evolving, but it will continue to play a central role in determining the Japanese-led Big Science. The budget community should also function as a gatekeeper to reduce unrealistic expectations.

48.3 International Linear Collider (ILC)

48.3.1 Reconstruction plan for recovery from the big tsunami in Iwate Prefecture and ILC.

The ILC has a long history, beginning with the Japan Linear Collider (JLC) proposal in 1986. However, it has gained increasing public attention in Japan since 2011 because it was included in the budget for the reconstruction plan for Iwate Prefecture following the Great East Japan Earthquake and Tsunami in 2011, which is called 3.11 in Japan. The damage caused by 3.11 was devastating, and the total number of dead and missing reached 25 949, of which 5794 were in Iwate Prefecture. A total of 26 077 houses in Iwate Prefecture were destroyed or partially destroyed, most of which were destroyed by the tsunami. Reconstruction to recover from 3.11 is a major issue for the local administration. The ILC first received national attention after 3.11, when the governor incorporated the ILC into the reconstruction budget.

In 2011, there was another candidate site: the Sefuri Mountains across Fukuoka and Saga Prefectures in Kyushu. However, in 2012, the physicists who promoted ILC decided that the Kitakami Mountains in Iwate Prefecture were more suitable because of there are no fault lines in the rock. These deliberations were reported in the national press and attracted much attention.

In Japan, discussions about importing a 250 GeV ILC into the country have continued. The construction cost of the ILC, which produces a large number of Higgs particles and performs precise measurements in particle physics, is estimated at 735.5~803.3 billion yen (635.0~702.8 billion yen for the accelerator and 100.5 billion yen for the detectors) [19]. When the ILC is obtained, the government is expected to bear nearly half the cost of the plan and most of the civil engineering budget.

48.3.2 History of discussions about ILC construction

The European Strategy for Particle Physics Report [20], the United States Particle Physics Project Prioritization Panel (P5) Report [21], and several statements by the

International Committee for Future Accelerators (ICFA) have supported the importation of ILC to Japan [22,23]. Japan's Ministry of Education, Culture, Sports, Science, and Technology (MEXT) and the Science Council of Japan (SCJ) have investigated the possibility of ILC in Japan several times.

The MEXT ILC Advisory Panel (Phase I) deliberated for four years before issuing a summary report on July 4, 2018 [19]. On December 19, 2018, the SCJ announced the results of its deliberation, saying that it could not support the plan [24]. The SCJ pointed out that it was unclear whether the size of the budget was sufficiently large compared with the scientific significance of the project, that the tax burden and international sharing were unclear, and that the ILC should be discussed in conjunction with other big projects in academia. On March 7, 2019, the MEXT stated that it would discuss cooperation with other countries.

On January 30, 2020, a committee of the SCJ published the Master Plan 2020, in which 161 large-scale bottom-up science projects were listed. The plan included a list of 31 high-priority projects. ILC was listed in the Master Plan 2020, but was not included in the high-priority projects list. Nonetheless, some local media coverage was extensive, which stimulated people to promote the project. The kind of activities used to increase the attractiveness of the ILC in Japan have been characterized as 'regional economic revitalization'.

In July 2021, the ILC Advisory Committee (Phase 2) was established. At the beginning of the meeting, the MEXT pointed out that they did not officially recognize Iwate Prefecture as a candidate site. Not only local governments but also Diet members have been actively involved in this. The local goverment's funds have already been invested in this project.

The MEXT ILC Advisory Panel reported as follows. ILC is as effective as other science and technology projects. However, this project had limited technical and economic benefits [25]. However, in today's Japan, economic and intellectual activities are concentrated in cities, and dwindling populations and lower economic activity in regional communities seem to be inevitable. As the ILC is being considered, there are still enormous expectations for regional revitalization, such as the economic effects on local companies that receive contracts for construction and the formation of towns to house overseas researchers.

Finally, in February 2022, the ILC Advisory Committee, while recognizing the scientific significance of the Higgs Factory, was encouraged to shift its discussion from the premise of attraction [26]. In particular, it was recommended that international discussions be held from the beginning, including whether or not the ILC should be selected for the next-generation project. In other words, it was pointed out that a promotion method in which a country that wants to attract a host country raises its hand and proactively promotes it is impossible on the scale of the ILC.

In response, Japan's central institute, High energy accelerator research organization (KEK), declared that it would follow this policy and deepen discussions with the international community while continuing to develop the technology [27]. The ILC discussion has entered a new era.

48.3.3 Is the Japanese public aware of the ILC?

The construction of accelerators for particle physics experiments is based on a bottom-up approach favored by the research community. Because large particle accelerator projects are part of Big Science and need a big budget, scientists have been asked about the 'usefulness' of the project beyond the research itself. This question has been asked many times in the past [28]. In 1969, Robert Wilson, Director of Fermilab, said, 'It has nothing to do directly with defending our country except to help make it worth defending' [29]. During the Cold War era, usefulness was perceived as 'the rise of national prestige.'

In the era of the Superconducting Super Collider (SSC), spin-off claims, such as development by a high-energy accelerator, have been criticized as exaggeration [30]. However, as a national project, increasing technical capabilities developed by the United States are expected [31]. The SSC era began in 1987 during the Reagan administration and ended in 1993 during the Clinton administration. During the era, the type of usefulness was 'science and technological development.'

In these days, ILC flags are flying in front of Ichinoseki and Mizusawa Station in Iwate Prefecture, and some media have enthusiastically reported on the attraction of the ILC. This type of usefulness seems to be 'regional economic revitalization.' If an ILC laboratory is to be built in the region, tunnel construction and many other construction projects will be initiated, and many researchers will be present, requiring housing, hotels, restaurants, and other services. The ILC may also have a positive effect on the education of children. Therefore, the ILC is a large-scale public project in the region, and activities involving local governments and the industrial community have been increasing. Of course, a win-win situation is desirable for both the ILC research project and the local community.

However, this was a huge project that had never been experienced, and the situation complicated the discussion because it did not follow the traditional decision-making process for large projects, and politics was primary. Therefore, the following research questions are addressed in this study:

RQ1: Does the general public know about ILC? Attracting a huge project should be a nationwide discussion, not just a regional one. Even if it is known in Iwate Prefecture, what is the level of awareness in major cities, such as Tokyo and Osaka and in Fukuoka and Saga, which were previous candidate sites?

RQ2: What does the public consider important in attracting an ILC? Do their priorities differ from the citizens of Iwate and citizens of other regions?

48.3.3.1 Method

To confirm differences in public awareness in Iwate Prefecture and other regions, we performed three online surveys, one in each of the three years: 2015, 2018, and 2019. One survey topic concerned the public's recognition of the ILC. Another topic was the themes that the public thinks are important in a dialogue with society. The second questions were included only in the 2018 survey.

We selected four prefectures and regions: Iwate Prefecture, Fukuoka, and Saga Prefectures where there was a candidate site, and two big cities, Tokyo and Osaka. We measured the responses for the first time from March 24 to 26, 2015, for the second time from June 5 to 7, 2018, and for the third time from March 12 to 15, 2019, using the Macromill, Inc. internet survey company. In total, 1674 people responded each time, and the respondents in each prefecture provided 25% of the answers. Gender and age were selected to reflect the Japanese population. In the second and third rounds of the survey, people who had responded in the past were not selected.

48.3.3.2 Results

The results showed that recognition of the ILC was high only in Iwate Prefecture; the percentage of people in that prefecture who had at least heard about the ILC was around 80%. However, in the other prefectures, the percentages were less than 40% (figure 48.1).

One question in the survey asked, 'Which themes are important for a dialogue with society?' (three items selected from eight) (figure 48.2). About 70% of the citizen respondents chose 'tax burden and international sharing' and 'scientific significance'. Both choices scored roughly the same across the four prefectures, and both were pointed out by the SCJ [24]. The third choice was 'economic effect,' even in Iwate Prefecture.

In the other prefectures, Japan leading in worldwide basic science was selected more often than economic effects. These findings suggest that the public in Iwate Prefecture recognize that 'tax burden and international sharing' and 'scientific significance' are more important than revenue generated in the region.

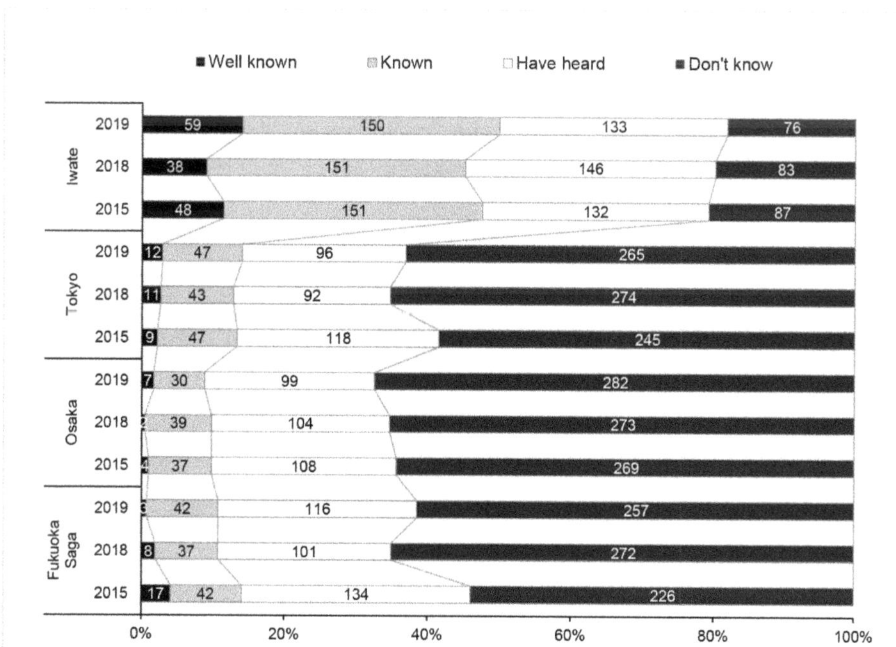

Figure 48.1 . Recognition levels in the four prefectures

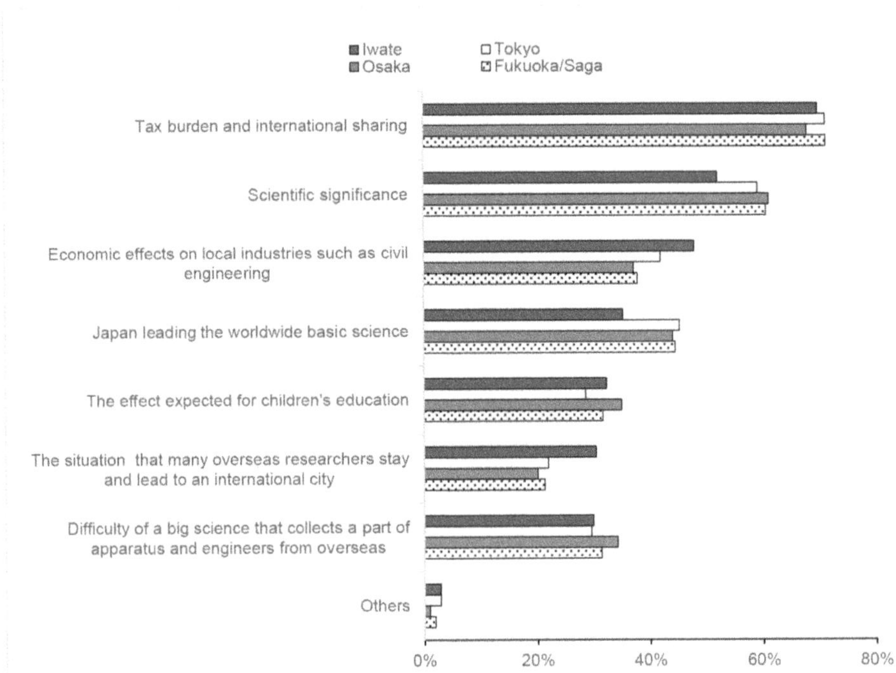

Figure 48.2. Responses to the question, 'Which themes are important for a dialogue with society?' (three items were selected).

48.3.3.3 RQ1 result

The results showed that the number of citizen respondents who had at least heard about the ILC was around 80% in Iwate Prefecture, and less than 40% in the other prefectures. There was a twofold difference in recognition between Iwate and other sites. Although this finding implied a high interest in ILC in Iwate prefecture, it suggested a lack of interest in the other prefectures.

48.3.3.4 RQ2 result

The results showed that 'tax burden and international sharing' and 'scientific significance' were among the top two selected in prefectures, including Iwate Prefecture. This finding supports scientific discussions.

48.3.4 Discussion

What is implied by the twofold difference between Iwate Prefecture and the other prefectures in terms of awareness of the ILC, while the importance of the ILC was similar in all four prefectures?

In Iwate Prefecture, many people are aware of ILC because it has been appealed. Flags and models have been erected in front of train stations, and there are educational activities about ILC in schools. Iwate Prefecture citizens are not only aware of ILC but also understand that it is an international project and that it requires a huge budget. Therefore, they perceive the importance of international sharing. They also understand that the ILC is a scientific project, and they also recognize its scientific importance. The findings suggest that Iwate Prefecture's citizens have calmly evaluated this huge scientific project, in contrast to the frenetic atmosphere induced by the media's promotion of the project.

In Japan, basic science is supported, and physicists are respected. This is because many of Japan's Nobel laureates and many physicists have communicated scientific knowledge through media, lectures, and books. They always behaved peacefully and humbly.

Scientific projects continue to grow in size. Among them, the future of high-energy physics is challenging. Because it consumes high amounts of electricity, it is not in line with green policies and therefore requires ingenuity. Furthermore, it is necessary to discuss future plans internationally, including the Future Circular Collider (FCC) of Centre Européen pour la Recherche Nucléaire (CERN). Japan should go beyond 'regional economic revitalization' and develop a new strategy for collaborating with the rest of the world.

Acknowledgments

We would like to thank Professor Kazuya Nakayachi, A/Professor Mia Takeda and A/Professor Saho Kawashima for their design, analysis, discussion for section 48.2.2. We gratefully acknowledge the funding received through the JSPS Kakenhi Grant (17K01020, 25350237). This work was supported by the World Premier International Research Center Initiative (WPI), MEXT, Japan.

References

[1] ESFRI 2006 (2006, September 29). Strategy report on research infrastructures. Roadmap 2006 https://www.esfri.eu/sites/default/files/esfri_roadmap_2006_en.pdf

[2] The UK Research Council 2008 Large Facilities Roadmap 2008

[3] Department of Energy. (2003, December).'Facilities for the Future of Science: A Twenty-Year Outlook.' https://fribusers.org/frib/docs/2003-DOE20YearScienceFacilities.pdf

[4] National Science Foundation. (2007, February 5). 'Major Research Equipment and Facilities Construction.' http://www.nanoscience.gatech.edu/zlwang/news/news/NSF-overview.pdf

[5] Science Council of Japan. (2010, March 17). *Proposal: Large-scale Academic Facility Plan / Large-scale Research Plan-Planning / Promotion Measures and Master Plan Formulation* (in Japanese) http://www.scj.go.jp/ja/info/kohyo/pdf/kohyo-21-t90-2.pdf

[6] Science Council of Japan. (2014, February 24). Proposal Master Plan for the 22nd Academic Large-scale Research Plan (Master Plan 2014) (in Japanese). http://www.scj.go.jp/ja/info/kohyo/pdf/kohyo-22-t188-1.pdf

[7] Science Council of Japan (2017, February 8) Proposal Master Plan for the 23rd Academic Large-scale Research Plan (Master Plan 2017) (in Japanese) http://www.scj.go.jp/ja/info/kohyo/pdf/kohyo-23-t241-1-0.pdf

[8] Ministry of Education, Science and Technology, Science and Technology Council, Academic Subcommittee, Research Environment Infrastructure Subcommittee, Working Group on Large-scale Academic Research Projects (2010, October 27). Promotion of Large-scale Academic Research Projects (Summary of Deliberation) - Basics on Promotion of Large-

scale Academic Research Projects' Formulation of the concept 'Roadmap' (in Japanese). https://www.mext.go.jp/b_menu/shingi/gijyutu/gijyutu4/toushin/1298714.htm

[9] Ministry of Education, Culture, Sports, Science and Technology. (2014, September 14). Formulation of Basic Concept Roadmap for Promotion of Large-scale Academic Research Projects—Roadmap 2014 (in Japanese). https://www.mext.go.jp/b_menu/shingi/gijyutu/gijyutu4/toushin/1351171.htm

[10] Ministry of Education, Culture, Sports, Science and Technology. (2017, July 28). Formulation of Basic Concept Roadmap for Promotion of Large-scale Academic Research Projects—Roadmap 2017 (in Japanese). https://www.mext.go.jp/b_menu/shingi/gijyutu/gijyutu4/toushin/1388523.htm

[11] MEXT Minutes of the 69th Meeting of the Research Environment Infrastructure Subcommittee (2014) https://warp.ndl.go.jp/info:ndljp/pid/11293659/www.mext.go.jp/b_menu/shingi/gijyutu/gijyutu4/010/gijiroku/1372163.htm (in Japanese)

[12] Popper K R 1959 *The Logic of Scientific Discovery* (London: Hutchinson)

[13] Merton R K 1942 The Normative Structure of Science In ed R K Merton *The Sociology of Science: Theoretical and Empirical Investigations* (Chicago, IL: University of Chicago Press) pp 267–78

[14] Kuhn T 1962 The Structure of Scientific Revolutions In ed M Ruse *Philosophy after Darwin* (Princeton: Princeton University Press) pp 176–7

[15] Gieryn T 1994 Policing STS: A Boundary-Work Souvenir from those of the Smithsonian Exhibition on 'Science in American Life *Science, Technology, & Human Values* **21** 100–15

[16] Fujigaki Y 2001 Interaction between the Science Arena and the Policy Arena in the Public Sphere: Validation Boundaries and Cultural Styles In ed S Dresner and N Gilbert *The Dynamics of European Science and Technology Policies* (Aldershot, UK: Ashgate) pp 165–81

[17] Ikkatai Y, McKay E and Yokoyama H M 2018 Science Created by Crowds: A Case Study of Science Crowdfunding in Japan *Journal of Science Communication* **17** A06

[18] Hallonsten O 2020 Research Infrastructures in Europe: The Hype and the Field *European Review* **28** 617–35

[19] MEXT, International Linear Collider (ILC) Advisory Panel (2018, 4 July) Summary of the ILC Advisory Panel's Discussions to Date after Revision. https://www.mext.go.jp/component/b_menu/shingi/toushin/__icsFiles/afieldfile/2018/09/20/1409220_2_1.pdf

[20] European Strategy Session of Council. (2013, 7 May). The European Strategy for Particle Physics Update 2013 (CERN-Council-S-0106). https://cds.cern.ch/record/1567258/files/esc-e-106.pdf

[21] Ritz S *et al* 2014 Building for Discovery: Strategic Plan for US Particle Physics in the Global Context, Report of the United States Particle Physics Project Prioritization Panel (P5). https://usparticlephysics.org/wp-content/uploads/2018/03/FINAL_P5_Report_053014.pdf

[22] ICFA Statement on the Progress towards an International Linear Collider (ILC) in Japan (2014, 20 February). https://icfa.fnal.gov/wp-content/uploads/ICFA_Statement.pdf

[23] ICFA Statement on the ILC Operating at 250 GeV as a Higgs Boson Factory (2017, November). https://icfa.fnal.gov/wp-content/uploads/ICFA-Statement-Nov2017.pdf

[24] Science Council of Japan. (2018, 19 December). Observations on the Draft Review of the International Linear Collider Plan (in Japanese) https://www.scj.go.jp/ja/info/kohyo/pdf/kohyo-24-k273.pdf

[25] Hirata K 1999 An Essay on Big-Device-Pure Sciences (in Japanese) *Japanese Journal for Science, Technology & Society* **8** 51–74

[26] Summary of Discussions on Issues Related to the International Linear Collider (ILC) Project (in Japanese) https://www.mext.go.jp/b_menu/shingi/chousa/shinkou/064/toushin/220214.htm
[27] Next Step toward the ILC Realization: MEXT Expert Panel Publishes Recommendations (2022, February 25) https://www.kek.jp/en/topics-en/202202251335/
[28] International Linear Collider (ILC) Advisory Panel. (2018). 'Summary of the ILC Advisory Panel's Discussions to Date after Revision.' https://www.mext.go.jp/component/b_menu/shingi/toushin/__icsFiles/afieldfile/2018/09/20/1409220_2_1.pdf
[29] Fermilab. *'Felmilab History and Archives Project. R.R. Wilson's Congressional Testimony, April 1969'* ; https://history.fnal.gov/historical/people/wilson.html
[30] Kevles D J 1995 Good-bye to the SSC: On the Life and Death of the Superconducting Super Collider *Engineering and Science* **58** 16–25
[31] Wojcicki S 2009 The Supercollider: The Texas Days—A Personal Recollection of Its Short Life and Demise ed A W Chao and W Chou *Rev. Accel Sci. Technol.* (Singapore: World Scientific) pp 265–301

IOP Publishing

Big Science in the 21st Century
Economic and societal impacts
Panagiotis Charitos, Theodore Arabatzis, Harry Cliff, Günther Dissertori, Juliette Forneris and Jason Li-Ying

Chapter 49

Pathways to economic and societal impact of large-sized facilities in China

Chen Guang and Ting Wang

Large scientific facilities are one of the characteristics of Big Science, with their high cost, huge size, and great complexity. Although they have contributed many scientific advances in many fields, there are many disputes surrounding their value for society. This essay suggests a conceptual framework to evaluate the benefits of large scientific facilities from the view of science, economy, and society (SES). In the case of the 500 m aperture spherical radio telescope (FAST), this large scientific facility contributes not only to the development of astronomy in China, but has also changed the region's economy profoundly and transformed the local society.

49.1 Introduction

Big Science is characterized by large facilities or instruments, mostly supported by public funds from government, and is an important platform for national innovation systems and an important foundation to support a country as a scientific and technological power. The initial purpose of building large facilities is to achieve breakthroughs in major scientific and technological fields, enhance scientific and technological competitiveness, and finally serve the all-round development of the economy and society in turn. However, large facilities have the characteristics of requiring substantial construction investment and a long operation life cycle, and the former in particular might have a great impact on public finances. Therefore, government departments need to properly evaluate and supervise the construction and operation process of Big Science facilities (Bezdek and Wending 1992, Martin and Tang 2007, Del Bo *et al* 2016b). With the increasing construction costs of large facilities, governments are not always able or willing to bear the relevant costs (Broad 1990). At the same time, policymakers pay more and more attention to the economic and social impacts of large facilities in addition to the output of scientific

doi:10.1088/978-0-7503-3631-4ch49

and technological achievements (Lee *et al* 2013). The comprehensive evaluation of the benefits of large facilities is helpful to guide policy decision-making and promote multiple value creation. Therefore, it has become a common concern of governments and administrative bodies all over the world.

The 500 m aperture spherical radio telescope (FAST) is very unique case among large facilities due to its location in an undeveloped area in China, in that we can observe more information on its impact other than science. FAST is also known as China's 'heavenly eye'. It is one of the nine national science and technology infrastructures reviewed and determined by the national science and education leading group, and a major national science and technology infrastructure construction project in the eleventh five-year plan. Its comprehensive performance exceeds that of the Bonn telescope in Germany and the Arecibo telescope in Puerto Rico (owned by the US National Science Foundation), and has high popularity and influence at home and abroad. FAST was conceived by Chinese astronomer Nan Rendong in 1994, approved by the National Development and Reform Commission in 2007, and started construction in Pingtang County, Qiannan Prefecture, Guizhou Province in 2011. It was completed on 25 September 2016, with a construction period of 5.5 years and a total investment of US$179 million (1.15 billion RMB[1]). On 11 January 2020, FAST passed the national acceptance criteria and was officially put into operation. On 1 April 2021 FAST was officially opened to domestic and international scientists. This paper mainly discusses the comprehensive scientific, economic, and social benefits brought by FAST.

49.2 Analytical framework

The research on the benefits of large facilities can be traced back to Schmied's (1975) evaluation of the benefits of the European Organization for Nuclear Research (CERN), and the follow-up research has mostly been carried out from the perspective of scientific, economic, social, and comprehensive benefits.

The research on the scientific benefits of large facilities is relatively rich, and most studies use the relevant indicators of research paper output. Hallonsten (2013, 2014) proposed that the scientific benefits of large facilities are not only reflected in the number of journal papers and publications, but also can give full play to their scientific benefits by supporting discipline development and promoting discipline intersection. Therefore, Hallonsten advocates evaluating their scientific benefits through citation analysis, and puts forward the concept of 'facility metrology'. Heidler and Hallonsten (2015) not only adopt the indicators such as the number of publications, timeliness index, and impact factors, but also combine with the method of citation analysis to measure the scientific benefits of large facilities. Del Bo (2016a) analyzed the scientific benefits of large facilities from an empirical perspective, discussed the impact of R&D investment on cost efficiency and the number of citations relying on the facility operation organization, and further considered the potential spillover effects and externality.

[1] In this article we calculate the exchange rate to be 1 US dollars = 6.424 RMB.

The research on the economic and social benefits of large facilities mainly focuses on technology spillover and knowledge transfer from the perspective of stakeholders (Gomez 2015, Nilsen and Anelli 2016) and the improvement of innovation ability. Castelnovo *et al* (2018) who used the balance sheet data of more than 350 suppliers from 35 countries and regions provided by CERN from 1991 to 2014 found that there is a significant positive correlation between CERN's procurement behaviors and the suppliers' R&D capability, knowledge creation, production efficiency, and profitability. Florio *et al* (2016) divided the economic and social benefits of large facilities into six categories, i.e. technology spillover, human capital accumulation, knowledge output, cultural impact, public goods, and third-party services, and constructed a set of analysis framework to measure the above mentioned effects of large facilities based on the cost–benefit analysis (CBA) method. Input–output analysis (IOA)[2] is also used in the economic benefit evaluation of large facilities, and can obtain more information on efficiency than CBA. Battelle (2011) used this method to evaluate the economic impact of the Human Genome Project. Due to the difficulty in obtaining data, case investigation methods are often used to analyze the economic and social benefits of large facilities in combination with semi-structured interviews and file access. Only a few such studies use large samples. For example, a large-scale survey by NASA (2012) shows that in the first 11 years of the twenty-first century, civil products derived from NASA aerospace technology have created 14 000 jobs, driving a direct economic income of US $5 billion and cost savings of US $6.2 billion. In addition, when the data acquisition is particularly unsatisfactory, the conditional value method (CVM)[3] can also be used to calculate the economic benefits. For example, Pung *et al* (2004) used the CVM to calculate the economic benefits of the British Library, which is found to be 4.4 times the total investment.

The research on the comprehensive benefits of large facilities requires building a more comprehensive evaluation index system such as the research carried out by Technopolis. Zuijdam *et al* (2011), Simmonds *et al* (2013), Reid *et al* (2015), and Farla *et al* (2016) have established an index system framework for the evaluation of the comprehensive benefits of large facilities, involving the three aspects of the economy, environment, and society, as well as 12 first-level indicators, such as knowledge spillover, market spillover, policy formulation, strategic impact, urban security, and national defense, etc. EvaRIO (2013) divides the comprehensive benefits into four types: direct impact, capacity impact, impact caused by the implementation of relevant activities, and indirect impact. RI-PATHS (Griniece *et al* 2018, 2019) divides the comprehensive benefits into four dimensions of direct economic, human resources, social, and policy impact. ESFRI (2013) proposed the pan-European relevance index system in 2013, and based on that further proposed

[2] Unlike the net profit emphasized by CBA, the IOA emphasizes the benefits.

[3] CVM, also known as the expected shortfall, is a risk assessment measure that quantifies the amount of tail risk an investment portfolio has. CVM is derived by taking a weighted average of the 'extreme' losses in the tail of the distribution of possible returns, beyond the value at risk cutoff point. Conditional value at risk is used in portfolio optimization for effective risk management.

an evaluation system including 21 key performance indicators, including available machine hours, number of registered users, number of publications, number of top 10% highly cited papers, training time of employees, etc (Kolar *et al* 2019). OECD (2019) proposed the SEIRI framework based on research in 2014, which lists 58 indicators from seven aspects, including becoming a national or world leading scientific research institution, becoming a favorable facility to support innovation, forming regional clusters, promoting educational development and knowledge diffusion, providing scientific support for public policies, providing high-quality scientific data and related services, and fulfilling social responsibilities. In addition, the comprehensive benefits of large facilities for the location have been summarized by Du and Shang (2011) through the aspects of informatization, resource development, technology transfer, and transformation, and by Chen (2014) using the aspects of promoting economic development, inducing technological innovation, strengthening human capital and social networks, and increasing social welfare.

Based on the previous research on the benefit evaluation system and indicators related to large facilities, most of which depend on statistical data focusing on the large facilities that cannot cover the whole impact including economy and society, this chapter suggests a conceptual framework of science–economy–society (SES), and selects the three dimensions science, economy, and society to evaluate the economic and social benefits generated by FAST. Specifically, scientific benefits include bringing new scientific discoveries and improving the basic conditions of scientific development; economic benefits are measured in terms of technology, industry, GDP and employment; and social benefits are analyzed from the perspective of the supporting capacity to government public services (Mu *et al* 2012), including infrastructure, transportation, public education, health care, public security, and city image.

49.3 Data and results

49.3.1 Scientific benefits

49.3.1.1 Bring new scientific discoveries and achievements

First, the total number of papers related to FAST published in leading academic journals at home and abroad is increasing (figure 49.1)[4]. As of 31 December 2020, 145 papers have been included in SCI, and the number of published papers has been increasing in particular after 2015. Second, as of early September 2019, FAST has found 138 high-quality pulsar candidates, among which 99 have passed certification, far exceeding the sum total found by multiple pulsar search teams in Europe and America in the same period, and the pulsar timing accuracy is more than four times higher than the international level. The large collecting area makes FAST the most sensitive telescope in the world and the precisely controlled adopt reflectors enables it to achieve excellent focus and track on celestial targets[5]. Third, the astronomical

[4] From CNKI and InCites databases, searches with 'FAST' or '500 m aperture spherical radio telescope' as the subject words.

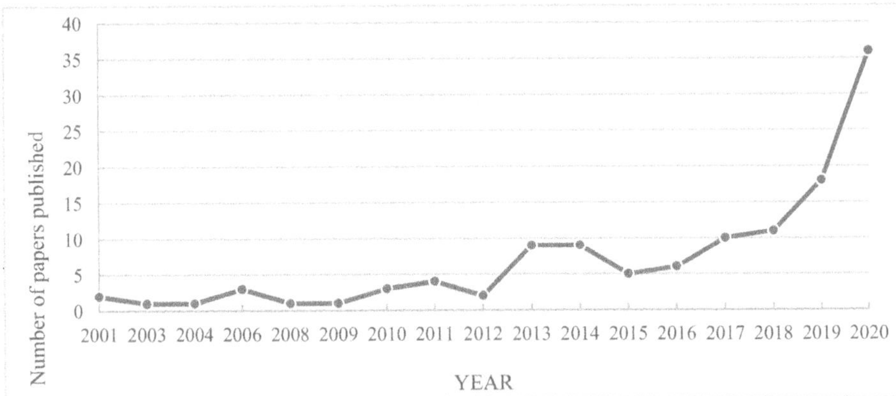

Figure 49.1. Trend of FAST-related SCI papers published over the years.

data output has increased significantly. It is expected that by 2027 the data storage of the FAST radio telescope data center will reach 100PB, and the computing capacity will reach 1000 TFLOPS[6]. The improvement of astronomical data storage and processing technology will provide data for supporting the study of the earlier moments of the Universe closer to the Big Bang.

49.3.1.2 Improving the basic conditions for discipline development

First, regarding scientific and technological talents, the FAST project construction has trained a group of outstanding scientists and engineers, of whom chief scientist Nan Rendong, chief technologist Wang Qiming, and deputy chief engineer Li Di are typical representatives. The FAST research and operational team has grown from the initial five members to about 100. Second, FAST has contributed to international cooperation in this field of astronomy. Relevant units of FAST have established new collaborations with many counterparts from other countries in the world, such as the Jodrell Bank Observatory in Manchester, UK, the Australia Telescope National Facility (ATNF), etc, and international astronomical experts at all levels have participated in a large number of substantive cooperations and exchanges. At the end of 2018 FAST's engineering commissioning team cooperated with Tianma's telescope team and successfully realized joint observation for the first time and obtained the interference fringes of the very long baseline interferometry (VLBI), indicating that FAST has the ability to participate in the joint observation of the VLBI network[7]. On 1 April 2021 FAST invited applications from the global community of astronomers and scientists who want to submit their projects. Third, FAST promotes higher education in astronomy in China with the development of

[5] 'FAST reveals mystery of fast radio bursts from the Universe' https://www.eurekalert.org/news-releases/799512.

[6] 'FAST radio telescope data center of the National Astronomical Observatory of the Chinese Academy of Sciences settled in Guizhou' http://www.cas.cn/cm/201705/t20170526_4602974.shtml (in Chinese).

[7] 'FAST and Tianma successfully obtained VLBI interference fringes for the first time' http://lssf.cas.cn/lssf/500mkjwyj/xwdt/201902/t20190212_4546064.html (in Chinese).

astronomical discipline construction in Guizhou University, Guizhou Normal University, and Qiannan Normal University for Nationalities. The National Astronomical Observatory has built the early data center of 'China's heavenly eye' in Guizhou Normal University, and is planning to build the Guizhou Radio Observatory and 'China's heavenly eye' data processing center there.

49.3.2 Economic benefits

49.3.2.1 Drive for influential high-tech development

First, a number of technological breakthroughs have been made during the construction of FAST. Among the technological challenges for building and operating this facility, one can mention the cable net structure with super-long span in ultra-high-precision and super-fatigue active displacement working mode; the high-strength, long-term bending fatigue resistant and low loss dynamic optical cable; the large-scale and high-precision real-time measurement system; the ultra-high standard electromagnetic compatibility design; and the 19-beam L-band cryogenically cooled receiver, etc. The required technological R&D for the construction of FAST has promoted and continues to promote scientific and technological progress and industrial upgrading in many high-tech fields in China[8]. For example, the Qingdao Oriental Tower has made a breakthrough in the manufacturing and installation of its large-volume feed-support tower based on the FAST-related technology (figure 49.2). At the same time, the construction experience of FAST will have a positive impact on the development of China's manufacturing technology[9]. For example, the flexible feed cable drive system developed by Dalian Heavy Industry & Dalian Crane Works adopts optical electromechanical integration technology to realize multi-angle, large-range, and high-precision spatial positioning movement. Some products based on such technology sell well[10]. Second is the improvement of the engineering process system. In the project construction, the unique Karst landform in this area is used as the station site of the FAST telescope. An actively deformed paraboloid reflector is adopted, and the operation mode of light-weight feed receiver is realized through the support of a six-tower and six-cable system (figure 49.2). The process system is conducive to the accurate construction of other large and complex projects and has good practical application value. The cable net structure with ultra-high fatigue performance has been applied to large stadiums and gymnasiums, and the high-precision cable net structure production system has been applied to the construction of the Hong Kong–Zhuhai Macao Bridge.

[8] 'Building the giant radio telescope FAST' by Jiang Peng, Research Professor of the National Astronomical Observatory of the Chinese Academy of Sciences https://www.cas.cn/cm/202105/t20210506_4786655.shtml?from=timeline (in Chinese).

[9] 'What will the FAST telescope bring us?' http://www.cas.cn/cm/201607/t20160705_4567465.shtml (in Chinese).

[10] China's 'heavenly eye'—the FAST telescope https://idea.cas.cn/viewdoc.action?docid=56176.

Figure 49.2. Panorama of the feed-support tower of the FAST telescope. Image credit: National Astronomical Observatories, CAS.

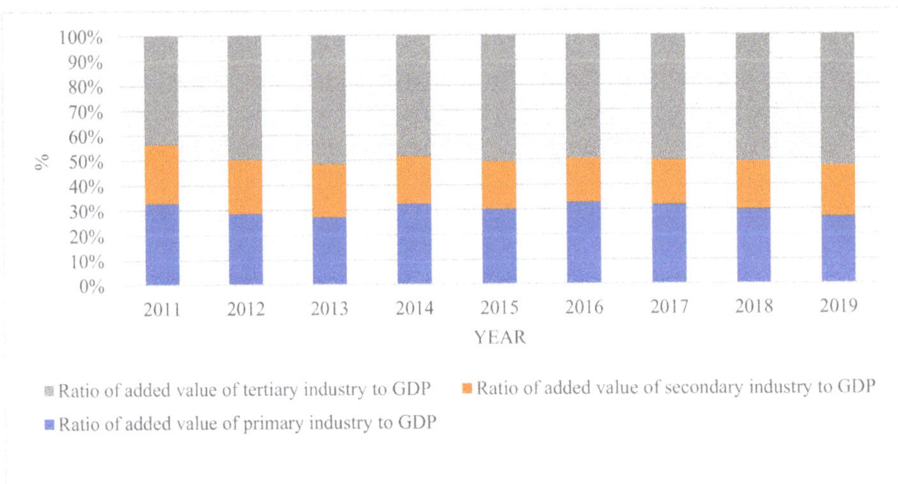

Figure 49.3. Changing trend of industrial structures in Pingtang county from 2011 to 2019.

49.3.2.2 Optimize the regional economy and industrial structure

First, during the FAST facility operation period of 2016–19, the tertiary industry in Pingtang county accounted for a larger proportion and grew steadily (figure 49.3). In 2019 the added value of the three major industries accounted for 27.0%, 20.4%, and 52.6% of the GDP, respectively. Compared with the previous year, the added value

[11] http://www.bao.ac.cn/xwzx/zhxw/201412/t20141203_4267089.html.

of the primary industry and the secondary industry decreased by 0.8% percentage, respectively, but the added value of the tertiary industry increased by 1.6%[12]. Second, FAST has improved the local investment environment and increased regional investment attraction. In 2019 the investment in real-estate development increased by 19.09% year-on-year, and in particular the investment in commercial premises increased by 26.15% year-on-year. In 2019 there are 47 new industrial projects attracting investment, with the actually received investment of $1.17 billion, a year-on-year increase of 7.51%. Third, FAST has greatly promoted the development of local tourism. In 2019 the number of domestic overnight tourists in Pingtang county was 487 000, a year-on-year increase of 35%, and the number of inbound overnight tourists was 3200, a year-on-year increase of 41.8%. FAST also offers more choices to tourists traveling in the region and allows their travel itinerary in Pingtang county have a broader turnover space. In 2019 the passenger turnover of the whole county increased by 15% year-on-year. Fourth, Guizhou Province has taken the opportunity of the FAST project to promote the integration of the big radio (telescope) and Big Data. Gui'an New Area, a new high-tech zone of Guizhou Province, has gathered the data centers of a large number of industry leaders such as Huawei and Tencent. After these data centers are officially put into operation, the total number of servers in the Gui'an New Area will reach 3.6 million. Under the new infrastructure trend, the Big Data industry in the Gui'an New Area is welcoming new opportunities[13].

49.3.2.3 Drive the growth of economic aggregate and labor employment

First, using the input–output analysis method, during the whole project construction period of FAST (2011–16), according to the input–output table of Guizhou Province in 2012, the simple GDP multiplier of the construction industry in the province was 1.403, the marginal propensity for consumption was 0.1015, the cumulative GDP multiplier was 1.774, the total GDP multiplier was 3.177, and the total investment of the FAST project was about $179 million. Excluding the factor of rising prices, FAST has created a cumulative GDP of $569 million during the construction period from 2011 to 2016. Since the construction of FAST, the growth rate of the regional GDP has remained above 12% from 2011 to 2018 (figure 49.4). In 2019 the regional GDP of the county was $1.34 billion, a year-on-year increase of 7.9%, and the total economic output jumped from the tenth to the eighth position in the prefecture.

Second, the employment multiplier of the construction industry in Guizhou Province is 463.87 excluding the factor of price increases and according to the above employment multiplier, FAST has directly created or retained about 5335 jobs during the construction period from 2011 to 2016. FAST has driven the growth of urban employment and promoted the employment and entrepreneurship in this impoverished region. Since the construction of FAST, by the end of 2018, 485 400

---

[12] Pingtang County Bureau of Statistics, Statistical Bulletin of National Economic and Social Development of Pingtang County, Guizhou Province, 2012–2017 http://www.gzpt.gov.cn/zfbm/xtjj_5770996/zfxxgk_5770966/fdzdgknr_5770969/tjsj_5121431/tjgb/ (in Chinese).

[13] The FAST scientific research and data processing center will be built soon http://www.cas.cn/cm/202004/t20200413_4740952.shtml (in Chinese).

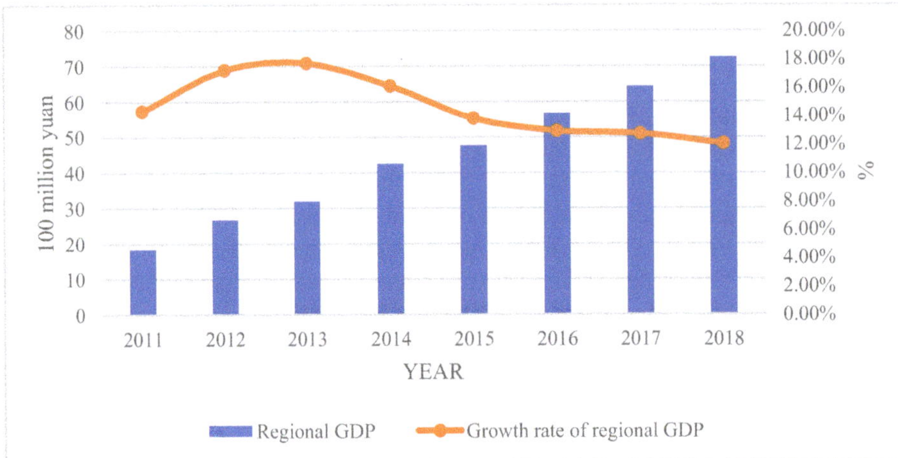

Figure 49.4. Changing trend of regional GDP and growth rate of Pingtang county from 2011 to 2018.

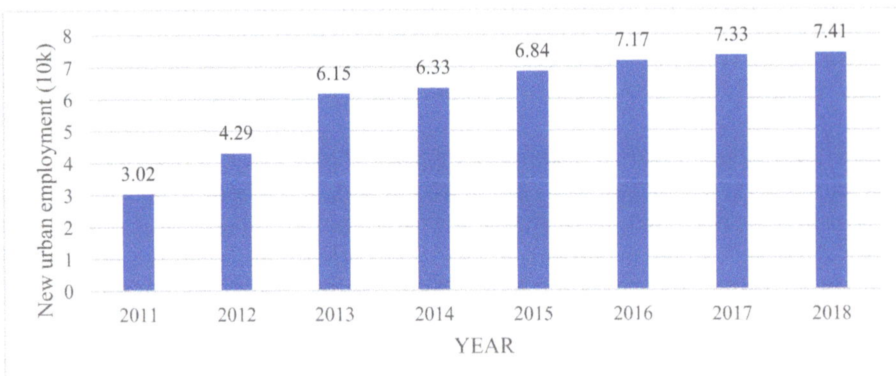

Figure 49.5. Changing trend of new urban employment in Pingtang county from 2011 to 2018.

urban jobs have been provided. From 2013 to 2018 the number of new urban jobs was more than 60 000 every year (figure 49.5).

49.3.3 Social benefits

During the construction and operation period of FAST, the social development level of Pingtang county where the project is located was generally good, the urbanization rate was significantly improved, and the living standards of residents were effectively improved[14]. The GDP of Pingtang was about $1 billion in 2017, compared with 0.41

[14] The infrastructure data and environmental protection data of the quiet area are from the reply of Pingtang County Development and Reform Commission; the police force allocation data are from the reply of Pingtang County Public Security Bureau; other data are from the Pingtang County Bureau of Statistics. Statistical Bulletin of National Economic and Social Development of Pingtang County, Guizhou Province from 2012 to 2017 http://www.gzpt.gov.cn/zfbm/xtjj_5770996/zfxxgk_5770966/fdzdgknr_5770969/tjsj_5121431/tjgb/ (in Chinese).

billion in 2012, increasing by about 141%. At the same time the average increase rate in the whole district was 94%. In 2017 the total number of registered residents of Pingtang County was 331 841, of which 92 311 were urban residents, an increase of 69 511 compared to 2012. As of 2017 the urbanization rate was 38.26%, an increase of 2.94 times compared with that of 2012 (7.06%); the floating population in the county increased by 6.5 times compared with that of 2015; and the per capita disposable income of urban residents increased by 78.1% compared with that of 2012, in which, the growth of per capita disposable income of urban residents was 18.3% higher than that of the prefecture average of 59.8% (table 49.1).

49.3.3.1 Promote infrastructure construction

First, progress has been made in the construction of roads, broadband networks, power grids, and water conservancy. By the beginning of 2018 more than 360 km of roads have been added in Pingtang county, the so-called 'last kilometer' of 'broadband access to villages' had gone through, and the rural broadband coverage rate of Pingtang county reached 84%; 555 km of rural power grid had been upgraded, two 110 kV transformer substations built, and four 35 kV to 110 kV transformer substations expanded. A total of $230 million has been invested in water conservancy infrastructure construction, which have effectively guaranteed the production and living electricity for people in remote areas and basically achieved full coverage of safe drinking water in the rural areas. Second, a quiet area was built to secure the operation environment for FAST. For this purpose, the local government relocated the residents within 0–3 km of the core area of FAST. By the beginning of 2018, 659 resettlement houses and 596 buildings (Ma'an resettlement site) have been built through the integration of funds, and 481 households relocated intensively. The government closed all the coal and stone production enterprises in the quiet area. Four coal production enterprises originally located in the quiet area were closed, with an annual loss of sales revenue of about $27.7 million.

49.3.3.2 Promote the development of transportation

First, the county's highway passenger transport, freight volume, and turnovers continue to increase rapidly. At the end of 2017, compared with that of 2013, the

Table 49.1. Changes in key indicators of social benefits (growth rate, 2012–17).

	GDP	Per capita disposable income of urban residents	Passenger turnover	Freight turnover
Pingtang County	141%	78%	3910%	4210%
Prefecture average*	94%	60%	1027%	1080%
Guizhou Province	99%	56%	2.4%	48.3%
The whole country	59.3%	48.1%	−2%	13.3%

Data source: Statistical Bulletin of National Economic and Social Development of China, and Guizhou Province and its counties and cities from 2012 to 2017.

*The average values are the county and city averages as per available data.

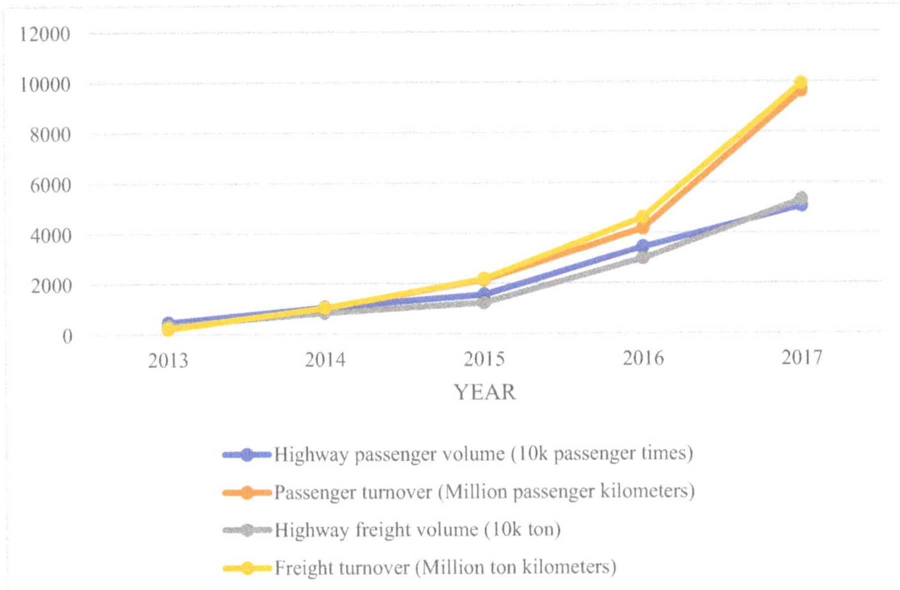

Figure 49.6. Variation trend of highway passenger transport, freight volume, and turnovers in Pingtang county.

road passenger volume increased by 9.5 times, the passenger turnover increased by 39 times, the road freight volume increased by 13.9 times, and the freight turnover increased by 42.6 times (figure 49.6), in which the growth of both the passenger and the freight turnovers were significantly higher than the prefecture average levels (table 49.1). Second, the amount of public transport operating vehicles, operating lines, and taxis in the county has increased dramatically. At the end of 2017, compared with that of 2012, the number of public transport operating vehicles increased by 2.8 times, the number of operating lines increased by 86%, and the number of operating taxis increased by 88.4%.

49.3.3.3 Promote the development of public science, technology, and education

First, the conditions and quality of preschool education and basic education have been further optimized and improved. The number of teachers in 2017 increased by 18% compared with that in 2013. The number of kindergartens increased from 20 to 52. Second, various forms of astronomical science popularization activities have been carried out in combination with the 'Science and Technology Activity Week' and 'National Science Popularization Day', etc to help more people understand FAST and popular astronomical knowledge, so as to further improve the scientific and cultural qualities of the people. Third, FAST brings more scientific research activities to the region, and the investment in scientific and technological research there has also increased significantly. In 2019 the annual local science and technology expenditure increased by 58.72% year-on-year.

49.3.3.4 Promote the development of medical and health care

First, medical service conditions in the region have been effectively improved. By the end of 2017, in order to better serve tourists and the FAST staff, the service capacity of the Kedu Town Central Health Center had been comprehensively improved. In order to further improve the medical service level and meet the internationalized needs of development, relevant official departments have planned to build a new Class 3A hospital capable of providing bilingual services. Second, it has directly promoted the development of medical and health care undertakings. By the end of 2017 the number of medical institutions had increased from 25 to 32, including township health centers, health and epidemic prevention stations, and maternal and child health centers (figure 49.7), and gradually realized the 'full coverage' of clinics in all the administrative villages in the county; however, during the same period, the number of medical institutions in most other counties and cities in the prefecture remained unchanged or decreased slightly.

49.3.3.5 Promote the development of public security

First, the police force has been strengthened, the number of criminal cases began to decline from 2016, and the case-solving rate increased as a whole. Second, the early warning and prediction of natural disasters have been strengthened and the management of the ecological environment improved. From 2011 to 2018 there were more than 156 early warnings and predictions of natural disasters and more than 50 prevention and control times, which has effectively provided a reliable guarantee for the construction and operation of FAST. The comprehensive index of

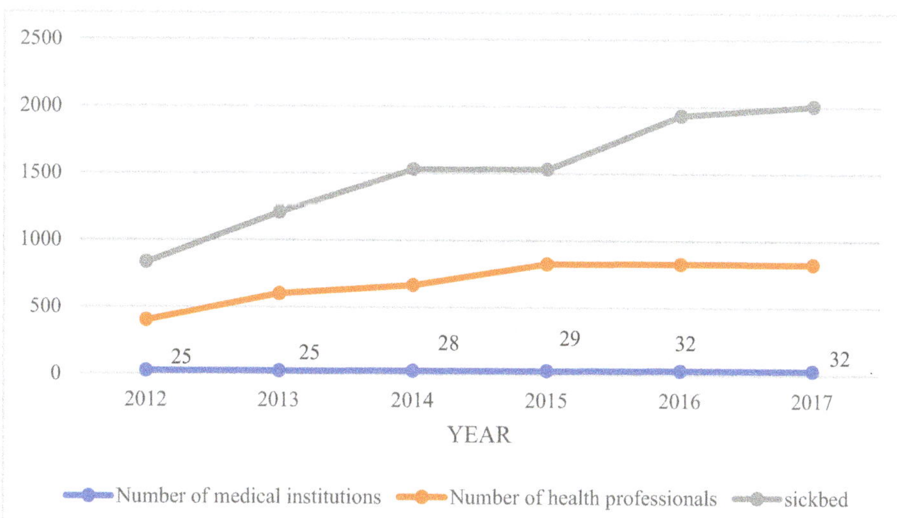

Figure 49.7. Changing trend of key medical and health indicators in Pingtang county.

forestry ecological construction, protection, and development reached 87.3[15], the atmospheric environment quality and management index reached 1.88[16], and a demonstration effect of ecological civilization construction and ecological function has appeared.

49.3.3.6 Create a new name card for the regional image shaping

The key words of 'astronomical science and technology' are the biggest driving force for FAST, and it will affect the regional economic development in the long term. As one of the first ten science and technology tourism bases in China, Pingtang county has built the Pingtang International Radio Astronomy Popular Science Cultural Park and the International Astronomy Town, which promise to become important astronomy science and technology tourism centers in the world and a national astronomy popular science education base. Pingtang county will also continue to attract astronomers from all over the world and become an important site for various conferences in the field of astronomy for a long time. At the same time, scientific researchers and tourists attracted by FAST will also generate a publicity effect, i.e. when they finish their research activities or sight-seeing trips at FAST, they may help publicize their FAST experience with a snowballing effect, creating an impressive image for Pingtang.

49.4 Discussion and prospects

This chapter analyzes systematically the positive impact of FAST from three aspects: scientific (and technological), economic, and social benefits. In general, the construction of FAST has made China's astronomical observation technology leap to the forefront in the world, turned the site location county into a world-class astronomical exchange center and one of the astronomical popular science tourism destinations, effectively promoting the rapid development of the local economy and the optimization of its industrial structure, has promoted the overall social development and effectively improved the living standards of the local residents, and helped in local poverty alleviation efforts. FAST has brought great achievements due to the correct selection of the site. The geographical location is not only scientifically suitable for the construction of specific facilities such as FAST, but also greatly promotes the regional development in terms of comprehensive economic and social impacts, in particular the new urban image design drives the rapid development of tourism. This provides an enlightening idea for the location of future large facilities.

---

[15] The comprehensive index of the forestry ecological construction, protection, and development is a composite index, which reflects the effectiveness of forestry development and protection in an economic region and is obtained from the synthesis of various indicators such as the forest coverage rate, the afforestation rate, the investigation and closure rate of cases of destruction of forest resources, and the ecosystem protection and restoration rate, as well as other relevant indicators.

[16] The atmospheric environmental quality and management index is a dimensionless index describing the comprehensive situation of urban ambient air quality. It comprehensively considers the pollution degree of SO_2, NO_2, PM10, PM2.5, CO, and O_3 specified in the technical regulations for ambient air quality index (AQI) (Trial) (HJ633-2012). The larger the value of the comprehensive index of air quality, the heavier the comprehensive pollution degree.

Different types of large facilities vary greatly, and evaluation of the site selection from the perspective of the comprehensive benefits of the large facilities can better promote the development of the related industries and the spillover effect to the region, and plays an important role in maximizing the comprehensive benefits of government public expenditure.

The construction of large scientific facilities in China is developing rapidly. The National Plan for Medium and Long Term Development of Major Scientific and Technological Infrastructure (2012–30) proposes that by 2030, China will basically build a major scientific and technological infrastructure system with complete layout, advanced technology, efficient operation, and strong support. In recent years, the investment in large scientific facilities has been increasing, and the total investment will soon exceed 10 billion US dollars. In China, these large facilities not only serve scientific progress, but also are required to provide better radiation and driving roles for economic and social development. The Chinese government pays particular attention to global openness and cooperation relying on the large scientific facilities. Chinese leaders have repeatedly mentioned the need to work with the global scientific community to do a great job in international Big Science plans and Big Science projects. Large scientific facilities not only belong to one country, but also the common infrastructure of the global scientific community, which will bring global scientists together to jointly carry out international scientific and technological cooperation and research.

Acknowledgments

This work was supported by the National Natural Science Foundation of China (grant number 71804179) and by the Chinese Academy of Sciences (grant number Y9X1691Q01).

References and further reading

Bezdek R H and Wendling R M 1992 Sharing out NASA's spoils *Nature* **355** 105–6

Broad W J 1990 Big science: is it worth the price? A periodic look at the largest new research projects. Heavy costs of major projects pose a threat to basic science *The New York Times* 27 May, Section 1, p 1

Castelnovo P, Florio M, Forte S, Rossi L and Sirtori E 2018 The economic impact of technological procurement for large-scale research infrastructures: evidence from the Large Hadron Collider at CERN *Res. Policy* **47** 1853–67

Chen G 2014 Economic and social impact of large scientific installations *Stud. Dialect. Nat.* **4** 118–22 (in Chinese)

Del Bo C F 2016a The rate of return to investment in R&D: the case of research infrastructures *Technol. Forecast. Soc. Change* **112** 26–37

Del Bo C F, Florio M and Forte S 2016b The social impact of research infrastructures at the frontier of science and technology: the case of particle accelerators: editorial introduction *Technol. Forecast. Soc. Change* **112** 1–3

Du C and Shang Z C 2011 *National Science Project* (Beijing: Beijing Institute of Technology Press) (in Chinese)

ESFRI 2013 Indicators of Pan-European Relevance of Research Infrastructures *ESFRI Roadmap 2021* https://www.esfri.eu/indicators-pan-european-relevance-research-infrastructures-october-2013

EvaRIO 2013 List of BETA-EvaRIO metrics and indicators *Final Report* http://evario.u-strasbg.fr/final-report

Farla K, Simmonds P, Rosemberg C and Rentel M 2016 Evaluating the economic and social impacts of cycling infrastructure: considerations for an evaluation framework *Report* https://www.gov.uk/government/publications/cycling-infrastructure-framework-for-evaluating-economic-and-social-impacts

Florio M, Forte S, Pancotti C, Sirtori E and Vignetti S 2016 Exploring cost–benefit analysis of research, development and innovation infrastructures: an evaluation framework *Working Papers* 201601 CSIL Centre for Industrial Studies

Griniece E, Cvijanović V, Kroll H and Zenker A 2019 Concept note of the modular impact assessment framework for research infrastructures (RI-PATHS Project) *Technical Report* https://researchgate.net/publication/337001994_Concept_note_of_the_modular_impact_assessment_framework_for_research_infrastructures_RI-PATHS_project

Griniece E, Reid A, Angelis J, Cvijanović V, Florio M, Kroll H and Zenker A 2018 Research Infrastructure Impact Assessment Pathways (RI-PATHS Project) *Project*

Gomez, A 2015 Techonological spillovers of research infrastructures, DEMM Working Papers

Hallonsten O 2013 Introducing facility metrics: a first review and analysis of commonly used measures of scientific leadership among synchrotron radiation facilities worldwide *Scientometrics* **96** 497–513

Hallonsten O 2014 How expensive is Big Science? Consequences of using simple publication counts in performance assessment of large scientific facilities *Scientometrics* **100** 483–96

Heidler R and Hallonsten O 2015 Qualifying the performance evaluation of Big Science beyond productivity, impact and costs *Scientometrics* **104** 295–312

Kolar J, Cugmas M and Ferligoj A 2019 Towards key performance indicators of research infrastructures arxiv:1910.00304

Lee S H, Peng M W and Song S 2013 Governments, entrepreneurs, and positive externalities: a real options perspective *Eur. Manag. J.* **31** 333–47

Martin B R and Tang P 2007 The Benefits from publicly Funded Research, SPRU Electronic Working Paper Series, Paper No.161, University of Sussex, Brighton

Mu R P, Kang X M and Fan Y G 2012 *Reflections on the Construction of Social Service Innovation Capacity in ChinaHigh-tech Development Report 2012* ed Chinese Academy of Sciences (Beijing: Science Press) (in Chinese)

NASA 2012 *SPINOFF2012* https://spinoff.nasa.gov/Spinoff2012/pdf/Spinoff2012.pdf

Nilsen V and Anelli G 2016 Knowledge transfer at CERN *Technol. Forecast. Soc. Change* **112** 113–20

OECD 2019 Reference framework for assessing the scientific and socio-economic impact of research infrastructures *OECD Science, Technology and Industry Policy Papers* https://oecd-ilibrary.org/science-and-technology/reference-framework-for-assessing-the-scientific-and-socio-economic-impact-of-research-infrastructures_3ffee43b-en

Pung C, Clarke A and Patten L 2004 Measuring the economic impact of the British library *New Rev. Acad. Librariansh.* **10** 79–102

Reid A, Griniece E and Angelis J 2015 Evaluating and monitoring socio-economic impact of investment in research infrastructures *Technical Report* https://researchgate.net/publication/

275037404_Evaluating_and_Monitoring_the_Socio-Economic_Impact_of_Investment_in_Research_Infrastructures

Schmied H 1975 A study of economic utility resulting from CERN contracts *CERN Yellow Reports, Monographs* 29 CERN

Simmonds P, Kraemer-Mbula E, Horvath A, Stroyan J and Zuijdam F 2013 Big science and innovation *Report* Technopolis https://assets.publishing.service.gov.uk/government/uploads/system/uploads/attachment_data/file/249715/bis-13-861-big-science-and-innovation.pdf

Zuijdam F, Deuten J, Meijer I, Boekholt P and Vermeulen N 2011 The role and added value of large-scale research facilities *Report* Technopolis https://academia.edu/3618621/The_role_and_added_value_of_large-scale_research_facilities

IOP Publishing

Big Science in the 21st Century
Economic and societal impacts
Panagiotis Charitos, Theodore Arabatzis, Harry Cliff, Günther Dissertori, Juliette Forneris and Jason Li-Ying

Chapter 50

Big Science in ASEAN in the twenty-first century?

Emmanuel Tsesmelis and Leong-Chuan Kwek

Focusing on Big Science, this chapter provides an overview of how science and the scientific community have been developing in Southeast Asia in the recent past and an outlook for science in the region for the twenty-first century. An active and dynamic scientific research landscape has been realised in academia, government, and industry, which can lead to exciting opportunities for the future. After a brief introduction to the historical development of science in southeast Asia, prospects for growth in the area of Big Science are discussed. Examples from other regions and countries are used to illustrate possibilities.

50.1 Introduction

Big Science plays an important role in today's science. Big Science typically refers to large-scale scientific research activity comprised of projects funded usually by several international entities. It also typically involves capital-intensive shared facilities and a large number of geographically distributed researchers. There are typically two camps: one that advocates Big Science as essential for societal progress and the other that objects to Big Science as a waste of resources [1–8].

Alvin Weinberg, who reputedly coined the term 'Big Science', compared the monuments of Big Science in the twentieth century to the construction of the pyramids, ancient temples, and churches [9]. He also saw 'the injection of a journalistic flavor into Big Science which is fundamentally in conflict with the scientific method' and yet concluded that Big Science was here to stay. Yet, due simply to their sheer size and complexity, Big Science projects could not avoid becoming embroiled in institutional, bureaucratic, and national politics [10].

doi:10.1088/978-0-7503-3631-4ch50

50.2 Basic science or Big Science?

In the mid-1960s, Weinberg formulated the rationale for Big Science as an issue of whether ‘governments should support basic science as a form of high culture or as an overhead charge on applied science and technology’ [11–13]. However, discovery comes at a cost. Indeed, the world spends more than US$1 trillion each year on R&D alone [14, 15]. Since World War II, the bulk of the capital for Big Science has been solicited from governments interested in supporting science for national interests, in particular national security [10].

As noted in [16], many countries that contribute to Big Science projects are interested not just in advancing the frontier of scientific knowledge but also in the economic impact on industry within their borders. It is from this perspective that the authors also remark that ‘any procurement for innovation by Big Science projects constitutes a public-policy intervention with a broad range of desirable effects’ [16].

Many countries are inevitably involved in these projects with significant resources, and the impact and benefit to society is evidently huge [17].

Science has expanded beyond boundaries of institutions, nations, cultures, and disciplines in the twentieth century. There is a shift across barriers from paper to digital and an accompanying shift from scarcity of knowledge to abundance through the Internet. One only needs to note that not very long ago scientific knowledge was regarded as a scarce and precious resource [18].

The recent COVID-19 pandemic evidently demonstrates the importance of doing Big Science and how poorer countries can leverage international collaboration for vaccine resources [19–22]. As Seth Berkley said in his recent article [19], ‘All of this (trials and development of vaccines) will require substantial funding, which is the big ask of Big Science. Late-stage clinical trials are not cheap, nor is vaccine manufacturing.’

50.3 Southeast Asia and ASEAN

The Association of Southeast Asian Nations (ASEAN) was formed in 1967 with five founding members—Indonesia, Malaysia, the Philippines, Singapore, and Thailand—to engage in joint efforts to accelerate the economic growth, social progress, and cultural development within the region. There was no consideration of science at the beginning of ASEAN. In 1967 there was little emphasis on science in all the founding countries. The priority at that time was to raise the economic level of all countries through cooperation. Three years later, in 1970, the Ad hoc Committee on Science and Technology was established to promote and intensify cooperation in the area[1]. With the formation of the Committee for Science and Technology in 1978, ASEAN has resolved to strive for the following objectives:

- Initiation and enhancement of regional cooperation in scientific and technological activities; and the generation and promotion of the development of scientific and technological expertise and manpower in the ASEAN region.

[1] https://asean.org/wp-content/uploads/2017/10/02-APASTI-2016-2025-Implementation-Plan-FINAL.pdf

- Facilitation and acceleration of the transfer of scientific developments and technologies among ASEAN countries. This facilitation includes the transfer of science and development from the more advanced industrialized countries to the ASEAN region.
- Provision of support and assistance in the application of the results of research and development (R&D), and in the more effective use of natural resources in the ASEAN region.
- Provision of support towards the implementation of present and future ASEAN (scientific) programmes.

Figure 50.1 shows the gross domestic expenditure on research and development (GERD) among the ASEAN countries. The data are extracted from different years, so they may not fully reflect the current status. However, it illustrates that there is a significant disparity in spending among these countries on R&D (figure 50.1).

Moreover, the total number of publications among the ASEAN nations from 1996 to 2011 as reported in [23] is given in table 50.1. It is noted that there is a significant disparity in the total number of scientific publications, reflecting the different values for each country. In fact, as seen in figure 50.2, there is a general linear log–log correlation between the amount of money invested in R&D with the total number of scientific publications, which is altogether not surprising. The result corroborates somewhat with the science, technology, and Innovation (STI) performance of each ASEAN member state by means of the composite indicator, the

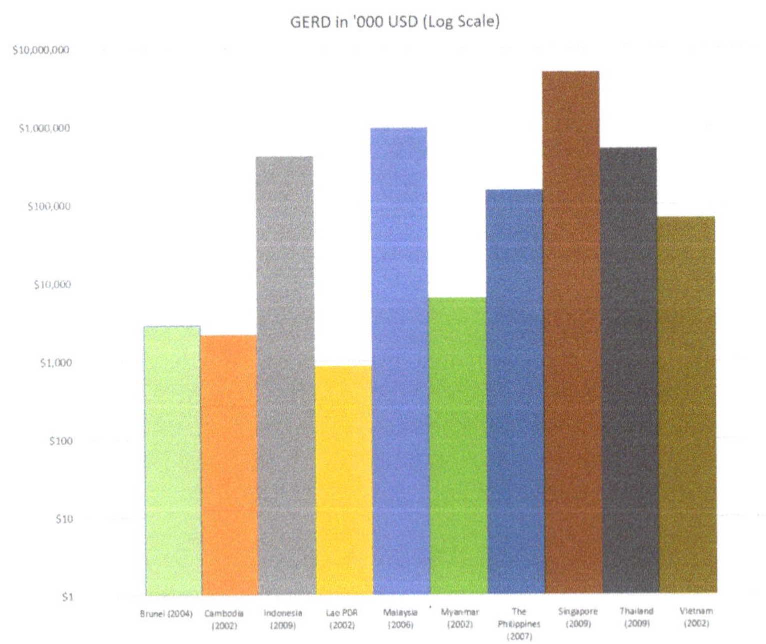

Figure 50.1. GERD amongst ASEAN countries. Note that the year in which the data were collected differs from countries to countries. The log scale has been chosen to reduce the disparity in the data. (Adapted from https://astnet.asean.org/structure/ with permission from the ASEAN secretariat.)

Table 50.1. Total publications between 1996 and 2011 for ASEAN countries [23].

Country	Total publications
Brunei	1064
Cambodia	1296
Indonesia	16 139
Laos	853
Malaysia	75 530
Myanmar	906
Philippines	11 326
Singapore	126 881
Thailand	69 637
Vietnam	13 172

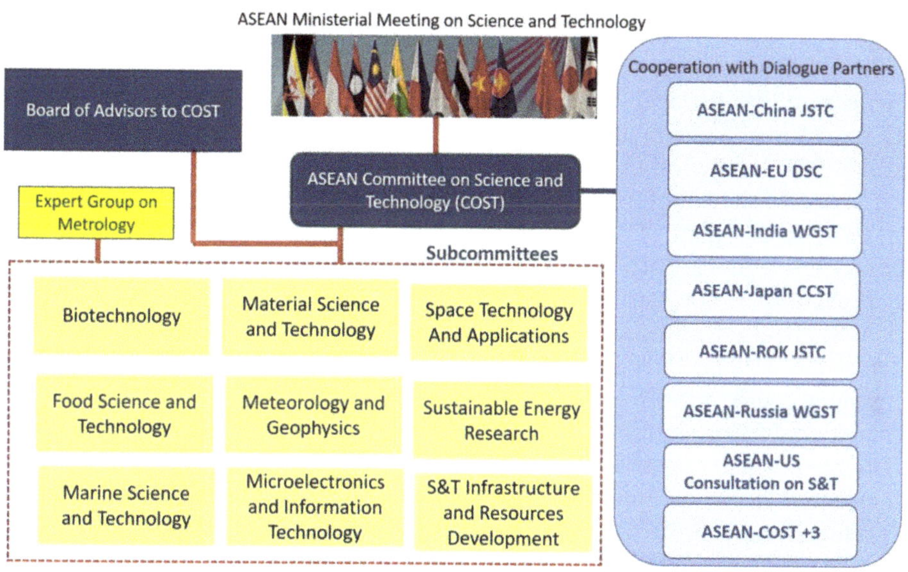

Figure 50.2. JSTC: Joint Science and Technology Committee; DST: Dialogue on Science and Technology; WGST: Working Group on Science and Technology; CCST: Cooperation Committee on Science and Technology; ASEAN COST+3: Committee on Science and Technology Plus Three. (Adapted from https://astnet.asean.org/structure/.)

summary science, technology and innovation index (SSTII) [24, 25]. Of course, at the finer scale, money is not the sole factor nor the most important one [26].

50.4 ASEAN science and technology

ASEAN leaders have recognised the importance of science and technology (S&T) as a key factor for sustaining economic growth. They have also long recognized the importance of enhancing community well-being and promoting integration in

ASEAN. ASEAN member countries need to move up the technology ladder, focusing on enhancing their capability for technological innovation in many areas by pushing forward the goal of building a strong S&T base for ASEAN as a single integrated economy. This goal will ensure that ASEAN remains globally competitive. This task, however, is not easy as ASEAN is a grouping of ten developing countries. With great variation in the level of scientific and technological capability, the group has yet to move the cooperation forward more vigorously to promote deeper integration. ASEAN must also promote greater awareness among its populations on the need for S&T activities and the roles and contributions of such activities towards sustainable economic development, and the need for communities such as the private sector in S&T undertakings, particularly research and technology development, and the commercialisation of technologies must also be promoted. The science, technology, and innovation cooperation in ASEAN focuses on nine programme areas, namely (i) biotechnology, (ii) food science and technology, (iii) marine science and technology, (iv) materials science and technology, (v) meteorology and geophysics, (vi) microelectronics and information technology, (vii) space technology and applications, (viii) sustainable energy research, and (ix) S&T infrastructure and resources development.

To ensure that the S&T cooperation in ASEAN remains relevant and supportive with regard to the directives of the ASEAN leaders and S&T ministers, a series of ASEAN Plans of Action in Science and Technology (APASTs) have been developed since the establishment of the Committee on Science and Technology (COST) in 1978. The most recent plan, The Action Plan of Action on Science, Technology and Innovation (APASTI) 2016–2025, was endorsed by the ASEAN Ministers for Science and Technology at the 16th ASEAN Ministerial Meeting on Science and Technology (AMMST) held on 6 November 2015 in Vientiane, Lao PDR. The APASTI 2016–25 provides the strategic direction for COST in implementing the programmes/activities focusing on public–private collaboration, talent mobility, people-to-people connectivity and inclusiveness, enterprise support, and public awareness and STI enculturation.

The plan, which comprises six goals, four thrusts, and fifteen supporting actions, will essentially provide appropriate guidelines for identification and formulation of programmes and projects to achieve better coordination and cooperation to strengthen the capabilities of STI in ASEAN.

Specifically, the four strategic thrusts are:

- Strengthen strategic collaboration among academia, research institutions, networks of centres of excellence, and the private sector to create an effective ecosystem for capability development, technology transfer, and commercialisation.
- Enhance mobility of scientists and researchers and people-to-people connectivity, and strengthen the engagement of women and young people in STI.
- Establish innovative systems and smart partnership to nurture STI enterprises to support micro, small, and medium enterprises (MSMEs), nurture knowledge creation and STI applications to raise competitiveness.
- Raise public awareness and strengthen STI enculturation to enhance ASEAN science and technology cooperation.

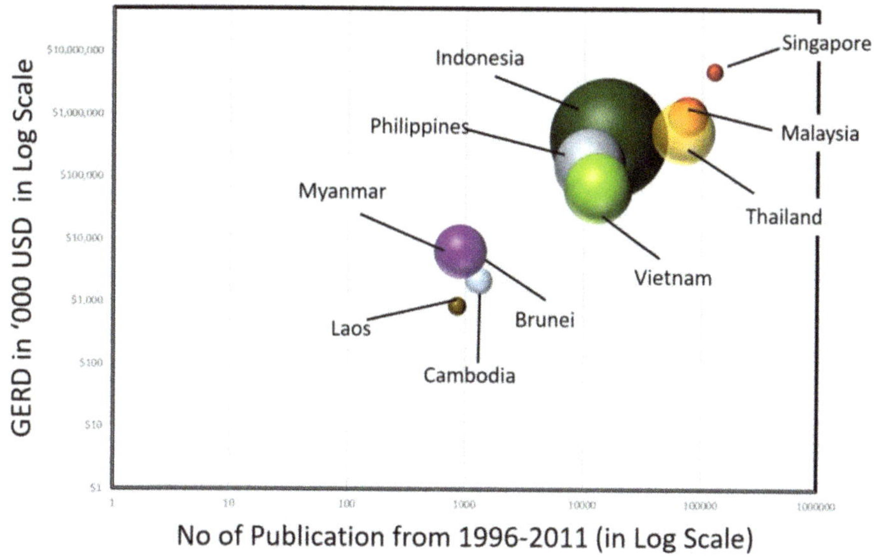

Figure 50.3. Log–log correlation plot of GERD ('000 USD) with publication numbers from table 50.1 from 1996 to 2011. Note that the sizes of the dots correspond to the populations in 2020.

The organisational chart of ASEAN S&T cooperation is shown in figure 50.3.

To provide better services to ASEAN public and government agencies in the fields of meteorology and seismology, two ASEAN centers have been established under COST, they are ASEAN Specialised Meteorology Center (ASMC) and ASEAN Earthquake Information Center (AEIC) which were established in 1993 and 2000, respectively.

ASEAN COST is actively pursuing close cooperation with its dialogue partners and sectoral dialogue partners. Several permanent bodies have been established with ASEAN dialogue partners such as China, the European Union (EU), India, Japan, the Republic of Korea (ROK), Russia, the United States (US), and also with the Plus Three (China, Japan, ROK). Regular meetings among S&T officials from ASEAN and dialogue partners are convened to promote closer cooperation. In addition, ASEAN also continues to forge close collaboration in S&T with international organisations such as UNESCO, UN-ESCAP, and the World Meteorological Organisation (WMO).

50.5 ASEAN and the European Union

In terms of population and land area, the European Union (4.2 million km^2 with 447 million people) and ASEAN (4.5 million km^2 with 661 million people) appear ostensibly similar. However, there is a large gap at the moment between the two regions in terms of scientific output and collaborations [27, 28], particularly in terms of Big Science. Several Big Science projects are hosted in Europe, such as the European Organization for Nuclear Research (CERN, hosted jointly by EU country France and non-EU country Switzerland), the International Thermonuclear

Experimental Reactor (ITER), the European Space Agency (ESA), the European Molecular Biology Laboratory (EMBL), the European Synchrotron Radiation Facility (ESRF), the European Organisation for Astronomical Research in the Southern Hemisphere (ESO), the European X-ray Free-Electron Laser Facility (European XFEL), and the Institut Laue-Langevin (ILL). Would it be possible for ASEAN to match the Big Science facilities in the European Union in the future?

Within ASEAN, there are currently few Big Science projects, some exceptions being in the domain of plant biotechnology [29]. In this chapter, the results of an analysis related to plant biotechnology in the ten year period 2004–13 to assess the research performance, impact, and collaboration of member states of ASEAN are elaborated. The analysis shows that plant biotechnology is currently one of the main areas of cooperation between ASEAN member states and among the research areas promoted to achieve regional food security and sustainable development. In general, the findings indicate increased scientific output, influence, and overall collaboration of ASEAN countries in plant biotechnology over time. The research performance and collaboration (domestic, regional, and international) of the region in plant biotechnology are linked to the status of the economic development of each member country. Thailand produced the most publications of the ASEAN member states while Singapore had the highest influence, as indicated by its citation activity in plant biotechnology, among the ASEAN countries. Domestic and international collaborations on plant biotechnology are numerous. Regional collaboration or partnership among ASEAN countries was, however, found to be very limited, which is a concern for the region's goal of economic integration and science and technology cooperation.

Certain similarities and differences between the European Union and ASEAN have been noted by Tommy Koh:

- Both are regional international organisations with legal entities. The EU had 28 members (27 following the exit of the United Kingdom in 2021), while ASEAN has 10 members, with Timor Leste expressing interest to join.
- Both were founded to promote peace. The EU was established, after two disastrous world wars, to prevent the recurrence of war in Europe and to institutionalize peace through economic integration. Likewise, ASEAN was founded to create a peaceful environment in Southeast Asia so that the ASEAN countries could focus on their economic development.
- Both organisations share a commitment to human rights. The EU has a Charter of Fundamental Rights and ASEAN has a Declaration of Human Rights. ASEAN has two commissions on human rights: the ASEAN Intergovernmental Commission on Human Rights and the Commission on the Rights of Women and Children. On the other hand, the EU has a European Court of Justice. ASEAN does not have a similar structure.
- Both ASEAN and the EU have concluded many free trade agreements or comprehensive economic partnership agreements with other countries. For example, the EU and Singapore have concluded a free trade agreement which is pending ratification. ASEAN has concluded such agreements with China, Japan, South Korea, India, Australia, and New Zealand, but not with the EU.

- Both organisations seek to integrate the economies of their member states into a single market and production platform. In the case of the EU, there is freedom of movement of goods, services, capital, and labour. In the case of ASEAN, the movement of labour is not free. However, the ASEAN Charter obliges the member states only to facilitate the movement of professionals, talents, and labour for business and academic purposes. Indeed, this is also a major difference between ASEAN and the EU.
- Both ASEAN and the EU hold regular political and economic dialogues with important external partners. The EU holds annual summits with many countries including the United States, China, Japan, and Russia.

It does not take long to see that nothing is explicitly mentioned about science and technology. In the following section we see another comparison which perhaps is closer to showing what ASEAN can achieve.

50.6 Areas of strength in ASEAN countries

50.6.1 Brunei

The main research focuses in science and technology in Brunei are biodiversity, energy, halal food, food security, agrotechnology, climate change, Asian studies, and Islamic banking and finance. Currently, its major contributions to GDP are oil, natural gas, garments, and agriculture.

50.6.2 Cambodia

Cambodia's S&T niches are agriculture, mines, and minerals, and possibly construction, based on existing infrastructure, human resources, actual production performance, and potential. Cambodia's major contributors to GDP are agriculture, garments, tourism, and construction.

50.6.3 Indonesia

Indonesia is the most populous country in Southeast Asia. Based on the total number of publications in international journals, public health, plant sciences, environmental sciences, ecology, multidisciplinary geosciences, tropical medicine, microbiology, pharmacology and pharmacy, infectious diseases, zoology, immunology, biology, biochemistry, molecular biology, medicinal chemistry, and agronomy are some of the main research fields in Indonesia. In particular, the publication in the fields of molecular biology and genetics, microbiology, immunology, space science, environment and ecology, and clinical medicine are mostly cited from Indonesian authored publications, with average citation indexes of over 11 citations per paper. The area of agriculture, medical sciences, and technological sciences are the fields with the biggest number of researchers in Indonesia, accounting for 22% of the total Indonesian researchers. The major contributions to GDP in 2009 were from agriculture, mining and primary resources, manufacturing, and services.

50.6.4 Lao People's Democratic Republic

Lao PDR is the only landlocked country in Southeast Asia. Agriculture has always featured prominently in the country's economy. Consequently, it is not surprising that agricultural R&D plays huge role in the development of the country. The estimated GDP in 2010 is $6.341 billion, with agriculture sector accounting for 29%, the industry sector contributing 26.5%, and services constituting 39%.

50.6.5 Malaysia

The Malaysian Science and Technology Indicators Report (2008) showed that the field of engineering science contributed the largest number of researchers (more than 3500 persons), followed by applied S&T (about 2000), medical and health science (1500), agriculture science (1500), ICT (about 1900), and social science (about 1500). The same report also showed that five highly ranked scientific publications in Malaysia by subject areas for a period from 1909 to February 2009 were medicine (18%), followed by engineering (14%), biochemistry, genetics, and molecular biology (9%), agriculture and biological sciences (9%), and physics and astronomy (8%). In line with the strategy to generate new sources of growth, the government has focussed significantly on targeted R&D to build competence and specialization in emerging technologies. Focus is directed into areas including biotechnology, ICT, and nanotechnology. Major contributors to GDP (2010) include agriculture, mining and quarrying, manufacturing, construction, and services.

50.6.6 Myanmar

Engineering is considered an area of high potential because of the number of higher educational institutions (HEIs) or universities that are dedicated to the engineering disciplines. There are 32 technological universities in Myanmar, all under the Ministry of Science and Technology (MOST). These universities offer a wide range of engineering disciplines including petroleum engineering, textile engineering, and mining engineering. Two of these technological universities offer MSc and PhD programs in 13 engineering fields or disciplines. The largest number of PhD students are in civil engineering and electrical power engineering. There are also 25 computer-focus universities, also all under MOST. Two well-known universities in these areas are the University of Computer Studies in Yangon and the University of Computer Studies in Mandalay. Both universities offer MScs and PhDs in computer studies. Computer science and information technology are considered an area of high potential. Other areas where Myanmar is trying to build S&T capability are in biotechnology, nanotechnology, nuclear technology, renewable energy, and materials science. The major contributors to GDP are agriculture, livestock and fishery, processing and manufacturing, construction, mining, forestry, transportation, and trade.

50.6.7 The Philippines

Major areas of strength in S&T/R&D are disaster science and the management of climate change, aquaculture, and biodiversity. Major contributions to GDP are:

business process outsourcing (including knowledge process outsourcing), tourism, electronics, mining, housing, agribusiness/forest-based industries (banana, pineapple, mango, papaya, and okra; preserved fruits; beverages; and processed marine products), logistics, and shipbuilding. These sectors identified in GDP could contribute significantly to inclusive growth and employment generation.

50.6.8 Singapore

Singapore's areas of strength or S&T niches can be gleaned from its research priorities as earlier mentioned under the S&T development strategy. Some key highlights of capabilities include the following:

- *Electronics.* Singapore has strong capabilities in data storage and semiconductors, which have contributed to the growth and development of the electronics sector. In data storage, Singapore has built up expertise in areas such as spintronics, media and interfaces, mechatronics and recording channels, optical materials and systems, and network storage technologies. In semiconductors, Singapore has built up strong capabilities in areas such as integrated circuit design and subsystems, packaging, complementary metal-oxide semiconductors (CMOS) and microelectromechanical systems (MEMS).
- *Biomedical sciences.* Singapore has strengths across the entire biomedical sciences spectrum, in areas such as bioprocessing technology, human immunology, eye diseases, stem cell, and cancer biology.
- *Infocomms and media.* Singapore has built capabilities in interactive and digital media, IT, computational science, communications and multimedia technologies, cloud computing, and security.
- *Engineering.* In precision engineering, Singapore research institutes will advance their strong manufacturing capabilities through research in advanced materials, microfluids, printed electronics, nano-manufacturing of multifunctional products, and sustainable manufacturing. In transport engineering, R&D is a key driver for growth in the aerospace and marine and offshore sectors. The A*STAR Aerospace Consortium engages companies across the entire value chain in pre-competitive research in maintenance repair and overhaul (MRO), engine-related technologies, avionics, and composite materials.
- *CleanTech.* Singapore leads in its expertise in water and solar research, and will develop capabilities in intelligent energy systems and energy efficiency.

Singapore's major contributors to GDP are electronics and chemicals manufacturing as well as finance and business services.

50.6.9 Thailand

On the basis of the human resources available, institutional capabilities, and actual industry experience, Thailand's S&T strengths lie in food biotechnology, life sciences, the agricultural sciences chain, energy technologies, and manufacturing technologies, particularly in the automobile and electronics industry. In the medical field, Thailand is strong in the medical hub concept and in system based public health. Of the total number of papers published in the Science Citation Index (SCI) by Thais in 2007, the biggest number (1439 out 4214) came from the medical sciences. On the basis of the top ten Thai exports, one can have a glimpse of Thai's areas of strength, which would logically include the science and technology components. The top ten Thai exports in 2011 were the following:

- Computers and parts.
- Automobiles and parts.
- Jewelry.
- Rubber.
- Plastic.
- Electrical circuits.
- Chemicals.
- Rubber products.
- Cooking oil.
- Rice.

Thailand's economy functions as an anchor economy for the neighboring developing economies of Laos, Burma, and Cambodia. Thailand's recovery from the 1997–98 Asian financial crisis depended mainly on exports, among various other factors. Thailand ranks high among the world's automotive export industries along with manufacturing of electronics goods.

50.7 ASEAN science and technology: policy and strategy

Finally, some recommendations are proposed in this study that are aimed at attaining the goal of building the S&T sector in the ASEAN Community. These recommendations cover the following:

(i) Setting an ASEAN vision for S&T that will contribute positively to the building process for the ASEAN Community.

(ii) Creating an active R&D network that will be useful for ASEAN member states. This collaboration will include ways to utilize the know-how and knowledge accumulated in universities and research institutions in every member state. Areas of collaboration will include socio-culture development and regional development. ASEAN member states will develop a security network in the areas of food and energy, water management, biodiversity, climate change, disaster management, and environmental preservation. This will evolve through a scheme called 'Open Innovation' that will provide researchers and scientists access to shared R&D facilities and information across ASEAN. Other segments of the R&D Network will share models of research consortia among universities, public research institutions, and corporate business units. Public–private partnerships are recommended for S&T utilization projects that target regional development and public services across the region such as food and energy security, disaster management, and environmental preservation and rehabilitation.

(iii) Ensuring human resource development among ASEAN member states through the AUN-SEED Program in collaboration with the Committee on Social Development. This means an expanded AUN-SEED Program for advanced education in engineering and science. Included here is the adoption of a 'sandwich' program and internship for researchers and technical personnel in R&D/S&T institutes ASEAN-wide.

(iv) Strengthening regional competitiveness through the development of innovative and cooperative strategies that will support SMEs in technological upgrading. This will be accomplished by setting up SME technology helpdesk centers and networks across ASEAN and promoting high-tech firm start-ups through programs of cross-ASEAN technopreneurships that combine ideas, socio-cultural uniqueness, highly skilled human resources, and capital utilization across ASEAN.

(v) Establishment of the ASEAN Fund for Innovation organized under ABA-PAST/ABASF and S&T utilization project funding organized under a COST sub-committee as well as venture capital fund to support technology-based SMEs and tech-firms start-ups across ASEAN. ASEAN should take advantage of international cooperation and the scheme of public–private partnership to establish such funding

50.8 Towards Big Science in ASEAN

ASEAN, like the EU, is not a single country. ASEAN is close to India, China, South Korea, and Japan in the northern hemisphere and to Australia and New Zealand in the southern hemisphere. ASEAN is therefore a strategically ideal place to host a Big Science project. ASEAN can also leverage countries in Asia such as China, Japan, South Korea, and India. These countries are within its geographical proximity. Moreover, it is ultimately of mutual benefit to these countries to invest some of their Big Science projects in Southeast Asia. Science strategy and policy should have an important focus on the key issues of science and technology.

In Southeast Asia, Singapore is a shining example of research in science, technology, and innovation. Singapore can bring up the research level in Southeast Asia through regional collaboration. Unfortunately, the country is small, in size and population. If bigger countries in ASEAN, such as Indonesia, Thailand, Malaysia, the Philippines, and Vietnam, can leverage on the scientific achievement of Singapore to scale new heights in science and technology, ASEAN can easily steer towards scientific excellence in the next ten to twenty years.

50.9 Concluding remarks

ASEAN is a rapidly growing economic region in the world. It has comparable land area and population as the European Union. Yet, in terms of scientific research and culture, it still lags behind the European counterpart. In this brief review, we attempt to discuss the possibility of 'Big Science' research in ASEAN. We are confident that the prospect is bright as long as the ASEAN countries work together to forge a common goal.

References

[1] de Solla Price D J 1963 *Little Science, Big Science* (New York: Columbia University Press)

[2] Wall T D and Wood S J 2005 The romance of human resource management and business performance, and the case for big science *Hum. Relat.* **58** 429–62

[3] Schatz G 2014 The faces of big science *Nat. Rev. Mol. Cell Biol.* **15** 423–6

[4] Stix G 2001 Little big science *Sci. Am.* **285** 32–7

[5] Cyranoski D 2006 'Big science' protein project under fire *Nature* **443** 382–3

[6] Petsko G A 2009 Big science, little science *EMBO Rep.* **10** 1282

[7] Westfall C 2003 Rethinking big science: modest, mezzo, grand science and the development of the Bevalac, 1971–1993 *ISIS* **94** 30–56

[8] Moshkova D M 2019 Public–private partnership in implementing 'mega-science' projects *J. Phys: Conf. Ser.* **1406** 012012

[9] Weinberg A M 1961 Impact of large-scale science on the United States *Science* **134** 161–4

[10] Capshew J H and Rader K A 1992 Big science: price to the present *Osiris* **7** 2–25

[11] Weinberg A 1963 Criteria for scientific choice *Minerva* **38** 159–71

[12] Weinberg A M 1964 Criteria for scientific choice II: the two cultures *Minerva* **3** 3–14

[13] Ziman J 2000 Criteria for scientific choice—commentary *Minerva* **38** 253–69

[14] Florio M and Giffoni F 2017 Willingness-to-pay for science as a public good: a contingent valuation experiment *Departmental Working Papers* Department of Economics, Management and Quantitative Methods at Università degli Studi di Milano

[15] Florio M, Giffoni F, Giunta A and Sirtori E 2018 Big science, learning, and innovation: evidence from CERN procurement *Ind. Corp. Change* **27** 915–36

[16] Eerme T and Nummela N 2019 Capitalising on knowledge from big-science centres for internationalisation *Int. Market. Rev.* **36** 108–30

[17] Nilsen V and Anelli G 2016 Knowledge transfer at CERN *Technol. Forecast. Soc. Change* **112** 113–20

[18] Wagner C S 2018 *The Collaborative Era in Science* (Berlin: Springer)

[19] Berkley S 2020 Covid-19 needs a big science approach *Science* **367** 1407

[20] Rochel de Camargo K Jr 2020 Trying to make sense out of chaos: science, politics and the covid-19 pandemic *Cad. Saude Publica* **36** e00088120

[21] Ros F, Kush R, Friedman C, Gil Zorzo E, Rivero Corte P, Rubin J C, Sanchez B, Stocco P and Van Houweling D 2021 Addressing the Covid-19 pandemic and future public health challenges through global collaboration and a data-driven systems approach *Learn. Health Syst.* **5** e10253

[22] Blau J 2020 EU scientists unite against corona *Res.-Technol. Manag.* **63** 4

[23] Gholizadeh H, Salehi H, Embi M A, Danaee M, Motahar S M, Ebrahim N A, Habibi F and Osman N A A 2014 Relationship among economic growth, Internet usage and publication productivity: comparison among ASEAN and world's best countries *Mod. Appl. Sci.* **8** 160–70

[24] Rodriguez V and Soeparwata A 2012 ASEAN benchmarking in terms of science, technology, and innovation from 1999 to 2009 *Scientometrics* **92** 549–73

[25] Nguyen T V and Pham L T 2011 Scientific output and its relationship to knowledge economy: an analysis of ASEAN countries *Scientometrics* **89** 107–17

[26] Fortin J-M and Currie D J 2013 Big science vs little science: how scientific impact scales with funding *PLoS One* **8** e65263

[27] Hassan S-U, Haddawy P, Kuinkel P, Degelsegger A and Blasy C 2012 A bibliometric study of research activity in ASEAN related to the EU in FP7 priority areas *Scientometrics* **91** 1035–51
[28] King D A 2004 The scientific impact of nations *Nature* **430** 311–6
[29] Payumo J G and Sutton T C 2015 A bibliometric assessment of ASEAN collaboration in plant biotechnology *Scientometrics* **103** 1043–59

Chapter 51

India's road to HEP

D K Srivastava and V S Ramamurthy

India has always considered the pursuit of knowledge as the highest form of human endeavor. However, as the pursuit of fundamental knowledge requires large-scale research infrastructures/tools, it raises questions about funding and the societal impact of investments in these tools. We discuss here India's participation in CERN activities in the field of experimental high energy and particle physics and in the creation of new facilities including the Large Hadron Collider. This participation has not only opened new opportunities for young researchers from India, but has also become a model for several other international collaborations in which India is taking part.

51.1 Prologue: India and its scientific traditions

Curiosity about natural phenomena has always characterized *Homo sapiens* since time immemorial. Unraveling these phenomena has led to an enormous intellectual and cultural development of humanity. The search for new knowledge has also gone beyond/traversed boundaries of all kinds, specifically geographical and political.

India has always played an active role in many of these intellectual quests, both at the individual and at the institutional levels. The University of Takshshila (or Taxila, now in Pakistan) is considered to be one of the earliest universities in the world and flourished from 600 BC to 500 AD, until it was destroyed by the invasion of the Huns. At its peak, it housed more than 10 000 students from across the world and in addition to religious philosophy, offered courses in astronomy, medicine, surgery, languages, warfare, economics, and creative arts. The other world-famous University at Nalanda operated from 427 AD to 1197 AD—until it was similarly destroyed by invaders, who also burned down its three world-famous libraries, which are believed to have housed 9 million books. Some of these books are known to have survived in copies and translations by Chinese scholars who had visited Nalanda for studies. It was also known to have housed more than 12 000 students,

doi:10.1088/978-0-7503-3631-4ch51

2000 teachers, and several laboratories for teaching creative arts, science, etc. Several other institutions of higher learning and individuals such as Charaka, Panini, Kautilya, Aryabhatta, and Bhaskara had also flourished in various parts of the country at different times and also educated students from Central Asia, the Far East, and Southeast Asia.

We have mentioned the above here only to emphasize that the search for knowledge beyond the known frontiers has always been a part of the Indian cultural tradition and India has gone beyond its borders in this search.

Coming closer to the present, the nineteenth century and the early part of the twentieth century were periods of very rapid developments in science and technology. Not only were very rapid strides made in our understanding of the material Universe, such as atomic theory, electromagnetism, quantum mechanics, the theory of relativity etc, the period also saw the growth of several technologies such as the combustion engines, electric motors, material and manufacturing technologies, wireless communication, etc.

Institutional support to modern science in India during the British Rule was however limited to field sciences such as cartography, weather, geography, geology, and botany, primarily driven by the political and economic agenda [1] of the colonizers. Curiosity driven individuals had to take the tradition forward mostly on their own during these decades. There were notable individuals such as P C Ray, who started several industries such as Bengal Chemicals to produce chemicals for household as well as industrial use, often based on his own work.

Sir J C Bose (1858–1937) was a legendary Indian scientist of those decades who carried out globally competitive scientific research in India. He laid the foundations of experimental science in India in biology, physics, and botany, working all by himself in Calcutta. He pioneered the investigation of radio and microwave optics, and remote wireless signaling and was the first to use a semiconductor junction to detect radio waves [2]. He made significant contributions to plant science using very sensitive instruments which he himself designed to measure their growth and the effect of stimuli. Incidentally, he was also the teacher of M N Saha and S N Bose and several others, who went on to make outstanding contributions to science.

M N Saha (1893–1956) laid the foundation of astrophysics by deriving the mathematical relationship between the observed spectra of stars and their temperatures [3]. He had also essentially envisaged the Compton effect as he estimated the radiation pressure of solar radiation to explain the fact that the tails of comets always pointed away from the Sun.

S N Bose (1894–1974) developed his quantum statistics of identical particles, now called bosons, which led to the formulation of Bose–Einstein condensates [4]. It is known that Bose essentially derived his statistics during a lecture he was delivering to students! He also envisaged the spin of photons as one, which was later confirmed by Sir C V Raman, in a series of brilliant experiments [5].

The discovery of Sir C V Raman (1888–1970) that won him the Nobel Prize in 1930 established that light that passes through a material is scattered and the wavelength of the scattered light is changed because it has caused an energy state transition in the material's molecules [6]. It went on to become a valuable tool for analysing the composition of liquids, gases, and solids.

While we celebrate the success stories, one should also not forget the contributions of Indian scientists who bore the brunt of resource constraints and lost opportunities.

D M Bose (1885–1975) (who had developed a cloud chamber earlier, to study radioactivity and cosmic rays) and his student Bibha Chowdhuri started high altitude cosmic ray studies in Sandakphu in 1937 in the Darjeeling Himalayas at a height of 12 000 feet using photographic plates ('Ilford plates'—which were glass plates coated on one side, halftone, with photographic emulsion). The 'full-tone' plates (coated on both sides), which would have been better, were not only more expensive, but all of them were also requisitioned for the war efforts of the Second World War and were not available to civilians. Studying their exposed prints, they observed several tracks, in addition to the usual tracks of protons. These they identified with the 'mesotron' (as early as in 1940)—the exchange particle of Yukawa, which we now call the pion [7]. They estimated its mass to be between 150 and 270 times the mass of the electron. It is now understood that the lower-end mass particles were muons and the higher-end mass particles were pions. However, with the resolution available using halftone plates, it was not possible for them to distinguish the two. After the end of the War, Cecil Powell used the same procedure, with high density full-tone emulsion plates to separate pions and muons (1948) and win a Nobel Prize (1950). The path-breaking efforts of Bose and Bibha Chowdhury were forgotten in the course of time.

We have highlighted these developments only to emphasize that even under colonial rule, individual scientists, working essentially by themselves with barest minimum (and often personal) resources, were making contributions to frontiers of physics [8].

51.2 Science in independent India

India emerged from nearly three centuries of colonial rule and became an independent nation on 15 August 1947. India is one of the few countries in the world that chose not only the path of democracy but also the path of science and technology for development right after becoming independent. Jawaharlal Nehru, who became the first prime minister of India, writing in his famous book *Discovery of India* (written in 1944 and published in 1946), argued that the scientific temper was a framework that one should adopt to overcome bias, prejudice, and irrationality. He grounded his vision of modern India in this and wrote [9]:

> The scientific temper points out the way along which man should travel. It is the temper of a free man. We live in a scientific age, so we are told, but there is little evidence of this temper in the people anywhere or even their leaders. [What is needed] is the scientific approach, the adventurous and yet critical temper of science, the search for truth and new knowledge, the refusal to accept anything without testing and trial, the capacity

> to change previous conclusions in the face of new evidence, the reliance on observed fact and not on pre-conceived theory, the hard discipline of the mind—all this is necessary, not merely for the application of science but for life itself and the solution of its many problems.

India is the first and only country to explicitly adopt scientific temper in its Constitution as a Fundamental Duty [10]:

> [It shall be the duty of every citizen of India] To develop scientific temper, humanism and the spirit of inquiry and reform.

There was no doubt that the country had an uphill task at the dawn of independence. The following challenges were quite apparent:

i. Can a country of perennial food shortages feed its citizens?
ii. Can democracy deliver with most of the population being illiterate and living in villages?
iii. Can an economically weak, under-developed country like India with poor educational and industrial infrastructure and a fledgling bureaucracy afford science and technology for development?

It is to the credit of the political and science leaders of those years that India today has a surplus in food production, a viable educational and industrial infrastructure, and is a global player in several areas of Big Science and high technology, such as nuclear technology, space technology, information and communication technologies, bio-technology, etc. In this chapter, we limit our discussion to India's journey into experimental high energy and particle physics and, briefly, observational astronomy.

As was mentioned earlier, Indian scientists have always been participating in emerging areas of nuclear and particle physics research, even before India's independence. Following independence, several small nuclear facilities were established to nurture nuclear science research in our universities and research laboratories. It was to the credit of Homi Jehangir Bhabha, who not only recognized early the importance of nuclear technology for satisfying the increasing energy needs of the country, but also of nuclear research as training ground for young students in emerging area of technologies. Bhabha came to India in 1939 for a vacation and could not go back due to the Second World War. With several path-breaking discoveries already to his credit [11] in particle physics and cosmic ray research at Cambridge, he started working at the Indian Institute of Science, Bangalore, and initiated studies on cosmic rays using balloons. He went on to establish the Tata Institute of Fundamental Research (TIFR), Bombay, where several experimentalists were trained to continue the research on high altitude cosmic rays. The group also went on to make the first observation of atmospheric neutrinos in the 2 km deep Kolar Gold Mines in a TIFR–Osaka–Durham Collaboration in 1965 [12]. TIFR also became the nursery to nurture the newly founded Indian Atomic Energy

Establishment, Trombay, which was later renamed the Bhabha Atomic Research Centre.

Vikram A Sarabhai, who had earlier worked with Bhabha, went on to establish his own Institute, the Physical Research Laboratory, in Ahmedabad, to study cosmic rays using rockets—an effort which ultimately blossomed into the Indian Space Research Organization, which went on to produce satellites for astrophysical studies, discover water on the Moon using its probe, and send a probe to Mars.

Starting with a few small charged-particle accelerators in our universities and in our National laboratories, the Indian Department of Atomic Energy moved on to set up a variable energy cyclotron and later a superconducting cyclotron at Calcutta, and a high voltage Pelletron accelerator for the study of heavy-ion induced nuclear reactions in TIFR. An Inter University Accelerator Centre was also set up in New Delhi with a similar heavy-ion accelerator. Both facilities were subsequently upgraded with superconducting LINAC boosters designed and fabricated in India. The Indian contributions to medium energy heavy-ion research with these facilities are well recognized.

In parallel, the Raja Ramanna Centre for Advanced Technology at Indore set up Indus-1, a 450 MeV, 100 mA electron storage ring, emitting radiation from mid-infrared to soft x-rays with a critical wavelength of ~61 Å, and Indus-2, a 2.5 GeV electron storage ring designed to produce x-rays. These experiences were to prove invaluable for further collaborations with high energy accelerator facilities abroad, as we shall see.

The emergence of several advanced high energy particle accelerators and detector systems in laboratories across the world did edge out cosmic ray experiments with nuclear emulsions and cloud chamber photographs. However, they also opened a new window of opportunity for scientists across the world to participate in studies using these accelerators in conjunction with nuclear emulsions and cloud chamber photographs, without being physically present for the experiments. In an early step towards Big Science, the research groups from universities and other institutions continued with their studies using nuclear emulsions exposed in accelerator-based experiments elsewhere. The researchers at Aligarh Muslim University, Benares Hindu University, and Jadavpur University had also started performing emulsion experiments. In the mid-1990s, researchers from Jammu University, Punjab University, and Rajasthan University joined an international collaboration led by the Lund Group to expose emulsion plates to proton and heavy-ion beams at JINR Dubna, AGS BNL, and CERN SPS at energies ranging from 4.5 GeV/nucleon to 200 GeV/nucleon. In addition to the usual procedure of exposing emulsion plates horizontally, an innovative procedure using thin emulsion chambers, equipped with thin target foils, exposed perpendicular to the emulsion surface was also utilized. The main aim of the research was to study the properties of nuclear matter under conditions of intense high temperature and density, as is believed to have existed a few moments after the creation of Universe through 'Big Bang'. There were some other groups from India which also participated in such emulsion experiments.

Over a period, the groups also matured to design and construct detector systems and perform measurements and analysis of the large body of data collected in these experiments. For example, in mid-1990s, Utkal University, Punjab University, and TIFR joined the Belle collaboration at KEK, Japan, to study CP violation in rare B-decays and the mystery behind the matter dominated Universe. Several other institutions have now joined that collaboration, and it is continuing with the Belle-II collaboration, with vastly improved capabilities to study rare decays of B and D mesons, and tau leptons. In the original Belle experiment, the Indian groups participated in measurements and analysis of the large body of data. In Belle-II, the group led the design, prototyping, and construction of layer 4 of the silicon strip-based vertex detector.

India's participation in CERN is an outstanding example of science driven collaboration. India's relationship with the organization dates back more than five decades, initially through cooperation with the Tata Institute of Fundamental Research, whose high energy physicists have been actively participating in experiments at CERN since the 1960s, as students, post-docs, and guest scientists joining different programs at CERN soon after its formation. This matured as participation of the experimental high energy physics group from TIFR when they joined the L3 experiment and contributed hardware for the endcap hadron calorimeter and made contributions to the development of software, measurements, and physics analysis for the study of Z boson and W^+W^- physics. It is indeed to the credit of this group that the Indian team leader eventually became the spokesperson of the LEP upgrade L3 group. In the late 1980s another group from Variable Energy Cyclotron Centre (VECC) joined the SPS/WA80 group to study high energy heavy-ion collisions. The group was entrusted with the design and fabrication of a photon multiplicity detector (PMD).

CERN had already planned to set up the Large Hadron Collider (LHC), as a successor to LEP, and installed in the same 27 km tunnel to look for the Higgs boson along with the other electro-weak searches, with a possibility to have heavy-ion beams to look for the so-called quark–gluon plasma. Brookhaven National Laboratory had canceled ISABELLE for a 200 GeV + 200 GeV pp collider and decided to pursue a Relativistic Heavy Ion Collider (RHIC) to have pp and nucleus–nucleus collisions to study quark–gluon plasma. Several theoreticians had started working on the formation, evolution, signals, and properties of quark–gluon plasma in the country as well as on NNLO contribution to the cross-section for the production of the Higgs boson in pp collisions.

By 1990–91, a group led by the Variable Energy Cyclotron Centre at Kolkata along with researchers from Jammu University, Punjab University, and Rajasthan University (the Institute of Physics, Bhubaneswar also joined for the WA98) designed, developed, and fabricated a pre-shower photon multiplicity detector (PMD) used in the WA93 Experiment at CERN SPS to study S+Au collisions and this led to the discovery of elliptic flow. A much larger detector using 55 000 pads was used to study Pb–Pb collisions at the WA98 Experiment (figure 51.1) [13]. The support for the two detectors had been provided by Department of Science and Technology and Department of Atomic Energy.

Figure 51.1. Photon Multiplicity Detector: WA98 (1994, 55K cells) [13] (left to right: T K Ghosh, B Nandi, Y P Viyogi, S K Badyal, and S Raniwala).

While the PMD was being redesigned for use in the ALICE experiment at the Large Hadron Collider—where the pads were to be replaced by honeycomb proportional counters [14], the experimenters joined the STAR experiment at BNL, with the same conceptual design [15] and obtained valuable results from the study of fluctuations, flow, limiting fragmentation, and limits on the formation of disoriented chiral condensates, etc at forward rapidities in Au+Au collisions at 200 GeV/nucleon.

The PMD emerged as a major work horse in all high energy heavy-ion collision experiments, not only at CERN but also in BNL and USA/STAR.

51.2.1 A large ion collider experiment (ALICE)

ALICE is a detector dedicated to heavy-ion physics at the Large Hadron Collider (LHC). It is dedicated to studying the physics of strongly interacting matter at extreme energy densities in nucleus–nucleus and pp collisions, where a phase of matter called quark–gluon plasma is produced, as indicated earlier. By the time India joined the ALICE experiment, the India–CERN collaboration for it had expanded to include about 50 research scholars and 50 scientists from 14 institutes and universities. The much larger and complex PMD, with 200 000 cells of varying sizes—complete with readout employing Multiplex Analog Signal Processor (MANAS) chips, designed, and produced in India, were installed in the ALICE detector (figure 51.2). ALICE also has a forward muon spectrometer to study the complete spectrum of heavy quarkonia (J/Ψ, Ψ', Υ, Υ', Υ'') via their decay in the $\mu^+\mu^-$ channel. The Indian collaboration, led by Saha Institute of Nuclear Physics,

Figure 51.2. Photon Multiplicity Detector: ALICE (2009, 200 000 cells) [14].

Kolkata, designed, fabricated, installed, validated, and went on to take data and analyse the data from the second station of the ALICE dimuon spectrometer (figure 51.3). The total requirement (with spares) of 88 000 MANAS chips with 16-channel readout for the muon spectrometer and 20 000 chips for PMD was also met by India. Participation in these experiments also gave Indian researchers access to all the raw data measured at the ALICE experiment, which put them at the forefront of the search and study of quark–gluon plasma, now also seen in high multiplicity pp collisions.

51.2.2 The Compact Muon Spectrometer (CMS) experiment

The goal of the CMS experiment is to investigate a wide range of physics, including the search for the Higgs boson, extra dimensions, and particles that could make up dark matter. The TIFR Group, followed its success with the L3 experiment, joined the CMS collaboration around the same time, and were soon joined by several other groups. The detector also has excellent capabilities to study heavy-ion collisions, which provide independent confirmation for the formation of quark–gluon plasma, studied in ALICE experiment, as well at ATLAS and LHCb experiments. Indian

Figure 51.3. ALICE-India Muon Spectrometer (left to right: S Bose, D Azmi, S Chattopahyay, H Gutbrod, and B Sinha). Image copyright 2007–2023 CERN.

scientists from Bhabha Atomic Research Centre, Delhi University, Indian Institute of Technology, Bombay, National Institute of Science Education and Research, Bhubaneswar, Panjab University, Saha Institute of Nuclear Physics, Tata Institute of Fundamental Research, and Vishva Bharati University, Shantiniketan, played a significant role in the CMS detector, which is one of the two large experiments that have led to the discovery of the Higgs boson. The CMS–India collaboration consists of about 40 scientists/faculties and more than 50 PhD research scholars. The experiment is now concentrating on physics beyond the standard model, matter–antimatter asymmetry, dark matter candidates, and sources of dark energy. As indicated earlier, CMS has excellent possibilities to study the quark–gluon plasma created in nucleus–nucleus and even pp collisions and the group at Bhabha Atomic Research Centre has contributed extensively to the study of quarkonia suppression, considered to be one of the cleanest and definite signatures of quark–gluon plasma formation.

The India–CMS collaboration [17] was involved in two major hardware fabrications—for the outer hadron calorimeter (HO) and silicon pre-shower detector (Si-PSD). The Si-PSD played an important role in distinguishing photons coming from the Higgs and from neutral pion decay. The HO part of the detector helps in the precise measurement of the transverse momentum of invisible particles and jets. The India–CMS collaboration also built a considerable number of resistive plate chambers, which measure the timing to a precision of pico-seconds, track the muons, and help in triggering of the Z boson, and thus directly helped in the discovery of the Higgs boson, as Higgs decays to ZZ* and Z and Z* decays to two muons each. Further, India has already supplied 17 GEM detectors (and will supply more) which

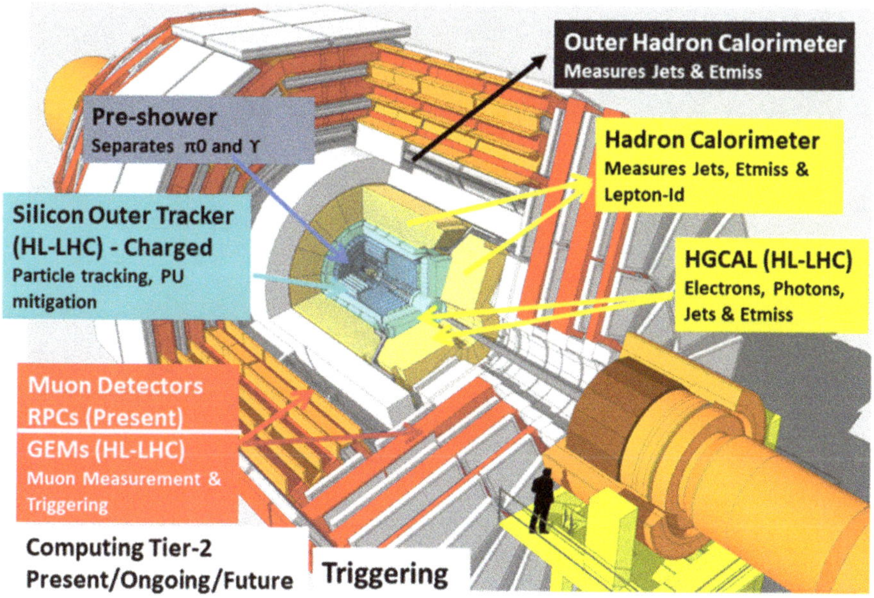

Figure 51.4. The main contributions of the India–CMS collaboration [16].

will help in understanding the properties of the Higgs boson and in the discovery of new particles and is participating in the High Granularity Calorimeter and Silicon Tracker Upgrade (figure 51.4).

The Delhi University researchers had earlier joined the E706 experiment at Fermilab to study direct photon production in the collision of protons and gluon structure function, supported by PL-480 and the Nuclear Science Foundation of the USA. In the late 1980s TIFR, Panjab University, and Delhi University joined the D0 experiment at Fermi Laboratory, which lasted for close to 20 years, and led to the discovery of the top quark and setting limits for the mass of the Higgs boson.

51.2.3 India–CERN collaboration in building the LHC

As the Indian high energy and particle physics community prepared to participate in major experiments at the LHC, the accelerator physics and technology community, led by the Raja Ramanna Centre for Advanced Technology (RRCAT), Indore, made significant contributions to the construction of the LHC accelerator, both in the areas of design, development, and supply of hardware accelerator components, and in software development and deployment in the LHC machine [17] (figure 51.5).

The most notable contributions included superconducting corrector magnets, precision magnetic positioning system jacks, accelerator protection systems, quench protection heater power supplies, quench detection electronics, control electronics for circuit breakers of energy extraction systems, vacuum system design for long

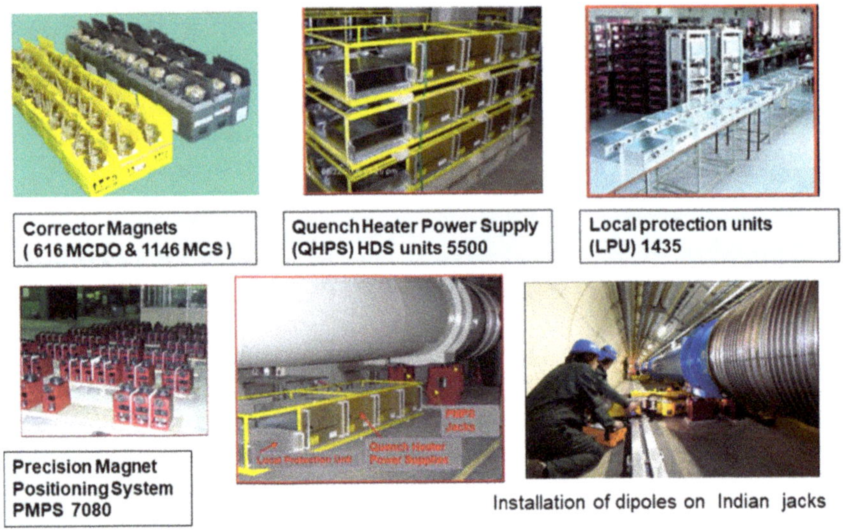

Figure 51.5. Main contributions of RRCAT Indore, Department of Atomic Energy to the LHC at CERN [17].

beam transport lines, and cryogenic systems. A team of operators and central control room supervisors worked at CERN for complete tests and measurements on the full series of magnets and gained valuable insight and experience. India actively participated in the development of LINAC4 for the luminosity upgrade of the LHC and in the Compact Linear Collider test facility project (CTF3) related to the post-LHC electron–positron linear collider program. Further, Indian industry has also taken new strides in the large-scale manufacturing and supply of high technology items meeting international standards to CERN.

These contributions of the RRCAT Indore, of the Department of Atomic Energy, were valued at about 47 million USD. In a unique arrangement, half of that amount was deposited in a CERN–India Fund, which was used to support the visits of Indian researchers to CERN and to buy equipment for their experiments. This proved to be valuable until the Department of Atomic Energy and the Department of Science and Technology, Government of India, started direct funding for CERN–India collaboration.

We have not covered several other collaborations, such as ISOLDE for radio-active ion beams, RD-51 for detectors, and medical imaging, etc, for want of space.

51.2.4 Computing

The ALICE and CMS experiments have several detectors and collect a vast amount of data (more than 100 Mbyte s^{-1} and requiring storage of 1.5 Pbytes per year) which needs to be analysed. Indian researchers contributed to designing, developing, and deploying software for the Worldwide Large Hadron Collider Grid (WLCG). Thus, 'Gridview' and 'MyWLCG' were developed to monitor the status of different components, services, and jobs in the WLCG. These software products are widely used by researchers at CERN as well as at 170 computer centers across 42 countries. Currently, India is engaged in contributing towards 'Cloud computing efforts' at CERN by developing Cloud management software and contributing to 'Openstack', in collaboration with CERN. 'Machine learning for job classification' is now under development. These efforts have contributions from various institutes of the Department of Atomic Energy, such as the Bhabha Atomic Research Centre, Indira Gandhi Centre for Atomic Research, Raja Ramanna Centre for Advanced Technology, Tata Institute of Fundamental Research, and Variable Energy Cyclotron Centre.

The Tier-2 Grid computing facilities at VECC and TIFR, and Tier-3 computing facilities at participating institutes, connected by the 10 Gbytes s^{-1} National Knowledge Network have met the pledged resources and have operated with 96% uptime, thereby facilitating the running of jobs by ALICE, CMS, and WLCG collaborations spread all over the world. The CERN–TIFR optical fiber link plays a key role in data dissemination from CERN to Indian institutes. The Tier-2 facility at VECC is one of the top super computers of India, which has been operating since 2002 and now provides 4000 cores of computing and 1.5 Pbytes of storage.

The activities discussed above were started with India and CERN signing a Cooperation Agreement in 1991. In 1996, the Atomic Energy Commission decided to take part in the construction of the LHC, and to contribute to the CMS and ALICE experiments and to the LHC Computing Grid with Tier-2 centers in Mumbai and Kolkata, which we have discussed briefly above. The significant contribution of Indian researchers in accelerator technology, computing, detector fabrication, data-taking, and analysis earned the status of Observer for India to the CERN Council in 2002.

By then India was firmly progressing to become one of the largest economies of the world and a powerhouse of young talent in information technology, sciences, and manufacturing involving high technology. India was being recognized across the world as an immense resource of experienced and high caliber professionals in all spheres of knowledge and technology. India has had a strong focus on science and technology, right from the days after independence, and had a strong realization that scientific research is a key element of economic growth. This was propelling India to a position among the topmost countries in the world in the field of scientific research. CERN collaboration is but one example of India contributing to the enterprise of the advancement of basic science via participation in the front-ranking worldwide efforts.

India became an Associate Member of CERN on 16 January 2017.

51.2.5 Astronomy and astrophysics

Any chronicling of Indian activities in high energy nuclear and particle physics will remain incomplete without a reference to associated Indian activities in observational astronomy and astrophysics. This was also greatly facilitated by the transfer of skills, and skills acquired in the above studies.

As was mentioned earlier, India has always had a tradition of research in astronomy. We have already mentioned some of them at the beginning of this chapter. During the early eighteenth century Sawai Raja Jaisingh-II (1688–1743) established five large observatories (Jantar Mantars) in different parts of India to accurately record the movements of celestial objects and to correct existing tables. Chandrashekhar Samanta (1835–1904), better known as Pathani Samanta [18], belonged to the tradition of naked eye astronomers and mathematicians of India. He lived in a remote, rural, hilly part of Odisha and had access to only ancient Sanskrit texts on astronomy and mathematics. With homemade instruments using bamboo and wood, he used to observe the motion of planets and went on to correct several existing tables.

Starting with a cylindrical paraboloid radio telescope in Ooty built in 1970, the National Centre for Radio Astronomy (NCRA) of TIFR built a unique facility, the Giant Metre wave Radio Telescope (GMRT) at Pune, during 1980s for radio astronomical research. The GMRT has been upgraded recently (and rechristened as uGMRT) and is run as an international facility to study the formation of galaxies in the early Universe. Astronomers from across the world congregate there to study galaxies, pulsars, supernovae, and Sun and solar winds. It went on to discover one of the most distant galaxies—which is about 12 billion light years away—and helped in the observation of the biggest explosion in the history of the Universe, the Ophiuchus Supercluster Explosion.

Another radio-observatory has been operated by the Raman Research Institute and Indian Institute of Astrophysics (IIA) at Gauribidanur, near Bengaluru since 1976.

A solar observatory was built at Udaipur in 1976 and has been studying solar oscillations. The Udaipur Observatory is a part of the Global Oscillation Network Group. Several other solar observatories have also been built in India in the last few decades. IIA runs solar observatories in Kodaikanal, Kavalur, and Hanle. The Hanle telescope (The Himalayan Chandra telescope) is a 2.01 metre class telescope operating in the optical and infrared wave bands. The HCT can be operated remotely from Bangalore using a dedicated satellite link.

The Aryabhatta Research Institute of Observational Sciences (ARIES) at Nainital, which started in 1950s, operates several optical telescopes, including one 3.6 metre optical telescope at Devasthali. ARIES carries out research on the formation and evolution of the Sun, stars, and galaxies, in addition to studying air pollution and climate change on Earth.

A High Energy Gamma Ray Telescope (HAGAR), capable of detecting 250 GeV photons has also been built in Hanle. A Major Atmospheric Cherenkov Experiment Telescope (MACE) has been built in Hanle, for studying astrophysical sources in energies of the 20 GeV to 10 TeV range by detecting Cerenkov radiation.

The research capabilities built by India over the last few decades in the areas of accelerators, detector systems, astronomical observatories in different wavelength regions, information and Computer technology capabilities to move, model, and analyse large volumes of data give the country unique opportunities to explore regions beyond the known horizons.

51.3 Spreading its wings

India of the twenty-first century is a country in transition—transition from a developing country to a developed country, transition from a poor country in which rich also reside to a rich country in which poor also reside. The growing participation of India in international initiatives in frontier areas of research amply reflects this ambition. As was mentioned earlier, Indian scientists have been participating in several international mega science projects for a long time. However, such participation was based on individual commitments. India's participation in CERN's Large Hadron Collider was the first national commitment towards an international mega science project [19].

The experience and confidence gained, not only by India but also the partnering countries, has led to India's participation in several mega science and technology projects beyond the CERN collaborations.

One of the most ambitious research and development collaborations that India has joined as a partner (in 2005), is the International Thermonuclear Experimental Reactor (ITER) to build and demonstrate the scientific and technological feasibility of fusion energy. India would provide the cryostat (termed as the largest thermos in the world and a 30 m tall and 30 m diameter outer vacuum shell), in-wall shielding, cooling water systems, cryo-lines and cryo-distribution systems, ion-cyclotron RF heating systems, electron cyclotron RF heating systems, diagnostic neutral beam systems, multi-megawatt power supplies, and several ITER diagnostic systems.

India is also one of the early partners to join (in 2010) the GSI Darmstadt Facility for Antiproton and Ion Research (FAIR) with a focus on basic investigations, not only in the field of nuclear physics and atomic physics, but also in several other application-oriented research areas in materials research, plasma physics, biophysics, and nuclear medicine. India's contributions will include costs towards building the accelerator and also experimental facilities such as the muon arm of the Compressed Baryonic Matter (CBM) experiment, Nuclear Structure, Astrophysics and Reactions (NUSTAR) experiment, and the plans for the Antiproton Annihilation at Darmstadt (PANDA) experiment.

To study objects in our own Solar System and stars throughout our Milky Way and its neighboring galaxies, as well as forming galaxies at the very edge of the observable Universe, near the beginning of time, India joined the Thirty Meter Telescope (TMT) Collaboration in 2010 as an Observer Member and, in 2014, as a Full Member.

After years of involvement in its design, India formally joined the Square Kilometre Array (SKA) project in 2015, to study gravitational waves, test the general theory of relativity, and investigate the evolution of the Universe.

The Laser Interferometre Gravitational-wave Observatory (LIGO)–India is planned to be an advanced gravitational-wave observatory as part of the worldwide network. With its addition, the number of observed events producing gravitational waves being captured is likely to increase by a large factor and their origin is likely to be determined with dramatically improved accuracy. It is envisaged as a collaborative project between a consortium of Indian research institutions (Institute of Plasma Research, Inter University Centre for Astronomy and Astrophysics, and Raja Ramanna Centre for

Figure 51.6. Understanding the Universe through Mega Science Projects [20].

Advanced Technology) and the LIGO Laboratory in the USA, along with its international partners. It is supported by Department of Atomic Energy and Department of Science and Technology, with formal approval being obtained in 2016.

A unique India-based Neutrino Observatory (INO) is to be constructed in South India to study neutrinos from various natural and laboratory sources using an iron calorimeter detector. It is envisaged that such an underground facility will develop into an international centre for other studies as well, in physics, biology, geology, etc, all of which will make use of the special conditions that exist deep underground. INO is also expected to benefit from the participation of Indian researchers in the Long-Baseline Neutrino Facility (LBNF) at the international Deep Underground Neutrino Experiment (DUNE) of Fermilab.

As can be seen from figure 51.6, the bouquet of scientific collaborations [20] is not a set of randomly chosen scientific projects but spans the entire time scale of the evolution of the Universe, from the Big Bang to the present.

51.4 Epilogue

One of us (VSR) was born when India was still under colonial rule. When India became independent in 1947 we were a fledgling democracy facing acute food shortages and weak infrastructure on all fronts. It was often said that India had to import everything—from a pin to a plane. The educational infrastructure was weak as well and a handful of researchers, fired by a missionary zeal, continued to carry out research with very limited resources. VSR still remembers his early days in nuclear research when the required instruments had to be rigged locally, often scavenging components from World War II rejects or scraps. It thus gives us immense satisfaction that India today not only participates as an equal partner in many projects at the forefront of scientific research, but also contributes cutting edge equipment designed, developed, and manufactured in India.

We have seen that the participation of Indian scientists in CERN programs started as early as the 1960s. By the 1990s it had expanded to building large detectors and contributing to the construction of the Large Hadron Collider, which culminated into India becoming an Observer to the CERN Council in 2002 and becoming an Associate Member in 2017. It is a fascinating story of scientific research in India, moving from strength to strength.

The exemplary contribution of Indian scientists and engineers at CERN gave them the confidence to face the world on their own terms and gave confidence to the world in the capabilities and commitment of Indian scientists, engineers, and governmental organizations to Big Science. This paved the way for India's participation at FAIR, ITER, TMT, SKA, and LIGO, and for planning the large India-based Neutrino Observatory.

CERN started as a European entity—the other collaborations are from all over the world. These have nurtured generations of brilliant students who are working successfully in a highly creative, competitive, and frontier area of science and technology in a multinational, multicultural environment. They have, by now, developed wonderful networks of their own of colleagues and collaborators across the nations of the world, which, after all, remains one of the greatest rewards of doing science.

These collaborations strengthened the interaction of Indian scientists and engineers with Indian industries which excelled and rose to the occasion and developed cutting edge technologies, best seen in the construction of components for the LHC mentioned above, the ITER cryostat—the largest stainless steel high-vacuum pressure chamber ever built (16 000 cubic meters)— to provide the high-vacuum, ultra-cool environment for the ITER vacuum vessel and the superconducting magnets, and the components of LIGO–India which will include enormous ultra-high vacuum chambers and very stable and precision laser sources and detectors. Several hundred precision power converters for FAIR have been made by Indian industry. FAIR will be one of the important international experimental facilities, partly owned by India, with extensive contributions to the Compressed Baryonic Matter (CBM) and Nuclear Structure and Reaction (NUSTAR) experiments, and the accelerator itself.

One aspect we have not mentioned so far is the stirring of emotions, excitement, and curiosity for science which CERN continues to evoke among Indian students and even common people. The excitement preceding the start of LHC and the announcement of the discovery of the Higgs boson, contributed a lot more to science outreach than years of efforts put together. This blossoming of love for basic research in young minds is perhaps one of the best outcomes of the engagement with CERN. We note with satisfaction that CERN has been welcoming summer interns from India for years. Several of these young students have realized their dream of working at CERN.

Participation in collaborations at CERN has also provided exposure to an entirely new culture of doing science, where several thousand scientists and engineers work for decades to build the equipment, perform the experiment, analyse the data, and examine every word of the manuscripts before they are finally submitted for publication. The research papers for the discovery of Higgs boson had more than 5000 authors each and those on the studies of quark–gluon plasma had more than 1000 authors from across the world.

These large number of scientists working so closely together are one of the best hopes for mankind, which now faces the extreme danger of global warming, which has been predicted for years by hundreds of scientists toiling together to prepare and regularly update the Intergovernmental Panel for Climate Change reports, which the world continued to take lightly in the early days, and which in turn has only exacerbated the situation. We hope that some of the concepts such as the Accelerator Driven Subcritical Systems developed at CERN can be realized soon and become a potent instrument in our fight against this crisis facing the entire mankind.

We add that, Indian traditions have always considered the 'entire world as a family' ('वसुधैव कुटुम्बकम्'—*vasudhaiva kutumbakam*), which truly represents the philosophy of cooperation and collaboration cutting across borders of all kinds. The artistic traditions of India have celebrated the Creation of the Universe in the 'Ananda Tandava', the Cosmic Dance of Creation by Nataraja. The installation of a statue of Nataraja performing the Cosmic Dance of Creation at CERN is a fitting tribute to this tradition (figure 51.7). It is also a matter of great satisfaction that the

Figure 51.7. The world's tallest free-standing statue of Nataraja, performing the Cosmic Dance of Creation, 'Anand Tandava', gifted to CERN by India, 18 June 2004 (left to right: Anil Kakodkar, Robert Aymer, and K M Chandrashkhar). Copyright CERN 2004-2020 - CERN Publications, IR-ECO-CO

names of several of the Indian researchers mentioned in the prologue appear on the sculpture at the Globe of Science and Inventions at CERN.

Acknowledgments

The authors would like to express their gratitude to Subhashis Chattopadhyay, Sukalyan Chattopadhyay, Brajesh Choudhary, Tapan Nayak, Sreerup Raychaudhuri, Vikas Singhal, Arun Srivastava, Purushottam Srivastava, and Yogendra Pathak Viyogi for sharing information while preparing this article. The support of the Department of Atomic Energy and Department of Science and Technology was crucial for these endeavors. We also thank the Department of Atomic Energy for their kind permission to use figure 51.6.

References

[1] Kochhar R K 1991 Science as a tool in British India *Econ. Polit. Wkly* **26** 1927–33

[2] Emerson D T 1997 The work of Jagadis Chandra Bose: 100 years of millimeter-wave research *IEEE Trans. Microw. Theory Res.* **45** 2267–73

[3] Saha M N 1920 Ionization in the solarchromosphere *Phil. Mag. Ser. 6* **40** 472–88
Saha M N 1921 On a physical theory of stellar spectra *Proc. R. Soc.* A **99** 135–53
Saha M N 1919 On radiation-pressure and the quantum theory *Astrophys. J.* **50** 220
See also; Ajayan P, Kim P and Banerjee S 2016 Two-dimensional van der Waals materials *Phys. Today* **69** 38

[4] Bose S N 1924 Plancks Gesetz und Lichtquantenhypothese *Z. Phys.* **26** 178

[5] Raman C V and Bhagavantham S 1931 Experimental proof of the spin of the photon *Indian J. Phys.* **6** 353
Raman C V and Bhagavantham S 1932 Experimental proof of the spin of the photon *Nature* **129** 22–3

[6] Raman C V and Krishnan K S 1928 A new type of secondary radiation *Nature* **121** 501–2
Raman C V 1928 A new radiation Indian *J. Phys.* **2** 387–98
Raman C V and Krishnan K S 1928 A new class of spectra due to secondary radiation. Part I *Indian J. Phys.* **2** 399–419

[7] Bose D M and Chaudhuri B 1941 Origin and nature of heavy ionization particles detected on photographic plates exposed to cosmic rays *Nature* **147** 240–1
Bose D M and Chaudhuri B 1941 A photographic method ofestimating the mass of mesotron *Nature* **148** 259–60
Bose D M and Chaudhuri B 1942 A photographic method of estimatingthe mass of mesotron *Nature* **149** 302
See also; Roy S C and Singh R 2015 D M Bose and cosmic ray research in India *Ind. Jour. Hist. Sci.* **50** 438
Lattes C M G, Muirhead H, Occhialini G P S and Powell C F 1947 Processes involving charged mesons *Nature* **159** 694–7

[8] Raychaudhuri S 2016 The early days of particle physics in India *Proc. of CU Phys.* 100

[9] Nehru J 1946 *The Discovery of India* (Calcutta: Signet) pp 452–3
See also; Mahanti S 2016 Nehru's vision of scientific temper *J. Sci. Temper* **4** 154–66

[10] Constitution of India, Article 51 A(h) https://www.constitutionofindia.net/constitution_of_india/directive_principles_of_state_policy/articles/Article%2051

[11] Bhabha H J 1936 The scattering of positrons by electrons with exchange on Dirac's theory of the positron *Proc. R. Soc. Lond.* **A154** 195–206

Bhabha H J and Heitler W 1937 The passage of fast electrons and the theory of cosmic showers *Proc. R. Soc. Lond.* A **159** 432

[12] Achar C V *et al* 1965 Detection of muons produced by cosmic ray neutrinos deep underground *Phys. Lett.* **18** 196

Achar C V *et al* 1965 Observation of a non-elastic cosmic ray neutrino interaction *Phys. Lett.* **19** 78

[13] Aggarwal M M *et al* 1996 A preshower photon multiplicity detector for the WA93 experiment *Nucl. Instr. Meth* A **372** 143

Chattopadhyay S, Ahammed Z and Viyogi Y P 1999 Application of neural network for photon–hadron discrimination in a preshower detector in high-energy heavy-ion collisions *Nucl. Instr. Meth.* A **421** 558

[14] Schukraft J *et al* 1999 *ALICE Technical Design Report: Photon multiplicity detector* CERN-LHCC-99-32, CERN-OPEN-2000-184 CERN

[15] Aggarwal M M 2003 The STAR Photon Multiplicity Detector *Nucl. Instr. Meth.* A **499** 751–61

[16] Banerjee S 2019 Collider experiments and India *International Meeting on High Energy Physics (Bubaneswar)*

Chaudhury B C 2020 India's participation in CERN-LHC program *Vigyan Samagam (New Delhi, January)*

[17] Sahni V C 2004 Indian contribution to the Large Hadron Collider under construction at CERN, Geneva *Curr. Sci.* **87** 441

Shrivastava P 2019 Indian contributions at CERN *Vigyan Samagam (Bengaluru, September)*

Singhal V 2020 Grid computing for CERN-LHC collaboration *India International Science Festival*

[18] P W E 1899 Siddhánta-Darpana; a treatise on astronomy *Nature* **59** 436–7

1899 Siddhánta-Darpana; a treatise on astronomy *Knowledge* 22 257–8

Naik P C and Satapathy L 1998 Samanta Chandra Sekhar: the great naked eye astronomer *Bull. Astr. Soc. India* **26** 33–49

[19] Ramamurthy V S 2011 Global partnerships in scientific research and international mega-science projects *Curr. Sci.* **100** 1783

[20] Sriavastava A, Bheesette S and Biswas A 2020 From micro to macro: understanding the universe through mega science projects *Vigyan Samagam 2020–2021* L-89320/2020

IOP Publishing

Big Science in the 21st Century
Economic and societal impacts
Panagiotis Charitos, Theodore Arabatzis, Harry Cliff, Günther Dissertori, Juliette Forneris and Jason Li-Ying

Chapter 52

LIGO-India in the global Big Science landscape

Sukanta Bose

52.1 Introduction

The objective of the LIGO-India project is to build and commission a Laser Interferometer Gravitational-wave Observatory (LIGO) (Aasi *et al* 2015) detector in India. Thereafter, the observatory will enter an operations and management (O&M) phase when it will conduct observation runs, often in coordination with the two LIGO detectors in the US and other kilometer scale-detectors such as Virgo in Italy (Acernese *et al* 2014) and KAGRA in Japan (Akutsu *et al* 2020). Four institutions in India have been charged with the responsibility of delivering the project in its first 5 years. These are the Raja Ramanna Centre for Advanced Technology (RRCAT), Indore, Institute for Plasma Research (IPR), Gandhinagar, Directorate of Construction, Services and Estate Management (DCSEM), Mumbai, and Inter-University Centre for Astronomy and Astrophysics (IUCAA), Pune. Through a bilateral memorandum of understanding (MoU) signed between the Indian funding agency the Department of Atomic Energy (DAE) and its US counterpart the National Science Foundation (NSF), these four institutes will collaborate with their American counterparts at Caltech and MIT. A second funding agency in India, namely, the Department of Science and Technology (DST) is a 50% partner with DAE in LIGO-India. In the O&M phase, it will be the responsibility of IUCAA, which is supported by the University Grants Commission of India, to exploit the observatory to support the science of stakeholders in India and abroad. While the project received in-principal approval from the government of India in 2016, soon after the first ever direct observation of gravitational waves was announced by LIGO and Virgo, it is awaiting budget sanction as of early-2022.

The site for the LIGO-India detector has been acquired in the town of Aundha—near Nanded city in the Indian state of Maharashtra. Like other LIGO detectors this one will also be a Michelson interferometer, with Fabry–Perot cavities in the

doi:10.1088/978-0-7503-3631-4ch52

two 4 km long arms perpendicular to each other, in the shape of the letter ‘L’. The choice of the site was principally governed by the requirement that it be seismically quiet.

The Indian team will build the detector and the complete civil and vacuum infrastructure. The latter includes building a vacuum system, with ultra-high vacuum chambers and tubes capable of achieving and sustaining a vacuum at the level of a trillionth of a bar. The optics and the laser will be contributed by the US. The aim is to reach a detector strain sensitivity of 1 part in a 100 sextillion. Once built, LIGO-India will carry out gravitational-wave observations in coordination with its US and international science partners.

52.1.1 LIGO-India: one of several Big Science projects in India

India is contributing to over half-a-dozen Big Science projects. The projects covered by the DAE-coordinated Science Fair in 2019 and 2020 list seven such projects (India’s Mega Science Exhibition n.d.). A good number of them will contribute to establishing the full-scale ecosystems in science and technology research that the population of a country such as India deserves. Among those the India-based Neutrino Observatory (INO) and LIGO-India will be constructed on Indian soil. INO is a particle physics project under construction with the aim of studying atmospheric neutrinos and measuring the neutrino mixing parameters. It is a multi-institute collaboration and one of the largest experimental particle physics projects undertaken in India.

LIGO-India is the other project being constructed on Indian soil. The tremendous advantage that this affords is that it provides the scope for training a variety of personnel in a wide spectrum of skills. Soon after the in-principal approval was granted, the DAE nurtured a technology development and capacity building project through RRCAT to carry out the required preparatory work at the three lead institutes RRCAT, IPR, and IUCAA. Under this project, IPR has completed the fabrication of prototype one-to-one size BSC and HAM ultra-high vacuum chambers for the LIGO detector’s vacuum system. A set-up for steel out-gassing measurement has also been set up by IPR. Other vacuum accessories such as cryo-equipment, gate valve, chambers, etc, are being procured. The design and development of the laser, optics, and vacuum system for setting up of a 10 m arm-length prototype interferometer at RRCAT for training purposes is progressing. The development of a narrow line-width laser system required for the 10 m prototype has been completed and the vacuum system has been fabricated. Construction of the LIGO Project Testing & Training facility at RRCAT has been concluded as well.

52.2 How does India’s participation help science and technology, for LIGO and beyond?

With multiple gravitational-wave (GW) detections already accomplished, our ability to discern signals from noise has also improved. Therefore, initial insistence on signals appearing loudly in at least two detectors before it can qualify as a detection is ebbing. The near identical orientation of the two LIGO detectors in the US was partly to ensure that their antenna patterns are similar so that the signals observed by them from a common source have similar strengths, which in itself is a signal consistency test that improves our confidence in them not being noise-related. With a third LIGO detector, in India, the antenna patterns are much different. At this stage in GW astronomy, this is a difference that we can not only afford but also demand: complementarity of the antenna patterns expands discovery potential by allowing for better sky localization, which aids in multi-messenger astronomy and measuring signal polarization. The last measurement is useful for testing general relativity (Saleem *et al* 2022).

Additional detectors in the global GW network allow for multiple types of redundancies. New technologies can be tested by staging them strategically so that the downtime of detector baselines is mitigated. One of the scientific targets of GW detectors is to observe the GW signal from a post-merger object that results from the collision of two neutron stars at the end of their binary inspiral. This signal occurs at frequencies determined by the neutron star equation of state but is expected to be at over a kilohertz where the detector strain noise is dominated by photon shot noise and worsens with increasing frequency (Bose *et al* 2018). There are proposals for using squeezing to reduce the noise at high frequencies so that the post-merger signal enjoys a better chance of showing up. However, any reconfiguration of detectors in this way comes at a cost. Ideally, we would like to see the inspiral part of the binary signal with a significant signal to noise ratio as well since observing it jointly with a post-merger signal will inform us more precisely about the equation of state. An additional detector in the network makes it much more compelling to get one of the detectors to be tuned to best catch the post-merger signal.

Expanded observatory footprints that are being limited by the size of the Earth are among the criteria that define Big Science in this century. By allowing the gravitational-wave observatory network to localize sources better, the addition of detectors such as LIGO-India enable multi-messenger astronomy, thereby allowing us to subject the same astronomical sources to multifarious analysis and extract maximal science from it. More detectors also increase the duty factor, or observation time, and improve the detectability of a large subset of the sources. Both of these factors will enhance our detection rates, which have not only triggered the development of smarter detection and source characterization algorithms and analysis strategies to keep up with bigger volumes of data, but have also spawned new dimensions of scientific pursuits—instead of being limited to understanding individual sources astronomers are already preparing to peer into large populations of binary black holes to understand a variety of problems, such as the stellar evolution history, the expansion rate of our cosmos (by using those binaries as standard 'candles'), and testing general relativity (since by smartly combining multiple events one can improve the sensitivity of our observations to any subtle deviations from its predictions).

LIGO-India will be a close replica of the LIGO detectors in the US. Even so, as the detector systems are being installed, a fresh start, a new location, and a new team allows for new ideas and possibilities to emerge. To wit, hitherto unknown sources of noise artifacts could emerge. Furthermore, there are instances of unbudgeted or unaccounted for excess noise in LIGO detectors that could arise from ageing detector components or suboptimally configured installations. Such noise may not show up in a new installation of detector components. Whenever such differences are found, they can help troubleshoot their causes and improve the overall network sensitivity to GW sources.

There are several ways in which a subset of technologies developed for LIGO-India can be transferred to other research groups and the industry. This process of technology incubation is expressly supported by the entities in the Indian government, as noted in the DAE-organized Mega Science Exhibition in 2019–20 (India's Mega Science Exhibition n.d.). Umpteen examples exist of such transfers in India. One relatively recent example is the commercialization of the lithium-ion battery technology developed by the Indian Space Research Organization (ISRO) for use in electric vehicles, as facilitated by a panel of the Indian government in 2018 (Gupta 2019). Another example is a green-light reflecting materials coating that was developed at the Central Scientific Instruments Organisation (CSIO) in Chandigarh, India, before being transferred to Bharat Electronics Limited for commercialization. This technology was originally used for the 'Head Up Display (HUD)' for India's indigenously developed combat aircraft Tejas (Tribune News Service n.d.) and allows pilots to see various flight parameters on the windshield without obstructing their view ahead or forcing them to look down at dashboard panels. It is now being developed further for applications in avionics and elsewhere, e.g. in visors, wherever see-through visual displays help mitigate disruption in primary viewing.

Coatings are important for LIGO too, where they are used to turn the test masses into highly reflective mirrors. However, the random atomic and molecular vibrations in the micron-thin coatings due to thermal noise interfere with the detectors' ability to measure the strain in the detector arms precisely and hurt its sensitivity to gravitational-wave signals (Harry *et al* 2002). Therefore, a thrust of LIGO research is to find coatings that have lower thermal noise but without sacrificing other attractive features of current coatings (Amato *et al* 2021, Vajente *et al* 2021). Some of these properties are desirable in many other experiments outside of LIGO, such as high-resolution microscopes and profilometers, where technology transfer can happen in India and abroad.

Apart from the above examples, there is a host of additional ones where this project can contribute towards India's self-reliance and to the world's supply chain. By its very nature of being a precision measurement system, with thousands of servos and control loops, LIGO-India will contribute to low-noise electronics, high-precision servo controls, ultra-stable high-power lasers, high-tech vibration isolation techniques, high-precision metrology, etc (Unnikrishnan 2013). Some of the ideas here may not be completely novel, but in many cases they have new elements that are tailor-made for GW observatories with their unique requirements, e.g. on broad-band noise mitigation.

LIGO-India can help spawn research in multiple experimental groups in India not only for the first phase of the projects but also for upgrades and next generation technology development. A recent example of this kind is GMRT (Gupta *et al* 2017), which got scaled up to uGMRT or upgraded-GMRT. Here it is important to be clear that at the moment there are no plans for any upgrades in LIGO-India beyond the 'A+' configuration. This is partly because it is important to ensure that we accomplish this leg successfully first. However, it does not preclude future improvements akin to uGMRT. Now, the worldwide GW community has investigated possible future generation detectors. Some would argue that the closest of these for LIGO-India would be the 'Voyager' configuration (Adhikari 2020), which entails cryogenics, changes in laser, optics, etc, but utilizing the same facilities and the vacuum tubes along the arms. India could contribute here in multiple ways. In regards to detector components, there are research groups in India with expertise in fabricating specialized optical fibers with low-noise and other desired characteristics for producing ultra-stable high-power lasers at wavelengths suitable for mirrors in cryogenic settings (CSIR-Central Glass and Ceramic Research Institute n.d.). These solutions will be useful worldwide for any project that decides to use Voyager-like upgrades.

Indian experts in wind-engineering are studying the optimal geometries of site structures that will reduce wind load on the observatory building, which is coupled to the various detector systems it houses. Wind shields and breaks are being studied and modeled for optimal porosity, height, and distance from observatory structures so that wind-induced noise in the detector can be mitigated. This work employs computational fluid dynamical modeling and wind-tunnel experiments. While in some cases these shields can be artificial structures, in other cases they can be rows of vegetation and trees planted with a planned grade in heights, e.g. to guide the wind over the observatory structures. A set of our findings can help other observatories—those that have already been constructed—as well, since many of these installations are external to the facility. In certain cases, small superficial changes to buildings can reduce wind noise. Wind shields and breaks will be easy to install in those cases too.

Wind is not the only source of environmental noise for us. The observation band of LIGO detectors is above 10 Hz, but high *seismic* noise at lower frequencies can affect the detector noise power at higher frequencies. While various technologies, such as multi-stage suspensions, are used to aid seismic isolation of the detector test masses, an idea that has not been implemented yet is the use of metamaterial pile-arrays embedded in the ground around the observatories to damp the low-frequency surface waves associated with ground vibrations. Some of this noise is produced by on-site heating, ventilation, and air-conditioning systems. Strategically selecting the pile material and embedding the piles can utilize their phononic band-gaps to effect such damping of the noise generated. Finite element modeling and simulations suggest a reduction of around 94% to 99% in incident wave magnitude in the 0–20 Hz band (Balasubramaniam and Rajagopal 2021, Thippeswamy *et al* 2021). Work is ongoing to probe its effectiveness in realistic situations. Similarly, it is yet to be explored how effective such structures would be in attenuating earthquakes and reduce loss of interferometer locks.

52.3 Human resource development: building the user community

For students who will be tomorrow's torch bearers in India's scientific progress, there is no substitute for hands-on experience in today's experiments that make exciting discoveries happen. Collaborative arrangements with laboratories abroad give valuable exposure but to a limited number of students and researchers. A large experiment in one's own country changes that radically. It allows for establishing ecosystems where year on year students and advisors from institutions nationwide can participate in sustained research programs once the observatory has been constructed and dedicated to the nation. Also, given the young average age of the Indian population, it is crucial that there are sizeable opportunities available for training our youth today in skills and knowledge relevant for jobs tomorrow. Avant-garde projects such as LIGO-India are among such crucibles.

The first question I heard a senior science policymaker ask me in 2014 in regards to funding LIGO-India was the disarmingly simple, 'Where are the users?'. As many in this business know, protesting how anyone could expect any users when the project was still awaiting approval would be understandable, but particularly unwise with such an audience. Encountering these 'chicken-and-egg' problems in jump-starting projects of this magnitude is not new. Imagine physicists in the eighties asking for US federal funding of hundreds of millions of dollars for building LIGO when it was not clear at all how frequent binary black hole or neutron star mergers are. Fortunately, there were takers for the idea that developing a completely new way of observing the Universe is compelling enough a reason to fund it. Thankfully, there is now a growing list of technology transfers from GW experiments that is making this noble support of the taxpayer an even more rewarding pursuit. As I describe below, starting around 2010 we had managed to grow a core group of a few dozen scientists—most already contributing to LIGO and Virgo. While that community would be important to engage the Indian policymakers further, we also had to identify what the scope of growth was in another 5 and 10 years, where would those people come from, and what other existing communities and industries would be interested in and benefit from LIGO-India? We were able to answer all these questions, but none of our responses would be convincing if we had not built that core group and used it to argue that it was the members of that group who would effectively act as a nucleation center by carrying out all of those human resource development activities.

Gravitational-wave research in India has been pursued since at least the late 1980s. Early on this was limited to mostly data analysis and binary waveform modeling research at IUCAA and the Raman Research Institute (RRI), and to a limited extent at RRCAT—on lasers and optics. A renewed Indian interest emerged in 2008 when it was recognized in LIGO (US) that it made much more scientific sense to relocate one of its three detectors—in particular, the second detector in Hanford—to somewhere across the globe so that the resulting long baselines could be utilized for better time-delay triangulation and, therefore, better localization of GW sources in the sky. This in turn aids multi-messenger efforts where by combining GW, electromagnetic (EM), and even particle (e.g. neutrino) emissions from the source we learn much more about the progenitor, its environment and, indeed, the cosmos itself (e.g. by utilizing both the GW distance measurement and EM redshift determination to measure the expansion rate of the Universe via the Hubble parameter).

These scientific prospects were very attractive to the fledgling but seriously interested community of Indian relativists and high-energy astrophysicists. Therefore, when the Australian gravitational-wave community was seeking funding for locating the third LIGO detector in that continent in 2009–10, India offered to support it as a minority international partner. The Australian tryst and the upcoming Advanced LIGO and Advanced Virgo commissioning had by now captured the attention of the science leaders in India. In parallel, the concerted effort stemming from the LIGO–Virgo collaborations for engaging with astronomers worldwide in preparation for electromagnetic follow-up of GW events expected in the Advanced LIGO–Virgo observation era, slated to begin soon after 2014, helped in galvanizing a section of the Indian community as well. When the Australian proposal did not meet with success, the Indian community already had the preparations required to make a bid for its own proposal. Several workshops and international conferences were hosted in India in the early 2010s to showcase the worldwide support we had for building a LIGO detector here. In the end, it was the first direct observation in September 2015 of gravitational waves from a binary black hole merger that led in February 2016 to an official announcement of in-principal approval for LIGO-India.

In the years between 2009 and 2014, the Indian participation in LIGO grew from a couple of authors to around a few dozen authors. While this is small compared to the LIGO author list in its first science run papers in 2003, it was still significant owing to the fact that this community was built with no formal recognition yet for the LIGO-India project. Growing it clearly benefitted from the fact that the LIGO–Virgo detectors were already acquiring and analyzing science data. This story highlights the fact that to make the first discoveries one has to take chances, which LIGO, followed by Virgo, took in good measure. Developing countries may not be able to afford it in the same numbers. Embarking on a project like the Superconducting Super Collider (Cramer 1997) and then terminating it midway is unthinkable in most countries. However, there are many examples of scientific pursuits where the risk-to-benefit ratio is quite small. Essentially all mega-science projects India is participating in, as noted above, would qualify in this category. In that sense, India's scientific and political leadership is less averse to taking calculated risks in its investments in science and technology. To that extent, it is creditable that scientists and science bureaucrats in this country had confidence in this emerging area and began investing in it even when not a single direct detection had been made of GWs.

52.4 The role of charity

Interestingly, LIGO-India has benefitted immensely from philanthropy. The author is aware of at least one charity trust, belonging to the Tatas (Tata Trusts n.d.), who provided funds that were critical to seed the bilateral engagement that was significant in building our community. It supported the travel of scientists to the LIGO sites in the US for hands-on training, conducting ‘town hall’ meetings for advertising the project, and hosting training workshops. The flexibility in what activities the funds can be legitimately used for (within the broad parameters of the funding agreement) is most useful, especially in the initial phase of this kind of a project.

Sometimes the extant rules and stipulations of funding bodies prove too restrictive for seeding a new project. Such scenarios arise naturally because it can be stifling to encumber a project with too many constraints while they are being conceived. Once formulated, funding and infrastructural (e.g. availability of land in the right location) bottlenecks are likely to come to light. A repartee one sometimes encounters to this predicament is that if the experiment is sound enough, these difficulties would not be unsurmountable. Even when this is true, it takes time for the bureaucracy to adapt to support a project. In that teething phase, even a fraction of the project budget proves extremely valuable to initiate critical activities, such as identifying potential sites for an observatory, growing the user community, or liaising with the industry to explore the scope for technological spin-offs or commercial applications. This is where philanthropic funding proved valuable for LIGO-India.

52.5 Multinational collaborations and cultural differences

It is not lost on anyone that cultures vary across institutions as well as the globe. Accounting for this in planning and executing international collaborations can prevent unnecessary miscommunication, frustration, and grief, especially, because culture-related variations can be subtle and are not always documented adequately. There are aspects of building an observatory that may go beyond objectivity and require cultural knowledge of how to get a workforce to get the job done. In some cultures, if a worker is of the opinion that a certain implementation cannot be done, he will be equally comfortable in communicating this straightforwardly to both his subordinates and his superiors. In other cultures, such an opinion is conveyed much more subtly to superiors. A lack of awareness of such differences runs the risk of delaying or even derailing project execution. This example highlights the value of long-standing collaborations among international partners for it helps people develop an appreciation of these cultural differences and adopt objective practices for accounting for them. It also emphasizes the importance of involving experts who can coach teams in addressing these differences, in particular because it can sometimes be domain specific—a data analyst may not readily anticipate cultural miscommunications that can occur between an engineer and an artisan or a millwright. (In fact, the last term, while prevalent in the US is not found in most English dictionaries in India!) Much of the cultural acclimatization in LIGO-India happened with several visits and interactions with workers across the spectrum—from site security people to district commissioners keen on how the project might help in improving the lives of the rural populace in the surrounding villages.

52.6 The necessity of education and public outreach

Apart from the 'greater good' rationale for pursuing Education and Public Outreach (EPO), whereby we engage in that activity for the benefit of the public as well as being accountable to the taxpayer, EPO has other values that are critical or even indispensable to projects of this magnitude. From its conception, it was clear to the leaders of the project that it will make a sizable commitment to education and public outreach in the communities around the site. To that end, we have been running EPO programs in their schools since the project obtained in-principal approval in 2016 and have received a lot of appreciation for this. It was obvious that for this project, charity should 'begin at home'. Of course, LIGO-India now supports EPO activities around the country and even contributes to international outreach via its members in the LIGO Scientific Collaboration (Inter-University Centre for Astronomy and Astrophysics n.d.).

EPO also works as the soft-power of a project in cultivating goodwill. One can imagine this being valuable in the first decade of observations of LIGO and Virgo when nothing was detected (as expected, and judiciously advertised as such). Projects such as LIGO-India and KAGRA have the benefit of riding on the goodwill already generated owing to the wondrous discoveries, but the soft-power of EPO can never be understated. Apart from directly influencing the learning of new generations of students—some of whom may even join a future avatar of the project—and improving the scientific awareness of the masses, it also prepares policymakers to take informed decisions on future scientific and technological investments. It also helps attract private donations to projects, which are often valuable in funding activities that the original (state) funds may not have the scope to support.

In some parts of the globe attracting promising bright minds to STEM disciplines has proved difficult. In our part of the world finding them opportunities for applying their talent to solving STEM problems within the country has been a greater challenge. In fact, losing them in large numbers to other shores has been a worry. Projects such as LIGO-India are bringing cutting-edge STEM research to their doorsteps where they can find the platform to learn and showcase their skills. Our EPO initiatives are highlighting the desirable aspects of skill diversification whereby scientific problems on the world stage can benefit from approaches to modeling and solving them that may be peculiar to our subcontinent today. National media has also contributed to these efforts by advertising the project as well as the achievements of our scientists in the LIGO–Virgo discoveries.

52.7 Why can't one wait to start?

Sometimes there is scientific compulsion for not waiting to start a project—e.g. some events, such as a supernova in our Galaxy, may not repeat frequently enough. Indeed, waiting for answers may actually end up wasting money by supporting ill-founded research. Also, collaborative projects help to make the best use of the current knowledge front in building bridges to the next. In the context of LIGO-India, a timely construction enhances the possibility that we will learn useful science about GW sources, e.g. by localizing them better. This in turn can help make informed choices about developing the technology for the next generation of GW detectors—those studies are already underway. Cost savings are multiplied when one broadens this discussion to include science that can be tapped by exploiting coordinated multi-messenger observations. Lagging parts of a collaboration can create jaggedness in the time-evolution of that knowledge front, which can result in lost opportunities. For example, the Hubble Space Telescope is one of those observatories that saw light from the kilonova associated with the binary neutron star merger GW170817 (Tanvir *et al* 2017, Kilpatrick *et al* 2022). The joint observations enriched our knowledge of these phenomena far more than physicists had imagined. At the same, it gives our community concern when any of these instruments glitch, just as when a payload computer on the Hubble Space Telescope (HST) halted unexpectedly on 13 June 2021. While that problem was fixed eventually, it reminded us that the HST has been making observations for over three decades and may not last much longer. Indeed, often science projects are founded with an ecosystem for discovery potential in mind. Delays or terminations of the components or participants in this system hurt scientific returns and ultimately waste money by prolonging ignorance. In the case of LIGO-India, one of the difficulties of having a gap of several years between the detector installation and commissioning in the US and that in India is that details that are not documented may not be readily available to us. Longer gaps also mean that people originally involved in the project may no longer be available now, forcing one to reinvent certain details. A timely start optimizes the return on investment.

52.8 Conclusion

How good a world we leave behind for future generations depends on the actions and investments we make now. This is true for addressing climate change and is true for seeding initiatives now that lead to technologies that make lives better tomorrow. Thankfully, scientists have learned to tap pure science for shorter-term returns as well. Big, multinational science projects have learned how to facilitate this process. First, collaborations such as LIGO, Virgo, and KAGRA have a high degree of transparency within them, while supporting competition, so that an idea that germinates in one institution is capitalized upon not only for GW science but also for spin-offs in technology and industry worldwide with relative efficiency. Outside observers have wondered (Blaustein 2020) how it is possible to compete while collaborating. Are not these terms antithetical, so much so that governments have enacted laws for incentivizing competition? In the case of LIGO-India, as indicated above, the collaborated-upon product will be of higher value than what is obtainable by individual groups. Such prospects elevate the whole group collectively and make collaborations attractive. This is the same reason why we work with our colleagues towards the betterment of our institution. Its improved stature is salutary for each one of us. All the same it is crucial that the institutional or collaboration culture is transparent enough to allow for individual excellence to be visible not just within the collaboration but also worldwide, and be rewarded. Here special attention must be devoted to junior and underrepresented colleagues who may be most prone to not getting their due—sometimes owing to existing or historical biases in a system that normalizes misplacing credit and at other times due to their positions or arrangements effectively disempowering them from claiming it. With the right reward and grievance redressal mechanisms and atmosphere a collaboration can ensure its success.

LIGO-India is being built after the other two LIGO detectors have already been installed and commissioned. Thus, we are in a position to benefit from the lessons learned in the US. In spite of the Hanford and Livingston detectors being touted as replicas, they harbor important differences in detail, which impact their observational sensitivities—given how fine the measurements are that they are built to make. There is no place for skimping in those details here in the engineering, installation, and commissioning—or it will not be possible to attain the same sensitivities here. Thankfully, the fundamentals of the site chosen for LIGO-India look good. Months- to year-long surveys of seismic noise and weather patterns (including, e.g., wind-induced noise) promise good environmental conditions for the observatory.

By participating in this worldwide collaboration members of LIGO-India will be privileged not only in bringing their insights into contributing to such a system but also take back to their institutions a working knowledge of what defunct systems to dissolve and what new twenty-first century solutions to replace them with. It is in this sense too that the diversity and inclusivity of a collaboration is so very enriching. With their inherent variety, they help bring new ways of tackling problems. Such an arrangement is also a two-way street—for good—since through its participants and practitioners, who are citizens of both the worldwide collaboration and their institutions in their own countries, it effectively induces our communities in various member nations to adopt these intrinsically higher values in our functioning.

While the article was in press, the much-awaited news was received that the Government of India has granted the final approval for building LIGO-India.

Acknowledgments

I would like to thank colleagues in the LIGO-India project and the LIGO Scientific Collaboration for providing multiple opportunities for several interactions over the years. In particular, thanks are due to Sendhil Raja, Subroto Mukherjee, Milind Gowardhan, Somak Raychaudhury, Tarun Souradeep, Fred Raab, David Reitze, Rana Adhikari, Michael Landry, Ajit Kembhavi, Suresh Doravari, Bala Iyer, C S Unnikrishnan, Sanjeev Dhurandhar, Sanjit Mitra, and Subhadeep De for useful discussions. This document has been assigned the LIGO Document Control Center No. LIGO-DCC-P2100384.

References

Acernese F *et al* (Virgo Collaboration) 2014 Advanced Virgo: a second-generation interferometric gravitational wave detector *Class. Quantum Grav.* **32** 024001

Adhikari R X 2020 A cryogenic silicon interferometer for gravitational-wave detection *Class. Quantum Grav.* **37** 16 165003

Akutsu T *et al* (KAGRA Collaboration) 2020 Overview of KAGRA: KAGRA science arXiv:2008.02921
Amato A *et al* 2021 Optical and mechanical properties of ion-beam-sputtered Nb_2O_5 and TiO_2–Nb_2O_5 thin films for gravitational-wave interferometers *Phys. Rev.* D **103** 072001
Aasi J *et al* (LIGO Scientific Collaboration) 2015 Advanced LIGO *Class. Quantum Grav.* **32** 074001
Balasubramaniam K and Rajagopal P 2021 Waveguide metamaterial rod as mechanical acoustic filter for enhancing nonlinear ultrasonic detection *APL Mater.* **9** 061115
Blaustein R 2020 Growing the gravitational-wave network: in conversations with David Reitze *Phys. World* July https://physicsworld.com/a/growing-the-gravitational-wave-network/ (Accessed: 28 March 2022)
Bose S, Chakravarti K, Rezzolla L, Sathyaprakash B S and Takami K 2018 Neutron-star radius from a population of binary neutron star mergers *Phys. Rev. Lett.* **120** 031102
Cramer J G 1997 The decline and fall of the SSC. The alternate view column *Analog Science Fiction and Fact Magazine* May http://npl.washington.edu/AV/altvw84.html (Accessed: 28 March 2022)
CSIR-Central Glass and Ceramic Research Institute n.d. *Research Division in Fiber Optics and Photonics* https://www.cgcri.res.in/research/research-divisions/fiber-optics-and-photonics/ (Accessed: 28 March 2022)
Gupta Y *et al* 2017 The upgraded GMRT: opening new windows on the radio universe *Curr. Sci.* **113** 707
Gupta U 2019 ISRO transfers space-grade Li-ion cell technology to BHEL *Photo-Voltaics Magazine—India* 29 November https://pv-magazine-india.com/2019/11/29/isro-transfers-space-grade-li-ion-cell-technology-to-bhel/ (Accessed: 28 March 2022)
Harry G M *et al* 2002 Thermal noise in interferometric gravitational wave detectors due to dielectric optical coatings *Class. Quantum Grav.* **19** 897–918
India's Mega Science Exhibition n.d. India's first global, Mega Science Exhibition 'Vigyan Samagam' begins today in Mumbai *Press Information Bureau of the Government of India* 8 May https://pib.gov.in/PressReleasePage.aspx?PRID=1571744 (Accessed: 28 March 2022)
Inter-University Centre for Astronomy and Astrophysics n.d. *LIGO-India Education and Public Outreach* https://ligo-india.in/activities/epo/ (Accessed: 28 March 2022)
Kilpatrick C D *et al* 2022 Hubble space telescope observations of GW170817: complete light curves and the properties of the galaxy merger of NGC 4993 *Astrophys. J.* **926** 49
Saleem M *et al* 2022 The science case for LIGO-India *Class. Quantum Grav.* **39** 025004
Tanvir N, Levan A and Jäger M 2017 Hubble observes source of gravitational waves for the first time *Science Release* heic1717 16 October https://esahubble.org/news/heic1717/
Tata Trusts n.d. https://tatatrusts.org/ (Accessed: 28 March 2022)
Thippeswamy M C, Kuchibhatla S A R and Rajagopal P 2021 Concentric shell gradient index metamaterials for focusing ultrasound in bulk media *Ultrasonics* **114** 106424
Tribune News Service n.d. CSIO develops indigenous avionics test rigs for IAF *Tribune New Service* 16 December https://sanjhamorcha.com/csio-develops-indigenous-avionics-test-rigs-for-iaf/ (Accessed: 28 March 2022)
Unnikrishnan C S 2013 IndIGO and LIGO-India: scope and plans for gravitational wave research and precision metrology in India *Int. J. Mod. Phys.* D **22** 1341010
Vajente G *et al* 2021 Low mechanical loss TiO_2:GeO_2 coatings for reduced thermal noise in gravitational wave interferometers *Phys. Rev. Lett.* **127** 071101

Chapter 53

Big Science in South Africa: a catalyst for human capital development and mission-driven innovation

Daniel Adams and Adrian Tiplady

South Africa (SA), as with many developing countries, suffers from capacity constraints with respect to highly skilled and trained graduates in science, engineering, technology, and innovation (SETI). In many instances, the global skills shortage within SETI aggravates a historical 'brain drain' phenomenon, where people that are highly skilled in next-generation technologies and research follow major projects and infrastructure.

South Africa has a small science system in terms of academics with PhDs and the number of PhDs per million of the population. Currently (i) only 43% of all academics have a PhD (the National Development Plan (NDP) [1] 2030 target is 75%); (ii) there are only 36 doctoral graduates per million of the population (the NDP 2030 target is 100 per million of population); and (iii) fewer than 50% of all doctoral graduates are in SET.

Investment in hosting and participation in large international research infrastructure provides a benign research and development environment with outputs such as human capital and research capacity development, and mission-driven innovation. South Africa has pockets of excellence in Big Science fields, but the critical mass of these is not sufficiently resilient, resulting in a loss of critical competency and expertise due to instability in the project life cycle or funding model. Participating in and hosting global research infrastructure enables (i) scientists and engineers to participate within a global network and be involved in cutting edge projects that do not require them to leave the country; and (ii) provides the opportunity to 'excite the youth' to pursue careers in science, engineering, and technology, hence stimulating the postgraduate pipeline.

doi:10.1088/978-0-7503-3631-4ch53

Over the last decade SA has invested significantly in the establishment of (locally hosted or in-bound) large international research infrastructure such as the Southern African Large Telescope (optical astronomy) and the MeerKAT/SKA projects (radio astronomy), as well as supporting researchers, scientists, and students to access (out-bound) international infrastructure such as the LHC at CERN. The primary benefits and returns on these investments to date were in the area of human capital and research capacity development, science awareness, and knowledge outputs. Over the last few years the shift in global research infrastructures to focus on their socio-economic impact has also found traction in SA. In this regard, significant gains were made in deriving socio-economic value from large research infrastructure hosting and participation. The SKA and participation in the CERN experiments are demonstrable examples of this.

In this chapter we will demonstrate how investment in Big Science in South Africa has served as a catalyst for human capital development and mission-driven innovation.

53.1 Introduction

Large-scale research facilities are becoming prohibitively expensive and challenging to build and maintain by one single country. The establishment of global research infrastructure has therefore become a norm for the modern research enterprise, where mutually beneficial cross-border partnerships can harness not only the sharing of resources, but collective knowledge and international collaboration.

For a developing country such as South Africa, access to global infrastructure is a vital component of it's research effort and policy framework—the benefits of which are manifested through a variety of avenues for both parties involved, including but not limited to:

- Human capital development, particularly with regard to student training and skills development through short- and long-term exchange programmes;
- Access to research infrastructure platforms that may not otherwise have been available in the country for use by the scientific research community; and
- Collaboration with the international scientific community based at, or connected to, the research infrastructure platforms.

The points above naturally lead to enhanced quality and quantity of scientific output.

From the perspective of global research infrastructure, and major global issues such as climate change, infectious diseases, food security, and natural disasters, international cooperation is vitally important. The fostering of international partnerships promotes efficiency and quality of research, whilst broadening the scope of diversity in the knowledge pool.

Access to major international research infrastructure is currently facilitated by a number of agreements brokered by the Department of Science and Innovation, and implemented through the National Research Foundation.

As a developing country, the decision to participate in global research infrastructure is often confronted with key local socio-economic and developmental challenges. Whilst investment in science, technology, engineering, and mathematics (STEM) research is seen as key to catalyse the development of a knowledge economy, it is important for South Africa to utilise a decision-making framework that ensures a high effective return on investment. This chapter gives an account of the significant return on the investments made to date through the interrogation of two major infrastructure investments (in South African terms), and hence provides evidence for success of such a decision-making framework, which is based on a two-component value proposition model.

53.2 South African investment in Big Science: the value proposition model

Strategic investments made by South Africa over the past few decades in Big Science infrastructure has been informed by, and anchored on, a two-component value proposition model, described in figure 53.1, which adopts an underlying principle of mutual benefit sharing. That is—how does South Africa benefit from investment[1] in Big Science, and what benefit does the relevant research infrastructure derive from South Africa's investment?

The return on investment resulting from delivery of the global and local value proposition ultimately creates value for South Africa's stakeholders, both internal and external. This has become an increasingly important consideration as

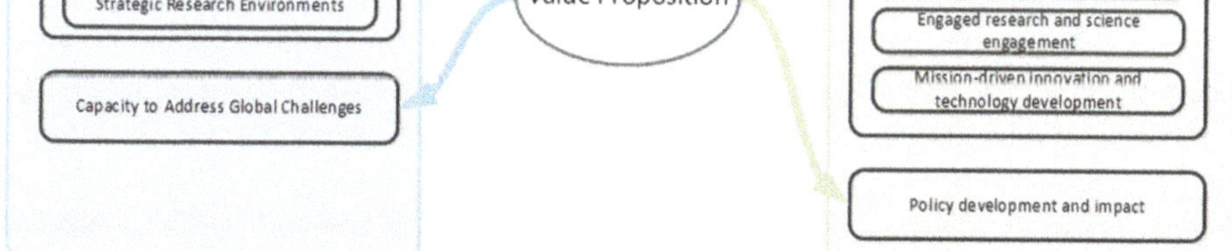

Figure 53.1. Two-component value proposition model, which informs South Africa's strategic investments in Big Science.

[1] In the context of this position paper, investment is executed through a variety of different forms from both financial investment to technological and research collaboration, to the hosting of research infrastructure platforms in South Africa.

investment decisions come under greater scrutiny from a globally connected society, requiring retention of a 'social license to operate' by funders and decision makers[2].

The first component of the value proposition model speaks to the delivery of value that contributes to the broader direct and indirect socio-economic development objectives that characterise a developing country such as South Africa (a local value proposition). This complex set of objectives is informed by various national policy documents that ultimately provide policy intent towards the delivery of the National Development Plan (NDP 2030) [2]—a key long-range plan adopted in South Africa for socio-economic development. The second component speaks to the delivery of value to the global scientific enterprise, not only through the making available of research capacity to address global challenges, but also the utilisation of existing strategic resources and advantages that South Africa has at its disposal. This can range from historical investments, geographic advantage[3], strategic and unique research environments[4], and even taking advantage of its demographic dividend. Such a global value proposition ultimately supports the positioning of South Africa, from a science diplomacy perspective, as a key trusted and influential partner that positively contributes towards addressing global challenges and relevant global discourse—a preferred destination for scientific cooperation and investment.

The framing of such investments therefore becomes critical for their long-term success and sustainability. Ultimately, this is interrogated through a 'stakeholder value' framework, which assesses derived value from a return on investment (ROI) model based on a local, and global, value proposition.

53.2.1 Unpacking the national or local value proposition

The local value proposition is delivered through national policy impact, and the delivery of direct, and indirect, socio-economic benefit. The pathways to impact resulting from investment in fundamental research are known to be unpredictable, long-term, and hence particularly difficult to model quantitatively. But there is sufficient anecdotal evidence within the science system at large that strengthens the case for a strong correlation between such investment, and broader socio-economic development.

Seen holistically, the development of the South African economy is driven, in part, by a recognition that its historical resource-based economy, upon which much of South Africa's economic growth and prosperity was based over the nineteenth

[2] Such an approach is increasingly gaining momentum within the commercial sector, where there is a migration from the traditional 'shareholder value maximisation' principle to that of 'stakeholder value maximisation'.

[3] Namely access to the southern oceans and southern skies, unique biodiversity, and the country's rich palaeontological heritage.

[4] In some cases, the 'misfortunes' plaguing the country, such as HIV/AIDS and tuberculosis, offer unique opportunities for collaboration from which foreign institutions could derive significant benefit.

and twentieth centuries, requires diversification[5] and strengthening of the local knowledge economy. Investment in access to science research infrastructure is a critical pillar towards the development of this knowledge economy—the development of which enables greater participation in the fastest growing sectors of global trade, typified by knowledge intensive products and services.

The development of human capital and research capacity, together with ongoing STEM skills development, supports the broader objectives codified within South Africa's National Development Plan 2030 (NDP 2030). There is reciprocal interaction between human capital development, and access to research infrastructure. To increase the number of skilled people, adequate and appropriate infrastructure for education and training is required. The establishment of more research infrastructure, on the other hand, requires highly trained people that can maximally derive benefit from this investment. As a response to the above-mentioned, the national Department of Science and Innovation (DSI) developed a white paper on science, technology, and innovation [3], which argues that research and development is a key driver for technological innovation, a critical requirement to speed up economic growth and improve livelihoods of citizens.

Research infrastructure has the ability to attract, develop, and retain an ensemble of a critical mass of highly skilled individuals, whilst providing a benign, enabling environment for mission-driven innovation and international research collaboration. A key benefit of such a critical mass is the development of transferable skills, which not only expresses itself through the strengthening and contribution towards the broader scientific knowledge (as measured through general research output), but can bring about technological innovation, and the production of products and processes that arise as a consequence of the research activity itself, or development of the research infrastructure.

In addition to the scientific and technological impacts of infrastructure, there is a materially significant economic impact (the medium- and long-term effects on the economy relating to technological advances induced by the establishment of research infrastructures), skills development and job creation, and social impact (the contributions to general well-being arising from progress made in science, which stems from the research process and its contribution to improving the quality of life of citizens). The utilisation of transferable skills has been clear in various case studies—the driving force of analysis and response to the recent COVID-19 pandemic in South Africa has been capacitated from the critical mass of expertise established through investment in fundamental science infrastructure.

Critical to South Africa's social development is the need to transform the legacy of a marginalised and disenfranchised research community, categorised by institutions that benefited from South Africa's apartheid legacy, and 'historically disadvantaged institutions'. Investment in access to research infrastructure at a national level contributes toward the establishment of an equitable landscape across the research enterprise.

[5] There are sufficient economic indicators, published annually, suggesting that significant progress has been made in this regard.

Finally, the recognition of a need for a 'social license to operate' has become a critical driver in ensuring the adoption of an 'engaged research' agenda—where broader public participation, engagement, and understanding is a necessary condition to ensure broader societal support and, in many cases, enhanced research impact[6].

53.2.2 Unpacking the global value proposition

The global value proposition delivers value to the global community in two parts: the provision of research capacity, expressed through enhanced internationalisation of science in South Africa, to address global scientific challenges[7]; and the provision of unique and strategic research resources that enhance global scientific output, making South Africa a preferred destination for scientific investment and collaboration, and enhancing South Africa's international scientific standing, visibility, and active participation in science diplomacy.

Mutually beneficial research infrastructure partnerships across borders can harness collective knowledge, assist in meeting global challenges faced by society, and enhance global science capacity[4]. Global science capacity is enhanced not only through the number of researchers, but more importantly, diversity in research and cultural backgrounds. One should ask that, without such diversity, do we ask the *right* questions to make progress in fundamental science? In responding to the issue, Dr Echkard Elsen (former Director for Research and Computing at CERN) said the following:

> An endeavour such as the future exploration of the structure of matter and the forces is a formidable task. It does require the brightest of minds and we simply cannot ignore that a very large fraction of the young population today is living in Africa. While education in Fundamental Physics is clearly not offered everywhere or not at the adequate level the curiosity—*a priori*—is in every child and student. We have to nourish this stamina for more understanding and allow the different cultural backgrounds to flourish and eventually to contribute. We also have to offer a path for young people to engage, become familiar with the scientific approach and develop and refine their own methods. This diversity in thinking is a key and becomes obvious in the following: Do we ask the right questions to make progress in fundamental science? It does not help to ask the same question over and over and never find an answer. The wrong question never has a good answer… Nature so far has been cleverer than us and clearly indicated New Physics (take Dark Matter) but we have no clue what it could be.

---

[6] An excellent example is in the dealing of matters of public health, where an informed society is more likely to adopt good public health measures to combat communicable disease.

[7] Including, but not limited at, those challenges impacting the Sustainable Development Goals, as well as broader scientific questions being addressed by the global scientific community.

Such a perspective reflects on the benefits that can arise from Africa's demographic dividend—a young population that can secure greater participation in the global research endeavour than its predecessors.

South Africa also has long standing strategic investments and capability, with a track record of excellent science outputs. Achievements in nuclear physics are well known, and rapidly emerging activities in large-scale astronomy, such as the Square Kilometre Array, have established South Africa as an influential partner to global astronomy.

South Africa holds strategic advantage in not only physical characteristics, but unique research environments. Areas have been declared, and protected through legislation (Astronomy Geographic Advantage Act, Act No. 21 of 2007), as Astronomy Advantage Areas due to the unique characteristics suitable for undertaking astronomical observations. With the ever more demanding requirements on the scientific performance of astronomical facilities, the scientific qualities of a possible site have a significant impact on the scientific output of the instruments. Population demographics, and the prevalence of communicable diseases such as tuberculosis and HIV/AIDs, also provide unique opportunities for research.

53.3 South Africa's strategic investments: two case studies

Over the last decade, South Africa has made significant investments in the establishment of large scale, international research infrastructures such as the Southern African Large Telescope (optical astronomy), and the MeerKAT and Square Kilometre Array projects (radio astronomy), whilst it has supported researchers, scientists, and students to access international research infrastructure such as the Large Hadron Collider (LHC) at CERN, the Joint Institute for Nuclear Research (JINR) in Dubna, and the European Synchrotron Facility (ESRF) in Grenoble.

During the last few years, the shift within global research infrastructures to focus on socio-economic impact has also found traction within South Africa. In this regard, significant gains were made in deriving socio-economic value from large research infrastructure hosting and participation. The Square Kilometre Array (SKA) project, and participation in the LHC experiments, are two clear demonstrable examples of where South Africa has invested in mobility, access to, and hosting of relevant research infrastructure (in the case of the SKA, drawing on its strategic geographic advantage)[8]. As an organising framework, we have classified the research infrastructures into two groups, namely: (i) out-bound research infrastructures, such as the LHC at CERN, JINR (Dubna), and ESRF (Grenoble); and

[8] In 2003 the South African Department of Science and Innovation decided to prepare and submit a competent bid, on behalf of South Africa and its eight African Partner Countries, to host the next-generation SKA radio telescope. Hosting of such a facility would be a natural realisation of the 1996 White paper on Science and Technology, and the 2002 National Research Development Strategy, which identified astronomy as being an area in which South Africa exhibits a natural geographic advantage. It is also a recognised driver of technological innovation—providing a benign research environment for the development of next-generation high-performance computing technologies, data storage systems, processing and utilisation of 'Big Data', artificial intelligence, and even new systems of work and high-tech project management.

(ii) in-bound research infrastructures, such as the MeerKAT/SKA and SALT projects. This section will present two case studies that characterise South Africa's strategic investments in both out-bound and in-bound research infrastructure: the LHC at CERN, and the MeerKAT/SKA project.

SA entered into a memorandum of understanding with CERN for collaboration on ATLAS, ALICE, ISOLDE, the LHC Theory Group, and the Worldwide LHC Computing Grid (WLCG)—the SA Centre for High Performance Computing hosts an LHC Tier 2 Data Node. The country's financial participation is determined by the proportional scientific participation in various experiments. SA is a member of the Review and Resources Board (RRB), the role of which is to: oversee the implementation of the collaboration agreements; monitor the projects and funds; and approve of new projects, upgrades and budgets for the different experiments. SA's participation at CERN is locally coordinated through an SA-CERN Collaboration Board—comprising iThemba LABS for financial management and secretariat support, and coordinators representing the science community for the different experiments (ATLAS, ALICE, ISOLDE, and Theory as well as CHPC representing the WLCG). This is funded by the DSI and implemented by the NRF.

SA was awarded the right to co-host the international SKA project in 2012, following an eight year competitive bid process. It was a founding board member of the SKA Organisation, when it was established as a company limited by guarantee in 2011 for the purposes of preparing for the construction and operation of the SKA telescope. SA became a founding member of the SKA Observatory, an inter-governmental treaty organisation established in January 2021 to construct and operate the SKA telescope, and as such is represented by the DSI on its council, the highest decision-making body within the SKA Observatory. As a host country, SA's financial participation attracts a premium over and above its proportional scientific participation. The SKA Organisation will be dissolved following acquisition of its assets and liabilities by the SKA Observatory through a business purchase agreement.

The benefits derived by South Africa from collaboration at the LHC, and the hosting of and participation in the SKA project, are in line with the emerging and accepted understanding of the benefits of research infrastructures based on a cost–benefit analysis approach[5], and align with the value proposition model as outlined earlier in this chapter.

53.3.1 Local value proposition

53.3.1.1 Enhanced scientific output

Between 2005 and 2018, South Africa's global positioning in respect of research publication output in physics improved from rank 41 to rank 32 in the world, with a fourfold increase in its journal publication output. A substantial driver of this improved scientific output has been South Africa's participation in the MeerKAT/SKA projects and CERN.

Figure 53.2 describes the improvement in overall scientific output and the impact of SA resulting from the key investments made by the DSI. It plots the mean

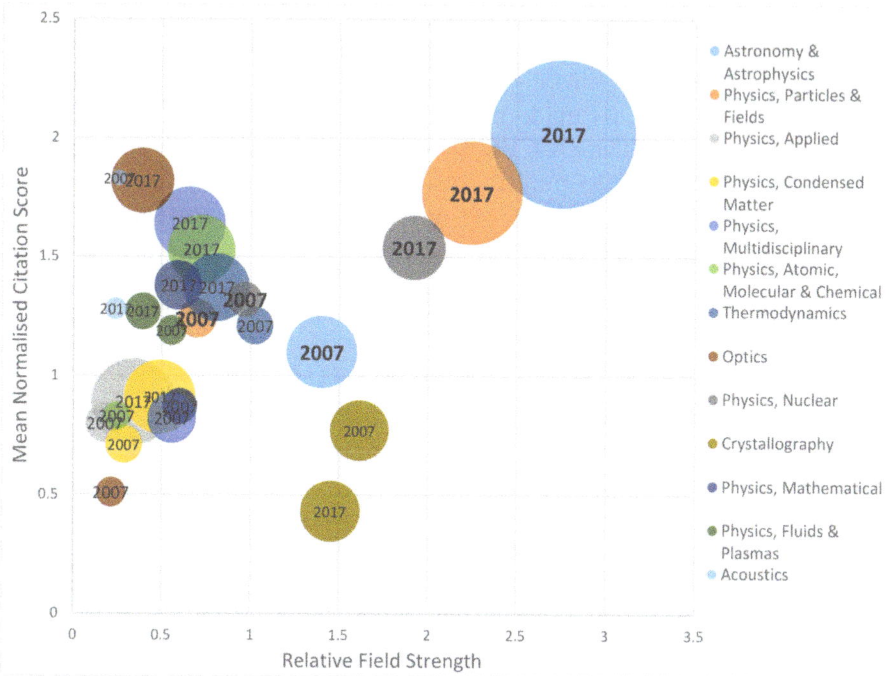

Figure 53.2. Positional analysis of physics in South Africa between 2005–07 and 2015–17 [8].

normalised citation score (MNCS[9]) against relative field strength (RFS[10]) for various physics related disciplines over the periods 2005–07 and 2015–17. The size of the bubbles is proportional to the publications produced during the relevant period. There is a clear relative improvement in astronomy and astrophysics, and particle physics related fields, as compared to the other physics related disciplines. From the period 2007 to 2017, the MNCS for astronomy and astrophysics has improved from 1.1 to 2.02—meaning that the visibility of astronomy and astrophysics output from SA is twice the global average. There has also been a doubling in the relative field strength, and fourfold increase in the number of papers published. This has likely improved even further in recent years—the MeerKAT radio astronomy facility, a pre-cursor to the Square Kilometre Array and one of the premier radio astronomy facilities in the world, was only commissioned in 2018. Anecdotally, research outputs in radio astronomy have increased substantially with higher impact, and broader international collaboration.

[9] The mean normalised citation score (MNCS) is an indicator of visibility and impact based on citations. This indicator is normalised to account for differences in the citation behaviour of different scientific fields as well as different publication years. A higher MNCS indicates more visibility and impact with an MNCS between 0.8 and 1.2 generally considered average.

[10] The relative field strength (RFS) measures what proportion of a country or organisation's total scientific output is in a specific field and compares that to what proportion of the world's total scientific output is in that same field. An RFS equal to 1 indicates that the country or organisation is just as strong or active in that field as the whole world is, while a higher value indicates more strength or activity.

This improvement is also seen in particle and nuclear physics. The increased normalised impact, and relative field strength (as compared to the global average), in the fields of astronomy and astrophysics, nuclear and particle physics, coincide with the key strategic investments made by South Africa in access to CERN and the MeerKAT/SKA projects. The SA-CERN collaboration was formalised in 2008, and SA researchers contributed to publications on the discovery of the Higgs boson, which peaked over the period 2011–13. Over the same period, substantial investment was made in SA to establish radio astronomy research infrastructure through human capital development programmes (initiated in 2005) and the MeerKAT project (commissioned 2018, with early development work starting in 2007).

53.3.1.2 Institutional transformation

Higher education institutions in SA have leveraged against the key investments made. The establishment of the SA-CERN collaboration contributed to institutional transformation of some of the smaller universities with no existing formal programme in particle and nuclear physics. For example, the High-Throughput Electronics Laboratory at the Institute for Collider Particle Physics at the University of the Witwatersrand, Johannesburg, partnered with the University of the Zululand to establish an electronics laboratory where students do their projects. Currently the University of Zululand have students and academics doing projects at ATLAS. The Wits Institute also co-supervises students at the University of Venda in the areas of artificial intelligence and machine learning.

The establishment of strong radio astronomy and astrophysics programmes at institutions not historically active in the field was also realised. For example, a very active radio astronomy group was established at the University of the Western Cape, whilst astronomy has been a driver for development of Sol Plaatje university (as a new university) into a research institution as a result of its proximity to the SKA telescope. Sol Plaatje is the only university in the country offering an undergraduate degree in data science. There has also been the establishment of the Inter-university Institute for Data Intensive Astronomy, and *ilifu* (a data intensive research cloud), as a consortium of universities undertaking research in astrophysics, bioinformatics, and energy.

53.3.1.3 Human capital and research capacity development

A significant impact of the SA-CERN collaboration is the attraction of respected scientists to take up senior academic positions at South African higher education institutions. Participation of South Africa in the LHC experiments has resulted in a nine-fold increase in the number of masters and doctoral students pursuing degrees in high-energy and particle physics between 2010 and 2019, whilst providing access to a unique learning environment that goes beyond what would be available within a university context. The complex environment of CERN provides exposure to learning and practical environments such as solving engineering challenges, managing large datasets, data analytics, machine learning and artificial intelligence, software and algorithm development, systems engineering, project management, and systems administration.

The growing pool of SA senior scientists participating in the CERN collaboration also provides excellent supervisory capacity for the growing student cohort in the field of high-energy physics. The growth in the supervisory capacity is a strong and definite sign of the partnering universities' support for the SA-CERN programme as an HCD investment vehicle to improve the output of high-end skilled students for academia and industry.

To fully exploit hosting of the international Square Kilometre Array radio telescope in South Africa, a targeted human capital development programme was implemented to establish a sustainable and diverse research community. In order to do this, a pipelined approach towards educational and research support was undertaken[11]. However, at the same time, South Africa identified an opportunity to significantly strengthen the local value proposition of its investment through direct participation in the scientific and technological design and construction of the SKA telescope. This was facilitated through the local development of the MeerKAT radio telescope (the premier radio astronomy facility of its type until the SKA is built). As a result, the human capital development programme was broadened to cover not only scientific competencies, but also technical and engineering, and in so doing established a critical mass of high skilled scientists, engineers, and project managers capable of delivering high-tech, next-generation scientific facilities.

Over the past 16 years the human capital development programme has been implemented, over 1400 bursaries and grants have been given out for study in astronomy and astronomy related engineering and technical disciplines. This has resulted in a near order of magnitude increase in the size of the active South African radio astronomy community. It has also, in addition, supported the critical mass of engineering and technical skills within the South African Radio Astronomy Observatory, a National Facility of the National Research Foundation and implementing agent for strategic investments in radio astronomy, for the delivery of next-generation radio astronomy facilities.

Anecdotally, a recent request for proposals to the international astronomy community garnered a substantial international response and South African-based astronomers were awarded, on a scientific competitive basis, approximately one third of the observation time available. Local participation in research infrastructure that is globally relevant speaks well of the human capital development programmes and initiatives underway.

53.3.1.4 Mission-driven innovation and technology development

Engagement on global research infrastructure provides an excellent model for mission-driven innovation and technology development. Coupled with the development of transferable skills that are highly relevant for next-generation technologies,

[11] One of the key challenges in growing the size of the community was lack of supervisory capacity for postgraduate studies. In order to alleviate this, the pipeline approach was implemented in a staged manner—beginning at post-doctoral fellowship and PhD level, before extending down to Masters, Honours, and undergraduate degrees. This was also supported through co-supervision with a number of professional astronomers abroad—which continues to this day.

Big Data, and the 4th Industrial Revolution, this provides an exciting innovation environment.

In the case of the MeerKAT and Square Kilometre Array projects, South Africa's low human capacity meant it developed its critical mass from the ground-up—drawing on experienced individuals from the defence industry, as well as highly competent but inexperienced graduates. Such a young team provided an opportunity to learn, but also develop new ways of doing things. As a result, processes and technologies developed for the MeerKAT make it a next-generation facility. The adoption of strong system engineering processes—common in the defence industry, but not widely known within the astronomy community—resulted in the delivery of a facility on time, within budget, and with four times the promised performance. The local development of stringent, highly specified requirements for research infrastructure created a demanding environment for local industry to push the innovation envelope. Significant achievements were delivered not only within the high-technology space, but even within the areas of civil engineering, where an underground radio frequency shielded facility to house onsite computing resources was established, logistics engineering, where a highly optimised logistics and maintenance operation was put in place to enable the construction and operation of very remote research infrastructure[12], and the steel industry, which participated in the fabrication and deployment of MeerKAT dishes[13].

As a consequence of the project, key technology developments took place that are positioned well for commercialisation. Key developments include:

- The development of a passive radar system, for use in the aviation industry;
- The development and deployment of supercomputing infrastructure at 30% the commercial cost; and
- The development of extremely high capacity data storage and processing systems.

The SA-CERN collaboration has led to the establishment of a High-Throughput Electronics Laboratory at the Institute for Collider Particle Physics at the University of the Witwatersrand, Johannesburg, South Africa, and its expansion leading to the inclusion of a Technology Transfer Pillar (TTP)—to be implemented through the Technology Innovation Platform (TTP) located the iThemba Labs. The value-add of the TTP and TIP is such that it enhances the capability of South Africa to develop high-throughput electronics components, required by the LHC experiments in South Africa. This paradigm shift for the country—moving from an in-cash to an in-kind contributor—in turn stimulates local innovation, industry development, job creation, and indigenous technology and skills development.

[12] Special Platinum Award at the 2017 South African Logistics Achiever Awards.

[13] Awarded the 2015 Overall Winner, and Tubular Category Winner, by the Southern African Institute of Steel Construction (SAISC) for excellence in the use of structural steel in the design and building of the MeerKAT antenna.

53.3.1.5 Catalysing growth of the knowledge-based economy

Growth of a knowledge economy is catalysed through investment in basic research infrastructure such as the MeerKAT. Florio (2019)[6] indicated that the empirical estimation of the correlation between infrastructure investment, and economic growth, has attracted noticeable research attention in economics for decades. Realisation of benefits arising from investment in research infrastructure was exploited in South Africa through the CERN collaboration and SKA/MeerKAT projects. The development of transferable skills is a critical component in catalysing growth of the knowledge-based economy—application of the highly specialised skills developed within a science research environment in other areas of the economy.

With the outbreak of the Covid-19 pandemic, the research team of the Institute for Collider Particle Physics at the University of the Witwatersrand under the leadership of Professor Bruce Mellado could use the knowledge, learning, and experience that they acquired through working on the complex data systems at CERN to assist the country with epidemiological analyses to track the spread of the virus, the identification of hotspots of infections, and management of the vaccine administration. These interventions were constantly showcased through public presentations, use of the results in decision-making around the lockdown restrictions, and publications, which brought public awareness of the value of science during disasters.

One of the key contributions by the Institute for Collider Particle Physics was the development of an early detection AI-based algorithm for the prediction of pandemic wave speaks. The advent of infection waves is driven by circumstances that are difficult to predict and therefore to control. In this complex environment, early detection algorithms can provide an early warning to policy-makers and the population. Early detection algorithms are able to issue an alert when the data displays a significant change that is consistent with the advent of a new wave. AI technology provides invaluable potential to develop early detection and alert systems that are highly needed for rapid and dynamic decision-making under risk and uncertainty under conditions such as the Covid-19 pandemic[7].

Likewise, the experience and learnings through the development of the MeerKAT radio telescope could be used to coordinate the National Ventilator Project, a key COVID-19 intervention to successfully supply 20 000 locally conceptualised and built ventilators to the national health system, at a cost level of less than 600 euros per ventilator.

53.3.1.6 Engaged research and science engagement

Ensuring informed public opinion is critical for a successful science system in South Africa. First, it ensures successful implementation, or adoption, of scientific results. Ranging from public policy development, to the adoption of preventative public health measures such as during the COVID-19 pandemic, ongoing science engagement has been a cornerstone of successful adoption.

Second, ensuring legitimacy in public opinion is critical for ensuring ongoing sustainable investment in research infrastructure. This ‘social license to operate’ manifests itself through public discourse, and ultimately at the ballot box. Through both the CERN collaboration, as well as the MeerKAT and Square Kilometre Array projects, science engagement has been critical to ensuring ongoing support from the public, private, and political spectrum.

The Square Kilometre Array project in particular occupies an enviable space within media, recognised as a 'good news story' for South Africa. The delivery of education and science engagement programmes has formed a critical pillar of the multi-pronged stakeholder management approach for the Square Kilometre Array in South Africa, structured to create a range of value propositions for different stakeholder groups within the vicinity of the site, and nationally across the various metropolitan areas. Initiatives have included:

- Extensive participation in, and partnership with, public science, local government, and industry events that establish the project as a key actor within the 'national system of innovation' and research enterprise.
- Development and distribution of science outreach materials, such as a comic book series targeted at primary and secondary schools.
- Development of a Science Centre in one of the local towns.
- Investment in educational infrastructure that supports the education pipeline within the towns surrounding the Square Kilometre Array site, including teacher and learner support programmes, laboratory equipment and facilities, educational programmes, and bursaries.

It is worth noting that the ownership of the aforementioned 'social license to operate' is seen as critical for the long-term success of establishment and operation of research infrastructure in the modern world[14]. Stakeholder engagement and participation is therefore key, along with a system-wide view of the potential unintended consequences of such a major investments. In addition to directed engagements and agreements with representative bodies, South Africa undertook a multi-year 'Strategic Environmental Assessment' for the Square Kilometre Array. This assessment, which resulted in a comprehensive Integrated Environmental Management Plan, was highly consultative through a number of public participatory processes, and considered the social, economic, and environmental factors arising from the construction and operation of the Square Kilometre Array. Implementation of the resulting plan includes various mitigation measures across the socio-economic-environmental spheres to support not only long-term environmental sustainability, but also social and economic sustainability of neighbouring communities. These measures included such diverse things as:

- Protection of resources of environmental (fauna and flora), cultural and heritage significance.

[14] This is mirrored in the corporate world, where long-term organisational sustainability was becoming increasingly dependent on the shift away from the traditional 'shareholder' model, where the primary objective is to make profit for its shareholders, to a 'stakeholder' model, where the primary objective is to create value for its stakeholders.

- Protection of agricultural activities on neighbouring farms through appropriate and sustainable land management practice.
- Establishment of a livestock feedlot to support the local, largely agricultural economy.
- Establishment of subsidised 'radio astronomy friendly' telecommunication services.

53.3.1.7 Policy influence and impact

Whilst an active scientific community will ultimately create an enabling environment for evidence-based decision-making and policy formulation, in the case of the MeerKAT and SKA projects it prompted the development and promulgation of globally unique legislation[15] for the establishment and protection of geographic areas suitable for the operation of astronomy facilities.

53.3.2 Global value proposition

Due to SA's world-class supercomputing, data management and storage, and high speed data transmission capabilities as well as its active participation in LHC experiments, a Tier 2 data node was established in South Africa that allowed the country to become part of a global network for hosting and managing CERN data. The Tier 2 node is located at the Centre for High Performance Computing (CHPC). Currently, the LHC Tier 2 data node at the CHPC processes over 1600 computing jobs per day for two CERN experiments, namely ATLAS and ALICE. A direct benefit of CHPC's participation in the Worldwide LHC Grid is that CERN agreed to host and train system administrators from South Africa. The long and established history of South African research competency in this field has made a positive contribution towards the global knowledge field.

Similarly, South Africa utilised its geographic advantage for the establishment of next-generation radio astronomy facilities—first, through the MeerKAT radio telescope, and second through hosting of the future international SKA project. Through SARAO, South Africa has executed strategic projects and programmes to establish globally competitive research infrastructure platforms at a world-class site, ensured the growth of a diverse, sustainable, and transformed globally respected technical and scientific research capacity, and derive socio-economic benefit from investment in research infrastructure. It is fair to say that there exists sufficient anecdotal, and auditable, evidence to support the claim that South Africa has emerged as an influential and global partner to radio astronomy, and the SKA

[15] The Astronomy Geographic Advantage (AGA) Act was adopted as a comprehensive long-term instrument to protect astronomy in all its forms in South Africa, in particular but not limited to the MeerKAT and SKA projects to be developed in the Northern Cape Province. The AGA Act provides for a wide range of requirements for the advancement and protection of astronomy including the development of skills, capabilities, and expertise; the identification and protection of areas in which astronomy projects can be undertaken; the provision of a framework for the establishment, protection, preservation, and maintenance of a national system of Astronomy Advantage Areas highly suitable for astronomy; and the regulation of activities which cause or could cause interference.

project in particularly. SARAO is recognised as a technical leader in the design and development of next-generation radio astronomy, and scientific infrastructure, as well as being an enabler of ground-breaking research through its locally designed research platforms. It has established itself as a credible actor within the national system of innovation, with the capability of delivering innovative socio-economic returns to the broader public.

As a result, South Africa has increased its scientific visibility and enhanced its standing with the world of science diplomacy. It has become a viable destination for investment in scientific and research infrastructure, which results in substantial positive direct and indirect local socio-economic benefits.

53.4 Conclusion and discussion

South Africa's approach in evaluating decisions regarding investment in research infrastructure has been informed, in part, by evaluation within the context of a two-component stakeholder value framework, which encompasses both a local and global value proposition. The development of this framework is informed by the long-term development aims of South Africa, expressed through its National Development Plan and other supporting policy statements, as well as global policy aims such as those codified through the Sustainable Development Goals. In the development of this framework, however, it has been critical to ensure that the value proposition framework reflects a multi-pronged stakeholder approach—that is, to ensure a 'social license to operate' that enhances the legitimacy of any investment from the perspective of each identified stakeholder group, including funders, decision makers, and directly affected local communities.

This approach has been largely successful in creating an enabling environment to optimise inclusive socio-economic return on investment in research infrastructure in South Africa. It has been able to deliver across a wide range of local and global value propositions, including:

- Enhanced scientific output.
- Institutional transformation.
- Human capital and research capacity development.
- Mission-driven innovation and technology development.
- Catalysing the growth of a knowledge-based economy.
- Engaged research and science engagement.
- Policy influence and impact.
- Strategic resource advantage.

Creating an enabling environment for the long-term scientific sustainability of the research infrastructure has necessitated investment in key interventions across broader socio-economic-environmental spheres.

In summary, the return in investment in research infrastructure is facilitated and catalysed through the adoption of two key principles:

- Development of a value proposition framework that reflects the value proposition of key stakeholder groups that are impacted by the investment.
- Recognition of ensuring a 'social license to operate' to ensure long-term organisational sustainability, and hence scientific success.

References

[1] National Planning Commission 2017 *Our Future—Make it Work: National Development Plan 2030* https://www.gov.za/sites/default/files/gcis_document/201409/ndp-2030-our-future-make-it-workr.pdf (Accessed: 12 May 2021)

[2] National Planning Commission 2017 *Our Future—Make it Work: National Development Plan 2030* https://www.gov.za/sites/default/files/gcis_document/201409/ndp-2030-our-future-make-it-workr.pdf (Accessed: 12 May 2021)

[3] Department of Science and Technology 2019 *White paper on Science, Technology and Innovation* https://www.dst.gov.za/images/2019/ WHITE_PAPER_ON_SCIENCE_AND_TECHNOLOGY_web.pdf (Accessed: 13 May 2021)

[4] Meeri K 2020 White paper on African-European Research Infrastructure Cooperation *RI-VIS* https://zenodo.org/record/4475595#.YIlArKGxV3g

[5] Florio M 2019 *Investing in Science: Social Cost–Benefit Analysis of Research Infrastructures* (Cambridge, MA : MIT Press)

[6] Florio M 2019 *Investing in Science: Social Cost–Benefit Analysis of Research Infrastructures* (Cambridge, MA : MIT Press)

[7] Wits University 2021 *AI Powered Algorithm Released to Detect the Third Wave in South Africa* https://www.wits.ac.za/news/latest-news/research-news/2021/2021–03/ai-powered-algorithm-released-to-detect-the-third-wave-in-south-africa.html

[8] Mouton J derived from ; Mouton J *et al* 2019 A scientometric assessment of physics in South Africa *Final report* Department of Science and Innovation

IOP Publishing

Big Science in the 21st Century
Economic and societal impacts
Panagiotis Charitos, Theodore Arabatzis, Harry Cliff, Günther Dissertori, Juliette Forneris and Jason Li-Ying

Chapter 54

Recent progress towards an African light source

Simon Connell, Katharina C Cramer, Edward Mitchell, Sekazi K Mtingwa and Prosper Ngabonziza

Africa is a rapidly changing continent, moving out of its colonial shadow to drive towards a vision of socio-economic development and prosperity. Its greatest ambition is to develop a knowledge-based economy for Africa. To achieve it, science, education, technology, and innovation play important roles. To strengthen all four and to reap the strong and many-fold returns on investment, the vast majority of countries from the Global North are investing in large-scale research infrastructures as a key element of their science programmes. Africa should be no exception to this. Of all large-scale research infrastructures, the advanced light source (AdLS) is clearly among the most transformative. Despite the importance of research infrastructures, in particular the AdLS, Africa is falling behind. All other habitable continents already have had AdLSs for decades, and many are busy with or planning the latest ten-thousand-fold upgrade. However, without doubt, Africa brings with it many of the requirements needed to become a global player. After all, Africa long ago passed the threshold of sufficient local and Diasporan AdLS user-experts. While an AdLS in Africa would be an ambitious flagship project, the benefits would be awesome. An AdLS would create a globally high-impact, multi-disciplinary, and transcultural hub of science and technology. It would represent a major step of science diplomacy towards the ideal Pan-African vision, and play an important role in stemming the African scientific Diaspora, and perhaps even reversing it as the Diaspora returns. In short, an African AdLS would be truly transformative, not only socio-economically, but also heralding progress in the transformation of the mind and of the culture. In Africa, there would of course also be the expected massive advance in trained human capacity, scientific and technological potential, and economic capacity. This transformation would inure to the benefit of all humanity as a symbol of progress of inclusivity and working together with a common drive and vision.

doi:10.1088/978-0-7503-3631-4ch54

54.1 Introduction

Exploring matter and unravelling its secrets are at the heart of humanity's quest, and subsequently are the main drivers spurring innovative developments in science and technology by industry and academia. Advanced light sources (AdLSs) offer unique opportunities to expand the boundaries for the exploration of new materials and living matter, thus often leading to scientific discoveries and industrial innovations, as well as to huge socio-economic benefits around the world in new sustainable technologies and ways of living.

Figure 54.1 depicts the locations of AdLSs, called synchrotron light sources and free-electron lasers (FELs), around the world. Currently, these AdLSs are playing a key role in globally stimulating science and innovation, and enhancing industrial competitiveness. What is striking is that Africa is the only habitable continent without an AdLS. Yet, there are attempts to remedy the situation. Indeed, in the past few years, African scientists with extensive synchrotron light source experience, in collaboration with their international colleagues, have come together to advocate for an African light source (AfLS); and to gather momentum towards its realisation on the African continent. An AfLS is expected to revolutionise fundamental, applied, and industrial sciences that are relevant to Africa. These include, but are not limited to, the following disciplines: structural biology, drug discovery, and vaccine development; medical science and biomedicine; palaeontology, archaeology, and cultural heritage; agriculture and resource exploration; competitive industry; energy and environmental sciences; mineral- and geo-sciences; nanoscience and nanotechnology; and many more (see figure 54.2).

This essay reports recent the progress towards the realisation of an AfLS. First, we discuss how an AfLS will contribute to achieving a stable Pan-African vision of science, technology, and innovation. Second, we elaborate on expected African

Figure 54.1. Location of advanced light sources (AdLSs) around the world. The map shows most of them, and currently Africa is the only habitable continent without an AdLS. (Figure adapted from *LAAAMP* (2020).)

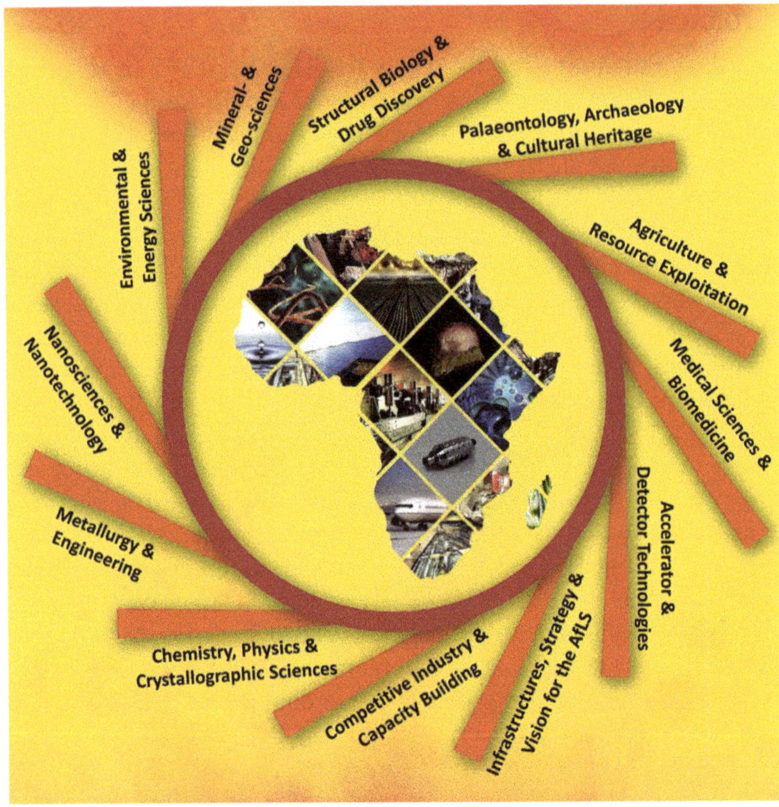

Figure 54.2. The African light source is expected to be a prominent mega-science techno-industrial and fundamental research facility, as well as an innovation hub in Africa. It would promote local and regional research platforms, as well as massive advanced human capacity development and employment in Africa.

socio-economic benefits and priority drivers for the AfLS. Next, we show the current growth of AdLS usage by African researchers. Finally, we conclude with a discussion of potential pathways for conceiving, constructing, and operating an AfLS; as well as ongoing efforts to produce the Conceptual Design Report (CDR) that will summarise the ideas; and a description of the AfLS project in detail. For a detailed discussion on the early history of the AfLS project, the interested reader is referred to (Connell *et al* 2018, 2019, Ngabonziza 2019).

54.2 A Pan-African vision of science, technology, and innovation—moving forwards beyond aid

Africa is a growing, rapidly modernising continent. In the coming few decades, Africa will become home to the bulk of the world's youth, with most of the African population currently under the age of 30 (UN DESA 2019). This aspiring and young population is highly motivated to be independent and productive, moving quickly into a sustainable and equitable future that is based on meaningful dialogue and participation. These visions drive the rationales for African investments in science, technology, and innovation. It is in this vibrant environment that advanced research infrastructures have a firm place as central scientific and technological, but also strategic opportunities to be seized for the African continent.

Such a new, large-scale scientific infrastructure on the continent needs to serve a diverse range of research needs, and it must also be able to compete with other cutting-edge research infrastructures around the world. The obvious candidate for such an endeavour is an AdLS, in particular the synchrotron-based AfLS, which has the ability to accommodate far more users than the FEL AdLS. A modern-day AdLS is a very large-scale research infrastructure that produces electromagnetic radiation—'brilliant light'—to investigate the properties of living matter and materials down to the atomic level in a variety of scientific disciplines. It embraces essentially all light-based spectroscopy, scattering, and imaging techniques, generating and using radiation from the infrared to the hard x-ray part of the electromagnetic spectrum. The radiation is considerably brighter (of the order of a billion times) than traditional laboratory-based infrared, ultraviolet, and x-ray sources, and each beamline at an AdLS is usually energy tunable and has a significant component of coherence, which can be utilised in (for example) high performance imaging techniques (e.g. Bilderback *et al* 2005).

Due to its high potential for scientific discovery and innovation, a synchrotron-based AfLS would contribute to regional development, providing an institutional baseline for further efforts to build capacity and competence. Moreover, the ability of such a facility to shape the contours of what we know would promote the establishment of independent and sustainable economic systems and societies. One reason hereof is that AdLSs are meeting points for researchers, and thus a location where valuable know-how is passed around. Indeed, an AdLS is arguably the leading research infrastructure that hosts multidisciplinary research activities that crisscross traditional disciplinary silos and collaborative boundaries. An AdLS provides parallel access to dozens of research groups simultaneously, around the clock, year in and year out. Thus, thousands of researchers visit an AdLS annually. Moreover, there is the ability for many more to acquire data remotely by sending experimental samples to the AdLS for data acquisition by in-house scientists. AdLSs are also available for industrial cooperation that ranges from open collaborations and joint developments to confidential research. Currently, Africa is missing out, because as a consequence of not having an AdLS, African researchers must travel to facilities on other continents, such as the European Synchrotron Radiation Facility (ESRF) in Grenoble, France, to perform their experiments. Thus, an AdLS in Africa would enable thousands of African scientists, engineers, and students to gain access to such a tremendous scientific and technological tool, without the need to travel to foreign facilities and be subject to foreign research priorities.

Many have asked whether Africa is ready for such a technologically sophisticated large-scale scientific infrastructure. The answer is a resounding YES! To this date, there exist approximately 50 AdLSs worldwide, including synchrotron radiation sources and free-electron lasers, most of them in Europe, Japan, and North America. However, South American and Southeast Asian countries also have constructed a number of national AdLS[1]. Moreover, roughly half of the world's operating AdLS have either implemented or plan to implement upgrades to provide a ten-thousand-fold upgrade in the figure of merit for the brightness of their beams in combination with detector technology enhancements. Thus, Africa has not only been left behind in the initial implementation, but also in the commissioning of subsequent generations of higher performing AdLSs. Africa is therefore long overdue for the acquisition of such a transformative facility.

Therefore, African countries are looking ahead, aiming to jump start technology and knowledge-driven economies, and leapfrogging ahead to catch up fast with the rest of the world. This ambition and dramatic change must be driven by instituting the educational framework and research infrastructures, both large and small, that will provide unprecedented opportunities for high-level training, skills growth, and sustainable careers in Africa for Africans. Individual countries are already embarking upon this road, such as South Africa with the world's largest radio telescope called the Square Kilometre Array (SKA), and setting targets for ambitious percentages of GDP to go towards funding science and technology, thereby putting them at the heart of post-colonial African priorities.

Certainly, an AdLS in Africa requires financial capacity, but can and should be made a priority due to its potential. The necessary financial capacity exists. An AdLS costs approximately one billion US dollars to construct, and a similar amount to operate, maintain, and upgrade every decade thereafter. The cost is similar to a football stadium, or a percentage of a large municipal railway system, or a large-scale power station; all of which are eminently affordable by the governments of Africa, especially if they pool their financial resources in a collaborative endeavour. They should make it a priority, because the significant cost would be worth it. For one, AdLS have enormous capacities, which contribute to solving the grand societal and economic challenges, such as climate change, health, and global pandemics. Moreover, AdLSs also contribute immensely to scientific advancement and knowledge creation. For example, since the early 2000s, several Nobel Prizes have relied considerably upon insights gained from experiments conducted at synchrotron radiation sources (Brooks-Bartlett *et al* 2015). Since early 2020, all synchrotron radiation facilities in Europe have become immersed in COVID-19 related research that explores, for example, the structure of the virus, vaccines, and other antiviral medications (Cramer 2020a, Kolar *et al* 2021). In sum, African governments must commit to acquiring an AdLS in order to transform their research landscapes, and strengthen the ability to deal with problems independently. Examples of specific challenges that AdLSs help addressing are: producing sustainable energy, environmental, and water resources; beneficiating Africa's mineral wealth; creating competitive industries; addressing Africa's disease burden with its own innovations and its own pharmaceutical industries; leading its own studies of its cultural heritages; mastering and developing information technologies; participating strongly in global

[1] Information retrieved from elightsources.org: https://lightsources.org/lightsources-of-the-world.

science; and finally, establishing a culture of learning and innovation, on the continent that is equal to all others.

If African countries do not jump on the train, they will fall behind even further, risking increasing dependence on other countries. In many countries from the Global North, visions of economic transition, prosperity, and welfare through knowledge creation and innovation have driven science policy agendas and investments in cutting-edge research facilities. Small and large research infrastructures became important strategic assets (Cramer *et al* 2020c). For example, the European Commission characterises these projects as no less than 'key drivers for European capacity building' (European Commission 2008, p 15). The sociologist Olof Hallonsten (Hallonsten 2020) argued that there has been a veritable 'political hype' around research infrastructures in Europe in the last few years. Almost all the European Union countries, as well as the United Kingdom, Australia, Japan, and China, have published their own research infrastructure roadmaps to drive coherent policy frameworks and the operation of powerful national and collaborative facilities that cater to the needs of their scientific communities (Bolliger and Griffiths 2020). China's long-term vision for research infrastructures entails large scientific facilities as key resources towards Chinese global leadership in science and technology (e.g. Chen 2011, Marcelli 2014). Thus, there is clearly an enormous imperative worldwide to host and enhance large-scale infrastructures.

54.3 African socio-economic and priority drivers for an advanced light source

The extremely high photon brightness and flux provided by AdLSs have empowered significant scientific and technological advances for the benefit of society. They include but are not limited to drug discovery and development; better understanding of diseases, based upon deciphering the protein structures of bacteria, viruses, and parasites; and advances in data storage through studies of electronic structures and magnetism in selected materials. The knowledge gained from bacterial, viral, and parasitic protein structures facilitates the understanding of their functions and the subsequent development of targets for medical interventions.

Ultimately, AdLSs provide countries with the technology that is the most important initial step for vaccine development, and all the benefits associated with such a capability. The enormous importance of these facilities is why for instance the structural biology beamlines at AdLSs remained open during the harshest shutdowns of the COVID-19 pandemic. Effectively, they were frontline health facilities. Furthermore, AdLSs show tremendous promise in other medical applications, allowing innovative imaging techniques of such organs as the heart, lungs, brain, and breast, leading to new diagnostic techniques and therapies.

More recently, AdLSs have been used in what has become a revolution in palaeontology and cultural heritage studies. For example, three-dimensional tomographic images have been made of fossils, rock art, and other artefacts. A key feature of AdLSs is their power to observe materials during their manufacture, processing, and end-use conditions, making AdLSs second-to-none in providing key insights into transient and dynamic processes, such as catalysis, fuel injection, and additive manufacturing.

An AdLS is not just about good science; it is also about empowerment, building communities, training young people (particularly in creating PhD-trained researchers, engineers, technicians and administrators who run international facilities), retaining innovators, and seeding local competitive industries. As such, an AfLS would contribute significantly to the African science renaissance, return of the African science Diaspora, enhancement of university education, training of a new generation of young researchers, growth of competitive African industries, enhancement of Pan-African science diplomacy, and advancement of research that addresses issues, challenges, and concerns relevant to Africa.

Establishing an AdLS on the African continent would empower African researchers to set their own agendas and pursue research of particular relevance for Africa. Recent scholarship has highlighted that big economic and societal challenges, such as climate change and health, are best addressed and solved by tight and globally spread collaborations of science-based knowledge regimes (e.g. Wagner 2008). However, it has also revealed that African scientific communities only act at the periphery of such knowledge regimes, which are dominated by researchers from the Global North. This has led to significant asymmetries regarding agenda setting, funding, and division of labour between collaborators from the Global North and South (e.g. Skupien and Rüffin 2020). Thus, topics that carry immense importance to Africa, such as the environment, health care, energy, agriculture, clean water, mineral extraction, human origins, and the origins of human culture, have not received enough attention.

An AfLS could change things for the better. The specific research opportunities at an AfLS, at the molecular level of using protein crystallography to drive drug discovery and vaccine development, would provide the insights needed to tackle diseases endemic to Africa, such as malaria and other neglected tropical diseases, that also have the potential to (re-)emerge in Europe and elsewhere. This would have significant benefits for Africa, by drawing upon the rich treasure chest of biodiversity and biomolecules available, and perhaps yet undiscovered, opening new therapeutic pathways under African research and ownership. There is a window of opportunity, as many African governments have committed themselves to African innovation in vaccine development and have partnered with new African pharma industries in order to play their parts in defeating future global pandemics. The COVID-19 pandemic clearly shows that the health of the entire planet requires every human to be healthy.

Moreover, an AfLS in Africa could raise the chances of African scientists being noticed within the scientific community, which could open doors for them and the continent. Regarding AdLSs in Europe and the United States, the scientists and researchers that use those facilities form a crucial part of the scientific elite, with several being Nobel Laureates. One could easily argue that access to those facilities raises one's level of recognition and credibility amongst the scientific communities and the likelihood of being awarded grants and prizes, which are important assets for further career advancement. Thus, an operable, competitive, and cutting-edge

AfLS would enable users to become experts in AdLS-based techniques at an early career stage, thereby maximising their productive time in cutting-edge research. It also would encourage the return of the African science Diaspora, establishing a global research community on the continent of Africa. We should not limit ourselves to thinking that an AdLS would only benefit one African country, the one in which it is sited. Indeed, it would belong to, and benefit, the whole African continent. Experience during the COVID-19 pandemic has shown that an AdLS is easily utilised remotely. The need for on-site use has been diminished. Indeed, the realisation of an AfLS entails the companion realisations of better developed local and regional research infrastructures that elevate each of Africa's 55 nations. No African state would be left behind.

AdLSs have made the leap from serving academia to serving industry. Most AdLSs worldwide have dedicated industrial liaisons and business development offices, with the mission to engage with industry for utilisation of AdLS facilities, skills, and intellectual property. They have progressed to become a premier tool that drives and supports industrial innovation (biotechnology, materials engineering, energy technology, and many more) through the advanced characterisation of materials and living matter, going far beyond the capabilities of conventional x-ray sources. Having such diverse applications and impacts in academic, applied, and commercial research and innovation, an AdLS would impact the Sustainable Development Goals (SDG) as defined by the United Nations (UN). Figure 54.3 illustrates in substantive terms how an AfLS would address all the 17 UN SDGs, as well as the African continent's own Grand Challenges.

Specifically, AdLSs can engage all actors relevant for innovation. A key ingredient for innovation is the collaboration with the private sector. Looking at existing AdLSs around the world, several facilities, such as the Advanced Photon Source (USA), SPring8 (Japan) and Swiss Light Source (Switzerland) operate

Figure 54.3. Informative chart illustrating how an African Light Source will address all 17 United Nations Sustainable Development Goals (UN SDGs), which are an urgent call for action by all countries.

beamlines that are wholly or partially owned by industry, and most AdLSs run significant commercial programmes with industry for proprietary access to the x-ray beamlines. Complementing this, it is important to note significant industry engagement through the public access programmes, wherein industry, usually in collaboration with academic teams, works on applied R&D with publishable results. This is an important engagement mechanism, supporting innovation via pre-competitive research, and light source reporting typically estimates that some 20%–40% of public programmes have industry relevance or engagement as part of the work being conducted. Not only does industry utilise AdLSs, but AdLS themselves utilise the high technology provided by industry. Industry-supplied accelerator components are particularly crucial during the main construction phase, but continues as the facility develops, renovates, and upgrades. Facilities may also create spin-off activities, including those involving advanced instrumentations, thereby creating value, jobs and industrial competitiveness.

Aside from the facility itself, an AdLS often forms the anchor for a combined science and industry park that springs up around it. Thus, it has the potential of becoming a clustered 'science-megalopolis' of globally competitive research, training, innovation, and inspirational outreach activities. To allow an AdLS to harness its full potential, it should therefore be highly accessible internationally, have a lot of space around it, be embedded within or close to a city, and also well connected to communication and energy grids. This may be a green-field site, or where there is space, on or close to an existing university or technology campus.

Not least, the impact of an international facility is also in lowering borders, enabling transnational mobility, understanding, and exchange, and working with science as a diplomatic tool in the context of science for diplomacy (e.g. Flink and Ulrich 2010, Ruffini 2017). The use of science in a diplomatic context is increasing with large-scale research infrastructures that are able to act as science and technology hubs, flying above or around political radars, and bringing together nations that may otherwise find relationships challenging.

There are many examples of how large-scale scientific collaborations have played a critical role in the politically delicate contexts of alliance-building and decision-making. CERN, which now plays a well-recognised role as a global hub for new scientific alliances and initiatives, started as a peaceful European collaboration after the end of the Second World War. It was financially and politically supported by the United States (e.g. Krige 2006). In another example, SESAME became the first AdLS in the Middle East. The most compelling aspects of this project are its members. They are Bahrain, Cyprus, Egypt, Iran, Israel, Pakistan, the Palestinian Authority, and Turkey. Even though Iran, Israel, and the Palestinian Authority, as well as Turkey and Cyprus, are mired in deep political conflicts, they have found common ground in SESAME (e.g. Schopper 2006, Schopper 2017). CERN, UNESCO, the EU and several influential scientists provide critical support for project. Their support takes various forms. For instance, they support it through various Framework Programme projects, such as CESSMAG, OPEN SESAME, and BEATS. They also provide organisational infrastructures and initial funding, which is one reason why the project is frequently named mini-CERN. Indeed, the founding histories of European collaborative AdLSs, such as the ESRF, and similar research infrastructures, and their tight connections to European politics and national science policy agendas are testaments to how large-scale research projects can be politics by other means (e.g. Hallonsten 2014, Cramer 2020b).

Today, there is a chance to bring Africa as a continent closer together through scientific collaboration. Africa is an interesting case politically, because Africa has country boundaries superimposed on it by former colonial powers. These boundaries do not coincide with the regional roots of similar African cultures. Put differently, Africa did not make its own internal boundaries, but inherited them. However, they have in some sense become *de facto* the established inter-country boundaries, and one does not see these easily changing.

The enunciation of the Pan-African vision has many authors, dating back to the times of the slave trade and early African colonial liberation movements (Makalani 2017). The vision is today enshrined in the African Union and many Pan-African institutions and projects. This gives the AfLS a special role to drive Pan-African science diplomacy. Africa must unite around this objective, plan for it, develop it, build it, and utilise it. The AfLS would then take its rightful place amongst the greatest of entities that facilitate the healing of the continent, and uniting it in peace and democracy towards solving the continent's greatest challenges.

54.4 The growth of advanced light source utilisation by African researchers

The likely first published measurements involving African scientists at an AdLS occurred around 1990 in the North of Africa (Bhat *et al* 1990) and around 1994 (Als-Nielsen *et al* 1994) in the South. From that time, the African userbase of international AdLSs has grown steadily. The first formal call for a Pan-African AdLS was recorded into the *2002 Strategy and Business Plan* of the aspirant African Laser Centre, which was then established the following year, based in Pretoria, South Africa (Mtingwa *et al* 2018). This early formal proposal for a synchrotron in Africa was something strongly encouraged by the Edward Bouchet-Abdus Salam Institute (EBASI). The EBASI in turn is supported by the Abdus Salam International Centre for Theoretical Physics (ICTP) in Trieste, Italy. The conversations, writings, and calls for an AdLS in Africa grew so that the number of both individuals and institutions involved became too numerous to mention explicitly. Likewise, the developments within each African country of smaller-scale local laboratory-based research, utilising similar techniques to those at large-scale AdLSs, grew, as did the use of external national and international AdLS facilities. It is not the purpose of this essay to chronicle this development at a national scale, but simply to recognise that these national-level developments occurred in parallel with conversations surrounding a joint Pan-African AdLS.

In this way, interested parties built the momentum towards an AfLS. It became necessary to draw all the participants together into a common, coherent, fully mandated vehicle. In building the momentum, colleagues scoured electronic media to increase the number of participants in this AfLS conversation, and they solicited as widely as possible for nominations to an AfLS Interim Steering Committee (AfLS-ISC). Finally, on 16 August 2014, the early AfLS motivators accepted all the nominations and constituted the AfLS-ISC, which inaugurated itself electronically on that date. It had representation from the following countries and regions: Egypt, Ethiopia, Nigeria, Rwanda, Senegal, South Africa, Zimbabwe, United States, Europe, and Japan. The AfLS-ISC had the limited mandate to initiate a transparent, inclusive, and democratic process, incarnated finally in the 1st African Light Source Conference and Workshop, hosted by the ESRF. At the conclusion of the latter, a duly elected, and more representative and inclusive, AfLS Steering Committee assumed leadership in the grand push for an AfLS. From that momentum, the vision for an AfLS has taken form, matured, and is constantly attracting new support.

There are many parallel efforts in Africa related to the development of science and technology, and their impact on socio-economic development and the strengthening of democracies. A significant one was the 1st African Higher Education Summit on Revitalising Higher Education for Africa's Future, which convened in Dakar, Senegal during 10–12 March 2015. Several key Pan-African organisations collaborated in convening the Summit, including the African Union Commission (AUC), TrustAfrica, Council for the Development of Social Science Research in Africa (CODESRIA), United Nations Africa Institute for Development and Economic Planning (IDEP), Association for the Development of Education in Africa (ADEA), Association of African Universities (AAU), African Development Bank, as well as other national and international partners. Two outcomes of this meeting were a Declaration and Action Plan, to bring about a shared Strategic Framework for inclusive growth and sustainable development, and a Global Strategy to optimise the use of Africa's resources for the benefit of all Africans. Establishing a synchrotron as a centralised African scientific facility was recommended, representing the second major Pan-African statement identifying the crucial significance of an AdLS for all Africa.

At a national level, significant work was also underway to join international facilities. On 21 May 2013, South Africa signed a medium- term arrangement with the ESRF at a level of 0.3% of the ESRF budget. In doing so, South Africa became the 20th country to join the European collaborative and international AdLS. The first South African experiments at the ESRF began as early as 1994, just as the facility commenced user operations. These and other early efforts grew, leading to the emergence of a vibrant and interdisciplinary South African synchrotron AdLS user community. Collaborations with the ESRF have continued to strengthen and flourish, and important breakthroughs have been achieved, notably in the domains of palaeontology, palaeoanthropology, materials science, and macromolecular crystallography. Palaeontology in Africa more generally has been one of the big success stories of AdLS science, with the involvement of some 12 African countries.

The number of African AdLS users is extremely difficult to ascertain, since there is no easily accessible system of user statistics. Highly trained potential African AdLS users have a limited possibility of using their skills from within Africa, or of obtaining AdLS-relevant employment within Africa. In fact, one purpose of this document is to highlight to African governments that this is the case, to the detriment of Africa. At this stage of the Roadmap towards an AfLS, it is vital that Africa retains the high-end skills that are being produced. African governments should commit to providing sufficient employment opportunities and local infrastructures that are relevant to AdLS research as their entry-level contributions to the AfLS project. Up to now, most of the interventions to build high-end capacity in Africa have simply fed the Diaspora, as the highly trained emerging researchers not only acquired skills, but also global science networks.

However, in recent years, much progress has evolved. Many African governments, and also the African Union (AU) and other organisations, are articulating the pre-eminent role of science, technology, and innovation in wealth beneficiation. Learned science bodies and academies now exist in many African countries, and they are networking in a Pan-African effort to enhance a large suite of disciplines. Moreover, national science funding agencies exist in many African countries. There is currently a call for a standard 1% of GDP in each country to be allocated for science R&D (STISA 2024). The international connections of African science are increasing at a good pace, both using the Diaspora and friends of Africa, as well as individuals and institutions that perceive that, for science to be effective, it must also be global. The AfLS has by now deep connections, many long-term, with most of the AdLSs worldwide. The same is true for its connections to the major learned societies of the world. Special mention should be made of the IUCr-UNESCO-*LAAAMP* OpenLabs, sponsored by the International Union for Crystallography (IUCr), UNESCO, and Lightsources for Africa, the Americas, Asia, Middle East and Pacific (*LAAAMP*).

The collaboration of those three organisations has led to the establishment of significant local and regional research infrastructures for x-ray-based sciences. The OpenLabs facilities are regional research and training laboratories, and they are also feeder infrastructures for a future AfLS. As an outgrowth of *LAAAMP*, the Benin government established X-TechLab, within the government's new innovation hub called Sèmè City in the city of Cotonou, to train students in the science of crystallography, which has many applications at AdLSs. Similarly, the East African Institute for Fundamental Research (EAIFR) located in Kigali, Rwanda functions as a regional infrastructure, as does the Sir Aaron Klug Centre for Imaging and Analysis in Cape Town, South Africa. There are now many such facilities, and a compendium of these facilities is collected and kept up-to-date on the Website of the AfLS (AfLS 2021), in the section on African science.

In summary, the userbase of African scientists (by origin/nationality) who have experience as AdLS users now has a footprint in more than 50% of Africa. If one includes the Diaspora, the African userbase may be on the order of 300 users. This number is unofficially taken by some AdLS Directors as a threshold to justify the construction of a new AdLS facility. Brazil did not use such a threshold, and the eminence in the youth was ably created by the need for emerging scientific leaders. From the perspective of established and potential scientific competence, we are confident that Africa is more than ready for an AfLS.

54.5 The roadmap to conceiving, building, and operating an African light source

An AdLS for Africa, in Africa, conceived, designed, constructed, commissioned, and operated by Africans, cannot wait for tomorrow. It is a challenging project, far beyond the scale and science capacities of any research infrastructure so far constructed in Africa. As such, it will take several years to scope, fund, and establish on the African continent. Nonetheless, steps have necessarily started along this road towards the goal of an operational AfLS by early the next decade.

Such a goal would mean construction (facility, infrastructures) to start before 2030, with funding and site secured and chosen within the next few years. This leaves about five years from today to build the case for an AfLS, and further, to build the skills, capacity, and active user community. Moreover, it will be essential to bring onboard the wide African scientific community, politicians, and funding bodies to achieve the goal of an AfLS. Indeed, the step of achieving the full support of African governments and the African Union will be indispensable. In the meantime, it will be essential to establish smaller, but fully research competent, key regional centres, such as Benin's X-TechLab and the African Laser Centre, which is a Pan-African network of laser research and training laboratories. Many students should pursue training in accelerator science and technology, as well as x-ray analytical techniques, and engineering and data science. Indeed, there will be a need to encourage young Africans to pursue science, technology, engineering, and mathematics (STEM) education and careers. Africa will not be alone in this endeavour, as many friends from around the world are ready to provide advice and support. However, African nations themselves must provide the principal funding for an AfLS.

Such pooled, collaborative approaches to fund, construct, and operate AdLSs have already been achieved in international research infrastructure consortia around the world, such as the ESRF in France, SESAME in Jordan, and the European XFEL project in Germany. That project, which is mainly driven by Germany and Russia as the two largest shareholders, originated under the auspices of the Hamburg-based German synchrotron facility called DESY and has close ties to Hamburg University and local and regional industry.

54.5.1 Building momentum and community

To date, the main AfLS activities have been coordinated within the African Light Source Foundation, which is a legal entity with a constitution and registered as a Trust in the Masters Office of the High Court in South Africa (AfLS 2021). A fully mandated (by the scientific community) AfLS Executive Committee (AfLS-EC), which emanated from the previous AfLS Steering Committee and is composed of established African and international scientists with extensive AdLS experience, coordinates the Foundation's activities. Over the past two years, the AfLS-EC has been in conversations with several African governments, Pan-African bodies (including the AU, learned societies and academies), and international organisations. The AfLS Foundation produced an AfLS Roadmap in 2015 and has updated and ratified it several times, involving both bottom-up and top-down approaches (AfLS 2021). There has been an extensive programme of training for young African scientists in the advanced research techniques available at AdLSs. These trainings are occurring in partnership with a number of AdLS facilities around the world. Moreover, those trainings are organised in collaboration with a few other organisations, including the Lightsources for Africa, the Americas, Asia, Middle East, and Pacific (LAAAMP 2020), X-TechLab in Benin (X-TechLab 2021), International Atomic Energy Agency (IAEA), African Synchrotron Network for Advanced Energy Materials (ASNAEM 2021), Synchrotron Techniques for African Research and Technology (START 2021), and many others. Relative to the partnering of South Africa with the ESRF, there has been a string of training and outreach activities, with young South African researchers and engineers benefitting from the ESRF environment for comprehensive and long-term training in AdLS research.

The AfLS Foundation has organised several international conferences and workshops to promote the AfLS project. The 1st AfLS Conference and Workshop (AfLS1) convened at the ESRF in Grenoble, France in 2015. The 2nd AfLS Conference and Workshop (AfLS2) convened jointly in November 2019 with the Pan-African Conference on Crystallography at the University of Ghana in the Accra suburb of Legon. Due to the COVID-19 pandemic, a somewhat smaller AfLS-2020 Workshop convened online in November 2020 jointly with the African Physical Society (AfPS). The 3rd AfLS Conference and Workshop (AfLS3) will also be a joint virtual meeting with the African Physical Society, Pan-African Conference on Crystallography, and several other Pan-African professional science societies, all hosted by X-TechLab. The 4th AfLS Conference and Workshop (AfLS4) is planned to convene during November 2022 as an in-person meeting in Kigali, Rwanda.

In efforts to progress along the roadmap towards the establishment of an AdLS on the African continent, the AfLS Foundation is establishing good relationships with the AU and various African governments, as well as key organisations inside and outside Africa. The AfLS Foundation is currently arranging in-person visits of key African leaders, government officials and potential supporters to existing AdLS facilities around the world. The prime goal of these visits is to show them the kind of amazing scientific facility that we are envisaging to establish in the near future in Africa.

The African Academy of Sciences (AAS), through its African Synchrotron Initiative (ASI), has also been advocating for an African synchrotron project. The AfLS Foundation and the AAS-ASI have been in discussions to define how best they can work together towards making an African synchrotron light source a reality. In addition, both the AfLS Foundation and the AAS have been involved in conversations with the AU and different African governments.

54.5.2 The Conceptual Design Report

The Conceptual Design Report (CDR) is one of the most significant documents in the Roadmap for an AfLS. This document will concretise the conversations with African governments, as well as Pan-African and international organisations. Its intended audience includes policymakers, politicians, academics, engineers, technicians, businesspersons, industrialists, funding bodies, strategic thinkers, all possible stakeholders and stakeholder organisations, and of course, the general public. For many large-scale multinational research infrastructures, the true day zero, from which the path to the actual commissioning of the facility can be dated, is in fact the date of completion of the CDR. This will represent the transition from a dream to a vision that has been concretised and described in detail, including its motivation, context, purpose, and content. However, the CDR will not specify the site or source of funds for a future AfLS facility. However, it will point the way to criteria for site selection. Furthermore, it will provide a reasonably accurate conception of what must be constructed, including site preparation, supporting infrastructure, and staff requirements. This means that the CDR will empower detailed conversations from the level of political concerns, to scientific goals, and all other levels, including engineering, technical, and financial.

The CDR will consist of five volumes as follows (AfLS 2021):

Volume I. Scientific, Socio-Economic, Educational, and Political Benefits
Volume II. Machine Design Concepts
Volume III. Scientific Capabilities and Beamline Technical Concepts
Volume IV. Technical Infrastructure and Building Design
Volume V. Multinational Project Governance and Finance

The production of the CDR will embody the African concept of *Ubuntu*, which means that it will be a document by the community and for the community. The procedure will be explicitly consultative, democratic, and inclusive. Each volume of the CDR will have its own sub-editor, and there will be an overall CDR editor. Each chapter, or sometimes each group of chapters, of the CDR volumes will have its own set of lead authors, who will have the dual roles of writers and convenors, taking the lead in arranging virtual online, well-advertised, and carefully planned Community Consultative Workshops (CCWs). The CCWs reach out to the community, on a per-discipline basis, inviting all to become co-authors. Each of the co-authors or groups of co-authors submit at the workshop a letter of interest (LoI), supported by a presentation. The LoIs and presentations define the research and science the community would like to perform, or any other aspects of the CDR where they would like to make input. The community will have the opportunity to discuss co-authors' ideas in the context of similar or related presentations. The lead authors then facilitate collaborations among co-authors and groups of co-authors. Ultimately, co-authors produce white papers that describe the collaborative inputs as succinctly as possible. These will be the final outcomes of the CCWs, and the white papers will be peer reviewed. This step assures that the (groups of) co-authors retain their rights for recognition of their roles in producing the CDR. It establishes the full authorship of the CDR, including the editor, sub-editors, lead authors, and co-authors. The AfLS CDR Team will edit, arrange the review of, and produce a set of journal articles for the publishing of the white papers. The particular journal for this part is still under discussion; however, it will be an Africa-based journal. The lead authors, under supervision of the sub-editors, will integrate the white papers smoothly into the content of the five volumes of the AfLS CDR. The CDR editor has overall responsibility for the CDR. The identification of the editor, sub-editors, and lead authors is nearly complete, and the plan is to launch the advertisement of the per-discipline CCWs early in 2022. The CDR is planned for completion by the end of 2022.

54.6 Conclusions

This essay argues the case for an AfLS and has examined the development and activities of the AfLS Foundation, which is driving the case forward and building momentum towards its realisation. There are many stakeholders in this endeavour, the multiplicity of which can work together in a coherent way, retaining their own branding but cooperating towards a common goal. A new feature of the progress along the Roadmap are examples where external funding sources partner well with local African entities, and even more significantly, where African governments take responsibility for the development of local infrastructures.

The AfLS Foundation's Roadmap involves training in AdLS accelerator and beamline advanced research techniques. This is happening, although the throughput is still low. Unfortunately, even at the low throughput we currently have, the poor availabilities of appropriate positions and sufficiently well-equipped local laboratories mean that the training programmes dominantly feed the African science Diaspora, bleeding talent away from Africa rather than retaining it.

An AfLS is becoming ever more firmly established on the agenda of the African Union. Benin's government firmly supports X-TechLAB as a contribution to the AfLS's feeder infrastructure, and the Ghanaian Government is positioning itself to champion the AfLS wherever it may ultimately be sited. An advanced light source on the African continent would create and harness world-leading science and technology capacities, thereby achieving African goals for poverty reduction, scientific competitiveness, job creation, and industrial growth, and African sustainability and resilience.

There has been significant progress to date; however, much remains to be done to realise an advanced light source in, by, and for Africa no later than the early 2030s.

References

AfLS 2021 *The African Lightsource* https://africanlightsource.org/ (Accessed: 7 October 2021)

Als-Nielsen J, Freund A K, Grübel G, Linderholm J, Nielsen M, Sanchez del Rio M and Sellschop J P F 1994 Multiple station beamline at an undulator x-ray source *Nucl. Instrum. Methods Phys. Res.* B **94** 306

ASNAEM 2021 *African Synchrotron Network for Advanced Energy Materials* https://asnaem.org/ (Accessed: 7 October 2021)

Bhat H L, Clark S M, El Korashy A and Roberts K J 1990 A furnace for *in-situ* synchrotron Laue diffraction and its application to studies of solid-state phase-transformations *J. Appl. Crystallogr.* **23** 545

Bilderback D H, Elleaume P and Weckert E 2005 Review of third and next generation synchrotron light sources *J. Phys.* B **38** S773–97

Bolliger I and Griffiths A 2020 *The Introduction of ESFRI and the Rise of National Research Infrastructure Roadmaps in Europe* Big Science and Research Infrastructures in Europe ed ed K C Cramer and O Hallonsten (Cheltenham: Edward Elgar) pp 101–27

Brooks-Bartlett J C and Garman E 2015 The Nobel science: one hundred years of crystallography *Interdiscip. Sci. Rev.* **40** 244–64

Chen H 2011 *Large Research Infrastructures Development in China: A Roadmap to 2050* (Berlin and Heidelberg: Springer)

Connell H *et al* 2018 The African light source project *Afr. Rev. Phys.* **13** 108–18 http://lamp.ictp.it/index.php/aphysrev/article/view/1610/586

Connell H *et al* 2019 Towards an African light source *Biophys. Rev.* **11** 499–507

Cramer C K 2020b *A Political History of Big Science: The Other Europe* (London: Palgrave Macmillan)

Cramer K C 2020a The science must go on. Working practices at single-sited European research infrastructures in times of COVID-19 pandemic *Elephant in the Lab* 14 May https://elephantinthelab.org/the-science-must-go-on

Cramer K C, Hallonsten O, Bolliger I and Griffiths A 2020c Big Science and research infrastructures in Europe: history and current trends *Big Science and Research Infrastructures in Europe* eds ed K C Cramer and O Hallonsten (Cheltenham: Edward Elgar) pp 1–26

European Commission 2008 Challenging Europe's Research: Rationales for the European Research Area (ERA). Report of the ERA Expert Group (Luxemburg: Office for Official Publications of the European Communities)

Flink T and Schreiterer U 2010 Science diplomacy at the intersection of S&T policies and foreign affairs: toward a typology of national approaches. *Sci. Public Pol.* **37** 665–77

Hallonsten O 2014 The politics of European collaboration in Big Science *The Global Politics of Science and Technology* ed M Mayer, M Carpes and R Knoblich vol 2 (Berlin: Springer) pp 31–46

Hallonsten O 2020 Research infrastructures in Europe: the hype and the field. *Eur. Rev.* **28** 617–35

Kolar J, Harrison A and Gliksohn F 2021 *Effect of the COVID-19 Pandemic on the Working Practices of Analytical Facilities II* https://ceric-eric.eu/wp-content/uploads/2021/01/PAPER_Effect-of-the-COVID-19-Pandemic-on-the-Working-Practices-of-Analytical-Facilities-II.pdf

Krige J 2006 *American Hegemony and the Postwar Reconstruction of Science in Europe* (Cambridge, MA: MIT Press)

LAAAMP 2020 *Advanced Light Sources and Crystallography: Tools of Discovery and Innovation* ed E Malamud pp 1–42 https://laamp.iucr.org/tasks/brochure (Accessed: 7 October 2021)

LAAAMP 2021 Lightsources for Africa, the Americas, Asia and Middle East Project https://laamp.iucr.org/home (Accessed: 7 October 2021)

Makalani M 2017 *Africana Age* http://exhibitions.nypl.org (Accessed: 16 February 2017)

Marcelli A 2014 The large research infrastructures of the People's Republic of China: an investment for science and technology *Phys. Status Solidi b* **251** 1158–68

Mtingwa K S and Winick H 2018 Synchrotron light sources in developing countries *Mod. Phys. Lett.* A **33** 1830003

Ngabonziza P 2019 The need for multidisciplinary scientific infrastructures in Africa: the importance of an African lightsource *Elephant in the Lab*

Ruffini P-B 2017 *Science and Diplomacy. A New Dimension of International Relations* (Berlin: Springer)

Schopper H 2006 SESAME. Synchrotron-light for experimental science and applications in the Middle East *Sharing Knowledge Across the Mediterranean Area. Towards a Partnership for Sustainable Management of Resources and the Prevention of Catastrophes* ed P Faugeras, A Hoummada and R Klapisch (Washington, DC: IOS) pp 89–92 https://ebooks.iospress.nl/volume/sharing-knowledge-across-the-mediterranean-area

Schopper H 2017 The light of SESAME: a dream becomes reality *Riv. Nuovo Cimento* **40** 199–239

Skupien S and Rüffin N 2020 The geography of research funding: semantics and beyond *J. Stud. Int. Educ.* **24** 24–38

START 2021 *Synchrotron Techniques for African Research and Technology* https://start-project.org/ (Accessed: 7 October 2021)

STISA 2024 *Science, Technology and Innovation Strategy for Africa 2024 Reporting on the AU Executive Council Decisions of 2006* https://au.int/sites/default/files/newsevents/workingdocuments/33178-wd-stisa-english_-_final.pdf

United Nations, Department of Economic and Social Affairs (UN DESA) 2019 *World Population Prospects 2019: Highlights* ST/ESA/SER.A/423 https://population.un.org/wpp/Publications/Files/WPP2019_Highlights.pdf

Wagner C S 2008 *The New Invisible College: Science for Development* (Washington, DC: Brookings)

X-TechLab 2021 *X-ray Techniques Lab in Bénin* https://xtechlab.co/en/ (Accessed: 7 October 2021)

IOP Publishing

Big Science in the 21st Century

Economic and societal impacts

Panagiotis Charitos, Theodore Arabatzis, Harry Cliff, Günther Dissertori, Juliette Forneris and Jason Li-Ying

Chapter 55

Latin America–Europe collaboration with HELEN and EPLANET: 2005–2015

Luciano Maiani and Veronica Riquer

Galileo Galilei, in the seventeenth century, begins the Scientific Revolution and inaugurates a new style of doing research: '*Misura ciò che è misurabile, e rendi misurabile ciò che non lo è* .'[1]

Galileo taught us to exploit all opportunities, to improve our instruments constantly, make use of our talent, and publish and divulgate our results. He made evident the value of this advice by creating and improving the telescope and by writing his fundamental findings in a beautiful Italian accessible to all, in *Dialogo Sopra i Due Massimi Sistemi.*

The lessons of Galileo prompted enormous progress in the resolution power of telescopes, i.e. the area where incoming light is collected. In the subsequent four centuries, the area of telescopes has increased at an exponential rate from the size of a coin to the order of hundreds of square meters, and will increase even more in the future with the diffusion of telescope arrays.

A similar exponential progress has occurred for the energy reached by particle accelerators, which regulates our capability to explore the structures of the microcosm of subnuclear particles and to discover new particles of increasing mass (following the energy–mass relation of Albert Einstein).

The latest leap in this field was the invention of *particle colliders*. An alternative to sending energetic particles toward a target made of a dense material (what in Anglosphere newspapers is called an 'atom smasher'), colliders produce head-on collisions between particles of two different beams which travel in opposite directions. It is a concept that was developed for the first time in Italy, in the Frascati Laboratories, by the Austrian physicist Bruno Touschek who had settled in

[1] Measure what you can measure and make it measurable what still cannot.

doi:10.1088/978-0-7503-3631-4ch55

Figure 55.1. The region near Geneva, where CERN is located. The LHC is located in an underground tunnel at about 100 m depth, indicated by the red circle. Lake Leman is visible above the circle, with Geneva in its right corner. In the background are the Alps and Mont Blanc. Copyright 2001-2023 CERN.

Rome after the war, with the first electron–positron collider called the Anello di Accumulazione (AdA). In his Italian with a charming Austrian accent, he describes the collision as '*treno contro treno*'[2].

The diameter of AdA was about one meter, but the concept was so revolutionary and efficient that electron–positron colliders of increasing size have since been constructed in laboratories all over the world, up to the Large Electron Positron (LEP) collider, built at CERN in the 1980s, which was installed in a circular ring 27 km in circumference.

The first collider of protons and antiprotons was constructed at CERN in the late 1970s, by Carlo Rubbia and Simon Van der Meer, to discover the carriers of the weak forces, the so-called intermediate vector bosons, in 1983. The largest proton–proton collider in the world, built at CERN in the LEP tunnel, is called the Large Hadron Collider (LHC) (figure 55.1). It has led to the discovery of the Brout–Englert–Higgs boson, the missing element of the theory of elementary particles, in 2012 by the ATLAS and CMS Collaborations.

[2] Train against train.

55.1 International collaboration in science

International collaboration in physics was born in Europe after the Second World War, from the need to build large and complex research infrastructure to explore the world of subnuclear particles.

This world had been unveiled by the reactions produced by cosmic rays in Earth's atmosphere and it could be explored only with big particle accelerators that no country in Europe could afford to build with her own resources.

The vision of a European laboratory dedicated to fundamental physics with large-scale infrastructures had been outlined by the Nobel Prize winner Luis De Broglie, one of the founding fathers of quantum mechanics, with a letter sent to a conference in Lausanne in 1949.

The vision of distinguished European scientists and state persons, among them Edoardo Amaldi, Pierre Auger, Leo Kowarski, Alcide De Gasperi, and Konrad Adenauer, led to the realization of CERN in 1954, only five years after the prophetical letter of De Broglie. Established by twelve member states, CERN is now led by 22 member states [1].

Similar organizations followed, dedicated to the study of outer space, such as the European Space Agency (ESA), of molecular biology, such as the European Molecular Biology Laboratory (EMBL), and of astronomy on a large scale, such as the European Southern Observatory (ESO).

55.2 The user community

An important transformation took place in CERN in the 1980s, following the construction of the Large Electron Positron collider (LEP).

Experiments at LEP required very large and sophisticated detectors and correspondingly large human and financial resources, which CERN itself could not make available.

Universities and the associated countries formed four large-scale collaborations, with extended funds for the construction and operation of detectors and also to support the travel to Geneva of professors and students, the CERN users, who would collect, interpret, and finally translate into new physics the large mass of data produced by LEP.

The phenomenon repeated itself, on a larger scale, with the Large Hadron Collider (LHC), which was installed in the LEP tunnel from the beginning of the year 2000. CERN users amount today to well over 10 000 physicists and engineers from around the world.

55.3 Bringing Latin America to Europe (2005–16 and beyond)

Cosmic rays and elementary particle physics in Latin America have been pioneered by outstanding personalities, starting from the years 1930s, including Manuel Sandoval Vallarta in Mexico, and Gleb Watagin, Mario Schenberg, and Giuseppe Occhialini in Brazil. After the War, Latin American physics acquired an international dimension due, among others, to Marcos Moshinsky in Mexico,

Roberto Salmerón and José Leite Lopes in Brazil, and Carlos Bollini and José Giambiagi in Argentina.

Relationships with the United States have been considerably reinforced by the historical teaching of Richard Feynman in Rio de Janeiro in 1951 and by the visit of Leon Lederman to Brazil in 1981. James Cronin, Nobel Prize winner in 1980 for the discovery of CP violation in K decays, has been among the promoters of the Pierre Auger Observatory for ultra-high energy cosmic rays, realized by a large international collaboration in Malargue, Argentina. In general, until 2005 physics in Latin America has remained closely connected to the United States and to FermiLab experiments, with essentially episodical links to European physics.

In May 2001 the first Latin American School of Physics was organized by CERN in collaboration with Centro Latino Americano de Fisica (CLAF). A this time, construction of the LHC and related detectors was in full development. The number of registered users at CERN from Latin American countries was slightly above 40 units, mostly related to collaboration with the DELPHI and L3 experiments at LEP.

Luciano Maiani, then Director General of CERN, participated in the school. Juan Antonio Rubio and Verónica Riquer organized a meeting on the possible evolution of Latin America–CERN relations, attended by physicists and by funding agency representatives from Argentina, Brazil, and Mexico. Following Riquer's suggestion, the meeting proposed to hold the next CERN–CLAF school in Mexico in 2003 and to initiate consultations on a possible Latin American collaboration in LHC experiments and computing GRID.

To explore the possibility of closer CERN–Latin America links, Juan Antonio Rubio prepared a report on the particle physics potential in Latin America [2]. In synthesis, Rubio reported of a community of about 800 investigators, 60% in theoretical and 40% in experimental physics, many working at Fermilab and a minority at CERN.

With Riquer and Rubio, we observed that a major obstacle for Latin American scientists to take part in the experiments at the LHC was the lack of funds for participation in the CERN experiments, in particular for students—*mobility was the key*. The outcome was Project HELEN [3] (High Energy Physics Latinamerican–European Network).

Following the suggestion made to us by the EU Commissioner Philippe Busquin, we submitted the project to América Latina Formación Academica (ALFA), a program created by the European Commission to facilitate the scientific interchange between Europe and Latin America.

In Europe, the project was focused on the large laboratories of CERN, DESY, and Gran Sasso.

On the other side of the Atlantic, the Pierre Auger Observatory for high energy cosmic rays in Malargue (Argentina), had already attracted broad European participation from CERN member states (e.g. France, Germany, Italy, Spain, Portugal). It naturally became the Latin American pole of HELEN and, later, of EPLANET. HELEN started in July 2005 and was concluded in April 2009 (figure 55.2).

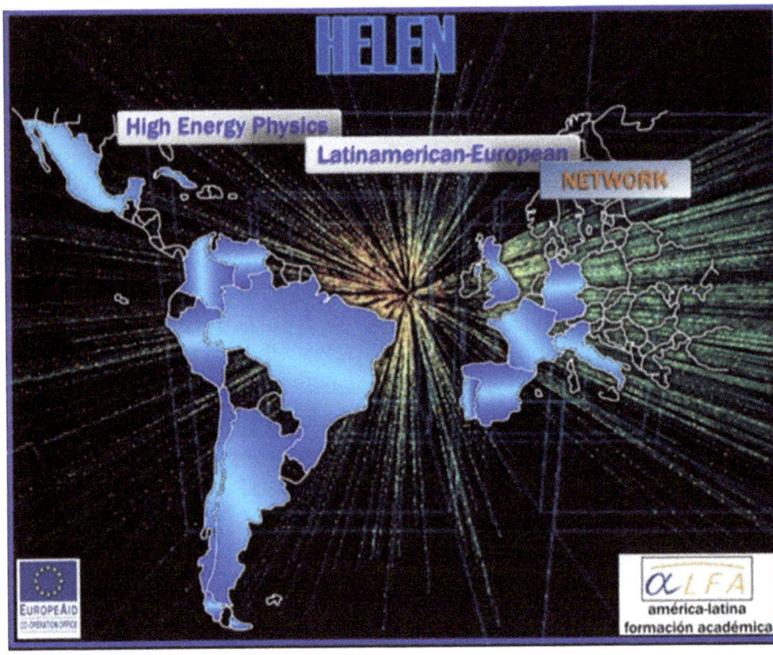

Figure 55.2. HELEN's logo. The countries of participating institutions are in blue.

Prompted by the success of HELEN, we proposed a new project that started in February 2011. This was EPLANET (European Particle Physics Latinoamerican Network), funded by the European Union in the Marie Curie-People action IRSES, a part of the 7th Framework Program. Supported by EPLANET, professors and graduate students from Latin America have been able to participate in the exciting exploration of the microcosm started in 2010, when LHC began its first physics run.

In the following, we shall illustrate the adventure of creating a Latin American community in the very scientific heart of Europe as we lived it, Verónica Riquer, Juan António Rubio, and myself, first with HELEN and continuing with EPLANET.

55.4 HELEN, July 2005–April 2009

HELEN' s declared objectives were:

- Train younger generations of physicists in high energy physics, thereby promoting fundamental physics in Latin American countries and contributing to the modernization of physics education there. CERN, DESY, and Gran Sasso facilities, in particular the LHC, HERA, and their experiments, as well as the experiments in the Pierre Auger Observatory for Cosmic Rays in Malargue (Argentina), were indicated as the fundamental tools for an advanced training program.

- Facilitate access of Latin American countries to the technological benefits in the domains of accelerator, detector, and information technology (the global LHC data GRID as an example).
- Strengthen the integration of the European and Latin American physics communities.

Institutions participating in HELEN formed a large network, and are listed in tables 55.1 and 55.2. The network consisted of:

- 22 universities/research institutions from eight Latin American countries.
- 16 universities/research institutions from six European countries.
- The European organization CERN (Switzerland).
- The Pierre Auger Observatory (Argentina).

Table 55.1. Latin American institutions participating in HELEN. Interlocutors elected to HELEN's first Executive Board (EB) are indicated.

Institution	Interlocutor	Country
Universidad de Buenos Aires	R Piegaia	Argentina
Universidad Nacional de La Plata	T Dova (EB)	Argentina
Universidad Nacional de Mar del Plata	M Benedetti	Argentina
Centro Brasileiro de Pesquisas Fisicas (CBPF)	J Dos Anjos	Brazil
Universidade de São Paulo—USP	A Szanto de Toledo	Brazil
Universidade do Estado do Rio de Janeiro—UERJ	A Santoro (EB)	Brazil
Universidade Federal do Rio de Janeiro—UFRJ	L De Paula	Brazil
Pontificia Universidad Catolica de Chile	M A Diaz	Chile
Universidad Técnica *Federico Santa María*	I Schmidt (EB)	Chile
Universidad de Antioquia	W Ponce	Colombia
Universidad Antonio Narinõ	M Losada	Colombia
Universidad de los Andes	C Avila (EB)	Colombia
Universidad Nacional de Colombia	R Martinez	Colombia
Instituto Superior de Tecnologías y Ciencias Aplicadas InSTEC	F Guzman (EB)	Cuba
Centro de Investigación y de Estudios Avanzados del IPN CINVESTAV	A Zepeda (EB)	Mexico
Universidad Nacional Autonoma de México (UNAM)	A Menchaca	Mexico
Universidad Michoacana de San Nicolas de Hidalgo (UMICH)	L Villasenõr	Mexico
Benemérita Universidad Autónoma de Puebla (BUAP)	A Fernandez	Mexico
Universidad Peruana *Cayetano Heredia*, Lima	J Castromonte	Peru
Pontificia Universidad Catolica del Peru	A Gago (EB)	Peru
Universidad Central de Venezuela	P Arias	Venezuela
Universidad de Los Andes	L Nuñez (EB)	Venezuela

Table 55.2. European institutions participating in HELEN. Interlocutors elected to HELEN's first Executive Board (EB) are indicated.

Institution	Interlocutor	Country
IN2P3		
Université de Paris VII—Denis Diderot, Paris	J Ocariz (EB)	France
Université Pierre et Marie Curie, Paris	M Boratav	France
CERN	J Salicio	Switzerland
DESY	F Schrempp (EB)	Germany
Istituto Nazionale di Fisica Nucleare (INFN)	U Dosselli	Italy
Universit'a degli Studi di Padova *Il Bo*	G Zumerle (EB)	Italy
Universit'a degli Studi di Roma *La Sapienza*	L Maiani	Italy
Universit'a degli Studi di Roma *Tor Vergata*	P G Picozza	Italy
Universit'a degli Studi di Torino	P Giubellino	Italy
Laboratório de Instrumentao e Física Experimental de Partículas (LIP)	G Barreira (EB)	Portugal
Centro de Investigaciones Energéticas, Medioambientales y Tecnológicas, CIEMAT	J A Rubio (EB)	Spain
Universidad Complutense de Madrid	F Arqueros	Spain
Universidad de Santiago de Compostela	E Zas	Spain
Universitat de Barcelona	D Espriu	Spain
Universidad de Valencia	A Ferrer	Spain
Imperial College of Science, Technology and Medicine	G Hall (EB)	United Kingdom

The mobility of personnel was used as a basis to:

(i) Provide specialized training for young Latin American scientists.
(ii) Contribute to the modernization of physics education in Latin America.
(iii) Foster the scientific collaboration in fundamental physics between Latin America and Europe by using existing or future large and expensive facilities.
(iv) Contribute to the technological development of Latin America through the HEP associated technologies.

55.5 The organization of HELEN

Mobility within the wide networks in tables 55.1 and 55.2 required efficient organization. This was based, in the first instance, on the *interlocutors*—scientists who had participated in the formulation of the HELEN proposal, on behalf of their institution.

The interlocutors coordinated the participation in HELEN of the professors and students of their laboratories. One interlocutor per country (indicated in tables 55.1

and 55.2) was elected to form the Executive Board (EB), the decision body of the project. The EB was chaired by the Project Coordinator (L Maiani) with V Riquer as Deputy-Coordinator.

The decisions of the EB were prepared by a Standing Executive Committee (SEC) formed by the Coordinator, Deputy-Coordinator, the CERN Interlocutor and four interlocutors elected by the EB, two for Europe and two for Latin America. In its first meeting, the EB elected J A Rubio and J Ocariz (for Europe) and A Santoro and A Zepeda (for Latin America). J Salicio completed the SEC as the CERN Interlocutor.

Finally, a Scientific Advisory Committee (SAC) was appointed in the first EB meeting, chaired by Luis Epele (Argentina) and Paolo Giubellino (Italy)[3].

HELEN's boards met twice a year. A general November Meeting of the Executive Board, alternating between Latin America and Europe, and a June Meeting of the Standing Executive Committee at CERN, for project monitoring and advice. Preliminary to the June Meeting, SAC met at CERN to input its recommendations and advice to the SEC of the following days.

The first EB took place in Rio de Janeiro in November 2005 (figure 55.3), where the first HELEN grants were assigned. Subsequently, SEC met at CERN in

Figure 55.3. First meeting of the Executive Board of HELEN, Rio de Janeiro, November 2005. The photo was taken by Alberto Santoro who hosted the meeting in UERJ.

[3] Other SAC members were: Nick Ellis, Elie Aslanides, Hans Hoffmann, Giorgio Matthiae (AUGER), Ornella Palamara (LNGS), Klaus Mönig (DESY); Luis Alvarez Gaumè (CERN TH), Marilena Streit Bianchi (CERN Techn. Transf.), and Pere Mato (CERN LHC GRID).

June 2006, preceded by the first SAC meeting. Two other EB meetings took place in Lisbon (2006) and Mexico City (2007). The Helen program ended formally in July 2008. It was prolonged to April 2009, to allow completion of the pending grants.

55.6 Results

Mobility was realized in HELEN with a large number of grants, totaling 1596 person-months over the full period, of which 1354 were from Latin America to Europe, 119 from Europe to Latin America, and 64 within Latin America. The latter grants were essentially a novelty for Latin American countries and have been quite successful in fostering international collaboration there, which had previously been very restricted.

The total cost of HELEN was about 3.0 million euros, of which about 2.7 million was provided by the EU.

The first HELEN grant holders from Latin America arrived at CERN in January 2006. In June of the same year, the *CERN Courier* could report: 'Now a small but active HELEN community is building up at CERN, and has established a HELEN club to allow the exchange of views and to help newcomers in the complex CERN environment'. Interesting events were organized by HELEN, in Argentina and in Mexico, to transfer CERN technologies in accelerator physics and computing.

To quote another important case, CMS researchers were able to travel to Brazil to help in setting up a CMS Computing Centre of the LHC Data GRID in Rio de Janeiro (at UERJ) and São Paulo (USP). The upgrade of the center was eventually approved in Brazil with financial support from FAPERJ and CNPq, which has allowed our Brazilian colleagues to increase their participation in CMS.

All in all, HELEN is perceived, in the high energy physics community, as an unprecedented and successful effort to integrate the particle physics communities of Europe and Latin America.

Peter Jenni, spokesperson of the ATLAS experiment, in a speech delivered in 2007 could speak of a 'before HELEN' and an 'after HELEN' for ATLAS in Latin America. In his words, 'HELEN has helped in a decisive way to foster and consolidate the very fruitful collaboration with Latin American colleagues and Institutions in the case of ATLAS. In particular the contributions from students and young researchers are a strong asset to the experiment, in addition to the institutional strengthening of the Collaboration and its resources.'

55.7 EPLANET, February 2011–January 2016

The main objective of EPLANET [4] was the formation of the scientific personnel in the beneficiary and partner institutions, through the participation in world-class experiments (see table 55.3). Experiments were to be performed at CERN (particle physics at the LHC) and at the Pierre Auger Observatory, Malargue, Argentina (the observation of ultra-high energy cosmic rays) (figure 55.4).

The rules of the framework program allowed the admission as Latin American partners from only four countries, namely Argentina, Brazil, Chile, and México.

Table 55.3. Participants in EPLANET. CIEMAT and University of Leeds retired from the project in 2012(*); ICTP-Trieste was admitted as a beneficiary in 2014(**).

Partner number	Institution	Country
1 Coordinator	Universita di Roma *La Sapienza*	Italy
2 Beneficiary	Centre National de la Recherche Scientifique—CNRS	France
3 Beneficiary	Istituto Nazionale di Fisica Nucleare—INFN	Italy
4 Beneficiary	Laboratorio de Instrumentacao e Fisica Experimental de Particulas—LIP	Portugal
5 Beneficiary	Centro de Investigaciones Energeticas Medioambientales y Tecnologicas—CIEMAT(*)	Spain
5a Beneficiary	International Centre for Theoretical Physics-Trieste-UNESCO(**)	Italy
6 Beneficiary	Universidad de Alcala de Henares—UAH	Spain
7 Beneficiary	Universidad de Barcelona—UB	Spain
8 Beneficiary	Universidad Complutense de Madrid—UCM	Spain
9 Beneficiary	Universidad de Santiago de Compostela—USC	Spain
10 Beneficiary	Universidad de Valencia—UV	Spain
11 Beneficiary	European Organization for Nuclear Research—CERN	Switzerland
12 Beneficiary	University of Leeds(*)	United Kingdom
13 Partner	Comision Nacional de Energia Atomica—CNEA	Argentina
14 Partner	Universidad de Buenos Aires—UBA	Argentina
15 Partner	Universidad Nacional de La Plata—UNLP	Argentina
16 Partner	Universidad Nacional de Mar del Plata—UNMP	Argentina
17 Partner	Centro Brasileiro de Pesquisas Fisicas—CBPF	Brazil
18 Partner	Universidade do Estado de Rio de Janeiro—UERJ	Brazil
19 Partner	Universidade Federal de Rio de Janeiro—UFRJ	Brazil
20 Partner	Universidade Estadual Paulista—UNESP	Brazil
21 Partner	Universidade Estadual de Campinas—UNICAMP	Brazil
22 Partner	Universidade do Sao Paulo—USP	Brazil
23 Partner	Pontificia Universidad Catolica de Chile—PUCC	Chile
24 Partner	Universidad Tecnica Federico Santa Maria—UTFSM	Chile
25 Partner	Benemerita Universidad Autonoma de Puebla—BUAP	Mexico
26 Partner	Centro de Investigacion y de Estudios Avanzados—CINVESTAV	Mexico
27 Partner	University of Guanajuato—UGTO	Mexico
28 Partner	Universidad Michoacana de San Nicolas de Hidalgo—UMSNH	Mexico
29 Partner	Universidad Nacional Autonoma de Mexico—UNAM	Mexico

CERN provided additional funds to continue the collaboration initiated with HELEN with Colombia, Peru, and Venezuela, that could not be continued in EPLANET for lack of specific agreements of these countries with the European Union.

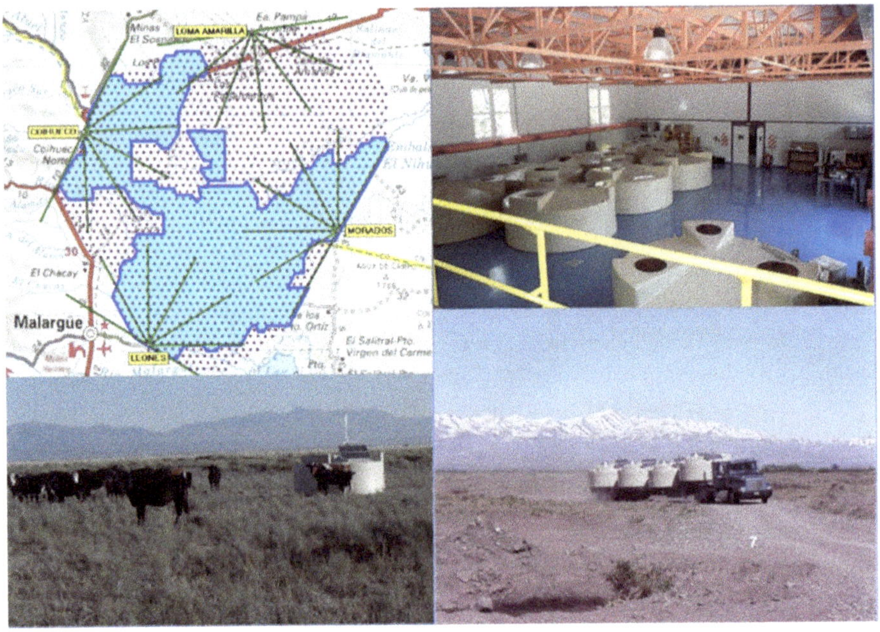

Figure 55.4. Surface Čerenkov detectors at the Pierre Auger Observatory in Malargue. From upper left, clockwise: (1) map of the site (dots indicate the locations of Č-detectors); (2) detectors in stock; (3, 4) detectors in the Argentinian Pampas. In the background are the Andes with the peak of Aconcagua.

Exchanges from Latin America to CERN and other European institutions and from European countries to the AUGER Observatory consisted of short exchanges (1–2 months) for senior and longer exchanges (2–12 months) for junior investigators, for a total of about 1500 months. The total support from the EU to cover accommodation (but not travel) expenses was about 3.2 million euros for four years. An extension to a fifth year, within the same budget, was approved in 2014.

EPLANET programs were organized around nine Work Packages (WPs):

- WPs 1–4: Mobility of early stage researchers (ESRs) and experienced researchers (ERs) to the experimental facilities of the LHC at CERN, namely ATLAS, CMS, Alice, and LHCb.
- WP 5: Collaboration between Europe and Latin America in theoretical physics and experimental subnuclear physics, other than LHC, at CERN.
- WP 6: Accelerator technologies, in view of the LHC upgrading foreseen to take place in the following years.
- WP 7: Medical applications, in particular to hadron therapy.
- WP 8: Scientific computing. In this framework, a series of lectures by CERN personnel in Latin American institutions, focused on ICT methods for data handling, was carried out.
- WP 9: Mobility of ESRs and ERs to the Pierre Auger Observatory in Argentina, operated by a large international collaboration. Activities consisted of the consolidation of the apparatuses, data taking, and preparation for further upgrading. The planned deliverables have all been achieved.

Table 55.4. Key values from the EPLANET—IRSES Marie Curie program, 2011–16. NOTES: (*) where EPLANET support is acknowledged; (**) source: http://inspirehep.net.

Secondments (months)	Researchers	Gender distribution	Workshops, schools	Seminars, conferences	Scientific publications (*)	Citations (**)
1448	509	F/M = 0.23	18	124	1132	55 128

The years 2011 and 2012 witnessed the coming into full operation of the LHC at CERN and the data collection and analysis of the four large experiments, resulting in a large number of technical reports and published papers. EPLANET participants have contributed largely to the discovery of the candidate Brout–Englert–Higgs boson, the new particle identified by ATLAS and CMS on 4 July 2012, with a mass of around 125 GeV.

During the years of EPLANET, three major discoveries have been made by the AUGER experiment: (i) the confirmation of a high energy cutoff in the spectrum of ultra-high energy cosmic rays, (ii) the first determination of the particle composition of cosmic rays in the cutoff region (i.e. proton versus nuclei abundances); and (iii) the first indication of point-like extragalactic sources.

EPLANET, like HELEN, has produced a remarkable number of theses and degrees of high quality, originating from research done in world-class facilities with highly qualified supervisors. The work of trainees has led to a large number of high quality publications (see table 55.4).

55.8 Ten years of science diplomacy

HELEN and EPLANET have established an important network of formal links between CERN and Latin American countries, extending to this new arena the tradition of 'science diplomacy' inaugurated by CERN more than 60 years ago.

The following are among the most important results:

- Cooperative agreements signed in 2007 for scientific and technological cooperation in HEP: CERN—CONICYT, Chile; CERN—CONACYT, México.
- ATLAS admitted to partnership with several Latin American institutions: UBA and UNLP in Argentina; USP in Brazil; UTFSM and PUCC in Chile; University A Narinõ in Colombia.
- ALICE: Two experimental detectors, V0 and ACORDE, were built in Latin America, specifically in Mexico, with HELEN support for teacher and student travel.
- CMS admitted UNIANDES, Colombia.
- CMS researchers helped in creating a CMS Tier2 center of the LHC Data Grid in UERJ and São Paulo.
- The possibility offered by HELEN to send students to the largest facilities in Europe and Latin America has greatly enhanced the activity between the two regions, changing the historical orientation of HEP groups in Latin America towards the United States.

- HELEN involved the most advanced institutions and the best research infrastructures in particle physics in the two continents.
- Promoting regional Latin American—Latin American integration by internal mobility has been one of the most remarkable results of HELEN.
- EPLANET has allowed scientists and students from Latin America to participate in the discoveries of the first LHC years, bringing new ideas and enormous drive.
- Teaching and outreach have been promoted energetically.
- Commitment to CERN and Europe have been consolidated, in view of the upgrades of the next years: the High Luminosity LHC and eventually, the energy increase.
- Experimental laboratories, in Mexico, Brazil, and Chile (and also in Colombia and Peru) have been created and will be the natural points of departure to develop technology transfer.
- *No brain drain*! We have seen postdoctoral students in HELEN becoming professors and group leaders during EPLANET, and some finding positions in industry, in their own countries.

Sustainability of the collaboration has been reached (almost!) by Brazil and Mexico.

55.9 Summary and future perspectives

Following Peter Jenni (section 55.6) we can well claim that the HELEN–EPLANET programs mark a *before* and an *after* for Latin American–CERN relations, underlined by the remarkable rise of the number of registered CERN users from Latin America versus time, shown in figure 55.5.

The decline after 2017 indicates that the resources provided by EPLANET have not been replaced by comparable in-house resources of groups participating in LHC experiments. CERN should pay attention to this alarming signal.

In 2015 several Latin American countries were almost at the point of obtaining regular funds to continue the collaboration with CERN, in particular to cover maintenance and operation expenses related to their participation in LHC experiments. The end of EPLANET and the economic crisis in the largest countries have frustrated most of these efforts. The economic crisis has drastically reduced the prospects of some Latin American countries, Brazil in particular, to become associate members of CERN.

Aiming to become a global laboratory, CERN cannot loose the scientific community of a whole continent. To include the the Latin American community in its future, positive actions from CERN are urgently needed. From our experience, we would suggest:

- To continue to organize and support the Latin America–CERN School.
- To create a liaison office for Latin America, reporting to the Director General and led by a scientifically recognized person with proven contacts to the Latin American scientific community, e.g. from the LHC experiments.

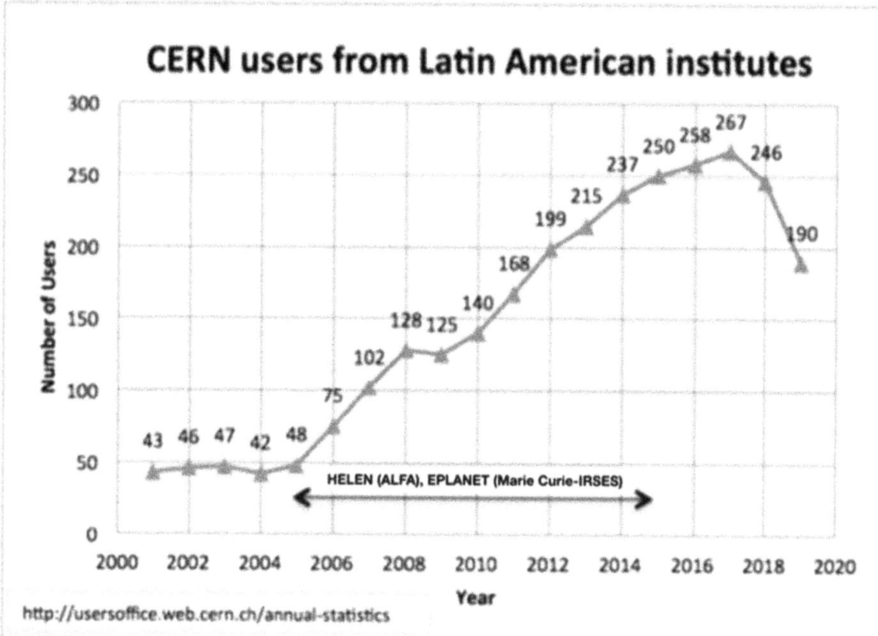

Figure 55.5. The number of CERN users from Latin America versus time since the year 2000. (I de Bediaga, CBPF [5], copyright CERN.)

- To include in the liaison office a similarly expert CERN administrator.
- The liaison office could, in particular, keep in contact with LASF4RI-HECAP, the initiative that is preparing for Latin American governments a plan of scientific facilities for research in high energy physics, cosmology, and astroparticle physics [6].
- Most importantly, in the absence of external funds CERN should consider investing from its own budget to encourage and support the mobility to CERN experiments of ESRs and ERs. In all, with Latin American institutes supporting travel costs, the HELEN and EPLANET programs have provided about 0.6 million euros/year over 10 years, much less than the added value to CERN of the work performed by the Latin American groups.

55.10 Conclusions: particle physics beyond the Pillars of Hercules

With the observation of the Higgs boson and of the cosmic ray high energy cutoff, we are now sailing beyond the Pillars of Hercules, in an uncharted sea where we hope to see new phenomena. Earth and satellite based astronomy and large distance cosmology are also on unexplored paths.

We are aiming at a more complete vision of the micro- and macro-cosmos, equipped with many prejudices and conflicting ideas. Certainly, we do not know where we will land, but we know it will be more strange and more fascinating than anything we are able to imagine today.

And, by the way, we are sure that the mobility of researchers and international collaboration will be essential elements for progress. The vision that De Broglie outlined in 1949 will again come back, this time at a global rather than regional scale.

Appendix A: List of acronyms

ALFA:	América Latina Formación Avanzada
CBPF:	Centro Brasileiro de Pesquisas Fisicas, Brazil
CERN:	Conseil Européen pour la Recherche Nucléaire
CIEMAT:	Centro de Investigaciones Energéticas, Medioambientales y Tecnológicas
CLAF:	Centro Latino Americano de Fisica
CINVESTAV:	Centro de Investigación y de Estudios Avanzados del Instituto Politecnico Nacional, Mexico
CNPq:	Conselho Nacional de Desenvolvimento Científico e Tecnológico
DESY:	Deutsches Elektronen-Synchrotron
EPLANET:	European Particle Physics Latinamerican Network
FAPERJ:	Fundação de Amparo à Pesquisa do Estado do Rio de Janeiro
HEP:	High Energy Physics
HELEN:	High Energy Physics Latinamerican–European Network
ICT:	Information and Communication Technology
IRSES:	International Research Staff Exchange Scheme, one of the Marie Curie actions
LASF4RI-HECAP:	Latin America Strategic Forum for Research Infrastructures for High Energy Physics, Cosmology and Astroparticle Physics
LHC:	Large Hadron Collider
LNGS:	Laboratori Nazionali del Gran Sasso, Italy
PUCC:	Pontificia Universidad Católica de Chile
UBA:	Universidad de Buenos Aires, Argentina
UERJ:	Universidade do Estado do Rio de Janeiro, Brazil
UFRJ:	Universidade Federal do Rio de Janeiro, Brazil
UNAM:	Universidad Nacional Autónoma de México
UNLP:	Universidad Nacional de La Plata, Argentina
USP:	Universidade de São Paulo, Brazil
UTFSM:	Universidad Técnica *Federico Santa María*, Chile

References

[1] CERN Our member states https://home.cern/about/who-we-are/our-governance/member-states
[2] Rubio J A 2003 Experimental Particle Physics in Latin America *Report* CERN-LA-4/6/01
[3] High Energy Physics Latinamerican–European Network https://www.roma1.infn.it/exp/helen/
[4] European Particle Physics Latin American Network https://www.roma1.infn.it/exp/eplanet/
[5] de Bediaga I Participação da América Latina no CERN *CLAF* https://www.claffisica.org.br/news-item/participao-da-amrica-latina-no-cern
[6] Latin American Strategy Forum for Research Infrastructure https://lasf4ri.org

IOP Publishing

Big Science in the 21st Century
Economic and societal impacts
Panagiotis Charitos, Theodore Arabatzis, Harry Cliff, Günther Dissertori, Juliette Forneris and Jason Li-Ying

Chapter 56

The rise of SESAME: a very personal point of view

Eliezer Rabinovici

Parts of this chapter have been reprinted from Rabinovici E 2017 SESAME: A personal point of view *International Journal of Modern Physics* A **32** 8. Copyright 2017 World Scientific Publishing Company.

This chapter is being written at the time when the idea of a Middle-Eastern scientific collaboration is nearly 30 years old. The idea ended up manifesting itself in a light source built for that very purpose. It is called SESAME. SESAME stands for 'Synchrotron-light for Experimental Science and Applications in the Middle East' and is a regional 'third-generation' synchrotron light source situated in Allan, Jordan. It is the Middle East's first major international research center, operating since 2017.

The current members of SESAME are Cyprus, Egypt, Iran, Israel, Jordan, Pakistan (Pakistan has taken a step back), Palestine, and Turkey. Active current observers include, among others, The European Union, France, Germany, Greece, Italy, Japan, Kuwait, Portugal, Spain, Sweden, Switzerland, the United Kingdom, and the United States of America (for a full current list, see the SESAME website https://www.sesame.org.jo/). The common vision driving the many people who have contributed to this project is the belief that human beings can work together for a cause that furthers the interests of their own nations and that of humanity as a whole.

I will describe from a very personal point of view how this project came about, the principles around which it was built, some of the hurdles it had to surpass, and its present status. It is known that each success has many parents and that failures tend to be orphans. In addition, a light source is mainly a home for experimentalists and I am a theoretical high-energy physicist, much of my research being in string theory. String theory does flirt with the idea of multiverses and some parallel universes. So does SESAME, as I will return to later.

That said, the story of SESAME is a story of many dedicated people all over the region and all over the globe. There are many without whom SESAME would not have reached this moment. Some have been very visible and others have contributed far from the limelight out of belief in the project. Some of these deeds I was well aware of and in other cases all I saw was the result of those efforts. It is to those unknown people that I wish to dedicate this chapter as a small token of my thanks to them.

56.1 Science as a bridge for understanding

Nobel Laureate Abdus Salam, as far back the 1950s/60s, identified the light source as a tool which could help thrust what were then considered as third-world countries (he was mainly concerned with Muslim countries) directly to the forefront of scientific research. The very same Abdus Salam joined our efforts in 1993!

Before entering into the history of the SESAME project I shall discuss, with the help of hindsight, some generalities. The questions to be addressed are: Why does science have the potential to be a bridge for understanding? Why should the scientists themselves be involved? What are the prerequisites needed to practice this? What should be the guiding principles involved? and last but not least, What should be the human environment needed for such an endeavor to have a chance of succeeding?

I am discussing science as a potential bridge for improving understanding, not as a bridge to 'peace'; the word peace and in particular 'just peace' means different things to different people and different nations. The term serves more often as an obstacle rather than as a bridge for understanding.

The necessary state of mind needed for anyone embarking on such a project should be to have an infinite, and nothing less than infinite, dose of optimism ingrained in them. They must learn to ignore the media reports and maybe also learn how to decouple from reality a little bit as they move along.

56.1.1 Why science?

Because when scientists meet they in most cases already have a common agreed upon language, that of science, they need not spend large amounts of time just to agree on the basic common terms for communication. This is a very crucial simplification which allows them to work side by side. I learned to appreciate this when, as director of the Israeli Institute for Advanced Studies at the Hebrew University Jerusalem, I hosted a group of multi-religion experts who tried to form a common view of the five first books of the Bible. They viewed the fact that they had found common ground on less than ten verses in ten months as a phenomenal success and I am sure they were right....

Finally, the professional appreciation developing from the common work of scientist has the potential to lead to more personal contacts and to the formation of some form of mutual trust.

56.1.2 Why scientists?

Scientists of all nations have a common language—science and many of them have a track record of participation in successful international collaborations such as at CERN. Scientists are neither better nor worse than other human beings but, given the privilege to pursue pure and applied knowledge, it is their duty to try to do their best to contribute back to society. One such way is to try and build a bridge of understanding between nations.

56.1.3 What kind of projects should one use as bridges?

The first prerequisite for a successful collaboration is that each side has something essential to contribute to the project and something essential to take out of the project. My own natural tendency is to encourage the bottom-up approach in scientific research and to strengthen small intimate collaborations. However, I had to accept in the process that for the time being in our region the viable approach is to focus on the rather large scale and top-down approach. This one can compromise on, but one must not compromise on the quality of the project. Only first-class science can serve a useful purpose in 'science for understanding' attempts. It is better not to have a project at all than to have a mediocre one! This theme follows SESAME throughout its history.

56.1.4 What does one need from the human environment in order to push such a project forward?

In retrospect one needs one 'good' person in an organization. What do I mean by a 'good' person? Well, it should be one who agrees with my vision....one who has the authority to decide essentially on their own whether to join the adventure, and, to be practical, the help they will agree to offer should be a small but not zero part of the funds they are responsible for. It is in large part thanks to such people, most of whose names never have and never will reach the headlines, people who have nudged the project at its many crucial moments, that SESAME has survived the arduous journey it has traveled.

56.2 From the CERN corridors through the Sinai beaches to the SESAME concept

For me SESAME starts just after the Oslo Accords when the outstanding scientist and my close friend and collaborator the late Professor Sergio Fubini from CERN/Torino (and MIT for a time) approached me in the corridor of the Theory Group at CERN telling me that it could now be the time to test what he called 'your idealism'. He was referring to my ideas on future joint Arab–Israeli scientific projects. Together with many others from the region and the world we founded the Middle Eastern Science Committee (MESC) with the idea of trying to forge meaningful scientific contacts in the region. CERN is a very appropriate venue for the inception of such a project. CERN itself was built after World War II in an effort to help heal Europe and European science and scientists in particular. One of its own missions was to promote understanding through science. Indeed, Abdus Salam was one of the eminent scientists who were willing to lend their name and endorse MESC. By joining our scientific committee Salam made public his belief in the value of Arab–Israeli scientific collaborations, something he had expressed several times in private.

In preparation for focusing our vision Sergio gave me some homework to do. He invited me to deliver two talks at a meeting in Torino to be held on the occasion of his 65th birthday. One talk was a mini-review on the status of string theory which, as mentioned, is my field of research, and the other was on the status of Arab–Israeli collaborations. Both are published in the book dedicated to that conference. I still find my mini-review on the status of string theory to be first-rate, but back to Arab–Israeli collaboration.

My conclusion, which appears in part of the principles outlined above, was that those collaborations that were successful were those in which each of the parties had brought into the project knowledge which was essential for its working and each party had a great deal to learn and gain by working on the project. A shining example of that were collaborations concerning an Egyptian–German–Israeli study of various features and properties of the Red Sea.

Following the meeting in Torino we traveled to Cairo to meet Professor Venice Gouda, the Egyptian minister for higher education, and other Egyptian officials. At that stage we were just self-appointed entrepreneurs.

We were told that President Mubarak had made a political decision to take politics out of scientific collaborations with Israel. As a result of that meeting we organized in close collaboration with the Egyptian authorities a high-quality scientific meeting in Dahab, the Sinai desert. It brought together about one hundred young and senior scientists, including Egyptian, Israeli, Jordanian, Palestinian, and Moroccan scientists from the region and outstanding researchers from all over the world. Amongst them, present and future Nobel and Fields Medal laureates (figure 56.1).

The meeting was held on 19–26 November 1995 in a large Bedouin tent (see figure 56.2), in the weeks after the murder of the Israeli prime minister Rabin. All of us stood for a moment of silence respecting those who fought for peace, this at the request of Professor Venice Gouda the Egyptian minister for higher education. The silence still echoes in my ears today.

The first day of the meeting was also attended by other representatives, including Professor Jacob Ziv, the president of the Israeli Academy of Sciences and Humanities. The academy had been supporting such efforts in general and the meeting in the Sinai and the SESAME project in particular. This support was expressed in words and in deeds, including financial support. Meetings and preparing for meetings do involve costs. It was thanks to the additional financial help of Professor Miguel Virasoro, a well-known physicist who was at the time the Director-General of ICTP (the institute founded by Abdus Salam), and also the help of Professor Daniele Amati, a well-known physicist and the director of SISSA, which is also based in Trieste, that the meeting was held. All three decisions of support were made at watershed moments and on the spur of the moment. It is gratifying to recall this invaluable and timely support.

SINAI MEETING ON
HIGH ENERGY, CONDENSED MATTER
AND ENVIRONMENTAL PHYSICS

19-26 November 1995
Dahab, Sinai Peninsula, Egypt

****

Under the auspices of:
ICTP
CERN, UNESCO
Egyptian Ministry of Scientific Research
The Israel Academy of Sciences and Humanities
Istituto Italiano per gli Studi Filosofici
Istituto Nazionale di Fisica Nucleare (INFN)
The Higher Council for Science and Technology, Amman
National Research Centre, Cairo
National Institute of Standards, Cairo
The Hebrew University, Jerusalem
Bethlehem University
International School for Advanced Studies (SISSA), Trieste
University of Cagliari, University of Napoli, University of Torino

****

This International Meeting has been planned with the aim of bringing together experts in these very active fields of research and putting them in contact with young researchers from the Middle East and from the whole Mediterranean area in a spirit of co-operation and friendship. The meeting is going to be a starting point for further similar encounters in this area. A limited number of young researchers from each country will be admitted. The participation of senior researchers is very important, so the only limitation on their number will be the available room at the meeting site.

The meeting will consist of a series of lectures, a set of seminars and demonstrations, and there will be ample time for discussions and for sessions of questions and answers.

Following is the list of the topics of the mini-courses and their organizers:

1) *Particle- and Astrophysics*, M. Jacob and G. Veneziano;
2) *Superconductivity (Physics and Applied Technology)*, A. Barone and F. Bonaudi;
3) *Spontaneous Symmetry Breaking*, R. Jackiw;
4) *Non-Abelian Gauge Theories*, E. Rabinovici;
5) *Environmental Physics*, G. Furlan.

There will be seminars by:

L. Alvarez-Gaumé, *Anomalies*; M. Assad Abdel-Rauf, *Theory of Four-body Systems: Rigorous and Variational Proofs of the Possible Coexistence of Atoms and Antiatoms*; M. Berry, *Geometric Phases*; E. Brézin, *1/N Expansion*; H.B. Ghassib, *Gauge Theoretic Description of Superconductivity*; R.B. Laughlin, *Particle Physics in Miniature: the Emulation of Quarks and Gluons by Quantum Antiferromagnets*; E. Rabinovici, *Dualities in Physics*; M. Virasoro, (to be announced); E. Witten, (to be announced).

****

International Scientific Advisory Committee

Abdus Salam and Victor F. Weisskopf (Honorary Chairmen)

L. Alvarez-Gaumé (CERN), M. Assad Abdel-Rauf (Ain Shams U., Cairo), J.J. Atick (Rockefeller U.), A. Barone (U. of Napoli), M. Berry (U. of Bristol), F. Bonaudi (INFN, Torino), E. Brézin (Ecole Normale Supérieure), N. Cabibbo (ENEA and U. of Roma), A. Devoto (U. of Cagliari), S. Fubini (Chairman - U. of Torino), G. Furlan (U. of Trieste), H.B. Ghassib (U. of Jordan, Amman), R. Jackiw (MIT), M. Jacob (CERN), R.B. Laughlin (Stanford U.), F. Nicodemi (U. of Napoli), E. Rabinovici (Hebrew U., Jerusalem), S. Sciuto (U. of Torino), G. Veneziano (CERN), M. Virasoro (ICTP), E. Witten (IAS, Princeton).

****

Local Organising Committee

M.M. El Halwagi (Ministry of Scientific Research, Cairo), M. El-Fiki (NIS, Cairo), M. El-Raey (U. of Alexandria), M.S. Shalan (NIS, Cairo), M. Fakhri (NRC, Cairo), M.S. El-Wahab (Ain Shams U.), M.A. Sadky (NIS, Cairo), A.I. El-Ibiary (Ministry of Scientific Research, Cairo), M. Tag El-Din Kamal (NRC, Cairo).

****

For further information please write to:
A. Devoto
Dipartimento di Scienze Fisiche
Via Ospedale, 72
09124 CAGLIARI (Italy)
FAX: (39) (70) 666126
E-mail: DEVOTO@CAGLIARI.INFN.IT **OR** E-mail: DIRECTOR@ictp.trieste.it

August 1995

Figure 56.1. The announcement of the SINAI Meeting on High Energy, Condensed Matter and Environmental Physics held in November 1995, which paved the way for the construction of the SESAME facility. The meeting led to the establishment of a Middle East Scientific Cooperation group, a precursor of SESAME. (Copy image of the official announcement from the author's personal archive.)

Figure 56.2. Photograph of the meeting that was held on 19–26 November 1995 in a large Bedouin tent in Dahab Sinai, Egypt. (Credit: From author's personal archive/Eliezer Rabinovici.)

The meeting at Dahab was very special—at moments it seemed we were following some over-the-top Hollywood script. We survived with no casualties a 6.9 magnitude earthquake (in other places in the Sinai there were casualties), some of the participants had seen Mt Sinai shake. A student of one nationality was rescued by a student of another nationality from drowning in the Red Sea. An admission committee of one Israeli institute convened under the Sinai moonlight to examine candidates for PhD and MSc studies. What clearer signals could one ask for to show us that we are on an interesting track? SESAME itself directly faced another such test from above when its roof collapsed in the winter of 2014 under piles of snow resulting from a highly irregular storm. The roof has since been reinstated. The 2020 Covid-19 pandemic was another such challenge but was of course a global challenge.

The meeting was followed by a very successful effort to identify concrete projects in which Arab–Israeli collaboration could be beneficial to both sides, but then the attempts to continue the project in the region itself were blocked by a turn for the worse in the political situation. Politics had been returned to the arena of scientific collaborations.

MESC then decided to retreat, with little dignity, from the region itself and return to Turin, Italy. During a meeting in November 1996 there was a section devoted to studying the possibilities of cooperation via experimental activities in high-energy physics and light source science. During that session the late German scientist Gus Voss suggested (on behalf of himself and Hermann Winnick from SLAC) to bring the parts of a German light source situated in Berlin called Bessy, which was about to be dismantled, to the Middle East. Herwig Schopper, a former Director-General of CERN, had attended this workshop.

The operation of MESC had resulted in building a sufficient amount of trust among the parties and thus had provided an appropriate infrastructure to make such an idea into something concrete. The idea of building a light source was very attractive thanks to the rich diversity of fields of science that can make use of such a facility, from biology through chemistry, physics, and many more to archeology and environmental sciences.

Such a diversity would also allow the formation of a critical mass of real users in the region. In fact, there was one virtue to the decades it took to build SESAME, it gave us time to build up, with the invaluable help of many scientists from all over the world, a significant community of potential users. The IAEA in Vienna has been a key financial supporter of these training activities. This training is one of the long term goals of the construction of a high-quality light source in the region. The major drawback of the Bessy proposal was that there was no way a reconstructed dismantled 'old' machine would be able to attract first-class scientists and science.

It was in the midst of these deliberations that Fubini had asked Schopper, who had a rich experience in managing complex experimental projects, to join us and take a leadership position. Fubini had also developed health problems and gradually left the center stage for others to carry on the flame.

56.3 From a name to a real structure

With Professor Herwig Schopper the focus of possible collaborations had been narrowed down to the large top-down project of constructing a light source (Herwig Schopper has written up his own account of the events in that period).

In a meeting convened in Uppsala by Professor Tord Ekelof it was decided to use the German machine as a nucleus around which to build the administrative structure of the project. That was the main purpose that the German machine would serve. It allowed one to address the administrative aspects of constructing a joint project given the (non-) relations among several of the members. Indeed, that was a very serious challenge on its own. It was overcome also by using the auspices of UNESCO as a place to deposit the instruments of joining the project. This was suggested by Schopper following the example of the way CERN was assembled in the 1950s. In fact the statutes of SESAME have also been, to a large extent, copied from the rather compact statutes of CERN. I was observing with quite some surprise our good luck as to how a band of self-appointed entrepreneurs had evolved into a self-declared interim Council of SESAME with Schopper as its president. The next major move was to choose a seat for the project. This was a complex process that involved choosing among many different candidate sites.

On 15 March 2000 I flew to Amman for a meeting on the subject. I met there with Professor Khaled Toukan and after studying a map sold at the hotel where we met, a map which effectively contained no information on the cities in Israel, we discussed what site Israel would support. We also asked at the time that a Palestinian be the Director-General. Due to various developments, none of which depended on Israel, this was not to happen. The decision on the site venue was taken at a meeting on

11 April 2000 at CERN. Jordan was selected as the host state. Jordan had and has diplomatic relations with all the parties involved.

The German machine was dismantled by Russian scientists, placed in 'Lego' boxes and shipped with assembly instructions to the Jordanian desert to be kept until the appropriate moment would arise. This was made possible also thanks to a direct contribution by the Director-General of UNESCO at the time, Professor Koichiro Matsuura. His help came at one of those watershed moments for SESAME. A good person doing what should and could be done on his own authority. His personal action did not pass unnoticed at UNESCO but he overcame the doubters.

A major effort was made by Dr Khaled Toukan from Jordan who at the time was the president of a Jordanian University and since has served in several ministerial capacities in Jordan and has become the Director-General of SESAME. The story of his recruitment to SESAME is one for Schopper to tell.

Now that the administrative structure was set in place it was time to address the engineering and scientific aspects of the project.

Technical committees had designed a totally new machine. The old machine would eventually be used as one boosting component of the whole new machine. Many scientists in the region were introduced by SESAME workshops to the scientific possibilities that SESAME could offer. Scientific committees considered and chose appropriate day-one beam lines, and yet, day one seemed very far in the future.

Technical and scientific directors from abroad helped define the parameters of the new machine and identified appropriate beam lines to be constructed. Administrators and civil servants from the members started meeting regularly in the finance committee. In Israel it was the Ministry of Science which sent delegates to these meetings and for a long time it was the Ministry of Education which covered the annual membership fees from its UNESCO desk. Jordan built the facility to host the light source and has made major additional financial contributions.

At this stage it became time for the SESAME interim Council to transform into a permanent council and in the process cut its umbilical cord from UNESCO. This transformation, required by legal aspects, presented new hurdles. It was required of every member of the interim council that wished to become a member of the permanent council that its head of state or someone authorized by the head of state sign an official document sent to UNESCO stating this wish. Given the nature of the project one can imagine that the process of obtaining these documents presented a serious challenge. Also this was overcome (Herwig knew only half of the story). In general in such cases it is required that there be a minimum number of members expressing this wish explicitly for the council to be official. The commitment of Israel was delivered to the Director-General of UNESCO Koichiro Matsuura as he was boarding the plane in Paris *en route* to Jordan. It was signed by the Israeli Foreign Minister at the time, Silvan Shalom. That statement was the one needed to reach the minimum quota. The UNESCO Director-General was flying to join the King of Jordan as well as representatives of the member countries in attending a heart-warming groundbreaking ceremony (see figure 56.3). That was followed several

Figure 56.3. King Abdallah of Jordan, Director-General of UNESCO, Herwig Schopper from CERN, Jordanian minister of education Khaled Toukan, and Eliezer Rabinovici are among those who attended the groundbreaking of Sesame. (Credit: SESAME/CERN.)

years later by a celebration dedicated to the completion of the building which would host the SESAME staff and later on the SESAME machine itself.

While the host building was indeed constructed it remained essentially empty (see figure 56.4). SESAME had received support from leading light source laboratories all over the world. This support was a source of strength to the members to continue with the project.

However, the challenge of building a new world-class machine was far from being overcome. In fact attempts to obtain significant funding support had failed time and again. It was agreed that the running costs of the projects should be borne by the members, however, the one-time large cost needed to construct a new machine was outside the budget parameters of most of the members. Most of them did not have a tradition of significant support for basic science. The European Union gave support to the project at that stage only through its bilateral agreement with Jordan and was not specific to SESAME. In the end several million euros of those projects did find their way to SESAME, but the coffers of SESAME and its building remained essentially empty (figure 56.5).

56.4 Putting together the pieces of the puzzle

After the building hosting the SESAME staff and planned to hold the SESAME light source itself was completed, Herwig Schopper asked to be allowed (yes, in this project you need to be allowed to seek some rest) to step down as president of the Council and he was replaced by another former Director-General of CERN, Professor Chris Llewellyn Smith from Oxford University, UK. Llewellyn Smith is

Figure 56.4. Top: Construction of the main building that today hosts the SESAME facility. Bottom: The interior of the SESAME building after the installation of the radiation shield—al though at that stage there was no source to radiate! (Credit: from the author's personal archive/E Rabinovici.)

a well-known theoretical physicist in his own right. Schopper himself attends most of the council meetings and is still heavily involved today, at the end of the year 2022.

As Llewellyn Smith stepped up to the plate his main challenge was to obtain the funding needed to construct a new light source and to remove from SESAME the

Figure 56.5. Top: A proposed design of the SESAME facility presented to the SESAME Council. Bottom: The main entrance of the SESAME facility (bottom). Credit: CERN.

black cloud of perception that it was simply a reassembled very old light source of little potential attraction to top scientists.

In addition to searching for sources of significant financial support there was an enormous amount of work still to be done in formulating detailed and realistic plans for the next year and for the subsequent years. A grinding systematic effort was made to begin to endow SESAME with the structure needed for a modern working accelerator. One also needed to build from scratch information material about the project.

Llewellyn Smith, like his predecessor, needed to also deal from time to time with political issues. For the most part the meetings of the SESAME council were totally devoid of politics. On the contrary, they were to me a parallel universe.

As I mentioned earlier, being a string theorist, I encounter more and more the need to consider the possibility of the existence of a multiverse or parallel universes. Many of those in the field would love to be able to travel to and marvel at other universes, but I am the one who gets to visit a parallel universe where administrators and scientists from the region get to work together, in the interests of their own people and in the interest of humanity as a whole, in a common project. Each one brings his/her own scars and prejudices and each one is willing to learn. That said, there were also moments when politics did contaminate the spirit forming in SESAME. In some cases this was isolated and removed from the agenda and in others some bitter taste remains. But these are just at the very margin of the main thrust of SESAME.

Another interesting idea emerging from black hole physics and string theory is holography. Superficially, it states that for some systems all the information is encoded on the surface enclosing a certain volume.

How nice it would have been if the walls of the SESAME building would be sufficient to encode such information and there would be no need for the financial resources to host a light source inside the building, but this was just a light distraction for me. There was a serious problem to solve.

The empty SESAME building was filled with radiation shields giving the appearance of a full building but the absence of the light source itself formed a void. This absence radiated all over. I observed that the morale of the local staff was in steady decline; in my opinion the project was in quite some danger.

At that stage I approached two persons in the ministry of finance in Israel. One should know that it is difficult to find more bitter enemies than some in the finance ministry and the university professors. They have battled over money, time and again, over decades accumulating their own scars. Nevertheless I was very well received and when I asked that Israel show an example by making a voluntary contribution to SESAME of five million dollars to build a new light source I was not shown the door. Instead they requested to come and see SESAME. After their visit they roped in the Budget and Planning Committee (BPC), which in Israel funds all higher education, and together agreed to indeed contribute the requested funds on the condition that others join them. The chair of the BPC was at the time Professor Manuel Trajtenberg.

Each member of the unlikely coalition consisting of Iran, Israel, Jordan, and Turkey pledged an extra five million dollars for the project in an agreement signed in Amman. Since then Israel, Jordan, and Turkey have paid up to their commitment, and Iran claims that they recognize their commitment but are obstructed by sanctions. Time will tell how that part will evolve. However, in any case this has encouraged the Director-General of CERN at the time, Rolf Heuer, to convince the EU to dedicate five million euros to the project, this in addition to the approximately three million euros which were directed earlier to the project from a bilateral EU–Jordan agreement. The essential EU support was led by the commissioner Carlos Moedas, as well as I Smith and B Fabiane. The director of the INFN Italy, Professor Fernando Ferroni, has also come on board by giving in 2015 almost two million euros for the project and seriously considering doubling the contribution. Many leading world laboratories, in a heartwarming expression of support to the project and its spirit, have donated possible equipment for future beam lines as well as fellowships for the training of young people.

With their help SESAME had crossed the point of no return. The undefined stuff dreams are made off turned into magnets and girdles made of real hard steel—I touched them at CERN where they were being assembled by joint teams of CERN and SESAME. It was a very emotional moment for me to feel an idea turning into reality. Eventually, we successfully reached the final stretch of construction.

The pace of events had finally accelerated, a star-studded inauguration (including among many others the Directors-General of CERN and UNESCO, the Israeli Science Ministry, the King of Jordan, the EU Commissioner of Research), this time it was a real inauguration as far as the machine is concerned. In the ceremony, amazingly, the political delegates of different member states listened to each other without leaving the room (as is the standard practice in other international organizations) and what was even more unique was that each member state delegate taking the podium gave essentially the same speech. As if for once they all instructed their speech writers to say the same thing, namely, 'We are trying here to achieve understanding via collaboration'.

At that moment the baton was passed from one ex-Director-General of CERN, Chris Llewellyn Smith, to another, Rolf Heuer (figure 56.6). The SESAME Council is the only world organization (except for CERN) which benefited from the fact that all its council presidents were ex-Directors-General of CERN.

56.5 To be judged by science

The high-quality 2.5 GeV light source started its operation during 2017. There are currently two beam lines in SESAME. One is a x-ray absorption fine structure/x-ray fluorescence (XAFS/XRF) spectroscopy beam line. The other is an infrared (IR) spectromicroscopy beam line for research. A third beam line is currently being added for materials science (MS) and plans for more are at advanced stages (figure 56.7).

If you are waiting for more details about them and future beam lines, forget it, I am a string theorist and I ran out of space in this essay, but the former and present scientific directors of SESAME, Giorgio Paolucci and Andrea Lausi, who oversaw the construction, are much better equipped to elaborate on this.

There is, however, concern that the only component used from Bessy, the Microtron, may break down, thus endangering the operation of the whole machine. We continue to operate on a shoe-string budget, the current approved 2022 budget is

Figure 56.6. Passing the baton between SESAME's Council Presidents. From left to right: Eliezer Rabinovicci, Chris Llewellyn Smith, Herwig Schopper, and Rolf Heuer. Credits: from the author's personal archive/E Rabinovici.

about $5.3 million US, much smaller than that of any other modern light source. I marvel at the ingenuity of the SESAME staff allowing SESAME to operate and I am sad about the seeming budget indifference of many of the parties involved. The world media was less indifferent: the BBC, *New York Times*, *Le Monde*, *Washington Post*, *Brussels Libre*, *The Arab Weekly*, as well as regional newspapers and TV stations did cover various aspects of SESAME. The AAAS has highlighted the significance of SESAME by recognizing the five founding fathers of SESAME and awarding them the AAAS Prize For Science Diplomacy. They are Chris Llewellyn Smith (UK, CERN), Eliezer Rabinovici (Israel), Zehra Sayers (Turkey), Herwig Schopper (Germany, CERN), and Khaled Toukan (Jordan).

SESAME has been inspired by CERN in many ways, yet it was a much more challenging task to construct. CERN was built after the war was over, it was clear who had won and who had lost. In my region the conflicts are not over, there are different narratives on who is winning and who is losing, as well as what to win or lose means. CERN took less than 10 years to set up the original construct. SESAME took about 25 years to do that. Thus, I would request an external reader to think of SESAME now, as CERN was in around 1960. Trying to find its place among the national, Soviet Union, and US giants.

As an Israeli I note with happiness that for the first time ever, Israeli scientists did high quality research at a facility established on the soil of an Arab country, Jordan. This unique facility started operating in uncanny normality. Well over a hundred proposals for experiments were submitted, refereed, and spots were allocated to the

Figure 56.7. Top: One of the SESAME magnets at CERN. SESAME magnets were planned and assembled by joint CERN/SESAME teams. Bottom: Photo of the components of the light source installed at SESAME. Photographer: Maximilien Brice/CERN. Top: from the author's personal archive. Bottom: Caraban Gonzalez, Noemi. Credit: CERN (CERN-PHOTO-201801-015-42).

experiments chosen. Data was gathered, analyzed, and the results were and are being published in first-rate journals (figures 56.8 and 56.9).

In addition, thanks to the help of the EU, SESAME became the first green light source, its energy being generated by solar power. This also had the added bonus of stabilizing the energy costs of the machine (figure 56.10).

Then came the outbreak of the COVID-19 pandemic and the challenge to operate a machine in such circumstances. SESAME's handling of the situation was second to none. As we are coming out of the present COVID period so has SESAME.

More experiments are being performed and more papers published in all fields of science. Myself, I was elected for the second time as a Vice-President of the SESAME council (it could be worth noting that I was unanimously approved for that position as was my co-Vice-President, an Egyptian, my predecessor, an Iranian, and my successors, a Palestinian and a Jordanian!). Under the Presidency of Heuer

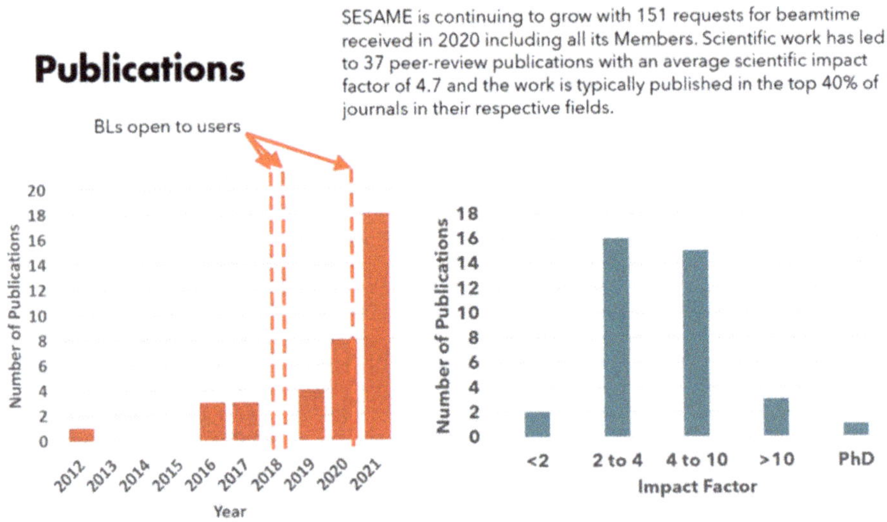

Figure 56.8. Plot showing the scientific output in terms of number of publications resulting from the experiments operating at the SESAME facility. Plot provided by the author. Credit: E Rabinovici/SESAME.

we streamlined the council meetings and started dealing in earnest with the issues that any light source faces. Science brings a semblance of normality even to our region. Thanks to the Italian contribution it was possible to build a hostel at SESAME, as in most light sources. The hostel was named 'The Sergio Fubini Hostel' a touching and most appropriate homage to Sergio.

Let me conclude this contribution with some general words: Many in the region and beyond have taken their people to a place their governments most likely never dreamt or planned to reach. These governments are aware now of where they are and as of now they did not blink.

I do not have here the place to give due credit to the many people without whose efforts SESAME would not have reached the point it is at now, I hope to be able to do it carefully elsewhere. Writing this in October 2022, I need to make a comment on the impact that the invasion of Ukraine by the Russian Federation had on CERN. Here I mean the impact it had on holding on to the vision and human mission of CERN.

In many ways SESAME is a very special child of CERN. Sometimes we find that our children can teach us important lessons. As President of the CERN Council in 2022, I can say that the way in which the member states of SESAME conducted themselves during the decades of storms that affect our region can now serve as a benchmark of how to keep bridges for understanding under the most trying circumstances. The SESAME spirit (so far) being a lighthouse even to the CERN Council.

Maintaining this attitude in a stormy political environment is very difficult and not enough appreciated and, yet, one miracle was achieved. One has shown that the

Three Beamlines in Operation

SESAME is continuing to grow with 151 requests for beamtime received in 2020 including all its Members. Scientific work has led to 37 peer-review publications with an average scientific impact factor of 4.7 and the work is typically published in the top 40% of journals in their respective fields.

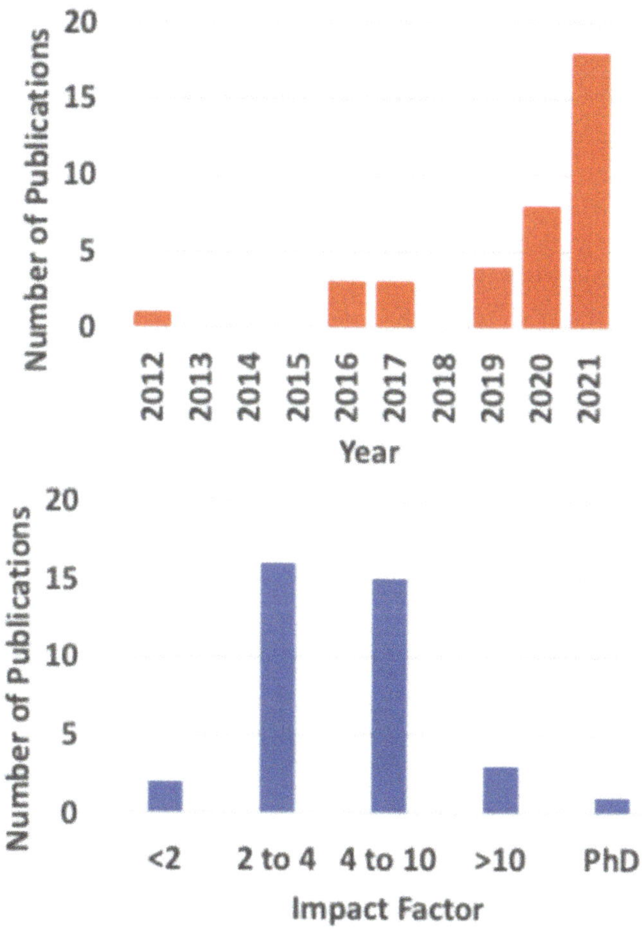

Figure 56.9. Plot showing the number of requests for beamtime reflecting a steady increase in its user community. A third beam line is currently under construction to enhance the research opportunities at SESAME. Plot provided by the author. Credit: E Rabinovici/SESAME.

Figure 56.10. SESAME the first accelerator in the world to be powered by renewable energy. Photograph: Caraban Gonzalez, Noemi. Credit: CERN (CERN-PHOTO-201811-345-7).

people in the region can work together for their own benefit and the benefit of humanity.

However, this saga ends, we have proved that the people of the region have in them the capability to work together for a common cause. Thus, the very process of building SESAME has become a beacon of hope to many in our region. The time is approaching to match this achievement with high-quality scientific research. This will be the responsibility of SESAME in the coming years.

High-quality science requires appropriate funding and help. SESAME needs them. SESAME is continuing very successfully its mission to train hundreds of engineers and scientists in the region. It is far too early to assess what industry can bring to and benefit from a light source. This issue is a serious challenge even for light sources placed in the most scientifically advanced nations.

My dream is that in the not-so-distant future people will be able to point to a significant discovery and say 'This was discovered at SESAME'.

Note from the author: Parts of my personal point of view were published in 2017 *Int. J. Mod. Phys.* A **32** 1741008 and in the Amaldi conferences 2015 and 2022 proceedings.

IOP Publishing

Big Science in the 21st Century
Economic and societal impacts
Panagiotis Charitos, Theodore Arabatzis, Harry Cliff, Günther Dissertori, Juliette Forneris and Jason Li-Ying

Chapter 57

Big Science collaborations—lessons for global governance and leadership

Mark Robinson

57.1 Global gridlock

Thomas Hale *et al* in *Beyond Gridlock* (2017) argue that the complexity, interdependencies, and interconnectivity of the modern world create new and grave global challenges that demand innovative collaborative solutions. Exasperatingly the world is beset by an inability of countries to cooperate via international institutions to address these policy problems that span borders. This international relations gridlock encompasses both a deadlock or dysfunctionality in existing organisations and the inability of countries to come to new agreements as issues arise.

Global gridlock research has never just been about describing the problems; many academics explore solutions and invite debate on proposals. This harder enunciation of potential pathways and mechanisms to solve global problems is what distinguishes this approach over more traditional international relations' frameworks that are rooted in safer descriptive territory. Disciples of the approach aim high by seeking to create and collectively build an evidence base for more effective management of global challenges in the twenty-first century. The inspiring message is that while gridlock is a dominant tendency within global governance, it not an inevitable one; there are ways for individuals, groups, and states to overcome it. The resultant theory identifies eight pathways through and beyond gridlock and their mechanisms. It advocates that it is through the combination of these pathways that international collaborations can be initiated, maintained, and delivered: (i) shifts in major powers' core interests, (ii) autonomous and adaptive international institutions, (iii) technical groups with effective and legitimate processes, (iv) multiple, diverse organisations and institutions coalescing around common goals/norms, (v) mobilisation of domestic constituencies for cooperation and compliance, (vi) civil society coalitions with reformist states, (vii) innovative leadership, and

doi:10.1088/978-0-7503-3631-4ch57

(viii) innovative funding (Hale *et al* 2017, Robinson 2018, p 10). This nomenclature (*Beyond Gridlock*, pathways 1–8) is used throughout this chapter.

57.2 Big Science communities

Big Science communities have been remarkably effective in evolving complex processes and mechanisms to enable international collaborations. Major powers who are fierce rivals in other domains, form lasting alliances. Communities of experts have found a way to work together to tackle global problems. The gist of *why* these collaborative efforts prevail is captured by the retired Head of Strategic Communications at ORNL, Mark Uhran when considering physics collaborations:

> For the first time in history, we find industrialized nations forming partnerships to design and build complex, technological assets for which no nation alone can bear the cost, or the risk. By partnering together, the effort and cost is shared; the risk distributed, and; the benefits accrue to all. (Uhran 2015)

It is the deeper question of *how* this has been achieved that this chapter will address through examining the interlinked pathways employed by science communities. In this context it is analytically valuable to consider three of the most relevant cases: the Conseil Européen pour la Recherche Nucléaire (CERN) community, the International Thermonuclear Experiential Reactor (ITER) nuclear fusion project community, and the International Space Station (ISS) community. CERN has attracted a great deal of attention through its pathfinding science results that have captured the public's imagination. It has spawned a succession of Nobel Prize winners for physics and several notable spin-offs (OECD 2014, pp 20–58). The ITER project is comparatively unknown outside of its nuclear fusion community although it is potentially of the greatest global significance from an environmental point of view. The ISS has entered modern culture with astronauts from 18 countries providing a continuous human presence in low Earth orbit since 2000.

Big Science project membership can be characterised by the concept of an epistemic community which Peter Haas (1992, p 3) defines as 'a network of professionals with recognised expertise and competence in a particular domain and an authoritative claim to policy-relevant knowledge within that domain'. The three case studies are all estimable examples of epistemic communities in their respective fields. For CERN it comprises the high energy physics community that has developed since the founding of the field in the 1950s. For ITER it comprises the nuclear fusion community and particularly those embracing magnetic confinement fusion. For the ISS it comprises all of those involved from the constituent national space agencies.

The boundaries of these communities are difficult to define and vary over time. 'The World-Wide Lab' is a term used in CERN's 62nd Annual Report (CERN 2016, p 30) to succinctly convey the confident camaraderie of purpose. The term naturally implies the prevalence of science at the heart of the enterprise; but this is not wide enough here. The CERN community is a product of complex social, scientific, technical, economic, and political interdependencies. It also reflects an ongoing relationship between shared aims and the everyday world constraints of its member states (Robinson 2018, p 3). The CERN community of nations has grown from its original 12 full member states to the 23 of today and has also seen an increase in the number of nations with associate or observer membership status.

For ITER the magnetic confinement fusion community and the central project are tightly bound; they will succeed or fail together. The number of members has not changed since the signing of the ITER Agreement in 2006. However, ITER's seven members already represent over half the world's population and over 80% of the world's GDP (OECD 2018): the People's Republic of China, Europe (28 EURATOM Treaty member states), India, Japan, the Republic of Korea, the Russian Federation, and the USA. Expansion of membership is unlikely due to the prospective buy-in costs of any new applicant and that the risk of disturbing the power balance within the ITER Council if a new major power joined.

There is also a continual variation in which national bodies interface with the central body. For example, at the start of the ITER project, the US State Department, due to security concerns, was opposed to the USA joining and the Department of Energy (DOE) was a solid supporter. A situation reversed today with the DOE, mainly due to budget pressures from competing national projects, struggling to support the project, whereas the State Department now fully backs it. It is a strength of the communities that they cope with this relentless boundary uncertainty. The Haas definition and the changing boundary conditions need to be kept in mind throughout this chapter wherever the term 'community' is used.

It is not anticipated that the three communities either singularly or collectively provide a panacea to global gridlock, but nevertheless their study is intended to provoke ideas for consideration in other domains.

57.3 Diplomatic, governance, and leadership mechanisms that have been used to implement the *Beyond Gridlock* pathways

To answer the question *how* global Big Science communities achieve their effective collaboration pathways with member states, we can consider to what degree each of the *Beyond Gridlock* pathways are employed.

Pathway 1 *shifts in major power core interests* are utilised by all three case studies. The collaborations that set them up were all long prepared and advocates rode timely waves of strategic opportunities. Pathway 2 *autonomous and adaptive international institutions* is true for CERN which is authorized by its members to be highly autonomous and adaptive but less so for ITER and the ISS where member states control and limit central autonomy. Pathway 3 *technical groups with effective and legitimate process* is achieved by CERN where infrastructure and experiment teams work seamlessly and again less so for ITER and the ISS where central design authority is continuously hard fought for, and interface management achieved despite difficulties. Pathway 4 *multiple diverse organisations coalesce around common goals and norms* is true for all three case studies that have high focality and legitimacy. Pathway 5 *mobilisation of domestic constituencies* only partly applies in the sense that central organs do encourage healthy domestic agencies which are beneficial to the relative communities, but this is not their focus. Pathway 6 *civil society coalitions with reformist states* does not apply to Big Science communities. Pathway 7 *innovative leadership* strongly applies to all the case studies: CERN exercises strong governance and light-touch leadership, the ISS demonstrates effective collaborative operational leadership and somewhat belatedly ITER has established effective central leadership. Finally, Pathway 8 *innovative funding* is utilised by all three case studies where different levels of contributions in-cash and in kind are used to fund the projects.

A pattern of successful combinations through gridlock is revealed: pathways 1, 4, 7, and 8 are used extensively, pathways 2 and 3 are used more by some than others, and pathways 5 and 6 are not applicable or used sparingly. Research has revealed that pathways 1 and 4 emerge as core requirements for global scientific international collaborations to be initiated, maintained, and delivered. They are supported in their implementation by pathways 7 and 8. Each pathway and the relationship between them will be examined in the four sections that follow, using the same structure: first CERN, then ITER, and finally the ISS. Each section ends with a summary, supported by figures and tables, that compares and contrasts the evidence.

57.3.1 Shifts in major powers' core interests

57.3.1.1 CERN

CERN was first conceived in the late 1940s as Europe was still recovering from World War II. Fascism and war had depleted most of the continent's resources and many of the top scientists had fled to the United States. Forward thinking scientists, such as Pierre Auger (French), Raoul Dautry (French), Lew Kowarski (Russian born, naturalised French), Edoardo Amaldi (Italian), and Niels Bohr (Danish), put aside national interests and strove to build bridges between nations in education, science, and culture.

CERN did not triumph over rival science laboratory concepts by chance; the fundamental physics community made their own luck. A consensus led by several Nobel laureates emerged for the need for a world-class facility dedicated to nuclear research of a pure scientific and fundamental character. French physicist Louis de Broglie first proposed a multi-national European lab in 1949. This call coincided with the 1950s European movement of reconciliation after World War II. At the same time the United States was emerging onto the global scene as a major leader in scientific research. European physicists wanted to compete, but no single country could match the resources of the emerging superpower.

After years of lobbying for a high energy physics 'global laboratory' spearheaded by Isador Rabi, a Nobel laureate and the United States envoy to the newly established UNESCO, the community settled on a facility to be based in Europe. The timely cooperation between the fundamental physicists and the politicians provided an opening and twelve European nations came together to sign the convention, officially forming CERN in 1954. Nations that 10 years earlier had been at war were bound together in peaceful research—a dramatic improvement in international relations that the community had worked together to achieve. The historian Professor John Krige, in an interview conducted for this research during a symposium at CERN to celebrate 25 years of operation of the flagship Large Hadron Collider, noted:

> If Europe wanted to do high energy physics, the only way was by having governments collaborating with each other because it's such an expensive field…pooling resources is the best way…. [indeed] the *only* way they were going to be able to afford it.

57.3.1.2 ITER

For ITER, the wave of political change was the mid-1980s rapid thawing of the Cold War between the two global superpowers of the period: the USA and USSR. US President Ronald Reagan had set the scene with his March 1982 message to the US Congress (Reagan Presidential Library 2019a):

> …we also recognize that, while the United States retains international pre-eminence in many areas of science and technology, we are no longer in a position to dominate each and every field. Nor do we hold a monopoly on the world's supply of scientific talent… There are areas of science, such as high energy physics and fusion research, where the cost of the next generation of facilities will be so high that international collaboration among… nations may become a necessity. We welcome opportunities to explore with other nations the sharing of the high costs of modern scientific facilities.

The catalyst for tangible progress was the appointment in 1985 of Mikhail Gorbachev as General Secretary of the Central Committee of the Communist Party and, by custom, Leader of the Soviet Union. Vast military expenditure coupled with bureaucratic central planning meant that by the time he took office the Soviet economy was in dire need of improvement. He quickly embarked on a radical reform programme and demonstrated a willingness for openness and dialogue with the West that his more elderly predecessors had shunned. His first international visit was to France to meet President Mitterrand. Accompanying Gorbachev was his chief scientific advisor and long-time friend, Evgeny Velikhov; they had met thirty years previously as students at Moscow State University. Velikhov was an ardent supporter of fusion research and was later universally accepted by the community as the 'godfather' of the ITER project. In a rare interview given in 2015 he recalled the political context of the times (Petrov 2015):

> It is important to remember that another collaborative project in fusion, called INTOR, had started in the Brezhnev times during a thaw in relations with the West. When war in Afghanistan started relations

chilled again. As a result of these complications, Europe and the USA moved forward on negotiations on fusion collaboration in the framework of the 1982 Versailles Summit without the participation of the Soviet Union. Negotiations had stalled again however; therefore, I decided to deliver this idea [of fusion] to Gorbachev who was going on a visit to Paris.... [he] took to the proposal and invited me to the meeting. Mitterrand immediately picked up on the idea of international cooperation in fusion.

Reagan was a complex personality and a paradox: publicly a Cold War warrior but privately someone who hated 'the bomb'. His Strategic Defence Initiative (SDI) and election in January 1985 to a second term, meant that the peacemaker Reagan could come to the fore and he considered the merits of a superpower summit. He faced opposition from hawks such as Defence Secretary Casper Weinberger and his deputy Richard Perle who both deeply distrusted the Soviets. Through careful management by US Secretary of State George Shultz and USSR Foreign Minister Eduard Shevardnadze, a reformer appointed by Gorbachev, the summit was kept on track (New York Times 1985).

The 19–20 November 1985 Reagan–Gorbachev meeting in Geneva entered Cold War folklore as the 'fireside' summit. The world's media were primed to pass verdict on who had 'won' the encounter. While little was achieved in terms of arms reduction or SDI (this was to take another two summits) historians later concluded that the two men started a rapport and trust-building that would ultimately lead to the ending of the Cold War some six years later (Reynolds 2007, p 178). In contrast to previous US–Soviet summits, no pre-Summit draft statement had been prepared due to US concerns that it could unduly reveal negotiation positions. This provided a golden opportunity for advocates to expertly nudge the long-term fusion vision into the mainstream national policy decision-making process. Although the idea of developing a large international collaboration was the last agenda item and the final entry to be agreed in the joint communiqué, it was the one with the farthest reach (Arnoux 2015). The text ending with the same evocative words that Eisenhower had used in his address to the UN General Assembly 30 years earlier (UN 1985, p 6, IAEA 2017):

> The two leaders emphasized the potential importance of the work aimed at utilizing controlled thermonuclear fusion for peaceful purposes and, in this connection, advocated the widest practicable development of international cooperation in obtaining this source of energy, which is essentially inexhaustible, for the benefit for all mankind.

For the fusion community this was the long-planned-for political breakthrough, reinforced when Reagan even cited 'fusion energy' in his address to a Joint Session of Congress on 21 November 1985 (Reagan Presidential Library 2019b). Two years later the IAEA, at its headquarters in Vienna, held the first meeting of the Quadripartite Initiative Committee comprising Europe (EURATOM nations),

Japan, USSR, and the USA where the concept and the name ITER, standing for International Thermonuclear Experimental Reactor, was agreed on (IAEA 1988).

57.3.1.3 ISS

US technological pre-eminence in space had been underlined by the Moon landing in 1969. For the ISS, contacts between US and USSR space exploration practioners steadily increased from the détente inspired, first joint Apollo–Soyuz mission in 1975 (Launius 2018, p 212). Although this was a one-off event, it did provide a tantalising glimpse to the respective national space agency communities of what might come. Personal relations between the specialists developed and trust between the two space agency camps steadily improved. This gave credibility to the idea of inviting Russia to take part when the opportunity came along and in January 1984 US President Reagan invited 'Friends and Allies' to join the challenge and share the costs, risks, and rewards. Prominent advocates in the US such as Daniel Goldin (NASA Administrator from 1992 to 2001) were eagerly ready when the Cold War Space Race finally ended to lead the effort for an international approach.

While the Russian experience in space was certainly valuable there was an undeniable lingering Cold War inspired mistrust, from both parties. ISS advocates successfully canvassed the cautious officials in the US State Department to, somewhat counterintuitively, collaborate with the Russians in a domain that had seen fierce Cold War rivalry. The strategic argument to involve the Russian space technological and science teams was partly to wrap them in a relationship that would allow little capacity for anything else. There was a desire to limit any science and technology transfer to former Warsaw Pact states and rogue states. In hindsight it was a masterful decision as it not only helped to achieve this restriction but also safeguarded the pace of the project when US launch capability was lost with the second Space Shuttle disaster in 2003 and Russia stepped in to fill the gap.

57.3.1.4 Summary of shifts in major powers' core interests

Figure 57.1 summarises the major waves of opportunity that shifted major powers' core interests (*Beyond Gridlock*, pathway 1) and helped provide the necessary surge to bring nations together:

- The establishment of UNESCO in 1946 and the birth of the first European movement in 1947 for CERN (convention signed in 1953).
- The period of US–USSR détente, 1967 to 1979, and US invitation to potential partners in 1984 for the ISS (IGA signed in 1998).
- The collaborative stance taken by the US and USSR leaders at the 'fireside summit' in Geneva in 1985 for ITER (agreement signed in 2006).

Figure 57.1 also shows the commendable longevity of the case studies that has been achieved despite the four categories of significant constraints that have beset the communities:

- **Financial:** such as the German funding crisis following reunification in 1990, the global financial crisis of 2008, and the Swiss franc crisis of 2015.

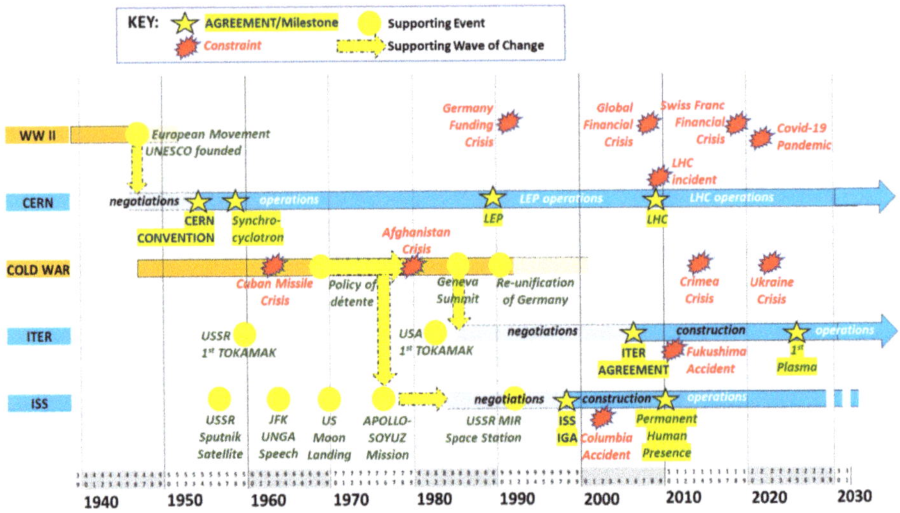

Figure 57.1. Timing of establishment and major milestones of case studies set against major constraints affecting the member states. Adapted from Robinson (2020) John Wiley & Sons. Copyright 2020 The Authors. Global Policy published by Durham University and John Wiley & Sons Ltd.

- **Political:** such as the Afghanistan crisis of 1979, the Crimea crisis of 2014, and at time of writing (March 2022) the Ukraine crisis of 2022.
- **Technical:** such as the Columbia Space Shuttle accident of 2003, the Large Hadron Collider (LHC) incident in 2008, and the Fukushima nuclear accident of 2011.
- **Health:** such as the 2020–22 COVID-19 pandemic affecting all member states and their supply chains.

57.3.2 Multiple, diverse organisations and institutions coalesce around common goals/norms

CERN is expanding human knowledge in fundamental physics that could help answer how our Universe was created and where we come from. ITER is endeavouring to provide a route to carbon-free, commercial power generation that could help abate global warming and change geopolitics. The ISS is carrying out unique micro-gravity science and building the knowledge mankind will need for space travel to Mars and beyond. Who would not be proud to part of these communities? While national agendas will always be present in projects of this magnitude, the driving factor that trumps these is the collective higher-level goal. There is a strong sense of purpose that nation states, non-governmental bodies, and people eagerly rally behind.

The enthusiasm generated in these endeavours surfaces in the way that the respective communities mark their project inaugurations. Like large engineering projects, they may plant a tree or lay a foundation stone; or in the case of the ISS deliver a first module in low Earth orbit! What sets them apart is that they also feel the need to emphasise and celebrate the meeting of minds between major powers. At the signature of the ITER Agreement on 21 December 2006, in the Elysée Palace in Paris, the host of the event, President Jacques Chirac of France, noted that the occasion marked a memorable moment in the history of science (Eurofusion 2006):

> Exceptional for its scientific ambition to harness the sun's power to take up the challenge of ecological energy, and exceptional for its international scale: the unprecedented association of seven major partners from the North to the South. It is the hand, held out to future generations in the name of solidarity and responsibility.

Rousing rhetoric that would have seemed embarrassingly out of place for something of less importance. President Kennedy in his 1961 address to the UN General Assembly calling for joint space missions with Russia (JFK Presidential Library 2019) is another striking example. These moments matter, since they help legitimise the projects, raise their profile beyond the ordinary, and are cited by members to this day.

The notion of high meaningfulness in the mission is also encapsulated by Maria Spiropulu, a physicist at Caltech, vice-chair of the American Physical Society Forum on International Physics and a member of the Compact Muon Solenoid (CMS) collaboration at CERN:

> Being a fundamental physics lab inherently played a big role in pushing competing nations to work together...the goal is the same, and it's very clear that it is fundamental research .... That alleviates a lot of the conflict...because we're all going towards the same fundamental goal and the same dream (Lucibella 2014, p 1).

This strong, inherent meaningfulness is important as the benefits of these endeavours are not necessarily seen early on; these are long-term considerations. This makes the projects harder to sell to a political class who crave short term results to bolster popularity and to a sceptical public who can easily point to other more pressing societal needs. The communities appreciate that *sustained* backing would not be achieved by the sudden seizing of political agendas by stunts or inducements. Effective communication of a long-term vision is needed to achieve and maintain stakeholder backing for these projects.

57.3.2.1 Summary of multiple, diverse organisations and institutions coalescing around common goals/norms

The communities distil the meaningfulness by establishing common goals/norms around which multiple diverse organisations and institutions can coalesce (*Beyond Gridlock*, pathway 4). Focality and legitimacy concepts are closely intertwined throughout the case study communities and are vital in gaining and maintaining the support of members. By measuring the degree with which community actors look to and converge around the central teams and the quantity and quality of the

Table 57.1. Attributes of focality and legitimacy across the case studies (adapted from Abbott and Hale 2014).

		Case Study 1 CERN	Case Study 2 ITER	Case Study 3 ISS
Focality	Actors look to and Converge Around the centre	VERY HIGH accepted global hub for fundamental physics	VERY HIGH Project and Community are united in achieving one aim	VERY HIGH self-evident unique global lab for micro-gravity research
	Connectivity of the centre with the community	VERY HIGH world-wide community share data through landmark comms network	HIGH Connectivity has improved since early days of the project	MIXED national operational communication paths work alongside central coordination
Legitimacy	Track record of Science Results	VERY HIGH several Nobel Prizes and meaningful spin-offs	LOW meaningful results only available in operations phase but potential is VERY HIGH	CONTESTED differing views of worth to date versus cost; commercialisation may bear fruit
	Moral Authority: Professional Community	VERY HIGH infrastructure and embedded experiment projects work seamlessly together	MEDIUM meaningful control of supply base rests with members; central authority in some areas is contested	MEDIUM limited central authority; national space agencies run operations
	Moral Authority: Public perception	VERY HIGH well developed community, education and public outreach programs	LOW broadly unknown to the public despite public outreach programs	VERY HIGH well developed community, education and public outreach programs

connectivity of the centre with the dispersed project elements, it is possible to assess the level of focality. By analysing the track record of science results, moral authority within the community and standing with the public, the sense of achievement in legitimacy can also be made. Table 57.1 provides a summary of both features.

57.3.3 Innovative leadership

An aspect of leadership that has emerged from the research is the concept of measured governance *by the lead entity* in the community. It concerns the approach taken to the other members by the central organisation leadership (for CERN) or by the principal member's leadership (Europe for the ITER project and the USA for the ISS).

A shared mission alone cannot sustain a laboratory for 65 years, and CERN's administration is designed to accommodate the member states' scientists and politicians as much as possible. When Germany during reunification in 1990 was forced to reduce its cash contribution to the central capital fund, just when the organisation needed more infrastructure funds for the flagship Large Hadron Collider, the decision reached was to reduce the cash contributions of *all* the member states by the same proportion. The central management team realised that the overriding need was to maintain fairness, a judgement that was then endorsed by the council who subsequently authorised innovative funding mechanisms to plug the cash shortfall. The ultimate authority for the laboratory rests in the governing council. Each member state gets two seats: one reserved for a scientist and the other for an administrator.

For ITER, the members are both stakeholders (through the council) and suppliers (through the in-kind contributions they are responsible for). This produces a tension in the council that must be sensitively handled if its legitimacy is to be preserved. The project delays that the council has also had to deal with have added to the pressure on them to collectively keep the community on track. All ITER interviewees agreed that the situation in this regard has improved since the early days of the project. Community consensus has prevailed, despite the inevitable tensions, as without it the project's critics would have exploited any major divisions between the members. The main member is Europe whose group of EURATOM states contribute most of the funds, enjoy the largest work-share, and host the project in France. An ongoing advantage that this has provided to the community is financial stability. Europe operates on a six or seven-year funding cycle rather than a one or two-year cycle as is the case in most of the other members.

For the ISS, governance of the project is inexorably linked to the leadership provided by the USA. The dominant role of the USA in the management arrangements was accepted by all the parties when they agreed to Paragraph 2 of Article 1 of the Intergovernmental Agreement (US State Department 1998): 'The Partners will join their efforts, under the lead role of the United States for overall management and coordination, to create an integrated ISS.' The international reputation of the USA has therefore been under scrutiny from the outset. Henry Hertzfeld is the Director of the Space Policy Institute at the Elliott School of International Affairs, George Washington University. In an interview for this research, he and several other interviewees attested that the US has consistently provided sound leadership. At the centre of NASA's approach to ISS operations throughout the project's life-cycle has been the will to reach consensus with the other four Space Agencies (the Canadian Space Agency (CSA), the European Space Agency (ESA), the Japanese Aerospace Exploration Agency (JAXA), and the Russian Federal Space Agency (ROSCOSMOS)).

A geo-political event that had the potential to wreak havoc across two of the case studies was the Crimea crisis of 2014. On the ITER project the Russians, when their currency devalued by a factor of three, boldly acted internally by making cuts to other national projects in order to maintain their cash contribution. In an interview for this research, the historian John Logsdon described the high-level US space policy thinking that was necessary to circumvent this threat:

> When the troubles with Russia started in Crimea and Ukraine there was a very conscious policy decision on behalf of the White House to build a wall around the station and not let our unhappiness with Russia in these other areas have an impact on the integrity of the station collaboration including the access needed to get there. The President agreed that it be excluded from any restrictions, reparations or sanctions that might be put in place. We had put ourselves in a position both in the architecture of the station itself and by our failure to have a shuttle replacement where we had become communally dependent on Russia.

At time of writing (March 2022), there has been extensive condemnation of the Russian invasion of the sovereign state of Ukraine. Severe and far reaching sanctions on the Russian leadership and state have been imposed by western governments and multi-national corporations. The leadership of Big Science communities world-wide have also strongly condemned the aggression. Given Russia's standing in global science, there has also been an impact on the three case studies.

On the 8 March 2022, the CERN Council agreed on a resolute position that had three elements: (A) CERN will promote initiatives to support Ukrainian collaborators and Ukrainian scientific activity in the field of high energy physics; (B) the Observer status of the Russian Federation is suspended until further notice; and (C) CERN will not engage in new collaborations with the Russian Federation and its institutions until further notice (CERN 2022).

As we have seen Russia has been intrinsically part of the ITER project since its inception. Russian deliverables form crucial parts of the in-kind contributions needed for the device construction, commissioning, and operation. As one of the seven members they also have a sizeable presence at the HQ in Cadarache, France; it is understood that the Russian staff there are as equally outraged as all the staff by their government's actions. The project's council has so far not commented on events. The principles in the ITER Agreement including at Article 10 the framework for sharing intellectual property, are inviolable and essential to the delivery of the results of this global project that are so needed by us all (IAEA 2006, Takano 2016).

For the ISS, the situation is different from 2014, in that the other partners are no longer solely dependent on Russia for re-supply or crew rotation. There are now commercial options and SpaceX has been transporting NASA astronauts and cargo to the space station since 2020 (SpaceX 2022). However, due to the load factors and general maintenance tasks there is still a reliance on Russian contributions as witnessed by the recent arrival of three ROSCOMOS cosmonauts in a Soyuz MS-21 spacecraft (Garcia 2022). Russian President Vladimir Putin appointed a close ally Dmitri Rogozin as head of ROSCOSMOS in 2018. Rogozin is noted for his bombastic nationalist statements, which have reached new levels in the recent crisis, including implying that any unscheduled ISS de-orbit would be more likely over the US or Europe than anywhere else (Wermus 2022). There have been swift re-assuring messages from the other ISS partners including NASA that promote normalcy of operations. NASA Administrator Bill Nelson said in remarks at a 1 March 2022, meeting of the NASA Advisory Council: 'NASA continues the working relationship with all our international partners to ensure their safety and the ongoing safe operations of the ISS' (Foust 2022, Gorman 2022). Undoubtedly, however, the severity of the Russian actions in Ukraine has heightened tensions and rapidly changed the state of international collaboration in space to become much more sensitive to politics here on Earth. It is perhaps only the limited operating life left for the ISS that is allowing the operational teamwork to continue.

Finally, consensus is maintained through an innate need for community survival. For ITER, there is a realisation that the set of circumstances that established this nuclear collaboration would be difficult to repeat in today's increasingly nationalistic political climate. There will not be a second chance for generations; members view failure as not an option for the planet or their community. National interests from non-European ITER members also arises in that they want to avoid any possibility of Europe taking over the remnants of the project should the international collaboration collapse. This feeling of profound responsibility for continuance is also prevalent within the CERN and ISS communities. Interviewees from all levels of management throughout all three case studies appreciated that together with the privileges of working on these iconic endeavours is the sense of duty to carry on the work of their predecessors.

57.3.3.1 Summary of innovative leadership

In endeavours of this scale, cost, and longevity, there will always be vocal sceptics ready to swoop on fall-outs between member states. It is not enough that the communities combat those who would wreck their collaborative spirit, they have to continually advance the cause (*Beyond Gridlock*, pathway 7). The leadership of the communities appreciate that they *have* to be shining exemplars of what can be achieved when national interests are trumped by common needs; they symbolise a nuanced practical side of neorealism beyond straightforward power theory (Buzan *et al* 1993). Each CERN Director-General has strived to advance, not just the research capability, but also the high energy community *esprit de corps* itself. Each ITER Director-General has had to balance the needs of the project with the needs to keep together a governing council whose members, in other fields, rarely sit around the same table. Each ISS crew has seamlessly passed on the metaphorical baton of collaboration since the first low Earth habitation in 2000.

57.3.4 Innovative funding

The time when any new intergovernmental organisation (IGO) would be established and maintained by solely cash contributions of the member states is long over. The need to sustain the institutions and industries in their own territories, bound by the project's detailed in-kind contribution commitments, is now an imperative consideration. The adoption of innovative funding is not undertaken lightly and the set-up phase for any new endeavour needs to take as long as necessary to stipulate the in-kind deliverables, time-scales, and relative value.

For CERN, members contribute to a central fund for the infrastructure of the laboratory as a whole. However, the embedded experiments are not funded from central cash funds but rather by in-kind contributions of those taking part in that research. This means that while much of the infrastructure was built by CERN using the money contributed to the central fund, the adjunct giant detector experiments were funded, designed, and built by independent coalitions of the willing. This way, if the development of one experiment falls behind it does not necessarily mean the entire research effort will suffer and thereby rewards innovation and risk-taking.

The task of providing for and promoting cooperation among the diverse group of ITER members was a challenge during the negotiations that continues to this day. Not least in the difficulties facing the central ITER organisation was the balancing act of keeping member's content with their respective work-share and project

Table 57.2. Big Science community, Central IGO Team, and member state factors positively and negatively affected by the level and blending of contributions provided in-kind and in-cash.

COMMUNITY	High level of contributions in-cash	Appropriate blending of contributions	High level of contributions in-kind
INTER-GOVERNMENTAL ORGANISATION/ CENTRAL PROJECT TEAM	More autonomy and nimbleness in decision making More potential for unauthorised [by the member states] mission creep	Sufficient authority to tackle problems and combat external constraints is delegated to the centre Sufficient project control levers are provided to the centre to work with member states to control the project	Increased project resilience; less likelihood that the project could be cancelled Less central autonomy Cumbersome decision making High administrative burden to manage changes
MEMBER STATES	Harder to receive national go-ahead and regular re-authorisation from home national funding agencies Simpler collaborative effort required Free riding easier [although responsible member states will not take up the opportunity	Primacy of the member states is sacrosanct protected in voting rights being retained in highest decision making body Improved project performance and thereby better and more timely accrued results to the member states	Easier to receive national go-ahead and regular re-authorisation from national funding agencies Increased benefits to home industry and institutions More interfaces with central organisation and other member states is challenging for configuration control and schedule adherence

progress while exercising actions to maintain that progress through the members own dispersed organisations.

For the ISS, the nature of the construction is that the members provide their respective modules to the central entity. It is a consummate in-kind jigsaw puzzle assembled and operated in space by remarkable coordinated efforts, not least exercised by the widely respected astronaut corps. There are central cash contributions for the Multilateral Coordinating Board (MCB) but only to the level necessary for its limited functions. There is no IGO, director-general, or large international staff to fund. It is to the ISS communities' credit that despite this imbalance they have made the collaboration work.

The skill of the negotiators is to agree the correct balance of contributions. There has to be enough cash contributions to let the central entity function effectively and enough in-kind contributions to allow the members to maintain backing for the project in their home territories. Table 57.2 shows how the level of autonomy that the community wishes the central entity to have is affected by the degree of one type of funding versus the other. This balance changes as the projects mature with the member states progressively granting more power to the centre while always retaining the final say. Many interviewees attested that over the past 60 years CERN has developed the optimum balance between the two types of funding.

57.3.4.1 Summary of innovative funding

The in-kind management system that the Big Science communities employ *demands* close cooperation between the member states and helps temper their rivalry. There is a need for constant dialogue and tight interface management. The three Big Science communities examined in this chapter have exercised consummate power relations at a distance that David Held highlighted as one of the requirements of managing globalisation (Held *et al* 1999, p 28). There is no room for ambiguous political positions or free riders. The member states involved are the power houses of the industrialised world who fiercely compete with each other in many other domains. The innovative funding design provides a way to reconcile the competition between members outside of the project community with the need to closely collaborate between the same members within it (*Beyond Gridlock*, pathway 8). The control of the in-kind management system is far from easy and adds a level of effort to the member states that they would not need with a simple delivery of cash contributions. However, once mastered it is an important reason for them staying bound together.

57.4 Conclusion

Today, the world news, amplified by social media, is dominated by continuing armed conflicts, migration, and environmental problems that span borders. The urgent and essential need to make progress in several domains means that advocates of progress run the risk of jumping at any solution or struggling through painful workarounds, both of which courses of action will no longer do. There is a desire from diplomats, subject matter experts, and the public to improve performance and to find a way through these impasses. There can be no more important question in international relations than what could work to improve global collaboration.

Robert Keohane in *After Hegemony* (1984, p 12) makes the point that, once the arguments have been won and the effort has been made to bind major powers and establish a collaboration, then hegemony is less important for the continuation of cooperation than for its creation. Keohane was also correct when he said ‘cooperation is not easy’, and this is certainly true for Big Science projects. The technical and engineering challenges of designing, constructing, and operating facilities that are at the outer limits of what today’s technology can achieve by multiple partners working together make them extraordinarily difficult to manage. This degree of difficulty is present even before factoring in the international relations considerations of major nuclear powers who comprise the core of the membership. All of this makes their self-evident success even more noteworthy.

Extensive field work in each of the three case study communities reveal winning combinations of pathways for Big Science collaborations. Shifts in major powers’ core interests and multiple, diverse organisations, and institutions coalescing around common goals emerge as the two core pathways. They are supported in their implementation by two enabling pathways: innovative leadership and innovative funding. The relationship between all four provides clues on mechanisms that can be used to good effect in other international relations domains that are caught up in the debilitating global gridlock phenomenon.

Looking up at the ISS provides a welcome respite to this gloom and offers inspiration to many as a beacon of higher-level collaborative human behaviour. Similarly, CERN inspires people that there is a better way to collaborate peacefully on the fundamental questions of nature. ITER provides hope of a carbon-free commercial power source for the benefit of mankind. This chapter has outlined that they also provide proven mechanisms for combatting global gridlock.

Two contributions emerged for consideration by others in the international relations field. The first shows that communities should be primed and ready to exploit shifts in the core interests of the major powers in order to launch new endeavours. Communities need to have allies of the necessary calibre in key positions in order to influence policy makers and maximise the timing advantage. Ingenious preparation by the Big Science communities included positioning key respected individuals within the influential circles of policy makers. It was high-level strategic lobbying by like-minded individuals who had personal relationships developed over decades that held the key. The second is how an ingeniously designed funding system that allows member states to commit to projects, permits the central IGOs to operate effectively and, at the same time, maintains support in the member states' homelands.

The applicability of the eighth pathway, concerning funding, to other domains outside of this research is unknown. A thorough systematic investigation of the advantages and disadvantages of the in-kind management system as it applies to intergovernmental collaboration is needed. For example, could academics and practioners combine to design an in-kind management system to help overcome constraints and improve collaborative performance for the United Nations?

A common feature of the formation of both CERN and ITER is that both sets of advocates were primed and ready to ride momentous waves of change as major powers shifted their core interests (pathway 1). Hindsight shows that the timing to step onto each wave was also flawless. For the ISS the take-up of shifts in major powers' core interests were subtler. Given the urgency of addressing the wicked problems that face the very existence of humankind, to simply await the next wave of change that may galvanise global collaboration is too complacent. Instead, research could be focused to determine how an artificial wave of change could be designed in each domain to provide a tailor-made plan to overcome global gridlock that major powers, expertly nudged by strategically placed advocates, could then rally behind.

Notes:

This chapter is an abridged and updated version of an article first published in *Global Policy* in 2020.

The data that support the findings in this chapter are available from the author upon reasonable request.

Author information

Mark Robinson is a tutor in Science Diplomacy at the Centre for International Studies and Diplomacy at SOAS, University of London. He has a PhD in Government and International Affairs from Durham University an MSc from Oxford University in Major Programme Management and an MA in Global Diplomacy from London University. Mark is also currently the Head of the In-Kind Management Division at the European Spallation Source in Lund, Sweden. Previously, Mark worked at three IGOs: the ITER Nuclear Fusion Project in

France, the European Southern Observatory in Germany/Chile, and a NATO Management Agency in Germany.

References

Abbott K W and Hale T 2014 Orchestrating Global Solution Networks: A Guide for Organizational Entrepreneurs *SSRN* 10.2139/ssrn.2431956 (Accessed: 12 August 2017)

Arnoux R 2015 Conceived in Geneva, born in Reykjavik, baptized in Vienna *ITER Newsline* 16 Nov https://iter.org/newsline/-/2323 (Accessed: 11 February 2019)

Buzan B, Jones C and Little R 1993 The logic of anarchy: neorealism to structural realism *Int. Affairs* **69** 743–4

CERN 2016 *2nd Annual Report of the European Organisation for Nuclear Research* CERN, Geneva, Switzerland http://library.cern/annual-reports (Accessed: 2 December 2017)

CERN 2022 *CERN Council Responds to Russian Invasion of Ukraine* https://home.cern/news/news/cern/cern-council-responds-russian-invasion-ukraine (Accessed: 10 March 2022)

Eurofusion 2006 ITER Organization Embarks on its Mission... 'Exceptional in every way' *Eurofusion Newsletter* 29 November https://euro-fusion.org/news/detail/detail/News/iter-organization-embarks-on-its-mission-exceptional-in-every-way/ (Accessed: 3 June 2020)

Foust J 2022 Roscosmos head again questions future of ISS while NASA emphasizes cooperation *Spacenews* 3 March https://spacenews.com/roscosmos-head-again-questions-future-of-iss-while-nasa-emphasizes-cooperation/ (Accessed: 18 March 2022)

Garcia M 2022 Soyuz trio docks to station, NASA astronaut nears departure *NASA Space Station* https://blogs.nasa.gov/spacestation/2022/03/18/soyuz-trio-docks-to-station-nasa-astronaut-nears-departure/ (Accessed: 18 March 2022)

Gorman S 2022 NASA shrugs off Roscosmos leader's rant over US sanctions and space station *Reuters* https://reuters.com/world/europe/nasa-shrugs-off-roscosmos-leaders-rant-over-us-sanctions-space-station-2022-02-26/ (Accessed: 17 March 2022)

Haas P 1992 Introduction: epistemic communities and international policy coordination *Knowl. Power Int. Policy Coord.* **46** 1–35 https://www.jstor.org/stable/2706951

Hale T *et al* 2017 *Beyond Gridlock* (Cambridge: Polity)

Held D, Goldblatt D, Perraton J and McGrew A 1999 *Global Transformations* (Cambridge: Polity)

IAEA 1988 *Establishment of ITER: Relevant Documents ITER Documentation Series* No. 1 IAEA, Vienna

IAEA 2006 Agreement on the Establishment of the ITER International Fusion Energy Organization for the Joint Implementation of the ITER Project *IAEA Information Circular* 702 25 April https://iaea.org/publications/documents/infcircs/agreement-establishment-iter-international-fusion-energy-organization-joint-implementation-iter-project (Accessed: 3 June 2017)

IAEA 2017 Address by Mr Dwight D Eisenhower, President of the United States of America, to the 470th Plenary Meeting of the United Nations General Assembly, Atoms for Peace Speech, 8 December 1953 *International Atomic Energy Authority, History* https://www.iaea.org/about/history/atoms-for-peace-speech (Accessed: 29 December 2017)

JFK Presidential Library 2019 *President J F Kennedy Address at UN General Assembly, 25 September 1961* https://jfklibrary.org/learn/about-jfk/historic-speeches/address-to-the-united-nations-general-assembly (Accessed: 3 December 2018)

Keohane R O 1984 *After Hegemony: Cooperation and Discord in the World Political Economy* (Princetown, NJ: Princetown University Press)

Launius R D 2018 *The History of Space Exploration* (London: Thames and Hudson)

Lucibella M 2014 CERN: 60 years of collaboration *American Physical Society News* vol 23, August/September 2014 https://aps.org/publications/apsnews/201409/cern.cfm (Accessed: 27 November 2017)

New York Times 1985 *Letter from Secretary of Defense Casper Weinberger to President Reagan, November 15, 1985* https://nytimes.com/1985/11/16/world/weinberger-letter-to-reagan-on-arms-control.html (Accessed: 22 August 2018)

OECD 2014 Impacts of large research infrastructures on economic innovation and on society *CERN Case Study* https://oecd.org/sti/sci-tech/CERN-case-studies.pdf (Accessed: 23 March 2018)

OECD 2018 *World GDP Statistics* https://stats.oecd.org/ (Accessed: 1 June 2018)

Petrov A 2015 Evgeny Velikhov: 'I decided to deliver this idea to Gorbachev.' *ITER Newsline* https://iter.org/newsline/-/2326 (Accessed: 10 October 2018)

Reagan Presidential Library 2019a President Reagan report to Congress on US international activities in science and technology, 22 March 1982 *Reagan Presidential Library Archives* https://reaganlibrary.gov/research/speeches/32282a (Accessed: 12 April 2017)

Reagan Presidential Library 2019b Address before a Joint Session of the Congress following the Soviet–US summit in Geneva, 21 November 1985 *Reagan Presidential Library Archives* https://reaganlibrary.gov/research/speeches/112185c (Accessed: 12 April 2017)

Reynolds D 2007 *Summits: Six Meetings that Shaped the Twentieth Century* (London: Allen Lane)

Robinson M 2018 The CERN community: a mechanism for effective global collaboration? *Global Policy* **10** 41–51

Robinson M 2020 Big Science collaborations; lessons for global governance and leadership *Global Policy* **12** 66–80

SpaceX 2022 DRAGON: Sending humans and cargo into space *SpaceX* https://spacex.com/vehicles/dragon/ (Accessed: 18 March 2022)

Takano A 2016 The framework for sharing ITER intellectual property *ITER Newsline* 5 December https://iter.org/newsline/-/2595 (Accessed: 4 March 2022)

Uhran M 2015 Mission accomplished! The role of systems engineering and integration in the International Space Station Program *EEE 26th Symposium on Fusion Engineering (Austin, TX)* pp 1–8

UN 1985 Text of US–USSR Joint Statement, issued in Geneva on 21 November 1985, on meetings held in Geneva from 19 to 21 November between the President of the USA, Ronald Reagan, and the General Secretary of the Central Committee of the Communist Party of the Soviet Union, Mikhail Gorbachev *UN Department for Disarmament Affairs Reference Library* https://unoda-web.s3-accelerate.amazonaws.com/documents/library/A-40-1070.pdf (Accessed: 30 December 2017)

US State Department 1998 Space Station agreement between the USA and Other Governments, Signed in Washington January 29, 1998 *Treaties and Other International Acts Series 12927* https://www.state.gov/wp-content/uploads/2019/02/12927-Multilateral-Space-Space-Station-1.29.1998.pdf (Accessed 11 March 2023)

Wermus K 2022 Russian Space Agency head suggests sanctions may lead to ISS plummeting *Newsweek* 24 February https://newsweek.com/russian-space-agency-head-suggests-sanctions-may-lead-iss-plummeting-1682474 (Accessed: 5 March 2022)

IOP Publishing

Big Science in the 21st Century
Economic and societal impacts
Panagiotis Charitos, Theodore Arabatzis, Harry Cliff, Günther Dissertori, Juliette Forneris and Jason Li-Ying

Chapter 58

From the National Laboratory to international collaborations (framing Big Science for the age of science diplomacy)

Simone Turchetti

This chapter offers a framing analysis of the term 'Big Science' to chart its transition in meaning from a notion advertising the national laboratory in the 1960s to one promoting international collaborations and, contingently, publicizing science diplomacy. To accurately chart this terminological shift, the essay first introduces a text-mining analysis of academic literature featuring the notion, and charting changes in frequency of words associated with Big Science, hence showing a change in meaning. It then presents a historical narrative that explains this change by focussing on the content of this academic literature. It therefore reveals that such a shift primarily results from the crisis of the national laboratory model of the 1980s and 1990s, when it became less viable financially and politically. In turn, this crisis propelled a different idea of Big Science as more collaborative and international. Indeed, it helped in re-branding Big Science collaborations not only as a promising model for large-scale scientific research, but also as a global governance exemplar instigated through science diplomacy.

Big Science's meaning has changed dramatically since the term's first introduction. Alvin M Weinberg's 1961 *Science* article that widely popularized the new notion identified Big Science with the rise of large-scale national laboratories distinctive of the post-WW2 era, positing its emergence firmly in the United States, the country then leading the global scientific enterprise. The facility that Weinberg directed at the time, the Oak Ridge National Laboratory (ORNL, Oak Ridge, TN), was an important example of scientific production on a large scale [1]. A recent article written exactly 60 years later associates instead Big Science with international scientific collaborations. Elaborated at the SOAS Centre for International Studies and Diplomacy, the article singles out recent examples of collaborative Big Science enterprises such as

doi:10.1088/978-0-7503-3631-4ch58

the European Laboratory for Nuclear Research (CERN), the International Thermonuclear Experimental Reactor (ITER), and the International Space Station (ISS) as core examples for what Big Science is today [2].

In the last 60 years the concept of Big Science has thus metaphorically travelled from the Tennessee prairies to 370 km above the Earth, conveying in its journey different ideas about the nature of the large-scale scientific enterprise. But how, why, and when did its association with the national laboratory morph into a new one framing it in the realm of 'global science' [3]?

In addressing this question, it is important to note that this transition in meaning has also informed changes in perspective on Big Science projects in the policy realm. In 1961 Weinberg drew on his experience as ORNL director to claim that if large-scale research products (rockets, accelerators, reactors) were the icons of twentieth century science, 'gigantism' made large-scale scientific enterprises corruptible to external economic and political forces due to the craving for sponsors able to support their costs [4]. The aforementioned SOAS article offers instead an antithetical reading. Not only do Big Science collaborations represent a virtuous model of science-making, in that they offer opportunities for sharing ideas across borders, but exactly because of this feature, they should be seen as a model of global governance too, since they enable for scientific projects to thrive at global level 'through major political and programmatic constraints' [5]. It is a model associated with the notion of science diplomacy, which emphasizes the merits of promoting collaborative research internationally, also on a large scale, as a vehicle to address global challenges and establish constructive relations between governments [6].[1]

This chapter aims to chart the transition in meaning that has led Weinberg's image of large-scale research as a national enterprise to morph into one focussing instead on international collaborations. To do so it begins from a different set of assumptions from those of previous historical analyses on Big Science. Those who have examined this historical phenomenon before have often displayed an eagerness to better define the term, arguing that it was actually too vague to be associated with only one aspect of the large-scale scientific enterprise [7]. While this scholarly work has undoubtedly been very useful, this article focuses instead, in line with communication studies, on the *framing* role that the term Big Science, as *bridging* concept, has played in the global science policy landscape distinctive of the last 60 years [8].[2] In particular, this article argues that the term originally featured in science publications eager to advertise the national large-scale laboratory, while over recent years it aimed instead to publicize international collaborations.

What follows highlights first, through a text-mining survey, how the set of terms that Big Science has been associated with has changed considerably over the last

[1] The molecular biologist N V Fedoroff defines it as 'the use of scientific collaborations among nations to address the common problems facing the 21st century and to build constructive international partnerships' [6].

[2] I rely here on the definition of framing as communication devices that help to give meaning to structures of the social world. Within it there are concepts that work as 'bridges' in terms of connecting between domains and areas of social life. See [8].

60 years in line with this shift in advertising. It then offers an historical narrative for this transition, and examines how changes in the framing of Big Science have informed the definition of new large-scale projects, especially in the USA. It focuses in particular on how during the 1990s the advocacy for opening up science in general, and Big Science projects more specifically, to international partnerships was particularly effective in pushing for a re-configuration of ongoing large-scale research enterprises in the USA, re-fashioning them as international collaborations. The article thus concludes that in its current version, the term Big Science better accommodates the growing interest in promoting these international collaborations in the globalized environment in which scientists now operate.

58.1 Big Science speech

The evolution of a concept's meaning can today more easily be charted through the novel tool of text-mining. Now more popular in humanities and social sciences research, it consists of statistically evaluating the recurrence of words and groups of words in a corpus of texts (or *corpora*) [9]. The study of these recurrences also allows charting associations between certain words in the context of a specific text as part-of-speech tagging. It is therefore useful to map variations over time of the 'constellation' of words associated with Big Science, especially those that reveal its association with national and international research enterprises.

Since a considerable number of mainstream science publications, such as for instance the journal *Science*, feature in the digital library JSTOR, I have used the library's text-mining platform Constellate to complete a survey of literature in which Big Science features as the keyword.[3] A corpus of 4520 articles and book chapters in total published between 1960 and 2010 has thus been examined to figure out the changing association of Big Science with other words.

It is important to note first of all that there is considerable variation in the usage of the word Big Science across the decades (figure 58.1). While the term easily entered the scholarly vocabulary in the 1960s, its usage was marked up until the 1990s. Since then its popularity has decreased somewhat, although by no means it is less relevant today.

Big Science's association with other words can be analysed through the compilation of n-grams—graphs charting deviations in their frequency as percentage of the entire corpora over time. Figure 58.2 displays the frequency of the adjectives national, international, and global in the Big Science literature (all non-capitalized). While evidently 'national' continues to be an inescapable referent for this literature, 'international' and 'global' are defining adjectives occurring more frequently in this literature at least from the 1980s. 'Global' was virtually absent from this literature (below 10%) in the 1960s, but by 2010 more than half the articles discussing Big Science use the term 'global' in some form. Of course, these adjectives may be too broad in meaning to tag Big Science literature accurately enough. For instance, one

[3] Within quotation marks as 'big science' (capitalization returns the same results). See: https://constellate.org/. Search tool last accessed and used for text mining purposes on 2 February 2022.

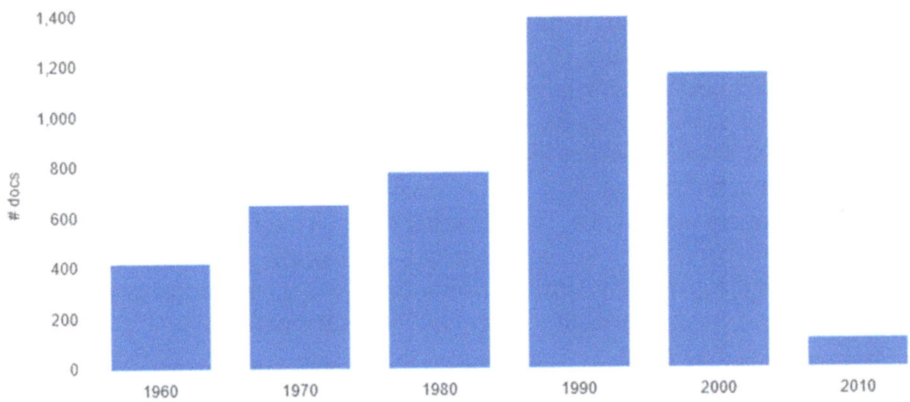

Figure 58.1. The use of 'Big Science' as a keyword in 5384 JSTOR documents between 1960 and 2010 (elaborated by the author using Constellate). Figure credit: ITHIKA, www.constellate.org.

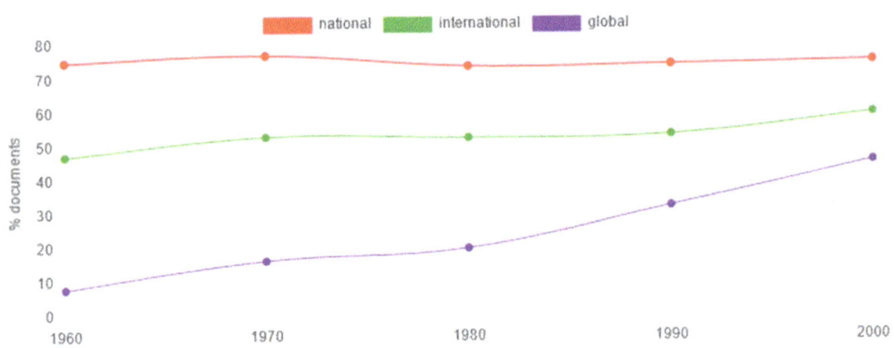

Figure 58.2. Frequency of 'national', 'international', and 'global' in the JSTOR corpora (elaborated by the author using Constellate). Figure credit: ITHIKA, www.constellate.org.

article may generically cite 'national statistics' and this would affect text analysis without having real relevance for the ways in which authors have framed Big Science as a national enterprise. Hence, we need to narrow the survey further by picking up more words that further validate what the search shows. One way to do so is to use nouns more directly geared towards national and international contexts, such as 'national laboratory' and 'international collaboration'.

Figure 58.3 displays their occurrence. While inevitably the pool of relevant Big Science articles using these terms is much lower than the previous one, the n-gram confirms the dissociation of Big Science from the concept of 'national laboratory' displaying its decreasing use (from 13% to 8%) over the 60 years considered in this survey. There is, notably, a parallel increase in the use of 'international collaboration' (from 1% to 4%) over the same period.

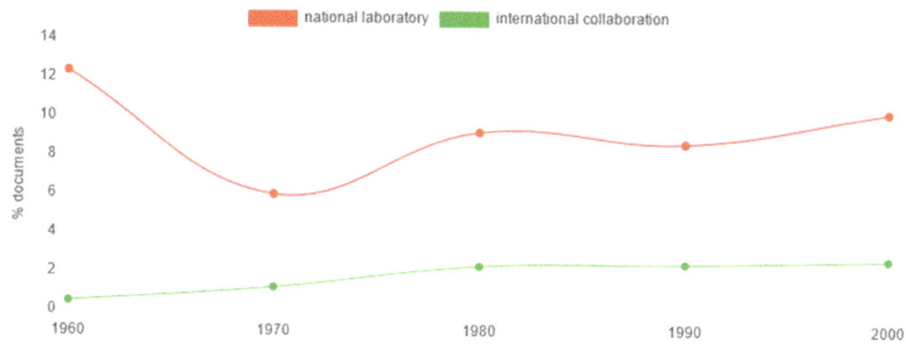

Figure 58.3. Frequency of the term 'national versus 'international collaboration' in the JSTOR corpora (elaborated by the author using Constellate). Figure credit: ITHIKA, www.constellate.org.

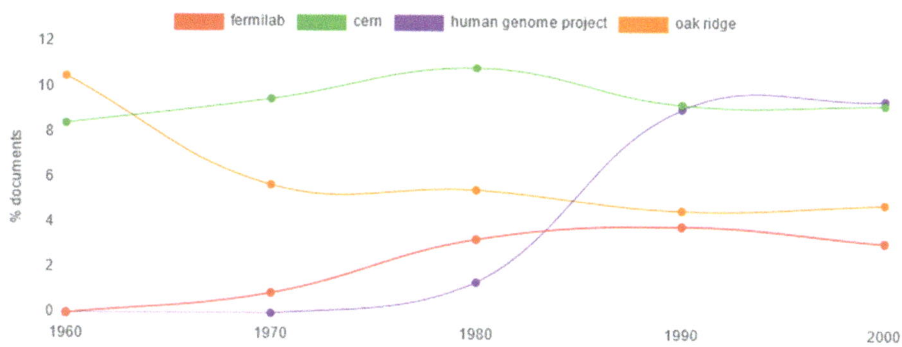

Figure 58.4. 'Fermilab', 'CERN', 'Human Genome Project', and 'Oak Ridge' in the JSTOR corpora (elaborated by the author using Constellate). Figure credit: ITHIKA, www.constellatc.org.

A sample of specific Big Science projects considered models of large-scale research either of a national or of an international kind have then been selected in the next n-gram (Fermilab, Oak Ridge, CERN and Human Genome Project) (figure 58.4). They confirm the trend highlighted by previous n-grams. Since the 1980s two leading US national laboratories (Fermilab and Oak Ridge) have been mentioned less in Big Science literature, whereas international collaborations such as CERN and the Human Genome Project have been associated with Big Science more frequently. Particularly significant is that since the mid-1960s the percentage of documents citing Oak Ridge within the literature fell below that citing the CERN and that, more generally, from the 1990s this literature features CERN and the Human Genome Project nearly twice as much as it features Oak Ridge and Fermilab.

Even more surprising is the stark contrast between two archetypal large-scale projects, one national and the other international, examined in the next n-gram (figure 58.5). These are the International Geophysical Year (IGY), namely the pioneering international collaborative project in the geosciences extending through

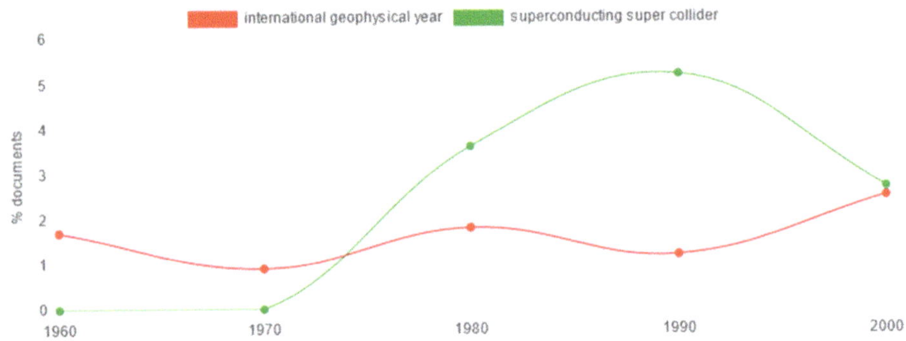

Figure 58.5. Occurrence of 'International Geophysical Year' and 'Superconducting Super Collider' in the JSTOR corpora (elaborated by the author using Constellate). Figure credit: ITHIKA, www.constellate.org.

18 months from 1 July 1957, and the Superconducting Super Collider (SSC), the largest US high-energy project ever planned and famously abandoned in 1992. While overall occurrences are again low in percentage, it is worth noting that from 1990 more authors associate Big Science with the IGY than the SSC. This confirms further the change in framing of the term Big Science as defining international collaborations rather than national laboratory work.

This preliminary text-mining exercise thus shows that over 60 years, a considerable number of scientific publications have identified Big Science with different entities and projects. Moreover, they have increasingly associated Big Science with international collaborative endeavours, casting aside its association with nationally based research enterprises—in particular through the distinctive configuration of the large-scale national laboratory. What follows offers an historical narrative seeking to explain these changes.

58.2 First came the national laboratory

Since its inception, scientists, policy experts, and science communicators wished to convey specific views about science policy by framing the word Big Science as a bridging concept promoting the large-scale national laboratory. When Weinberg introduced the term he was a science administrator busy seeking a way out of a policy conundrum about the trajectory of the laboratory that he directed. A former assistant of nuclear physicist Eugene Wigner during the Manhattan (atom bomb) Project, he went on to lead the ORNL after WW2, when the facility was used primarily to put together the fissile material for the nuclear weapons dropped on Japan. Weinberg witnessed its expansion during the Cold War years, when the US Atomic Energy Commission (AEC) funded its programmes [10].

The 1961 *Science* article that popularized the term Big Science drew on Weinberg's experience as director to emphasize the need for scientists, and especially high-energy physicists, to acquire more powerful and expensive machines in order to lead in their fields. Weinberg therefore conveyed the message that only expensive research equipment allowed to achieve the results needed, especially in high-energy physics. But he also manifested an anxiety about how these vital research resources would make scientists, and in particular those administering large-scale research, more vulnerable to external pressures because of the need to gain considerable financial support to build and run them [11].

Those who re-used Weinberg's new term did show, however, more interest in his argument about the need for scientists to 'scale up' in order to be ahead, than that about how the search for larger research budgets could make science unprincipled. They thus went on to emphasize the large-scale national laboratory as the last frontier in the realm of modern science. Indeed a powerful advertising machine now adopted the notion while overlooking Weinberg's concern about the drawbacks of 'gigantism'. For instance in 1970 a *Science* reporter enthusiastically emphasized Japan's entry into the Big Science era, by presenting a new set of national facilities in that country and emphasizing that they represented a 'lucrative business' [12]. Similar pledges for national Big Science projects featured in other articles describing recent developments in disciplines such as chemistry, oceanography, and computer science. When plans for Big Science facilities failed, such as in the case of the Intense Neutron Generator (ING) that the Canadian atomic agency never built, commentators pointed their fingers at inept policy-makers, effectively portraying a clash between inherently progressive large-scale national scientific endeavours and backward political habits [13]. Thus in its early days the term Big Science publicized the large-scale laboratory, indirectly seeking to persuade more investment in individual national projects.

This appreciation of Big Science as a nationally focussed enterprise often entailed belittling international collaborative projects as less endowed alternatives for those countries that could not develop similar large-scale national facilities. For instance, while today we consider CERN as a chief example of Big Science, Weinberg's contemporaries portrayed it as a cheaper alternative for poorer aspirants to large-scale research. The ORNL director framed it as an instrument 'set up to enable several countries to share costs that are too heavy for them to bear separately' [14, p 167]. And this framing effectively gave leeway to its understanding as a means to help poorer countries to complete large-scale projects in particle physics. Indeed, a UK official explained at one point that his country funded the CERN mainly because it was an 'internationally shared machinery' [15, p 349]. The implication, quite evidently, was that sharing ended once research proper started within national laboratories.

This framing also entailed portraying what we now see as a pre-eminent example of Big Science collaborations, the International Geophysical Year (IGY), as something different from Big Science proper. A recent article by Elena Aronova, Naomi Oreskes, and Helen Baker contends that the IGY pioneered a new type of Big Science connected with the sharing, internationally, of scientific data. This is because the IGY introduced a system of World Data Centres scaling up research data collection and sharing [16]. While there are obvious merits in thinking about Big Science in this way, it is important to recognize that very few scientists did so at the time. Actually, specialists, such as for instance the president of the American Geophysical Union, stressed at one point that Big Science was not to going to be as relevant in geophysical studies as it was in high-energy physics, exactly because of its international orientation and collaborative methodology [17].

The president's anxiety about associating international collaborations and Big Science possibly derived from the Cold War, since the conflict between superpowers urged to prioritize the sponsorship of large-scale national research projects; especially those with implications for national defence. This unease may have also derived from the understanding that the association between Big Science and international collaborations heavily featured instead in the proceedings of organizations seeking a way out of the Cold War by promoting international exchanges more, even across the Iron Curtain. Weinberg, for instance, discussed the potential of Big Science collaborations at the Seventh Pugwash Conference on Science and World Affairs organized at Stowe (VT) devoted to 'International Cooperation in Pure and Applied Sciences'. Together with other leading American scientists he then went on to recommend the creation of a 'globular cluster of big science centers', managed directly by international organizations such as the United Nations [18]. No 'globular big science cluster' emerged under the UN aegis any time soon [19]. Actually, there was resistance within national governments to promoting international collaboration on a large scale. For instance, a 1967 proposal for a World Health Organization Big Science research centre (dubbed as 'Operation Megamouse') never saw the light of the day as the five countries responsible for its sponsorship (West Germany, USSR, US, UK, and France) unanimously rejected it [20].

So when did the portrayal of Big Science as a nationally focussed endeavour begin to change (and why)? From the late 1960s, the unrest in university campuses in Western nations, and the growing opposition to the Vietnam conflict, fuelled greater scrutiny of modern science, including its large-scale manifestations. In turn, the growing dissatisfaction for the ties between scientists and military patrons tarnished the widely advertised image of Big Science as a progressive system of large-scale national laboratories. Indeed, especially in the US, this system featured heavily in the exposés of radical scientists given its contribution to military operations, especially in Vietnam [21]. This helped to re-frame Big Science in science publications, in particular since Weinberg's early concerns about its being corruptible could no longer be overlooked.

58.3 Good and bad press

By the 1970s many scientific articles recalled that Big Science catalyzed funds for research leaving little for studies with a social impact, such as for instance those focussing on pollution and under-development. These publications indirectly contended that smaller research groups producing research with greater social impact were penalized by the growth of Big Science. For instance in 1972 the US National Academy of Science's Committee on Science and Public Policy concluded that: 'the advent of these facilities (National Accelerator Laboratories) will hasten the trend to "big science" and large user-groups *at the expense, if no other money can be found, of smaller research group* [my italics]' [22].

Conversely, some international projects received praise exactly because they addressed transnational social and environmental challenges without large budgets. For instance in 1970 the director of a cancer research international agency based in Lyon (France) advocated an international investment on the environmental causes of cancer because 'environmental pollution does not respect national boundaries' [23]. University of Washington atmospheric scientist Richard J Reed portrayed Big Science as an opportunity to scale up research in his field but emphasized the disadvantages of this transition to large scale in the prospect of greater social impact for atmospheric sciences [24].

The advertising of other branches of the Earth sciences, such as oceanography, now considerably changed, as these were depicted (in contrast with the IGY), as examples of a Big Science with a social twist. For instance, in 1972 University of Miami oceanographer F G Walton Smith indicated large-scale oceanographic projects as more viable than traditional national laboratory work in high-energy physics exactly because science 'must not be merely practical, it must be socially relevant' [25].

It is quite telling that one of the main IGY successors in the realm of internationally coordinated research, the International Biological Programme (IBP), was, in contrast with its predecessor, plainly presented as a novel example of collaborative Big Science. The IBP, which started in 1964, actually represented a solution to Weinberg's conundrum about how to harmonize the scaling up of laboratory research. In particular, as science studies scholars Chunglin Kwa and Stephen Bocking have separately showed, Weinberg cleverly used the IBP to funnel ORNL funds towards radiation ecology studies, so to give to Big Science projects at his organization a more visible environmental dimension [26].

This new emphasis chimed more generally with the promotion of international collaborative Big Science as a device for development; hence to address imbalances between richer and poorer regions of the world. In 1971 the head of Organization for Economic Co-operation and Development (OECD) science policy division, Jean-Jacques Salomon, enthusiastically described the emergence of a new *internationale* of science. Cooperation, he contended, was a fundamental aspect of modern science stifled by the ongoing Cold War. Big Science, if understood as distinctive of the national laboratory, was instead a phenomenon about to end, given that Western Europe was now building capacity 'on the same scale [...] sustained by the bipolar powers', and soon intergovernmental organizations would have the budgets needed to compete with their national research agencies [27].

The new rhetoric of scientific internationalism on a large-scale also instigated a new framing of CERN. Scientists and science communicators alike portrayed it as a more sustainable model of Big Science, due to its shared costs, also catering for international interactions. In 1973 the incoming CERN executive director, British particle physicist John Bertram Adams, recalled that to be 'at the frontier of [their] subject', nuclear physicists ought to travel regularly to the international laboratory to exchange ideas [28]. Quite a different stance from that of the fellow national who had previously described CERN as little more than a shared piece of machinery.

With the 1980s approaching, a new narrative thus emerged in the scientific presses portraying international collaborative projects on a large-scale as more viable, economically and politically, than their national equivalents. In 1981 *Science*, for instance, positively reported on the decision taken by the Deutsches Elektronen-Synchrotron (DESY) laboratory in Hamburg to build a new \$265 million accelerator (HERA) in the context of an international consortium [29]. Two years later the boson discoveries at the CERN set the collaborative Big Science organization on a leadership path in high-energy physics, setting off the competition with US national laboratories such as Fermilab and Brookhaven. Conversely, scientists viewed traditional national laboratory-based Big Science as more problematic. In 1977 biostatistician Irwin D J Bross wrote, for instance, that the new generation of scientists 'felt threatened by the technology of Big Science' [30].

58.4 Big Science studies

The advocacy for reforming Big Science to make it less tied to national defence imperatives received further support in the 1980s through historians and social scientists enquiring into its circumstances and this examination soon became an area study in its own right. The 1963 book *Little Science, Big Science*, written by British physicist turned historian Derek De Solla Price, had actually offered a very positive image of Big Science as a national research enterprise, effectively framing its gigantism as a positive development (in contrast with Weinberg). De Solla Price elaborated the content of his book as a set of presentations for the prestigious Pegram Lectures annually delivered at the Brookhaven National Laboratory. It is thus plain to see that his views chimed with the organizations' promotion of the large-scale national laboratory as a form of Big Science. Through the use of scientometric analysis, De Solla Price framed it as a healthy evolution of modern science, and definitely not a pathological development [31].

However, during the 1970s and 1980s a new generation of scholars embedded the radicalism distinctive of university protest movements into novel historical analyses, hence strengthening the critiquing of Big Science in its national manifestations. Some, like historian Paul Forman, gave foundation to Weinberg's early concerns about Big Science as a pathological development. As noted by Julia Harriet Menzel and David Kaiser, Forman replaced 'the quantitative idiom [of Solla Price] with a language of morality' to denounce how the physicists involved in the large-scale national laboratory had lost their autonomy in the process of searching for ways to be ahead in the field [32]. Forman's portrayal of Big Science sparked an intense debate within history of science studies, but it also crossed disciplinary domains [33]. In addition, during the 1980s many other studies confirmed further Forman's negative portrayal. For instance, Robert Seidel showed that while the AEC's laboratory system represented a home for Big Science, it equally defined a 'conflict between patronage based upon meritocratic and upon geopolitical considerations' [34]. A few years earlier, a study of the Jet Propulsion Laboratory exposed the intellectual and organizational problems internal to the growth of Big Science [35]. A history of fusion research openly described it as a 'Faustian bargain' [36]. Importantly, historians did not portray international scientific collaborations in the same way.

Sociologists of science were equally influential in casting Big Science under a negative light, in particular by exploring its output. Nation-based high-energy physics laboratories were less productive in terms of publications than internationally based networks [37]. Gender studies scholars contributed to the mounting critique by exposing what they identified as 'the embeddedness of big science in a "culture of death"' due to its, often undisclosed, ties to military agencies [38].

These efforts paralleled historical research tracing the legacy of Big Science. Robert Seidel and John Heilbron, for instance, wrote the history of the Lawrence Berkeley Laboratory, which opened in the 1930s and represented an important precursor for many national laboratories of the post-WW2 era [39]. An economic history study on the post-1945 California sardine depletion posited it as a study in the 'institutional development of post-war "Big Science" [...] to cope with complex environmental issues in resource-extractive industries' [40, p 403]. Yet, the impressive set of articles and books exploring the large-scale national laboratory presented a largely critical portrayal of its main facets, especially from an ethical and political viewpoint. In addition, this depiction echoed the growing concern of science policy advisers and commentators about the sustainability of large-scale national research projects from an economic viewpoint, lending further support to the search for (international) alternatives.

58.5 Big Science enters the age of scientific diplomacy

Since the 1990s science publications have more frequently associated Big Science with internationally collaborative schemes in a world research system undergoing an important transformation due to the end of the Cold War and the rise of globalization. One of the key drivers for this transition was the advocacy of scientific collaborations as a foreign policy device, which coupled with the aforementioned criticism of nation-based Big Science schemes, an important economic downturn, and a reduced budget for projects tied to a defence agenda. In turn, anchoring the bridging concept of Big Science to international collaborations offered an opportunity to promote its viability at the dawn of the new century, especially by emphasizing more its relevance to international relations.

Already from the mid-1980s national Big Science projects struggled to gain the same support they had received in previous years. For instance, a US National Resource for Computation in Chemistry (NRCC) was phased out following a joint consultation of its patrons, the National Science Foundation (NSF) and the Department of Energy (DOE, ex-AEC). Promoted as the 'Chemists' First Try at Big Science', the NRCC ended up being negatively framed as a project shoring up funding needed for equally meritorious research exercises on a smaller scale [41]. Moreover, in 1986 *Science* gave resonance to the shortcomings of US expenditure in large-scale research. As a consequence, in contrast with Japan, Europe, and the Soviet Union, US space agencies had been unable to send a probe to track the comet Halley [42].

In order to avoid more criticism (and, possibly, the NRCC fate), some scientists deliberately advertised government-funded large-scale research as Big Science of a different variety. For instance, they framed one of the largest project in gravitational wave research, the Laser Interferometer Gravitational-Wave Observatory (LIGO), as Big Science tied to small science practices. Although the telescope construction represented the most expensive single project that the NSF had ever undertaken, its instigators at Caltech and MIT claimed that they 'envision the work itself to still be in the style of small science' [43].

Large-scale international projects were as burdensome for public finances as national projects, so the economic downturn affected them too. Yet, what helped international Big Science collaborations to avoid the criticism then hitting their national equivalents was the realization of their importance for foreign relations. By then, US politicians had already engaged in debates on the role of science and technology in foreign policy, often fuelled by policy scholars underlining the need to use science as a device to build constructive relations between countries. In 1985 the director of the MIT Centre for International Studies, Eugene Skolnikoff, openly advocated for re-thinking the merits of international collaborations during hearings at the US House Committee on Science and Technology. On that occasion, he also criticized 'the overwhelming domestic orientation of the American R&D enterprise' [44].

This advocacy went hand in hand with a shift in thinking in sectors of the US administration, and especially State Department, about the role of science and technology in foreign policy. Instrumental to its definition was political scientist and diplomat Joseph Nye, who not only recognized the 'complex interdependence' of world countries, especially in the ongoing process of globalization, but vouched support for the need of managing foreign relations through 'soft power' devices. He thus contended rooting US foreign policy in projects based on persuasion rather than coercion [45]. Nye's thinking implicitly cast a positive light on the setting up of collaborative schemes, in the sciences, as a key instrument to propel US relations with other countries.

In the early 1990s Skolnikoff and Nye's advocacy received further support in the US Congress. Key figures, such as the Chair of the US Science and Technology House Committee George Brown Jr, explicitly called for a new international orientation of Big Science projects as: ' [...] international cooperation will prove essential for the success of "big science" projects in the future. [...], international burden-sharing on large, complex and costly scientific and technological projects will be necessary, if not sufficient, for their survival [...]' [46].

This new thinking, also catering for the popularization of the notion of scientific diplomacy informed the trajectory of ongoing Big Science projects in the US, shifting their orientation. They were decisive, for instance, in re-considering that on NASA's space station and the planned mission to Mars, which until that point had been the archetypal Big Science project in the US space research programme. In 1992 the US administration cancelled funding for the station, while the end of the Cold War propelled negotiations between NASA and the Russian Federation space agency, Roscosmos, on a similar collaborative enterprise. The talks paved the way to several agreements between the two agencies, leading, in 1998, to the signing of a memorandum of understanding on jointly setting and running an International Space Station (ISS), which was both a scientific and foreign policy achievement in the post-Cold War climate.

As Daniel Kevles and others have shown, only US Big Science projects with a clear international dimension survived the economic recession of the 1990s, when only funding available for research sponsors expanding their international programmes, such as the NSF and the Environmental Protection Agency, could increase [47]. What enabled the Human Genome Project, then framed as marking 'biomedicine's entry into the big science league' [48],[4] to survive and overcome considerable financial pressures, was the web of international collaborations that its main executor, the National Institute of Health, instigated. In particular, the 1996 Bermuda principles allowed the project to thrive as a collaborative exercise through the international sharing of pre-publication data on sections of the human genome [49].

This development contrasted with what was then happening in high-energy physics. In 1992 US authorities cancelled the largest scheme in this study area, the Superconducting Super Collider (SSC), notwithstanding the large sums put aside in previous years for its completion. The economic recession that hit the US persuaded both Republican and Democrat representatives in the US Congress to vote against its future development [50]. In taking this decision they were also, presumably, aware that competing projects in the same area with an international collaboration component offered greater prospect of success with a lighter burden for federal funds. For instance, plans for an international laboratory on fusion research had reached by then a turning point. Talks for an International Thermonuclear Experimental Reactor (ITER) had developed from 1987, always in the context of a post-Cold War rapprochement between superpowers. They coupled the explicit ambition to offer diplomatic support to the scientists' plan to harness fusion energy with the implicit one of showing the foreign policy merits of pursuing Big (Energy) Science through an international collaborative scheme. The plans evolved through a partnership under the aegis of the International Atomic Energy Agency including the atomic energy organizations of Russia, Japan, Spain, and France [51]. By 2003 the five billion dollars ITER received priority in the list of new projects funded through the US Department of Energy (formerly the chief SSC sponsor).

A parallel initiative, the building of a high-energy physics laboratory in the Middle East, more vividly displayed the foreign policy advantages to be derived from pursuing Big Science through international scientific collaborations. Housing the German synchrotron BESSY, the new laboratory aimed to benefit particle physics research in the region, but it indirectly displayed the importance of this Big Science collaboration for the peace process in the Middle East by showcasing the potential for a research partnership between Israeli and Arab physicists [52].

Historical work on Big Science further injected into the science policy debate more analytical elements, now inviting to re-think large-scale research as entailing

[4] Doubling in funding from \$87 million to \$154 million [48].

more than large-scale equipment, and including relational elements distinctive of collaborative enterprises. For instance, drawing on a study of the high-energy accelerator Bevalac, US historian Catherine Westfall, emphasized the significance of these relational aspects [53]. David Munns reconstructed the history of the US National Radio Astronomy Observatory along similar lines, emphasizing its alterity to the national laboratory model [54]. The pitfalls of national Big Science projects and the emphasis on the collaborative dimension were also at the origins of the introduction of Lilian Hoddeson's notion of 'megascience', where economic constraints are instrumental in the search for external collaborations [55].[5] And, finally, John Krige could think anew about the history of CERN focussing on its implications for US foreign policy in the post-WW2 years, and emphasizing its significance for transatlantic relations [56]. Drawing on Krige's study, recent research has appraised Big Science projects in Cold War Europe as decisive in shaping the relations between European countries, often across the Iron Curtain [57].

In 1992 science historian Peter Galison acknowledged that Big Science 'inevitably requires collaboration' [58]. Yet over the following 30 years, scholars and commentators alike have focussed more on this collaborative element to unearth its significance in transforming Big Science into an international, and at times global, enterprise, also re-considering its significance for international relations. Indeed Big Science is actually a component of what Michael E Peters defines as 'global science', namely an 'emerging geography of scientific knowledge and collaboration as an aspect of globalization and its interconnectedness within a globalized world'. 'Big Science is [now] Global'—he summarizes [59].

58.6 Conclusions

In 2010 Oreskes, Aronova, and Baker appropriately pointed out that we should no longer think of Big Science as 'this' or 'that', but rather as both an analytical and framing category equally serving diagnostic and promotional purposes during the Cold War and beyond [60]. This paper adds to their useful remark an evaluation of how, since its conception, Big Science has pivoted around competing policy perspectives regarding the scientific enterprise. And these standpoints have advertised, on the one hand, a geographically localized and identifiable large-scale research facility (the national laboratory), and, on the other a diffused complex of research structures either interconnected through data flows or sharing research facilities through collaboration.

While more sustained and detailed research will fully detail the articulation of the Big Science debate around these two competing perspectives, this paper has first charted its trajectory through an analysis of the changing meaning of the word Big Science as detected by the terms it associated with in the relevant literature. We have seen that over the last 60 years, the word Big Science has been connected more with international collaborations which were alien to its original make-up as a notion

[5] 'Forming invisible mini-institutions within the programs of large physics laboratories' [55].

intertwined with the large-scale national laboratory. Contingently we have seen how projects such as CERN or the IGY that gravitated at the periphery of this concept when it was first introduced, eventually featured more heavily within its literature, while archetypal projects (Oak Ridge, Fermilab) have been framed less in Big Science literature in recent decades.

This paper explains this transition by charting how scientists, policy advisers, and science communicators sought to portray Big Science and what aspects they sought to emphasize more. In the 1960s, their narratives underscored the role of the large-scale national laboratory as a foundational element in the pursuit of Big Science, in itself seen as the ultimate frontier in the pursuit of modern scientific research. From the late 1960s voices critical of modern science and its large-scale manifestations cast a shadow on these promotional perspectives, recalling the political, ethical, and social shortcomings of large-scale investments in research. In turn, while for the following 20 years Big Science continued to be an inescapable referent for national research, international alternatives gained support in science publications, often due to their social, developmental, and environmental ambitions. Moreover, even when, such as in the case of gravitational wave research, national agencies continued to pay for expensive large-scale facilities, practitioners wished to frame their study differently to shy away from critical narratives targeting Big Science. Historical work on large-scale research gave foundation to these critical stances by displaying the ethical and political drawbacks.

From the late 1980s and through the 1990s, Big Science entailed the promotion of international collaborations due to the folding of key Big Science projects such as the SSC, the NRCC, and the NASA Space Station. The search for international synergies helping to realize collaborative projects such as ITER and the ISS demonstrates further the ascendancy of collaborative Big Science. While there are multiple reasons for these historical developments (including the end of the Cold War and the ensuing economic recession), this chapter has displayed the importance of a new way of thinking about science, particularly in the US, recalling its implications for foreign affairs. While Skolnikoff did not single out Big Science in positing the centrality of scientific collaborations to US external relations, George Brown Jr openly campaigned for the important shift which led to the ISS being pursued as a collaborative rather than US (NASA) project.

Presumably, to some of these advocates for international collaboration, CERN appeared as a model. However, we have seen that the portrayal of the European nuclear research laboratory changed considerably over time. Initially presented as a cheaper alternative to far more illustrious examples of Big Science in the USA, it evolved into a prominent example of Big Science collaborations defining the uses of scientific diplomacy in the post-war years.

This reminds us of the importance of not taking science diplomacy advertising at face value and uncritically following portrayals of international Big Science project as being inherently progressive, or taking over from antiquated Cold War ones. The history of CERN shows that its implications as a science diplomacy device eventually emerged, but only gradually. That of ITER and the ISS shows instead that the foreign policy merits of their implementation overlapped the economic downturn and the end of the Cold War. Hence, what today many science diplomacy enthusiasts wish to consider examples of global governance are in fact just the result of intricate historical processes that have ultimately helped in re-framing the promotion of Big Science as an instrument of twenty-first century diplomacy.

Acknowledgements

Research for this article has received funding from the European Research Council (ERC) under the European Union's Horizon 2020 research and innovation programme (Grant agreement No. 101021098).

References

[1] Weinberg A M 1961 Impact of large-scale science on the United States *Science* **134** 161–4 For an historical analysis see also Johnston S F 2018 Alvin Weinberg and the promotion of the technological fix *Technol. Cult.* **59** 620–51

[2] Robinson M and Hallonsten O 2021 Big Science Collaborations; lessons for global governance and leadership *Glob. Policy* **12** 66–81 To chart some of the changes distinctive of Big Science see ed Hallonsten O 2016 *Big Science Transformed: Science, Politics and Organization in Europe and the United States* (Springer: Berlin)

[3] See for instance Peters M A 2006 The rise of global science and the emerging political economy of international research collaborations *Eur. J. Educ.* **41** 225–44

[4] Weinberg A M 1961 Impact of large-scale science on the United States *Science* **134** 162 See also ed Weinberg A 1968 *Reflections on Big Science* (Cambridge, MA: MIT Press)

[5] Robinson M 2020 Big Science collaborations; lessons for global governance and leadership *Global Policy* **12** 66–80

[6] Fedoroff N V 2009 Science diplomacy in the 21st century *Cell* **136** 9–11 On science diplomacy, see also; Turchetti S and Rispoli G 2020 Science Diplomacy *Histor. Stud. Natur. Sci.* **50** (special issue)

[7] See for instance Capshew J H and Rader K A 1992 Big Science: price to the present *Osiris* **7** 2–25 Galison P and Hevly B (ed) 1992 *Big Science: The Growth of Large-Scale Research* (Palo Alto, CA: Stanford University Press) Kevles D 1997 Big Science and big politics in the United States: reflections on the death of the SSC and the life of the human genome project *Hist. Stud. Phys. Biol. Sci.* **27** 269–97

[8] Reese S D 2007 The framing project: a bridging model for media research revisited *J. Commun.* **57** 148–4 On framing more generally see Chong D and Druckman J N 2007 Framing theory *Annu. Rev. Polit. Sci.* **10** 103–26

[9] Wiedemann G 2016 *Text Mining for Qualitative Data Analysis in the Social Sciences: A Study on Democratic Discourse in Germany* (Wiesbaden: Springer)

[10] See ed Weinberg A M 1994 *The First Nuclear Era: the Life and Times of a Technological Fixer* (New York: AIP Press) pp 198–200 See also Johnston S F 2018 Alvin Weinberg and the promotion of the technological fix *Technol. Cult.* **59** 620–51

[11] Weinberg A M 1961 Impact of large-scale science on the United States *Science* **134** 164

[12] Boffey P 1970 Japan (I): on the threshold of an age of Big Science? *Science* **167** 31–5

[13] Doern G B 1970 Big Science, government and the scientific community in Canada: the ING Affair *Minerva* **8** 357–75

[14] Weinberg A M 1963 Criteria for scientific choice *Minerva* **1** 159–71

[15] Rose H 1967 The rejection of the WHO Research Centre: a case study of decision-making in international scientific collaboration *Minerva* **5** 340–56
[16] Aronova E, Oreskes N and Baker H 2010 Big Science and Big Data in biology: from the International Geophysical Year through the International Biological Program to the Long Term Ecological Research (LTER) network, 1957–present *Hist. Stud. Nat. Sci.* **40** 183–224
[17] Malone T 1964 A golden age for geophysics? *AGU Trans.* **45** 295–300
[18] 1961 Conferences on science and world affairs *Science* **134** 984–91
[19] (eds) Kraft A and Sachse C 2020 *Science, (Anti-)communism and Diplomacy: The Pugwash Conferences on Science and World Affairs in the Early Cold War* (Leiden: Brill) p 114
[20] Rose H 1968 The Rejection of the WHO Research Centre *Minerva* **6** 263–4
[21] For an overview see ed Turchetti S 2016 Looking for the bad teachers: the radical science movement and its transnational history *Science Studies During the Cold War and Beyond* ed E Aronova and S Turchetti (New York: Palgrave Macmillan) pp 77–101
[22] Hammond A 1972 Physics in perspective: a new report *Science* **177** 597–9
[23] Higginson J 1970 International research: its role in environmental biology *Science* **170** 935–9
[24] Reed R 1971 The effects of GARP and other future large programs on education and research in the atmospheric sciences *Bull. Am. Meteorol. Soc.* **52** 458–62
[25] Smith F G W 1972 What the oceans means to man *Am. Sci.* **60** 16–9 See also Hammond A L 1974 Physical oceanography: big science, new technology *Science* **185** 247
[26] Kwa C 1987 Representations of nature mediating between ecology and science policy: the case of the international biological programme *Soc. Stud. Sci.* **17** 413–42
[27] Salomon J-J 1971 The Internationale of Science *Sci. Stud.* **1** 23–42
[28] Adams J B 1973 Some problems of a Big ScienceDedalus **102** 111–24
[29] Robinson A L 1981 DESY looks to an international future *Science* **213** 530
[30] Bross I D J 1977 A coming battle? *Science* **195** 933–4
[31] De Solla Price D J 1963 *Little Science, Big Science* (New York: Columbia University Press)
[32] Menzel H and Kaiser D 2020 Weimar, Cold War, and historical explanation: re-reading Forman *Hist. Stud. Nat. Sci.* **50** 31–40
[33] See on this debate ed Oreskes N 2014 Introduction *Science and Technology in the Global Cold War* ed N Oreskes and J Krige (Cambridge, MA: MIT Press) pp 1–10
[34] Seidel R W 1986 A home for Big Science: the Atomic Energy Commission's laboratory system *Hist. Stud. Phys. Biol. Sci.* **16** 135–75
[35] Tobey R C 1984 Big Science in the national security state *Rev. Am. Hist.* **12** 135–8
[36] Lakoff S 1983 Pitfalls and benefits of Big Science *Minerva* **21** 320–7
[37] Dickson D 1983 Study of Big Science Groups hits raw nerve *Science* **220** 482–3
[38] Jansen S C 1990 Is science a man? New feminist epistemologies and reconstructions of knowledge *Theory Soc.* **19** 235–46
[39] Heilbron J and Seidel R 1990 *Lawrence and His Laboratory: A History of the Lawrence Berkeley Laboratory* (Berkeley, CA: University of California Press)
[40] McEvoy A F and Scheiber H N 1984 Scientists, entrepreneurs, and the policy process: a study of the post-1945 California Sardine depletion *J. Econ. Hist.* **44** 393–406
[41] Robinson A 1980 Pulled plug on chemistry computer center *Science* **209** 1504–6 See also Robinson A 1976 A computational chemistry: getting more from a minicomputer *Science* **193** 470–2
[42] Logsdon J M 1989 Missing Halley's comet: the politics of big science *ISIS* **80** 254–80
[43] Waldrop M M 1990 Of politics, pulsars, death spirals-and LIGO *Science* **249** 1106–8

[44] Holden C 1985 Global cooperation in Big Science *Science* **229** 32–3 Since the late 1960s had advocated a more sustained commitment of the US administration to using science and technology as a diplomacy device, see ed Skolnikoff E 1967 *Technology and American Foreign Policy* (Cambridge, MA: MIT Press)
[45] Keohane R O and Nye J S 1989 *Power and Interdependence: World Politics in Transition* (New York: Harper) Nye J 1990 Soft power *Foreign Policy* **80** 153–71 On Nye's circumstances, see Turchetti S 2020 The unflinching Mr Smith and the nuclear age *Ber. Wiss.* **43** 521–41
[46] Brown G 1994 Changing rationales: foreign policy and the space station *Harv. Int. Rev.* **16** 34–71
[47] Hamilton D P 1991 Hard times for Big Science *Science* **252** 1058
[48] Norman C 1990 Bush budget highlights R&D *Science* **247** 517–9
[59] Marshall E 2001 Bermuda rules: community spirit, with teeth *Science* **291** 1192
[50] Kevles D 1997 Big Science and big politics in the United States: reflections on the death of the SSC and the life of the human genome project *Hist. Stud. Phys. Biol. Sci.* **27** 284 In 1999 Steve Fuller pointed out that the SSC defeat 'may actually awaken American science and technology policy from the spell cast by the Cold War', see Collier J H 2000 Divining the Oracle of Big Science: steps on the path to a new republicanism *Minerva* **38** 109–20
[51] Malakoff D *et al* 2003 ITER Tops DOE's list of next big science projects *Science* **302** 1126–7 See also Åberg A 2021 The ways and means of ITER: reciprocity and compromise in fusion science diplomacy *Hist. Technol.* **37** 106–24
[52] Masood E 1999 Middle East synchrotron facility could bring regional research cooperation *Nature* **39910** 507 Morton O 1999 Finding a new home for BESSY in the Middle East *Science* **284** 2077–9
[53] Westfall C 2003 Rethinking Big Science: modest, mezzo, grand science and the development of the Bevalac, 1971–1993 *ISIS* **94** 30–56
[54] Munns D 2003 If we build it, who will come? Radio astronomy and the limitations of 'National' Laboratories in Cold War America *Hist. Stud. Phys. Biol. Sci.* **34** 95–113
[55] Bodnarczuk M and Hoddeson L 2008 Megascience in particle physics: the birth of an experiment string at Fermilab *Hist. Stud. Phys. Biol. Sci.* **38** 508–34
[56] Krige J 2006 *American Hegemony and the Postwar Reconstruction of Science in Europe* (Cambridge, MA: MIT Press)
[57] See for instance Lalli R 2020 Crafting Europe from CERN to Dubna: physics as diplomacy in the foundation of the European Physical Society *Centaurus* **63** 103–31 On NATO Big Science projects, see Turchetti S 2018 *Greening the Alliance: The Diplomacy of NATO's Science and Environmental Initiatives* (Chicago, IL: University of Chicago Press)
[58] Hwang K 2008 International collaboration in multilayered center-periphery in the globalization of science and technology *Sci. Technol. Hum. Values* **33** 101–33
[59] Peters M A 2006 The rise of global science and the emerging political economy of international research collaborations *Eur. J. Educ.* **41** 226
[60] Aronova E, Baker K S and Oreskes N 2010 Big Science and Big Data in biology *Hist. Stud. Nat. Sci.* **40** 183–224

www.ingramcontent.com/pod-product-compliance
Ingram Content Group UK Ltd.
Pitfield, Milton Keynes, MK11 3LW, UK
UKHW051911100926
8556IPUK00005B/851